BIOLOGY
Today and Tomorrow

SECOND EDITION

BIOLOGY
Today and Tomorrow

SECOND EDITION

Jack A. Ward

Howard R. Hetzel
Illinois State University

West Publishing Company

St. Paul   New York   Los Angeles   San Francisco

Library of Congress Cataloging in Publication Data

Ward, Jack A.
    Biology, today and tomorrow.

    Bioliography
    Includes index.
    1. Biology. I. Hetzel, Howard R. II. Title.
QH308.2.W37 1984   574   83-23425
ISBN 0-314-69684-9

COPY EDITING: Jo-Anne Naples
ILLUSTRATION: John Foster, Sue Sellars,
                        James Teason,
                        Evanell Towne
COMPOSITION: The Clarinda Company
COVER: Al Satterwhite/The Image Bank

A study guide has been developed to assist you in mastering the concepts presented in this text. The study guide reinforces concepts by presenting them in condensed, concise form. Additional illustrations and examples are also included. The study guide is available from your local bookstore under the title *Study Guide to Accompany Biology: Today and Tomorrow* prepared by George S. Arita.

If you cannot locate it in the bookstore, ask your bookstore manager to order it for you.

Portfolio Photo Credits

The Organization of Living Organisms: p. 1 (a) Manfred P. Kage, © Peter Arnold, Inc., **(b)** Oxford Scientific Films, Animals Animals; **p. 2 (a)** Jane Burton, Bruce Coleman Incorporated, **(b)** Oxford Scientific Films, Animals Animals, **(c)** © Hugh Spencer/ Photo Researchers, Inc.; **p. 3 (a)** Runk/Schoenberger, Grant Heilman Photography, **(b)** Peter Scoones, Seaphot Limited, **(c)** Roger Klocek, Shedd Aquarium, **(d)** Richard S. Funk; **p. 4 (a)** Grant Heilman Photography, **(b)** Oxford Scientific Films, Earth Scenes, Animals Animals, **(c)** John Shaw, Bruce Coleman Incorporated; **p. 5 (a)** Richard S. Funk, **(b)** Richard S. Funk, **(c)** copyright Werner H. Müller, Peter Arnold, Inc.; **p. 6 (a)** Peter Scoones, Seaphot Limited, **(b)** Stephen Dalton/Photo Researchers, Inc., **(c)** Hans Reinhard, Bruce Coleman Incorporated; **p. 7 (a)** W. Clifford Healey/Photo Researchers, Inc., **(b)** Karl H. Maslowski/Photo Researchers, Inc.; **p. 8 (a)** Manfred P. Kage, Peter Arnold, Inc., **(b)** Manfred P. Kage, Peter Arnold, Inc., **(c)** Carolina Biological Supply Company, **(d)** Carolina Biological Supply Company.

Animal Reproduction and Development: p. 1 (a) Ingeborg Lippman, © Peter Arnold, **(b)** Hans Richter, T.F.H. Publications, Inc.; **p. 2 (a)** H. Hansen, T.F.H. Publications, Inc., **(b)** © 1978 Nancy Sefton/Photo Researchers, Inc., **(c)** G. I. Koorman, © Animals Animals, **(d)** Richard S. Funk; **p. 2 (a)** Grant Heilman Photography, **(b)** Carolina Biological Supply Company, **(c)** Carolina Biological Supply Company, **(d)** Carolina Biological Supply Company, **(e)** Carolina Biological Supply Company, **(f)** Carolina Biological Supply Company; **p. 4 (a)** copyright Bill Wood, Robert Harding Associates, **(b)** Runk/Schoenberger, Grant Heilman Photography, **(c)** Gene Wolfesheimer/Photo Researchers, Inc.; **p. 5 (a)** copyright Hans Pfletschinger, Peter Arnold, Inc., **(b)** Stephen Dalton/Photo Researchers, Inc., **(c)** Runk/Schoenberger, Grant Heilman Photography; **p. 6 (a)** Manfred P. Kage, Peter Arnold, Inc., **(b)** Martin Rotker/Taurus Photos, **(c)** Donald Yaeger, Camera M.D.; **p. 7 (a)** Donald Yaeger, Camera M.D., **(b)** Donald Yaeger, Camera M.D.

Plant Reproduction: p. 1 (a) Clyde H. Smith, © Peter Arnold, **(b)** Runk/Schoenberger from Grant Heilman; **p. 2 (a)** Photography by Grant Heilman, **(b)** Runk/Schoenberger from Grant Heilman, **(c)** Photography by Grant Heilman, **(d)** Photography by Runk/Schoenberger; **p. 3 (a)** and **(b)** Jack Wilburn, Earth Scenes **(c)** Photography by Runk/Schoenberger; **p. 4 (a), (b),** and **(c)** John Colwell, Runk/Schoenberger, Grant Heilman Photography.

The Environment: p. 1 (a) Jacques Jangeux, © Peter Arnold, Inc., **(b)** Raymond A. Mendez, © Animals Animals; **p. 2 (a)** © M. P. Kahl/Photo Researchers, Inc., **(b)** Clem Haagner, Bruce Coleman Incorporated, **(c)** N. Myres, Bruce Coleman Incorporated; **p. 3 (a)** © Tom McHugh/Photo Researchers, Inc., **(b)** Roger C. Anderson, Illinois State University, **(c)** Richard S. Funk, **(d)** Jack A. Ward; **p. 4 (a)** Joseph E. Armstrong, **(b)** Richard S. Funk, **(c)** Grant Heilman Photography; **p. 5 (a)** Breck P. Kent, Earth Scenes, Animals Animals, **(b)** © W. E. Ruth, Bruce Coleman Incorporated; **p. 6 (a)** © Len Rue, Jr., Bruce Coleman Incorporated, **(b)** David C. Fritts, Earth Scenes, Animals Animals; **p. 7 (a)** Jack A. Ward, **(b)** © Karen Lukas/Photo Researchers, Inc., **(c)** R. Mahony, Bruce Coleman Incorporated; **p. 8 (a)** Gene Ahrens, Bruce Coleman Incorporated, **(b)** Jack A. Ward, **(c)** U. S. Environmental Protection Agency.

Animal Behavior: p. 1 (a) Jerry Cooke, © Animals Animals, **(b)** Jack A. Ward, **(c)** Boyce A. Drummond III, Illinois State University; **p. 2 (a)** © Joe Branney, Bruce Coleman Incorporated, **(b)** San Diego Zoo, **(c)** © Tom McHugh/Photo Researchers, Inc.; **(d)** Grant Heilman Photography; **p. 3 (a)** Rudolph Zukal, T.F.H. Publications, Inc., **(b)** Irven De Vore, Anthro-Photo, **(c)** © Animals Animals, **(d)** © Animals Animals; **p. 4 (a)** © J. Foott, Bruce Coleman Incorporated, **(b)** Boyce A. Drummond III, Illinois State University, **(c)** © Leonard Lee Rue III/Photo Researchers, Inc., **(d)** L. L. T. Rhodes, © Animals Animals.

*This book is dedicated
to
the memory of Jack A. Ward*

CONTENTS IN BRIEF

SECTION ONE

INTRODUCTION 1

Chapter One
Science and the Living World 2

The Next Decade: The Continuing Search for
Life 11

Chapter Two
The Chemical Basis of Life 13

The Next Decade: Your Chemical Environment 32

SECTION TWO

THE BIOLOGY OF CELLS 35

Chapter Three
Cell Structure 36

The Next Decade: The Cells of Henrietta Lacks 54

Chapter Four
Cell Functions 56

The Next Decade: The Continuing Exploration
of Cells 75

Chapter Five
Cell Reproduction 77

The Next Decade: The Fight Against Cancer 90

SECTION THREE

**ANIMAL AND PLANT STRUCTURE
AND FUNCTION 93**

Chapter Six
Human Organ Systems: Intemgumentary, Skeletal,
and Muscular 94

The Next Decade: Fighting Paralysis and Replacing
Lost Limbs 110

Chapter Seven
Human Organ Systems: Digestive, Respiratory,
and Circulatory 112

The Next Decade: Will A Body Scanner Replace
Your Doctor? 144

Chapter Eight
Human Organ Systems: Excretory, Nervous,
and Endocrine 147

The Next Decade: Your Aging and Aching
Body 173

Chapter Nine
Plant Cell Structure and Functions 175

The Next Decade: Future Products From Plants 193

SECTION FOUR

**REPRODUCTION AND
DEVELOPMENT 195**

Chapter Ten
Animal and Plant Reproduction 196

The Next Decade: Cloning 220

Chapter Eleven
Human Sexuality and Reproduction 249

Chapter Twelve The Development
of Vertebrates 253

The Next Decade:
Fetal Therapy 277

SECTION FIVE

GENETICS 279

Chapter Thirteen
Mendel and the Principles of Genetics 280

The Next Decade: Future Food 289

Chapter Fourteen
Types of Inheritance 291

The Next Decade: Genes and Cancer 312

Chapter Fifteen
Molecular Genetics 315

The Next Decade: New Applications of Genetic
Engineering 331

Chapter Sixteen
Human Genetics 333

The Next Decade: Genetic Counseling 345

SECTION SIX

EVOLUTION 347

Chapter Seventeen
Darwin and the Evidence of Evolution 348

The Next Decade: Social Darwinism 369

Chapter Eighteen
The Mechanisms of Evolution 371

The Next Decade: Unnatural Selection? 389

Chapter Nineteen
Our Primate Heritage 391

The Next Decade: Primate Predictions 401

Chapter Twenty
Human Evolution 403

The Next Decade: Lucy's Legacy Shakes the Human
Family Tree 416

SECTION SEVEN

ECOLOGY AND ETHOLOGY 419

Chapter Twenty-One
The Biosphere 420

The Next Decade: Further Biome Destruction 435

Chapter Twenty-Two
Ecosystems and the Physical Environment 437

The Next Decade: Overloaded Cycles 450

Chapter Twenty-Three
Communities and the Biotic Environment 452

The Next Decade: Integrated Pest Management 463

Chapter Twenty-Four
Populations and Human Ecology 465

The Next Decade: Population of the United
States 488

Chapter Twenty-Five
The Biology of Behavior 490

The Next Decade: Sociobiology 509

APPENDIX 1   The Metric System 513
APPENDIX 2   The Major Phyla of Living
Organisms 515
GLOSSARY 523
INDEX 545

CONTENTS

SECTION ONE

INTRODUCTION

CHAPTER ONE
Science and the Living World 2

SCIENCE AND SCIENTISTS 3
CONCEPT SUMMARY: Scientific Method 5
BASIC CHARACTERISTICS OF LIVING
  ORGANISMS 5
Cells and Organization 5
Metabolism 5
Homeostasis and Death 5
Reproduction and Heredity 7
Irritability 7
Evolution 7
CLASSIFYING THE LIVING WORLD 7
KINGDOMS OF LIVING THINGS 8
SUMMARY 10
THE NEXT DECADE: The Continuing Search
  for Life 11
FOR DISCUSSION AND REVIEW 12
SUGGESTED READINGS 12

CHAPTER TWO
The Chemical Basis of Life 13

THE STATES OF MATTER 13
THE STRUCTURE OF MATTER 14
Covalent Bonds 17
Hydrogen Bonds 17
CONCEPT SUMMARY: Six Elements Comprise 99
  Percent of Living Matter 14
BOX 2-1: Chemical Symbols 17
CHEMICAL REACTIONS 21
BOX 2-2: Acids and Bases 21
ORGANIC COMPOUNDS 21
Proteins 23
BOX 2-3: Enzymes 23
Carbohydrates 24
Lipids 26
Nucleic Acids 26
THE CHEMICAL ORGINS OF EARTH
  AND LIFE 27
Primordial Soup 27

BOX 2-4: Spontaneous Generation 30
THE NEXT DECADE: Your Chemical
  Environment 32
FOR DISCUSSION AND REVIEW 33
SUGGESTED READINGS 33

SECTION TWO

THE BIOLOGY OF CELLS

CHAPTER THREE
Cell Structure 36

STUDYING CELLS 37
ANIMAL CELL STRUCTURE 38
The Cell Membrane 40
The Endoplasmic Reticulum 40
The Golgi Complex 41
Lysosomes 41
Mitochondria 43
Microfilaments, Microtubules, and Associated
  Organelles 43
The Nucleus 44
PLANT CELL STRUCTURE 46
The Cell Wall 46
BOX 3-1: Abnormal Cells 46
Plastids 48
Vacuoles 50
Bacteria 50
Viruses 51
CONCEPT SUMMARY: Cell Organelles and Their
  Functions 50
SOME OTHER DESIGNS 50
CONCEPT SUMMARY: Differences Between Plant
  and Animal Cells 52
SUMMARY 53
THE NEXT DECADE: The Cells of Henrietta Lacks 54
FOR DISCUSSION AND REVIEW 55
SUGGESTED READINGS 55

CHAPTER FOUR
Cell Functions 56

ENERGY 57
PHOTOSYNTHESIS 57

BOX 4-1: The Power Molecule 58
The Light Reaction 60
The Dark Reaction 61
HIGHLIGHT: C4 Plants 63
CELLULAR RESPIRATION 63
Aerobic Respiration 63
Glycolysis 64
The Krebs Cycle 64
The Electron Transport System 64
Anaerobic Respiration 67
CONCEPT SUMMARY: Comparison of the Processes
  of Photosynthesis and Aerobic Respiration 68
CELL PERMEABILITY AND ACTIVE
  TRANSPORT 68
Water and Its Transport 68
Diffusion and Osmosis 69
Cell Walls 70
Contractile Vacuoles 71
Environments 71
Membrane Transport of Large Molecules 71
Phogocytosis 72
Pinocytosis 72
Facilitated Diffusion 72
Active Transport 73
SUMMARY 73
THE NEXT DECADE: The Continuing Exploration
  of Cells 75
FOR DISCUSSION AND REVIEW 76
SUGGESTED READINGS 76

CHAPTER FIVE
Cell Reproduction 77

THE CELL CYCLE 77
Mitosis 79
Prophase 79
Metaphase 80
Anaphase 80
Telophase 80
Cytokinesis 81
Plant Cells 81
HIGHLIGHT: Chromosome Numbers 82
MEIOSIS 82
Meiosis I 82
Meiosis II 85
BOX 5-1: Random Assortment 86
Meiosis in Plants 87
CONCEPT SUMMARY: Comparison of Mitosis
  and Meiosis 87
CELLULAR DIFFERENTIATION 88
CELL DEATH 88
SUMMARY 89
THE NEXT DECADE: The Fight Against Cancer 90
FOR DISCUSSION AND REVIEW 92
SUGGESTED READINGS 93

SECTION THREE

ANIMAL AND PLANT STRUCTURE AND FUNCTION

CHAPTER SIX
Human Organ Systems: Integumentary, Skeletal,
and Muscular 94

TISSUES 94
Epithelial Tissue 94
Connective Tissue 95
Muscle Tissue 96
Nerve Tissue 96
The Integumentary System 96
THE SKELETAL SYSTEM 97
HIGHLIGHT: Suntan or Sunburn 98
Bone 99
Bone Development 100
Joints 101
HIGHLIGHT: Aching Joints 102
THE MUSCULAR SYSTEM 103
Muscle Contraction 104
CONCEPT SUMMARY: Comparison of Smooth,
  Cardiac, and Skeletal Muscle 105
Types of Movement 106
BOX 6-1: Muscles and Skeletons of Animals Other Than
  Humans 107
THE NEXT DECADE: Fighting Paralysis and Replacing
  Lost Limbs 110
FOR DISCUSSION AND REVIEW 111
SUGGESTED READINGS 111

CHAPTER SEVEN
Human Organ Systems: Digestive, Respiratory and
Circulatory 112

DIGESTION AND NUTRITION 112
The Digestive System 112
HIGHLIGHT: Ulcers and Gallstones 116
Nutrition 117
Carbohydrates 117
BOX 7-1: Feeding and Digesting in the Animal
  Kingdom 118
Proteins 120
Fats 120
Vitamins 122
Minerals 122
CONCEPT SUMMARY: Digestion and Absorption
  of Basic Foods 122
THE RESPIRATORY SYSTEM 123
Breathing 126
BOX 7-2: Respiratory Devices in Aniamls Other Than
  Humans 127

External Respiration 128
HIGHLIGHT: Smoking and Your Lungs 129
CONCEPT SUMMARY: Comparison of Arteries
 and Veins 130
THE CIRCULATORY SYSTEM 132
The Heart 132
The Pacemaker 133
Blood Pressure 134
The Systemic Circulation 134
BOX 7-3: Heart Disease 136
BOX 7-4: The Circulatory Systems of Animals Other
 Than Humans 137
Blood 138
The Lymphatic System 138
THE IMMUNE SYSTEM 140
Immunity 142
SUMMARY 142
THE NEXT DECADE: Will A Body Scanner Replace
 Your Doctor? 144
FOR DISCUSSION AND REVIEW 145
SUGGESTED READINGS 146

CHAPTER EIGHT
Human Organ Systems: Excretory, Nervous,
and Endocrine 147

THE CONCEPT OF HOMEOSTASIS 147
BOX 8-1: Endotherms and Ectotherms 148
THE EXCRETORY SYSTEM 149
HIGHLIGHT: The Artificial Kidney 151
CONCEPT SUMMARY: Events During Urine
 Formations 152
BOX 8-2: Excretion and Osmoregulation for Organisms
 Other Than Humans 153
THE NERVOUS SYSTEM 155
Central Nervous System 155
Peripheral Nervous System 158
BOX 8-3: The Nervous Systems of Organisms Other
 Than Humans 159
HIGHLIGHT: Drugs and the Nervous System 160
The Neuron 161
Sensory Neurons 161
Interneurons 161
Motor Neurons 161
The Senses 163
BOX 8-4: Brain Chemistry 164
Taste 164
Smell 165
Vision 165
Hearing and Balance 167
THE ENDOCRINE SYSTEM 168
Hormone Activity 169
BOX 8-5: Chemical Coordination 171
Prostaglandins 171

SUMMARY
THE NEXT DECADE: Your Aging and Aching
 Body 173
FOR DISCUSSION AND REVIEW 174
SUGGESTED READINGS 174

CHAPTER NINE
Plant Structure and Functions 175

PLANT TISSUES 175
Meristematic Tissue 177
Permanent Tissues 177
Epidermis 177
Parenchyma 177
Sclerenchyma 177
Collenchyma 178
Cork 178
HIGHLIGHT: Hemp 179
Xylem 180
Phloem 181
Celery 181
CONCEPT SUMMARY: Functions of Plant Tissues 181
PLANT ORGANS 182
Leaves 182
Stems 183
Roots 184
PLANT SYSTEMS 186
Photosynthesis and Respiration 186
Transpiration 186
BOX 9-1: EDIBLE PLANT PARTS 187
Digestion 189
Translocation 190
Hormones 190
Reproduction
SUMMARY 191
THE NEXT DECADE: Future Products From
 Plants 193
FOR DISCUSSION 194
SUGGESTED READINGS 194

SECTION FOUR

REPRODUCTION AND DEVELOPMENT

CHAPTER TEN
Animal and Plant Reproduction 196

ASEXUAL REPRODUCTION
Asexual Reproduction of Unicellular Organisms 196
Asexual Reproductions of Multicellular Organisms 197
Asexual Reproduction of Plants 198
HIGHLIGHT: Food For Throught 198
Limitations of Asexual Reproduction 199

Economic Importance of Asexual Reproduction   199
SEXUAL REPRODUCTION IN ANIMALS   201
External Fertilization   202
HIGHLIGHT: INTERNAL OR EXTERNAL
    FERTILIZATION?   203
External Fertilization with Amplexus   203
Internal Fertilization without Copulation   203
Internal Fertilization with Copulation   204
Hermaphroditism   204
Parthenogenesis   205
Alternating Sexes   205
SEXUAL REPRODUCTION IN HIGHER
    PLANTS   205
The Mosses   206
The Ferns   206
Seed Plants   208
Angiosperms   209
BOX 10-1: Hay Fever   209
BOX 10-2: Fruit or Vegetable   213
Gymnosperms   214
CONCEPT SUMMARY: Comparison of Sporophyte
    and Gametophyte Generations of Mosses, Ferns, and
    Seed Plants   216
SEXUAL REPRODUCTION IN SINGLE-CELLED
    ORGANISMS   216
SUMMARY   218
THE NEXT DECADE: Cloning   220
FOR DISCUSSION AND REVIEW   221
SUGGESTED READINGS   221

CHAPTER ELEVEN
Human Sexuality and Reproduction   222

THE MALE REPRODUCTIVE SYSTEM   222
External Anatomy   223
The Penis   223
The Scrotum   224
Internal Anatomy   224
Sperm Production, Semen, and Ejaculation   225
Hormonal Control of Male Sexuality   227
THE FEMALE REPRODUCTIVE SYSTEM   228
External Anatomy   228
HIGHLIGHT: Scents and Sex   229
Internal Anatomy   230
BOX 11-1: Estrus   231
Vagina   232
Cervix   232
Fallopian Tubes   232
BOX 11-2: Twenty-Eight Days   233
Egg Production and Ovulation   233
Puberty   234
Menstrual Cycle   234
Menopause   237
BIRTH CONTROL   238

CONCEPT SUMMARY: Events in The Menstrual
    Cycle   238
Mechanical Methods of Birth Control   238
Surgical Methods of Birth Control   238
Male Sterilization   238
Female Sterilization   238
Abortion   238
Chemical and Hormonal Methods of Birth Control   239
Natural Methods of Birth Control   239
Rhythm, or Calandar, Method   242
Basal Body Temperature Method (BBT)   242
Mucus (Ovulatory) Method   242
Coitus Interruptus   243
Abstinence   244
SEXUALLY TRANSMITTED DISEASES   244
Gonorrhea   244
Syphilis   245
Herpes   246
AIDS (Aquired Immune Deficiency Syndrome)   246
SUMMARY   247
THE NEXT DECADE: Alternate Routes to
    Reproduction   249
FOR DISCUSSION AND REVIEW   252
SUGGESTED READINGS   252

CHAPTER TWELVE
The Development of Vertebrates   253

VERTEBRATE EMBRYO DEVELOPMENT   253
Fertilizatoin   254
Cleavage   255
Gastrulation   255
BOX 12-1: OVISTS AND SPERMISTS   255
Neurulation   256
Morphogenesis   256
CONCEPT SUMMARY: Primary Germ Layers
    and Their Fates   256
TERRESTRIAL VERTEBRATE
    DEVELOPMENT   256
BOX 12-2: Cellular Differentation and Inducers   257
HUMAN EMBRYONIC DEVELOPMENT   258
Fertilization   258
Cleavage   259
Gastrulation   259
Implantation   260
Neurulation   261
Placenta   261
Extraembroyonic Membranes   261
SEQUENCE OF DEVELOPMENT   262
First Trimester   262
Second Trimester   263
Third Trimester   264
HIGHLIGHT: Homologies   266
CHILDBIRTH   266

HIGHLIGHT: Methods of Childbirth 269
TWINNING 270
HIGHLIGHT: Siamese Twins 271
DEVELOPMENTAL ANOMALIES 272
Teratology 272
BOX 12-3: QUESTIONS ABOUT CHILDBIRTH 272
Hormonal Anomalies 275
SUMMARY 275
THE NEXT DECADE: Fetal Therapy 277
FOR DISCUSSION AND REVIEW 278
SUGGESTED READINGS 278

SECTON FIVE

GENETICS

CHAPTER THIRTEEN
Mendel and the Principles of Genetics 280

GREGOR MENDEL 281
MENDELIAN GENETICS 282
The Law of Segregation 282
BOX 13-1: Mendel's Peas and Genetic Symbols 283
Principle of Independent Assortment 285
CONCEPT SUMMARY: Mendel's Conclusions 284
GENETICS AFTER MENDEL 286
SUMMARY 288
THE NEXT DECADE: Future Food 289
FOR DISCUSSION AND REVIEW 290
SUGGESTED READINGS 290

CHAPTER FOURTEEN
Types of Inheritance 291

MENDELIAN INHERITANCE 291
Fruit Fly Inheritance 291
BOX 14-1: Solving Genetic Problems 293
Human Inheritance 295
BOX 14-2: Pedigrees 296
MECHANICS OF NON-MENDELIAN
  INHERITANCE 298
Linked Alleles 299
Incomplete Dominance 299
X AND Y CHROMOSOMES 300
BOX 14-3: More Genetic Symbols 301
Sex Determination 301
Dosage Compensation 302
SEX-LINKED INHERITANCE 303
Hemophilia 303
Color-Blindness 304
SEX-LIMITED INHERITANCE 305
SEX-INFLUENCED INHERITANCE 306

MULTIPLE-ALLELIC INHERITANCE 307
POLYGENIC INHERITANCE 308
Skin Pigmentation 309
GENE INTERACTIONS 310
Albinism 310
Pleiotrophy 311
SUMMARY 311
THE NEXT DECADE: Genes and Cancer 312
FOR DISCUSSION AND REVIEW 313
SUGGESTED READINGS 314

CHAPTER FIFTEEN
Molecular Genetics 315

DNA 315
Replication 319
DNA and Chromosomes 320
RNA 321
PROTEIN SYNTHESIS 321
The Genetic Code 321
Transcription 322
Translation 322
CONCEPT SUMMARY: Events of Protein
  Synthesis 325
The One-Gene, One-Enzyme Principle 325
MUTATIONS 325
Sickle-Cell Anemia 326
GENETIC ENGINEERING 327
BOX 15-1: Another Type of Sickle-Cell Anemia? 328
SUMMARY 329
THE NEXT DECADE: New Applications of Genetic
  Engineering 331
FOR DISCUSSION AND REVIEW 332
SUGGESTED READINGS 332

CHAPTER SIXTEEN
Human Genetics 333

POINT MUTATIONS
Cystic Fibrosis 333
CONCEPT SUMMARY: Causes of Some Important
  Human Genetic Disorders 334
HIGHLIGHT: Genes and Heart Disease 334
Muscular Dystrophy 335
Tay-Sachs Disease 335
Phenylketonuria 336
BOX 16-1: Amniocentesis 337
CHROMOSOMAL ANOMALIES 338
Autosomal Trisomics 338
Down's Syndrome 338
Other Autosomal Trisomics 342
BOX 16-2: The Intelligence Debate 340
Sex Chromosomal Trisomics 342

Klinefelter's Syndrome 342
Trisomy-X 342
XYY Males 342
Sex Chromosomal Monosomy 343
Turner's Syndrome 343
SUMMARY 344
THE NEXT DECADE: Genetic Counseling 345
FOR DISCUSSIONS AND REVIEW 346
SUGGESTED READINGS 346

SECTION SIX

EVOLUTION

CHAPTER SEVENTEEN
Darwin and the Evidence of Evolution 348

DARWIN'S DISCOVERY 348
The Beagle 350
HIGHLIGHT: Galapagos Islands 352
The Origin of Species 353
The Scopes Trial 354
THEORIES OF EVOLUTION 355
CONCEPT SUMMARY: Major Points of the Darwin
and Wallace Theory on the Origin of Species by Natural
Selection 356
THE EVIDENCE OF EVOLUTION 356
Fossils 356
Rock Strata 358
BOX 17-1: Dating Fossils 360
Contenental Drift 360
The Geological Time Scale 362
Compartive Studies 362
Comparative Anatomy 362
Comparative Embryology 364
Vestigial Structure 365
Comparative Protein Chemistry 366
THE NEXT DECADE: Social Darwinism 369
FOR DISCUSSION AND REVIEW 370
SUGGESTED READINGS 370

CHAPTER EIGHTEEN
The Mechanisms of Evolution 371

NATURAL SELECTION 371
Directional Selection 372
Stabilizing Selection 372
Disruptive Selection 373
Divergent Evolution 373
Parallel Evolution 373
Convergent Evolution 374
SPECIATION 374
Species-Isolating Mechanism 375

Geographic Isolation 375
Seasonal Isolation 376
Ecological Isolation 377
Ethological Isolation 377
Mechanical Isolation 378
Gametic Isolation 378
Hybrid Iviability 378
Hybrid Sterility 379
The Origin of Species 380
Hybrid Speciation 380
Allopatric Speciation 380
Sympartric Speciation 381
THE HARDY-WEINBERG LAW 382
CONCEPT SUMMARY: The Hardy-Weinberg
Law 382
BOX 18-1: Mutations 383
EVOLUTION IN ACTION 385
THE HUMAN SPECIES 387
SUMMARY 387
THE NEXT DECADE: Unnatural Selection? 389
FOR DISCUSSION AND REVIEW 390
SUGGESTED READINGS 390

CHAPTER NINETEEN
Our Primate Heritage 391

EVOLUTION OF THE PRIMATES 392
CONCEPT SUMMARY: Classification of Humans 393
Opposable Thumbs 393
Locomotion 393
Perception 396
The Cerebrum and Behavior 397
EVOLUTION OF THE HOMINOIDS 398
SUMMARY 400
THE NEXT DECADE: Primate Predictions 401
FOR DISCUSSION AND REVIEW 402
SUGGESTED READINGS 402

CHAPTER TWENTY
Human Evolution 403

ARCHEOLOGICAL EVIDENCE 403
Olduvai Gorge 404
BOX 20-1: Naming Fossils 404
Skull 1470 406
THE EVOLUTION OF HOMINIDS 406
BOX 20-2: The Rest of the Story About the Evolution of
Hominids 408
Homo Erectus 409
CONCEPT SUMMARY: Major Points of Richard
Leakey's Views of Human Evolution 411
Migrations 411
BOX 20-3: Cave Dweller Capers 411
Neanderthals 411

CONTENTS

Cro-Magnons 413
Dispersion 414
SUMMARY 414
THE NEXT DECADE: Lucy's Legacy Shakes the Human
   Family Tree 416
FOR DISCUSSION AND REVIEW 417
SUGGESTED READINGS 417

SECTION SEVEN

ECOLOGY AND ETHOLOGY

CHAPTER TWENTY-ONE
The Biosphere 420

AN OVERVIEW 421
BOX 21-1: Climate 422
TERRESTRIAL ENVIRONMENTS 424
Tropical Rainforests 425
Grasslands 425
Deserts 425
Temperate Deciduous Forests 426
Coniferous Forests 426
Tundra 426
AQUATIC ENVIRONMENTS 427
CONCEPT SUMMARY: Characteristics of Terrestrial
   Biomes 428
Oceans 428
BOX 21-2: Ocean Water and Fresh Water 429
Estuaries 432
Freshwater Habitats 432
BOX 21-3: Water, Density, and Life 432
SUMMARY 434
THE NEXT DECADE: Further Biome Destruction 435
FOR DISCUSSION AND REVIEW 436
SUGGESTED READINGS 436

CHAPTER TWENTY-TWO
Ecosystems and the Physical Environment 437

WEATHER, CLIMATES, AND
   MICROCLIMATES 438
SUBSTRATES AND MEDIA 438
ENERGY AND ECOSYSTEMS 439
Producers, Consumers, and Decomposers 439
Food Pyramids 441
BOX 22-1: The Life of a Diatom 441
BIOGEOCHEMICAL CYCLES 443
Water Cycle 444
CONCEPT SUMMARY: Trophic Relationships Among
   Living Organisims 444
Carbon Cycle 445
Oxygen Cycle 446
Nitrogen Cycle 446
Phosphorus Cycle 448

SUMMARY 449
THE NEXT DECADE: Overload Cycles 450
FOR DISCUSSION AND REVIEW 451
SUGGESTED READINGS 451

CHAPTER TWENTY-THREE
Communities and the Biotic Environment 452

COMPETITION AND COEXISTENCE 453
TROPHIC RELATIONSHIPS 454
Prey-Predator Interactions 454
Herbivores and Plant Populations 455
Symbiosis 456
Parasitism 456
Mutualism 458
Commensalism 459
CONCEPT SUMMARY: Forms of Symbiosis 459
COMMUNITY STABILITY 459
BOX 23-1: The Gypsy Moth Plague 461
SUMMARY 462
THE NEXT DECADE: Integrated Pest
   Management 463
FOR DISCUSSION AND REVIEW 464
SUGGESTED READINGS 464

CHAPTER TWENTY-FOUR
Populations and Human Ecology 465

HUMAN POPULATIONS 467
CONCEPT SUMMARY: Factors That Affect Population
   Size 467
HUMAN ECOLOGY 469
Food Supply and World Nutrition 470
Irrigation 470
BOX 24-1: Marasmus and Kwashiorkor 471
Soil Erosion and Conservation 472
Green Revolution 473
Aquaculture 473
HIGHLIGHT: Americans, Meat, and the World Grain
   Supply 473
Nutrition 474
Energy and Nonrenewable Resources 474
Fossil Fuels 475
Nuclear Power 476
Solar Energy 477
Hydroelectric Power 477
Wind Power 477
Geothermal Energy 477
Energy from Garbage 477
Conservation 478
Pollution 478
Air Pollution 478

Water Pollution 482
Pesticide and PCB Pollution 484
Other Infamous Toxins 486
Radiation 486
Thermal Pollution 486
Noise Pollution 486
Food Pollution 486
The Future of Homo Sapiens 487
Box 24-2: Acid Rain 481
SUMMARY 487
THE NEXT DECADE: Population of the United
   States 488
FOR DISCUSSION AND REVIEW 489
SUGGESTED READINGS 489

CHAPTER TWENTY-FIVE
The Biology of Behavior 490

MODAL ACTION PATTERNS 490
RELEASING STIMULI 492
Visual Releasers 492
Auditory Releasers 494
HIGHLIGHT: Human Visual Releasers 495
Pheromones 496
HIGHLIGHT: Human Pheromones 497
CONCEPT SUMMARY: Types of Releasing
   Stimuli 498
THE UMWELT 498
Stimulus Filtering 498
Motivation 498
Nature versus Nurture 499

Imprinting 499
TERRITORY 499
Territorial Marking 500
Functions of Territories 501
HIGHLIGHT: Human Territoriality 502
AGGRESSIVE BEHAVIOR 503
Threat 503
Confrontation 503
Appeasement 504
Diversion 504
Displacement 504
CONCEPT SUMMARY: Responses to Threat
   Behavior 505
DOMINANCE HIERARCHIES 505
SEXUAL BEHAVIOR 506
Pair Formation 506
Courtship 506
Precopulatory Behavior 507
Copulatory Behavior 507
SUMMARY 507
THE NEXT DECADE: Sociobiology 509
FOR DISCUSSION AND REVIEW 510
SUGGESTED READINGS 510

APPENDIX 1   The Metric System 513
APPENDIX 2   The Major Phyla of Living
   Organisms 515
GLOSSARY 523
INDEX 545

Introductory biology books fall into distinct categories. Some follow the traditional approach to biology, and contain lengthy surveys of the plant and animal kingdoms. Others abandon the traditional approach, with the idea that no one book can fit every professor's approach to a general biology course. The latter type concentrates on "selected issues" and reflects the training and discipline of the authors. Some years ago, for example, relevance became the fad: Earth Day rallies and increased concern for our environment were followed by a raft of introductory biology textbooks that were environmentally or ecologically oriented. Newer on the scene are the human biology textbooks, which have a strong human orientation but often neglect the important principles and concepts of biology.

This introductory book is a compromise. It is not an encyclopedia of current facts and statistics nor a definitive reference work. Instead, it has been written for students enrolled in one- or two-semester biology courses, often for general education credit. Students who read this book probably will not become biologists or even scientists; most are taking their first biology course. Nevertheless, they need to become more aware of their role in a complex biological system and to better understand the world in which they live. If the students who read this book grasp only the suggestion that *Homo sapiens* (the human being) is not the master manipulator of the environment, but rather only a single species, depending on and interacting with thousands of other organisms, our efforts will have been rewarded.

The goal, then, is not to provide overall coverage of biological phenomena; even a feeble attempt would create an immense volume. Therefore, some aspects of biology are not treated in this book, and other topics are tersely treated. But some topics receive full, detailed coverage.

Organization of the Book

In most cases, sections are autonomous, meaning that the professor can assign the sections in any order that suits the organization of a particular course. As far as possible, the chapters of each section are autonomous as well. Autonomy, however, does not necessarily mean that the information in one chapter is independent of information in another chapter. Each chapter contains numerous cross-references to specific pages. These references are signaled by a small box in the text with corresponding page numbers appearing in a larger box in the margin. They signal that the referenced material is important in understanding the full implications of the concept being discussed. Students can quickly consult the cross-references to determine whether this material has already been covered in their course.

The chapters of each section follow a regular format. In most cases, the section begins with a brief chapter about the historical background of the principle or concept to be discussed. For beginning students in particular, it is important to understand that biology is not a discipline in isolation but one molded by thinking in many other areas. This introductory chapter also contains or is followed by a general overview of the section. Following chapters present many plant and animal examples—some can readily be observed, others will be new to students, unusual, and possibly verge on the bizarre. Each section concludes with a reexamination of concepts or principles that have special relevance to daily human existence. In some cases, both animal and human examples appear, so that students will not miss obvious parallels in biology.

The Preview Questions located at the beginning of each chapter are a new feature in this edition. These questions serve to interest and orient the student by indicating the most important topics in each chapter.

The chapters contain a number of special features. The terms that appear in boldface are defined in the glossary or where they appear in the text. Each chapter is profusely illustrated with diagrams, charts, and photographs. Chapters also contain information that is set off from the main body of the text as two types of boxed material. Information set off in a series of numbered boxes expand upon topics discussed in the text and give the topics expanded special emphasis. The second series of boxed information called Highlights contain information that may not have been mentioned in the test, but is of high-interest and will serve to motivate students. Concept Summaries are another new feature in this edition which graphically provide a clear review of one or two major topics within each chapter.

At the end of the text of each chapter is a point-by-point summary, so students can quickly review the contents. The summaries should make excellent outlines for students who take notes while reading. Summaries are followed by sections called The Next Decade, which attempt to forecast the future on the basis of our current biological knowledge. These sections are based on factual information even though the narratives sometimes are whimsical. They are followed by sections labeled For Discussion and Review which should help students pull together the major concepts of each chapter and provide useful topics for oral discussion. Finally, each chapter ends with Suggested Readings. The references for further reading are few in number but have been selected with students' convenience in mind. The Suggested Readings are the more interesting popular books and paperbacks found in most campus bookstores. References to *The American Scientist, Science,* and other scientific journals are at a minimum, because most beginning students have difficulty comprehending the articles in them. The Suggested Readings, however, should stimulate interest.

Other devices included to stimulate interest are the sections of colored photographs which are called Portfolios. Each Portfolio develops and illustrates an important theme or topic in the text.

Two appendixes are found at the end of the text. The first deals with the metric system of measurements with which students should become familiar, and the second briefly surveys the major phyla of living organisms.

The index includes all major principles and concepts, their component parts, and references to illustrations. The entries are differentiated so the student can quickly determine whether page references contain expanded coverage, word reference only, or illustrations.

The text is accompanied by a laboratory manual, a student study guide, and instructor manuals.

It is hoped that this format helps make biological concepts more accessible to those who are not biologists.

Finally, thanks are do to the many people who helped in the completion of this text and without whose assistance the task would have been far more difficult, if not impossible. The following reviewers have provided encouragement and pertinent, insightful and useful comments.

George S. Arita
Ventura College, California

Richard A. Boutwell
Missouri Western State College

Professor Mario Caprio
Suffolk County Community College, New York

Mel Cundiff
University of Colorado, Boulder

Professor Paul Desha
Tarant County Junior College, Texas

Professor Eugenia DeWitt
Los Angeles Valley College

Professor Wallace Dixon
Eastern Kentucky University

Professor Charles W. Gaddis
University of Arizona

Professor Mel Gorelik
Queensborough Community College, New York

Paul H. Gurn
Mattatuck Community College, Connecticut

Professor Lynn M. Hansen
Modesto Junior College, California

J. Rogert Harkrider
Cypress College, California

Professor Joseph Hindman
Washington State University

Professor Martin Ikkanda
Los Angeles Pierce College

Professor Arnold J. Karpoff
University of Lousville, Kentucky

Professor Robert A. Koch
California State University, Fullerton

William H. Leonard
University of Nebraska, Lincoln

Jon R. Maki
Eastern Kentucky University

Professor Marilyn Neulieb
Urbana College, Ohio

Aryan I. Roest
California Polytechnic State University, San Luis Obispo

Professor Donald Scoby
North Dakota State University

Burton Staugaard
University of New Haven, Connecticut

Professor M. L. Stehsel
Pasadena City College

Charles L. Vigue
University of New Haven, Connecticut

Fred H. Whittaker
University of Louisville, Kentucky

With a great deal of sadness I wish to report the sudden and unexpected death of the senior author, Jack A. Ward, in early December 1982. It was his enthusiasm for teaching and deep interest in students that led to the development of this text. Without his leadership and drive the completion of this revision has been more than doubly difficult. Without the help, hard work, and good spirits of my friends and colleagues, Marie Knight, Carol Ober, Mark Lamon, and G. Richard Hogan, the task would have taken months longer and would have been exceedingly more difficult. Finally, I am indebted to William Stryker, Administrative Editor, West Publishing Company, for his good sense of humor, high degree of professionalism and great patience.

Howard R. Hetzel

SECTION ONE

INTRODUCTION

The illustration above is a piece of baked clay, fashioned by a Greek artist in the fourth century B.C. The four fish and two shells represented comprise what is perhaps the first attempt to classify living forms—taxonomy.

1

SCIENCE AND THE LIVING WORLD

After reading this chapter, you should be able to answer the following questions:

1. What are the primary steps in the scientific method?

2. What are the unique features of living things?

3. Why are living organisms classified into groups and given scientific names?

4. What are the five kingdoms of living organisms? Which kinds of organisms are included in each one?

You are probably taking this introductory biology course for many reasons. One reason, we hope, is that you are interested in the living world and in yourself as a living organism. The living world envelops you, and you are part of it. You see a great variety of living organisms: trees, shrubs, flowers, birds, insects, pets, other people. You are also surrounded by microscopic organisms that are invisible to the naked eye: bacteria, protozoa, fungi, yeasts (not to mention the viruses you harbor in your own cells). If you search the environment carefully, you can find dozens of other organisms that you never knew existed.

People have always been interested in the world around them. At first, their interests centered on ways to get food from it. As culture and knowledge expanded, people began to appreciate the beauty of nature and to use living and nonliving resources as efficiently as they could. Agricultural techniques grew more sophisticated, crafts and industries developed, and science emerged as a major factor. Today, science plays a highly visible and dynamic role in human cultures. Magazines and newpapers contain articles on such topics as test-tube babies, cloning, pollution, famine, carcinogens and mutagens, endangered species, genetic engineering, the energy crisis, birth control, abortion, and on and on. As a citizen, you make judgments about these issues. We believe that a knowledge of the principles of biology will help you make more intelligent judgments about many complex human problems.

By studying biology, you should also gain an understanding of what science is and what scientists do. People often fear and distrust science, blaming it for many of the world's problems. Such feelings stem from a failure to understand how science works.

SCIENCE AND SCIENTISTS

Unlike the fields of religion and philosophy, which deal with abstract concepts, science focuses solely on phenomena that can be observed (directly or indirectly), measured, and tested. Conclusions are drawn from these observations and data, but no ethical or moral judgments are made about them.

For example, scientists have learned the causes of certain birth defects and have developed techniques to test human fetuses before birth to determine whether they have these defects. This is useful information. If parents find out that their unborn child has a serious birth defect, they may decide on an abortion or they may decide to allow the pregnancy to continue. The moral or ethical decision concerning the fate of the fetus is the parents'—and it is not scientific. Science cannot prove that abortion is desirable in the case of birth defects. It can, however, determine what causes a birth defect and predict the parents' odds of producing children with the defect. The facts are determined through objective scientific observation. The predictions are based on the established laws of genetics.

People who become scientists do so for a variety of reasons. Common to all these individuals are a deep curiosity about some aspect of nature and a strong motivation to expand knowledge about the subject (see Figure 1–1). By establishing cause-and-effect relationships among phenomena, scientists generate the knowledge that allows them to create guidelines for predictions about related phenomena. Science means knowledge, but it is knowledge gained in a special way. Scientists make their observations and conduct their studies in a systematic and logical way called the **scientific method.** The important steps in the scientific method can be illustrated with the following example.

Step 1. *Making observations.* The scientific method begins with objective observations of natural phenomena.

Example: A microbiologist observes that cultures of a certain species of bacteria grow more rapidly under constant illumination than under partial illumination.

Step 2. *Formulating the hypothesis.* During hypothesis formulation, the scientist uses inductive reasoning to create a possible explanation for the phenomenon observed. In this case, a cause-and-effect relationship be-

Figure 1–1

Four famous scientists. (a) Sir Alexander Fleming (1881–1955), the discoverer of penicillin, won the Nobel Prize for Medicine and Physiology in 1945. (b) Madame Marie Curie (1867–1935), pioneer in the field of radioactive chemicals and discoverer of the elements radium and polonium, won the Nobel Prize for Physics in 1903 and for Chemistry in 1911. (c) James D. Watson (b. 1928) and Francis H. C. Crick (b. 1916) shared the Nobel Prize for Medicine and Physiology in 1962 with Maurice H. F. Wilkins for discovering the molecular structure of the genetic material DNA. (The Bettmann Archive.)

(a)　　　　(b)　　　　(c)

tween the duration of illumination and the rate of culture growth could be proposed. This relationship, apparently the best explanation of the available facts, is called a **hypothesis.**

Example of a hypothesis: The length of time a culture receives illumination is positively correlated with the rate of culture growth. The longer the period, the more rapid the growth rate.

A hypothesis about the relationship between two aspects of an observation should be stated in a way that makes it possible to test its degree of validity. The testing, which often takes the form of an experiment, may produce evidence that either supports or disproves the hypothesis. However, a hypothesis is not something that can be proved. It can only be demonstrated under specific experimental conditions. In the example here, to fully prove the hypothesis, the scientist would have to observe the entire world population of this species of bacteria under the given cultural conditions—an impossible task. As more and more evidence accumulates in support of a hypothesis, we come closer and closer to the truth.

Step 3. *Testing the hypothesis.* Using deductive reasoning, the scientist devises a means to test the hypothesis. As is commonly done, in this case the scientist designs a **controlled experiment** to test the relationship between illumination and growth rate. Such an experiment makes it possible to test only one factor or variable at a time, holding all others constant. In our example, all but one of the factors that might affect the growth rate of the bacteria are held constant. The variable to be tested is the duration of illumination.

Example of controlled experiment: If the duration of illumination affects the growth rate, then cultures maintained in the dark should grow slowly or not at all. Two groups of cultures are set up. Both have the same number and type of bacteria, the same type of culture dish, the same type of growth medium, and so on. Both are incubated for the same length of time and at the same temperatures. But one set of cultures, the **controls,** is incubated under constant illumination, while the second set, the **experimental group,** is incubated in total darkness. Presumably, the only difference between the experimental group and the control group is the illumination. At the end of the incubation period, both sets of cultures are examined. If the cultures under constant illumination show

greater growth than do the cultures kept in the dark, the hypothesis will be supported and further experiments can be designed to gain more support for it.

The further experimentation may be done by setting up groups of cultures under different durations of illumination, such as three, six, twelve, and eighteen hours in each twenty-four hour period. The growth rates of these cultures can then be compared with each other and with the rates for cultures under constant illumination. But what happens if there are no differences in growth rate between the experimental and control cultures? If both grow at the same rate, the hypothesis will be disproved and an alternative hypothesis, based on further observations, will be sought.

Step 4. *Further observations and an alternative hypothesis.* If the original hypothesis is disproved, the scientist will reexamine the situation, seeking new information on which to base an alternative hypothesis. Maybe the cultures under constant illumination are slightly warmer than those under partial illumination. Therefore, maybe temperature is the variable that affects growth rate.

Example of alternative hypothesis: Higher temperatures are positively correlated with higher growth rates.

Step 5. *Testing the alternative hypothesis.* Again, an experiment is designed to control all the variables that may affect the growth rate of the bacterial cultures, except temperature. Three identical sets of cultures are set up. All are incubated for the same length of time under the same amount of illumination. However, each is incubated at a slightly different temperature. If, at the end of the incubation period, the culture at the highest temperature shows a growth rate greater than that of the culture at the next lowest temperature, and the culture at the lowest temperature shows the lowest growth rate, the alternative hypothesis is supported and additional experiments to study culture growth in relationship to temperature can be planned. One possible proposal is that there is an optimal temperature for maximum culture growth and that temperatures above and below it retard growth.

As hypotheses are formulated, tested, modified, expanded, and retested, scientists obtain ever more refined and accepted explanations of the phenomenon

SCIENTIFIC METHOD

Step 1   Make objective observations.

Step 2   Using inductive reasoning, formulate a hypothesis.

Step 3   Test the hypothesis under carefully controlled conditions—for example, use a controlled experiment.

Step 4   On the basis of the outcome of the test, either reject, accept, or modify the hypothesis.

Step 5   If the hypothesis is rejected, formulate an alternative hypothesis for testing.

under study. Ultimately, the body of knowledge becomes strong enough for the scientists to make predictions about related situations and observations. For example, you may propose that if temperature affects the growth rate of one bacterial species, it may affect the growth rate of other species in the same way. This may or may not be the case, but the thought establishes the basis for further hypotheses. A hypothesis that receives wide acceptance by the scientific community is often referred to as a **theory** and sometimes a law. (One of the most important and unifying theories in biology is Darwin's theory of evolution by natural selection.)

Scientific inquiry is usually hard work, and the constant repetition of experiments often proves dull. What, then, motivates scientists? The answer is satisfaction with their new knowledge. New knowledge is not always immediately applicable to human life. Many scientists study such odd groups of animals or plants and such unusual phenomena that their efforts may seem a waste of time. For instance, studying molds and fungi may seem silly. However, early studies of these organisms led to the discovery of antibiotics, which improved human health and life expectancy throughout the world. Studies that do not apply directly to humans are called **basic research.** Studies that have obvious human application are called **applied research.** The principles of biology are derived from basic research, which also forms the bases on which applied research can build. Thus, applied research would be impossible without basic research.

This semester will involve concentration on biological science. But keep in mind that all science is a single endeavor that includes anything and everything for seeking new knowledge through the use of the scientific method. Science is commonly broken down into what may seem like different disciplines: biology, chemistry, physics, and geology, for example. But each discipline is interrelated with all the others. A knowledge of chemistry and physics is important for understanding many biological principles. Biology is also divided into subdisciplines such as zoology, the study of animals; botany, the study of plants; genetics, the study of heredity; cytology, the study of cells; parasitology, the study of parasites; and mammalogy, the study of mammals.

BASIC CHARACTERISTICS OF LIVING ORGANISMS

Think for a moment about the great diversity of life. Giant trees form forests, and worlds of small animals and microbes exist among the leaves and at the tree bases. Coral reefs contain colorful kaleidoscopes of hundreds of species of fish, invertebrate animals, and plants. The great savannas of Africa abound with grazing animals and their predators, the carnivorous cats and others. Although living organisms seem to take an infinite variety of forms, all have a number of common characteristics.

Cells and Organization

Regardless of their size or form, all organisms are composed of one or more living units, called cells. **Cells** assume a variety of forms, but their basic structures and chemical compositions are remarkably similar in all living things. Organisms composed of a single cell are called **unicellular organisms.** Those of more than one cell are called **multicellular organisms.** Complex multicellular organisms may be composed of billions of cells organized as **tissues**—groups of cells with common functions and similar structures. Tissues may combine to form **organs,** and organs may combine to form organ systems. The heart, for example, is an organ within the circulatory system that is composed of cardiac muscle tissue.

Metabolism

All of the biochemical and physical processes of cells and organisms are called **metabolism.** Cells and organisms require nutrients and energy to sustain these processes and to constantly renew and repair themselves. Cells also produce products that must be excreted. Metabolism is controlled by the cellular genetic material known as **deoxyribonucleic acid (DNA)**□.

The ultimate source of energy required for metabolism is the sun. Plants transform light energy to chemical energy in the form of nutrients through the process of **photosynthesis**□. The nutrients, or energy molecules, provide energy and sustenance for plants and for all other organisms. Organisms release the energy in food for use by cells through the process of **cellular respiration**□.

Organisms that can convert solar energy to chemical energy in the form of nutrients are called producers, or **autotrophs.** Plants are autotrophs. Organisms that depend on plants or other animals for food and energy are called consumers, or **heterotrophs.** Ani-

mals are heterotrophs. **Herbivores** are animals that feed on plants. **Carnivores** are animals that feed on other animals and that derive energy (food) molecules from the tissues of what they feed on. Thus the energy from the sun is transferred through a series of organisms as chemical energy. Each series of such organisms is referred to as a **food chain**□ (see Figure 1–2).

All organisms die. Soon after death, their bodies become the source of nutrients and energy for the last link in the food chain—the **decomposers,** which include bacteria and other microorganisms. Through decomposition, the complex chemical structures of organs, tissues, and cells are reduced to simpler compounds. Plants then reuse these compounds to produce new plant tissues, thereby recycling them once again through the living world.

Figure 1–2
Recycling matter and energy through a small portion of the living world. Plants convert light energy into chemical energy in the nutrients they synthesize. Herbivores such as caterpillars utilize plant material for a source of nutrients and energy and are themselves used by carnivores such as the shrew. Other carnivores such as the fox feed on lower level carnivores. Upon the death of an organism decomposing organisms utilize the dead body for nutrients and energy and in so doing break down complex organic substances into simple inorganic compounds that are utilized again by plants during the synthesis of new nutrients.

Homeostasis and Death

Organisms continually use energy to maintain a more or less constant internal environment that differs from the external environment. This process is called **homeostasis.** The constant internal environment is the one in which the cells that comprise the organism function best. However, the distinction between the internal environment and the external environment cannot be maintained forever. When homeostasis is not maintained the life processes of cells are disrupted. As a result, all living things die.

Reproduction and heredity

All organisms have the ability to reproduce themselves. This fundamental ability comes from the nature of the hereditary material DNA, which is in all cells. Each molecule of DNA has the remarkable ability to reproduce itself. It also has a unique configuration, containing a variety of chemical codes that dictate the particular life form of the organism. DNA ensures that maple trees produce maple trees, beetles produce beetles, and humans produce humans. However, the organization of DNA in cells and organisms also allows for variation among members of the same species. Except in the case of identical twins, triplets, and so on, each human is genetically unique.

Irritability

All organisms respond to other organisms and to environmental variations in temperature, light intensity, and so on. These adaptive responses are only to the changes that are important to survival. They contribute to the well-being of the organisms. Inappropriate responses or a lack of response may lead to death.

Evolution

In the millions of years that life has existed on earth, environments have changed. The populations of organisms have successfully responded to these changes through various adaptations brought about by **natural selection**. Populations that do not adapt become extinct. Those that do adapt may give rise to new kinds of organisms. However, these changes occur very slowly. Often, hundreds of thousands of years are required for a new species to evolve. Nonetheless, the millions of species that exist on earth today are the result of natural selection, which continues to produce changes in living organisms.

CLASSIFYING THE LIVING WORLD

Some biologists, called taxonomists or systematists, devote their lives to identifying, naming, and classifying the different species of organisms. During their studies, they often discover species that have not been described before. But these scientists do not just compile long lists of names for plants, animals, and microbes. They also search for evidence indicating evolutionary relationships among the groups with which they work. Their efforts lead to an understanding of how different species are related through genetics and evolution. Organisms with similar characteristics and structure are grouped together.

The basic system of classifying organisms was developed during the eighteenth century by Carolus Linnaeus. It is still used today, although it has been refined considerably. As scientists learn more and more about organisms, it often becomes necessary to revise or rearrange taxonomic groupings. In addition, different references sometimes present different classifications of organisms. These differences reflect the opposing hypotheses of taxonomists. The basic system, however, remains essentially the same.

Every organism belongs to a distinct **species** that possesses certain traits. A species, most simply, is a group of organisms that is reproductively isolated ▫. Members of a particular species breed with one another—but not with members of other species—to produce fertile offspring. Similar species that are closely related through evolution are grouped together as genera (more than one genus). **Genera** with common characteristics are grouped into families, families are grouped into orders, and orders are grouped into classes. Classes with major similarities are grouped into **phyla,** and phyla are grouped into **kingdoms.** As an example of this system, consider humans:

Species:	*Homo sapiens*
Genus:	*Homo*
Family:	Hominidae
Order:	Primates
Class:	Mammalia
Phylum:	Chordata
Kingdom:	Animalia

Every species has a unique **scientific name** that is recognized by biologists throughout the world. Common names for animals and plants differ from country to country and even from region to region in the same country. For example, the word for dog is different in French, German, Russian, Spanish, and Chinese. In some areas of the United States, crows are called blackbirds and some blackbirds are called grackles (a completely different species). The North American cougar is also called a puma or a mountain lion. Standardized, internationally accepted scientific names eliminate the confusion created by the use of common names.

The scientific name of each species consists of two words. The first word is the genus. It is always a noun, and it is always written in italics (or underlined), with the first letter capitalized. The second word is an adjective and is not capitalized, but it does appear in italics (or underlined). For example, the domestic dog is *Canis familiaris*. *Canis* is the Latin word for dog; *familiaris* is an adjective meaning "familiar" or "common." Scientific names are always written with Roman letters, regardless of the native language. Table 1–1 gives the scientific names for some common plants and animals.

KINGDOMS OF LIVING THINGS

The phyla are grouped into a few major kingdoms. Traditionally, only two kingdoms, Animalia (the animals) and Plantae (the plants), were recognized be-

cause common organisms were easily placed in one or the other. Plants were generally viewed as being stationary, green in color (because of chlorophyll), and capable of photosynthesis. Animals were generally regarded as being able to move about and as feeding on plants or other animals (because they cannot synthesize their own food). However, as biologists became more familiar with microscopic creatures, many found that having only two kingdoms was unsatisfactory. Some unicellular creatures contain chlorophyll, carry on photosynthesis, and move by themselves. When they live in the dark, however, they lose their chlorophyll and depend on organic substances in the water for nourishment. Should such creatures be classified as animals or plants?

Recently, ecologist R. H. Whittaker proposed a five-kingdom system for classifying the living world. His system (shown in Figure 1–3) has been widely accepted by biologists. The kingdom **Monera** contains bacteria and blue-green algae. Cells of these organisms lack a membrane-enclosed nucleus and other structures found in the cells of all other organisms. The blue-green algae and some bacteria are autotrophs. The remaining members of the kingdom are heterotrophs. The kingdom **Protista** contains unicellular and simple multicellular organisms. The cells of multicellular organisms are not differentiated into well-defined tissues. Protozoa and most algae belong to this kingdom. Protista includes both autotrophs and heterotrophs. The kingdom **Fungi** contains yeasts, molds, mushrooms, and the like. All are heterotrophs that absorb dissolved nutrients from their environments. Tissues are not well developed. The kingdom **Plantae** contains the organisms easily recognized as plants, with well-defined tissues. The

Table 1–1
Common and Scientific Names of Some Familiar Organisms

COMMON NAME	SCIENTIFIC NAME
Human	*Homo sapiens*
Dog	*Canis familiaris*
Cat	*Felis domestica*
African elephant	*Loxodonta africana*
Blue whale	*Balaenoptera musculus*
Sugar maple	*Acer saccharum*
White oak	*Quercus alba*
Giant sequoia	*Sequoiadendron giganteum*

Figure 1–3
The five kingdoms of living organisms proposed by R. H. Whittaker. Kingdom Monera (bacteria and blue green algae) contains organisms whose cells lack membrane-bound nuclei. Kingdom Protista (protozoans and many of the algae) includes organisms that possess cells with nuclei, but multicellular forms are not organized into tissues, as is the case with most other multicellular organisms. Kingdom Fungi (molds, yeasts, mushrooms, and so on) includes organisms that absorb their nutrients from their environment. Kingdom Plantae (the true plants) includes organisms with tissues typical of plants, and they are associated with the process of energy transformation and food synthesis known as photosynthesis. Kingdom Animalia (the animals) includes organisms that actively feed on plants and other animals and possess nerve, muscle, epithelial, and connective tissues.

NUTRITION BY ABSORPTION

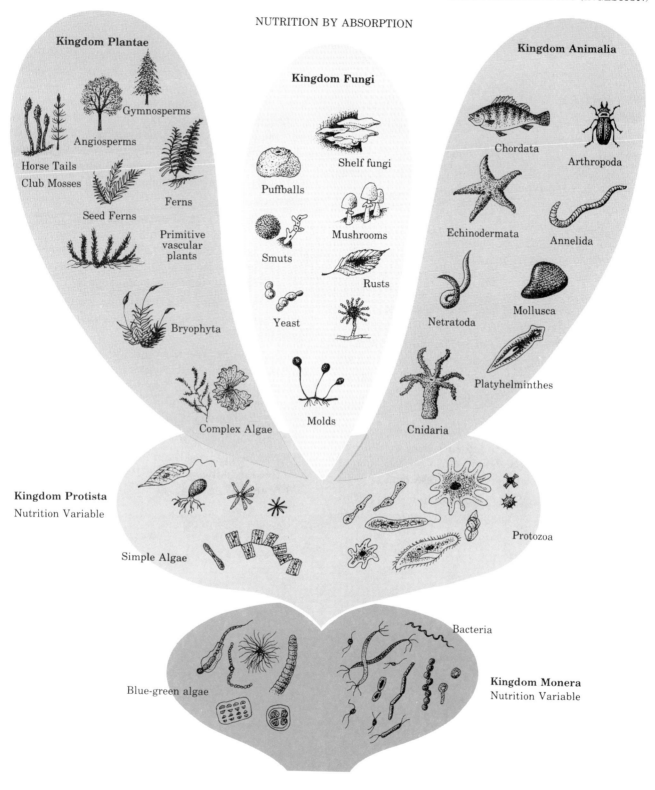

Kingdom Plantae

Gymnosperms

Angiosperms

Horse Tails
Club Mosses

Seed Ferns

Ferns

Primitive
vascular
plants

Bryophyta

Complex Algae

Kingdom Fungi

Shelf fungi

Puffballs

Smuts

Mushrooms

Yeast

Rusts

Molds

Kingdom Animalia

Chordata

Arthropoda

Echinodermata

Annelida

Netratoda

Mollusca

Platyhelminthes

Cnidaria

Kingdom Protista
Nutrition Variable

Simple Algae

Protozoa

Bacteria

Blue-green algae

Kingdom Monera
Nutrition Variable

kingdom **Animalia** contains the animals, which have well-defined tissues and a common pattern of embryological development□. Appendix 2 describes the basic characteristics of the members of each of these kingdoms.

SUMMARY

1. Science, including biology, plays a highly visible and dynamic role in modern society.

2. Basic curiosity drives scientists to expand knowledge, to establish cause-and-effect relationships among phenomena, and to create guidelines for predicting natural phenomena.

3. Science is concerned with making observations and collecting facts, which are systematically organized into a body of knowledge by the scientific method.

4. In the scientific method, hypotheses are formulated to tentatively explain observed phenomena.

5. The hypotheses are tested by controlled experiments that include experimental and control groups.

6. When a hypothesis is supported by the scientific community, it becomes a theory, the most acceptable explanation of a phenomenon at the time.

7. Basic research does not necessarily have immediate or direct application to humans. Applied research is intended to have obvious human applications. It would be impossible without basic research.

8. Although an amazing variety of living organisms exist, they all have common characteristics: (a) All are composed of either one (unicellular) or more (multicellular) living units called cells, which have common chemical and structural features. (b) In multicellular organisms, the cells may be organized into tissues and organs. (c) Cell metabolism (biochemical reactions) convert energy and utilizes nutrients to maintain the integrity of the organism through homeostasis. (d) When homeostasis is not maintained, death results. (e) Organisms reproduce their own kind because of the unique features of their hereditary material. (f) All organisms show irritability and respond to environmental change. (g) Changes in populations may lead to the evolution of new species through natural selection.

9. The basic source of energy for living organisms is sunlight, which photosynthetic plants convert into chemical energy and, eventually, into food molecules. Because animals are incapable of photosynthesis, they rely on plants for their basic food supply.

10. When organisms die, their bodies become sources of nutrients and energy for other organisms, called decomposers.

11. Taxonomists identify, classify, and name species of living organisms, attempting to organize them into groups that reflect their genetic and evolutionary relationships.

12. Organisms that interbreed comprise a species. They do not breed with members of other species.

13. The living world can be conveniently divided into five kingdoms: Monera, Protista, Fungi, Plantae, and Animalia.

THE NEXT DECADE

THE CONTINUING SEARCH FOR LIFE

For centuries, naturalists and other biologists have explored the far corners of the earth, seeking new forms of life to describe and preserve in the world's museums, zoos, and botanical gardens. Few areas remain to be explored, but the search continues. Expeditions are still sent to the least-explored pockets of remaining jungle, such as those of New Guinea. Divers still brave the icy polar waters, seeking life beneath the great ice floes. Scientists in self-contained submarines explore the ocean bottoms to ever-increasing depths.

In 1977, geologists explored the Galapagos Rift, a deep canyon located about four hundred miles northeast of the Galapagos Islands. At a depth of 2,500 meters, a series of geothermal hot springs were found. Around them were communities of organisms unknown until then. These densely crowded communities contained new species of sea worms up to three meters long, clams twenty centimeters long, and crabs, barnacles, fish, and a variety of other worms. These communities have now been studied in some detail, and others like them have been found in the Eastern Pacific Ocean, where there are similar hot water springs. Many of the animals are directly or indirectly dependent for food and energy on bacteria that form visible clouds in the warm water around the springs. The water from the springs is rich in hydrogen sulfide and carbon dioxide, from which the bacteria derive their en-

ergy. Here scientists have discovered not only a group of new species but a complex community of organisms capable of existing in a sunless environment because of a very unusual food chain.

The discovery of the Galapagos Rift community was unexpected. Sometimes searches are organized to find living creatures that may or may not really exist or that have been seen only rarely. People today still search for the fabled Loch Ness Monster and Bigfoot (both probably do not exist). A recent expedition to Africa searched for a dinosaur that natives had reported as still surviving in a remote jungle swamp (it was not found). On the other hand, a 1982 scientific article described an expedition to the mountainous region of New Guinea in search of a species of bower bird that bird specialists had never seen alive. All anyone knew about these yellow-fronted gardener bower birds was based on three preserved specimens in a museum. The expedition members not only found the living birds but were able to study their behavior and compare it to the behavior of other bower birds. These beautiful birds build elaborate platforms on which the males court the females.

Discoveries of new species of large multicellular organisms have become less and less frequent. However, scientists discover new species of insects, microbes, and a host of other small organisms every day. There are probably still hundreds of

thousands of living species on earth yet to be described for the first time. Nevertheless, some scientists are turning their attention to other planets and even to outer space in a search for living creatures. These scientists, who call themselves exobiologists, believe that our own galaxy, the Milky Way, is so vast that other planets similar to ours must not only exist but also support life, probably even intelligent life. Other scientists disagree about the existence of life beyond planet earth.

During the coming decade, scientists plan to develop new techniques that might allow them to communicate with other intelligent life, if it exists. Shortwave radio signals have been beamed from the earth into space. But if they have been heard, they have not been answered. Astronomers and astrophysicists also search for signals that might indicate intelligent life forms trying to locate us.

In 1977, Voyager I and Voyager II, two U.S. space probes, were launched to explore the outer realms of our solar system and then drift into outer space. Both contain recorded messages in many languages, photographs of the earth and humans, and recordings of music. In 1983, Voyager I was in outer space and Voyager II was approaching the planet Uranus. What they discover might make the biggest headlines in the next decade.

FOR DISCUSSION AND REVIEW

1. Name several great scientists. What special personal characteristics did they possess? Are such characteristics unique to scientists?

2. Outline the basic steps of the scientific method.

3. Design a controlled experiment to test one of the following hypotheses: (a) The growth of plant species X increases as the temperature increases. (b) Drug X increases the heart rate of mice. (c) The newly discovered vitamin Z is essential for the normal growth of such animals as rats.

4. List and explain the general characteristics of all living organisms.

5. Why is a uniform and internationally accepted method of naming and classifying living organisms essential?

SUGGESTED READINGS

THE GAME OF SCIENCE, 4th ed., by Garvin McCain and Erwin M. Segal. Brooks/Cole: 1982.
Examines the attitudes, rules, and concepts of science. Written to help the layperson appreciate and understand the workings of science.

FIVE KINGDOMS: AN ILLUSTRATED GUIDE TO THE PHYLA OF LIFE ON EARTH, by Lynn Margulis. W. H. Freeman: 1981.
A unique, well-illustrated catalogue of all the major groups of life on earth.

THE TWO CULTURES AND THE SCIENTIFIC REVOLUTION, by C. P. Snow. Cambridge University Press: 1959.
Explains how science does and does not relate to other disciplines.

The Organization of Living Organisms

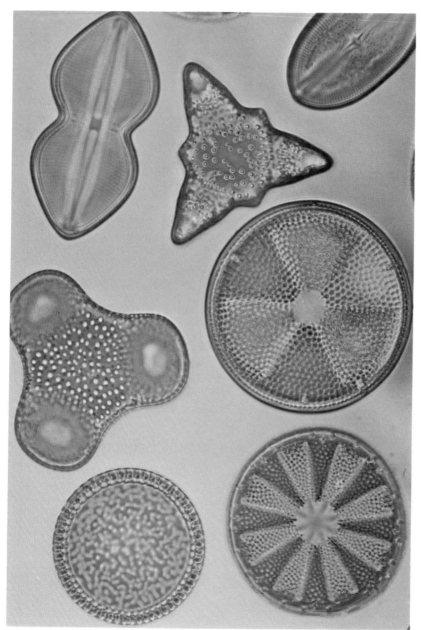

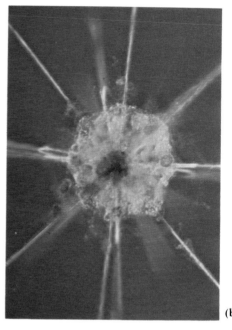

(b)

(a) Cell walls of diatoms photographed through an interference microscope which produces the brilliant colors. Diatoms are examples of unicellular organisms—creatures whose bodies are a single cell. They represent one of the most numerous kind of unicellular organisms. They are abundant in the oceans (and in fresh water) and are frequently called the "grass of the sea". Because they are photosynthetic organisms, they occur in the sunlight surface waters of the open ocean and form the base of many marine food chains. **(b)** A living radiolarian represents another group of unicellular organisms, the Protozoa. Radiolarians are also aquatic but are not photosynthetic. Most unicellular organisms are found in water, or moist soil, or in the body fluids of other organisms because few can withstand drying in their active forms. However, many can form inactive cysts or spores that can survive drying.

(a)

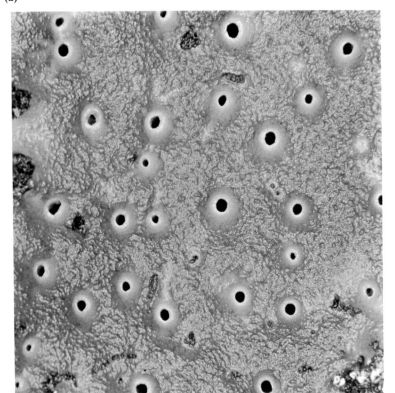

(a)

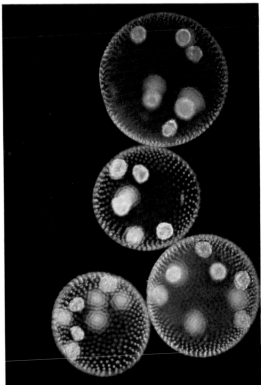

(b)

(a) An encrusting sponge, *Halichondria panicea.* Sponges are a unique group of multicellular animals whose cells are not differentiated into the basic tissues found in other animals. Some species represent loose aggregations of cells that if separated from one another will join with other cells to produce an entirely new sponge. (b) Four *Volvox* colonies each containing several daughter colonies. Each colony is composed of several hundreds of unicellular individuals each bearing two flagellae. The coordinated beating of the flagellae moves each colony through the water with one specific half always oriented anteriorly. Biologists believe multicellular organisms evolved gradually from colonies of unicellular organisms as the individuals in the colony gradually became more specialized and dependent upon the whole colony. (c) A slime mold. Slime molds exist during part of their life cycle as creeping unicellular forms which come together into slimy masses. In some types, the cells fuse together forming a single continuous mass of cytoplasm; in other types, the cells maintain their individuality. In either case, the masses produce reproductive bodies that produce spores, each of which becomes a new individual cell. Are slime molds unicellular or multicellular organisms?

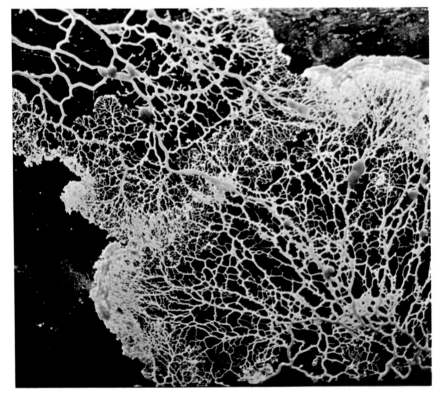

(c)

(a)

(a) This Portugese man-of-war consuming its fish meal is actually a complex colony of numerous individuals, each modified for a special function—some sting and capture prey, others digest the food, and so forth. All are coordinated to form what appears to be a single organism. **(b)** In this giant African Termite mound lives another kind of animal colony: a colony of social insects composed of thousands of individuals. Their activities are coordinated by chemical and behavioral controls for the good of the total colony. **(c)** Different species of organisms may interact in a variety of ways. Clown fish live unharmed among the stinging tentacles of sea anemones. The sea anemones provide protection for the fish which may lure food organisms to the anemones' deadly tentacles. **(d)** These brightly colored lichens are another example of the intimate relationships that have evolved between different species. Lichens are partly photosynthetic algae and partly non-photosynthetic fungi. The algae produce nutrients for the fungus, but it is less clear what benefits the alga derives from the association (photo by Richard S. Funk).

(b)

(d)

(c)

(a)

(a) Mushrooms and toadstools are among the most complex members of Kingdom Fungi which includes both unicellular and multicellular forms. However, none of the Fungi possesses the specialized tissues of the vascular plants. The multicellular fungi are composed of cytoplasmic strands, called hyphae, which exist separately as in the bread mold. Some form the underground mycelia that can produce spore-producing reproductive structures that are mushrooms, puffballs, and their relatives. **(b)** The hyphae and black spore producing structures called sporangia of the common bread mold, *Rhizophus*. **(c)** The yellow mycelia of a fungus found beneath a rotting log.

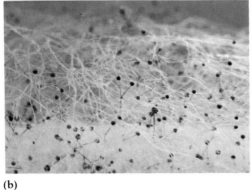

(b)

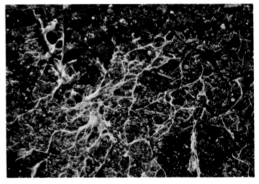

(c)

(a) A tulip is an example of a regular or radially symmetrical flower. The floral parts are arranged around a central axis in a way that makes it possible to divide the flower into two similar halves, by more than one longitudinal plane (photo by Richard S. Funk). (b) This Jacobean lily flower is an example of an irregular or bilaterally symmetrical flower. It can be divided into two similar halves only by a single longitudinal plane (photo by Richard S. Funk). (c) The sunflower is a member of the largest family of flowering plants. Although the flower may at first appear similar to those of other flowering plants, it is quite different. Each "flower" is composed of dozens to hundreds of small specialized flowers grouped together in a compact head. In the sunflower, sterile ray flowers that look like petals surround small tubular shaped disc flowers that produce the seeds.

(a) Like starfish and numerous other benthic animals, this sea anemone appears to have radial symmetry. Their major body parts are arranged around a central axis in a way that makes it possible to divide each animal into two similar halves by any of several planes passing through the central axis. Actually, one or more inconspicuous structures may destroy the perfect radial pattern, and then the animal is said to have biradial symmetry; this is true of the sea anemone. **(b)** The salamander, like most other animals, has bilateral symmetry. It has a definite anterior and posterior ends, an upper or dorsal surface, a lower or ventral surface, and left and right sides. Only one longitudinal plane can divide a bilaterally symmetrical animal into two similar left and right halves.

(c) A milliped also has bilateral symmetry, but it illustrates yet another form of animal anatomical organization, metamerism or segmentation. Arthropods such as insects, annelid worms such as earthworms, and, to a less obvious degree, chordates such as ourselves are metameric. Posterior to the head of the milliped, the body is divided into sections called segments or metameres. In this case, each metamere bears two pairs of appendages. In many animals, each body metamere is similar to every other one, but in many other animals, groups of similar specialized metameres appear, or even fuse, to form a distinct body region. The head of the milliped is a group of fused metameres.

(b)

(a) The larva, a caterpillar, of the cecropia moth. Many animal species have juvenile stages, called larva, that are very different in form and life style from the adult. When they reach a certain stage of development, they change to the adult form through a process called metamorphosis. (b) An adult male cecropia moth.

(a)

(b)

(a)

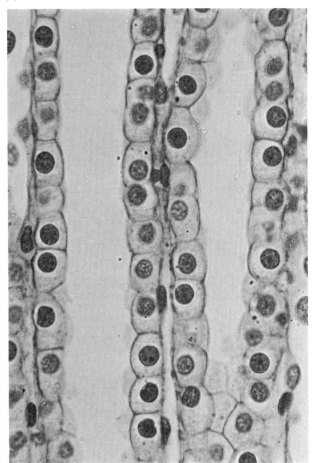

(b)

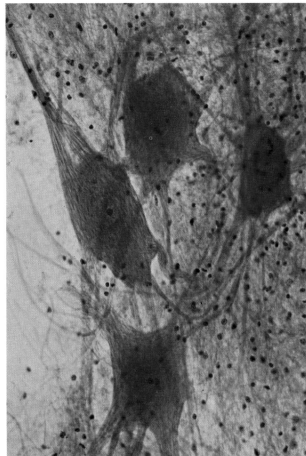

(a) Cuboidal epithelial cells of human kidney tubules. Complex multicellular organisms, such as animals and the seed plants, have bodies composed of millions of cells. These cells are of different types with specialized functions. These specialized cell types are called tissues. Epithelial tissues of animals and plants cover surfaces or line cavities, or tubes or ducts. (b) Neurons (nerve cells) from a spinal cord. One of the unique characteristics of animals is they possess nervous tissue that permits them to respond rapidly to changes in the environment. (c) Human blood cells. Blood is a type of connective tissue. Connective tissues are composed of cells suspended in a nonliving matrix. This photo shows many red blood cells and three white blood cells. These cells float in the fluid portion of blood, the plasma; the plasma is the matrix (courtesy Carolina Biological Supply Company). (d) Striated muscle tissue. The contraction of muscle fibers result in animal movements. The striations seen on these muscle fibers are caused by the patterns of overlapping filaments of muscle proteins (courtesy Carolina Biological Supply Company).

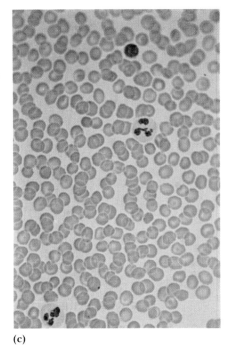

(c)

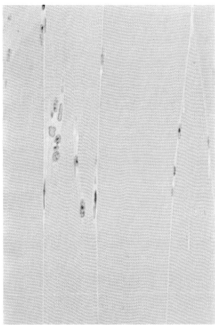

(d)

2

THE CHEMICAL BASIS OF LIFE

Professional biologists must know the details of chemistry as well as biology, because the physiological processes of cells and organisms are chemical in nature. This close link between chemistry and biology means that even students in introductory biology courses should understand some basic principles of chemistry. Such knowledge will help them appreciate the efficient, complex, and beautifully ordered biochemical basis of life. It will also help them understand the articles in magazines and newspapers on such topics as pollution, chemical induction of cancer, mutations, genetic engineering, and food and drug testing.

The basic laws of chemistry are not difficult to learn if they are approached in a step-by-step fashion. Therefore, they are presented in this chapter as a numbered series of statements and explanations.

THE STATES OF MATTER

1. All substances **(matter)** found in the universe are composed of chemicals and exist in one of three possible states—gas, liquid, or solid. For example, water exists as a solid when frozen into ice, as a liquid in its drinkable form, and as a gas when boiled into steam.

2. Matter in various states may be mixed together. For example, dust particles (solids) may be suspended in water (liquid) or in air. Such mixtures are called **suspensions.** They are not permanent mixtures, because the particles are so large and heavy that they settle out in time.

3. Another type of mixture is called a **solution.** It occurs when the particles or molecules of the solid are small enough to become uniformly distributed among the molecules of a liquid, such as water. They

PREVIEW QUESTIONS

After reading this chapter, you should be able to answer the following questions:

1. What are some of the chemicals you're exposed to every day?

2. What is the chemical structure of matter?

3. What are the theories about how life on earth originated?

4. Which chemicals are most abundant in the human body?

form a homogeneous mixture that does not separate out. Sugar mixed in water is a good example of a solution. The sugar, called the **solute,** is said to dissolve in the water, called the **solvent.** Water can dissolve thousands of chemicals, so it is no wonder that it forms the basic medium of living matter. Liquids and gases such as oxygen can also form solutions in water or in other liquids.

4. Yet another kind of mixture is called a **colloid.** In such mixtures, the particles of the solid are large and do not dissolve. They remain dispersed in a liquid medium because of their size and their electrical charges. Living matter is a complex mixture of thousands of substances in solution or in colloidal dispersion in water.

THE STRUCTURE OF MATTER

1. All matter, both living and nonliving, is composed of **elements**—pure substances that cannot be altered by natural processes but that can combine to form molecules and chemical compounds. There are ninety-two naturally occurring elements and a small but growing number of unstable human-made ones. Nonetheless, 99 percent of all living matter is composed of the following six elements: hydrogen (H), carbon (C), oxygen (O), nitrogen (N), phosphorus (P), and sulfur (S).

2. The smallest natural indivisible units of elements are **atoms.★** They are composed of subatomic particles. The three most important types of subatomic particles are **protons,** which carry a positive charge; **electrons,** which carry a negative charge; and **neutrons,** which carry no charge (see Figure 2-1).

3. Protons and neutrons make up most of an atom's mass and form the dense atomic nucleus. Electrons, which are much smaller than either protons or neutrons, rotate around the nucleus. The atoms of different elements have different numbers of protons, but the same number of electrons as protons. Because the number of positive charges on protons is equal to the number of negative charges on electrons, atoms are electrically neutral.

4. The number of protons in the atomic nucleus is

★We have learned to split atoms and release their contained energy. But when atoms are subdivided, the characteristics of the element are destroyed.

CONCEPT SUMMARY

SIX ELEMENTS COMPRISE 99 PERCENT OF LIVING MATTER

ELEMENT	IMPORTANCE TO LIVING ORGANISMS
Hydrogen	Component of proteins, carbohydrates, lipids, all other organic substances, and water
Carbon	Component of proteins, carbohydrates, lipids, and all other organic substances
Oxygen	Component of many organic substances and a requirement for respiration
Nitrogen	Component of proteins and nucleic acids
Phosphorus	Component of many proteins, other important organic substances, and bone
Sulfur	Component of some amino acids and necessary in the structuring of complex proteins

---

the element's atomic number. Thus each element has a different atomic number.

5. Like everything else, atoms have weight. The weight of an atom is primarily the combined weights of the protons and neutrons, because electrons are so small that their weight is negligible. Actually, a proton or a neutron weighs about 1,836 times as much as an electron. A carbon atom has six protons and six neutrons and therefore is given the atomic weight of 12 (6 + 6 = 12). The weights of atoms are measured in comparison to the weights of other atoms, not in units such as grams or ounces. Carbon is now the standard element against which the others are compared. Thus a hydrogen atom, containing a single proton and no neutrons, weighs one-twelfth of a carbon atom. An atom of the element magnesium, with twelve protons and twelve neutrons, weighs twice as much as a carbon atom.

6. The number of neutrons in an atom varies, although commonly it is equal to the number of protons. Variations in the number of neutrons do not change the chemical characteristics of the element but do produce various **isotopes** of it. Each isotope has a characteristic number of neutrons. Many isotopes are unstable. As they revert to a stable condition, they give off radiation.

For example, two isotopes of the carbon atom are carbon = 12 (written as ^{12}C) and carbon = 14 (written

Proton   Electron

Neutron

A HYDROGEN atom has:
1 proton
1 electron
0 neutrons

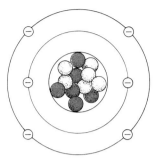

A CARBON atom has:
6 protons
6 electrons
6 neutrons

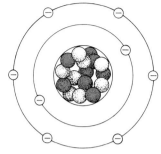

An OXYGEN atom has:
8 protons
8 electrons
8 neutrons

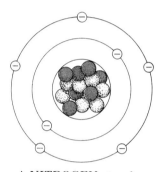

A NITROGEN atom has:
7 protons
7 electrons
7 neutrons

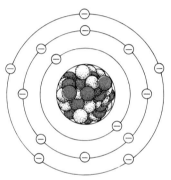

A PHOSPHORUS atom has:
15 protons
15 electrons
15 neutrons

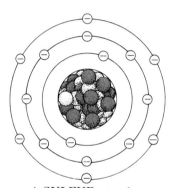

A SULFUR atom has:
16 protons
16 electrons
16 neutrons

Figure 2–1
Simplified diagrams of atoms of the elements hydrogen, carbon, oxygen, nitrogen, phosphorus, and sulfur.

as ^{14}C). The latter has two extra neutrons. In nature, most carbon exists as ^{12}C, with six protons and six neutrons in each atomic nucleus. However, ^{14}C is constantly being formed in the upper atmosphere through cosmic radiation of nitrogen atoms. Like many other isotopes, ^{14}C is unstable. It changes to ^{12}C at a slow but steady rate (a process called radio-active decay). As ^{14}C is converted to ^{12}C, radiation is given off. Therefore, ^{14}C is considered a radioactive substance.

7. Electrons are arranged in a number of orbitals, or energy-level "shells," that differ in their average distance from the nucleus (see Figure 2–1). The distance of an orbital from the nucleus predicts the energy

state of the electrons. The greater the distance, the higher the energy state.

8. The maximum number of electrons in each orbital is predictable: two in the orbital closest to the nucleus, eight in the next orbital, eighteen in the third, thirty-two in the fourth, and so on. However, eight electrons in the outermost orbital beyond the first form a particularly stable situation. If the outer orbital has more or fewer than eight electrons, the atom will have a certain degree of instability. This instability causes the atoms to react with other atoms by forming chemical bonds.

9. **Chemical bonds** tend to make the reacting at-

oms stable by completing the number of electrons in the outermost orbital. They also represent a store of potential energy. When two or more atoms chemically bond, the combination is called a **molecule.** For example, a molecule of water (H_2O) consists of two atoms of hydrogen bonded to a single atom of oxygen. (See Box 2–1).

10. A variety of chemical bonds exist. Three common and important types are ionic bonds, covalent bonds, and hydrogen bonds.

Ionic Bonds

An atom with only a few electrons in its outermost orbital may lose some to the outermost orbital of another atom that lacks only a few electrons to complete its stable number. The atom giving up electrons is called the electron donor, and the atom receiving the electrons is called the electron acceptor. Usually, atoms with fewer than four electrons in their outermost orbital are electron donors, and those with more than four are electron acceptors.

An electron donor ends up with fewer electrons than protons and becomes positively charged ($+$). An electron acceptor ends up with more electrons than protons and becomes negatively charged ($-$). The charged atoms are called **ions.** A positively charged donor is electrically attracted to a negatively charged acceptor, and their attraction creates an ionic bond. An example of ionic bonding is the formation of table salt from sodium (Na^+) and chloride (Cl^-) ions (see

Figure 2–2
The formation of an ionic bond.

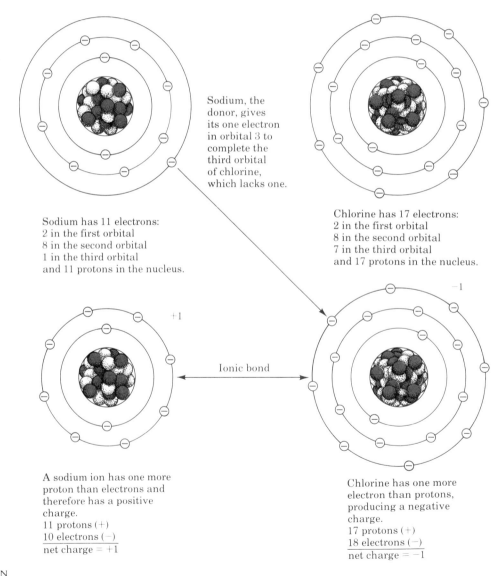

Sodium, the donor, gives its one electron in orbital 3 to complete the third orbital of chlorine, which lacks one.

Sodium has 11 electrons:
2 in the first orbital
8 in the second orbital
1 in the third orbital
and 11 protons in the nucleus.

Chlorine has 17 electrons:
2 in the first orbital
8 in the second orbital
7 in the third orbital
and 17 protons in the nucleus.

Ionic bond

A sodium ion has one more proton than electrons and therefore has a positive charge.
11 protons ($+$)
10 electrons ($-$)
net charge = $+1$

Chlorine has one more electron than protons, producing a negative charge.
17 protons ($+$)
18 electrons ($-$)
net charge = -1

CHEMICAL SYMBOLS

Each element is represented by a symbol: C for carbon, O for oxygen, Fe for iron, and so on. These symbols are used to designate the elements contained in molecules, which are substances composed of two or more atoms. For example, H_2O (water) is a molecule formed of two hydrogen atoms and one oxygen atom. Glucose, a simple sugar, is designated by the molecular formula $C_6H_{12}O_6$. The numerical subscripts give the number of each type of atom in the molecule: six carbons, twelve hydrogens, and six oxygens in glucose. However, molecular formulas do not provide biologists and chemists with as much information as do structural formulas.

The structural formula for glucose appears in Figure 2–13. It shows not only the quantity of each type of atom in the molecule but also how the atoms are arranged and which are bonded together. The structural formula of the fructose molecule is next to that of the glucose molecule. Clearly, the fructose molecule is different from the glucose molecule; it is a different sugar. However, the molecular formula for fructose is the same as that for glucose: $C_6H_{12}O_6$. Thus, we can see that it is possible for two substances that are structurally different to have the same molecular formula.

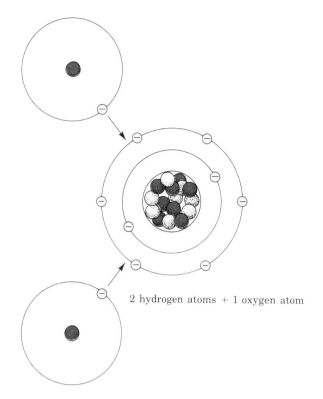

2 hydrogen atoms + 1 oxygen atom

Figure 2–2). Molecules that are ionically bonded have a tendency to disassociate (break apart) when in solution.

Covalent Bonds

Atoms forming covalent bonds do not give up or receive electrons; they share them. An example of covalent bonding is the formation of a molecule of water from two atoms of hydrogen and one atom of oxygen (see Figure 2–3). The single electron in the orbital of each hydrogen atom completes the outer orbital of the oxygen atom, raising the number of electrons from six to eight. A molecule of water (H_2O) is the result. Molecules of substances formed by covalent bonding do not disassociate into ions in solution.

Hydrogen Bonds

Water is a special compound, with many unique features important to living systems. Many of these fea-

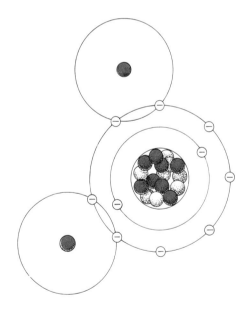

Water molecule

Figure 2–3
The formation of covalent bonds.

tures can be attributed to the hydrogen bonds that exist between water molecules. A single water molecule is polarized; that is, although the oxygen and hydrogen atoms share electrons, the oxygen atom attracts the electrons toward itself and away from the hydrogen atoms (see Figure 2–4). Therefore, the area of the water molecule occupied by the oxygen atom is slightly more negatively charged than the area occupied by the hydrogen atoms. Water molecules form lattices, or networks, when the negatively charged oxygen area of one water molecule is linked to the more positively charged hydrogen area of another water molecule (see Figure 2–5). This occurs because of the mutual attraction of unlike charges.

In effect, a hydrogen atom forms a bridge between two water molecules. This bridge is a hydrogen bond. Because water molecules are linked by hydrogen bonds, water is a very cohesive substance and has a high surface tension. A large amount of energy is necessary to heat water or to change it from a solid to a liquid or from a liquid to a gas. The polarized nature of water molecules makes water a nearly uni-

Figure 2–4
Water is a polarized molecule because the oxygen atom attracts the electrons, giving a portion of the water molecule a negative charge. The portion of the molecule formed by the hydrogen atoms is thus more positively charged.

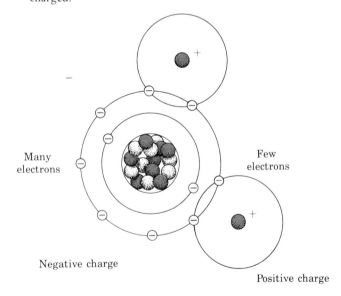

Many electrons

Few electrons

Negative charge

Positive charge

versal solvent, and the heat capacity of water allows life to exist over a wide temperature range.

Hydrogen bonds also form bridges between different parts of large, complex molecules, such as proteins and deoxyribonucleic acid (DNA)▫, thereby contributing to their complex shapes and stability. If hydrogen bonds are broken, the shapes of complex molecules are changed, and their biological activities are altered.

CHEMICAL REACTIONS

1. During chemical reactions, chemical bonds are formed or broken, and energy is either taken up or given off. When a chemical reaction gives off energy, it is called an **exergonic reaction.** When energy from an outside source is needed for a chemical reaction, it is called an **endergonic reaction.** In living systems, energy given off by exergonic reactions is commonly used to fuel endergonic reactions (see Figure 2–6).

2. During a chemical reaction, three possible events may occur.

— A molecule may be split into two or more molecules or atoms. For example, when hydrochloric acid (HCl) dissolves in water, free hydrogen and chloride ions are released (see Box 2–2):

$$HCl \rightarrow H^+ + Cl^-$$

— New molecules may be formed from two or more molecules or atoms. For example, when two molecules of hydrogen gas (H_2) react with a molecule of oxygen (O_2), two water molecules are formed:

$$2\,H_2 + O_2 \rightarrow 2\,H_2O$$

— A molecule may react with another molecule to produce two or more different molecules. For example, when sodium hydroxide (NaOH) reacts with hydrochloric acid (HCl), molecules of water (H_2O) and sodium chloride (NaCl) are formed:

$$NaOH + HCl \rightarrow H_2O + NaCl.$$

3. Oxidation and reduction reactions, called **redox reactions,** are common in metabolism. When an atom or molecule gives up an electron to another one,

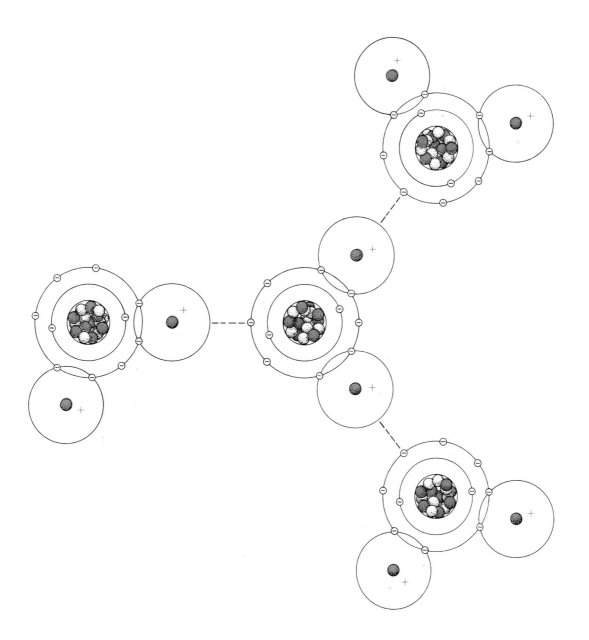

Figure 2–5

Water is a cohesive substance because hydrogen bonds form between water molecules. The negatively charged part of one molecule is linked to the more positively charged portion of another molecule by hydrogen bonds.

it is oxidized. The atom or molecule accepting the electron is reduced. The reason the terms *oxidation* and *oxidized* are used is that oxygen is an atom (although certainly not the only one) that commonly accepts an electron and is reduced. An atom or mole-cule cannot be reduced unless another atom or molecule is oxidized, and vice versa. The two processes are always linked. Redox reactions are common in such biological processes as photosynthesis▫ and cellular respiration▫

57
63

CHAPTER 2 *The Chemical Basis of Life*

BOX 2-2

ACIDS AND BASES

Substances like hydrochloric acid (HCl) and vinegar (acetic acid-CH_3CO_2H) produce hydrogen ions (H^+) when in solution. (A hydrogen ion is a proton without its single electron.) Because of this property, they are called acids—substances that produce protons, or proton donors. Bases are substances that accept protons.

When sodium hydroxide (NaOH) is dissolved in an acid solution, it disassociates into sodium ions (Na^+) and hydroxyl ions (OH^-). The hydroxyl ions have strong affinities for hydrogen ions (protons) and combine with them to form water molecules. This example illustrates the rule that acids and bases tend to neutralize each other and in doing so produce a salt and water:

$$HCl + NaOH \rightarrow NaCl + H_2O$$

acid   base   salt   water

Chemists have devised a scale, called the **pH scale**, to express how acidic or basic (alkaline) a solution is. The scale is a measure of the concentration of hydrogen ions in a liter of solution, expressed in chemical units called moles. The pH scale ranges from 0 to 14, and it is a logarithmic scale; that is, each whole number on the scale represents ten times more acidity or alkalinity than the next whole number. The midpoint of the scale, pH 7, represents neutrality, a state where the number of hydrogen ions equals the number of hydroxyl ions. Solutions with pH values lower than 7 are acids; they have more hydrogen ions than hydroxyl ions. Solutions with pH values greater than 7 are basic; they have fewer hydrogen ions than hydroxyl ions. The accompanying chart shows the pH values of some common substances.

A solution with a pH of 7 contains a ten-millionth (1/10,000,000, or 10^{-7}) of a mole of hydrogen ions. The concentration of hydrogen ions in a solution can be derived quickly when the pH is known. Converting the number of the pH scale to a negative exponent gives the measure of hydrogen ions in moles.

pH Values
of Some Common Substances

Normal human blood	7.4
Average human urine	6.0
Human saliva	5.8–7.1
Human stomach contents	1.5–1.7
Pure water	7

Acid   Alkaline

10^{-12}

10^{-2}

pH 2

10^{-7} 10^{-7}

MOLES H^+ IONS   pH 7   MOLES OH^- IONS

10^{-12}

10^{-2}

pH 12

	CONCENTRATION OF H^+ IONS (MOLES PER LITER)		pH	CONCENTRATION OF OH^- IONS (MOLES PER LITER)	
Acidic	1.0	$= 10^{-0}$	0	10^{-14}	
	0.1	$= 10^{-1}$	1	10^{-13}	
	0.01	$= 10^{-2}$	2	10^{-12}	
	0.001	$= 10^{-3}$	3	10^{-11}	
	0.0001	$= 10^{-4}$	4	10^{-10}	
	0.00001	$= 10^{-5}$	5	10^{-9}	
	0.000001	$= 10^{-6}$	6	10^{-8}	
Neutral	0.0000001	10^{-7}	7	10^{-7}	Neutral
Basic		10^{-8}	8	10^{-6}	$= 0.000001$
		10^{-9}	9	10^{-5}	$= 0.00001$
		10^{-10}	10	10^{-4}	$= 0.0001$
		10^{-11}	11	10^{-3}	$= 0.001$
		10^{-12}	12	10^{-2}	$= 0.01$
		10^{-13}	13	10^{-1}	$= 0.1$
		10^{-14}	14	10^{-0}	$= 1.0$

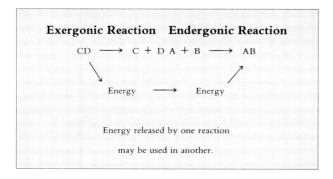

Figure 2–6

Exergonic reactions often furnish energy for endergonic reactions in living cells.

4. Many chemical reactions seem to occur spontaneously and rapidly, but most require some energy input—called **energy of activation**—to get them going (see Figure 2–7). For example, hydrogen gas (H_2) and oxygen (O_2) mixed together in a container will not automatically react with one another to form water. However, if an electrical spark is introduced into the mixture, providing energy of activation, the gases will react explosively to form water.

5. Many chemical reactions proceed slowly. For example, rust on an iron nail is actually the result of an oxidation reaction between iron and oxygen, a slow process under ordinary conditions. Many slow chemical reactions can be speeded up by increasing the temperature of the reactants.

6. A **catalyst** is a substance that reduces the amount of energy needed to initiate a reaction and that maintains the reaction at a specific rate. It combines with the reactants but is freed, unchanged, at the end of the reaction, so it can be used again. **Enzymes** are proteins that act as catalysts for reactions occurring in living organisms. (See Box 2–3.)

7. Certain enzymes require a second nonprotein substance, called a **coenzyme,** in order to function as catalysts. Coenzymes are often derived from vitamins. Figure 2–8 summarizes the relationship between an enzyme and its coenzyme in a reaction that splits one molecule into two.

ORGANIC COMPOUNDS

Carbon is a versatile element. Carbon atoms have four electrons in their outermost shells. These atoms readily form covalent bonds with other carbon atoms and with a variety of other elements, including hydrogen, oxygen, nitrogen, phosphorus, and sulfur.

Figure 2–7

Energy of activation. (a) An electric spark introduced into a mixture of hydrogen and oxygen molecules provides the energy of activation. Then water molecules form explosively. (b) Activation energy is necessary to overcome the energy barrier of the reactants before a chemical reaction can take place.

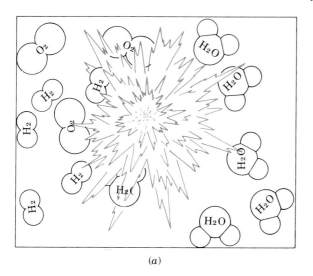

(a)

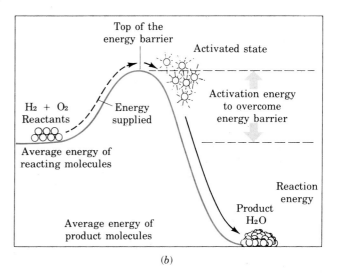

(b)

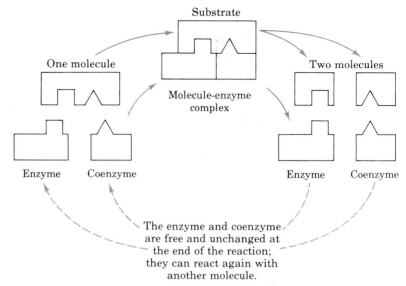

Figure 2–8
An enzyme and a coenzyme may combine in order to catalyze a reaction.

Substrate

One molecule

Molecule-enzyme complex

Two molecules

Enzyme   Coenzyme

Enzyme   Coenzyme

The enzyme and coenzyme are free and unchanged at the end of the reaction; they can react again with another molecule.

Thousands of substances contain large amounts of carbon; examples are coal, kerosene, gasoline, and plastics. Compounds containing various numbers of carbon and hydrogen atoms are called **organic substances.** The molecular structures of a few simple organic compounds are shown in Figure 2–9. Living matter contains large amounts of organic compounds; the most important are proteins, carbohydrates, lipids, nucleic acids, and vitamins.

1. **Proteins** are large, complex molecules composed mainly of carbon, hydrogen, oxygen, and nitrogen. Biochemists know the complete structure of very few of them. Proteins are important because they are part of the structure of cells and because many of them serve as catalysts for chemical reactions in cells. In the latter capacity they are called enzymes.

Protein molecules are made up of long chains of chemical units called **amino acids,** which are linked

Figure 2–9
Six organic molecules formed by atoms linked by covalent bonds.

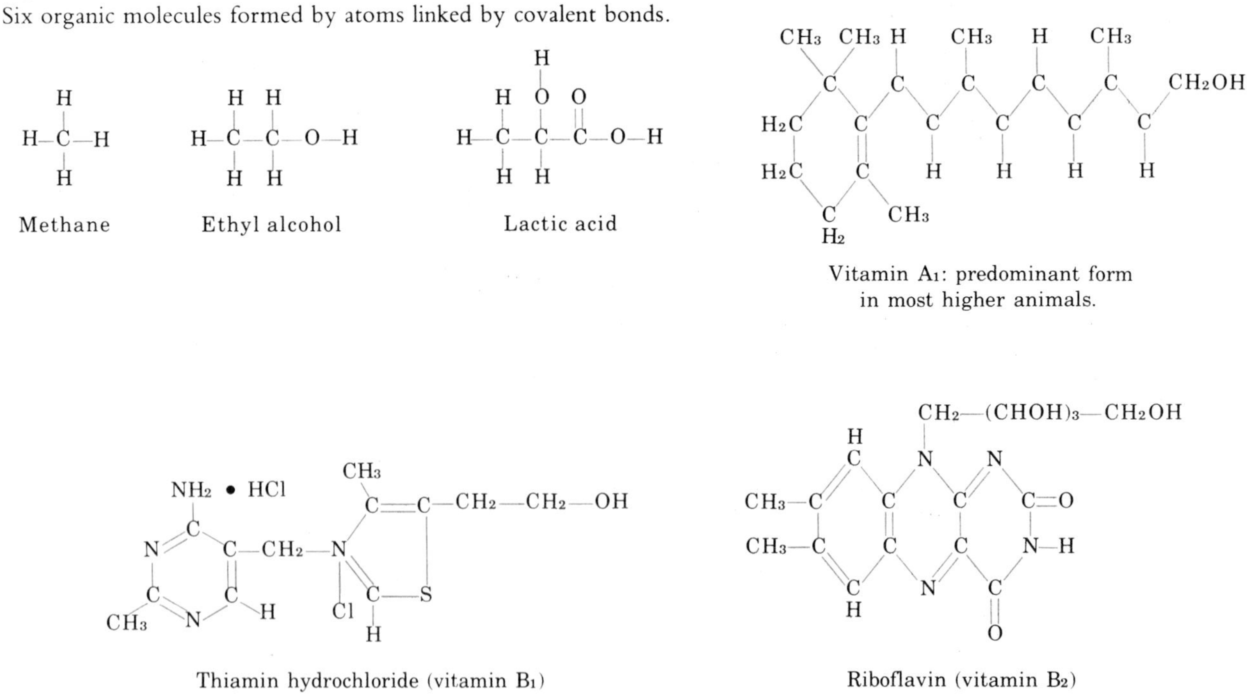

Methane

Ethyl alcohol

Lactic acid

Vitamin A_1: predominant form in most higher animals.

Thiamin hydrochloride (vitamin B_1)

Riboflavin (vitamin B_2)

ENZYMES

Enzymes are the protein catalysts of chemical reactions in living cells and organisms. They possess several unique characteristics, including the following:

1. Enzymes are highly specific. Few of them can catalyze more than one kind of reaction. Therefore, the body's cells contain countless different enzymes, which regulate hundreds of thousands of biochemical reactions. Some of these reactions are involved with food digestion, cellular respiration, and synthesis of new living matter.

2. Enzymes function efficiently only under highly specific conditions. For example, human enzymes function best at a temperature of 37 degrees Celsius (98.6 degrees Farenheit), normal body temperature. Enzymes of other organisms function best at the normal body temperatures of those organisms or at the optimal environmental temperatures for species that have no means of regulating their body temperatures. Also, enzymes function best in environments with specific pH values. The optimal pH for most enzymes is between 5 and 9. Above and below this range most enzymes are inactive.

3. Enzymes are easily inhibited. Certain substances may reduce or totally inhibit enzyme function. For example, carbon monoxide and cyanides inhibit enzymes important in cellular respiration and are therefore toxic to living organisms. Some enzyme inhibitors combine directly with the enzyme and render it inactive. Others combine with the substrates with which the enzyme would have combined.

together by covalent bonds. All life utilizes the same pool of about twenty different amino acids to form an incredible number of different proteins.

To understand the structure of a protein, it is necessary to understand the structure of an amino acid. Each type of amino acid is structurally and chemically different from all the others, but all amino acids have certain features in common. All contain carbon, hydrogen, oxygen, and nitrogen. Some contain additional elements, such as sulfur.

All amino acids in proteins have a carbon atom bonded to the following four groups of atoms: a single hydrogen atom (H); an amino group containing one nitrogen atom and two hydrogen atoms (NH_2);

a carboxyl group consisting of one carbon atom, two oxygen atoms, and one hydrogen atom (COOH); and an R group that represents the remainder of the amino acid molecule, the part that is different in each type of amino acid. Figure 2–10 shows what a generalized amino acid looks like and gives some specific examples.

During the formation of proteins, the amino group of one amino acid forms a **peptide bond** with the carboxyl group of the next amino acid in the chain (see Figure 2–11). In the process, one molecule of water is formed. Therefore, the process is called **dehydration synthesis.** A peptide bond is a type of covalent bond. Short chains of amino acids are called **polypeptides.** As these chains grow longer, they become proteins. For example, the protein myoglobin contains 153 amino acids.

The arrangement of amino acids in a protein molecule is not random. The sequence that forms the primary structure of the molecule is precise and genetically dictated. However, the chains of amino acids in a protein tend to coil like springs or spiral staircases. They give the molecules an additional three-dimensional form, called the secondary level of protein structure. Adjacent coils are linked to one another by hydrogen bonds, which contribute to the stability of these complex molecules. Complex proteins may reach a tertiary structural level when the coiled protein molecule folds on itself in complex patterns that are maintained by chemical bonds between R groups of amino acids in adjacent loops of the protein. Also, in some cases, several molecules with tertiary structural levels can be linked together to form an extremely complex macromolecule with a quaternary structural level.

Because so many amino acids can combine in so many different ways, it is easy to imagine the great variety of proteins that can exist, each with unique properties and functions determined by its specific structure. However, despite their complexity, proteins are easily altered by heat or by treatment with a wide range of chemicals—for example, acids. These agents alter the shape of the protein molecule, and this change alters or destroys the protein's properties and functions. Consider as an example the change in appearance and texture of egg white protein (albumin) when it is heated.

Most proteins are too large to enter or leave cells through cell membranes. They require digestion (chemical breakdown into amino acid units) before

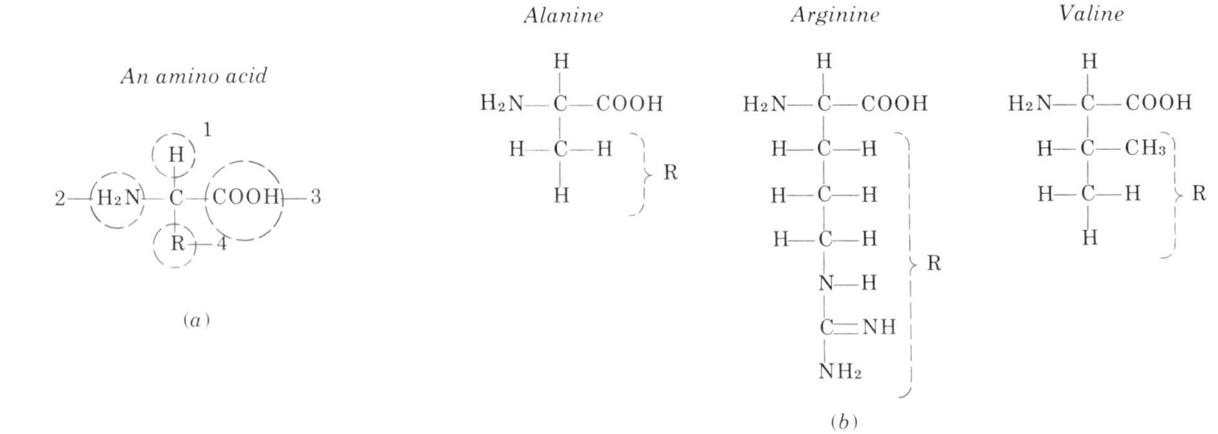

Alanine   Arginine   Valine

An amino acid

$$
\begin{array}{c}
\text{H} \\[2pt]
| \\[-2pt]
2-\text{H}_2\text{N}-\overset{\displaystyle 1}{\text{C}}-\text{COOH}-3 \\[-2pt]
| \\[2pt]
\text{R}-4
\end{array}
$$

(a)

Alanine
$$
\begin{array}{c}
\text{H} \\
| \\
\text{H}_2\text{N}-\text{C}-\text{COOH} \\
| \\
\text{H}-\text{C}-\text{H} \\
| \\
\text{H}
\end{array} \Biggr\} \text{R}
$$

Arginine
$$
\begin{array}{c}
\text{H} \\
| \\
\text{H}_2\text{N}-\text{C}-\text{COOH} \\
| \\
\text{H}-\text{C}-\text{H} \\
| \\
\text{H}-\text{C}-\text{H} \\
| \\
\text{H}-\text{C}-\text{H} \\
| \\
\text{N}-\text{H} \\
| \\
\text{C}=\text{NH} \\
| \\
\text{NH}_2
\end{array} \Biggr\} \text{R}
$$

Valine
$$
\begin{array}{c}
\text{H} \\
| \\
\text{H}_2\text{N}-\text{C}-\text{COOH} \\
| \\
\text{H}-\text{C}-\text{CH}_3 \\
| \\
\text{H}-\text{C}-\text{H} \\
| \\
\text{H}
\end{array} \Biggr\} \text{R}
$$

(b)

Figure 2–10
Amino acids. (a) The structure of an amino acid can easily be remembered by noting that a carbon atom in the molecule is bonded to four groups: a hydrogen atom, an amino group, a carboxyl group, and the remainder of the molecule, called the R group. (b) Three amino acids with their R groups indicated.

they can pass through the membranes. This is accomplished by the process of **hydrolysis,** which means "splitting with water." During hydrolysis, enzymes break peptide bonds, thereby freeing amino acids. The amino acids complete their structures by taking on H^+ or OH^- ions (see Figure 2–12).

2. **Carbohydrates** are another class of organic molecules that are much simpler than proteins. Sugar, starch, and cellulose are examples of carbohydrates. Cellulose and certain other carbohydrates are important structural components in a variety of cells, but one of the most important functions of carbohydrates is their role in energy transformations in cells. Carbohydrates are formed by plant cells during photo-

Figure 2–11
The formation of peptide bonds between amino acids, an example of dehydration synthesis.

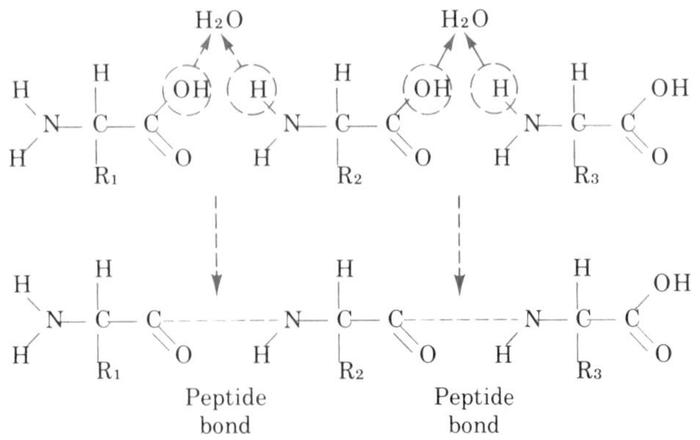

synthesis□. They store energy that can be released by the process of cellular respiration□ to fuel cell functions. Many carbohydrates, including starch and cellulose, are large, complex molecules. Their structure is easily deciphered, however, because they are composed of simple sugars—monosaccharides—linked together. Glucose is a monosaccharide that plays a pivotal role in cellular energy transformations.

Each **monosaccharide** is composed of carbon, hydrogen, and oxygen atoms and in which there are always twice as many hydrogen atoms as oxygen atoms $(CH_2O)_n$—where n represents any number. Another characteristic of monosaccharides is that an oxygen atom may form a double bond with one of the carbon atoms. Thus the oxygen and carbon atoms share two electrons rather than one. Monosaccharides containing three-carbon atoms are called trioses (*tri* means "three"). Trioses and pentoses (five-carbon sugars) are biologically important, but hexoses (six-carbon sugars) are the building blocks of the more complex carbohydrates.

In general, the carbon atoms form a chain to which atoms of hydrogen and oxygen are bonded, but the arrangement of the hydrogen and oxygen atoms differs from monosaccharide to monosaccharide. Although monosaccharides may exist as straight chains, they commonly exist in the form of rings. This molecular arrangement eliminates the double bond between the oxygen and the single carbon. Figure 2–13 shows the structures of two monosaccharides in both forms—as straight chains and as rings.

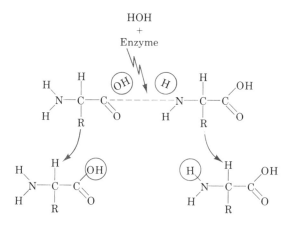

Figure 2–12
Hydrolysis of a peptide bond.

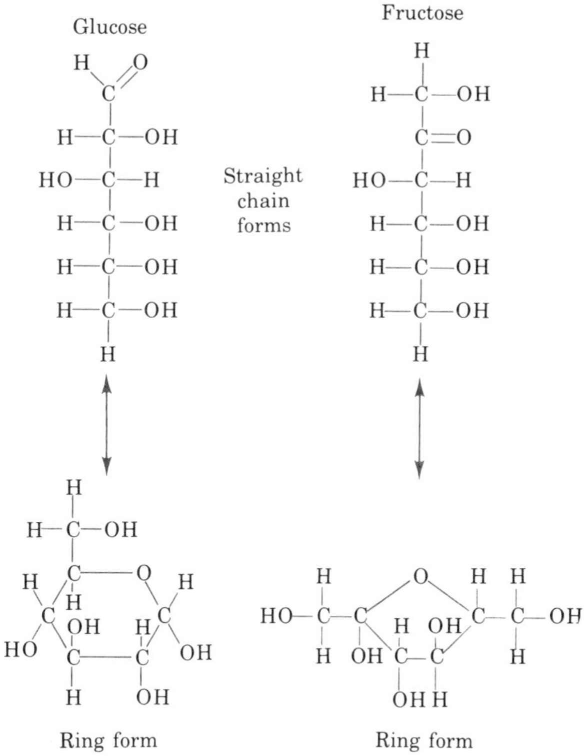

When two monosaccharides bond together, a **disaccharide** is formed. When many monosaccharide units are linked, **polysaccharides,** such as starch and cellulose, are formed. A molecule of water is formed whenever two monosaccharide molecules bond together. This is an example of dehydration synthesis. Two other examples of this type of reaction are illustrated in Figure 2–14. One example shows how the disaccharide maltose is created from two molecules of glucose. The other shows how the disaccharide su-

Figure 2–13
Glucose and fructose, two monosaccharides, in their straight chain and ring forms.

Figure 2–14
Carbohydrate bonding. (a) The formation of a molecule of the disaccharide maltose from two molecules of glucose. (b) The formation of a molecule of the disaccharide sucrose from a molecule of glucose and a molecule of fructose.

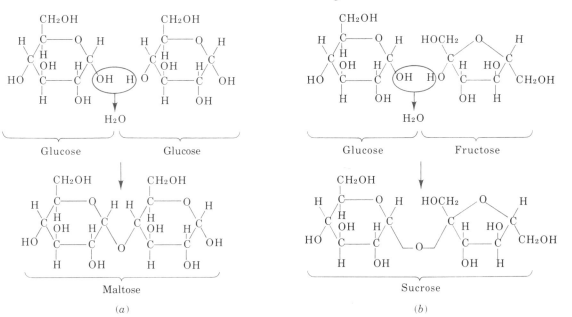

crose (table sugar) is created from one molecule of glucose and one molecule of fructose. Conversely, disaccharides or polysaccharides can be converted to monosaccharides through the process of digestion or hydrolysis. Water is added in this reaction to break the bonds between adjacent monosaccharides.

3. Like carbohydrates, **lipids** are composed primarily of carbon, hydrogen, and oxygen, but there is less oxygen in lipids than there is in carbohydrates. The simplest lipids are the fats; other lipids include the waxes and oils. All lipids dissolve in organic solvents but not in water. Lipids are important structural components of cells, and they may also serve as storehouses of chemical energy in cells. A molecule of a simple fat is composed of four building blocks: one molecule of **glycerol** bonded to three **fatty acid** molecules. Again, the bonds are formed by removing water molecules in dehydration synthesis.

Glycerol is a simple three-carbon molecule, but fatty acids usually consist of chains of twenty-four or more carbon atoms. The formation of a fat molecule is diagramed in Figure 2–15. The long chains of car-bon atoms are bonded to hydrogen atoms, and the carboxyl group (COOH) at one end of the chain bonds with the glycerol molecule. The digestion of fats by hydrolysis separates the molecules of fatty acids from the glycerol molecule through the addition of three molecules of water.

The difference between saturated and unsaturated fats (see Figure 2–16) has become an important nutritional consideration. **Saturated fats** solidify at room temperature, and every carbon atom in the carbon chain of the fatty acids is bonded to four different atoms. **Unsaturated fats,** commonly called oils, are liquid at room temperature. In some instances, two carbon atoms in the unsaturated chain may share two electrons, forming a double bond between them. Saturated fats are easily converted to the complex lipid **cholesterol,** a compound implicated in human heart and artery disease. Atherosclerosis, the accumulation of cholesterol in the walls of arteries, reduces blood flow and increases blood pressure▫.

4. **Nucleic acids** are among the most important compounds in living systems. They represent the ge-

Figure 2–15

The formation of a fat molecule from a molecule of glycerol and three molecules of fatty acids.

Glycerol

Fatty acid

Fat molecule $+ 3H_2O$

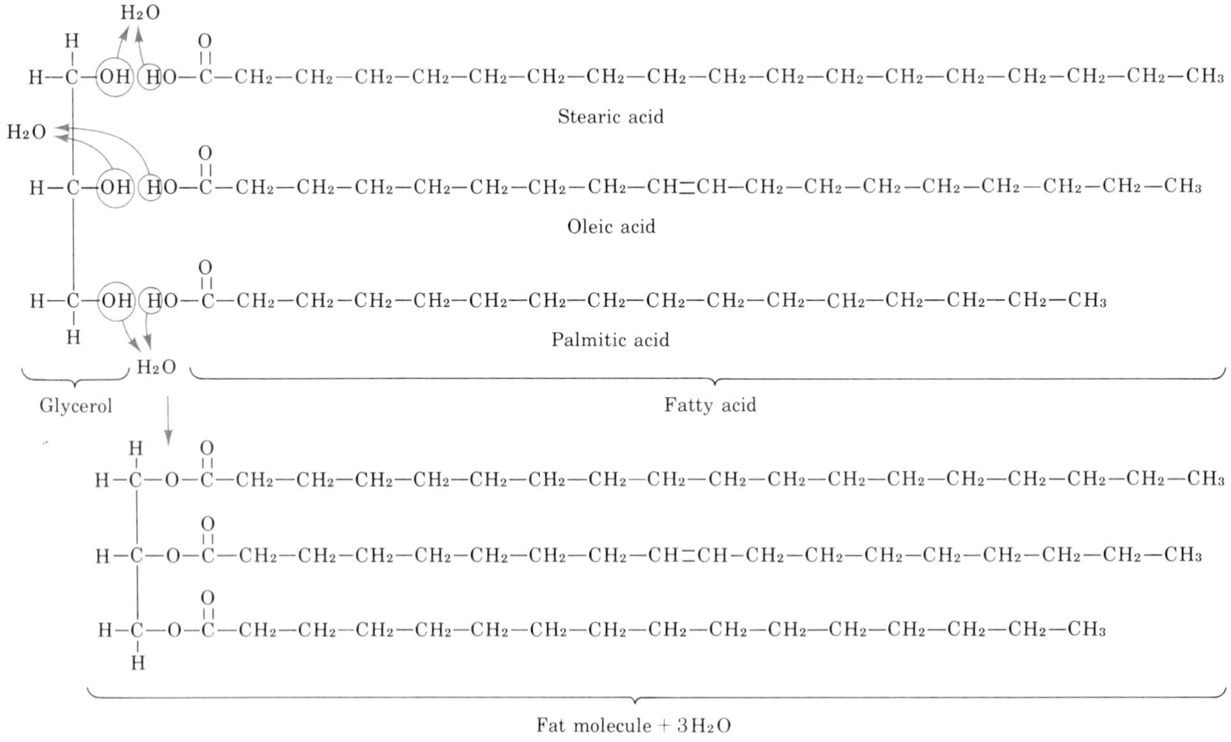

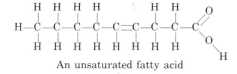

An unsaturated fatty acid

A saturated fatty acid

Figure 2–16

Unsaturated and saturated fatty acids. Fats containing saturated fatty acids are easily converted to cholesterol, a compound implicated in human heart and artery disease.

netic or hereditary material of the cell and are the key to translating genetic information into cellular function. Because these substances are so important, a detailed discussion of them appears in the section on heredity and genetic mechanisms▫.

THE CHEMICAL ORIGINS OF EARTH AND LIFE

Several realistic theories have been offered to explain the origin of the universe, but the most widely accepted theory today is the big bang theory, proposed in 1951.

According to the big bang theory, 20 billion years ago all of the universe was one big ball of **neutrons.** The movement of these particles became greater and greater until the big ball generated unbelievable amounts of heat (estimated at over a billion degrees Celsius). The increase in temperature caused a parallel increase in pressure. Finally, the big ball exploded and created the biggest bang ever known. Neutrons were flung everywhere.

As the neutrons moved farther from their point of origin, they began to cool and to produce negative charges, or **electrons.** The production of electrons left behind **protons,** and the attraction of electrons to protons created hydrogen. (Hydrogen is still the most abundant element in the universe.) This process continued until the newly formed particles began to aggregate into small balls. Each small ball became a galaxy. Our galaxy is the Milky Way. Within each galaxy, the process continued to form smaller balls, creating solar systems like our own.

These balls can be thought of as clouds of gases, which astronomers call dust clouds. As time passed, each dust cloud became cooler. Many developed temperatures that hovered near absolute zero. However, as the particles of the dust clouds slowed down and moved closer, heat once again was generated. The heat became so intense that it caused the fusion of hydrogen, forming helium and releasing energy in the form of light and heat.

The acceleration of this process caused dust clouds to throw off groups of particles, creating eddies of smaller clouds. (The process is analagous to throwing a stone into water. The energy of the stone as it strikes the water creates whorls of movement outward from the energy source.) The hot and illuminated central masses became the stars of the universe. The less-hot eddies of dust radiating around them became planets. Today these processes continue. New stars and planets are constantly being created, while others are being lost or destroyed.

In our own solar system, the earth was formed by this process about 4½ billion years ago. (Some estimates suggest that the earth could be 10 billion years old.)

Like all other planets, the earth was at first a hot, molten mass (cloud) of materials. However, as the mass cooled, hydrogen became the basic building block from which all other elements were made▫. The core of the earth today is still a hot molten ball. Volcanic eruptions demonstrate the existence of the molten core and provide a glimpse of what the earth was like much earlier, when volcanoes that dotted its surface were continually erupting.

Primordial Soup

Life as we know it did not begin until nearly 1 billion years after the earth was formed, when the earth's surface solidified and water accumulated in basins. But the precursors of life were formed as soon as the surface began to cool. Hydrogen was commonplace, and various chemical reactions near the molten surface produced small quantities of oxygen. The oxygen and hydrogen then fused to form water. Because the earth was so hot, the first water was in the form of steam, which rose continuously from the surface of the earth.

Other compounds were being formed at this time. Those important as precursors to life were the combination of hydrogen and nitrogen to form ammonia, hydrogen and carbon to form methane gas, and hydrogen and cyanide to form hydrogen cyanide (see Figure 2–17).

As the earth continued to cool, steam rising from its still-molten crust began to condense into droplets of water, the most important ingredient to life. At first, no water fell to the surface of the earth. Droplets forming in the early atmosphere were converted back to steam as they fell. When the earth cooled sufficiently, the droplets became rain. The rainy season of our young planet was tens of thousands of years long. Rain brought with it molecules of methane, ammonia, hydrogen, cyanide, and nitrogen, which dissolved in the first seas. The early ocean was a primordial soup that contained all these building blocks of life.

It is hypothesized that life did not originate at one specific spot at one specific time in the early oceans of the earth. Forerunners of living organisms originated again and again, wherever the necessary precursors accumulated. Various kinds of molecules became concentrated in bodies of water. With the energy of lightning, volcanic eruptions, and ultraviolet radiation, molecules fused together to form macromolecules. The molecules of major importance to life were **amino acids,** which fused to form **polypeptides** and finally **proteins**▫. (Rocks more than 3 billion years old that were excavated in South Africa contain twenty-two amino acids.) Amino acids and proteins may possess electrical charges, and the attraction of opposite charges is thought to have created increasingly larger molecular aggregates.

A. I. Oparin, a famous Russian biochemist, proposed that the first forms of life were protein aggregates he called coacervates. During their formation, the protein complexes incorporated other compounds, such as sugars▫, fats▫, and even nucleic acids▫. Coacervates had an outer boundary, a thin membrane much like the membrane that encloses cells today. The membrane made it possible for coacervates to control the flow of materials into and out of the aggregate.

Coacervates were subject to the workings of natural selection. Some were better at feeding on organic molecules in the primordial soup and assimilating materials than others were. They could more favorably compete because of the arrangement of molecules within them, and these arrangements became dominant. As soon as amino acids and proteins formed **enzymes**▫, the assimilation process accelerated. These characteristics, coupled with the inclusion of nucleic acids, allowed some coacervates to reproduce themselves. This first form of life probably resembled a virus or a gene encased in a thin membrane. Natural selection determined which form of primitive life perpetuated itself most efficiently.

22
24 26 315
23

Figure 2–17
The earth was dotted with many explosive volcanoes early in its development. Eruptions of these volcanoes released hydrogen, carbon dioxide, oxygen, and nitrogen, which made up the primitive atmosphere. Chemical combinations among these gases eventually formed water, ammonia, methane, and hydrogen cyanide.

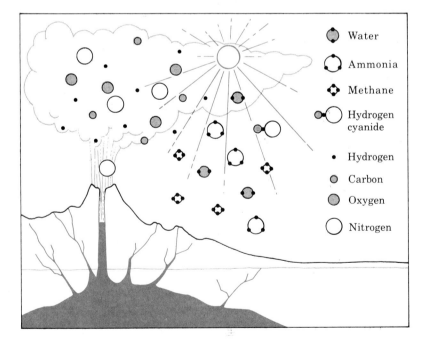

Oparin's proposal is not entirely speculation. Similar series of events have been created in the laboratory. In 1953, Stanley Miller demonstrated the formation of amino acids from similar precursors. Miller's experimental apparatus (Figure 2–18) heated water to form steam; an electrical charge simulated lightning; and the gases ammonia, methane, and hydrogen were introduced. The gases were carried by the steam and activated by the spark. The condensation contained newly formed amino acids as well as sugars and other organic compounds. The experiment has been repeated many times, sometimes with ultraviolet light or ultrasound instead of lightning.

Coacervates have also been made in laboratory test tubes. Sidney Fox, a U.S. biochemist, has heated coacervates and formed what he has called proteinoid microspheres. Microspheres are more stable than coacervates. After they reach a certain size (increasing by the uptake of amino acids), they reproduce by **budding.** It has also been suggested that microspheres contained the primitive precursors of **DNA** and **RNA**.

For millions of years, the coacervates and their descendants were heterotrophs using up the preformed organic food molecules in the environment. Early life would have become extinct if this had continued until the food was gone. Eventually, however, some of the early life forms evolved into autotrophs capable of synthesizing organic molecules from inorganic precursors and light energy. Thus they were the ancestors of photosynthetic green plants. It is also significant that the by-product of this biochemical conversion was oxygen. Autotrophs created an atmosphere rich in oxygen.

In the upper layers of the atmosphere, oxygen formed a layer of ozone (O_3), a filter that reduced the amount of ultraviolet radiation reaching the earth's crust. Ultraviolet radiation provided the energy for

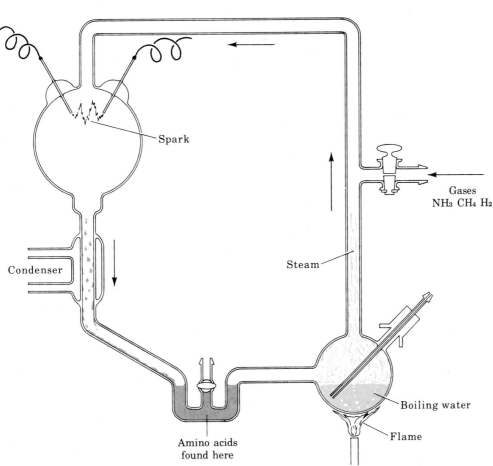

Figure 2–18
The experimental apparatus used by Stanley Miller in 1953 to demonstrate that steam plus the gases ammonia, methane, and hydrogen would produce amino acids in the presence of an energy source (the electric spark). This experiment suggested that similar events may have occurred on earth when life began.

Spark

Condenser

Steam

Gases
NH₃ CH₄ H₂

Boiling water

Flame

Amino acids
found here

SPONTANEOUS GENERATION

Since earliest recorded history, various individuals have theorized about the origin of life. A popular and long-lived idea was that life originated by **spontaneous generation**. The Egyptians believed that crocodiles in the Nile were generated from the sun-baked mud along its shores. Aristotle, who was aware that certain species of birds were found only at specific times of the year, proposed that they were spontaneously generated from the swamps and bogs that they inhabited. Today we know that the birds Aristotle witnessed were migratory species.

Somewhat later, in fifteenth century Europe, it was thought that leaves falling from trees turned into fish if they fell into water and turned into birds if they fell on land. Lambs were thought to originate from plant stalks that their mothers ate, the lambs being attached to the stalk by their umbilici. Two centuries later it was commonly supposed that maggots arose spontaneously from decaying flesh. One early physician even published a recipe on how to generate mice, if and when needed. A few dirty old shirts and a handful of grain stored together in a dark, quiet place would generate mice within a few weeks. When microorganisms were discovered, it was concluded that they too came about spontaneously from water, meat broths, milk, and the like.

The theory of spontaneous generation died during the last half of the nineteenth century, when Louis Pasteur demonstrated before the Academy of Sciences in France that nutrient broth boiled in flasks with curved necks would not "generate" microorganisms. Invading bacteria stuck on the walls of the curved necks and could not reach the broth. However, if the neck was broken off, the broth quickly became infested with minute organisms. Pasteur's process of purification by boiling was named pasteurization after him.

synthesis of the first molecules. A reduction in the levels of ultraviolet radiation reaching the earth stopped the creation of new organic molecules and other kinds of primitive living systems. Ultraviolet radiation of the magnitude that existed before the ozone layer formed would be lethal to life as we know it today.

Life was created in a relatively short time in comparison to the age of the earth. The conditions nec-essary for creating life will never again be present here. However, although the creation of life depended on a series of chance events, it is unlikely that it is a rare event in the universe. It has been estimated that the universe has at least 10^{17} stars with planets. It is within the realm of statistical probability that approximately 10^8 of those planets, 100 million of them, had or have conditions favorable for the creation of life.

In summary, the earliest forms of life were heterotrophic, depending on organic molecules and energy from the primordial soup. Autotrophs evolved from these forms of life. The by-products of the autotrophs' self-sustaining biochemical processes changed the earth's atmosphere. They made possible the evolution of more complex, multicellular heterotrophs and autotrophs that were capable of using oxygen for cellular respiration. The coexistence of these organisms has led to the delicate balance in nature that has lasted for millions of years—a balance now threatened by a dominant form of life, *Homo sapiens*.

SUMMARY

1. The basic substance of living organisms is a complex mixture of thousands of chemicals in solution or in colloidal dispersion in water.

2. All matter in the universe, both living and non-living, is composed of elements.

3. The smallest naturally indivisible units of elements are atoms, which are composed of the subatomic particles protons (positively charged particles), electrons (negatively charged particles), and neutrons (particles bearing no charges). The protons and neutrons make up most of the mass of the atom, forming the dense atomic nucleus.

4. The electrons rotate in orbitals, or energy-level shells, around the nucleus. Each orbital contains a predictable number of electrons, and the energy levels of the shells increase with distance from the atomic nucleus.

5. Atoms of elements can react (form molecules) with other atoms by losing, gaining, or sharing electrons in their outermost shells. Such atoms are then chemically bonded to each other. The most important types of chemical bonds are ionic bonds, covalent bonds, and hydrogen bonds.

6. Chemical reactions usually require some form of energy input (energy of activation) to begin, and they often proceed slowly unless the reactants are heated or unless a catalyst is present.

7. A catalyst is a substance that temporarily combines with the reacting chemicals to facilitate the reaction but that is released unchanged at the end of the reaction, free to function again as a catalytic agent.

8. Organic compounds are substances containing various numbers of carbon and hydrogen atoms. Among the most biologically important organic compounds are proteins, vitamins, carbohydrates, lipids, and nucleic acids.

9. Vitamins combine with protein catalysts (enzymes) as coenzymes in cellular biochemical reactions.

10. Proteins (complex, coiled, and folded chains of amino acids linked by covalent peptide bonds) are important structural components of cells. Many proteins also serve as enzymes.

11. Carbohydrates are structural components of cells and energy stores for cellular respiration. Most carbohydrates are chains of simple sugars bonded together.

12. Lipids exist in a variety of forms. The simplest, fats, are composed of three fatty acids bonded to a molecule of glycerol. Like carbohydrates, lipids serve as structural components of cells and as energy stores.

13. Nucleic acids are the genetic or hereditary material of a cell.

14. The big bang theory is currently the most widely accepted theory about the origin of the universe.

15. Life on earth began after the crust cooled and rains steadily fell. The water containing the precursors of macromolecules has been called the primordial soup.

16. The earliest forms of life were heterotrophic, dependent on organic molecules and energy from the primordial soup. The autotrophs that evolved from these earlier forms produced oxygen, which was added to the earth's atmosphere.

THE NEXT DECADE:

YOUR CHEMICAL ENVIRONMENT

Living matter and nonliving matter are composed of chemicals (atoms and molecules). Therefore, we exist in a chemical environment that changed slowly as life evolved. However, the changes have become more rapid as human populations have expanded. Since the Industrial Revolution, we humans have fouled our environment with human and technological wastes at an ever-increasing rate. Now, changes in the chemical environment may be so rapid that some living systems may no longer be able to keep pace.

Some of the chemicals and materials that we have developed have immensely improved the living conditions and life expectancies of many people—although certainly not all. But all of us will suffer the consequences of our altered chemical environment. All organisms have been exposed to such chemicals as the pesticide DDT, and certainly all will suffer any consequences resulting from increased levels of carbon dioxide in the atmosphere. During the coming decade, however, much attention will be focused on the dangers of polluting our environment with radioactive and other chemical wastes. We will undoubtedly be concerned with decisions about the extent to which nuclear power plants should be used and where chemical wastes can be safely stored.

Radioactive compounds such as those containing the isotope carbon-14 are frequently used in medicine and biological research. Atoms of carbon-14 can become part of the molecules in living organisms, so they can be used as markers or tracers to follow the progress of chemical reactions in cells and tissues. Carbon-14 and other radioactive isotopes can be beneficial if used properly and in small amounts. However, the amount of radiation given off by many isotopes is large, even though the rate of breakdown is slow. If living organisms are exposed to high levels of radiation or to low levels of radiation for long periods of time, cell components such as the genetic material DNA can be structurally altered causing mutations. The organisms may also take up and concentrate radioactive substances in cells, a process that can cause radiation illness.

Until recently most experts on atomic energy maintained that nuclear power plants were safe. These authorities emphasized that the plants were carefully constructed to retain the radioactive fuel and that they had cooling systems with large amounts of circulating water to absorb the great heat given off during radioactive decay. As long as the cooling systems operated efficiently and the structural integrity of the nuclear reactor was maintained, little or no radiation would be released into the environment.

But if the cooling system were to fail, the heat produced by radioactive decay would increase, causing the fuel core to melt (a meltdown). If the molten fuel and water from the cooling system were to come in contact, the steam could cause explosive ruptures in the reactor system. The result would be the release of radiation and radioactive material into the environment.

The unexpected occurred at the Three Mile Island nuclear power plant near Harrisburg, Pennsylvania, on 28 March 1979. Because of mechanical failures and human error (perhaps negligence), the cooling system failed to operate efficiently. The reactor's fuel core overheated and came dangerously close to a meltdown.

If a meltdown had occurred in the highly populated Harrisburg area, the results would have been disastrous.

Since Three Mile Island, several other accidents at nuclear power plants have occurred. Most of them have involved ruptures of corroded steam generator tubes. In 1981 the Ginna power plant in Ontario, New York, was shut down after about 7,200 liters (1,900 gallons) of water containing radioactive substances were released. Serious doubts about the safety of nuclear power plants have increased, and the opponents of nuclear energy have gained in credibility.

Another problem of great concern is where to dispose of nuclear and other chemical wastes safely. In the thirty-five years since the development of nuclear weapons, the United States alone has accumulated 335 million liters of radioactive defense waste. And since the development of nuclear power plants, the United States has accumulated 7,000 metric tons of radioactive spent fuel rods. All of this material is in temporary storage, waiting for someone to decide the best way to store it safely and permanently where it will not leak radioactivity or contaminate groundwater. Some isotopes take thousands of years to become safe.

Toxic wastes of many sorts have been stored in special containers or in specially constructed landfills in the earth. Although many of these storage systems were once regarded as safe, leaks eventually developed. They caused toxic substances to be released into soil and water, often contaminating entire communities. One such contamination was of the Love Canal community in Niagara Falls, New York.

Can we safely store nuclear wastes for thousands of years? Many think not, although ingenious ideas on the subject have been proposed. One plan is to solidify nuclear wastes in glass, place the glass in containers, and bury the containers in salt deposits where little or no water exists and where no earthquakes occur.

During the next decade, the debate about how and where to store chemical wastes will continue. The solution, if there is one, must come soon. Our chemical environment is rapidly deteriorating.

FOR DISCUSSION AND REVIEW

1. List the three states of matter. Give an example of a substance other than water that illustrates the three states.

2. Name the six elements that make up most living matter. What is the state of these substances in pure form?

3. In simple terms, define *atom* and describe its structure.

4. List three types of chemical bonds, and describe how each is formed. Name a substance that is a part of living systems that has each of these chemical bonds.

5. What kinds of events can occur during a chemical reaction?

6. What is the difference between an acid and a base?

7. Define or explain each of the following: (a) oxidation-reduction (redox) reaction; (b) energy of activation; (c) catalyst; (d) coenzyme.

8. What are vitamins?

9. Describe the structure of a simple protein and a complex protein. Name two proteins. What functions do proteins have in living systems?

10. Distinguish between hydrolysis and dehydration synthesis.

11. Name several carbohydrates. Describe the structure of starch. What important roles do carbohydrates play in living systems?

12. Describe the structure of a simple fat. What important roles do lipids play in living systems?

13. Distinguish between an organic compound or substance and an inorganic compound or substance.

14. Describe the probable origin of the earth and life.

15. Differentiate between coacervates and proteinoid microspheres.

SUGGESTED READINGS

CHEMICAL BACKGROUND FOR THE BIOLOGICAL SCIENCES, by Emil White. Prentice-Hall: 1969.
A good review of the basic chemistry used in the study of biology.

THE WORLD OF CARBON, by Isaac Asimov. Collier Books: 1962.
Everything you ever wanted to know about carbon.

THE WORLD OF NITROGEN, by Isaac Asimov. Collier Books: 1962.
Everything you ever wanted to know about nitrogen.

THROUGH THE MOLECULAR MAZE, by Allen Breed, Thomas Rodella, and Ronald Basmajian. William Kaufmann: 1975.
A helpful guide to the elements of chemistry for beginning life science students.

SECTION TWO
THE BIOLOGY OF CELLS

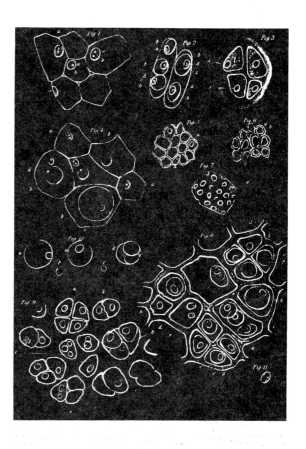

The drawing above was made by Theodor Schwann in 1838. These sketches of animal and plant cells represent the first attempt made to show the fundamental similarity between the structures of animal and plant cells.

3

CELL
STRUCTURE

PREVIEW QUESTIONS

After reading this chapter, you should be able to answer the following questions:

1. How do biologists study the structure of cells?

2. What structures are contained in cells? What are their functions?

3. What differences are there between plant cells and animal cells?

4. What differences are there between bacterial cells and other kinds of cells?

5. What is a virus?

One of the basic concepts in biology is **cell theory.** In general terms, this theory states that (1) all living organisms are composed of one or more cells, (2) all cells come from preexisting cells, (3) cells contain hereditary material, and (4) the metabolic (chemical) processes of organisms take place in cells.

Cells, the basic units of life, come in many sizes and shapes. Their shapes vary almost infinitely, but their sizes are usually between one and a hundred micrometers in length or diameter. A micrometer is one-millionth of a meter. Scientists throughout the world use the metric system of measurement. Students who are unfamiliar with the system should consult Appendix 1.

Some organisms consist of only a single cell; they are called **unicellular organisms**. Those with bodies composed of more than one cell are called **multicellular organisms**. Large organisms may be composed of millions or billions of cells, many of which are specialized to carry out some primary function that contributes to the functioning of the total organism. Groups of cells of similar type and function are called **tissues**.

Each human body is composed of approximately 60,000 billion cells, all descendants of the individual's original cell, the fertilized egg. The fertilized egg divides into two cells, then four, eight, and so on, until several billion cells have been arranged into a recognizable human embryo. All embryos, human and otherwise, develop in a similar fashion. The first cells are generalized cells; they give rise to increasingly specialized cells. This process, called **cellular differentiation**, accounts for the muscles, nerves, blood cells, and all the other specialized tissues and cells of multicellular organisms. Specialized cells have many unique characteristics, but they possess the basic structural components characteristic of all cells.

STUDYING CELLS

Most cells are too small to be seen without a microscope. Therefore, nobody actually saw a cell until the first microscopes were developed, during the second half of the seventeenth century. Robert Hooke of England made the first report of cellular microscopic structure in 1665. (See Figure 3–1.) Examining thin slices of cork with one of the first microscopes, he saw that the cork was composed of very small units separated by walls. He called these units *cellulae* which means "little rooms." Hooke was actually observing not the living cork cells but rather the spaces where living cells had once existed. Nevertheless, his cellulae became known as *cells*, the term still used to describe these basic units of life.

About the same time, in Holland, Anton van Leeuwenhoeck was constructing some of the finest microscopes available. In search of things to observe with his instruments, he discovered many tiny organisms that were previously unknown. He was the first to see human sperm and blood cells. Although he re-ported the existence of many kinds of cells, he never called them cells and was unaware of the significance of his findings.

With the development of better microscopes, more biologists described the microscopic anatomy of various organisms and tissues. Matthias Schleiden, a German botanist, in 1838, and Theodor Schwann, a German zoologist, in 1839, came to the same conclusion: All living things are composed of cells. Both men are credited with the development of the cell theory. In 1858 Rudolf Virchow added to cell theory with his proposal that all cells came into existence from preexisting cells. Later, Louis Pasteur proved conclusively that all living organisms originate from preexisting organisms. This evidence forced the scientific community to abandon its notion that cells and organisms arise spontaneously.

Today, the study of cells is known as **cytology**. Beginning less than a century ago, cytology advanced rapidly as better microscopes were developed and techniques for preserving and staining cells were improved (so more living detail could be deciphered). The clarity and detail of the cells pictured in this

Figure 3–1

(a) The microscope used by Robert Hooke, who first described cells in 1665. (The Bettman Archive.) (b) One of the types of electron microscopes commonly in use today (Courtesy Illinois State University—Photo Services.)

(a)

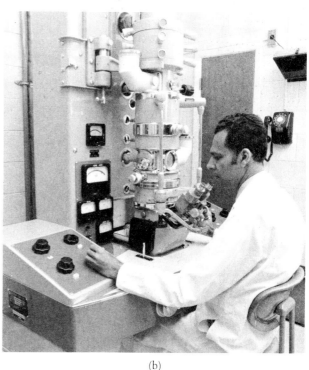

(b)

chapter are the result of many sophisticated techniques.

One of the prerequisites for observing cells is to preserve them in such a way that their structure and inner arrangement are retained as in life. This is accomplished as follows. The cells are treated with a variety of chemical substances called fixatives. Then the cells and tissues are sliced as thinly as possible for microscopic examination. However, before the tissues are sliced, they are frozen or infiltrated with paraffins or plastics to give them strength and support. Specialized machines then cut slices as thin as one or two micrometers. Finally, special dyes are added to enhance the microscopic picture of cells and tissues, creating contrasts and coloring the various structures. Special techniques are also available for examining living cells and tissues.

The electron microscope was developed in the 1930s. Although it has been in common use for only the past twenty years or so, its impact on concepts of cells and cellular components has been dramatic. Hooke first observed cells through a microscope that had a magnifying power equivalent to today's dimestore magnifying glasses. The microscope in any biology laboratory is capable of magnifying cells and tissues several hundred times. The light microscopes used by researchers can magnify objects a thousand times or more. The electron microscope, however, magnifies objects up to three hundred thousand times (see Figure 3–1).

Magnification power is not the only factor important for cellular studies. Microscopes must also have high resolving power—the degree to which two close points can be distinguished. Good light microscopes have resolving powers of slightly less than one micrometer, but electron microscopes can distinguish between two points as close as one nanometer or less. (A nanometer is one-billionth of a meter.)

ANIMAL CELL STRUCTURE

Whether you scrape a cell from the lining of your mouth and observe it under a microscope or examine a unicellular organism such as an amoeba (see Figure 3–2), you will see some of the most obvious struc-

Figure 3–2

An amoeba and its cellular structures are easily seen with an ordinary light microscope.

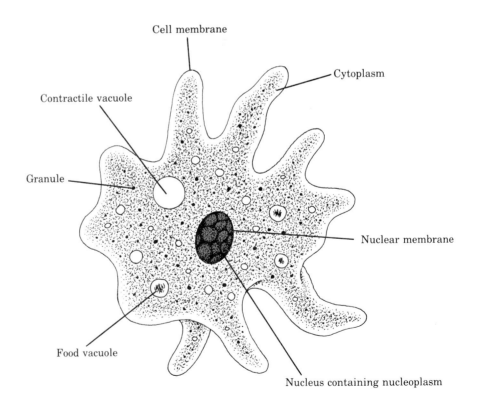

Cell membrane

Cytoplasm

Contractile vacuole

Granule

Nuclear membrane

Food vacuole

Nucleus containing nucleoplasm

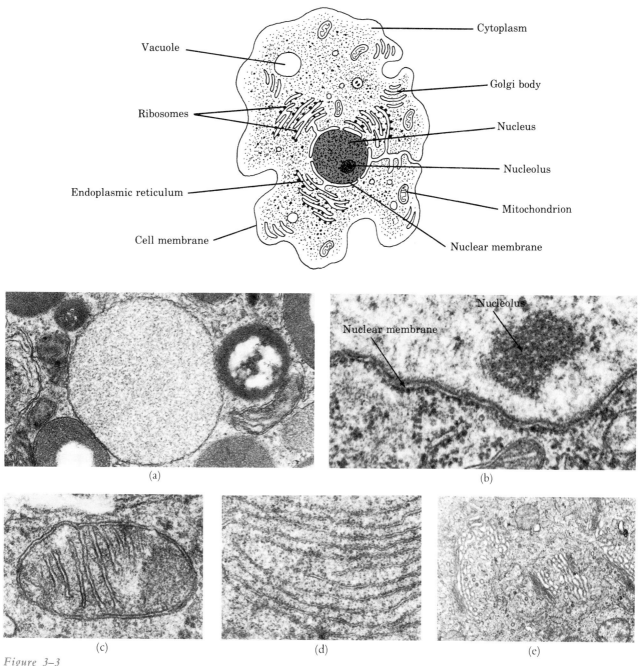

Figure 3–3

A simplified diagram of a thin section through an animal cell, with photomicrographs of (a) a lysosome, (b) a portion of a cell nucleus and its membrane, (c) a mitochondrion, (d) endoplasmic reticulum, and (e) Golgi complex. The cytoplasm of a eukaryotic cell such as this contains a complex system of membranes, the endoplasmic reticulum, that are continuous with the surface cell membrane and the nuclear membrane. The system forms channels that lead to the exterior of the cell and to pores in the nuclear membrane. These pores and channels allow the transfer of materials between the nucleus and the cytoplasm. The endoplasmic reticulum is involved with the synthesis of protein and other cellular components. Many cell organelles are also composed of membranes similar to those of the endoplasmic reticulum. For example, lysosomes are membrane-enclosed vacuoles containing lydrolytic enzymes; each mitochondrion or organelle associated with cellular respiration is composed to a smooth outer membrane and an inner membrane. These are convoluted to form inward projecting folds, called cristae. A Golgi complex is a complex series of concave membrane-enclosed cavities, called cisternae; they are the sites of the synthesis of a variety of cellular sections. (Photo by M. Nadakavukaren, Illinois State University.)

tures of cells. One part of each cell is too thin to be seen: the external **cell membrane**, which separates the contents of the cell from the environment and maintains the integrity of the cell. The membrane and the contents of the cell are collectively referred to as **protoplasm**. Within the protoplasm are structures called **organelles** ("little organs"). The **nucleus** is the most prominent organelle. Its contents, called **nucleoplasm**, are separated from the rest of the protoplasm by a **nuclear membrane**. The nucleus is the command center, because it contains the hereditary material that dictates the form and function of the cell. The contents between the nucleus and the cell membrane are referred to as **cytoplasm**. Thus the protoplasm has two distinct portions—the nucleoplasm and the cytoplasm.

If you examined the cytoplasm of a cell with the typical microscope in a biology laboratory, it would appear to contain granules, bubbles, and specks in a homogeneous medium. No other detail would be discernible because of the low resolving power of the microscope. The true complexity of cellular structure and organization can be appreciated only if the cell is observed with an electron microscope. In fact, electron microscopes show that each cell is a dynamic, ever-changing entity enclosed by a complex cell membrane that is also ever-changing (see Figure 3–3).

The number and location of organelles varies from cell to cell, depending on the primary function of the particular cell. Also, the number and type of organelles within a single cell may change, depending on the cell's activity. We will now examine the various organelles of a generalized animal cell.

The Cell Membrane

Often referred to as the plasma membrane, the cell membrane maintains the shape of the cell and regulates the passage of substances into and out of the cell. Simple materials move quickly and easily through the cell membrane. More complex materials are transported through it by carrier molecules. As already mentioned, a cell membrane is so thin that it cannot be seen through an ordinary microscope. In fact, until the electron microscope appeared, scientists could only assume that it existed.

The cell membrane appears to be composed of three layers—an inner layer and an outer layer separated by a less dense layer in the middle (see Figure 3–4). The denser layers are composed of protein, and

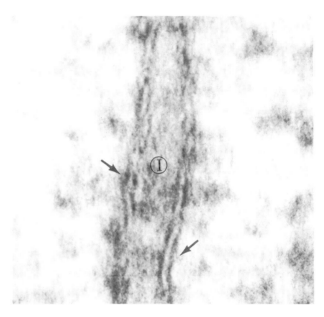

Figure 3–4

A photomicrograph of the cell membranes of two adjacent cells (indicated by arrows) separated by an intercellular space (I). Each membrane is composed of two dark, electron-dense layers separated by a less dense, lighter layer. (Photo by M. Nadakavukaren, Illinois State University.)

the middle layer is composed of fats, or lipids. The cell membrane is sometimes referred to as the **unit membrane**, because its structure is repeated throughout the cell, differing from organelle to organelle only in thickness and amounts of protein or lipids (see Figure 3–5a).

The uniform structure of the cell membrane has puzzled some cytologists. They wonder how the membrane can act so selectively when its structure is so uniform, and they propose an alternative view of the cell membrane, called the **fluid mosaic model** (see Figure 3–5b). This view is that proteins also extend through the middle layer. The arrangement of proteins in the inner and outer layers, in league with the proteins in the middle layer, permits selected substances to pass through the entire cell membrane. Evidence in favor of this new theory has been accumulating.

The Endoplasmic Reticulum

Indentations in the cell membrane penetrate the cytoplasm, sometimes as far as the nuclear membrane.

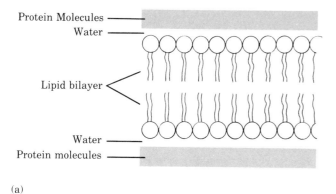

Protein Molecules ——
Water ——
Lipid bilayer ——
Water ——
Protein molecules ——

(a)

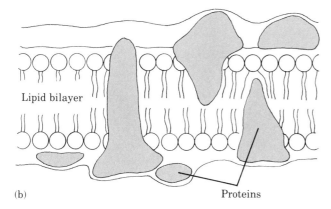

Lipid bilayer

(b)   Proteins

Figure 3–5

(a) the proposed structure of the unit membrane of a cell. (b) The fluid mosaic model of the cell membrane. Note that some protein masses extend into and through the lipid layer.

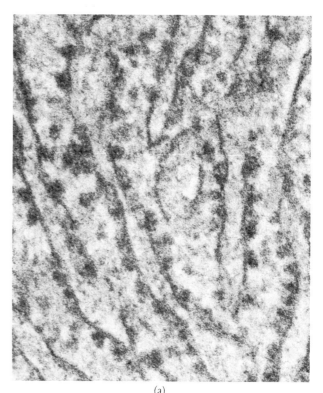

(a)

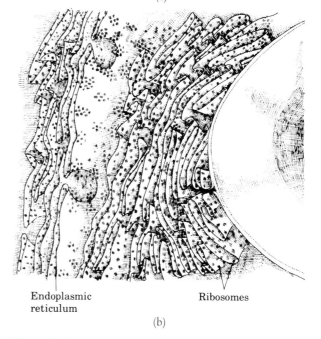

Endoplasmic   Ribosomes
reticulum
(b)

Figure 3–6

(a) Electron micrograph of rough endoplasmic reticulum with ribosomes. (Photo by M. Nadakavukaren, Illinois State University). (b) Three-dimensional representation of the endoplasmic reticulum.

This series of membranes, called the **endoplasmic reticulum (ER),** creates a vast array of channels within the cell that serve as lines of chemical communication among the nucleoplasm, the cytoplasm, and the external environment.

There are two types of ER: rough and smooth. **Rough ER** is studded with tiny granules that are easily seen with an electron microscope. These granules, called **ribosomes**, play an important role in the synthesis of proteins▫. (Ribosomes may also appear freely in the cytoplasm.) Rough ER is most abundant in growing cells or in cells that secrete proteins. For example, the cells in the pancreas that secrete insulin (a small protein) have large amounts of rough ER (see Figure 3–6).

Smooth ER is so named because it has no ribosomes. It is abundant in cells that secrete steroid or lipid substances, such as sex hormones. Smooth ER probably plays a role in the synthesis and storage of a variety of lipids.

The Golgi Complex

An organelle composed of a series of parallel membranes that enclose flattened, fluid-filled spaces called **cisternae** is known as the **Golgi complex**. The cisternae are often slightly curved, making the entire complex appear concave. Each Golgi complex has from five to thirty cisternae. The complex is involved in secreting substances from the cell. Some of the substances may be formed by the complex, but most are probably synthesized by the rough ER. The rough ER then transmits the substances to the Golgi complex, where they are processed and packaged before being secreted. When Golgi complexes are active, small sacs, called vesicles, break off from the ends of the flattened cisternae and slowly move toward the cell membrane. These vesicles become part of the membrane, and their contents are passed outside the cell. Golgi complexes are also involved in the formation of new cell membranes during cell reproduction□ and of the membranes that comprise the endoplasmic reticulum. They also produce lysosomes.

77

Lysosomes

Although commonly formed at the ends of the cisternae of Golgi complexes, some **lysosomes** develop from the endoplasmic reticulum. Lysosomes are essentially membranous bags that contain enzymes capable of digesting a wide variety of substances. When a cell engulfs viruses, bacteria, or other large particles, these substances become enclosed in a membrane-lined organelle called a vacuole. The vacuole then fuses with one or more lysosomes, which furnish the enzymes for digesting the foreign substance. Human white blood cells contain many lysosomes, which effectively destroy foreign matter and bacteria engulfed by these cells. This ability makes white blood cells a major defense against infectious disease.

Lysosomes may also digest cytoplasmic organelles and other membranes, allowing the cell to remodel and replace old organelles. This function may be involved in the aging process and the death of cells. In fact, lysosomes are sometimes called the suicide bags, because cell death is often associated with the rapid digestion of cell contents by lysosomes. Most of the destroyed cells are replaced, so cell destruction can have a positive and rejuvenating role in the biology of organisms□.

88

Some of the functional relationships among the endoplasmic reticulum, the Golgi complexes, and lysosomes are summarized in Figure 3–7.

Microbodies

The tiny membrane-enclosed vacuoles containing a mixture of enzymes are **microbodies.** A variety of microbodies, including peroxisomes and glyoxisomes, have been identified in plant and animal cells.

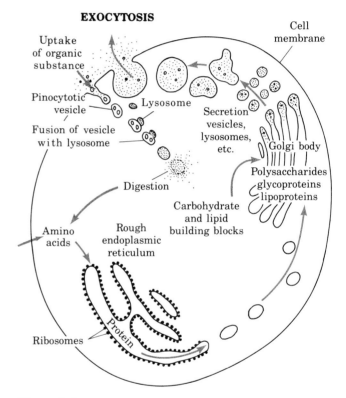

Figure 3–7
The functional relationships of various cell organelles. Amino acids are taken into the cell through the cell membrane. Some proteins are taken into the cell in small vacuoles called pinocytotic vesicles. After digestion, amino acids are liberated into the cytoplasm. They are incorporated into proteins by ribosomes. Some of the proteins are transferred to Golgi bodies, where they are combined with carbohydrates or lipids, forming complex compounds such as lipoproteins or glycoproteins. Some of these compounds are secreted by the cell.

The enzymes contained in microbodies are involved in a number of cell processes, including the breakdown of amino acids and lipids□.

Mitochondria

The powerhouses of cells are the **mitochondria.** They contain essential enzymes that release energy from nutrients, a process called cellular respiration□. The cells use the energy released by mitochondria in many ways—for movement, active secretion of substances, cell division, and so on. As might be expected, more active cells contain more mitochondria than do less active cells. For example, muscle cells have many more mitochondria than do gland cells.

Mitochondria show a distinct and easily identified structure when viewed with an electron microscope. As Figure 3–8 indicates, a mitochondrion is essentially a membranous sac composed of an outer membrane and an inner membrane. The inner space of the mitochondrion, the lumen, contains a semisolid substance, the matrix. The outer membrane is stretched smoothly over the mitochondrion and appears unwrinkled.

However, the inner membrane is folded into projections that extend into the lumen. These folds, called **cristae**, are the most distinctive feature of the mitochondrion. They increase the surface area of the inner membrane many times. A number of the important enzymes involved in cellular respiration are located on the surface of the mitochondrion's inner membrane. Mitochondria change shape readily and move through the cytoplasm, but the cristae always make them easy to identify.

Microfilaments, Microtubules, and Associated Organelles

The organelles of living cells occasionally move about within the cells. Cells are also capable of changing their shapes. These activities are made possible by the presence of extensive networks of highly contractile protein **microfilaments**, called the **microtrabecular system**. The microfilaments extend throughout the cell and act as a scaffold in supporting the organelles. The contraction and elongation of the microfilaments enable the organelles to change positions or the cells to change shapes. The actions of the microfilaments are evident in the constriction of the cytoplasm in dividing cells and in the contraction of muscle cells.

Many cells are enabled to move by the microscopic, hair-like **cilia** or **flagella** that extend from their surfaces. Unicellular organisms, such as certain groups of protozoa and algae, swim about because of the beating of cilia or flagella. Human sperm cells are also capable of swimming because each bears a single flagellum. Cilia and flagella flex and bend because of the microtubules they contain. **Microtubules** have a more complex structure then do microfilaments, as Figure 3–9 shows. Each is an unbranched cylinder composed of thirteen columns of globular protein molecules. Microtubules bend because of changes

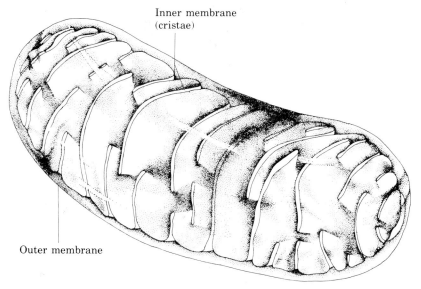

Inner membrane (cristae)

Outer membrane

Figure 3–8
A mitochondrion, showing the outer, smooth membrane and an inner membrane whose folds form projections, called cristae, that extend into the matrix of the structure.

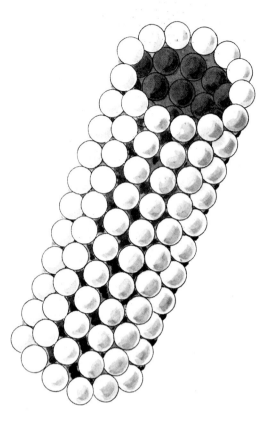

Figure 3–9
A microtubule is composed of thirteen columns of globular proteins.

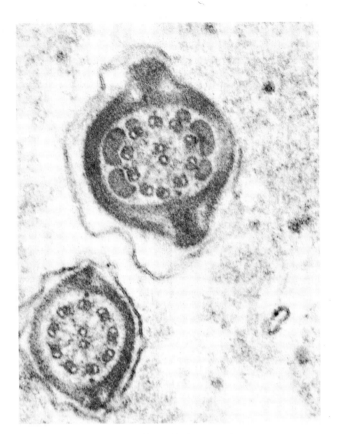

Figure 3–10 (Continued on facing page)
(opposite) Diagram of the arrangement of microtubules in a cilium and its basal body which contains the same arrangement of microtubules as does a centriole in an animal cell. *(above)* Electromicrograph of the cross sections of cilia, showing the arrangement of the microtubules. (Photo by M. Nadakavukaren, Illinois State University)

within this protein structure. The arrangement of microtubules in cilia and flagella is precise and always the same (see Figure 3–10). Extending the length of each cilium or flagellum is a pair of microtubules. This pair is surrounded by nine other pairs. Embedded in the cytoplasm at the base of each cilium is a structure appropriately called a basal body. It too contains microtubules, but in an arrangement different from that within the cilium itself. In basal bodies, microtubules are arranged in groups of three rather than in pairs. There is no central group; instead there are nine groups of three each, extending the length of the basal body and arranged in a radial pattern (see Figure 3–10).

Another cell structure that contains microtubules is the **centriole.** This structure is found in animal cells but not in cells of higher plants, and it is involved in the process of cell division. The microtubules in centrioles are arranged as they are in basal bodies.

The Nucleus

The largest and most obvious structure in a cell is the nucleus (see Figure 3–11). It contains the genetic material that codes and controls the cell's structure and general activity. This nucleus is separated from the cytoplasm by a double nuclear membrane, which is continuous with the endoplasmic reticulum of the cell. The nuclear membrane has pores that allow material to pass between the nucleoplasm and the cytoplasm.

The nucleoplasm contains numerous fine granules and threads, which are parts of unbroken **chromatin threads** (*chromo* means "color"; chromatin threads

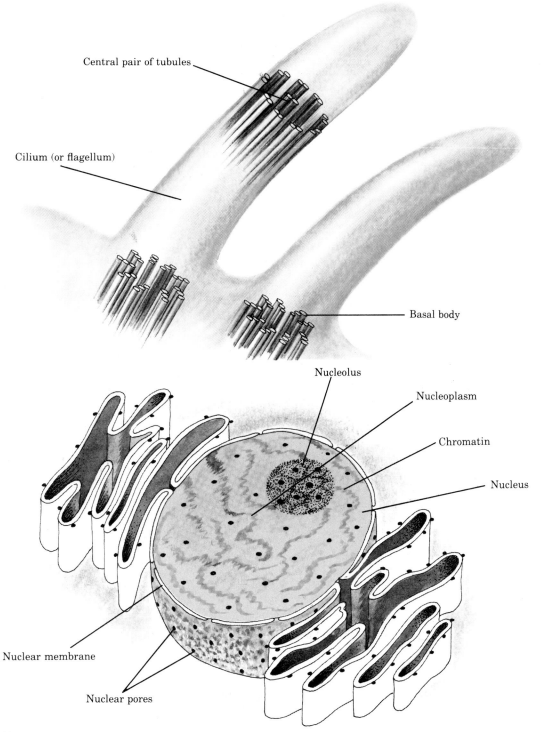

Central pair of tubules

Cilium (or flagellum)

Basal body

Nucleolus

Nucleoplasm

Chromatin

Nucleus

Nuclear membrane

Nuclear pores

Figure 3–11

The nucleus, the control center of the cell, contains the genetic material that regulates all cellular activities. Note that it is separated from the rest of the cell by a double nuclear membrane but that pores in the membrane provide communication channels between the nucleus and the cytoplasm.

become colored when treated with certain dyes). When a cell reproduces, the long, irregularly spaced chromatin threads are transformed into the more conspicuous **chromosomes**.

Chromatin threads are actually the genetic material **deoxyribonucleic acid (DNA)** in association with specific proteins called nucleoproteins. Unfortunately, the threads are too fine to study in detail, even with an electron microscope. An occasional large clump of chromatin, called **heterochromatin**, appears in the nucleus. The patches of heterochromatin are attached to the finer chromatin threads. Current theory favors the idea that the patches of heterochromatin are masses of inactive DNA.

The nucleus also contains one or more structures called **nucleoli** ("little nuclei"). Nucleoli, which are often spherical, are most obvious in actively growing cells or in cells engaged in protein synthesis. They are always associated with specific regions of certain chromosomes, called nucleolar organizing regions. Nucleoli are aggregates of **ribonucleic acid (RNA)** and granular and fibrous proteins. RNA that will later combine with proteins to become ribosomes is produced in nucleoli. Cytologists have observed a chain of biochemical events that begins with the DNA of chromosomes, involves nucleoli, and is linked with protein synthesis□ in the cytoplasm. (See Box 3–1.)

321

PLANT CELL STRUCTURE

The body of any multicellular plant contains a variety of cell types and tissues, as does the body of any multicellular animal. Figure 3–12 shows a generalized plant cell. Such cells contain almost all the structures common to animal cells; they lack only centrioles and, usually, lysosomes. In addition, some structures are unique to plant cells.

The Cell Wall

Plant cells have a rigid **cell wall**, external to the cell membrane (see Figure 3–13). The wall is a nonliving substance, but it is secreted by the living cell contained within it. Cell walls are most commonly composed of the polysaccharide **cellulose** and can be complex. The rigidity of plant tissues is regulated by

---

BOX 3–1
ABNORMAL CELLS

Cells can become abnormal both biochemically and structurally. Some of the abnormalities arise spontaneously, and some are inherited. Some are induced by foreign agents, such as viruses, or by other environmental factors.

Human cancer cells are good examples of abnormal cells; they are abnormal biochemically, structurally, and functionally. Pathologists can recognize them by their shape and the size of their nuclei. The nuclei are usually large and irregular, and they have deeply indented surfaces. Their nucleoli are also abnormal in size and shape.

Cancer cells grown in laboratories look different from normal cells grown under similar conditions. Cultured cancer cells are usually spherical; normal cells are not. It has been suggested that cancer cells are spherical because many of the microfibrils in their microtrabecular systems are less stabilizing than are microfibrils in normal systems.

Another feature of cancer cells is that they usually contain a greater number of chromosomes than do normal cells. For example, HeLa cells (see "The Next Decade" for this chapter) have between seventy and eighty chromosomes in their nuclei instead of forty-six, the normal number for humans. Also, when normal cells taken from a young human are cultured, they divide and continue to live for only fifty generations or so, but cancer cells go on dividing indefinitely. Finally, cancer cells reproduce more rapidly than normal cells. The study of abnormal cancer cells is one of the largest and most intense areas of biological and medical research today.

---

the thickness of cell walls. Wood is composed of extremely thick cell walls of dead plant cells.

Thick cell walls have two layers. A plant cell first secretes the thin primary cell wall, which is composed of loosely organized cellulose fibers. It is elastic and stretches as the plant cell grows. It attains maximum thickness when the cell reaches maturity and its maximum size. Then the cell secretes the secondary cell wall underneath the primary cell wall. The secondary wall is thicker and more rigid. It is composed of dense layers of cellulose fibers oriented at about sixty degrees to each other. It is frequently strengthened by the addition of hard materials, such as lignin. The

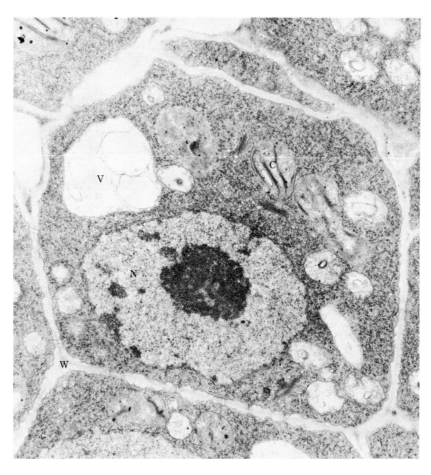

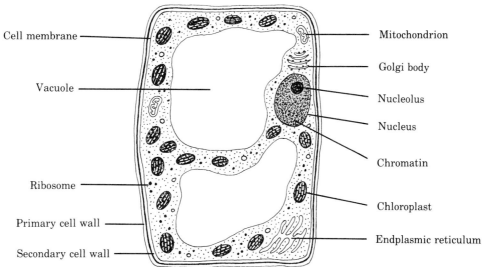

Figure 3–12
Electron micrograph *(above)* and a drawing *(below)* of a plant cell, showing a portion of the plant cell vacuole (V), the nucleus (N), chloroplasts (C), the position of the plant cell wall, and other structures. (Photo by M. Nadakavukaren, Illinois State University)

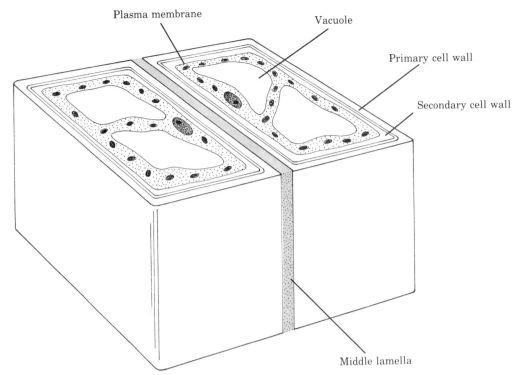

Figure 3–13

Plant cell walls. An immature plant cell first secretes a primary cell wall outside its cell membrane and later, as it matures, a secondary cell wall between its cell membrane and the primary cell wall. Plant cells are separated by a middle lamella containing pectin.

Plasma membrane

Vacuole

Primary cell wall

Secondary cell wall

Middle lamella

cells of flexible, soft plant parts often have only a primary cell wall.

Adjacent plant cells are glued tightly together by a layer outside their primary cell walls, called the **middle lamella**. It is commonly composed of pectin, a complex polysaccharide. (Pectin is extracted from citrus fruit peel and added to fruit juices to thicken or jell them.) The cells of all plant tissues are held together by middle lamellae, whether the cells have both primary and secondary cell walls or only a primary cell wall. Plant tissues with well-developed middle lamellae tend to be firm or hard; unripe fruits are an example. However, in some tissues, the middle lamellae are partially dissolved, and the tissues become soft. This is what happens when fruits ripen.

Cell walls and middle lamellae are not impervious barriers. Water and other important biological compounds pass through them rapidly. Small molecules move with ease among the cellulose fibers of cell walls and thus gain access to cells. More complex molecules are less able to do so. However, cell walls may bear depressions, called pits, that permit the direct transfer of materials from one cell to another through a continuous stream of cytoplasm.

Plastids

Plant cells commonly possess **plastids**, pigment storage organelles that are not found in animal cells. There are three types of plastids: **Chloroplasts** contain the green pigment **chlorophyll; chromoplasts** contain red, orange, or yellow pigments; and **leucoplasts** contain starch. All three types of plastids originate from a common organelle, the **proplastid**, and one type of plastid may change to another. For example, potatoes are composed almost entirely of leucoplasts, but if you cut them into pieces and expose the pieces to the sun, they will develop green edges. The leucoplasts become chloroplasts.

Chloroplasts are of great interest because they are the sites of photosynthesis, perhaps the most important of all biological processes. Green plants have the ability to convert light energy into chemical energy that can be used by organisms incapable of photosynthesis, such as humans.

With the help of electron microscopes, biologists have found that chloroplasts are essentially the same in all green plants (see Figure 3–14). Like mitochondria, they have a double membrane. However, the

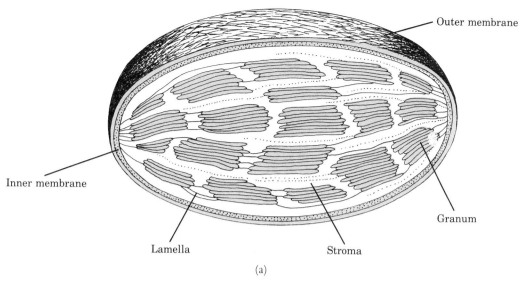

Outer membrane

Inner membrane

Lamella

Stroma

Granum

(a)

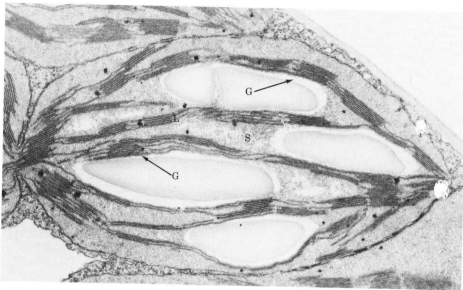

G

L

S

G

(b)

Figure 3–14
A chloroplast. In both the diagram (a) and the electronmicrograph note the lamellae (L), grana (G), and stoma (S) (photo by M. Nadakavukaven, Illinois State University).

inner membrane does not fold as it does in a mitochondrion. The center of the chloroplast contains a complex system of double membranes called **lamellae**. Frequently, the lamellae become concentrated and stack up like coins or plates; they are then called thylakoid discs. These regions of the chloroplast, known as **grana**, are interconnected by lamellae. Chlorophyll molecules are bound to the membranes of the lamellae and grana. The spaces within the chloroplast that lack double membranes and appear unstructured are the **stroma**.

Chromoplasts contain red, orange, or yellow pigments and give the familiar c s to flowers and fruits. The display of chromo s is never so obvious as in the fall foliage of deciduous t ees. Cool weather destroys the chlorophyll in chloroplasts that previously masked the other pigments.

Leucoplasts are abundant in the storage organs of plants. They contain starch and oils rather than pigments. The cells in the tubers of potatoes, for example, contain starch-filled leucoplasts.

CONCEPT SUMMARY
CELL ORGANELLES AND THEIR FUNCTIONS

ORGANELLE	FUNCTION
Cell membrane	Regulation of the passage of substances into and out of the cell
Endoplasmic reticulum	Synthesis of proteins, lipids, and other substances
Ribosome	Synthesis of proteins
Golgi complex	Synthesis of complex cellular substances
Lysosome	Digestion of substances taken into cells
Mitochondrion	Cellular respiration
Microfilaments and microtubules	Association with cellular movements
Nucleus	Housing of hereditary material (DNA)
Plastids	Housing of plant pigments such as chlorophyll (some are centers for photosynthesis)

Vacuoles

Fluid-filled vacuoles occur in both plant and animal cells, but they are much larger in older plant cells. In immature plant cells, many small vacuoles develop in the cytoplasm. They fuse together as the cells mature. A mature plant cell may contain a single vacuole that takes up more than 50 per cent of the cell's volume. The space occupied by the vacuole compresses the cytoplasm and nucleus into a relatively thin layer against the cell wall. Imagine that a box is the cell wall of a plant cell. A balloon inflated until it occupies at least half of the middle volume of the box represents the vacuole. The nucleus, nucleoplasm, cytoplasm, and organelles would have to be squeezed between box and balloon.

Vacuoles are filled with a fluid known as **cell sap**, which is a complex solution of sugars, amino acids, organic acids, and pigments. Blue or purple flowers and fruits (for example, grapes) owe their color not to chromoplasts but to the purple pigment authocyanin, which is contained in cell vacuoles.

The membrane enclosing the vacuole limits the passage of some materials. Osmotic pressure may cause water to move into the vacuole from the external environment. When the vacuole takes in water, its volume increases and the entire cell becomes rigid, or turgid. Under certain conditions, the vacuole losses water, the cell loses turgor, and the plant wilts.

SOME OTHER DESIGNS

The cells of all organisms of all kingdoms except Monera are called **eukaryotic cells** because the nucleus is isolated from the cytoplasm by a nuclear membrane or envelope. However, the cells of bacteria and blue-green algae, called **prokaryotic cells**, do not contain a membrane-enclosed nucleus. Prokaryotic cells may have existed before eukaryotic cells evolved, because they are much simpler structurally.

Bacteria

Almost everywhere on earth there are bacteria. Some bacteria cause the most dreaded human diseases (for example, cholera, typhoid fever, plague, tuberculosis, pneumonia, rheumatic fever, syphilis, and gangrene). Other bacteria are of enormous benefit. For example, the role of bacteria as decomposers is of great importance in recycling the essential raw materials in ecosystems.

Bacteria are unicellular and range in length from less than one micrometer to five micrometers. They exist in three common forms, as shown in Figure 3–15: **cocci** (spherical cells), **bacilli** (rod-shaped cells), and **spirilli** (spiral-shaped cells). But bacteria are classified by their physiological activities and abilities rather than by their shape and size.

The structure of bacterial cells is deceptively simple. A bacterium has a cell wall outside its cell membrane. However, the wall is composed of substances derived from amino acids and related compounds—not from cellulose, as in plants. Some bacteria also secrete a mucous capsule outside the cell wall. Disease-causing bacteria that produce these capsules are especially resistant to the defense mechanisms of the infected organism.

The cytoplasm of a bacterium, densely structured and rich with ribosomes, contains a clear area called a **nucleoid**. Within the nucleoid is a single circular strand of DNA, but the nucleoid is not enclosed by a

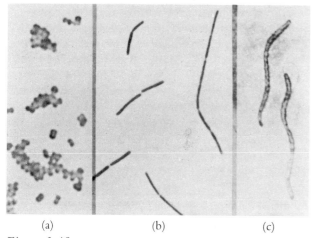

Figure 3–15

Three common forms of bacteria: (a) Cocci are tiny spheres. (b) Bacilli are rod-shaped. (c) Spirilli are spiral-shaped.

membrane. Other than the cell membrane, bacteria possess few, if any, membrane systems. Nor are there membranous structures such as mitochondria, lysosomes, Golgi bodies, vacuoles, or endoplasmic reticula. Despite this structural simplicity, bacteria carry on most of the metabolic activities common to other living organisms, including respiration and protein synthesis. Most bacteria are **heterotrophic** (feeding on organic substances), but some are **autotrophic** (able to manufacture their own organic nutrients). Some autotrophic bacteria are capable of photosynthesis. The chlorophyll in photosynthetic bacteria is associated with membrane systems.

The reproduction of bacteria is a simple process called binary fission (see Figure 3–16). When the circular DNA strand has replicated itself, the two strands separate. New cell membranes and cell walls form between them, dividing the cytoplasm in half. The halves separate to become two individuals.

Viruses

Biologists do not generally agree whether viruses (see Figure 3–17) should be considered living organisms. Viruses lack cells and the cellular machinery to respire, to absorb and use nutrients, and to synthesize protein. When outside living cells, viruses are inert. But when they infect living cells, they induce the infected cells to manufacture more viruses. Viruses are very small, ranging in size from 10 to 275 nanometers (millimicrons).

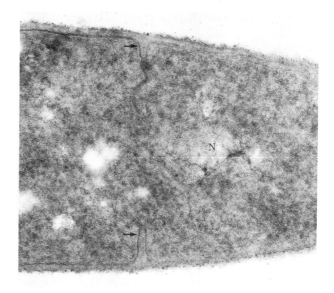

(a)

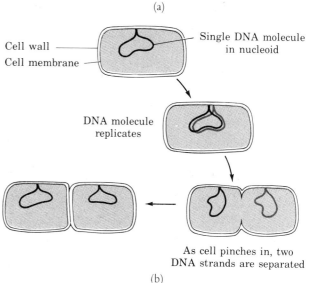

Cell wall
Cell membrane — Single DNA molecule in nucleoid

DNA molecule replicates

As cell pinches in, two DNA strands are separated

(b)

Figure 3–16

Bacterial cell division. (a) In this electron micrograph, arrows indicate where the cell wall is extending inward. The light area (N) is the location of the nucleoid of one of the future daughter cells. (b) The daughter strands of DNA replicate and separate during the division of a bacterium. (Photo by M. Nadakavukaren, Illinois State University.)

A great variety of viruses have been classified. Some infect only bacterial cells, others only plant cells, and others only animal cells. Viruses cause a number of human diseases, including the common cold, poliomyelitis, mumps, measles, chicken pox,

CONCEPT SUMMARY

DIFFERENCES BETWEEN PLANT AND ANIMAL CELLS

PLANT CELLS	ANIMAL CELLS
Possess cell walls	Lack cell walls
Commonly possess pigment containing plastids	Lack plastids
Older cells possess large vacuoles	If vacuoles are present, they are seldom large
Lack centrioles	Possess centrioles

or DNA, surrounded by a protein coat or covering. The protein coat protects the nucleic acid and determines what type of cell a particular virus can infect. During infection, the nucleic acid enters the cell's cytoplasm and provides the cell with the genetic information for producing viruses instead of its own cellular material (see Figure 3–18). After large numbers of viruses have been produced, the infected cell re-

viral pneumonia, smallpox, yellow fever, influenza, and rabies. Viral infections may also initiate certain types of cancer, such as leukemia.

Viruses vary in shape and size, but each one is composed of a molecule of nucleic acid, either RNA

Figure 3–17

An electron micrograph of a virus. (Photo by M. Nadakavukaren, Illinois State University.)

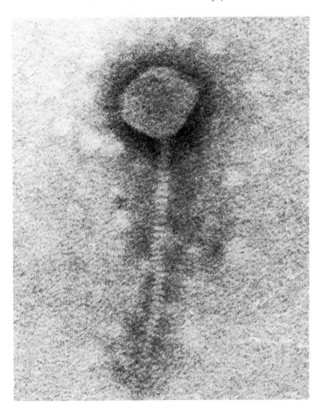

Figure 3–18

A bacterial virus injects its DNA into a bacterium. The viral DNA directs the bacterium to synthesize and assemble viral components into new viruses, which are released from the host cell.

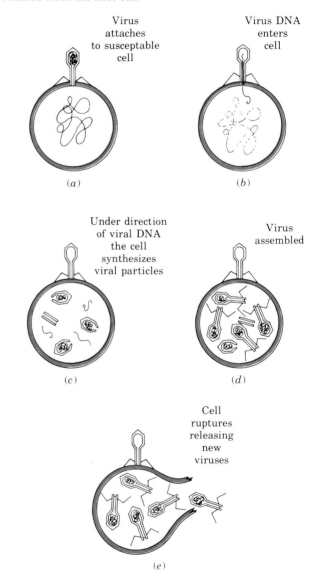

Virus attaches to susceptable cell *(a)*

Virus DNA enters cell *(b)*

Under direction of viral DNA the cell synthesizes viral particles *(c)*

Virus assembled *(d)*

Cell ruptures releasing new viruses *(e)*

leases them into the environment. In some cases, the infected cell bursts. In others, the viruses pass through the cell membrane without destroying the cell.

SUMMARY

1. The bodies of living organisms are composed of one or more structural units—cells. Unicellular organisms are composed of a single cell; multicellular organisms are composed of many cells.

2. Each cell is bordered by a cellular membrane that separates the cell's cytoplasm from the external environment.

3. Cells contain organelles composed of membranes similar in structure to the external cell membrane.

4. The largest and most conspicuous organelle is the nucleus, a spherical or oval structure that houses the genetic or hereditary material (DNA) of the cell. The nucleus is separated from the cytoplasm by a double membrane.

5. The endoplasmic reticulum is composed of membranes that extend from the cell membrane to the nuclear membrane. Some endoplasmic reticula (rough ER) bear small granules, the ribosomes, which are associated with protein synthesis. Some (smooth ER) lack ribosomes and function in the synthesis of other cellular products, especially lipids.

6. Golgi complexes are parallel curved membranes enclosing fluid-filled spaces. Alone or in association with rough ER, Golgi complexes synthesize, process, package, and secrete substances from cells.

7. Lysosomes are membrane-enclosed vesicles or sacs that contain digestive enzymes. They digest other cell organelles and foreign substances taken into cells.

8. Mitochondria have a smooth outer membrane and a convoluted inner membrane enclosing a fluid-filled space. They are the powerhouses of cells, and they contain enzymes for cellular respiration, which releases energy from nutrients.

9. Cells also contain microfilaments and microtubules, which contribute to cell movement. Other organelles containing microtubules are cilia, flagella, and centrioles.

10. Plant cells differ from animal cells in the following ways: (a) they lack centrioles; (b) they secrete a cell wall, composed largely of cellulose, outside the cell membrane; (c) they often contain large fluid-filled vacuoles that push the cytoplasm to the edges of the cell; and (d) they may possess pigment-containing organelles called plastids.

11. The most common plastids are chloroplasts, which contain complex double membranes (lamellae and grana) that house the molecules of chlorophyll essential for photosynthesis.

12. Bacteria and other prokaryotic cells are structurally simpler than cells of eukaryotes. The most notable difference is that prokaryotic cells lack the centrally located nuclei that eukaryotic cells have.

13. Viruses are not cells. They can reproduce only after introducing their DNA or RNA into the cytoplasm of a living cell, inducing the invaded cell to manufacture new viruses.

THE NEXT DECADE

THE CELLS OF HENRIETTA LACKS

Biologists have learned much about cell structure and function by studying cells and fragments of tissue removed from multicellular organisms. They nurture and maintain these cells and tissues in special culture chambers under carefully controlled conditions. The first truly successful cell cultures were established in 1912 from heart muscle and fibroblast cells taken from chicken embryos. They grew and reproduced for thirty-four years before they were allowed to die. Today many human cell lines have been cultured continuously through thousands of cell generations.

The most famous of the human cell lines is that of HeLa cells, established in 1951 after a young black woman had been admitted to Baltimore's Johns Hopkins Hospital with suspected cervical cancer. As an essential part of the diagnosis, a small sample of living tissue was removed from her cervix. Some of this tissue was given to Dr. George Gey, a researcher at the hospital who was trying to culture various kinds of cancer cells. Until then, no one had been successful in culturing cells from human malignancies, but this time the cells flourished. They reproduced a new generation of cells as often as once a day. Because of this great breakthrough, research on

these and similar cells began in earnest. HeLa cells were eventually sent to laboratories throughout the world, where they became the basic cell line for many kinds of cellular research.

What happened to the young woman whose cells were distributed to laboratories all over the world? Unfortunately, she died shortly after the diagnosis of cervical cancer was made, and she was soon forgotten. For many years even her name was a mystery. Many thought it was Helen Lane, others Helen Larson. HeLa is an acronym for her name. Today we know that her name was Henrietta Lacks, and although she is dead, her cells provide her with a unique type of immortality.

When carefully cultured, HeLa cells cling tenaciously to life. They have served as culture systems for numerous types of viruses, including the polio virus. If not for HeLa cell cultures, the development of the polio vaccine might have been long delayed. HeLa cells have been subjected to detailed chemical analyses and careful studies of nutritional requirements. They have been used to test the potential effects of radiation on human cells. They have been recipients of foreign nuclei, such as mouse nuclei, in studies of nuclear control of cell function. (In the latter

instance, HeLa cells synthesized mouse proteins.) Recently, biologists around the world have found many human cell cultures contaminated and taken over by the highly competitive HeLa cells.

The use of HeLa cells for experimental purposes continues to increase, and special facilities have been established to grow them in large numbers. The National Science Foundation supports two such major centers, one at the University of Alabama in Birmingham, the other at the Massachusetts Institute of Technology in Cambridge. Both institutions have received unbelievable numbers of requests for HeLa cells.

The University of Alabama Center has received and filled single orders for as many as a trillion HeLa cells (approximately 1.5 kilograms). The Massachusetts Institute of Technology receives weekly orders for 2 billion cells from one scientist alone.

As cellular research continues to expand, during the next decade, additional culture centers may need to be established, and the cells of the young black woman will continue to be used in research on cancer, viral infections, developmental biology, and genetics. Few, if any, scientists expect to ever clone humans. But if they do succeed, maybe the first clone will be Henrietta Lacks.

FOR DISCUSSION AND REVIEW

1. Draw a typical animal cell that includes the following structures: nucleus, nucleolus, chromatin, plasma membrane, mitochondria.

2. Draw a typical plant cell that includes the structures listed in question 1, as well as chloroplasts and a cell wall.

3. Describe the functions of the following: endoplasmic reticulum, ribosomes, Golgi complexes, lysosomes, mitochondria, microtubules.

4. How many centimeters are there in a yard? In a foot? How many liters are there in a gallon? In half a gallon? How many inches are there in a meter? How many centimeters in a meter? How many meters in a kilometer?

SUGGESTED READINGS

THE CELL, by Carl P. Swanson and Peter L. Webster. Prentice–Hall: 1977.
Well-written account of the biology of cells.

THE CELL: ITS ORGANELLES AND INCLUSIONS, by Don Fawcett. Saunders: 1966.
Outstanding electron photomicrographs of cell organelles, with brief descriptions of their functions.

THE CENTER OF LIFE: A NATURAL HISTORY OF THE CELL, by L. L. Larison Cudmore. Quadrangle Books: 1978.
An introduction to the biology of cells for the general reader.

MOLECULES TO LIVING CELLS: READINGS FROM THE SCIENTIFIC AMERICAN. W. H. Freeman: 1980.
Twenty-six excellent articles written by leading authorities in the field of molecular cell biology.

4

CELL
FUNCTIONS

PREVIEW QUESTIONS

After reading this chapter, you should be able to answer the following questions:

1. What is the source of energy needed by all living things?

2. What is photosynthesis? What important events are associated with it?

3. Why do most living things die if they are deprived of oxygen?

4. How is the energy in the nutrients you eat converted to a form your cells can use?

5. How do substances enter and leave cells?

To maintain cellular integrity and normal functions, living cells require three things: water, nutrients, and energy. Energy occurs in a variety of forms, such as heat and light. Living organisms use light and the chemical energy stored in chemical bonds to meet their needs. On the basis of their energy sources and nutritional requirements, both unicellular and multicellular organisms can be divided into two basic types: autotrophs and heterotrophs. An autotrophic organism (*auto* means "self"; *trophic* means "to feed") that is photoautotrophic uses light energy to synthesize its own organic nutrient compounds from simple inorganic chemicals. A chemoautotrophic uses energy from inorganic compounds to do the same job.

The majority of plants and many unicellular organisms are photoautotrophic—capable of converting light energy into chemical energy in the chemical bonds of organic molecules. In this process, called **photosynthesis,** the light energy is locked in chemical bonds of chemical energy in new organic compounds. Later this energy can be used by the autotrophs and heterotrophs. Only a few microorganisms, such as iron- and sulfur-reducing bacteria, are chemoautotrophs.

Heterotrophic organisms (*hetero* means "other") cannot use light energy directly. Instead, they depend on the chemical energy converted by autotrophs to supply their energy needs. Many microorganisms and all animals are heterotrophs. Therefore, the fundamental support of all life comes mainly from the process of photosynthesis. In this process, photoautotrophs convert light energy to chemical energy, producing organic molecules that autotrophs and, subsequently, heterotrophs use (see Figure 4–1). These organic molecules are used for the production

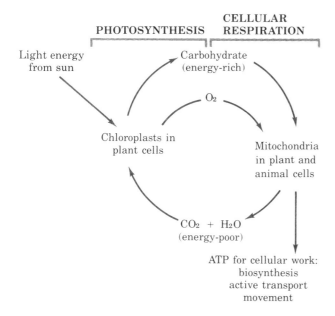

PHOTOSYNTHESIS   CELLULAR RESPIRATION

Light energy from sun

Carbohydrate (energy-rich)

O_2

Chloroplasts in plant cells

Mitochondria in plant and animal cells

$CO_2 + H_2O$ (energy-poor)

ATP for cellular work: biosynthesis active transport movement

Figure 4–1

The relationship between photosynthesis and cellular respiration. Both plant and animal cells respire, but only plant cells have the power of photosynthesis. During photosynthesis, plant cells use light energy from the sun to convert CO_2 and water into carbohydrates, which are a store of chemical energy (converted light energy). Mitochondria in both plant and animal cells can release the energy from carbohydrates and incorporate it into molecules of ATP for use in cell processes.

and repair of cell structures. Equally important is their use in furnishing the energy needed for cellular activities.

Cellular respiration is the process that transfers the chemical energy in organic compounds (such as glucose, a simple sugar◻) into molecules that make it more readily available to the cell. One such molecule is the power molecule, **adenosine triphosphate (ATP),** discussed in Box 4–1. The process of cellular respiration will be described later in the chapter.

ENERGY

A variety of energy forms exist: light, heat, electricity, mechanical energy, and chemical energy (the latter contained in the bonds that unite atoms into molecules). The first law of thermodynamics, the law of conservation of energy, states that energy can be converted from one form to another, but it is neither de-

stroyed nor created during the process. However, no process of energy conversion is completely efficient. For example, when a candle is burned to convert its chemical energy to light, much of the energy is converted instead to heat, which dissipates.

The second law of thermodynamics states that usable energy tends to dissipate. After an energy transformation, there is less energy in the system than was originally present. However, the dissipated energy (usually in the form of heat) plus the converted energy must total the amount of energy present in the original system. Another way of stating this law is: In a closed system—one in which neither energy nor matter enters or leaves—there is a continuing trend toward disorder and disorganization, called entropy.

Luckily for us, the earth is not a closed system. The sun continuously provides energy. Thus life can maintain its organization and even continue to increase in complexity. This is an important concept. We will soon see that when light energy is converted into chemical energy during photosynthesis, much energy is lost as heat. This is also the case when chemical energy in nutrients is converted into the chemical energy that is usable in cellular processes. Living systems on earth will therefore exist only as long as the energy input from the sun continues. We do not have to be too concerned, however, because the sun is predicted to last at least another 10 billion years.

PHOTOSYNTHESIS

Photosynthesis is the fundamental process that supports life on earth. Two important events occur during the process:

1. Photosynthesis converts light energy into chemical energy in the form of organic nutrients needed by both autotrophs and heterotrophs. These nutrients form the basis of all food chains ◻. It is estimated that photosynthesis by plants produces 90 billion tons of glucose each year.

2. Photosynthesis produces the by-product oxygen (O_2), which autotrophs and most heterotrophs require for cellular respiration. The process renews the earth's supply of oxygen, which is constantly being consumed (see Figure 4–1). In fact, photosynthesis produced the free oxygen in the atmosphere in the first place.

BOX 4–1

THE POWER MOLECULE

Adenosine triphosphate, or ATP, is present in all cells. It is one of the first molecules formed during the light reaction of photosynthesis, and it is one of the energy stores that fuels the dark reaction. ($NADPH_2$ is the other). ATP is also used in the synthesis of new carbohydrates. It is produced during cellular respiration, when molecules such as glucose are broken down, releasing their large energy store. However, ATP is relatively unstable and, unlike glucose or starch, is not used to store chemical energy in cells for long periods of time.

Each ATP molecule is composed of three parts: an adenine group, a ribose group and a chain of three phosphate groups. (See the diagram.) When the chemical bonds that link the second and third phosphate groups to the rest of the molecule are broken, exceptionally large amounts of energy are released. (Chemical bonds that release large amounts of energy are called high-energy bonds; they are designated by wavy lines.) When a cell needs energy, the terminal phosphate group is split from the ATP molecule and energy is released. Molecules of **adenosine diphosphate** (ADP) and PO_4^{-3} (P_i) are formed:

$$\text{ATP} \underset{\leftarrow}{\overset{\text{enzyme}}{\rightarrow}} \text{ADP} + P_i + \text{energy}$$

The two arrows pointing in opposite directions show that this reaction is reversible. During cellular respiration, energy released during the breakdown of

Adenine group

Ribose group

Chain of 3 phosphate groups

sugar molecules is used to link P_i to ADP molecules, and molecules of ATP are formed:

$$\begin{array}{c}\text{Respiration}\\\text{Photosynthesis}\end{array} \rightarrow \text{energy} \rightarrow \begin{array}{c}\text{ATP}\\\uparrow \downarrow\\\text{ADP}\\+\\P_i\end{array} \rightarrow \text{energy} \rightarrow \text{cell activity}$$

In a similar fashion, ADP may be degraded to **adenosine monophosphate** (AMP) and P_i, but only if ATP is in short supply. This reaction, also is reversible.

$$\text{ATP} \underset{\rightarrow}{\overset{\text{enzyme} \;\; \leftarrow}{}} \text{ADP} + P_i + \text{energy}$$

$$\text{ADP} \underset{\leftarrow}{\overset{\text{enzyme} \;\; \rightarrow}{}} \text{AMP} + P_i + \text{energy}$$

The process of photosynthesis can be summarized in chemical terms as follows:

$$12\ H_2O\ \text{(water)} + \text{light energy}$$
$$+\ 6\ CO_2\ \text{(carbon dioxide)}$$
$$+\ \text{chlorophyll} \xrightarrow{\text{enzymes}} C_6H_{12}O_6\ \text{(glucose)}$$
$$+\ 6\ H_2O + 6\ O_2\ \text{(oxygen)}$$

However, this summary is an oversimplification of the entire process, which actually involves a multitude of chemical reactions. Some of these reactions

are not fully understood even by experts in plant physiology and biochemistry. Also, each reaction is catalyzed by a specific enzyme. The equation shows that two essentials for photosynthesis, besides water and carbon dioxide, are light and the green plant pigment chlorophyll.

Visible light is part of what physicists call the electromagnetic spectrum. The spectrum includes radio waves, microwaves, infrared radiation, visible light, ultraviolet radiation, X-rays, and gamma rays (see Figure 4–2). When visible light passes through a prism, it separates into the colors of the spectrum—

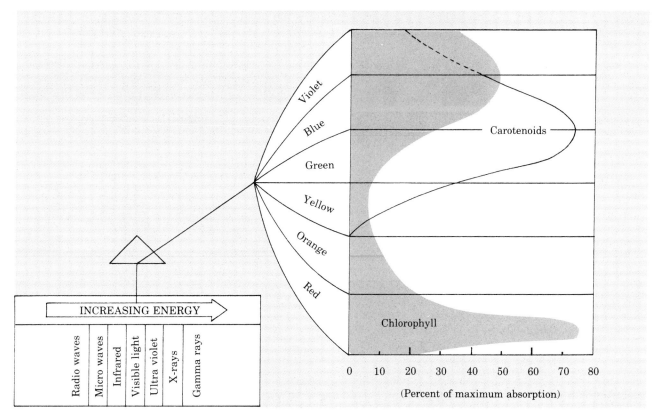

INCREASING ENERGY

| Radio waves | Micro waves | Infrared | Visible light | Ultra violet | X-rays | Gamma rays |

0 10 20 30 40 50 60 70 80

(Percent of maximum absorption)

Figure 4–2
When visible light passes through a prism, it separates into the colors of the spectrum. The green plant pigment, chlorophyll, reflects light in the green range of the spectrum while absorbing light in the other areas of the spectrum.

red, orange, yellow, green, blue, and violet. Each color has a different wavelength and a different energy value. The wavelength of light that an object reflects to the human eye determines its color.

Grass is green because it reflects green wavelengths and absorbs others. To be even more specific, grass is green because its chloroplasts contain large amounts of chlorophyll, which absorb and use the energy of blue, violet, and red light but reflects green light. However, plants do absorb and use the energy of much of the visible light spectrum because of other pigments. In addition to chlorophyll, chloroplasts contain the yellow to red **carotenoid pigments.** These pigments absorb the energy of green and blue light and transfer it to chlorophyll.

Light consists of tiny particles, or packets, of energy, called photons. The process of photosynthesis begins when an electron in a molecule of chlorophyll absorbs a photon of light. Photosynthesis takes place in two phases, the light reaction and the dark reaction, which cannot be shown in the general equation.

The light reaction occurs within the thylakoid discs that comprise the grana in chloroplasts (see Figure 3–13). The **dark reaction** occurs in the stroma. During the light reaction, electrons in chlorophyll molecules gain energy by absorbing light energy. The electrons then pass from the chlorophyll molecules through a series of oxidation and reduction reactions ▫, which release energy for the formation of ATP and $NADPH_2$, a coenzyme. These two molecules transfer energy to fuel the dark reaction. The electrons lost by chlorophyll molecules during the light reaction must be replaced. For this purpose, water molecules are split (a process called photolysis), the electrons are replaced, and oxygen and $NADPH_2$ molecules are formed. Figure 4–3 summarizes these events and relationships.

During the dark reaction, energy carried in chemi-

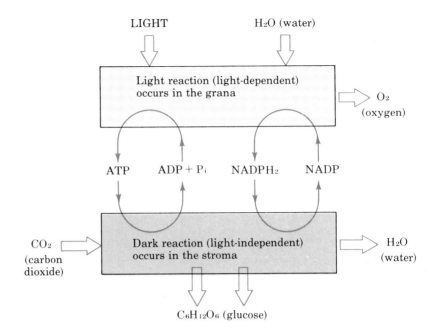

Figure 4–3
The major events occuring during the light and dark reactions of photosynthesis. The light reaction converts light energy to chemical energy in ATP and $NADPH_2$ which fuels the dark reaction.

LIGHT H_2O (water)

Light reaction (light-dependent) occurs in the grana

O_2 (oxygen)

ATP   ADP + P$_i$ $NADPH_2$ NADP

CO_2 (carbon dioxide)

Dark reaction (light-independent) occurs in the stroma

H_2O (water)

$C_6H_{12}O_6$ (glucose)

cal bonds of ATP and $NADPH_2$, formed during the light reaction, is used to combine carbon dioxide from the atmosphere and hydrogen from water molecules into simple organic compounds, such as glucose. Glucose chemical bonds represent stores of transformed light energy (see Figure 4–3).

Why are the two phases of photosynthesis called the light and dark reactions? The light reaction is so called because it is a photo event. That is, it requires light (is light-dependent); it will not occur in the absence of light. The dark reaction is so called because it is light-independent (it can occur in either the dark or the light). It is fueled by the energy in ATP and by the hydrogen from $NADPH_2$ formed during the light reaction.

The Light Reaction

Two different photosystems are used by the light reaction. **Photosystems** are organizations of pigments, proteins, and enzymes bound in the membranes of the thylakoid discs in the grana of chloroplasts (see Figure 4–4). Each system contains a different mix of chlorophyll and other light-absorbing pigments. The different mixes cause a peak absorption of light at slightly different wavelengths.

One photosystem has its peak absorption at wavelengths of about 680 nanometers and is therefore called the P680 system. The other shows peak ab-

sorption at wavelengths of about 700 nanometers and is therefore called the P700 system. In both systems, when a chlorophyll molecule absorbs two photons of light, two electrons increase in energy and leave the chlorophyll molecule. The two systems do not function at the same time, however. The P680 fires first.

When the electrons leave P680, they must be replaced. This is accomplished by the splitting of water molecules **(photolysis).** When a water molecule is split, the electrons from the hydrogen atoms replace those lost from the chlorophyll. At the same time, the hydrogen ions combine with NADP to form $NADPH_2$. Oxygen from the water molecule is released as molecular oxygen (O_2). The electrons released from the chlorophyll molecule are now enriched with the energy from the light. They are passed to a molecule that acts as an electron acceptor. This acceptor then passes the electrons to another acceptor, which passes them to yet another acceptor, and so on. During this series of oxidation and reduction reactions □, energy is released at each transfer of electrons. It is used to produce ATP molecules that will fuel the dark reaction. Figure 4–4 shows the path of the electrons given off by P680. The electrons go up in energy content and then through a set of oxidation and reduction steps. During which, the energy is released to form ATP, a process called **photophosphorylation.**

The same electrons finally reach P700, in time to

18

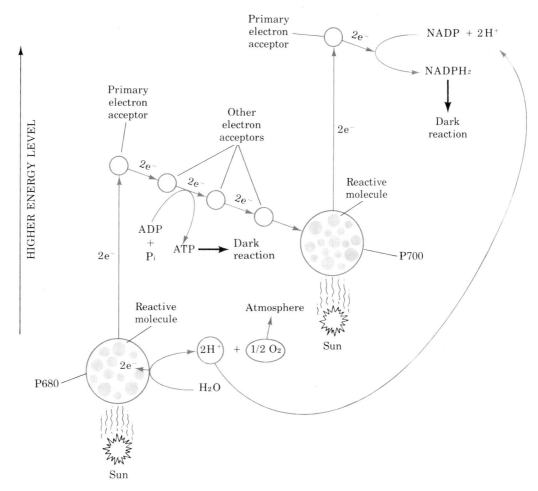

Figure 4-4
Summary of the major events in the light reaction of photosynthesis.

replace the two electrons given off when the system absorbs two photons of light. The electrons released by P700 are also passed through a series of carriers. They eventually enter the reaction between NADP and the hydrogen ions released from the water molecule during photolysis. This reaction forms $NADPH_2$, which (like the ATP molecules formed by the P680 events) passes on to the dark reaction. At this point, the light reaction has converted light energy into chemical energy in ATP and has produced $NADPH_2$ and free oxygen.

The Dark Reaction

During the dark reaction, carbon dioxide and hydrogen released from water and carried by $NADPH_2$ are combined by a series of reactions to form organic molecules. The type of organic molecule first produced by the dark reaction depends on the kind of plant. **C3 plants** produce molecules containing three carbon atoms (phosphoglyceraldehyde—PGAL). **C4 plants** produce molecules containing four carbon atoms (for example, malate). In both cases, the organic molecules are converted into numerous other molecules, including glucose. Corn and sugarcane are examples of C4 plants. Wheat, barley, and rice are examples of C3 plants.

Figure 4–5 shows the details of the dark reaction as it occurs in C3 plants. At the top of the diagram, free carbon dioxide taken from the air combines with a five-carbon sugar, ribulose diphosphate, to form an unstable six-carbon compound. This compound splits

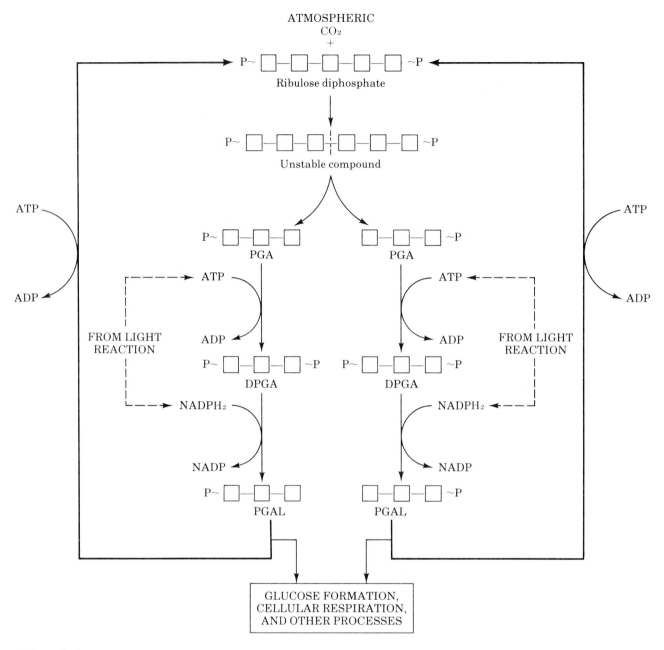

Figure 4–5
Summary of the major events in the dark reaction of photosynthesis.

to form two three-carbon molecules of phosphoglyceric acid (PGA). Each PGA molecule in turn is converted to another three-carbon compound, diphosphoglyceric acid (DPGA). This conversion is
accomplished with energy input from ATP that was
produced during the light reaction. Each DPGA molecule is then converted to PGAL at the cost of

$NADPH_2$ produced by the light reaction. ADP (produced when the ATP is utilized) and the NADP (produced during the last reaction) go back to the light
reaction and are converted again to ATP and
$NADPH_2$.

Many things may happen to the PGAL produced
by the dark reaction. Mostly, though, PGAL is con-

Botanists have identified over a hundred species of C4 plants. Most of them are monocots, and most are plants of tropical and arid climates. C4 plants are much more efficient than C3 plants at fixing CO_2 into organic compounds, and they reach a greater overall photosynthetic efficiency in terms of producing organic matter.

C4 plants use two types of cells during photosynthesis. First, CO_2 is fixed in loose mesophyll cells and incorporated into malate or other four-carbon compounds. These compounds are then moved into bundle sheath cells, which surround the veins of leaves. Here the CO_2 is again released and fed into the same series of reactions described in the text for C3 plants.

It would be exciting and profitable if agriculturally important C3 plants could be converted into C4 plants, which would be more efficient and more highly productive than C3 plants are. In this era of genetic engineering and sophisticated plant breeding, anything is possible. Someday we may see another agricultural revolution.

verted, through a series of complex chemical events, back to ribulose diphosphate, a process which requires more energy from ATP molecules. Ten out of every twelve PGAL molecules are recycled in this way, leaving only two out of twelve to be converted to glucose or used in other ways. This may not seem very efficient. In fact, the process of photosynthesis is far from efficient. Many C3 plants convert less than 1 percent of the light energy they absorb into chemical energy in organic compounds such as glucose. But C4 plants are much more efficient. Sugarcane, for example, converts as much as 8 percent of its absorbed light energy.

Although photosynthesis seems to be an inefficient system, it is the only system for converting light to the chemical energy that makes all life on earth possible. It provides the energy that is passed along all food chains from plants to herbivores, such as cattle, to our dinner plates. As we will see later, the conversions along the food chain are also inefficient. It's a good thing sunlight is abundant and free for the taking.

CELLULAR RESPIRATION

During cellular respiration, molecules of glucose are broken down by a series of chemical reactions. Energy from their chemical bonds is used in the formation of ATP from ADP and inorganic phosphate (see Box 4–1). Most cells require oxygen for this process to continue. Consequently, these cells are regarded as carrying on **aerobic respiration** (*aeros* means "air").

Some cells and organisms, such as microorganisms found in sewage and in oxygen-poor sediments, can survive with little or no oxygen for varying periods of time. When oxygen is not involved in cellular respiration, the process is known as **anaerobic respiration.** For both aerobic and anaerobic respiration, chemical energy in nutrients is transferred to ATP, which in turn furnishes the energy for cellular processes. Aerobic respiration provides much more energy than does anaerobic respiration; thus, it is the more efficient process.

Aerobic Respiration

The following equation is a summary of cellular aerobic respiration:

$$C_6H_{12}O_6 + 6O_2 + ADP + P_i \xrightarrow{\text{enzymes}}$$
(glucose)   (oxygen)   (inorganic phosphate)

$$6CO_2 + 6H_2O + ATP \text{ (stored energy)}$$
(carbon dioxide)   (water)

Even though the equation is only a superficial summary, we can see that oxygen is required in the process and that carbon dioxide and water are formed. However, the equation disguises the fact that a large number of different chemical reactions, each catalyzed by a specific enzyme, occurs during the process. Many of the enzymes are located in the mitochondria.

There is debate among specialists about the number of ATP molecules produced for every molecule of glucose respired. The numbers most commonly given are thirty-six or thirty-eight, but estimates as high as forty-four exist; and in certain microorganisms, the number may be lower than thirty-six. Most biochemical reactions require some energy input. This is equally true for respiration. For every molecule of glucose respired, two molecules of ATP are required just to get the process going. Then a maximum of only 40 to 50 percent of the energy in glucose is used in the formation of ATP; the rest is lost as heat.

For discussion, it is convenient to divide aerobic respiration into three parts: glycolysis, the Krebs cycle, and the electron transport system.

Glycolysis

It is in the fluid portion of the cytoplasm that **glycolysis** occurs. The enzymes that catalyze the reactions are not structurally associated with any cell organelle During glycolysis, a six-carbon molecule of glucose is split into two three-carbon molecules of pyruvic acid. The details of the process are complex. Figure 4–6 shows the major events of glycolysis.

During the initial phases of glycolysis, a molecule of glucose is chemically transformed into a molecule of fructose with a phosphate group attached to each end (it is called fructose-1,6-diphosphate). The phosphate groups and the energy to accomplish the transformation come from two ATP molecules. The fructose-1,6-diphosphate, which is not as stable a molecule as glucose, is then split. Eventually, it becomes two three-carbon molecules of phosphoglyceraldehyde, each of which is then converted to diphosphoglyceric acid. During the latter transformation, two important things happen. First, hydrogen atoms are transferred to an important cellular hydrogen acceptor, NAD, which carries them on to the electron transport system. This is the last series of reactions in aerobic respiration. Second, each molecule of PGAL gains another phosphate group, creating the two molecules of diphosphoglyceric acid.

During the next three steps in glycolysis, the phosphate groups are removed from the three-carbon diphosphoglyceric acid molecules. Along with the energy released from the molecules, the phosphate groups are used to form ATP. These reactions result in the formation of four molecules of ATP and two molecules of pyruvic acid. Only two ATP molecules are used to initiate the glycolysis of one molecule of glucose. Therefore, cells realize a net gain of two ATP molecules for every molecule of glucose processed. Obviously, then, only a small portion of the potential energy in glucose is released up to this point.

Before the Krebs cycle begins, each of the two three-carbon pyruvic acid molecules formed during glycolysis is converted to a molecule of acetyl-Coenzyme A. This conversion, which occurs inside the mitochondria, is diagrammed in the upper-left corner of Figure 4–7. During the conversion of each pyruvic acid molecule to acetyl-CoA, two hydrogen ions and two electrons are transferred to NAD and passed on to the electron transport system, and a molecule of carbon dioxide (CO_2) is formed. This is part of the total CO_2 produced as a by-product of respiration. The remaining two-carbon molecule, acetic acid, combines with a special carrier molecule, coenzyme A, to form acetyl-CoA. It then enters the next stage of aerobic respiration, the Krebs cycle.

The Krebs Cycle

As Figure 4–7 indicates, the **Krebs cycle** is a cyclic series of chemical reactions. It begins when a molecule of acetyl-CoA transfers the two-carbon acetic acid to a four-carbon compound, oxaloacetic acid, to form a six-carbon compound, citric acid. Through a series of nine reactions catalyzed by different enzymes, citric acid is converted back to oxaloacetic acid, which is then free to react with more acetyl-CoA. For every two molecules of acetyl-CoA that enter the Krebs cycle, eight atoms of hydrogen are passed to the electron transport system (six atoms combine with NAD and two with FAD, another electron acceptor), two molecules of ATP are produced, and four molecules of CO_2 are released. The enzymes essential to the Krebs cycle are located in the matrixes of mitochondria. [43]

The Electron Transport System

Located in the inner membranes of mitochondria, the **electron transport system** consists of a series of compounds that function sequentially as electron acceptors and donors. Most of these compounds are called **cytochromes.** The electrons used in the chain are from hydrogen atoms fed into the system, and the final acceptor is oxygen. As a result, water (H_2O) is formed. The great numbers of hydrogen ions and electrons that pass through the electron transport system come from reactions during glycolysis, the conversion of pyruvic acid to acetyl-CoA, and especially the Krebs cycle.

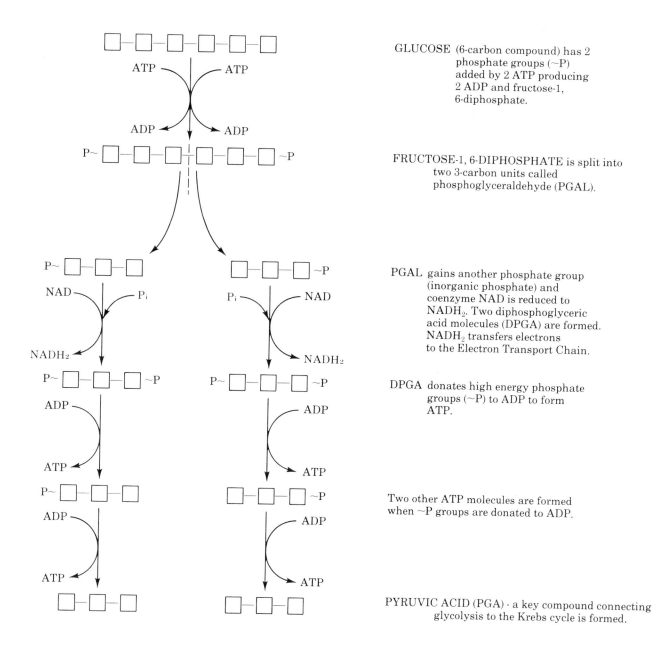

GLUCOSE (6-carbon compound) has 2
phosphate groups (~P)
added by 2 ATP producing
2 ADP and fructose-1,
6-diphosphate.

FRUCTOSE-1, 6-DIPHOSPHATE is split into
two 3-carbon units called
phosphoglyceraldehyde (PGAL).

PGAL gains another phosphate group
(inorganic phosphate) and
coenzyme NAD is reduced to
$NADH_2$. Two diphosphoglyceric
acid molecules (DPGA) are formed.
$NADH_2$ transfers electrons
to the Electron Transport Chain.

DPGA donates high energy phosphate
groups (~P) to ADP to form
ATP.

Two other ATP molecules are formed
when ~P groups are donated to ADP.

PYRUVIC ACID (PGA) - a key compound connecting
glycolysis to the Krebs cycle is formed.

Figure 4–6
A summary of the chemical events that occur during
glycolysis. Although four ATP molecules are produced by
the process, two are required to initiate it. Therefore, the
net gain is of only two ATP molecules.

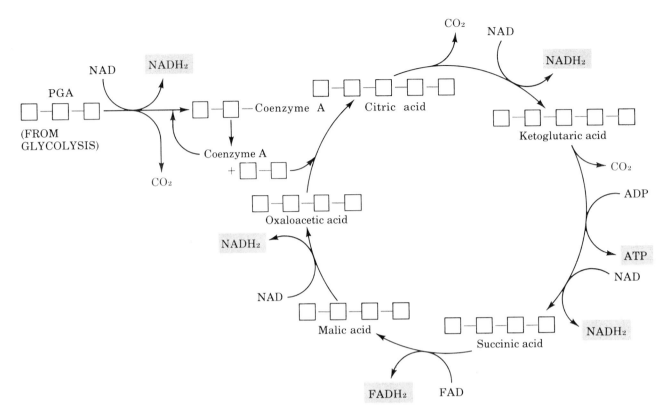

Figure 4–7

Conversion of pyruvic acid to acetyl CoA and summary of the reactions occurring during the Krebs cycle. One molecule of ATP is produced for each molecule of acetyl-CoA that enters the cycle, but two molecules of acetyl-CoA are produced for every molecule of glucose.

As Figure 4–8 shows, when an electron is transferred to the first electron acceptor, energy is given off. The resulting acceptor complex thus has a lower energy state. The transfer of the electron from the first to the second acceptor results in a further release of energy. This release continues until the end, when oxygen, the final electron acceptor, combines with two hydrogen ions. The result is water, which has the lowest energy level in the entire system.

Some of the energy released during the electron transfers is trapped when ATP is synthesized from ADP and P_i. The rest of the energy is lost as heat. Three ATP molecules are generated for each pair of electrons fed to the system by the $NADH_2$. Only two ATP molecules are formed for every pair of electrons passed to the chain by $FADH_2$.

NAD, FAD, and coenzyme A are essential for cellular respiration. All are derivatives of vitamin B. Biochemists call these essential cofactors of the respiratory process niacin, riboflavin, and pantothenic acid.

Cellular respiration is efficient. The oxaloacetic acid that combines with acetyl-CoA at the beginning of the Krebs cycle is regenerated again at the end of the cycle. Coenzyme A is released at the first step in the cycle and is then free to react with acetic acid to form new acetyl-CoA. When $NADH_2$ and $FADH_2$ give up hydrogen ions and electrons to the electron transport system, they are ready to accept more hydrogen and electrons. As each electron donor in the system passes its electron to the next carrier in the chain, it becomes an electron acceptor again. Some of the ATP generated in respiration is used for glycolysis of another sugar molecule to fuel the Kreb cycle. Glucose and oxygen are the only compounds that must constantly be provided anew to keep the system going.

Glucose isn't the only compound that can fuel cellular respiration. Fatty acids, glycerol (a component

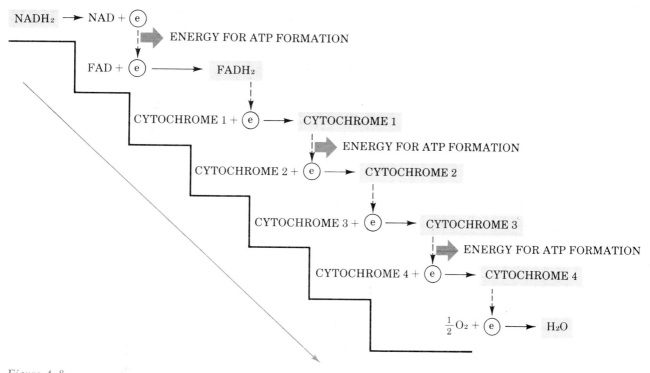

Figure 4–8

The electron transport system. The majority of the ATP molecules generated by aerobic respiration are produced when electrons are transferred from electron acceptor to electron acceptor. As each acceptor in the sequence is reduced, it represents a lower energy state. Some of the energy released by each transfer is incorporated into the high-energy bonds of ATP molecules. O_2 acts as the final electron acceptor.

of fats), and amino acids can enter the respiratory process at specific points. After chemical conversion, glycerol can enter the chain of reactions that comprise glycolysis. Fatty acids can be converted to acetyl-CoA molecules and enter the Krebs cycle. Amino acids can enter the Krebs cycle at several points after the amino group is removed.

Once these additional compounds enter the respiratory process, they too provide energy to produce ATP molecules. In fact, one of the principles of dieting is based on the role of fats in cellular respiration. If you restrict your intake of carbohydrates (which later are broken down into glucose molecules), your body will begin to mobilize its fat reserves. These reserves become the body's major energy source. As a result, you substitute your own body fats for glucose, and you lose weight.

Anaerobic Respiration

Many organisms and cells can live without oxygen for a certain period of time. They continue to respire but are restricted to anaerobic respiration. Without oxygen, the Krebs cycle and the electron transport system do not operate. Because most of the ATP molecules are formed in association with the electron transport system, anaerobic respiration is much less efficient than aerobic respiration.

In anaerobic respiration, as in glycolysis, a net gain of only two ATP molecules is realized. Glucose still undergoes glycolysis and is converted to two molecules of pyruvic acid, as in aerobic respiration. But in the absence of oxygen, the next step is modified. Pyruvic acid molecules act as hydrogen acceptors. Instead of being converted to acetyl-CoA, they are converted to other substances, such as lactic acid and ethyl alcohol.

In the absence of oxygen, yeast cells convert pyruvic acid to ethyl alcohol and carbon dioxide in a process called **fermentation.** Since at least the beginning of recorded history, humans have taken advantage of fermentation to produce alcoholic beverages such as wine and beer. Fermentation is also the process by

CONCEPT SUMMARY

COMPARISON OF THE PROCESSES OF PHOTOSYNTHESIS AND AEROBIC RESPIRATION

PHOTOSYNTHESIS	AEROBIC RESPIRATION
Converts light energy into chemical energy in the form of chemical bonds in organic substances, such as glucose	Converts chemical energy in organic substances, such as glucose, into chemical energy in other organic substances, such as adenosine triphosphate (ATP)
Forms new organic substances, such as glucose, from the inorganic substances water and carbon dioxide	Breaks down organic substances, such as glucose, into carbon dioxide and water
Requires carbon dioxide and produces oxygen	Requires oxygen and produces carbon dioxide
Performed only by photoautotrophic organisms	Performed by most living oganisms, including photoautotrophs

which bread and other doughs rise. The carbon dioxide produced by fermentation creates gas-filled spaces that make the dough rise and increase its volume. Although alcohol also is produced, we don't get drunk when we eat bread because the small amount of alcohol evaporates during baking.

The muscle cells of animals can respire anaerobically during periods of low oxygen supply. For example, if you run hard for a long time, your circulatory system will be unable to deliver sufficient quantities of oxygen to the active muscle cells. When this happens, pyruvic acid is converted to lactic acid. Lactic acid accumulates in the muscles, causing the pain of muscle fatigue. When sufficient amounts of oxygen are again available, the blood carries the lactic acid to the liver, where it is converted to pyruvic acid.

Unfortunately, brain cells don't live long without oxygen (about four minutes). If you stop taking in oxygen, your brain soon suffers irreversible damage. When you run or jog, you involuntarily respire faster as your body attempts to fuel aerobic respiration. The more you exert yourself, the more tired you feel, a sign of lactic acid accumulation and a switchover to anaerobic respiration. By breathing much harder dur-

You may be wondering why muscle cells don't produce alcohol or why yeast cells don't produce lactic acid. The answer is simple. Yeast cells and muscle cells have different enzymes. These enzymes are specific in function, catalyzing only certain reactions.

CELL PERMEABILITY AND ACTIVE TRANSPORT

If cells are to carry on life processes, they must maintain an internal environment that is significantly different, physically and chemically, from the external environment. If the internal environment is not maintained within narrow limits, normal cell function may be drastically impaired. For example, every cellular enzyme functions best within a narrow temperature range and at a specific pH. If these conditions are not [20] maintained, the enzyme will not function properly, and the biochemical reaction catalyzed by the enzyme will not occur. In extreme situations, enzymes and other proteins may be destroyed, and the cells may die.

The cell membrane plays a critical role in main- [40] taining the optimal internal environment. It keeps some substances out and allows others, such as oxygen and nutrients, to enter at a programmed rate. Some substances temporarily in short supply may actually be concentrated by cells. Accumulated cellular secretions or excretory products may be actively excreted. The discussion that follows describes some of the ways that cells absorb or eliminate substances through the cell membrane.

Water and Its Transport

Most living organisms and cells are 60 to 80 percent water, a substance that makes life as we know it possible. Some organisms may be more than 90 percent water. One of the special properties of water is its ability to dissolve a wide range of chemicals. We know the ease with which salt, sugar, and a number of other substances dissolve in water. Many important chemical reactions cannot take place unless the reacting substances are in solution.

Another special characteristic of water is its ability to help maintain temperature. A great deal of heat is necessary to increase the temperature of water even one degree. Water also holds heat and gives it off slowly as it cools. This is why coastal areas tend to have moderate climates and areas within continents tend to have more extreme ones. Water is equally important in regulating the temperature of living organisms. Perspiration is one mechanism by which water helps regulate the temperature of humans.

Water molecules are small enough to pass through the spaces between molecules of the cell membrane. Therefore, water enters and leaves cells freely. With a microscope, it is easy to see how readily water can enter cells. When human red blood cells are placed in pure distilled water, they rapidly swell. They continue to absorb water until their cell membranes disintegrate. Obviously, cells must have mechanisms to protect them from such excessive flooding. They do. But before this topic is discussed, we need to look at the processes of diffusion and osmosis.

Diffusion and Osmosis

All molecules are in constant random motion because of their heat energy. Molecules in liquids and gases, however, have greater freedom of motion than those in solids. As the molecules move about, they collide with one another and are deflected in new directions. High concentrations of molecules have a high frequency of collision. In time, the molecules move to areas of lower concentration, where the frequency of collision is lower. These molecules slowly become uniformly distributed within the space available to them. The movement of molecules from regions of higher concentration to regions of lower concentration is called **diffusion.**

Many people find the odor of Limburger cheese strong and objectionable. If someone opened a package of Limburger near you, it would take only a few seconds for you to smell the odor molecules as they diffused from the area of high concentration near the cheese to the area of lower concentration in your nostrils.

Water molecules, like any other kind of molecules, diffuse from place to place, even through cell membranes. Cell membranes are semipermeable; that is, they allow certain molecules, such as water, to enter or leave the cell while keeping others inside or outside it. **Osmosis** is the term used to describe the diffusion of water through a semipermeable membrane. As water moves by osmosis through a cell membrane, it still obeys the basic law of diffusion: It moves from the side with the higher concentration of water toward the side with the lower concentration.

As indicated earlier, when a red blood cell is placed in pure water, it swells (see Figure 4–9). This is because the contents of the cell include some water but a large amount of other substances, such as proteins. Fewer molecules of water than of other substances exist in the cell, which creates a lower concentration

Figure 4–9

Reactions of red blood cells when placed in hypotonic, isotonic, and hypertonic media. The results are due to osmosis—the movement of water through a semipermeable membrane from a more concentrated medium to a less concentrated medium.

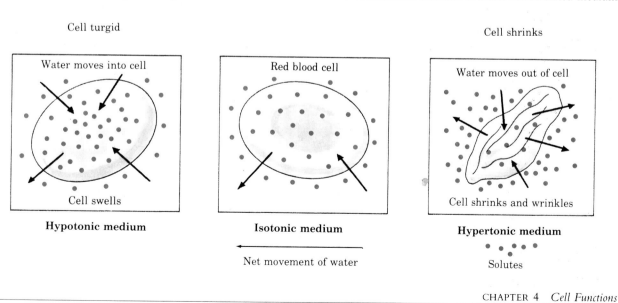

of water. Outside the cell, where there is no dissolved material, a higher concentration of water molecules exists in an equivalent space or volume. Therefore, water molecules will move across the cell membrane from the region of higher water concentration outside the cell to the region of lower concentration inside the cell.

This movement of water molecules across the semipermeable cell membrane is an example of osmosis. The cell swells as the cell membrane stretches to accommodate the entering water. In this situation, the water external to the cell is described as being a **hypotonic solution** (*hypo* means "less"). There is less dissolved material (and more water) in the solution than there is inside the cell.

If the solution exterior to the cell contains the same amount of dissolved substance as does the interior of the cell, the same concentration of water will exist on both sides of the membrane. The exterior solution here is described as an **isotonic solution** (*iso* means "same"). A red blood cell placed in an isotonic solution will not swell, because over time the same amount of water will enter the cell as will leave it.

One more combination exists: a **hypertonic solution** (*hyper* means "more"). In this situation, the solution surrounding a cell contains a greater amount of dissolved molecules (and therefore less water) than does the solution inside the cell. In a hypertonic solution, a red blood cell would shrink because of the loss of water from the interior to the exterior of the cell.

If the cell membrane were strong enough, the pressure of the water inside the cell would slow the movement of water into the cell. Eventually the system would come into equilibrium. The internal pressure that would slow and eventually stop the movement of water into the cell is called **osmotic pressure.** (Osmotic pressure is also a measure of the tendency of water to move from a hypotonic solution into a cell.) In pure water, most cells burst before they reach an osmotic equilibrium. Cells have evolved special adaptations, such as cell walls and contractile vacuoles, to prevent their oversaturation with water. Many live in an isotonic environment—a medium that is in osmotic equilibrium with the cytoplasm. Multicellular organisms—humans, for example—have excretory systems that play the major role in maintaining the proper balance of water in the body fluids around the cells.

Cell Walls

Plant cells have rigid cell walls; animal cells do not. Cell walls are strong enough to retain the swelling cells as water diffuses into them. The internal pressure in a plant cell builds up, and the plant cell becomes turgid, or rigid (see Figure 4–10). After a certain point, no more water accumulates. If water leaves a plant cell faster than it is replaced, the cyto-

Figure 4–10

Reaction of plant cells to hypotonic, isotonic, and hypertonic media.

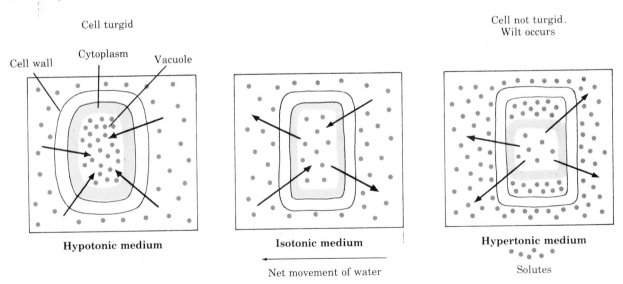

Hypotonic medium   Isotonic medium   Hypertonic medium

Net movement of water   Solutes

plasm shrinks and pulls away from the cell wall. The cell loses its turgidity, and the plant wilts. Without osmotic pressure, the cell walls are not strong enough to support the total weight of plant tissues.

Contractile Vacuoles

Freshwater protozoa commonly have **contractile vacuoles,** special cytoplasmic organelles that act as cellular sump pumps. These vacuoles drain water from the cytoplasm and discharge it through openings in the cell membrane (see Figure 4–11). Because the water is being discharged into the environment, where water molecules are more concentrated, the cell must forcibly extrude the water molecules. This requires the expenditure of energy derived from cellular respiration□.

Environments

Many plants and animals living in the ocean are bathed in seawater. This water contains enough dissolved salts and other materials to be an isotonic medium, one similar to the body fluids of these organisms. Therefore, the concentration of water in the ocean is essentially the same as that in the cytoplasm of the organisms. The net amount of water entering cells is thus about equal to the amount leaving, and the cells are never subjected to osmotic stress. In fact, the balance is so perfect that even fragile cells are unaffected by stress. For example, many ocean animals release eggs and sperm directly into seawater, and that is where fertilization occurs. However, for other organisms, such as fish, seawater is not isotonic. (See Box 21–2□.)

Many simple molecules move into and out of cells the same way that water does. For example, oxygen molecules diffuse into a cell from an outside region where they are more concentrated. However, oxygen molecules do not accumulate in cells, because they continuously react with hydrogen to form water during cellular respiration□. As a result, oxygen in cells is usually at a low level, which encourages the diffusion of more oxygen into cells.

For animals, the movement of carbon dioxide molecules is in the opposite direction. The concentration of carbon dioxide is great in the cells, so it diffuses into the environment. In human bodies, carbon dioxide is picked up by the blood and carried to the lungs, where it is exhaled. But because carbon dioxide is used up in the process of photosynthesis, plant cells absorb it from the surrounding environment.

Membrane Transport of Large Molecules

Many of the substances that cells need are too large to diffuse through cell membranes. How, then, are substances accumulated in the cytoplasm of cells?

Figure 4–11
Some cells—for example, the common amoeba—eliminate water with a contractile vacuole.

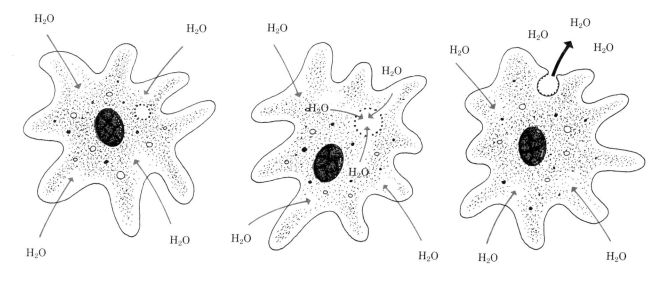

Phagocytosis

"Cell eating" is the literal meaning of **phagocytosis.** Cells eat when fingerlike projections of the cell membrane grow out and around a food particle, finally enclosing it in a bubble-like food vacuole (see Figure 4–12). A common example of phagocytosis is the eating activity of the amoeba. Human white blood cells engulf foreign particles and disease organisms in the same way that an amoeba engulfs food.

Phagocytosis is much more complex than it may seem. After phagocytosis, the engulfed particle is in a vacuole whose membrane has the properties of the cell membrane that gave rise to it. Therefore, the complex molecules of the object in the vacuole must be digested into smaller molecules before they can permeate the vacuolar membrane. This is essentially the same series of chemical steps that takes place in the digestive system of humans. Our digestion starts in the mouth and is completed in large cavities (stomach and intestines). When the larger, more complex molecules have been digested into simpler ones, they are absorbed by the cells that line these cavities. Even after foods are digested, though, the molecules may still be too large to enter cells by simple diffusion. If so, they are taken into the cells by pinocytosis.

Pinocytosis

Literally "cell drinking," **pinocytosis** is similar to phagocytosis, but pinocytotic vacuoles are much smaller than food vacuoles. The material that becomes enclosed within them is usually in solution rather than solid.

The process begins with a thin channel from the cell membrane into the cytoplasm, through which the cell drinks. Once solution fills the channel, the membrane pinches it off, creating a small vacuole in the cytoplasm. Almost immediately, the vacuole fuses with a lysosome◦, which carries digestive enzymes to break down the substances in the vacuole. After digestion, the small molecules pass through the vacuolar membrane into the cytoplasm. In some cases, the vacuolar membrane breaks down, and the vacuole's digested contents enter the cytoplasm. 42

Facilitated Diffusion

Some large molecules may move into or out of cells by a more complicated process, called **facilitated diffusion** (see Figure 4–13). Molecules move from regions of high concentration to regions of low concentration, just like molecules moved by simple diffusion. However, the actual passage across the cell membrane is "facilitated" by special protein molecules, called carriers. The molecules moved by carriers either are too large to diffuse through the layers of the cell membrane or are insoluble in the lipid layers of the cell membrane.

The carrier has enzyme-like properties, and it binds with the molecule that is to be transported at the outer surface of the membrane. The entire complex then glides through the lipid layer. On the other side of the membrane, the carrier breaks away from the molecule. The molecule moves into the cytoplasm, and the carrier goes to transport another molecule. Materials may be transported out of the cell in a reverse process.

Facilitated diffusion gives the cell membrane another property: selectivity. Like enzymes, each carrier protein is highly specific to certain kinds of mole-

Figure 4–12
Food particles are taken into the food vacuoles of cells such as amoebas by the process of phagocytosis. However, the food must be digested before any nutrients can pass through the membrane of the vacuole into the cytoplasm of the cell.

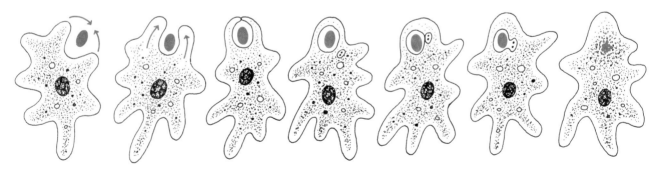

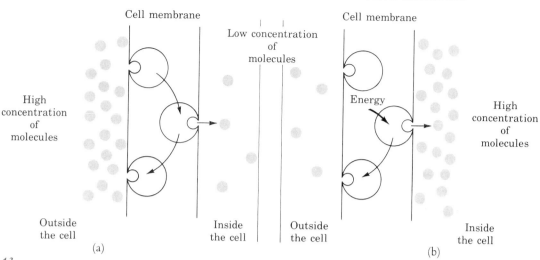

FACILITATED DIFFUSION

Cell membrane

High
concentration
of
molecules

Low concentration
of
molecules

Outside
the cell

Inside
the cell

(a)

ACTIVE TRANSPORT

Cell membrane

Energy

High
concentration
of
molecules

Outside
the cell

Inside
the cell

(b)

Figure 4–13

A comparison of facilitated diffusion and active transport shows that in both cases the molecule taken into the cell first forms a complex with a specific carrier molecule in the cell membrane. However, in the case of facilitated diffusion, the concentration of the molecule inside the cell is low. Thus the movement is along a diffusion gradient and does not require an expenditure of energy. In the case of active transport, the concentration of the substance is greater inside the cell than outside it. The substance is transported against a concentration gradient, and energy is expended during the process.

cules. The membrane selects for specific classes of molecules on the basis of the kinds of carriers present.

Active Transport

Cell membranes also have the ability to pass molecules from regions of lower concentration to regions of higher concentration, the reverse of diffusion and facilitated diffusion. Sometimes cells need to accumulate specific substances because cellular processes require them in higher concentrations than occur in the environment. The process that makes such movement of molecules possible is **active transport** (see Figure 4–13). It requires the expenditure of energy and frequently involves an exchange system whereby one molecule is actively transported and then exchanged for another.

One of the simpler examples of active transport is the **sodium/potassium exchange pump.** Many cells require higher concentrations of potassium ions than are present in the environment. The concentration of potassium within some cells may be twenty to fifty-five times greater than the concentration outside the cells. The active transport system of the cells pumps potassium into the cells and pumps sodium ions out in exchange.

SUMMARY

1. Photosynthesis is the biochemical process that converts light energy into chemical energy that is then used to fuel cell metabolism.

2. Organisms capable of photosynthesis, called autotrophs, synthesize the organic compounds that are essential nutrients for heterotrophs, which are incapable of synthesizing these compounds.

3. Oxygen, which is essential for cellular respiration, is produced during photosynthesis and released into the atmosphere.

4. Photosynthesis is divided into light and dark reactions. During the light reaction, light energy is captured and converted into the chemical energy that fuels the dark reaction. The dark reaction synthesizes organic substances from carbon dioxide and hydrogen.

5. Cellular respiration of organisms may be aerobic (requiring oxygen) or anaerobic (not requiring oxygen). The former is more efficient.

6. Aerobic respiration has three phases: glycolysis, the Krebs cycle, and the electron transport system.

Glycolysis splits glucose molecules. The smaller molecules are accepted into the Krebs cycle. During these biochemical conversions, ATP is formed, carbon dioxide is produced as a by-product, and electrons are released. The electron transport system accepts electrons from both glycolysis and the Krebs cycle. The energy it releases is involved in the formation of most of the ATP molecules formed during aerobic respiration. Oxygen is the final electron acceptor. These relationships are shown in the diagram.

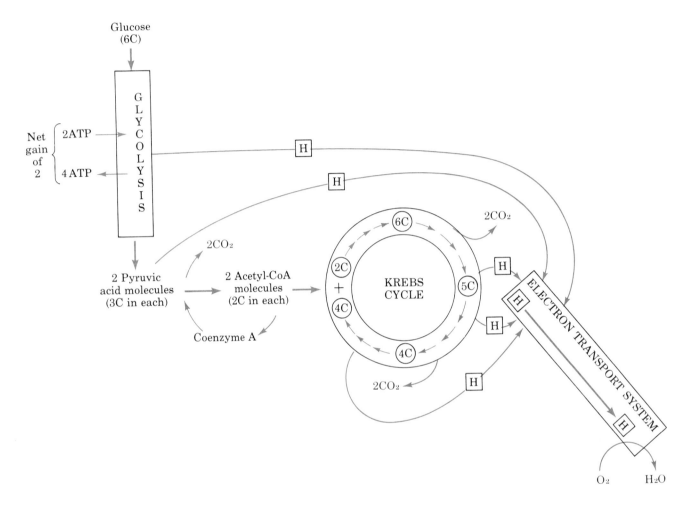

7. During anaerobic respiration, the Krebs cycle and electron transport system do not operate. Glycolysis produces alcohol or, in the case of muscle cells, lactic acid. Little energy is released during these conversions.

8. Small molecules may move through cell membranes by simple diffusion. Other substances enter cells by facilitated diffusion or active transport. Because the cellular membrane is selective, it is called a semipermeable membrane.

THE CONTINUING EXPLORATION OF CELLS

During the next decade, scientists will have new tools for exploring the fabric of our universe. Some will be reaching toward the planets and stars, but others will be focusing on the molecular universe within cells.

Not too long ago, biologists pictured cells as simple sacs filled with a semifluid substance in which floated the nucleus and some smaller, more obscure structures. With good light microscopes and special staining techniques, biologists began to study the structure and function of some of these smaller organelles. But it was not until electron microscopes were used that cytologists really began to appreciate the complex architecture of cells.

Now, million-volt electron microscopes are used to explore the finest details of cell structure. Just a few years ago, they were used in discovering the extensive fine filamentous cell skeletons, the microtrabecular system. Today, they are used to investigate the details of how microtubules form and function. But we cannot learn as much as we would like from photographs alone.

Other techniques have been developed to isolate cell organelles from living cells so their activities can be studied under controlled and monitored conditions. Cells are suspended in solutions that prevent osmotic shock and then are torn apart in machines similar to kitchen blenders. Next, the suspensions of cell parts are centrifuged. The heavier parts, such as nuclei, settle out during the slow rates of centrifugation and are collected in the sediment at the bottom of the centrifuge tube. The smaller, lighter structures, such as mitochondria and ribosomes, remain suspended in soluton and settle out when the suspension is centrifuged for longer periods of time at higher speeds. By careful control of the centrifugation rates, nearly pure samples of different types of cell organelles, such as mitochondria, can be collected. In fact, much of what we know about cellular respiration and the enzymes that regulate it was learned by isolating mitochondria under carefully controlled conditions.

To test the role of nuclei in cells and to understand their relationships to cytoplasm, biologists developed techniques for removing the nuclei from cells and transplanting them to other cells, sometimes even to cells of different species. In 1962, a new and exciting technique for studying cell function was developed, and it is widely used today. The technique is cell fusion. Two cells are induced to fuse together into a single cell, called a hybrid cell, that is capable of reproduction. The fusion is induced by treating the cells with solutions of polyethylene glycol.

Depending on the nature of the experiment, the nucleus of one of the cells may be destroyed or removed before fusion, or both nuclei may be left intact. Different kinds of cells can be fused together. Normal and cancerous human cells from the same individual or from different individuals may be fused, or cells from two different species may be fused. Human cells have been fused with bird and mouse cells, and mouse cells have been fused with rat and hamster cells. Cells of different plant species have also been fused.

Cell fusion studies have added to our knowledge about how normal cells become cancerous, how cells differentiate, and where hereditary factors appear on chromosomes. During the next decade, cell fusion experiments will enable us to learn much more about the intricate workings of cells. Also, cell fusion techniques using plant cells may lead to the development of more productive and disease-resistant plants or even of types of plants capable of growing in seawater.

Plant cells fuse more readily than do animal cells. In the procedure for fusing them, the cell walls are removed by enzymes. Then the cells to be fused (called protoplasts) are mixed and given the necessary chemical stimulus to fuse them. Following fusion, the new hybrid cells are stimulated to produce new cell walls and are cultured in glass tubes in special media that encourage their growth and reproduction. Undifferentiated masses of new cells are formed. Individual cells from these cultures are given a special hormone treatment that induces them to develop into new plants with features of both parents. As many as a thousand cloned plants can be produced from as little as one gram of hybrid cells.

Some examples of plants that have been produced by fusion are tomatoes that will grow in semi-arid conditions, tobacco that will grow in water with a salt content nearly that of seawater, and potatoes that are at least partially resistant to potato blight disease. Who knows what kinds of plants may be developed during the next decade and what benefits they will bring to the world?

75

FOR DISCUSSION AND REVIEW

1. What is the difference between autotrophic and heterotrophic organisms?

2. Write a generalized chemical formula for the process of photosynthesis.

3. Explain the difference between the light and dark reactions of photosynthesis. How are they connected?

4. What is the difference between aerobic and anaerobic respiration?

5. Write a generalized chemical formula for aerobic respiration.

6. Explain glycolysis, the Krebs cycle, and the electron transport system. How are these processes related to aerobic and anaerobic respiration? Which is more efficient in energy production?

7. Compare and contrast osmosis, diffusion, facilitated diffusion, and active transport. What are the processes that allow cells to be selectively permeable?

8. How are pinocytosis and phagocytosis similar? How are they different?

SUGGESTED READINGS

THE CELL, by Carl P. Swanson and Peter L. Webster. Prentice-Hall: 1977.
A well-written account of the biology of cells.

THE LIVING CELL: READINGS FROM THE SCIENTIFIC AMERICA, edited by Donald Kennedy. W. H. Freeman: 1965.
A collection of articles on cell structure and function, written for the general reader.

MOLECULES TO LIVING CELLS: READINGS FROM THE SCIENTIFIC AMERICAN. W. H. Freeman: 1980.
Twenty-six excellent articles, each written by leading authorities in the field of molecular cell biology.

PHOTOSYNTHESIS, by Eugene Rabinowitch. Wiley: 1969.
An excellent introduction to photosynthesis and energy conversion.

5
CELL REPRODUCTION

The ability to reproduce is one of the unique characteristics of living things, and the key to reproduction lies in the cell. Every cell in existence was involved at some time with cellular reproduction. In fact, one of the most fundamental principles of biology is that all cells come from preexisting cells.

During cellular reproduction (often called **mitosis** or **cell division**), each "daughter" cell receives an identical copy of the genetic material contained in the mother, or stem, cell.

Molecules of **deoxyribonucleic acid (DNA)** are the genetic material that enables cells to make exact copies of themselves. The nucleus of every normal human cell (except sperm or ova) houses forty-six **chromosomes**, which appear as twenty-three pairs. The amazing DNA molecules, coated with nucleoproteins, form the chromosomes. One member of each chromosome pair comes from the father, and the other comes from the mother. They form what biologists call **homologous pairs** of chromosomes. At the time of conception, the father's sperm and the mother's ovum join, bringing single chromosomes together to form pairs. When the cell that begins a human existence divides for the first time, the two resulting cells each receive a complete set of forty-six chromosomes. These cells continue to divide until the human becomes a reality.

THE CELL CYCLE

When cells reproduce, they follow a distinct four-phase cycle: gap one (G_1) phase, synthesis (S) phase, gap two (G_2) phase, and mitosis (M) phase. Figure 5–1 shows how these phases relate. The G_1, S, and G_2

PREVIEW QUESTIONS

After reading this chapter, you should be able to answer the following questions:

1. Why are daughter cells genetically identical?

2. What steps are involved in the production of male sex cells?

3. How does the production of female sex cells differ from the production of male sex cells?

4. List the main differences between mitosis and meiosis.

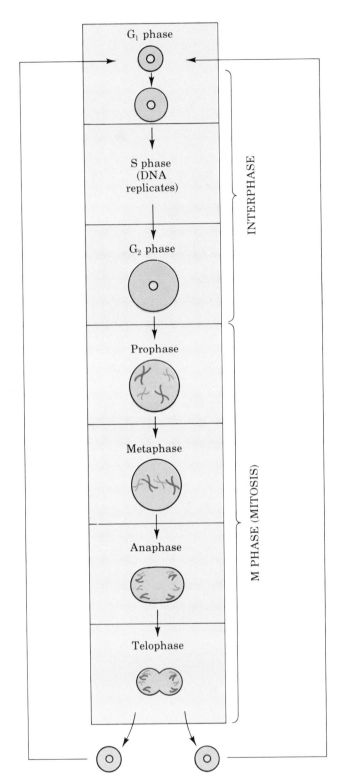

Figure 5–1

The cell reproductive cycle. When a cell completes cell division, it enters the G_1 phase, a period of cell growth. If the cell is to reproduce again, the G_1 phase is followed by the S phase, the period when the genetic material (DNA) is replicated. This is followed in turn by the G_2 phase, during which the cell is further readied for the period of cell division—the M (mitosis) phase.

phases together are the longest phase of the cell cycle. They used to be known as the interphase, and they were thought of as a resting stage between divisions. It is now known that DNA synthesis and most cellular metabolism occur at this period.

During the so-called interphase, cells have a characteristic appearance□. The nucleus is prominent and contains one or more nucleoli. The nuclear membrane is intact. Chromosomes cannot be seen because they exist as long, thin strands or threads of DNA that are often intertwined and entangled. These strands are called **chromatin threads**, and are so fine they are not clearly visible. Where they become more dense, they are known as **heterochromatin**. Cytoplasmic organelles are obvious at this phase. They have a characteristic appearance and distribution. All metabolic activities of the cell function during interphase.

The earliest interphase activity of a new cell involves a period of increased ribonucleic acid (RNA)□ synthesis, protein synthesis, and growth. This is the **G_1 phase**. Many cellular organelles, such as mitochondria, also increase in number during this time. When the cell reaches its maximum size, it may continue its vital functions in support of a particular tissue or organism instead of reproducing. However, if the cell begins a reproductive cycle, it enters the S phase.

During the **S phase**, the DNA in the cell replicates itself from precursors synthesized during G_1. The S phase is the key to cellular reproduction; during this phase, the duplicate copy of the genetic information is manufactured.

The **G_2 phase** follows. It is characterized by protein synthesis, and it paves the way for actual cell division, which occurs next.

The **M phase** is composed of two separate events: **mitosis** (sometimes referred to as nuclear division), in which the DNA molecules that replicated during the S phase separate into two equal groups; and **cytokinesis**, the process that cleaves the cytoplasm of the dividing cell into nearly equal portions.

Mitosis

Mitosis is thought to occur in four phases: prophase, metaphase, anaphase, and telophase. It is a continuous process; one phase moves smoothly into the next without any interruption. Figure 5–2 details the events of mitosis.

Prophase

One of the obvious events of prophase, the "before phase," is the shortening and thickening of the elongated chromatin threads into discrete, rodlike chromosomes. It is believed that the protein-coated strands of DNA, known as chromatin, become chromosomes by coiling. The discrete chromosomes appear as double, or twin, chromosomes. Each half of a twin chromosome is called a **chromatid**. The chromatids are connected at a common spot, called the **centromere**, at some point along their length. (See Figure 5–3.) The DNA molecule in each chromatid is continuous; it is not interrupted by the centromere.

As chromatids continue to coil and become more obvious, the nuclear membrane and nucleoli begin to disintegrate. Eventually, they disappear, leaving the chromosomes free in the cytoplasm of the cell.

Animal cells have two **centrioles**, which separate and move to opposite ends of the cell. (Plant cells lack centrioles. This is a major distinction between animal and plants cells.) As the centrioles migrate, fibers composed of microtubules□ form around each one.

Figure 5–2

Mitosis. The major events of animal cell mitosis: During prophase, the chromsomes become evident as they coil and appear to become shorter and thicker. Simultaneously, the nuclear membrane disappears, the centrioles move to opposite poles of the cell, and the asters and spindle fibers form. Some spindle fibers attach themselves to the centromeres of the chromosomes, which gradually move toward the metaphase, or equatorial, plate in the center of the cell. Metaphase is the brief period when the chromosomes are aligned with their centromeres on the metaphase plate. Anaphase begins when the centromere of each chromosome splits. The two now-separated chromatids move toward opposite poles of the cell. During telophase, the new single-stranded chromosomes at each of the two poles of the cell uncoil as a new nuclear membrane develops, enclosing them. Simultaneously, the cell divides by cytokinesis.

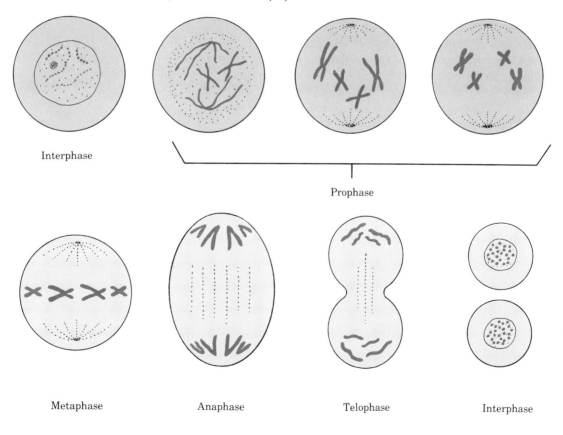

Interphase

Prophase

Metaphase   Anaphase   Telophase   Interphase

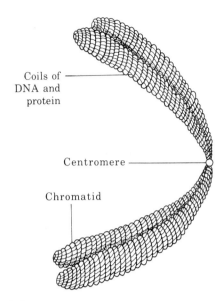

Coils of
DNA and
protein

Centromere

Chromatid

Figure 5–3
An idealized chromosome coiled to a maximum extent.

The fibers create a starlike structure called the **aster**. (*Aster* literally means "star.") When the centrioles lie at opposite ends of the cell, they become connected by **spindle** fibers. These fibers are filaments that form a pattern of two cones set end-to-end. Not surprisingly, this biconical structure is called the **spindle**. (See Figure 5–2.)

Late in prophase, the well-defined chromosomes begin to move—sometimes sporadically, sometimes in a smooth glide—toward the widest part of the spindle. The reason for this movement is unknown. The movement does not begin, however, until a spindle fiber from each centriole attaches itself to the centromeres.

Metaphase

The "middle phase" (metaphase) is brief in comparison to the other parts of the M phase of the cell cycle. It occurs when the centromeres of all the chromosomes come to lie on exactly the same plane in the middle of the spindle. (The arms of the chromatids may extend in any direction.) This brief alignment of centromeres is on an imaginary **equatorial plane**. It is during metaphase that the chromosomes are most distinct and easily seen, counted, and described.

Anaphase

The "after phase" (anaphase) begins when the cen-

tromeres divide and the chromatids of each twin chromosome separate and move toward opposite ends of the spindle. The separated chromatids are now known as **daughter chromosomes**. The centromeres of the daughter chromosomes lead the way to the opposite poles of the cell, and the arms of the chromosomes trail behind. Again, the mechanism that causes this movement is unknown. One widely accepted theory is that chromosome movement is some function of the movement or shortening of the spindle fibers attached to the centromeres. Anaphase ends when a complete set of chromosomes congregates at each pole of the spindle.

Telophase

The "end phase" (telophase) essentially repeats the events of prophase, but in reverse order. The daughter chromosomes at each pole lengthen and begin to uncoil; they lose their distinct appearance as chromosomes and begin to look like tangled masses of chromatin threads. The spindle and aster fibers disintegrate. A new nuclear membrane encloses the chromatin, and nucleoli are resynthesized. While these events are occurring, cytokinesis begins. Try to identify the phases of mitosis shown in Figure 5–4.

Figure 5–4
Cells in a developing whitefish blastula go through various stages of mitosis: M = metaphase, A = anaphase, T = telophase.

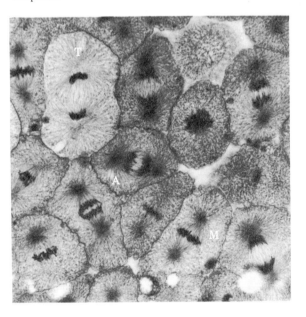

Cytokinesis

During cytokinesis, the cell membrane begins to constrict about midway between the reforming nuclei. The constriction continues until two separate masses of cytoplasm are obvious. Each mass contains about half of all the cytoplasmic organelles—mitochondria, ribosomes, and so on. Eventually, the two cytoplasmic masses with separate nuclei become completely separate **daughter cells**. The reason for the movement of the cell membrane during cytokinesis remains a mystery, but the movement is believed to be involved with a ring of microfilaments□ beneath the membrane at the site of the initial furrow. Figure 5–5 illustrates cytokinesis.

Plant Cells

During the M phase, the events in plant cells are basically the same as in animal cells. Although the cells of higher plants lack the centrioles that are characteristic of animal cells, a bioconical spindle forms and functions just the same. The major difference in the M phase of plant and animal cells is the manner in which cytokinesis occurs. During telophase, the plant cell membrane does not begin to constrict. Instead, a new **cell wall** begins to form midway between the newly formed daughter nuclei. (See Figure 5–5.)

The manufacture of the new cell wall begins in the center of the cell and proceeds outward. The first step is the formation of a series of compressed membranes derived from Golgi complexes and the endoplasmic reticulum□. These membranes aggregate along the midline of the cell, forming a visible structure called the **cell plate**. The cell plate continues to form until the two daughter cells are separate compartments of cytoplasm. Each of the daughter cells then produces a new cell membrane, which in turn secretes a new primary cell wall between the cell membrane and the cell plate. The cell plate remains between the two adjoining cell walls as the **middle lamella**.

Figure 5–5

Cytokinesis: (a) of an animal cell; (b) of a plant cell. There is a significant difference between the two processes. In animal cells, the cytoplasm constricts inward, gradually dividing into two daughter cells. In plant cells, the cytoplasm does not constrict during cytokinesis. Instead, a cell plate forms within the cytoplasm and expands outward, dividing the cell.

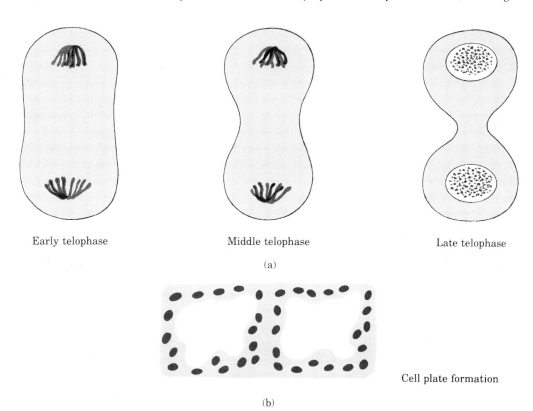

Early telophase   Middle telophase   Late telophase

(a)

Cell plate formation

(b)

During the cell cycle, a mother cell gives rise to two daughter cells, each with a complete complement of genetic material (DNA) identical to that in the original mother cell. Mitosis ensures equal distribution of chromosomes. The replication of DNA occurs not while the cell is dividing but before division, during the S phase of interphase.

MEIOSIS

The distribution of chromosomes is different during the production of sperm or ova. The process, called **meiosis,** causes a reduction in chromosome numbers (*meio* means "less") in daughter cells. This reduction is important, because if gametes (sperm or ova) contained the full complement of chromsomes typical of the species (forty-six in humans), the newly fertilized egg would have twice the number it needed (ninety-

two in humans). Each subsequent generation would have double the chromsomes of the preceding generation. If species are to be perpetuated, the chromosome number must be the same generation after generation.

Meiosis creates **gametes,** or sex cells (sperm or ova), that contain half the number of chromosomes— one member of each homologous pair. Thus human sperm and ova each possesses only twenty-three chromosomes. When an ovum and a sperm unite, the typical number of chromosomes, forty-six for humans, is restored. The number of chromosomes in gametes is called the **haploid number** (represented by *n*). All other cells contain the **diploid number** (represented by 2*n*). Figure 5–6 shows the difference between mitosis and meiosis in terms of chromosome number.

Meiosis has another, equally important function: It provides for new genetic combinations, which cause variations among individuals. The important ramifications of genetic variation are described elsewhere.

Meiosis will be described here as it would occur during the formation of sperm cells in the testes of the human male. This meiotic process, called **spermatogenesis** (*genesis* means "beginning") is illustrated in detail in Figure 5–7. Meiosis in human females creates ova and is called **oogenesis.**

A specific class of cells, called **spermatogonia,** lines the seminiferous tubules of the testes, where the sperm are produced. These diploid cells enlarge and give rise to **primary spermatocytes,** where "twin chromosomes" are produced during the S phase of the cell cycle. As a result of meiosis, each primary spermatocyte becomes four sperm cells. Meiosis involves two consecutive cell divisions, often referred to as meiosis I and meiosis II.

Meiosis I

During **meiosis I,** a primary spermatocyte divides into two **secondary spermatocytes.** The chromosome number is reduced from the diploid condition (primary spermatocyte) to the haploid (secondary spermatocyte). Meiosis I is often called the **reductional division** because the chromosome number is halved.

As was the case for mitosis, meiosis I is characterized by prophase, metaphase, anaphase, and telophase. During prophase I, the chromosomes shorten and become well-defined. Eventually, they appear as

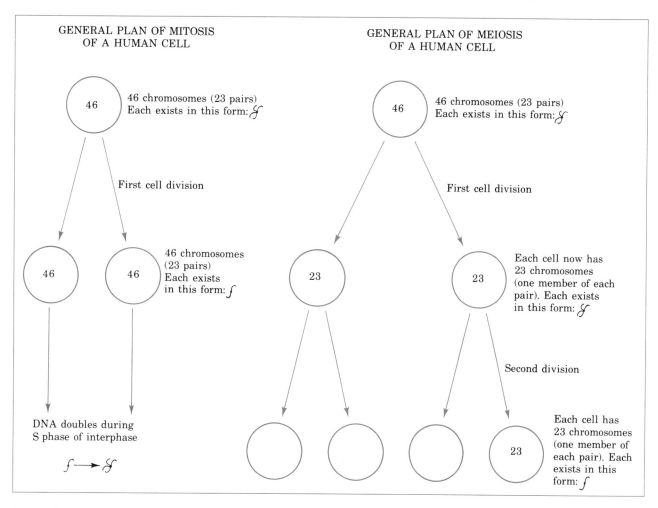

GENERAL PLAN OF MITOSIS OF A HUMAN CELL

46 chromosomes (23 pairs) Each exists in this form: ∮

First cell division

46 chromosomes (23 pairs) Each exists in this form: ∫

DNA doubles during S phase of interphase

∫ ⟶ ∮

GENERAL PLAN OF MEIOSIS OF A HUMAN CELL

46 chromosomes (23 pairs) Each exists in this form: ∮

First cell division

Each cell now has 23 chromosomes (one member of each pair). Each exists in this form: ∮

Second division

Each cell has 23 chromosomes (one member of each pair). Each exists in this form: ∫

Figure 5–6
During mitosis, each parent cell gives rise to two daughter cells, which are genetically identical to the parent. That is, they have the same complement of chromosomes as the parent cell. During meiosis, which is completed in two consecutive cell divisions, four daughter cells are produced, and the chromosome complement is reduced from the diploid to the haploid number.

twin chromosomes, or chromatids joined by a common centromere. However, before the chromosomes align on the spindle, homologous chromosomes (one of maternal origin, the other of paternal origin) come to lie in physical contact with each other, a pairing called **synapsis**. The alignment of the homologues is so exact that the genes on one chromosome touch the corresponding genes on the other. Frequently, the homologues are so closely associated that they appear to be a single chromosome. Actually, they form a bundle of four chromatids, referred to as a **tetrad**.

During synapsis, the homologues often exchange segments of DNA, a process called **crossing over**

(see Figure 5–8). This process, which is one of the mechanisms that create new genetic combinations, is a common occurrence. The result is that pieces of paternal and maternal chromosomes are exchanged, causing a mixing of genetic information.

After synapsis, homologous chromosomes begin to separate. As they do so, they align on the equatorial plane in pairs (metaphase I). In some cases the homologues separate completely; in others the arms of the chromatids remain entangled.

After brief alignment on the equatorial plane, the pairs separate and the chromosomes move to the opposite poles of the cell (anaphase I). However, the

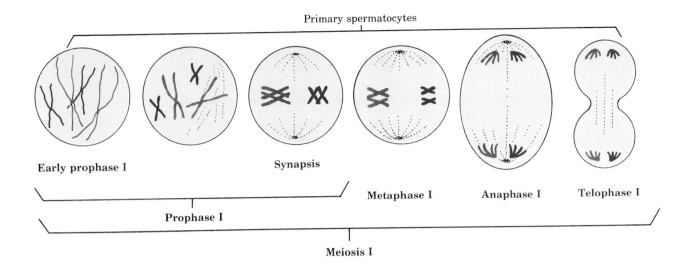

Primary spermatocytes

Early prophase I   Synapsis

Metaphase I   Anaphase I   Telophase I

Prophase I

Meiosis I

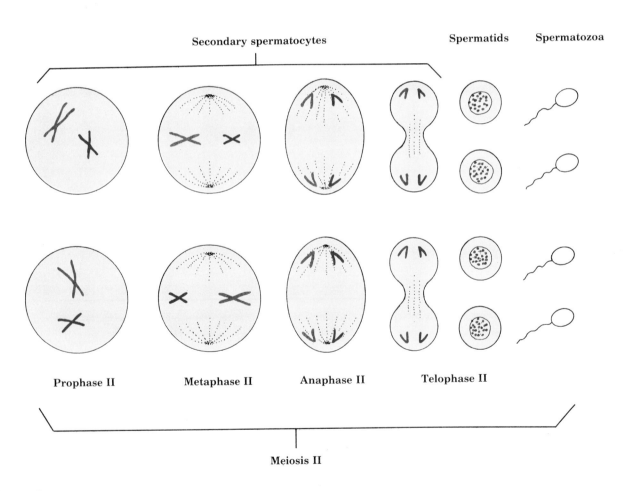

Secondary spermatocytes   Spermatids   Spermatozoa

Prophase II   Metaphase II   Anaphase II   Telophase II

Meiosis II

Figure 5–7

The major events of meiosis during spermatogenesis. Meiosis is a sequence of two cell divisions: meiosis I and meiosis II. Each division includes a prophase, a metaphase, an anaphase, and a telophase. During prophase I, the chromosomes in primary spermatocytes shorten by coiling and become obvious. A unique event, synapsis, occurs during the prophase of meiosis I. During synapsis, the two members of each pair of homologous chromosomes come together and pair so intimately that they appear to be one chromosome with four chromatids. At this time, crossing over also occurs. Following synapsis, the paired chromosomes begin to part, but they still move in pairs to align themselves on the metaphase plate. Metaphase occurs when all the chromosomes are at the plate. Metaphase I ends as the members of homologous pairs of chromosomes separate. During anaphase I they move to opposite poles of the primary spermatocyte. During telophase I, the primary spermatocyte divides, producing two secondary spermatocytes, each containing the haploid number of chromosomes. Often, these chromosomes remain partly coiled during telophase I. Therefore, prophase II of meiosis II may be brief. Prophase II is completed when the chromosomes (one member of each homologous pair) align themselves on the equatorial, or metaphase, plate in a secondary spermatocyte (metaphase II). During anaphase II, the centromeres of each chromosome split, and the new, separate chromatids move toward opposite poles of the cell. During telophase II, the new single-stranded chromosomes uncoil as nuclei re-form. Each secondary spermatocyte divides, producing two spermatids that soon differentiate into spermatozoa.

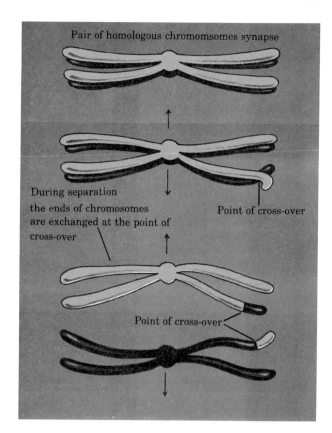

Figure 5–8

Crossing over. During prophase I of meiosis, homologous chromosomes pair so exactly and so tightly that they look like a single chromosome composed of four chromatids. During this pairing, and as the members begin to separate again after synapsis, corresponding sections of chromatids of homologous chromosomes are exchanged. This event, crossing over, results in greater genetic variation among the gametes produced by meiosis and therefore also among the offspring of sexually reproducing organisms.

centromeres holding chromatids together do not divide at the beginning of anaphase in meiosis I (as they did during mitosis). Therefore, the two entire chromatid complexes (twin chromosomes) move to opposing poles—the mostly maternal homologue to one pole, the mostly paternal homologue to the other.

The distribution of maternal and paternal homologues is random. In fact, it is referred to as **random assortment** (see Box 5–1). The number of random combinations of paternal and maternal chromosomes in the resulting two cells is $2n$, where n is the haploid number of chromosomes characteristic of the species. In humans, the number of possible chromosome combinations is 2^{23}, or 8,388,608—a very large number. Random assortment is another mechanism for providing genetic variation in offspring.

Meiosis I ends when the chromosomes have moved to opposite poles and cytokinesis has created two new cells, the secondary spermatocytes. Each chromosome in a secondary spermatocyte consists of two joined chromatids. The number of chromosomes, however, is exactly half the number of chromosomes in the cell that produced them. The twin chromosomes remain distinct, rodlike structures until meiosis II. However, there is no S phase of DNA replication between meiosis I and meiosis II.

Meiosis II

During **meiosis II**, the two secondary spermatocytes formed during meiosis I divide to form a total of four

BOX 5–1

RANDOM ASSORTMENT

Random assortment, or independent assortment, is the basis of Mendel's second law of genetics (discussed in Chapter 11). Here it will be explained using only three homologous pairs of chromosomes—A and A′, B and B′, and C and C′.

During anaphase of meiosis I, chromosome A might move toward one pole of the cell and chromosome A′ toward the opposite pole. However, chromosome pair BB′ need not follow the same pattern as AA′. The distribution of AA′ has no effect on the distribution of BB′. Thus the two events are random and independent. There is an equal chance that chromosome B will move toward the same pole as A did or toward the opposite pole.

Likewise, the distribution of chromosome pair CC′ is not influenced by the distribution of either AA′ or BB′. Chromosome C′ has an equal chance of moving toward either pole. Thus, one pole of the primary spermatocyte may receive any of the following combinations of chromosomes:

ABC
A′B′C′
A′BC
AB′C
A′B′C
ABC′
A′BC′
AB′C

These combinations occur purely by chance. With large numbers of sperm produced, there is an equal probability that any of these combinations will be present.

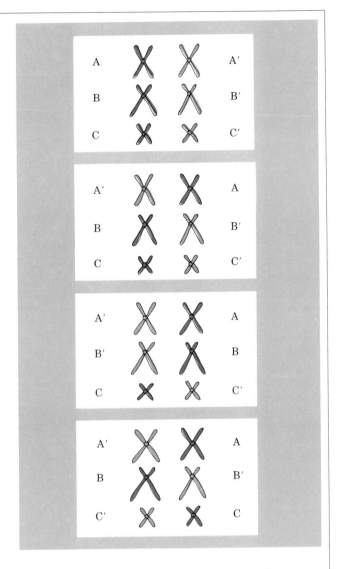

Possible chromosome alignment during metaphase I of meiosis

cells, called **spermatids**. The events in meiosis II are essentially the same as those in mitosis. There is an equal distribution of genetic information during meiosis II. For that reason, this stage is often called the **equational division**.

Prophase of meiosis II is brief. Twin chromatids, still connected by a common centromere, are already present. They move toward the center of the spindle, and their centromeres align on the equatorial plane (metaphase II). The centromeres divide at the begin-

ning of anaphase II. Then the chromatids separate into daughter chromosomes. During telophase and cytokinesis, the nuclei of the resulting cells are re-established and the chromosomes uncoil to form chromatin threads. The resulting cells are haploid spermatids. In human spermatogenesis, each of the secondary spermatocytes entering meiosis II possesses twenty-three double chromosomes. After the centromeres divide and the chromatids separate, there are two cells with twenty-three daughter (single-stranded) chromosomes each, the haploid number. Spermatids undergo additional differentiation, eventually becoming functional **sperm cells**.

During the formation of ova (oogenesis), the events of spermatogenesis are essentially repeated (see Figure 5–9)—with one major difference. In most species, oogenesis creates only a single haploid ovum for each **primary oocyte**, not four. During cytokinesis of meiosis I, each primary oocyte produces one large cell, the **secondary oocyte**, and one tiny cell, the **polar body**. In humans, each cell has twenty-three double chromosomes. During cytokinesis of meiosis II, the single secondary oocyte divides, producing one large cell, an **ootid**, and another polar body. Each of these cells contains twenty-three daughter (single-stranded) chromosomes.

The polar body formed during meiosis I may divide to form two additional polar bodies during meiosis II. These polar bodies are almost devoid of cytoplasm and cytoplasmic organelles; they disintegrate and die. The events of oogenesis thus create one

large cell, the ovum, with most of the cytoplasm of the primary oocyte. The cytoplasm of the ovum contains the nutrients that support the development of an embryo▫.

253

Meiosis in Plants

Higher plants are characterized by two distinct generations. The **gametophyte generation** (literally a "gamete plant") produces eggs and sperm. The **sporophyte generation** produces spores. These generations alternate, a phenomenon referred to as **alternation of generations**▫. Meiosis in plants is identical to 205 meiosis in animals, except that spores rather than gametes are produced. Specific cells of the sporophyte generation undergo meiosis to produce haploid spores. Each spore is capable of germinating into a haploid gametophyte generation that produces eggs and sperm cells directly by mitosis and cellular differentiation.

Unlike mitosis, meiosis reduces the diploid number of chromosomes in cells to the haploid number con-

Figure 5–9

The major events of oogenesis. During the first meiotic division, one small cell, a polar body, is formed. The other, larger cell, the secondary oocyte, receives most of the cytoplasm. During the second meiotic division, the ootid receives most of the cytoplasm. As a consequence, the ovum, or egg, retains most of the cytoplasmic resources, such as stored nutrients. These resources are used by the cells of the early embryo following fertilization of the ovum.

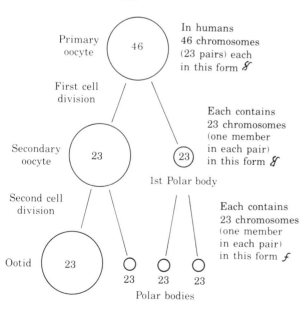

CONCEPT SUMMARY	
COMPARISON OF MITOSIS AND MEIOSIS	
MITOSIS	MEIOSIS
Occurs during a single cell division	Occurs during two sequential cell divisions
Results in the production of two diploid cells	Results in production of four haploid cells
Daughter cells all genetically identical	Daughter cells not all genetically identical
During metaphase, chromosomes are randomly aligned on the equatorial plate	During metaphase of first meiotic division, chromosomes are aligned in homologous pairs on equatorial plate
At beginning of anaphase, the two chromatids of each chromosome separate	At beginning of anaphase of first meiotic division, whole homologous chromosomes separate from each other (this is the reductional division); chromatids of chromosomes separate during the second equational meiotic division
In animals, does not result in production of gametes	Involved in production of gametes in animals

tained in sperm and egg cells. The process involves two sequential cell divisions, a reductional division (meiosis I) and an equational division (meiosis II). Meiosis also creates new genetic variations by synapsis, crossing over, and random assortment. This reordering of genetic material creates offspring that are genetically different from their parents and from one another.

CELLULAR DIFFERENTIATION

New cells created during cell reproduction usually undergo some structural change as they assume their gentically programmed activities. For example, spermatid cells produced through the second meiotic division of secondary spermatocytes look nothing like mature sperm cells, or spermatozoa (see Figure 5–10). A spermatid is a spherical cell much larger than a sperm cell. During its transformation, the cytoplasm of the spermatid becomes reorganized into a head region containing the nucleus and an acrosome (a structure involved in penetration of the ovum), a smaller middle piece containing mitochondria, and a terminal tail piece, or flagellum, used to propel the sperm cell.

During the early development stages of a multicellular organism, all the cells in the **embryo** appear to be structurally similar. As the embryo develops, cells and tissues begin to assume the form and function they will possess in the new individual. In the human embryo, some cells form epithelial cells, others connective tissue cells, nerve cells, muscle cells, and so on.

Cellular differentiation is under genetic control. But even though every cell in a human embryo carries the same set of genetic information, generalized cells become nerve cells, muscle cells, and various other kinds of cells. In order to differentiate, some of the genetic information must be suppressed (and thus is not expressed). Geneticists are currently studying the mechanisms that determine which part of the total genetic blueprint cells follow as they differentiate. They have found that the first few cells of a human embryo, if isolated, can develop a complete human being▢. Cells from later embryonic stages lose this potential.

270

However, some lower animals and many plants are capable of developing entirely new individuals from small fragments of tissue or parts of the organism.

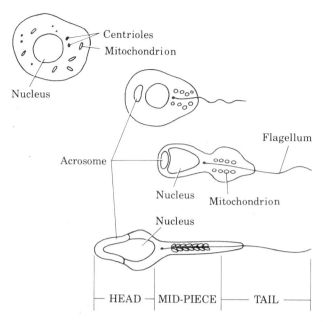

Figure 5–10
A spermatid undergoes considerable transformation while developing into a sperm cell. Much of the cytoplasm is lost, a flagellum is formed, and the cell becomes organized into a head region containing the nucleus, a middle piece with a mitochondrion, and a tail section. This is an example of cellular differentiation, the change in cell form and function from a generalized to a more specialized form.

This process is called **asexual reproduction▢**. Animals as complex as amphibians have the ability to regenerate lost limbs. At the site of the amputation, cells rapidly reproduce, differentiating into new muscles, bone, and so on. Regeneration occurs only if the remaining limb stump contains sufficient nerve supplies.

196

Humans and other mammals cannot regenerate structures as complex as limbs. In 1972, however, a scientist induced a degree of limb regeneration in rats by electrically stimulating the stump. Also, under certain circumstances, very young children have been able to regenerate lost tips of fingers down to the first joint. It is apparent that certain aspects of a cell's genetic capabilities can be turned on or off.

CELL DEATH

All organisms eventually die. The longest-lived organisms are conifers. The bristlecone pine trees of

California, for example, may live for four thousand years or more. Unicellular organisms, such as the amoeba, may come closest to eternal life, because the body of the mother cell becomes the bodies of the daughter cells.

The different kinds of cells in multicellular organisms have different life spans. Some types of cells survive until the death of the entire organism. Others, such as human red blood cells, have a more limited life span and must be replaced. But all cells eventually age and deteriorate. Some accumulate pigment granules and other inert substances as they age. Others may self-destruct by the release of lysosomal enzymes into the cytoplasm. The process of cellular aging is not well understood, but it is an area of extensive investigation.

Cell death in multicellular organisms occurs all the time and is often a necessary and desirable event. For example, cell death plays an important role in the transformation of many animals from the larval to the adult form (metamorphosis). During the metamorphosis of a tadpole to a frog, the tadpole's tail is reabsorbed by the destruction of muscle and other associated cells (see Figure 5–11).

During human development, the lips, which are initially fused to the gums, are freed from their attachment because of the death of cells. When human fingers and toes first develop, they are webbed by bridges of tissue that are also destroyed by the genetically programmed death of cells. In some infants, however, this tissue does not self-destruct during development. These infants are born with two or more fingers fused together. This condition, called syndactyly, is genetically inherited.

SUMMARY

1. The genetic material (DNA) is organized as pairs of structures, called homologous chromosomes, housed in the nucleus. During cell reproduction (mitosis), each daughter cell receives an exact copy of every pair of chromosomes.

2. The life of a cell consists of G_1, S, G_2, and M phases. These phases are called the cell cycle.

3. Chromosomes are duplicated during the synthesis (S) phase, which occurs before mitosis. Chromosomes then are double, each one being composed of two chromatids.

Figure 5–11
A tadpole during the process of metamorphsis into an adult frog. As the four limbs develop, tail cells die and are reabsorbed. Eventually, the tail is completely reabsorbed. (Photo Researchers, Inc.)

4. Mitosis is divided into prophase, metaphase, anaphase, and telophase—specific events in the distribution of chromosomes to daughter cells.

5. Cytokinesis is the division of the cytoplasm. In plant cells, it involves the formation of a cell plate and a new cell wall from the center outward. In animal cells, it occurs by constriction of the cell membrane.

6. During meiosis, gametes (sex cells) are formed. The diploid chromosome number is reduced to the haploid number.

7. Meiosis consists of two cell divisions, which produce four haploid cells. The first is the reductional division (reducing the chromosome number). The second is the equational division (mitosis).

8. Cellular differentiation of young cells is marked by growth and structural changes as the cells assume their genetically programmed functions.

9. No organism is immortal. Cells have a programmed period of time they can live.

THE FIGHT AGAINST CANCER

During the next decade, someone you know well probably will die, and the chances are one in six that the person will die of cancer. In the United States, cancer is second only to heart disease as a killer. In 1981, approximately 700,000 individuals here were diagnosed as having cancer. Although susceptibility to cancer increases with age, the disease affects all age groups. It is one of the top four causes of death of children and young adults.

Lung cancer, the most common form of cancer in males, accounts for almost 20 percent of all cancers, and its incidence is increasing. Lung cancer is strongly linked with smoking. Other common types of cancer in males are leukemia and cancer of the colon, rectum, prostate, bladder, esophagus, stomach, and pancreas. Breast cancer is the most common form of cancer in females. Leukemia and cancer of the stomach, uterus, lung, ovary, and pancreas are the leading types of cancer in females. However, cancer can strike almost any organ of the body.

Cancer is not considered a single disease. Rather, it is a large number of diseases, all with a common cause—uncontrolled cellular reproduction. Cellular reproduction and differentiation are programmed and controlled by genetic material (DNA) housed in the nucleus. If the genetic controls are lost or disrupted, cells may reproduce rapidly and repeatedly, and they may change their appearance and degree of differentiation. This happens in the case of cancer cells, which seem incapable of controlling their cell cycles.

Cancerous, or malignant, cells compete with normal cells for nutrients and other necessary raw materials, causing the death of normal cells. Masses of malignant cells, called neoplasms or malignant tumors, form and increase in size, crowding and disrupting normal tissues.

If such tumors are encapsulated by connective tissue and kept localized, they are called benign tumors. Such tumors are relatively easy to remove surgically. The great danger is that a malignant tumor will undergo metastasis. In this case, the uncontrolled cancerous cells are carried in the blood and lymph throughout the body and establish secondary tumors. When a malignancy has metastasized, the cancer is difficult to treat.

The numerous forms of cancer can be grouped into four general categories: carcinoma, lymphoma, leukemia, and sarcoma.

Carcinoma is cancer of the epithelial tissues that cover or line body organs, including the human breast and glands such as the pancreas. Carcinoma, by far the most common form of human cancer, includes lung cancer.

Lymphoma is cancer of the lymph nodes and the spleen. It results in the production of large numbers of lymphocytes, a type of white blood cell. Only about 5 per cent of all human cancers are of this type.

Leukemia is cancer of the bone marrow and other blood-forming organs. Although we hear a lot about it, leukemia actually accounts for only about 4 per cent of all human cancer.

Sarcoma is cancer of the connective tissues. It is the least common form of human cancer.

Most forms of cancer are considered to be neither inherited nor contagious. How, then, are they induced? Several hypotheses are being intensively studied in laboratories throughout the world and are gaining increasing support.

One hypothesis is that the genetic regulators that control cell cycles may alter spontaneously, perhaps by some type of duplicating error made during the S phase of the cell cycle. Genetic material (genes) may be added or deleted, disrupting the control mechanism. If this occurs, one should expect some malignant cells to revert to normal ones. Although this happens only rarely, such reversals have been documented.

Many cancer victims have had heavy exposure to pollutants and other chemicals. Thus an alternative hypothesis is that some of the chemicals in our environment, called mutagens, induce changes (mutations) in the genetic regulators of cell reproduction. Chemicals in cigarette smoke, automobile emissions, certain cosmetics, hair dyes, and many other substances have induced mutations in microorganisms and experimental animals. Exposure to X-rays and other forms of radiation also is known to induce mutations. Environmental agents thought to cause cancer are called carcinogens.

Evidence supporting the hypothesis that enviromental factors induce cancer is growing stronger every day. For example, only a few decades ago, lung cancer was rare. But as air pollution increased and humans, especially males, began smoking cigarettes, this form of cancer has become a common killer. Cancer of the colon shows a correlation with diet. It is far more common among humans whose diets are rich in red meat than among humans who consume less meat and more fiber-rich foods, such as grains, fruits, and vegetables. The World Health Organization reports that 80 to 90 percent of all cancers may be related to enviromental factors.

Still another hypothesis relates cancer to viral infections. A few forms of cancer in animals have been linked concluively to viruses. There is no direct evidence of viral-induced cancer in humans as yet, although circumstantial evidence suggests a link in some forms of the disease. It is even proposed that viruses may enter human cells and remain dormant for long periods, after which they become activated and somehow induce their host cells to become malignant. Perhaps carcinogens act indirectly on cells by stimulating viral activity.

Another view is that cancerous cells are being produced all the time in humans, whose immune systems recognize and destroy them. However, when the immune system becomes defective or loses its efficiency, as it does during the aging process, cancer cells survive and reproduce. The chances of a person with an immune-deficient disease getting cancer are as high as one in ten. People whose immune systems have been suppressed because of organ transplants have an even higher risk of cancer. Those who have received kidney transplants have a hundred times the risk of developing cancer as others of the same age.

The various forms of cancer may be caused in different ways or in combinations of ways. All this makes the likelihood of discovering a single cure or preventive measure for cancer rather unlikely. If cancer is caused by induced or spontaneous alterations of the reproduction-regulating systems of cells, it will be especially difficult to cure.

Currently, people whose cancers cannot be removed surgically are treated with radiation or drugs (chemotherapy), or both. Radiation can be focused directly on the cancer, destroying its cells. Drugs either interrupt the cell division of rapidly dividing cells, such as cancer cells,

or for other reasons are more toxic to cancer cells than to normal cells. Many new drugs have been developed, but often they work more effectively in some people than in others. A new cell culture technique allows a patient's cancer cells to be cultured in a short time. The cultured cells can be used to test the effectiveness of a variety of anticancer drugs. The most effective drugs can then be used to treat the patient.

Another new approach that may be effective in certain cases is heat therapy. Scientists have discovered that cancer cells are more sensitive to heat than are normal cells. If the affected part of the body can be heated to a temperature that kills the cancer cells but does not damage normal cells, a cure may be realized.

Other researchers hope to find a way to sensitize a patient's immune system to cancer cells. The immune system would then attack the cancer cells in the same way it attacks infectious bacteria and other foreign agents.

The fight against cancer continues. In the next decade, we will see many significant breakthroughs in the treatment and cure of cancer.

FOR DISCUSSION AND REVIEW

1. Describe the events that take place during prophase, metaphase, anaphase, and telophase of mitosis. (This exercise will be even more helpful if you can draw the nuclear or chromosomal events.)

2. Describe the typical cell cycle. What is the significance of mitosis in this cycle?

3. Explain the differences between cytokinesis in a plant cell and in an animal cell.

4. Describe the nuclear events during meiosis (gamete production), including the following: meiosis I, meiosis II, homologous chromosomes, synapsis, random assortment. (The exercise will be more helpful if you can chart the nuclear events.)

5. Explain the difference between diploid and haploid chromosome compliments. Which cells are haploid? Diploid?

6. Outline the major similarities and differences between spermatogenesis and oogenesis.

7. Define *cellular differentiation* and *cell death*. What are the mechanisms in each case?

SUGGESTED READINGS

THE CELL, by Carl P. Swanson and Peter L. Webster. Prentice-Hall: 1977.
An extremely well-written account of the biology of cells.

LOOKING AT CHROMOSOMES, by John McLeish and Brian Snood. St. Martin's Press: 1958.
A beautifully illustrated description of mitosis and meiosis in the lily.

MOLECULES TO LIVING CELLS: READINGS FROM THE SCIENTIFIC AMERICAN. W. H. Freeman: 1980.
Twenty-six excellent articles, each written by leading authorities in the field of molecular cell biology.

REPRODUCTION IN MAMMALS I: GERM CELLS AND FERTILIZATION, by Colin R. Austin and R. V. Short. Cambridge University Press: 1972.
Contains a short but excellent description of the processes of meiosis and gamete formation.

SECTION THREE

ANIMAL AND PLANT STRUCTURE AND FUNCTION

The drawing above illustrates a very early myth—leaves which fell into the water became fish, and those that fell on the land became birds.

6

HUMAN ORGAN SYSTEMS: Integumentary, Skeletal, and Muscular

PREVIEW QUESTIONS

After reading this chapter, you should be able to answer the following questions:

1. Your skin, or integument, has several functions. Name as many as you can.

2. Identify as many of the major bones in your skeleton as you can.

3. Other than support, what are some of the functions of the human skeleton?

4. How is a muscle cell contracted and shortened?

5. How is a muscle cell lengthened following contraction?

Specialized cells that work together to perform a specific function form **tissues**. The human body contains four basic types of tissue: epithelial, connective, muscle, and nerve. The tissues combine to form organs and organ systems. An **organ system** is a complex of several organs that fulfill some functions necessary for the survival of the organism and its component cells and tissues. There are ten organ systems in the human body: integumentary, skeletal, muscular, digestive, respiratory, circulatory, excretory, nervous, endocrine, and reproductive.

TISSUES

Epithelial Tissue

The tissue that covers the surface of organs or that lines their cavities is **epithelial tissue**. It always has one free surface. Three general types of epithelial tissue are recognized: simple, stratified, and pseudostratified (see Figure 6–1). **Simple epithelium** consists of a single layer of epithelial cells. **Stratified epithelium** is thicker and consists of several cell layers. **Pseudostratified epithelium** consists of a single layer of cells, but the cells are of different sizes and heights. They are mixed with taller cells that overhang smaller cells, and this gives them the appearance of being more than one cell layer.

Simple epithelium is further divided into tissue types based on the shape of the individual cells. If the

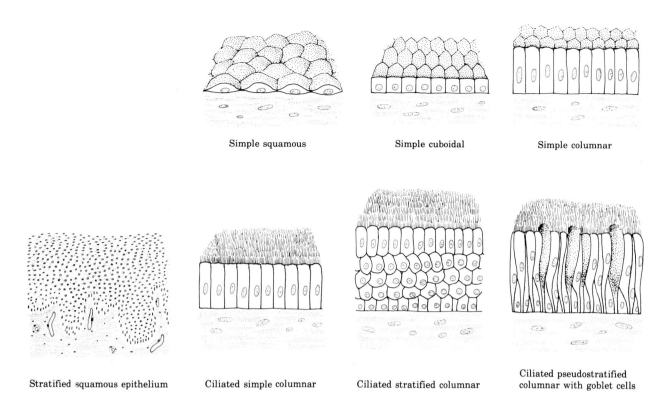

Simple squamous

Simple cuboidal

Simple columnar

Stratified squamous epithelium

Ciliated simple columnar

Ciliated stratified columnar

Ciliated pseudostratified
columnar with goblet cells

Figure 6–1
Several types of epithelial tissue. The primary differences among them are the shape of the epithelial cells and the number of cell layers.

cells are flat, they form **squamous epithelium**. If they are shaped like tiny cubes, they form **cuboidal epithelium**. If they are tall, they form **columnar epithelium**. If the cells of an epithelium bear cilia, that epithelium is also referred to as a ciliated epithelium.

As noted earlier, epithelial tissue forms the external covering of all organs and the lining of all cavities, passages, and ducts within the organs. In humans, for example, the ducts within kidneys and the smallest ducts of many glands are lined with simple cuboidal epithelium. The lining of human veins and arteries is composed of simple squamous epithelium. Nearly the entire length of a human intestine is lined with columnar epithelium. Ciliated epithelium lines human respiratory passages, and the outer layer of human skin is stratified epithelium.

Connective Tissue

The human body contains a large amount and variety of **connective tissues**, including blood▫, bone▫, car-

tilage, fibrous connective tissue, adipose tissue, and bone marrow. What makes it possible to put these diverse tissues into a single category? In each case, the living cells of the tissue are suspended in a large amount of noncellular material, called the matrix or ground substance. It is currently believed that the matrix is secreted by the living cells of the tissue as the tissue develops. The matrix is usually composed of some mix of carbohydrates and proteins. Its consistency ranges from liquid to gelatin to solid. For example, blood cells float in the noncellular fluid matrix portion of the blood, called **plasma**. Cartilage cells are embedded in **collagen**, a stiff, proteinaceous substance that is like gelatin. Bone cells are interspersed with calcium and magnesium deposits, which form the matrix and give bone its hardness and rigidity.

Two types of **fibrous connective tissue**, loose and dense, bind the cells together. In loose connective tissue, such as the dermal layer of skin, the connective tissue cells are scattered among a loose network of protein fibers, and both are suspended in a gel-like matrix. (See Figure 6–2) In dense connective tissue,

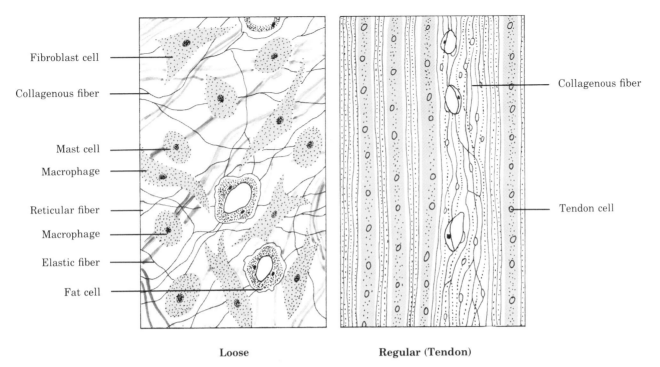

Fibroblast cell

Collagenous fiber

Mast cell

Macrophage

Reticular fiber

Macrophage

Elastic fiber

Fat cell

Collagenous fiber

Tendon cell

Loose

Regular (Tendon)

Figure 6–2

Loose and dense (regular) fibrous connective tissues.

such as the connective tissue surrounding muscle, the connective tissue fibers are more abundant. In ligaments (which hold bones together at joints) and tendons (which attach muscles to bone), the connective tissue fibers lie side by side in parallel fashion, giving extra strength to these important structures.

The final type of connective tissue to be described here is **adipose tissue**, or fat. Adipose tissue cells are called fat cells because they are specialized for the storage of fat, or lipids. Each fat cell usually contains a large droplet of lipid that pushes the cytoplasm and nucleus of the cell to the edges of the cell. When viewed through a microscope, a fat cell looks empty. Actually, the large empty space is the space occupied by the lipid droplet. Adipose tissue is always associated with the other kinds of connective tissue described here. It also occurs around joints and in a substantial layer below the skin. The skin layer reduces heat loss from the body and is therefore important in temperature homeostasis.

Muscle Tissue

Cells that are able to contract or shorten make up **muscle tissue.** The basic function of muscle is movement. There are three types of muscle tissue: smooth, cardiac, and skeletal. All are described later in this chapter.□

Nerve Tissue

The nervous system is composed of **nerve tissue**, which in turn is composed of neuroglia and neurons. The neuroglia cells are the supportive cells of the nervous system. They hold together the nerve cells, or neurons. The neurons are the major functional cells of the nervous system. They transmit coded messages, called nerve impulses, throughout the body. Their structure is described in Chapter 8.□

The Integumentary System

A surprise for many people is that skin is considered an organ. In fact, it is the largest organ in the body. It covers an average surface area of about three by seven feet, the size of a table top. The **integumentary system** is composed of the skin, all associated glands (such as sweat glands), and hair. Skin has two layers, an outer epidermis and an inner dermis (see Figure 6–3.

The **epidermis** consists of many cell layers. The

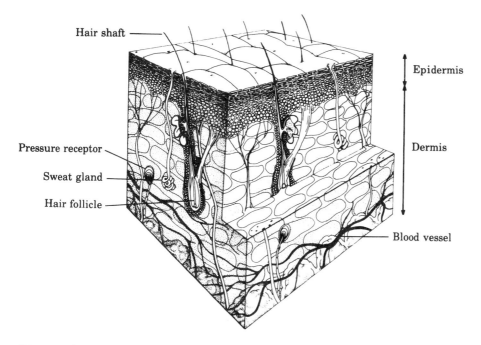

Hair shaft

Pressure receptor

Sweat gland

Hair follicle

Epidermis

Dermis

Blood vessel

Figure 6–3
A section of human skin.

innermost layer contains two types of cells, epidermal cells and pigment cells (melanocytes). The epidermal cells reproduce constantly and give rise to new outer layers of the epidermis. The **melanocytes** produce a dark pigment called **melanin**. The color of skin is related to the amount of melanin produced by the melanocytes. If you examine the cell layers of the epidermis, you can see that the shape and appearance of the epidermal cells change markedly as they are moved toward the surface of the skin. As new cells form, the old ones are pushed farther from the blood supply and become more flattened. They also accumulate an inert, insoluble protein, **keratin**, which waterproofs and protects underlying cells. The outer cell layers eventually die and wear away from the surface of the skin.

The underlying **dermis** is composed of loose fibrous connective tissue and fat cells, among which course small blood vessels and nerves. The dermis also houses the bulk of the tubular sweat glands, the hair follicles, and the sebaceous glands. The latter produce a fatty substance to lubricate the hair and skin. Sweat glands, hair follicles, and sebaceous glands are invaginations of the epidermis that extend into the dermis. The dermis also contains a number of sensory receptors that record the sensations of touch, deep pressure, heat, cold, and pain.

The thickness of skin differs from one part of the body to another. Skin is very thick (up to five millimeters) on the palms of the hands and the soles of the feet and even thicker where callouses develop. In most places, however, it is only one to two millimeters thick.

Skin has many functions. It separates the internal environment of the human body from the external environment. It protects the body against invasion by microorganisms such as bacteria. Its pigmentation screens out some ultraviolet radiation but admits enough to permit the synthesis of vitamin D. Skin also plays an important role in the maintenance of normal body temperature. Although the outermost layers of skin cells are dead, skin is thin enough for the nerves in the dermis to be stimulated by various forms of tactile stimuli, heat, and cold. This is a protective measure as well as a source of pleasure.

THE SKELETAL SYSTEM

The sizes and shapes of humans are to a large extent correlated with the development of their **skeletons**. The major function of the skeletal system is to support the body, but it also has other important functions. Parts of the skeleton, such as the long bones of

Many people aspire to a deep, rich suntan. When spring comes, they expose as much skin as possible to the sunlight. Unfortunately, sunburns are as common as suntans.

The component of sunlight that causes suntans and sunburns is ultraviolet radiation. Most of this radiation is absorbed by the melanin pigment in the melanocytes of the epidermis. Thus the underlying tissues are protected. Careful, limited, and repeated exposure to sunlight induces the melanocytes to gradually produce more melanin, and the skin gradually darkens.

All humans have about the same number of melanocytes in their skin. However, the melanocytes of people with dark skin produce more melanin than those of people with light skin. The difference is genetic.

If skin is exposed to the sun for long periods before sufficient melanin has been built up, epidermal cells in the inner layers are damaged. As a result, more blood flows to the dermis, making the skin look pink. If damage is severe, the blood vessels begin to leak a fluid that fills blisters in the skin. Following a severe sunburn, the epidermis eventually peels away. New epidermal cells replace this layer.

Prolonged exposure to sunlight over many years causes thickening of the skin. The skin of even young people may appear aged. Excessive exposure to sunlight may induce some forms of skin cancer. Obviously, suntanning should be approached carefully.

the arms and legs, are levers moved by the muscles attached to them. Systems of muscles and bones make movement possible. If either is abnormal or deformed, movement is limited. Parts of the skeleton also serve to protect delicate organs. The skull and vertebral column protect the brain and the spinal cord. The ribs protect the heart and lungs.

Bones also store important minerals, such as calcium and phosphorus. Calcium is necessary for normal blood clotting, muscle contraction, and the functioning of cell membranes. Phosphorus is an important component of DNA□ and ATP□. If these minerals are in short supply in the diet, the body uses the minerals stored in bones.

The skeletal system has two main divisions, the axial skeleton and the appendicular skeleton. The **axial skeleton** is the part of the skeletal system that is con-

fined to the body's central longitudinal axis, an imaginary line from the top of the head to the space between the feet. It therefore includes the skull, the bones around the larynx (voice box), the bones of the inner ear□, and the vertebral column, sternum, and

Figure 6–4
The axial and appendicular skeletons.

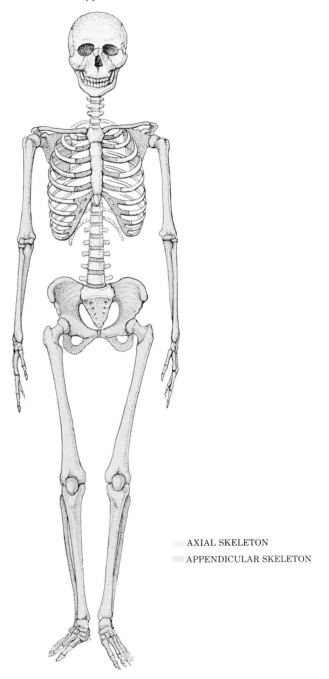

AXIAL SKELETON
APPENDICULAR SKELETON

associated ribs that surround the thoracic (chest) region (see Figure 6–4). The **appendicular skeleton** contains the bones that make up the arms and legs (the appendages), shoulders, hands, pelvic region and hips, and feet (see Figure 6–4).

Bones are often classified according to their shape. There are, for example, long bones, short bones, flat bones, and irregularly shaped bones. Even though bones may seem different, they have the same microscopic structure.

Bone

The hardness of **bones** enables them to be the major support structures of the body. Bones are hard because they accumulate compounds of calcium and phosphorus. They may seem inert, but they are made up of living, growing tissues. These tissues are contained in complex systems which include canals that run through the mineralized parts of bones. These systems look something like the concentric rings on a dart board and are called Haversian systems. Figure 6–5 shows the details of one Haversian system. In the center is a Haversian canal which is a passageway for nerves and blood vessels that supply bone cells with nutrients and oxygen. Haversian canals run longitudinally in the bone.

In concentric circles radiating away from each Haversian canal are layers of hard, calcified non-cellular substance, the matrix, which was secreted originally by living bone cells called osteoblasts. The matrix is

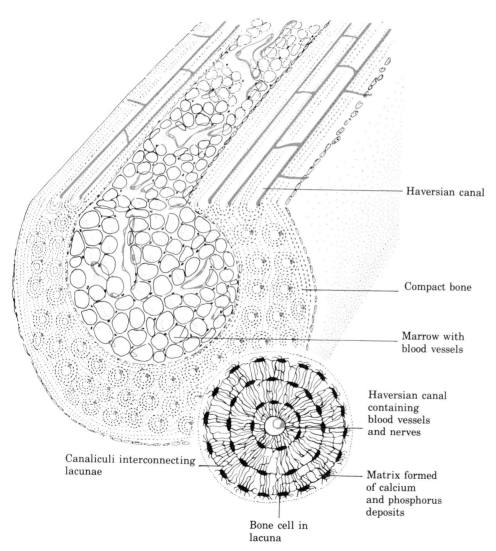

Figure 6–5
The structure of bone. Bone cells (osteocytes) located in spaces in the matrix called lacunae occur in rings around the canals, called Haversian canals, that course through the bone. Blood vessels and nerves pass through these canals.

Haversian canal

Compact bone

Marrow with blood vessels

Haversian canal containing blood vessels and nerves

Matrix formed of calcium and phosphorus deposits

Bone cell in lacuna

Canaliculi interconnecting lacunae

what we call bone. Between the calcified layers are circular layers that have small chambers or spaces; these chambers are called lacunae. The living bone cells (now called osteocytes) are housed in lacunae. (In mature humans, the majority of bone cells do not produce new bone tissue.) Still smaller radiating spaces, called canaliculi, connect the lacunae of in the same and in different layers, and ultimately connect the lacunae to the Haversian canal. Thus each Haversian system is composed of a central canal and circular layers of mineralized bone alternating with layers containing lacunae containing the living bone cells; all lacunae are interconnected by canaliculi. A single bone is composed of hundreds of thousands of Haversian systems.

Many bones contain large internal cavities filled with **bone marrow**, a type of connective tissue. The primary function of bone marrow is the production of certain types of blood cells.

Bone Development

The skeleton of a human embryo is preformed as cartilage. At about six weeks of age, the embryo's bone cells begin to deposit bone. They continue to do so to some extent throughout adulthood. This is the process that heals the broken bones of bodies of all ages (although healing takes longer with age).

The process of bone formation, or ossification (see Figure 6–6), begins near the center of a long bone, such as a leg bone. At this first site, called the center of bone formation, the living cells of the cartilage change and produce enzymes that cause the flexible and plastic-like matrix of the cartilage to calcify, or turn to bone. Bone is slowly formed outward from the original center toward the ends of the bone. While this is occurring, the cartilage at the ends of the bone continues to grow, causing the bone to elongate as the body ages. At adulthood, the bone has reached full size because of cartilage growth and is hard because of bone formation. Some bones retain a small layer of cartilage at each end. This layer creates the smooth surfaces that allow adjacent bones to move easily at joints (see Figure 6–7).

Flat bones are produced in a similar manner, but there is generally no precise direction of growth. Fingers of bone are produced everywhere, as cartilage

Figure 6–6
The development of a long bone.

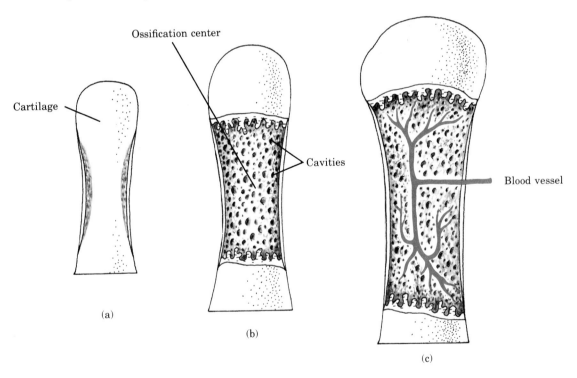

growth causes the bone to enlarge in size. The result is bone composed of a honeycomb of bone fragments and that is therefore less rigid and more fragile than other bones. This type of bone is called spongy bone.

Joints

Adjacent bones are connected by **joints**. There are three kinds of joints: fibrous, cartilaginous, and synovial.

Fibrous joints allow little or no movement between neighboring bones. The bones of the skull are held together by sutures, which are fibrous joints. Teeth are held to the jawbone by other fibrous joints.

In cartilaginous joints, two bones are held together by cartilage. The bones involved are capable of only slight movement. The joints of the vertebral column are a good example of this type of joint. We have all heard of slipped or ruptured discs. The disc is a cartilaginous pad between two vertebrae.

Most of the active body movements are possible because of synovial joints, such as those of the elbow and knee. Structurally complex, these joints are ac-

tually small cavities called synovial cavities. Because the bones at synovial joints do not actually touch, they have extreme freedom of movement. The ends of bones at these joints are often covered with smooth connective tissue, such as cartilage, which reduces friction during movement. The synovial space between the ends of the bones is filled with a small amount of synovial fluid (about one and a half teaspoons in the knee joint).

Many synovial joints also contain special fluid-filled saclike structures called **bursae**. These structures further reduce friction and contribute to the smooth, painless operation of joints. (The synovial joint of the knee, shown in Figure 6–7, has thirteen bursae.) These joints are strengthened by tough connective tissue bundles, called **ligaments**. Muscles that move bones are connected to specific sites on bone surfaces by **tendons**. Although joints work well when bodies are young, they are common sites of disease among the middle-aged and elderly. Some examples of joint disease are rheumatism, arthritis (of which there are several types), gout, bursitis, tendonitis (tennis elbow is an example), and sprains.

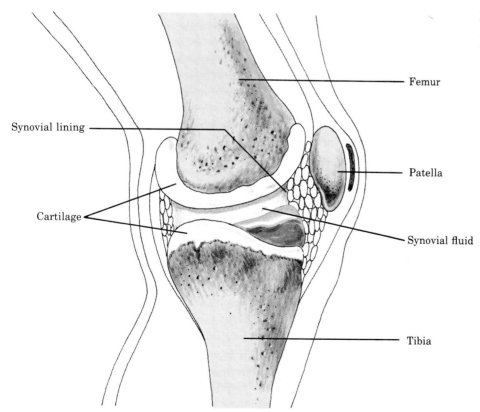

Figure 6–7
Articulation and cartilage of the knee.

Femur

Synovial lining

Patella

Cartilage

Synovial fluid

Tibia

Joint disorders are among the most common of all human ailments. Every student who reads this essay will suffer from some minor or major ailments of joints. There is a good biological reason for joints to be our Achilles' heels (so to speak). Joints receive much abuse. They are constantly being moved, rotated, and twisted—all of which places them under continual stress. Many of them support tremendous amounts of weight, which increases their stress. Physical abuse, usually unintentional, pulls joints apart; pulls and injures tendons, ligaments, and muscles; damages cartilage; and destroys bursae. Many kinds of illness and disease cause mineral reabsorption from bone, which weakens the joints, adds new mineral deposits to them, or reduces the amount of their synovial fluid. All this makes joints painful and sometimes immovable.

There are many disorders of the joints—far too many to mention them all here. These are the ones heard about most often: tennis elbow, trick knee, bursitis, and arthritis.

Tennis Elbow

Not too long ago, a new epidemic swept the United States: tennis elbow. The technical name for this disorder is tendinitis—inflammation of a tendon. Tendons attach muscles to bone. In the case of tennis elbow, the tendons that become inflamed are those that attach the muscles of the forearm to the bone of the upper arm at a point near the elbow joint.

Tennis elbow is more common among amateur tennis players than among professionals. It occurs when the muscles of the forearms are not well-developed and not well-prepared for a weekly afternoon of abuse. The abuse includes the unusual force the forearm takes as the tennis ball strikes the racket and the unusually rapid return movements that involve the wrist as well as the muscle of the lower arm. Tennis elbow does not develop because the elbow is bent too much. It comes about because the rapid and often unnatural movements of the forearm place considerable stress on the tendons near the elbow.

Those who develop tennis elbow are advised to discontinue their game for a few weeks. Aspirin or anti-inflammatory drugs are often prescribed. In serious cases, cortisone injections are given.

"Tennis elbow" can develop anywhere in the body that tendons attach muscles to bone, especially at the origins of muscles. Therefore, the inflammation can occur in wrists, shoulders, ankles, and even fingers (where it is called "trigger finger").

Trick Knee

In the case of trick knee, either the cartilage or the ligaments of the knee joint are damaged (see Figure 6–7). The ligaments in the knee area connect the two bones of the lower leg with the single bone of the upper leg. The knee joint is a perfect example of a hinged joint. It allows for only a single movement, that of flexing the lower leg backward and then extending it forward to its original position. Normally, no side, rotational, or forward movements are possible. However, if the ligaments are damaged (often referred to as being torn or ruptured), the hinge weakens and becomes wobbly. At this point, side-to-side or rotational movement becomes possible.

The trick knee is a knee with a new set of painful movements. These movements cause the joint to lock or to pop out temporarily. If the cartilage is damaged, clicks can often be heard when the person is walking. The trick knee usually acts up when the gait changes from walking to running, when stops are quick, when stairs are navigated, or when fast rotational movements of the body are made.

Trick knee can be painful and serious enough to interfere with normal activities. Experts in sports medicine have developed therapeutic routines that strengthen leg muscles so they will give added support to the knee joint. However, if the injury is serious, surgery is commonly recommended.

Bursitis

The bursae of joints often become inflamed, a condition known as bursitis. The inflammation can be caused by a physical injury or by constant pressure to the same joint over a long period of time. "Housemaid's knee" is a form of bursitis that is supposed to have occurred when the housemaid spent many laborious hours on her knees scrubbing floors. Although not a problem for the modern housekeeper, it can affect carpet layers, roofers, and so on.

Arthritis

We have probably heard more about arthritis than about any other joint disease. That is because arthritis is not just one disease but a general term that can be applied to as many as twenty-five malfunctions of the joints. The causes of arthritis are unknown at the present time. Many suggestions have been made, however. They range from local bacterial infection, to injury, to allergy, to hormone imbalance. One of the newer theories is that arthritis is an auto-immune disease. That is, the damage to the structure of joints is said to be caused by cells and cellular secretions that normally prevent infection and that instead attack the body.

Rheumatoid arthritis is the most common arthritic disease. Although often viewed as a disease of the elderly, it is, in fact, more common among adults who are thirty to fifty years old. It is also fairly common among younger adults, teenagers, and children. Rheumatoid arthritis is an inflammation of the synovial membrane in synovial joints. When this membrane, which is the source of synovial fluid, becomes inflamed, it produces too much fluid. The joint swells and becomes extremely painful. In response to the inflammation and swelling, a hard tissue forms over the cartilage articulations. This tissue makes the joint stiffen. Movement then becomes more painful. In time, the new tissue can grind away the entire cartilage. When this happens, the two bones fuse and the joint becomes totally immovable.

Osteoarthritis (*osteo* means "bone") is an affliction of the elderly. It affects only the cartilage at synovial joints. From years of use, the cartilage erodes and new bone is deposited in lumps. The lumps make movement difficult and, finally, impossible. Many people who suffer this form of arthritis have no pain, but their fingers may curl and permanently arch and their wrists and other joints may display bumps of bone formation.

Gout is another form of arthritis. It was once considered a disease of aristocrats, because in the Middle Ages only aristocrats developed the disorder. They achieved this distinction because they were the only people who had the opportunity to eat excessive amounts of protein. What this suggests is that gout is related to diet. In fact, gout is the accumulation of uric acid crystals in synovial joints. The accumulation makes movement both difficult and painful. Uric acid, a by-product of protein metabolism, is normally excreted in the urine.

Gout develops when no uric acid is excreted, when more than normal amounts are produced, or when both events occur. The uric acid results from the metabolism of nucleic acids that are consumed with animal flesh. Gout sufferers should avoid meat, particularly liver, kidney, brain, thymus gland, and other glands. There are still some unexplained facts about gout. One such fact is that women rarely develop the disease. Another is that the disease runs in families, which suggests some inherited disorder in either the metabolism of nucleic acids or the elimination of uric acid.

There are no cures yet for arthritis, although a number of new drugs are effective in relieving much of the pain associated with it. Most of the drugs do nothing for the diseased joints, however. As a consequence, one can move more easily because the pain is less, but the movements continue to destroy the joints. Bone deposits in arthritic joints are often removed surgically. At the present time, it is possible to replace nearly every synovial joint in the body with an artificial joint. The most frequent replacements are those of the knee and hip.

THE MUSCULAR SYSTEM

The human body contains three kinds of muscle tissue: smooth, cardiac, and skeletal. **Smooth muscle cells** form the muscle layers of internal organs such as the digestive tract. Each cell contains a single nucleus and is shaped like an elongated spindle, tapering at either end (see Figure 6–8). Smooth muscles are involuntary muscles, which means that we have no conscious control over their movements. They contract slowly.

The heart, a muscular pumping organ, is composed of **cardiac muscle**. Although the muscle of the heart is fast-acting (contracts quickly), its movement is involuntary. The muscle cells (fibers) of cardiac muscle frequently branch and produce complex systems of fibers containing several nuclei (see Figure 6–9). Like skeletal muscle, cardiac muscle is striated. That is, when viewed under an ordinary light microscope, the muscle fibers appear to have stripes or bands at regular intervals.

Skeletal muscle forms the muscle masses that are attached to the bones (see Figure 6–10). About 40 percent of the body weight of humans is skeletal muscle. If a skeletal muscle is dissected and viewed under a microscope, it can be seen that the muscle is composed of many separate muscle cells (commonly referred to as muscle fibers). (See Figure 6–11). Muscle fibers are generally very thin. They can also be quite long. Some are as long as thirty centimeters (twelve inches).

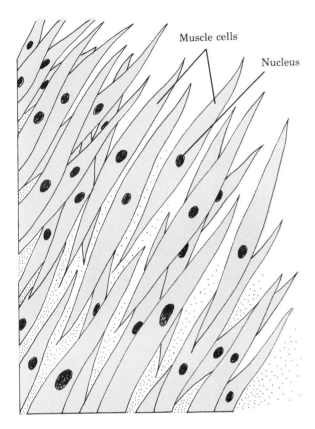

Figure 6–8
Smooth muscle.

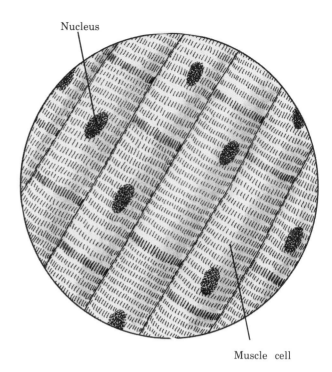

Figure 6–9
Cardiac muscle.

The cell membrane of a muscle fiber is called the sarcolemma. *Sarco* means "flesh." The term is used to denote muscle structure because the flesh we eat is almost entirely skeletal muscle. The cell membrane of muscle fibers is slightly different from the cell membranes described earlier□. At regular intervals, it forms tubes, called T-tubes, that penetrate into the cell. These tubes play a role in the muscle contraction□.

The cytoplasm of the fiber, often called sarcoplasm, may contain many nuclei and always contain a large number of mitochondria□. The magnifying powers of the electron microscope show that the sarcoplasm contains a complex series of small, threadlike structures called myofibrils (*myo* means "muscle"). Each myofibril is composed of yet smaller myofilaments. There are two kinds of myofilaments, thick and thin. The thick ones are composed of the protein **myosin**, and the thin ones are composed of the protein **actin**.

The myofilaments of a muscle fiber do not extend the entire length of the fiber. They are much shorter than muscle fibers and are arranged in a precise order in units called sarcomeres. A discrete line where one sarcomere ends and another begins can be seen through a microscope. This characteristic line and its regular repetition from one end of a muscle fiber to another produce striations. Because of the striations, skeletal muscle is often called **striated muscle**. The striations of cardiac muscle come about in the same way. Smooth muscle lacks striations because the fibers have fewer and more randomly ordered.

Muscle Contraction

A muscle contracts because of a stimulus from a neuron. The neurons that stimulate muscles are referred to as motor neurons. A single motor neuron branches many times and may connect to as many as 150 muscle fibers. Each branch that reaches a muscle fiber ends in a motor end plate. The motor neuron and all the muscle fibers it stimulates are referred to as a **motor unit**. All the muscle fibers of a motor unit con-

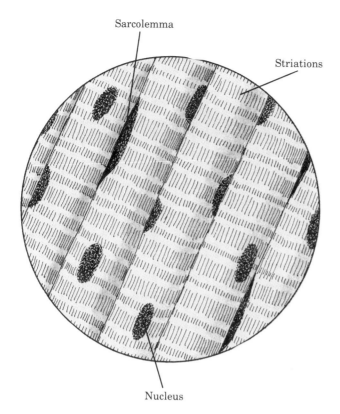

Sarcolemma

Striations

Nucleus

Figure 6–10
Skeletal muscle.

CONCEPT SUMMARY

COMPARISON OF SMOOTH, CARDIAC, AND SKELETAL MUSCLE

SMOOTH MUSCLE	CARDIAC MUSCLE	SKELETAL MUSCLE
Involuntary	Involuntary	Voluntary
Slow-contracting	Rapid-contracting	Very rapid-contracting
Found in internal organs such as the digestive tract	Found in the heart	Attached to bones
Cells spindle-shaped	Fibers cylindrical and branching	Fibers long and cylindrical
Containing a single nucleus	Containing several nuclei	Containing several nuclei
Without striations	Striated	Striated

tract and relax in perfect unison. Either there is or there is not sufficient stimulation for the contraction of a motor unit. This is the **all or none principle** of muscle contraction.

However, we know that muscles may shorten just a little or may shorten completely when contracted. We can demonstrate this by looking at the palm of the hand and curling the fingers inward. If we compare this curling to making a tight fist, we can see that the muscles used for both actions are the same. The movements and final response are different because of the amount of muscle shortening. In the first instance, only a few motor units were ordered to shorten. In the second instance, probably all the motor units of the muscles involved were stimulated. This property of muscle contraction makes it possible for the same muscles to produce what appear to be different actions.

When the motor neuron fires, a chemical transmitter substance is released from the motor end plate. This substance acts as the stimulus for the muscle contraction. It causes changes in the sarcolemma that are transmitted to the T-tubes and to the endoplasmic reticulum□ (in the case of muscles, usually called sarcoplasmic reticulum). The sarcoplasmic reticulum in turn releases calcium ions (Ca^{++}). The calcium is important for the changes it causes in the actin and myosin myofilaments.

The explanation for how a muscle contracts is called the **sliding filament model**. The thick myofilaments of mysoin have little extensions, or knobs, around the entire filament. These knobs are called cross bridges. When the motor unit is stimulated and calcium is released, the calcium moves to the cross bridges. There it promotes the breakdown of ATP□ and creates active sites for attachment to the thin actin myofilaments.

The first phase of muscle contraction, then, is the binding of actin to myosin. This binding occurs many times in rapid succession. The cross bridges attach and detach again and again in a ratchet-like fashion. This causes the myofilaments of actin to slide toward the myofilaments of myosin (see Figure 6–11). The movement of the actin myofilaments causes the sarcomere to shorten. All the sacromeres of the muscle fiber shorten in unison.

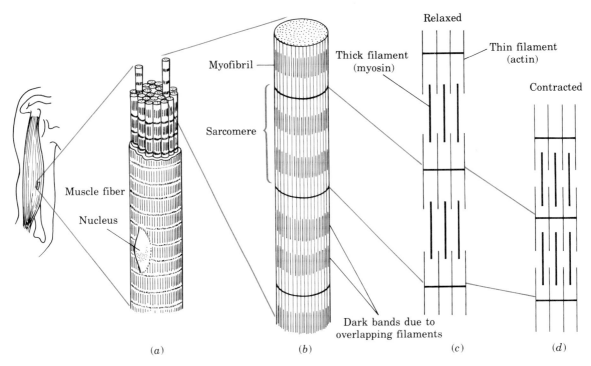

Relaxed

Thick filament
(myosin)

Thin filament
(actin)

Contracted

Myofibril

Sarcomere

Muscle fiber

Nucleus

Dark bands due to
overlapping filaments

(a)　　　　　　　　　(b)　　　　　　　　　(c)　　　　　　　　　(d)

Figure 6–11

The sliding-filament theory of muscle contraction. (a) Individual skeletal muscle fibers contain numerous myofibrils. (b) A myofibril is composed of contractile units, called sarcomeres, that contain overlapping filaments of the proteins actin and myosin. The arrangement of these filaments produces the banded appearance of each sarcomere, which in turn causes the striated appearance of skeletal muscle fibers. During muscle contraction, each sarcomere shortens when actin and myosin filaments actively slide along each other, modifying the pattern of banding. Notice the difference between the position of the actin and myosin filaments in (c) a relaxed sarcomere and (d) a contracted sarcomere.

An analogy to this movement is the following. Stretch your arm (which is thick, as is myosin) stiffly in front of you. Then have someone hold a shirt with one sleeve extended, ready for your arm (the shirt sleeve represents the thin actin myofilaments). The length of your arm plus the length of the shirt sleeve represents the length of the sarcomere when it is not contracted. The hairs on your arm (cross bridges) are activated and pull the sleeve over your arm. Your arm in the sleeve represents the contracted length of a sarcomere. This analogy works only if you imagine that the sleeve is pulled onto your arm by the hairs. It wouldn't work if you pushed your arm into the sleeve.

After the contraction of a muscle fiber, the calcium ions are quickly returned to storage and all reactions are reversed. Suddenly, all the cross bridge connections between actin and myosin are broken, and the actin myofilaments slide back to their original position.

Types of Movement

The blend of bones of many shapes, joined together in various ways, and pulled in many planes by muscles of differing lengths and bulks creates an orchestration of several types of movement. Consider a few examples. Open your hand with the palm facing you, and bend your fingers toward you. This is the movement called flexion. (Can you also flex your forearm? Your lower leg?) When you bring your fingers back to their original position, the movement is called extension. If you continue to extend your fingers until they begin to point downward, they become hyperextended. When you spread them like a fan, they are abducted. When you pull them together again, they are adducted. People are also capable of partial rotation of both the upper and lower arms and the upper leg, but not the lower leg. (The arrangement of the bones in Figure 6–4 suggests the reason.) You can protract your jaw by jutting it outward and retract it by the reverse movement.

These and all other complicated movements are possible because of the synovial joints, including hinges, pivots, and ball and socket joints. But movement also relies on **antagonistic muscles**. When one muscle group contracts, another relaxes. This "antagonism" not only facilitates movement in general but also makes quick movements possible.

The most commonly used example of antagonistic muscles is that of the muscles used in flexing and extending the forearm (see Figure 6–12): the biceps brachii and the triceps brachii. These muscles are attached to fixed (immovable) points in the upper arm and shoulder, their points of origin. Each muscle also has a point of insertion on the bones in the forearm When the brain issues the order to flex or bend th

Figure 6–12

The biceps and triceps function as antagonistic muscles. When the biceps contracts, the triceps relaxes and the arm flexes at the elbow. When the triceps contracts, the biceps relaxes and the arm is extended or straightened.

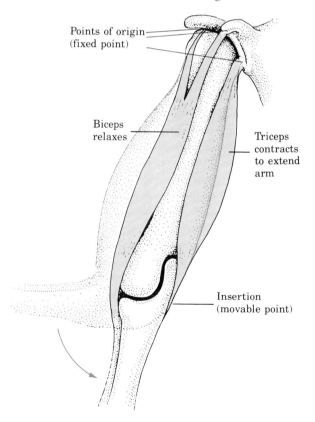

Points of origin (fixed point)

Biceps relaxes

Triceps contracts to extend arm

Insertion (movable point)

A basic characteristic of animals is their ability to move from place to place or, if they are sessile (fixed in one place), at least to move parts of their bodies. Rapid movements depend on well-developed skeletons and muscle systems. But most animals do not possess both bony internal skeletons and efficient sets of individual antagonistic muscles. For example, sponges are supported by internal skeletons composed of spicules, but they have no muscles and therefore cannot move. On the other hand, many tiny, almost microscopic, animals with no skeleton and little in the way of muscles move by means of rapidly beating cilia, which are borne on their epidermal cells. Of course, the use of such tiny appendages for locomotion is limited to only the very smallest of creatures.

Most of the simpler animals, such as worms, do not have separate, clearly defined muscles. They do, however, have more or less continuous bands of circular and longitudinal sheets of muscles that extend the length of their bodies. At first glance, it also appears that the simpler animals lack skeletons, but in fact they possess what are called hydrostatic skeletons. These skeletons are a kind of fluid-filled cavity.

How can sheets of muscles and fluid-filled hydrostatic skeletons work to produce movement? Well, first let's remember that a given volume of water cannot be compressed into a smaller volume under ordinary pressures. For example, if you squeeze a balloon filled with water, the volume of water in the balloon will stay the same, but the shape of the balloon will change. The change will depend on how and where you apply the pressure. The movements of soft-bodied animals with hydrostatic skeletons are similar. The general muscle tone of the two muscle layers maintains pressure on the fluid in the body cavity, and this pressure gives the animals a degree of firmness. If the longitudinal muscle cells on one side of the animal contract and those on the other side remain at rest, it is easy to see that the animal will bend toward the side of the contracting muscle cells. If the longitudinal muscle cells on the opposite side of the bend contract and the cells that caused the bending in the first place relax, the animal will straighten again.

For the body to lengthen, the longitudinal muscles

(Continued Next Page)

BOX 6–1 (CONTINUED)

MUSCLES AND SKELETONS OF ANIMALS OTHER THAN HUMANS

relax and the cells in the circular muscle layer contract. The force decreases the diameter of the animal. But because water is incompressible, the pressure on the fluid in the hydrostatic skeleton causes the animal to elongate as the longitudinal muscles relax. (Think of the water-filled balloon.) For the body to shorten, the longitudinal muscles contract, and the circular muscle layer relaxes, becoming shorter and fatter. Thus, the two muscle layers, circular and longitudinal, act as antagonistic sets working against the hydrostatic skeleton to bring about coordinated movements.

Most of the less simple animals do have hard skeletons, and most of these skeletons are external rather than internal. For example, the arthropods (insects, spiders, crustaceans, and their relatives, which together constitute 70 percent of the animal kingdom) all have exoskeletons that are continuous over their entire body surfaces. The exoskeletons are for the most part hard and rigid. However, to allow for various degrees of movement, they are constructed so that joints and lines of flexure occur at certain places on the body.

Like vertebrates, arthropods have individual muscles. However, their muscle systems differ from those of vertebrates. Arthropods' muscles are attached to the inside of the skeleton rather than to the outside. The flexure of an arthropod leg joint is brought about by contraction of individual muscles attached on inward projections of the exoskeleton. So is extension by the arthropod's antagonistic set of muscles. Also, the tube-like skeleton of **arthropod** appendages cannot support great weight. Therefore, arthropods are limited in size. The stronger, more solid internal skeletons of vertebrates have allowed vertebrates to evolve into the largest animals on earth.

arm, nerve impulses stimulate the striated muscle fibers in the biceps to contract. Because the biceps is attached at its origin and point of insertion, contraction causes the bones in the forearm to move toward the shoulder. As this happens, the muscle fibers in the triceps are relaxed. Thus the triceps extends while the biceps contracts. When the brain issues the order to extend the flexed arm, the biceps relaxes and the triceps contracts. This causes the upper and lower portions of the flexed arm to move away from each other, thereby straightening the arm.

The biceps and triceps work as a pair, each performing an opposite, or antagonistic, movement. Almost all human body movements are caused by coordinated contractions of antagonistic pairs or sets of skeletal muscles. The skeletons and muscle systems of some other organisms are described in Box 6–1.

SUMMARY

1. Organisms are composed of cells. These cells form four basic types of tissue: epithelial, connective, muscle, and nerve.

2. The three basic types of epithelial tissue are simple, stratified, and pseudostratified.

3. The major types of connective tissue are blood, bone, cartilage, fibrous connective tissue, adipose tissue, and bone marrow.

4. The three types of muscle are skeletal, cardiac, and smooth.

5. Nerve tissue is composed primarily of supporting cells, called neuroglia, and nerve cells, called neurons.

6. The integumentary system consists of the skin, sweat glands, sebaceous glands, and hair. The skin contains a number of receptors that record sensations of touch, pressure, heat, cold, and pain.

7. Functions of the integumentary system include separation of the body's internal environment from the external environment; protection against abrasions, invasion of the body by bacteria and other microorganisms, and ultraviolet radiation; regulation of body temperature; and synthesis of vitamin D.

8. The two main divisions of the skeletal system are the axial skeleton and the appendicular skeleton. The axial skeleton includes the skull, vertebral column, sternum, and associated ribs. The appendicular skeleton includes the bones of the shoulders, arms, hands, pelvis, hips, legs, and feet.

9. Functions of the skeleton include support of the body; formation of levers so muscles can cause body movements; protection of internal organs such as the brain, spinal cord, heart, and lungs; storage of minerals such as calcium and phosphorus; and production of blood cells in the bone marrow.

10. Bone is composed of Haversian systems, which are composed of a central canal (the Haversian canal) and circular layers of mineralized bone alternating with layers with lacunae that contain the living bone cells. These layers are connected by smaller spaces called canaliculi.

11. Bone marrow produces certain types of blood cells.

12. Bone is preformed as cartilage. Bone formation (mineralization) begins at the center of each long bone and extends outward in both directions toward the ends of the bone. Bones continue to grow because of increases in the growth of cartilage.

13. Bones are connected by joints. There are three kinds of joints: fibrous, cartilaginous, and synovial.

14. Muscles are composed of muscle fibers. The fibers are composed of small, threadlike myofibrils, which are composed of yet smaller myofilaments.

15. Myofilaments do not extend the full length of a muscle fiber. Instead they are housed in discrete units called sarcomeres.

16. There are two kinds of myofilaments: actin (which is thin) and myosin (which is thick).

17. When muscles contract, the actin myofilaments slide toward the myosin myofilaments and cause the sarcomeres to shorten. This is referred to as the sliding filament model.

18. The synapse between a motor neuron and a muscle fiber is called the motor end plate. The motor neuron may branch many times and may stimulate many muscle fibers. The motor neuron and all the muscle fibers it stimulates are referred to as the motor unit.

19. Because of its antagonistic muscle groups, the human body is capable of many complex movements.

FIGHTING PARALYSIS AND REPLACING LOST LIMBS

Because of motor vehicle and sports accidents, between 10,000 and 15,000 people in the United States become paraplegics each year. A paraplegic is a person whose lower body and extremities are paralyzed. Thousands more lose one or more limbs. Paralysis results from spinal injuries. It is especially tragic that the average age of spinal injury patients is only nineteen.

Until recently, not much could be done to restore movement to paralyzed limbs, and artificial limbs were little more than stiff crutches. However, biomedical engineers are changing the situation. In the next decade, restoration of movement to paralyzed limbs may become commonplace. New, computerized artificial limbs (and maybe even muscles) will be developed. They will give amputees some freedom of movement.

Using the technology developed in the areas of aerospace engineering, microelectronics, polymer chemistry, and biofeedback, engineers are developing the prototypes of many amazing devices. For example, a complex artificial leg has been tested on a young man who lost his leg in a motorcycle accident. The leg is operated by subconscious and conscious brain signals sent to the muscles of the thigh. When the man thinks of taking a step, the electrical activity generated in his thigh muscles is recorded and fed into a computer. The computer relays the appropriate signals to the pneumatic pistons that operate the artificial limb. The device, called the Drexel-Moss computer knee, allows the user not only to walk but also to perform conscious maneuvers such as avoiding obstacles and recovering from stumbling. The next step in development is to miniaturize the computer system so it can be strapped to the patient's belt. This will allow the patient to leave the hospital and gain complete mobility.

Similar systems have been developed to relieve paralysis. Recently, a paraplegic regained the use of his hands when he was attached to a device that translates the electrical signals in the shoulder and upper arm muscles into electrical signals that stimulate the nine key muscles in the hands. The patient eventually developed enough dexterity to feed himself, use the telephone, and comb his own hair. Using similar devices, patients whose legs were completely paralyzed are now riding stationary bicycles and lifting weights with their legs. As further development of such systems continues during the next decade, the future will look even brighter for paralysis victims.

More lifelike artificial limbs may be developed if it becomes possible to build artificial muscles that allow more natural movement and more freedom of movement. Physicists at the Massachusetts Institute of Technology are experimenting with soft, gel-like substances, called ionic gels, that decrease in volume by as much as five hundred times when stimulated by small electric currents. When the current is turned off, the gels return to their former volume and shape. Although a cylinder of gel an inch long requires about ten minutes to regain its original size and shape, scientists believe gels the size of muscle fibers would return to original size as quickly as muscles contract. Therefore, although the immediate use of these gels is for switches in various types of sensors, the MIT scientists envision that gel fibers may someday be engineered into artificial muscles that will be induced to contract by electrical signals originating in the patient's brain.

FOR DISCUSSION AND REVIEW

1. What is the difference between a tissue and an organ?

2. Where in the human body are the basic kinds of epithelial tissues?

3. Where in the human organism is melanin pigment? What is its function?

4. List the different kinds of joints that are in the axial and appendicular skeletons.

5. Describe the growth of a long bone. Of a flat bone.

6. What are the major differences between skeletal, smooth, and cardiac muscle? Where is each found?

7. Describe the sliding filament model of how a muscle contracts.

8. Describe the flexing of your leg in terms of antagonistic muscles.

SUGGESTED READINGS

BEHOLD MAN, by Lennart Nilsson. Little, Brown: 1978.
A remarkable color atlas of human anatomy and physiology, with many photographs of real organs and details of how they work.

A CHILD'S BODY: A PARENT'S MANUAL, by The Diagram Group. Bantam Books: 1977.
An illuminating look at what you should know about the anatomy and physiology of a growing child.

DISCOVERING THE HUMAN BODY, by Bernard Knight. Lippincott and Crowell: 1980.
An intriguing description of how the organs and organ systems were discovered and by whom. The book also contains anatomical detail not included in this chapter.

MAN'S BODY, by The Diagram Group. Bantam Books: 1976.
Male anatomy and physiology, from acne to cardiovascular problems.

WOMAN'S BODY, by The Diagram Group. Bantam Books: 1977.
Everything you wanted to know about female human anatomy and physiology.

7

HUMAN ORGAN SYSTEMS: DIGESTIVE, RESPIRATORY, AND CIRCULATORY

PREVIEW QUESTIONS

After reading this chapter, you should be able to answer the following questions:

1. How is the food you eat digested?

2. How well-nourished are you? Do you eat a well-balanced diet? Describe your diet.

3. Do you have complete voluntary control over your breathing rate?

4. What kinds of substances and cells are in your blood, and what are their functions?

5. How do you acquire immunity to various diseases?

DIGESTION AND NUTRITION

The Digestive System

A tube nearly nine meters long that extends from the mouth to the anus is the basic digestive system. (See Figure 7–1 for a drawing of the major parts of the digestive system.) A primary function of the digestive tract is to take in foods (nutrients) and chemically break them down into simpler molecules, such as amino acids, simple sugars, and fatty acids. This is the process of digestion. The simple molecules are absorbed through the intestinal lining into blood or lymph vessels. They are then either transported throughout the body to provide nourishment to the cells or brought to storage sites where they can be saved for later use. Undigested material passes through the digestive tract and is expelled from the body during defecation or elimination.

The digestive tube is modified into specific organs, such as the stomach, small intestine, and large intestine. These organs perform specialized functions that contribute to the total processes of digestion and absorption. Attached to the digestive tube are two glands: the liver and the pancreas. The liver secretes bile, and the pancreas produces enzymes that contribute to digestion.

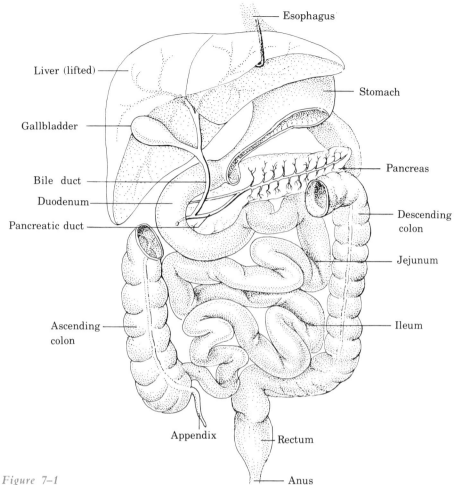

Esophagus

Liver (lifted)

Stomach

Gallbladder

Pancreas

Bile duct

Duodenum

Descending colon

Pancreatic duct

Jejunum

Ascending colon

Ileum

Appendix

Rectum

Anus

Figure 7–1
The human digestive system.

Digestion begins when a bite of food is taken into the mouth and chewed into small pieces. During chewing, the food is mixed with **saliva**, which is secreted into the mouth cavity by three pairs of **salivary glands**. These glands are the parotids, the submaxillaries, and the sublinguals. Saliva contains mucus, which helps moisten food, making it slippery and easy to swallow. Saliva also contains large amounts of the enzyme salivary amylase, which begins the chemical breakdown, or hydrolysis, of starch into disaccharides (maltose). (Table 7–1 explains the functions of various digestive enzymes.)

The voluntary muscles of the tongue and the back part of the mouth cavity (the pharynx) allow for swallowing the food after it has been chewed and mixed with saliva. The act of swallowing is partly voluntary and partly involuntary. After food is swal-

lowed, it is pushed through the digestive tract by wave contractions of the involuntary smooth muscle layers in the wall of the digestive tube. These slow waves of muscular contraction are called **peristalsis**.

The **esophagus** is a muscular tube that connects the pharynx to the stomach. Food passes through the esophagus in only a few seconds, so little or no digestion takes place there. The esophagus and the stomach are separated by a muscular ring, the gastroesophageal sphincter, that opens to allow the passage of food into the stomach.

When empty, the **stomach** is a tubelike organ. Its walls are elastic, however, so they can be greatly expanded to store food during the early phases of digestion. Except during regurgitation, food in the stomach does not pass back into the esophagus, because the sphincter remains closed. However, if some of the

TABLE 7–1
Digestive Enzymes

ORIGIN	ENZYME*	SUBSTANCE ACTED ON	ACTION
Saliva	Amylase	Starch—bread, potatoes, pasta, and so on	Produces maltose sugar
Stomach	Pepsin	Protein—meat, egg white and yolk, vegetable protein such as beans, and so on	Produces peptides—small units that make up proteins; splits the bonds between the amino acids phenylalanine and tyrosine
Liver via gallbladder to small intestine	Bile salts (although not enzymes, bile salts contribute to the digestion of fat)	Fats from meat, cooking oil, butter, margarine, and so on	Emulsifies fats so they become more soluble in water (much like the action of detergent on greasy dishes); emulsified or liquified fats are then digested further by other enzymes
Pancreas to small intestine	Amylase	Starch	Produces maltose sugar; completes job begun in the mouth
	Trypsin	Proteins	Produces peptides; splits the bonds between the amino acids arginine and lysine
	Chymotrypsin	Proteins	Produces peptides; splits the bonds next to the amino acids leucine, methionine, phenylalanine, tyrosine, and tryptophan
	Exopeptidase	Peptides	Strips off any terminal amino acids from polypeptides
	Lipase	Liquified fats	Produces glycerol and fatty acids
Glands of small intestine	Amylase	Starch	Produces maltose sugar
	Maltase	Maltose sugar	Produces glucose sugar
	Lactase	Lactose (milk sugar)	Produces glucose and galactose
	Sucrase	Sucrose (table sugar)	Produces glucose and fructose
	Exopeptidases (various ones)	Peptides	Produces amino acids
	Lipase	Liquified fats	Produces glycerol and fatty acids
	Enterokinase	Chymotrypsinogen, trypsinogen	Produces chymotrypsin and trypsin, both created by the pancreas in inactive forms that are activated by enterokinase

*Enzymes are usually named after their action. For example, the enzyme that breaks sucrose (a double sugar) into two single sugars is called *sucrase*, the suffix denoting enzymatic action.

acid contents of the stomach leak through the sphincter into the esophagus, heartburn occurs.

Muscular contractions of the stomach wall mix food in the stomach with gastric juices secreted by the stomach lining. These juices contain mucus for lubrication, hydrochloric acid, and pepsinogen. Pepsinogen is converted to the enzyme pepsin by the hydrochloric acid. Hydrochloric acid also lowers the pH of the stomach contents to the level where pepsin functions optimally. The low pH also kills foreign organisms. **Pepsin** converts complex proteins into shorter, simpler proteins, called peptides. Food is retained in the stomach from two to six hours. During this time, it is converted into a semifluid called **chyme**, which passes through the pyloric sphincter into the small intestine.

The highly coiled **small intestine** is approximately seven meters long. It is divisible into three regions: the **duodenum**, the **jejunum**, and the **ileum**. The final digestion of all food is completed in the small intestine. Therefore, as might be expected, a large number of digestive enzymes are secreted into the small intestine. These enzymes are secreted by the pancreas, a complex gland connected to the duodenum by a duct and by glands in the wall of the small intestine. Table 7–1 lists these enzymes, their sources, and the types of food they hydrolyze.

Fats are more difficult to hydrolyze than are

starches, sugars, and proteins. Before fats can be efficiently digested, they must be emulsified (separated into tiny droplets). The droplets produced by emulsification have a large surface area, which facilitates enzymatic action. The emulsifying agent is **bile**, which is produced by the liver and stored in the **gallbladder**, where it is concentrated by resorption of water. When food passes from the stomach into the small intestine, the intestine secretes a hormone into the bloodstream. When this hormone reaches the gallbladder, the bladder contracts and bile flows down the common bile duct into the duodenum. There it mixes with chyme and emulsifies fat.

The contents of the small intestine are moved along by peristalsis. The amino acids, simple sugars, and fatty acids are absorbed through the lining of the small intestine. The amino acids and sugars are taken into the circulatory system and transported in blood to the liver. The liver adjusts the levels of the various nutrients and other chemical components found in blood. The fatty acids are absorbed in lymph ducts. (They will be described later in the chapter.)

Figure 7–2 is a cross section of the small intestine. It shows the tissue layers that make up the wall of the digestive tract. From the outside inward are the following layers: (1) an outer layer called the serosa; (2) a layer of smooth muscle, with muscle fibers that extend along the length of the intestine, called the longitudinal muscle layer; (3) another layer of smooth muscle, with the long fibers extending around the intestine instead of along it, called the inner circular muscle layer; (4) a layer of loose fibrous connective tissue containing numerous small blood vessels and nerves; and (5) a mucous membrane lining the digestive tube. The mucous membrane is composed of a thin outer muscle layer, a middle layer of connective tissue containing blood and lymphatic vessels, and a layer of glandular and epithelial cells lining the digestive cavity.

The mucous membrane is convoluted. Many tiny

Figure 7–2

A section through a portion of the small intestine.

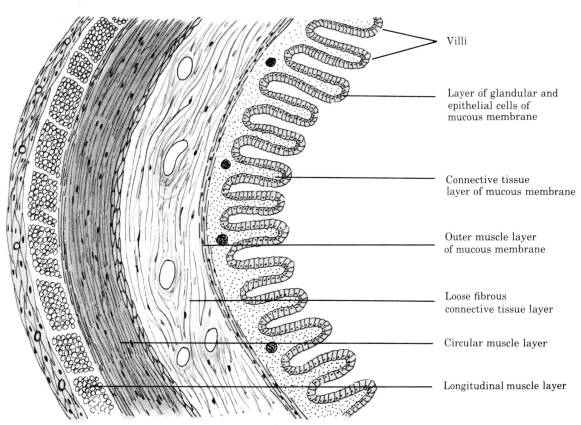

Villi

Layer of glandular and
epithelial cells of
mucous membrane

Connective tissue
layer of mucous membrane

Outer muscle layer
of mucous membrane

Loose fibrous
connective tissue layer

Circular muscle layer

Longitudinal muscle layer

ULCERS AND GALLSTONES

Two common disorders associated with the digestive tract are ulcers and gallstones. At least 20 million people in the United States suffer from ulcers, and at least 15 million have gallstones.

Ulcers are sores in the lining of the stomach or duodenum. The contents of the stomach and the top part of the duodenum are highly acidic and contain enzymes that digest proteins. The linings of the stomach and duodenum can be irritated by such abrasive chemicals. Ordinarily, they are protected by mucous secretions. However, many people secrete unusually large amounts of gastric juice. Such people tend to develop ulcers.

It is difficult to say what induces ulcers, although certainly nervous tension and stress are implicated. Once an ulcer has formed, stomach acid and acids in such foods as citrus fruits and tomatoes contribute to pain and distress. Coffee and alcohol also aggravate ulcers. Antacids, foods that reduce the concentration of acid, and medications are used to control and heal ulcers.

Gallstones form in the gallbladder, where bile is concentrated by the resorption of water. If the bile becomes too concentrated, crystals of cholesterol may form. The solidified cholesterol becomes gallstones. (See the illustration.) Although gallstones may be composed of other substances, cholesterol is the major component in 90 percent of the cases.

A gallstone passing from the gallbladder down the common bile duct to the intestine causes extreme pain. Sometimes the stone may become lodged in the duct, preventing the passage of bile into the duodenum. When this happens, fats can no longer be digested normally. The person with this condition feels pain and becomes ill. Bile salts spill into the blood, and the skin takes on a yellow coloration. When gallstones become serious, both the gallbladder and the stones are surgically removed. It is estimated that over $1 billion per year is spent on the treatment and removal of gallstones in the United States alone.

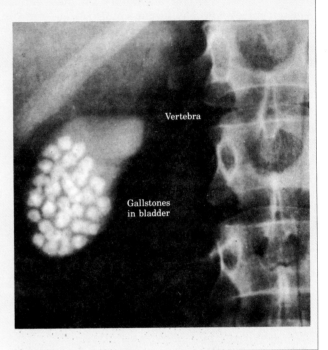

Vertebra

Gallstones
in bladder

fingerlike projections extend from it into the small intestine. There may be as many as 5 million of these structures, called **villi**. When a villus is enlarged, as in Figure 7–3, one can see that it is covered with still tinier projections, called microvilli. Villi and microvilli greatly increase the surface area of the small intestine, providing more area for absorption. Each villus has a tiny artery and vein, connected by capillaries that collect nutrients from the villus. Each also has a small branch of a lymphatic duct that contains lymph. Fatty acids pass into the lymph ducts, not into the blood vessels.

The blood that carries the nutrients goes directly from the small intestine to the **liver**, the largest single organ in the body and one of the most important.

The liver has more than a hundred functions. Besides producing bile, it regulates the level of nutrients in the blood. If the blood contains large amounts of nutrients, as would be the case following a large meal, liver cells remove and store some of them. During periods of fasting, the cells release their stores.

Material that is not digested in the small intestine passes into the **large intestine**. Near the juncture of the small and large intestines is a small, tubelike sac. This sac, the **appendix**, seems to have no vital function in humans. However, it may become infected (appendicitis) and require surgery.

The large intestine is approximately two meters in length. Most of that length is termed the **colon**. Water is absorbed by the lining of the colon. When water

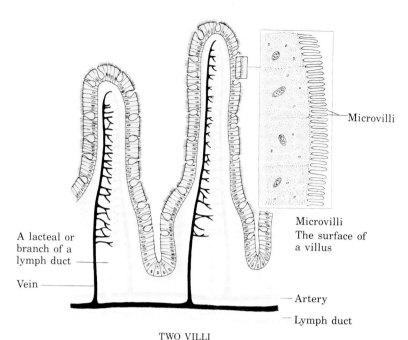

Figure 7–3
Section through a villus. The lining of the small intestine forms millions of tiny projections, called villi. Each bears countless smaller projections, called microvilli. The villi and microvilli increase the surface area of the intestinal lining for greater absorption of digested food.

Microvilli

Microvilli
The surface of
a villus

A lacteal or
branch of a
lymph duct

Vein

Artery

Lymph duct

TWO VILLI

is removed from indigestible material, the matter that is left forms feces. Constipation (the formation of hard, dry feces) results if indigestables are passed along the large intestine too slowly and too much water is absorbed. Diarrhea results if material is passed along too quickly and not enough water is removed.

Only 50 percent of the mass of feces is made up of undigested material. The remainder is composed of the millions of bacteria that flourish in the large intestine and of cells that are sloughed off the lining of the digestive tract. The bacteria get their nutrients from the undigested food and produce intestinal gases (flatus) as a by-product. However, these bacteria also play a more positive role. Some of them synthesize vitamins B and K, which are absorbed by the colon to help fulfill daily vitamin requirements. Feces pass from the colon into the terminal portion of the large intestine, the **rectum**, and then are eliminated through the **anus** during defecation.

Digestion and the digestive systems of some animals other than humans are discussed in Box 7–1.

Nutrition

It may seem naive, but we are going to assume here that some people do not know how to eat properly. Eating properly does not mean knowing how to use a knife and fork. It does mean knowing how to choose among foods so that the result is a well-balanced, nutritionally correct diet. It is painfully obvious that some people do not have a balanced diet because of a shortage of food. However, research has shown that many people with sufficient resources still do not have good diets.

For example, a real student who lived on junk food reported to a university health center because he felt ill. Tests showed that he had scurvy, a serious disease that is caused by a deficiency of vitamin C. Although it may seem impossible that anyone on your campus could have scurvy, a recent survey indicated that in the United States and Canada 15 percent of all age groups suffered some vitamin C deficiency.

We all strive for good health, but how do we tell if we are healthy? The main thing we associate with good health is feeling good. The thing that makes us feel good physiologically is good nutrition. Proper nutrition means eating foods that supply sufficient amounts of the five nutrient classes—carbohydrates, proteins, fats, vitamins, and minerals—as well as fluids (water being the most important).

Carbohydrates

The chemistry of **carbohydrates** has been dealt with elsewhere▫. Here carbohydrates will be discussed as one of the essential food groups.

24

CHAPTER 7   *Human Organ Systems: Digestive, Respiratory, and Circulatory*

BOX 7-1

FEEDING AND DIGESTING IN THE ANIMAL KINGDOM

All animals are heterotrophs. They need to consume organic substances such as proteins, carbohydrates, lipids, and vitamins in order to grow and reproduce. However, the types of foods and the manner in which they are ingested and digested vary from one animal group to another. A few examples will be given here.

Many marine animals—even such large, complex creatures as starfish—supplement their diets by absorbing dissolved nutrients from seawater directly through their body walls. Also, many parasitic animals, such as tapeworms, that live in the digestive tracts of their hosts, absorb nutrients in a similar manner. Tapeworms, which are flatworms belonging to the phylum Platyhelminthes, have no digestive system whatsoever. They depend entirely on the food already digested by their hosts.

Most animals, however, ingest particulate matter. If the food particles are large, they are mechanically broken down before digestion begins. The breaking down is accomplished by a variety of tooth-like structures and mouth parts. It also occurs in a special region of the digestive tract called the gizzard. Many aquatic animals feed on tiny particles of specific sizes that they strain out of the water. These animals are called filter feeders. Other aquatic animals feed on the dead organic litter, called detritus, that accumulates on the substrate. These scavengers are called detritus feeders. Yet other aquatic animals burrow through and ingest the substrate itself. These animals, called deposit feeders, digest the usable nutrients and eliminate the indigestible residues as fecal castings. Some animals, the herbivores, feed exclusively on vegetation. Others, the carnivores, eat other animals. Still others, the omnivores, feed on a wide variety of both plant and animal foods. Humans are omnivores.

The sponges (phylum Porifera) are a good example of filter feeders. They are primitive animals that lack any kind of digestive system. Instead, their bodies contain extensive systems of canals and chambers. The systems are lined in part by flagellated cells called choanocytes. Water containing microscopic food particles is circulated through the canals by the beating flagella. The choanocytes take up the food particles into food vacuoles, as do many single-celled creatures, such as protozoa (see Chapter 4). However, little digestion occurs in these vacuoles. The particles are later passed

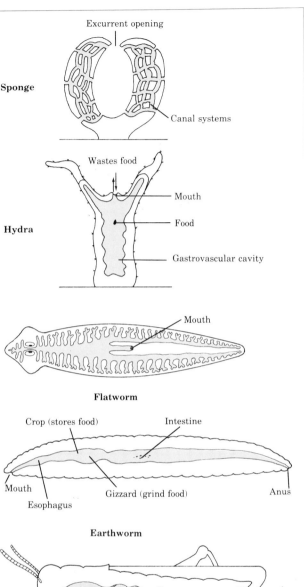

along into wandering amoeboid cells, where digestion is completed. From these cells, nutrients are distributed to the other cells in the sponge.

Simple incomplete digestive tracts are found in members of the phylum Cnidaria (jellyfish, hydras, sea anemones, and corals, for example) and in the free-living, nonparasitic flatworms (phylum Platyhelminthes). These digestive tracts are simple sacs that open to the outside of the animal at the mouth. They are called incomplete digestive tracts because there is no anal opening, only the mouth. Food taken into the sac is partially digested by enzymes secreted by cells in the sac lining. This process is called extracellular digestion. Ultimately, the food is broken down into microscopic particles. These particles are brought into food vacuoles by other cells in the lining of the digestive cavity. In the vacuoles, digestion is completed intracellularly. Indigestible material is ejected through the mouth.

Cnidarians are carnivores that feed on prey that they capture with the tentacles surrounding the mouth. The tentacles bear stinging cells called cnidocytes. Each cell ejects a long, hollow thread that penetrates the body wall of the prey and injects a deadly toxin. Free-living flatworms are either scavengers or predators. They suck food into the digestive cavity by an extendible proboscis. In many species of both phyla, the digestive sac is enlarged and the surface area for the uptake of nutrients is further increased by partitions or branching.

Most animals have complete digestive tracts. The simplest form occurs in roundworms (phylum Nematoda) and in segmented worms (phylum Annelida). For these worms, the digestive tract is basically a straight tube that opens to the outside at the mouth (the anterior end) and at the anus (the posterior end). Food is taken into the tract through the mouth and is digested completely in the intestinal tract (extracellularly). The digested nutrients are absorbed

through the wall of the digestive tract. Indigestible matter is eliminated through the anus. These worms are often described as having a tube-within-a-tube construction, because the tubular digestive tract is surrounded by a tubular body wall.

The same basic design is continued in the phylum Arthropoda, which includes crustaceans, spiders, scorpions, and insects. The digestive tract for this phylum is more highly specialized. Different regions are modified for specific functions. The crop is used for storage, the gizzard is where the food is broken down into smaller particles, and the midgut region is where digestion and absorption occur.

The arthropods are the largest phylum—over 70 percent of all animal species. They show great diversity in diet and feeding habits. Almost all arthropods have their head and sometimes other body appendages modified in different ways for grasping and manipulating food. For example, all insects have the same set of head appendages, but the appendages are modified for each insect group that has different feeding habits and diets. Herbivores have chewing mouth parts. Insects with liquid diets have sucking mouth parts. Insects (such as horseflies and mosquitoes) that feed on the blood of other animals have cutting or piercing mouthparts.

If we were to examine the entire animal kingdom, we would find many modifications of digestive tracts. A common theme, however, is the evolution of modifications for increasing the surface area of the intestine for more extensive and efficient absorption of digested nutrients.

Carbohydrates, the major source of energy for cells, are broken down into simple sugars by digestion▫. Of the many simple sugars, glucose is the major one required in cellular metabolism for the formation of ATP molecules▫. Many cells can also convert proteins and fats into usable fuels for producing ATP molecules. Brain cells, however, cannot. For cells in the nervous system to function normally, a continuous supply of glucose is necessary. It is obvious, then, that we must constantly eat carbohydrates. But we often hear that carbohydrates are bad for us. That is what is taught by many advocates of high protein diets. In fact, though, the high protein diets can be more dangerous to health than high carbohydrate diets are. Carbohydrates are not bad. Indeed, they are essential for good nutrition. About the

only negative thing that can be said about them is that people probably eat too many of them. Carbohydrates ingested in too great a quantity are converted to fats, which are stored in adipose tissue until they are called up from the reserves. (About the only ways they are called up are by exercise or supervised diet, or both.)

The amount of glucose in blood plasma is called the blood glucose level. Under normal conditions, there are between 80 and 120 milligrams of glucose per 100 milliliters of blood. However, shortly after carbohydrates are eaten, this value may increase as much as 20 percent. Certain cells of the pancreas are sensitive to the amount of glucose in the blood. When the amount is too high, say after a carbohydrate-rich meal, the pancreas releases insulin, which facilitates the transfer of glucose from the blood to the liver and muscles, where it is stored. Glucose molecules are chained together, forming a large storage molecule called glycogen. If the amount of blood glucose remains high, fat cells absorb the excess, which is then converted into more fat.

The reverse happens when insufficient amounts of carbohydrates are eaten. When blood glucose falls below the levels just mentioned, another hormone from the pancreas—glucagon—stimulates the breakdown of glycogen in the liver and the release of glucose back into the bloodstream. This hormone works only in the liver. When glycogen is converted to glucose in muscle tissue, it is used by the muscle tissue at the time of conversion, during strenuous exercise, for example. Fats are not recruited during these reactions.

Some people have an abnormal carbohydrate metabolism, and their blood glucose levels may always be low. This condition is known as hypoglycemia. These people often take medications and follow special diets. Other people have blood glucose levels that are always too high. Most often, this condition is the result of a failure of the pancreas such that insufficient insulin is produced. People with this condition are said to be hyperglycemic and to suffer from diabetes.

How can we avoid the excessive storage of carbohydrates, particularly as fat? Cutting back some on carbohydrates will help. However, eating balanced meals is the only real solution. A balanced meal contains some carbohydrates but not a lot. It also contains fats and proteins. Fats slow down the digestion and absorption of carbohydrates, and proteins enhance the release of glucagon. Thus part of the needed glucose is obtained from stores rather than from what was just eaten.

Proteins

The building blocks of all cells, tissues, and organs are proteins. (Their chemistry is discussed elsewhere.) Enzymes, which are proteins, are the catalysts necessary for all chemical processes in cells. There is a constant turnover of protein in human bodies, because little protein is actually stored. Some extra protein molecules are converted to carbohydrates and fats. The rest are converted to urea, which is excreted in the urine. Because of this turnover, it is recommended that each person consume a gram of protein per kilogram of body weight per day. That amounts to fifty-six grams for the average male col-

lege student and about forty-four grams for the average female college student. These recommendations and others made for all the other known nutrients are called the recommended daily allowance (RDA). The next time you eat your favorite breakfast cereal, note how much of the RDA it supplies. The amount is printed on the box.

Proteins are composed of many amino acids. Of the twenty-two known amino acids, nine cannot be synthesized by the cells of the human body. However, metabolism and growth require them. These amino acids are called the **essential amino acids** because they are required for good health. They must be included in the foods people eat because the body cannot make them.

Proteins are classified as complete or incomplete. A complete protein contains all the essential amino acids. An incomplete protein lacks some of them. In general, complete proteins are animal proteins, and incomplete proteins are plant proteins. For this reason, those who become vegetarians without researching their diets could suffer from amino acid deficiencies. However, vegetarians who carefully select their foods can fulfill all of their protein requirements. Most vegetarians should include beans (particularly soybeans) and dairy products in their diets.

A serious worldwide problem is protein deficiency. Sufficient proteins are particularly important for growing children. But in many countries, children subsist on diets composed almost exclusively of grains, with little or no animal protein. The grains may provide kilocalories sufficient for normal existence, but the lack of protein causes a severe deficiency, known as **kwashiorkor**. If the protein deficiency continues for a long time, mental retardation is one of the usual results.

Fats

Technically, fats are better known as lipids, although the words are commonly interchanged. There are two major categories of lipids: fats and oils. Therefore, fats are really a special class of a larger group of similarly constructed compounds. By general definition, fats are solid at room temperature; butter is an example. Oils are liquid at room temperature; vegetable oils used in cooking are examples. Chemically, fats are composed primarily of saturated fatty acids, whereas lipids are composed primarily of unsaturated fatty acids. This distinction is described in more detail

22

elsewhere . There are many fatty acids. The human body can manufacture all but one—linoleic acid. Thus there is only one essential fatty acid.

People in the United States and some other Western countries have a fat phobia. They are told by the media about the dangers of being too fat and eating too much fat, about avoiding cholesterol, and about fat as the cause of major diseases. Most of this publicity is not based on fact. Like the other major food groups, fat is an important nutrient. It is the basis of body oils that are needed by the skin and hair. It insulates us from cold. It is a source of energy. The fat layer around many organs insulates them from the physical shock experienced in car accidents or bad falls. Fat is also an important structural compound of many tissues and a particularly important element of many nerve cells.

When the consumption of carbohydrates is excessive, fat accumulates in the body in discrete layers. Only so much glycogen can be stored in the liver and muscles. Any excess glucose is stored as lipids in adipose tissue. Fat cells do not seem to limit the amount of lipids they can store. They simply increase in size as the lipid content rises. Fat reserves can be an important source of fuel for the production of energy. However, the fuel molecule that comes from fat is acetic acid, not glucose. Thus fat cannot fuel the tissues and cells of the nervous system□.

Fat is an important reserve energy source for muscular activity. The members of professional football teams, for example, rarely weigh under two hundred pounds. Some of their weight is stored fat. This fat provides fuel for the strenuous and continuous muscular activity involved in practice and in competitive games. The same is true for weight lifters and boxers. Most of the energy for sustained activity comes from fat because the stored glycogen in the liver and muscles is quickly used up.

The fats we eat become mixed with the rest of the stomach contents. In the stomach, they behave in the same way as do the oils. The fats and oils ultimately separate from the other compounds in the stomch and rise to the top. This occurs because lipids do not dissolve in water. Digestive enzymes cannot break down lipids in this state. The lipids must first be made water-soluble during a process called emulsification.

Bile salts secreted from the liver emulsify the lipids. Enzymes in the small intestine then break the lipids down into glycerol and fatty acids, which are absorbed and transported by the lymphatic system. The intestinal cells that absorb the fatty acids reassemble them and coat them in protein for transport.

The transport molecules are called lipoproteins. The liver cells recognize these lipoproteins. They remove the protein from them and utilize the fatty acids in a variety of ways, including the manufacture of cholesterol. If there are excessive amounts of lipoproteins, the liver cells repackage them as triglycerides and protein in what is called low-density lipoprotein (LDL). The LDL is sent to all parts of the body and serves many purposes, including the production of hormones. If the body needs energy, the lipids stored in fat cells are called into action. Whatever is not used for fueling muscle cells is packaged and sent through the circulatory system as another lipoprotein, called high-density lipoprotein (HDL).

One of the lipids that has received a great deal of publicity recently is **cholesterol**. A different kind of lipid from the triglycerides just mentioned, cholesterol is a sterol. It is also implicated as a potential cause of cardiovascular disease. Cholesterol is found in many foods, including egg yolks and animal fat (where the level is high). However, there is no cholesterol in plant foods.

The digestive system deals with cholesterol differently from the way it deals with most lipids. Cholesterol is absorbed intact by the cells of the intestine. Once in the lymphatic system, it is transferred directly to the liver. There it can be stored or excreted, or it can become mobile in the blood, where it may accumulate in arteries. If excreted, it combines with the cholesterol that is normally made by liver cells and becomes an important component of the bile salts that emulsify fat. After being excreted, much of it is reabsorbed and recycled back to the liver. When it becomes mobile in the blood, it is frequently associated with LDL and HDL.

If the amounts of these lipids in the blood become too high, they often accumulate as plaque in the walls of arteries. This plaque also contains calcium deposits. The formation of plaque is a disease called atherosclerosis. The complete causes of cardiovascular diseases are unknown. However, such diseases are related to atherosclerosis, high blood pressure, and arteriosclerosis (hardening of the arteries). The combination of the three reduces the inside diameter of blood vessels and restricts blood supply to vital organs, such as the heart.

For a long time it was thought that the amount of cholesterol in the blood was directly related to the

DIGESTION AND ABSORPTION OF BASIC FOODS

FOOD TYPE	DIGESTION PRODUCT	ABSORPTION
Proteins	Amino acids	Through villi in the small intestine directly into the bloodstream
Carbohydrates	Simple sugars	Through villi in the small intestine directly into the bloodstream
Lipids	Fatty acids and glycerol	Through villi in the small intestine into lacteals of the lymphatic system

amount eaten. There was considerable publicity about low-cholesterol foods and low-cholesterol intake, and people were told to eat only three eggs a week. Recent experiments, however, have suggested that the cholesterol to be feared is what the liver makes, not what is eaten. It appears that people make different amounts of cholesterol, depending on their genetic inheritance. It is also known, though, that heavier people have more LDL, HDL, and cholesterol in their blood than do thin people. The cholesterol controversy has not been settled. Cholesterol is obviously a complex biological problem.

Despite the pros and cons about fats in the diet, we know that fats are essential nutrients. The current recommendation (RDA) is that fats should contribute 30 percent of the kilocalories in human diets. It is estimated that the average intake in the United States exceeds 40 percent. Therefore, we probably should reduce our overall intake of fats.

Vitamins

Like the other classes of nutrients, **vitamins** are organic compounds. Generally, though, their molecules are smaller than those of other nutrients. Vitamins in general are **coenzymes**. The activity of coenzymes [22] has been described elsewhere□.

The two major classes of vitamins are water-soluble vitamins and fat-soluble vitamins (see Table 7–2).

What the body does with these two classes of vitamins is quite different. Intestinal cells absorb the water-soluble vitamins from food directly into the blood. The fat-soluble vitamins must first be emulsified, like lipids. Then the intestinal cells eventually absorb them into the lymphatic system. One of the largest vitamins, B_{12}, is so difficult to transport across the cell membranes of the intestine that it is wrapped in a special mucus from the stomach before it is actively transported□. Once vitamins are in general circulation, the fat-soluble ones are stored in the fat of many tissues and organs. Any excess of water-soluble vitamins is excreted in the urine. [73]

If large overdoses of fat-soluble vitamins are consumed, the vitamins may reach toxic levels in the body. For example, excessive amounts of vitamin D can induce nausea or diarrhea and, over a period of time, can adversely affect the heart, circulatory system, and kidneys. Too much vitamin intake may be as detrimental as too little. People who eat well-balanced diets probably get their daily requirements of all essential vitamins.

Minerals

Elements□ such as calcium, sodium, iodine, and iron [14] are **minerals**. They usually occur with other elements as simple compounds in foods. For example, table salt is sodium chloride. Table 7–3 lists the important minerals. It includes their sources and functions and explains what happens when deficiencies of them occur.

By themsleves, minerals do not play a role in body functions because they float inactively in the blood or are stored as inactive elements. They are, however, important building blocks, becoming incorporated in a large number of organic compounds through the activity of enzymes. Enzymes, then, are responsible for the ordered array of minerals in our bodies.

There are two types of minerals. Some minerals are water-soluble and are absorbed directly into the blood from intestinal cells. Other minerals must have a carrier for absorption. When they are absorbed, they can be stored in the same way as the fat-soluble vitamins are. An oversupply of such minerals can be toxic. Among the storable minerals are copper, fluorine, iodine, iron, magnesium, and manganese.

Minerals are also classified as either major or trace. This classification is based on how much of the min-

TABLE 7–2
Vitamins

VITAMIN	SOURCE	FUNCTION	RESULT OF DEFICIENCY
Water–Soluble Vitamins (cannot be stored)			
B_1 (thiamine)	Whole-grain cereals, nuts, pork, liver, eggs	Coenzyme in oxidation of carbohydrates and synthesis of ribose	Beriberi, which causes paralysis of smooth muscles
B_2 (riboflavin)	Beef, veal, lamb, eggs, many vegetables	Coenzyme in oxidation of glucose and fatty acids	Dermatitis
Niacin (nicotinic acid)	Meats, breads, cereals, peas, beans, nuts, many vegetables	Part of NAD and FAD	Pellagra, dermatitis, diarrhea
B_6 (composed of three compounds)	Spinach, corn, tomatoes, yogurt, cereals, liver, meats	Coenzyme in synthesis of proteins and nucleic acids	Slow growth, dermatitis, convulsions
B_{12}	Milk, eggs, cheese, most meats	Coenzyme for synthesis of DNA; red blood cell production	Pernicious anemia
Pantothenic acid★	Green vegetables, cereals, liver	Component of coenzyme A in Krebs cycle	Not known
Folic acid (folacin)★	Leafy green vegetables, liver	Coenzyme in synthesis of DNA; hemoglobin production	Megaloblastic anemia
Biotin★	Eggs, liver, yeast	Coenzyme in synthesis of nucleic acids and metabolism of amino acids	Dermatitis, fatigue, depression
C (ascorbic acid)	Citrus fruits, juices, leafy green vegetables, tomatoes	Involvement in synthesis of protein; collagen of bone and cartilage; promotion of iron absorption	Anemia, scurvy
Fat–Soluble Vitamins (can be stored)			
A	Yellow and green vegetables, fish, milk, butter	Synthesis of visual pigments; maintenance of epithelial tissues	Nightblindness; increased susceptibility to infection
D	Fish, egg yolk, milk, exposure of skin to ultraviolet light	Promotion of calcium and phosphorus absorption	Rickets
E	Nuts, wheat germ, green vegetables	Prevention of oxidation of vitamin A; possible help in maintaining stability of cell membranes	Sterility, kidney problems
K★	Spinach, cauliflower, liver	Blood-clotting factor	Delayed blood clotting

★Also produced by the bacteria in the intestinal tract.

eral is needed. Major minerals are important nutrients in our diets. It is suggested that we consume 0.1 gram of each of these minerals per day. Trace minerals, as their name implies, are needed in only small amounts. The suggested consumption is 0.01 gram of each trace mineral per day.

THE RESPIRATORY SYSTEM

Every cell in the human body requires oxygen for **internal (or cellular) respiration** and produces carbon dioxide as a result of the process. The function

TABLE 7–3

Minerals

MINERAL	SOURCE	FUNCTION	RESULT OF DEFICIENCY
Major Minerals			
Calcium	Dairy products, eggs, dark green vegetables	Formation of bone, teeth, blood clots; keeping muscle and nerve activity normal	Rickets, muscle spasms
Chlorine	Table salt, seafood, vegetables	pH balance of body fluids, tissues, part of stomach acid	pH imbalance, which may result in death
Magnesium	Dairy products, cereals, green leafy vegetables, seafood, chocolate	Helping in release of energy; catalyst; muscle relaxation	Convulsions, hallucinations
Phosphorus	Meat, fish, eggs, dairy products, whole grains	Formation of bone, teeth, cell membranes; keeping muscle and nerve activity normal; part of many compounds, such as ATP and nucleotides	Loss of bone minerals, many metabolic disorders
Potassium	Most fruits and vegetables	Normal muscle and nerve activity	Muscle disorders, including cardiac muscle and nerves
Sodium	Table salt, most foods	pH balance, normal muscle and nerve activity	Cramps, diarrhea, dehydration
Sulfur	Most meats, dairy products, eggs	Part of many proteins and vitamins; part of skin, hair, nails	Unknown
Trace Minerals			
Chromium	Meats and animal proteins (except fish), whole grains, kidneys, most other foods	Promotion of insulin action	Diabetes-like problems
Cobalt	Most foods	Part of vitamin B_{12}	Pernicious anemia
Copper	Most foods	Promotion of iron utilization in hemoglobin	Anemia
Fluorine	Fish, tea, most city water supplies	Prevention of tooth decay	Weak teeth prone to decay
Iodine	Iodized salt, seafood	Part of the thyroid hormones	Goiter
Iron	Most meats, dried nuts, cereals, beans	Part of hemoglobin	Anemia
Manganese	Liver, kidney, nuts, legumes	Cooperation with many enzymes; synthesis of hemoglobin	Infertility, spongy bones, menstrual problems, impaired fat metabolism
Molybdenum	Probably most foods	Part of several enzymes; synthesis of hemoglobin	Unknown
Selenium	Most foods	Antioxidation to prevent destruction of other molecules	Unknown; possible atrophy of heart lining
Zinc	Oysters, herring, milk, eggs, whole grains	Involvement in DNA; component of more than eighty enzyme systems; wound and burn healing; protein synthesis; immune reactions; general growth	Poor growth, slow sexual development, impaired wound and burn healing, loss of taste, retarded growth

of the respiratory system (see Figure 7–4) is to draw air into the lungs. Oxygen in the lungs diffuses into the blood, and carbon dioxide in the blood diffuses into the lungs and is exhaled. This process is called **external respiration.** The blood carries the oxygen to the cells and carries the carbon dioxide from the cells to the lungs. Inhaling air into the lungs and exhaling it out is called breathing, or ventilation.

When one inhales, air is drawn through the nostrils into the nasal cavities, where it is warmed and moistened. Hairs in the nasal passages filter out particulate matter, such as pollen grains and dust. Air then passes into the back part of the mouth cavity, the **pharynx.** The back of the pharynx houses the opening to the **larynx,** or voice box, which contains the vocal cords. When food is swallowed, this opening is covered and protected by the **epiglottis,** a flap of tissue near the rear of the pharynx. Air passes through the larynx and into the **trachea,** or windpipe. Cartilage keeps the larynx and trachea from collapsing because of changes in air pressure. You can feel the larynx, or Adam's apple, and the cartilage rings in the trachea by rubbing the front of your neck.

The trachea branches in the chest cavity to form

Figure 7–4

The human respiratory system. Note that the bronchus entering each lung branches repeatedly to form as many as 6 million bronchioli, each of which ends in a cluster of thin-walled sacs—the alveoli. Oxygen from the air in the alveoli diffuses into blood passing through capillaries associated with them. Simultaneously, carbon dioxide in the blood diffuses into the alveoli.

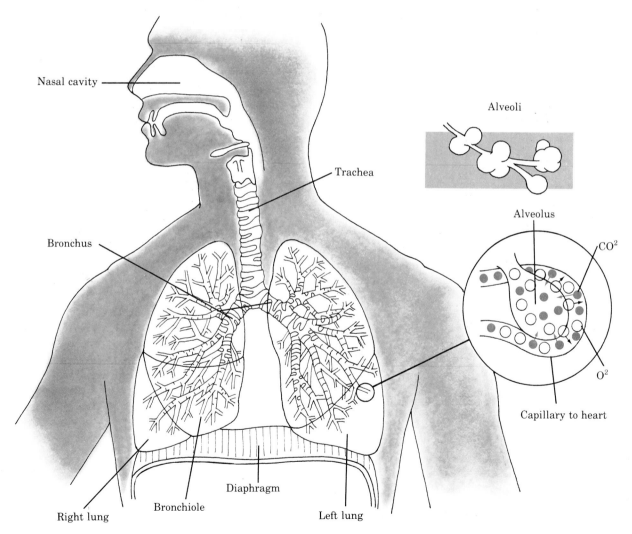

two **bronchi.** Air passes from the bronchi into the lungs. In the lungs, the bronchi branch into as many as 600 million small tubes, called **bronchioli.** Each bronchiolus ends in tiny, saclike structures, called **alveoli.** The thin walls of the alveoli are close to tiny blood-carrying capillaries. Oxygen in the alveoli diffuses⁰ into these capillaries, and carbon dioxide diffuses from the capillaries into the alveoli.

Each **lung** is located in a separate chamber, a **pleural cavity,** in the chest. The pleural cavities change size during breathing (see Figure 7–5). They are enlarged by elevation of the rib cage and lowering of the **diaphragm,** a sheet of muscle that forms the floor of the chambers. As the chambers enlarge, a partial vacuum is created, sucking air into the lungs. When the rib cage is lowered, the diaphragm returns to its former position. The chambers compress, forcing air out of the lungs. The respiratory systems of some organisms other than humans are described in Box 7–2.

At rest you probably inhale and exhale between fourteen and twenty times per minute, but while exercising, your rate of breathing increases dramati-cally. Although you can consciously control your breathing rate to a certain degree, it is impossible for you to hold your breath indefinitely.

Breathing

As you begin to read this section, try to increase your breathing rate. Then try to decrease it. What you have just demonstrated is that you can voluntarily control the breathing rate. However, most of the time, you pay no attention to it. That is because your normal rate is controlled by a breathing center in the region of your brain called the medulla oblongata.

The breathing center produces a regular rhythm that fires neurons and controls the breathing rate. Its control is an example of a feedback loop. The usual rhythm is a second or two for inhaling and two or three seconds for exhaling. This rhythm creates the average breathing rate: fourteen or fifteen times per minute.

During inhalation, the breathing center stimulates neurons that send messages to the **diaphragm** and the rib cage. When relaxed, the diaphragm is dome-

Figure 7–5

A model commonly used to explain the mechanics of breathing. Lungs (the balloons) are elastic sacs within separate pleural cavities (the jars). (a) When the rubber diaphragms at the bottom of the jars are stretched increasing the space in the jars, a partial vacuum is created. The balloons expand as air enters them through the glass tubes. Likewise, when the muscular diaphragm is contracted and the rib cage raised during inhalation, air passes into the lungs. (b) When the rubber diaphragms return to their original position, the pressure in the jars increases, and the balloons collapse as air is forced from them. Air is forced from the lungs during exhalation, when the diaphragm is raised and the rib cage lowered.

(a)

(b)

BOX 7–2

RESPIRATORY DEVICES IN ANIMALS OTHER THAN HUMANS

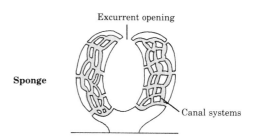

Sponge

Excurrent opening

Canal systems

Animals remove oxygen from the environment and transfer it to individual cells, where it is utilized in cellular respiration. Respiratory systems are the structures and organs used in the uptake of oxygen. Many aquatic animals do not have specialized respiratory structures. Instead, oxygen diffuses (see Chapter 4) through their body surface and into the tissues, where it is utilized.

Because diffusion is a rather slow process, one would expect only small animals with large surface areas and small total volumes to be able to get along without specialized respiratory structures. This is not always the case, however. Large sponges, jellyfish, sea anemones, flatworms, roundworms, and even earthworms all lack respiratory systems and obtain oxygen through their general body surfaces. These animals are all slow-moving and relatively inactive, and they have lower oxygen demands than more active animals.

In some aquatic animals, body construction is such that no body tissue is very far from oxygen-bearing water. For example, sponges constantly circulate oxygen-rich water through their extensive canal systems. Sea anemones and flatworms circulate water through their digestive tracts. In fact, the digestive tract of a sea anemone is called a gastrovascular cavity. It not only functions in the digestion of food, but it also circulates water with oxygen and wastes in and out of the cavity. In these organisms, no tissue is far removed from either an external or an internal source of oxygen.

Earthworms are animals that have solved the problem of oxygen distribution in a more complex way. Their digestive tracts do not contain significant amounts of oxygen, and their internal tissues are not located near enough to the external surface to receive oxygen by diffusion. However, they have evolved fairly efficient circulatory systems. These systems circulate blood, which contains hemoglobin, a red pigment with a strong affinity for oxygen. Beds of tiny capillaries lie just beneath the thin epidermis. Oxygen diffuses through this thin layer into the blood, where it binds with hemoglobin and is transported to all the internal organs. In order for the oxygen to diffuse efficiently through the epidermis, the surface of the earthworm must be kept moist. Earthworms live in

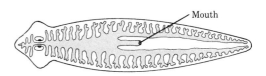

Mouth

Flatworm

Insect

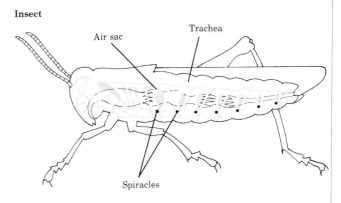

Air sac

Trachea

Spiracles

moist soil, and they constantly secrete fluids that bathe the epidermis. Animals as large and complex as frogs also can take up oxygen through their skin, but in all cases the skin must be kept moist and the animals must therefore live in moist habitats.

Many aquatic animals absorb oxygen only through certain external structures that contain numerous blood

Continued Next Page

CHAPTER 7   *Human Organ Systems: Digestive, Respiratory, and Circulatory*

BOX 7–2 (CONTINUED)

RESPIRATORY DEVICES IN ANIMALS OTHER THAN HUMANS

vessels and/or expand to provide a large surface area. Regardless of position, form, or shape, these specialized structures are called gills. Many worms possess long, extendable feeding tentacles that double as gills. In many animals, the respiratory structures are internal rather than on the surface. Often these structures are called internal gills or lungs. In fish, the gills are located behind the head. Fish gills work in the following manner. Water is taken in through the fish's mouth and passes over the extensive surface area of the gills and out through the gill slits between the gills. Deoxygenated blood coming from the heart passes through extensive capillary beds in the gills, where oxygen is taken up by the red blood cells. Sea cucumbers (phylum Echinodermata) draw water through their anal openings into two large, highly branched lung-like structures that are attached near the end of the intestine. Sea cucumbers literally breathe water in and out of the anus.

Land-dwelling vertebrates, such as humans, breathe air in and out of lungs. Insects and other land-dwelling arthropods have a unique system of air tubules, called trachea, that open by a number of pores, called spiracles, on the surface of the animal. The trachea branch and rebranch, extending air ducts to every body tissue, so these tissues have direct access to oxygen.

diaphragm relaxes and pushes upward into the chest cavity. At the same time, the rib cage contracts inward and downward. The elastic lungs push the air outward. Soon the breathing center sends another wave of neural stimulation to the diaphragm and ribs, and the cycle is repeated.

The breathing rate is not constant. Excitement and exercise increase it tremendously. When the body is asleep, the breathing rate is at its lowest. The rate is increased and decreased by special receptors, (chemoreceptors) in the aortic and carotid arteries. These receptors respond to the amount of carbon dioxide (CO_2) in the blood. They are sensitive not to the CO_2 itself but rather to hydrogen ions (H^+). When CO_2 is in the blood, it combines with water, forming carbonic acid (H_2CO_3), which quickly creates H^+ and carbonate (HCO_3^-) ions. Therefore, the H^+ ions are a direct measure of the amount of CO_2 in the blood.

During exercise, the cells are actively converting energy for this activity. As a by-product, CO_2 concentration increases in the blood. The chemoreceptors sense the increased concentration and send messages to the breathing center, which speeds up breathing until the CO_2 level is lower than what the receptors can detect. Then the normal breathing rate is restored. When these receptors are not stimulated, the breathing center operates at its usual pace.

Pressure and oxygen receptors are also involved in regulating breathing rates. The oxygen receptors respond only when the oxygen level of the blood is very low. If the blood pressure rises rapidly, the breathing rate decreases. If the blood pressure falls (which happens during shock), the breathing rate increases.

shaped. Nervous stimulation causes it to contract and pull downward. At the same time, stimulation of the rib cage muscles causes the rib cage to move outward and upward. The internal size of the chest cavity increases, and the air pressure within decreases. Because of the difference in pressure between the chest cavity and the atmosphere, air rushes into the lungs (see Figure 7–5). The lungs enlarge as the air reaches the alveoli.

The alveoli have stretch receptors in their walls. When these receptors are expanded, they send messages back to the breathing center. The center is momentarily inhibited, and the nerve impulses to the diaphragm and ribs are interrupted. As a result, the

External Respiration

Gas exchange, or external respiration, occurs across the walls of the alveoli. In the walls of the alveoli (which are one cell layer thick), the air just inhaled comes in close contact with circulating blood in capillaries (see Figure 7–4). The blood circulated to the lungs has a high concentration of CO_2, carried primarily as carbonate ions circulating in the plasma. The CO_2 concentration of the blood here is much higher than is the concentration of air in the alveoli. As a result of simple diffusion□, the CO_2 moves across the alveoli cell wall into the alveoli themselves.

Because the concentration of oxygen is just the reverse of the CO_2 concentration, oxygen crosses into

There have been more and more warnings about smoking. The risks are high. The constant irritation of the mouth, pharynx, trachea, and lungs can cause a multitude of problems and diseases. Cancer is the most feared. Smokers have more cancers of the structures listed here than do nonsmokers. They also have a higher incidence of cancer of the bladder and pancreas, stomach ulcers, bronchitis, emphysema, and cardiovascular disease.

Lung cancer is a disease that seems to have become much more common during the past fifty years. In 1912, it was something of a medical curiosity. The cases diagnosed at the time were among miners who were constantly exposed to a variety of metal ore particles suspended in the air they breathed. In 1939, the first link between smoking and lung cancer was suggested. The evidence mounted, and in 1962 the College of Physicians in England formally concluded that smoking does increase the risk of developing lung cancer. In 1964, the Surgeon General of the United States released an extensive report linking smoking and lung cancer. The warnings are everywhere today—on cigarette packages, on television and radio, in newspapers and magazines—but people still smoke.

Lung cancer usually develops slowly. It may take years of constant abuse of lung tissue before a carcinoma appears. The lining of a healthy lung is composed of ciliated columnar cells interspersed with mucus cells. The mucus cells produce a thin sheet of mucus that is swept upward by the beating of the cilia. This sweeping is a filter for the air that is breathed. When dust and other particles enter the lungs, they are trapped in the mucus and eventually eliminated when they reach the pharynx.

But with constant irritation by air pollutants, particularly cigarette smoke, the normal tissue lining of the lung changes dramatically. The glandular goblet cells begin to produce excessive mucus. At the same time, smaller cells, which are produced below the epithelium invade the layer of columnar and mucus cells. They reduce the number of cilia. Mucus begins to accumulate, and a "smoker's cough" develops. The changes up to this point are reversible if the smoking is stopped. However, if it continues, mucus begins to fill the tiny alveoli. This reduces the total surface area for external respiration. Breathing becomes difficult and labored. Coughing becomes more persistent. The symptoms of emphysema begin to appear. As the condition worsens, the smaller cells proliferate and become flattened squamous cells. This is cancer.

The cells divide uncontrollably, replacing the normal columnar and mucus cells. As the sheet of cancerous cells slowly covers the surface of the lung, the lung loses its elasticity and its ability to exchange gases. If it is not too late, the lung can be surgically removed. However, at this stage it is probable that many of the cancer cells have been transported to the blood or lymph (metastasis) and that new cancerous growths have begun to appear elsewhere in the body.

Emphysema is another slow, progressive disease caused by constant irritation to lung tissues. The irritants cause cellular changes in bronchioles and alveoli. The alveoli may develop thick fibrous coverings or may even burst. The bronchioles collapse, trapping air in the alveoli. As a result, the lungs become partially inflated—permanently. *Emphysema* literally means "full of air." The word was chosen because of the condition it describes. The first symptom of emphysema is labored breathing. The difficult part is exhaling, which takes a voluntary effort. This symptom may be associated with a cough. As the disease progresses, more and more bronchioles collapse and more and more respiratory surface is lost. Breathing becomes more difficult. Any form of strain or exercise is nearly impossible because the individual cannot obtain enough oxygen for sustained activity. The condition worsens as the lack of respiratory surface increases the CO_2 level of the blood to the point where the blood pH is lowered. The lowered pH interferes with the breathing center in the brain, and breathing difficulties increase. Certain tissues, particularly those of the brain and nervous system, become oxygen starved. As a result, the heart beats more vigorously and sends larger volumes of blood to the lungs for oxygenation. The overload on the heart usually produces serious heart problems.

There is no cure for the progressive changes caused by either lung cancer or emphysema. There is evidence, however, that the progression can be halted in the early stages. In many cases, even the damage can be repaired. Many smokers have saved their lives by quitting. The message is clear. If you never have smoked, don't; if you smoke, stop. If you are in your early twenties and smoke two packs a day, your life expectancy is eight years less than that of your classmates who do not smoke. Smoking makes you twenty times more likely to develop cancer of the respiratory system or other organs, emphysema, and artery and heart disease.

the blood. As the blood gives up CO_2, a chemical change occurs in hemoglobin molecules in red blood cells. This change causes a rapid attraction of oxygen to hemoglobin. Oxygenated blood, then, has a much lower CO_2 concentration, and nearly all hemoglobin molecules carry oxygen.

When the red blood cells with oxygen-rich molecules of hemoglobin reach the tissues, the reverse of external respiration occurs. The tissues are low in oxygen and high in carbon dioxide. Oxygen is released from hemoglobin molecules, and it diffuses into the tissues. The blood has little or no CO_2, so these molecules diffuse from the tissues to the blood. The CO_2 is quickly absorbed by the blood, where it combines with water to form carbonic acid. Thus H^+ ions and carbonate are formed. The CO_2 is carried as carbonate ions back to the lungs, and the H^+ combines with hemoglobin for the return trip. The attraction of the H^+ ions to hemoglobin is important for maintaining the pH balance of the blood.

THE CIRCULATORY SYSTEM

Every cell of the human body requires nutrients and oxygen, and every cell produces carbon dioxide and harmful waste materials. Blood in the circulatory system transports these materials, as well as substances such as hormones, to and from the cells. The blood is moved by a pump, the heart. Within a continuous system of blood vessels, the blood is circulated through the body and back again to the heart.

There are several kinds of blood vessels. **Arteries** carry blood away from the heart. They are composed of three layers: an inner layer of epithelial cells, a thick central layer of smooth muscle and elastic fibers, and an outer layer of connective tissue. Large arteries branch into smaller ones, the smallest of which are called **arterioles.** The arterioles eventually branch into numerous **capillaries,** many of which are the width of only a single red blood cell. Many substances from the blood cross capillary walls into the tissues, and many others cross capillary walls from the tissues into the blood.

Capillaries lead to small veins, called **venules,** which eventually unite into larger blood vessels, the **veins.** The structure of veins is similar to that of arteries, except that the middle layer of veins is thinner

and has fewer elastic fibers. Many of the larger veins (the leg veins, for example) have one-way valves so blood can move in only one direction. The blood pressure in veins is low, and the valves are a way to prevent the backflow of blood.

Figure 7–6 shows the distribution of some of the major blood vessels. If you study it (and Figure 7–9), you can see that the heart actually acts as a pump for two separate circulatory pathways. In the first, called **pulmonary circulation,** the heart pumps blood through the pulmonary arteries to the lungs. The blood flows through the capillary systems in the lungs, where it receives oxygen and gives off carbon dioxide. It then flows back to the heart through the pulmonary veins.

The newly oxygenated blood is then pumped through the second circulatory pathway, the **systemic circulation,** to the rest of the body and back to the heart again. During systemic circulation, the blood becomes depleted of oxygen and laden with carbon dioxide. When it arrives back in the heart, it is pumped again to the lungs. Obviously, if this system is to work, the oxygen-rich and oxygen-poor blood should not mix in the heart. To understand how these pathways are separated, we need to understand the structure of the heart.

The Heart

The hard-working human heart is one of the first organs to function in the fetus and it continues beating until the person dies. It is about the size of a closed

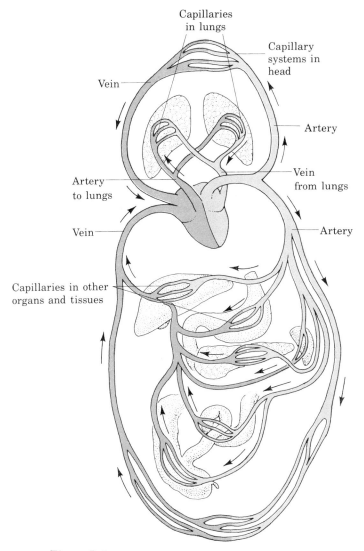

Capillaries
in lungs

Capillary
systems in
head

Vein

Artery

Artery
to lungs

Vein
from lungs

Vein

Artery

Capillaries in other
organs and tissues

Figure 7–6

A generalized view of the human circulatory system. Arteries carry blood away from the heart to the lungs and other structures, where they give rise to beds of capillaries.. Capillaries distribute nutrients and oxygen to the tissues and collect wastes that enter the blood. Blood in the capillaries passes into veins, which conduct the blood back to the heart.

fist, and it beats about seventy times per minute in a normal resting adult. The heart has a tough, muscular wall (cardiac muscle) that is lined on the inner surface by a layer of epithelial cells, called the **endocardium.** It is covered on the outside by a smooth layer of connective tissue covered by squamous epithelium called the **epicardium.** It lies in a cavity, the **pericardial cavity,** where it is free to beat.

The heart is divided into four chambers. Two of the chambers, the **atria,** have walls thinner than the walls of the other two chambers, the **ventricles.** The atria form the upper portion of the heart, and the ventricles form the lower, more massive portion (see Figure 7–7). The atria receive blood from veins that carry the blood back to the heart. The right atrium receives venous blood from the systemic circulation. The left atrium receives blood from the pulmonary veins. Therefore, the blood entering the right atrium contains low levels of oxygen and high levels of carbon dioxide. The left atrium receives blood directly from the lungs, so it is rich in oxygen and low in carbon dioxide. The two atria are separated, but each connects with the ventricle beneath it. The ventricles are also separated.

The flow of blood through the heart is as follows (see Figure 7–7 for details). Oxygen-poor blood enters the right atrium from the general circulation. Simultaneously, oxygen-rich blood enters the left atrium from the lungs via the pulmonary veins. As the atria contract, blood from the right atrium is forced into the right ventricle. At the same time, blood from the left atrium is forced into the left ventricle. When the ventricles contract, the oxygen-poor blood in the right ventricle is forced into the pulmonary arteries to the lungs. The oxygen-rich blood in the left ventricle is forced into the aorta and into the systemic circulation.

It may sound confusing that oxygen-poor blood goes to the lungs in the pulmonary arteries and returns oxygen-rich to the left atrium in the pulmonary veins. The confusion occurs because the arteries and veins are named not for the type of blood they contain but for whether the blood is moving toward or away from the heart. Arteries are always part of the outward circulation that begins at the heart, and veins are always part of the inward circulation that leads back to the heart.

The circulation of blood through the heart is a one-way system because of a series of valves. The valve between the right atrium and the right ventricle is called the **tricuspid valve.** It is composed of three flaps of fibrous tissues, called cusps. From each cusp, tiny inelastic tendons project toward the walls of the ventricle, where they attach themselves to the papillary muscles. The valve between the left atrium and the left ventricle is similarly constructed. It consists of only two cusps, however, and is called the **bicuspid valve.** When the ventricles relax and the atria

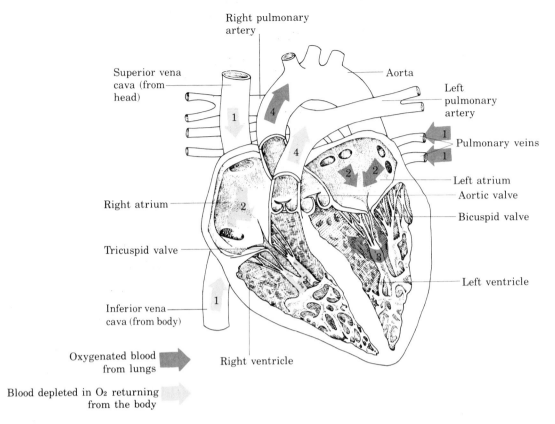

Figure 7-7

The double-pump structure of the human heart. Blood from the systemic circulation flows (1) into the right atrium, (2) and the left atrium receives blood from the lungs. When the ventricles relax, blood flows from the right atrium into the right ventricle and from the left atrium into the left ventricle. (4) When the ventricles contract, valves between them and the atria close, and blood is forced from the right ventricle into the pulmonary artery (3) toward the lungs and from the left ventricle into the aorta and systemic circulation.

contract, the tricuspid and bicuspid valves are forced open. Their cusps fold into the chambers of the ventricles. When the ventricles contract and the atria relax, the pressure of blood against these valves forces them closed. Without some check, the cusps would be forced upward into the atria and a backflow of blood would occur. To prevent this, the papillary muscles contract, exerting force on the tiny tendons. These tendons in turn pull at the cusps, preventing them from collapsing under the pressure. The valves then allow blood to flow in only one direction.

The valve between the left ventricle and aorta and the valve between the right ventricle and the pulmonary artery are called the **semilunar valves.** These valves are composed of three cusps that are shaped like half-moons. When blood is forced against them by the contraction of the ventricles, the valves open. When the ventricles relax and the pressure decreases, the blood in the arteries begins to surge backwards. As it does, the valves are forced closed.

When the valves open and close, they produce sounds. The closing of the valves makes the most distinct sound. When the tricuspid and bicuspid valves snap shut, they produce a characteristic sound called the "lub" sound. When the semilunar valves snap shut, they produce a louder but lower-frequency sound, the "dub" sound. The sounds of the heart are thus often referred to as the lub-dub sounds. Between these sounds can be heard the "swooshing" circulation of blood.

Physicians can also hear sounds that warn about a lack of cardiac homeostasis. These sounds are commonly called heart murmurs. Several different sounds are possible. One set of abnormal sounds is produced when one of the valves is faulty and blood seeps backwards. Other abnormal sounds are produced when the valves become hardened and blood is shot through them like water through a nozzle. Heart murmurs have many causes. Some are inherited. Others are the result of a disease such as rheumatic fever.

The Pacemaker

The contraction of the heart involves a series of rhythmic contractions that make up the cardiac cycle. The heart has its own natural pacemaker that paces each cardiac cycle: the **sinoatrial (SA) node,** located in the upper portion of the wall of the right atrium. At the beginning of a cardiac cycle, stimuli flash from the SA node through the walls of both of the atria. Ultimately, they trigger a second node, the **atrioventricular (AV) node,** located in the right atrium near where the four chambers converge. The AV node relays the stimulation down a series of modified muscle fibers located in the wall between the right and left ventricles. At the base of the ventricles, these fibers branch into numerous smaller fibers that create a net reaching to all of the ventricular muscles.

A stimulus sent by the SA node is traveling 60 centimeters per second as it reaches the AV node. From the AV node, it slows to 5 centimeters per second. But when it reaches the ventricles, it increases to 210 centimeters per second. The high speed in the ventricles ensures that all the ventricular muscles will contract at once.

The rate of heartbeat fluctuates. Nerves to the heart adjust it to meet variations in body activity. The muscles of an active person use oxygen at a faster rate than do the muscles of an inactive person, and the active person's heart beats faster too. This occurs because of the increased numbers of hydrogen ions in the blood, which trigger receptors in specific arteries that affect the breathing center. (This complex mechanism has been described elsewhere▫.) However, if the nerves to the heart were cut, the heart could continue to beat because it has its own pacemaker.

These electrical events of the pacemaking system can easily be recorded. The recording is called an **electrocardiogram (ECG**—formerly abbreviated EKG). The ECG of the normal beating heart has several predictable characteristics (see Figure 7–8). There

are three major peaks with valleys in between them, including the P wave, the QRS complex, and the T wave. The P wave represents the SA impulse and the contraction of the atria. The space between the end of the P wave and the beginning of the QRS complex represents the time it takes the SA impulse to reach the muscles of the ventricles. The QRS complex represents the electrical events of the ventricle contractions. The T wave represents the recovery of the ventricular muscles (that is, the time when they are ready to contract again). The distance between S and T is the time required for recovery. (The downward deflection of the Q part of the QRS complex is the recovery of the atrial muscles.) Physicians can tell a lot about a particular heart by comparing its ECG to the average ECG. For example, a person who has had a heart attack often suffers permanent damage to the ventricular muscles. The recovery time of the damaged muscles (the distance between S and T) is longer than average. Recovery may never appear complete (shown by various odd shapes of the T wave). A person with heart murmurs also has a characteristic ECG.

Figure 7–8
An electrocardiogram (ECG).

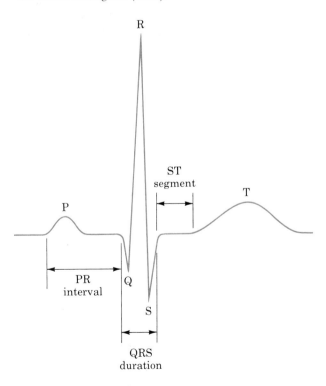

Blood Pressure

Contraction of the heart muscles is called **systole;** relaxation is called **diastole.** When the ventricular muscles contract, blood is forced into the arteries. The volume of blood and the forceful contraction create an internal pressure. As the blood is pumped into the arteries, which have a lesser volume, the pressure increases and the semilunar valves close. This height of pressure generation is called the **systolic pressure.** The blood is then squeezed through the arteries by the rhythmic contraction and relaxation of the artery walls. As this occurs, there is some loss of pressure. The drop in pressure is most marked when the ventricles are relaxed and therefore giving the lowest pressure, the **diastolic pressure.**

If you place your forefinger and middle finger over the vessels on the inside of your wrist, you will feel your **pulse.** What you are feeling is the surge of blood sent into the circulatory system from the heart during systole. This surge stretches the arteries for a moment. During diastole, when the ventricles are relaxed, the pressure drops and the arteries recoil. The combination of surge and recoil is the pulse. At rest, the average pulse rate is seventy to eighty beats per minute. When you are very active—for example, when you jog—your pulse rate may double.

Blood pressure is easily measured with a special instrument called a blood pressure cuff. The instrument allows a physician to determine both the systolic and the diastolic pressures. The blood pressure reading is expressed as one number over another, for example, 120 over 80 (120/80)—an average reading. The larger number is the systolic pressure, the smaller the diastolic pressure. An old rule of thumb is that the systolic pressure should be 100 plus your age. That rule is not a bad way of estimating the pressure. Blood pressure does increase with age. Arteries lose their elasticity, and various diseases of the circulatory system (such as atherosclerosis and arteriosclerosis) increase the blood pressure.

After about age forty, however, the 100-plus-age rule no longer applies. That is because blood pressure should rarely be higher than 140/90. If it is higher, the person suffers from high blood pressure, or hypertension. Physicians worry more about the lower number because it indicates an increased resistance to blood flow, which can eventually lead to artery and heart disease. On the other hand, increased systolic pressure (the higher number) can cause strokes. Many available medications are effective in controlling blood pressure.

The Systemic Circulation

When the ventricles contract, blood is forced into two major circulatory pathways: the **systemic circulation** and the **pulmonary circulation.**

Blood exiting the right ventricle is carried into the large pulmonary trunk. Near the heart, this vessel branches into the left and right pulmonary arteries. When these arteries reach the lungs, they branch again and again until the blood flows into the capillaries of the alveoli. After external respiration□ takes place, the reverse occurs, and the blood is transported to the left atrium by the pulmonary veins.

Blood exiting the left ventricle is carried by the large single **aorta.** As it leaves the heart the aorta arches to the left. Three major arteries branch from the upper portion of the arch. The first is an artery that branches quickly into two separate arteries, one of which supplies blood to the right shoulder and arm. The other supplies blood to the right side of the neck, throat, face, and brain (the right carotid artery). The second artery to branch from the aorta is the left carotid artery, which supplies blood to the left side of the neck, face, and brain. The third artery to branch from the aorta supplies the left shoulder and arm (see Figure 7–9).

After the branching of these major arteries, the aorta bends downward and passes into the abdominal region. Here various branches supply the diaphragm and liver, the muscles of the trunk and legs, and the digestive, excretory, and reproductive systems.

The returning venous circulation is composed of many veins, which eventually connect to the large superior (from the head) and inferior (from the body) vena cavae. The vena cavae return oxygen-poor blood to the right atrium (see Figure 7–9).

Two other important subdivisions of the systemic circulation are the coronary circulation and the hepatic portal circulation. The **coronary circulation** supplies oxygenated blood to the cardiac muscles of the heart. Right and left coronary arteries branch from the aorta just outside the heart, before the branching of the other major arteries. One of the coronary arteries branches into three major arteries, which branch into other smaller arteries. These arteries supply cardiac muscle with oxygen and nutrients.

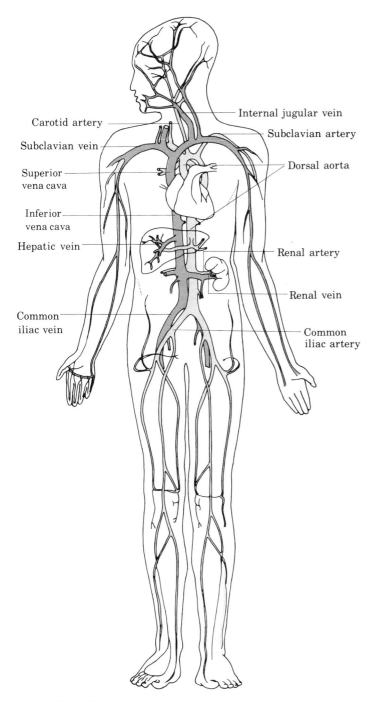

Figure 7–9
The major blood vessels of the human circulatory system.

Oxygen-poor blood and by-products of cardiac muscle metabolism are passed into veins. The veins eventually drain into the coronary sinus, which empties into the right atrium.

The narrowing of the coronary arteries because of plaque leads to coronary heart disease, one of the most common of all diseases in the United States. When cardiac muscle is deprived of oxygen-rich blood, the muscle dies. This death interrupts nerve impulses and creates an infarction, commonly known as a heart attack. After a heart attack, new vessels may develop to supply the areas of cardiac muscle where circulation was reduced by arterial blockage. However, in most instances the heart attack was caused by one or more of the major coronary arteries having a reduced blood flow. Coronary bypass surgery is used to correct the deficiency (see Box 7–3).

Hepatic portal circulation is a special circulation of oxygen-poor (venous) blood from the digestive system to the liver before the blood is returned to the heart. The veins leaving the digestive system are rich in the products of digestion and any harmful substances absorbed. The liver is like a giant sponge. It absorbs all the important molecules that are derived from digestion (for example, glucose) and that will be either stored or transferred to the blood for general circulation. The liver cells can store glucose as glycogen or fat, and they can produce glucose by metabolizing carbohydrates, fats, and proteins. They can also detoxify harmful substances. For example, liver cells convert to urea the poisonous ammonia compounds that result from protein metabolism. The urea is less poisonous than the ammonia compounds, and it is eventually eliminated in the urine. Liver cells can also engulf bacteria and other foreign matter that pass in the blood. In fact, it is the liver cells that engulf and destroy worn-out red blood cells

The liver receives oxygen-rich blood from the hepatic artery. All blood drained from the liver collects in the hepatic vein, which empties into the inferior vena cava.

The circulatory systems of some organisms other than humans are described in Box 7–4.

Blood

A complex substance, blood is composed of blood cells and platelets suspended in a chemically complex liquid called **plasma.** Blood cells and platelets constitute approximately 45 percent of the four to six liters of blood contained in each human body. Plasma represents the remaining 55 percent of the blood.

Plasma contains dissolved gases, various ions that help maintain pH, and a wide variety of proteins,

BOX 7–3

HEART DISEASE

If a coronary artery is suddenly blocked, a sudden heart attack may occur. Often death results. But coronary arteries and their branches often become occluded with plaque over a period of months or years. During this period, a person may suffer angina (chest pain) or exhibit other symptoms of occlusion.

If coronary arteries are becoming dangerously occluded, as in the photograph, a coronary bypass operation may be recommended. During the operation, sections of veins removed from the patient's legs are used to replace the blocked coronary arteries (see the drawing). Then blood can again flow to the heart muscle.

This type of operation is becoming more and more common, but it is major surgery and therefore entails considerable risk. A newer technique is proving successful, and with less risk. The patient's chest cavity is not even opened. The physician places a tube, called a catheter, into a major vein. At the end of the catheter is a small balloon-like device. With the view provided by X-rays, the surgeon moves the catheter until it reaches the coronary arteries. If the X-rays reveal blocked or nearly blocked arteries, the surgeon moves the balloon into these areas and expands it, pushing the plaque into the arterial walls and opening the arteries to near-normal size.

Today, every valve in the heart can be replaced with an artificial one. Many people have holes in their hearts that cause one chamber to leak blood into another. These holes can be patched by the use of various surgical techniques. Even the major blood vessels leaving the heart can be replaced in part or whole by synthetic textile tubing.

Many people suffer irregular heartbeat rhythms because of various abnormalities of the natural pacemaking system of the heart. Artificial pacemakers are commonly installed today. The pacemaker itself is placed just under the skin of the chest or abdomen. Wires lead to specific sites on the heart muscles. Some pacemakers stimulate the heart only if the heart begins contracting arhythmically. Other pacemakers fire at a constant rate, stimulating the heart all the time.

Today, however, this type of pacemaker is adjustable, so the stimulation rate can be changed to suit various activities.

Although the hearts of many people are able to be repaired, anywhere from 15,000 to 66,000 Americans each year suffer irreparable heart damage and of course, many of these people die. In many cases the only hope is a heart transplant or the implantation of an artificial heart. Artificial hearts have been tested on animals during the past several years, but an artificial heart was first implanted in a human on December 2, 1982. A 61 year old Seattle dentist who was near death of heart disease was selected to be the first recipient. His name was Barney Clark. Dr. Clark was an extremely ill man when the operation was performed, but he lived until March 2, 1983 when he died of medical complications not associated with the artificial heart itself.

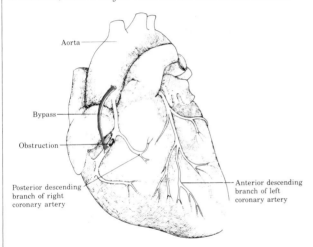

Aorta

Bypass

Obstruction

Posterior descending branch of right coronary artery

Anterior descending branch of left coronary artery

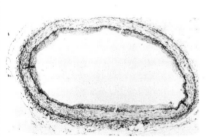

Normal coronary artery cross section

Deposits forming in inner lining

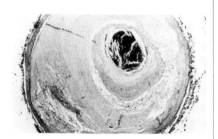

Narrowed channel blocked by blood clot

BOX 7–4

THE CIRCULATORY SYSTEMS OF ANIMALS OTHER THAN HUMANS

Large animals usually have circulatory systems to transport nutrients and oxygen to tissues and to carry off waste products. Various designs for circulatory systems have evolved in different animal groups. Some of them are efficient; others are not. Of course, if an animal is small enough, simple diffusion can handle the transportation. But relatively few phyla are made up of small organisms. Some groups of even large animals have no heart or veins or arteries—nothing that even vaguely resembles the kind of circulatory system humans possess.

For sponges, the canal systems that circulate water with oxygen and food particles act as circulatory systems. The water also carries away wastes and even gametes. Further distribution of nutrients and elimination of particulate wastes is accomplished by wandering amoeboid cells that creep through the tissues of the sponges. In both the Cnidarians and the flatworms, wandering amoeboid cells also contribute to the distribution of nutrients. But it is the circulation of water through the digestive cavities that brings oxygen within easy reach of internal tissues and that assists in waste elimination.

The groups mentioned so far do not possess any sort of true body cavities. However, when such cavities do exist, enclosing the internal organs of animals and serving as hydrostatic skeletons, the fluids they contain can be circulated by cilia and flagella or by the muscular movements of the body wall. The moving body cavity fluids act as an additional medium for the transport of materials from one body location to another. Also, the body cavities provide a space in which organs are free to move independently of the animal's body wall. The space may be used for the beating hearts, the muscular pumps necessary for efficient circulatory systems.

Two general types of circulatory systems have developed: open and closed. Mollusks and arthropods have open systems. In both cases, the heart pumps blood out through blood vessels, which lead to open blood sinuses. There the blood leaves the vessels and slowly flows around the organs and through the tissues. Finally, the blood makes its way back to the heart for recirculation. This kind of system is inefficient in large animals because the blood pressure drops drastically when the blood leaves the vessels, and circulation and

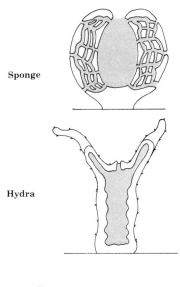

Sponge

Hydra

Flatworm

Lateral hearts   Earthworm

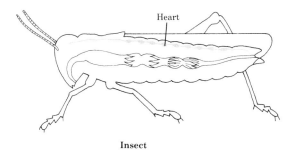

Heart

Insect

distribution of oxygen is relatively slow. Consequently, most mollusks are slow-moving. Squid and octopuses, however, are large and active. They have evolved additional hearts to pump blood quickly through the

(Continued Next Page)

THE CIRCULATORY SYSTEMS OF ANIMALS OTHER THAN HUMANS

gills before it is returned to the main heart for recirculation, and they possess closed circulatory systems. Insects and many other land arthropods are also active, but their blood does not transport oxygen. Instead, these arthropods have evolved air tubes, called trachea, that conduct oxygen directly to the tissues from the outside, thereby avoiding the problem of slow circulation of blood.

Vertebrates, including humans, have closed circulatory systems. The blood pressure is maintained at levels high enough for reasonably rapid circulation because the blood remains within arteries, capillaries, and veins during circulation. The simplest closed circulatory systems are those of segmented annelid worms. They have two major blood vessels, one above and one below the digestive tract, extending the length of the worm. The major blood vessels are connected to each other in each segment by lateral blood vessels (see illustration). In some annelids, part of the upper vessel is contractile and serves as the heart. In others, such as earthworms, some of the lateral blood vessels in the anterior region become hearts.

commonly called plasma proteins. Some of these proteins are important in certain immune responses, and others make the blood thick, or viscous. Blood is about four or five times thicker than water. The thickness of plasma causes it to move more slowly than water, thereby allowing time for the interaction of blood, body fluids, and cells. Plasma is the major transport medium for nutrients, hormones, and waste products.

Carried in the plasma are the **platelets**—cytoplasmic fragments of a particular class of blood cells. Platelets are necessary for the clotting of blood. When you cut yourself, two reactions quickly occur. First, because of the trauma of the cut, the smooth muscles of the blood vessels involved in the injury contract, decreasing the blood flow. However, this reaction is not sufficient to seal the wound. At the point of vessel injury, blood platelets begin to accumulate and then to disintegrate. Their disintegration releases into the blood several factors that eventually help form the

blood clot. The chemical steps in blood clotting are complex and involve many reactions. For example, four platelet chemical factors react with thirteen chemicals in blood plasma before a clot is formed. To keep it simple we can say that the major reactions are the formation of thromboplastin, the thromboplastin presence converting prothrombin to thrombin, and the thrombin presence converting fibrinogen to fibrin. The final product is **fibrin,** a fiber-like thread that is insoluble in water and that adheres to the wounded area and to many other platelets that have not yet reacted to the trauma. The clotting process is a chain reaction. More and more platelets get into the act until there are sufficient fibrin threads to form a clot and seal the wound.

Blood cells include red blood corpuscles (erythrocytes), which are the most abundant cell type, and a variety of white blood cells (leucocytes). **Erythrocytes** contain no nuclei (the nuclei are lost during cell differentiation) and are simple, concave, disk-shaped sacs that hold an iron-containing respiratory pigment called **hemoglobin.** Hemoglobin binds with and releases oxygen molecules. Oxygen transported in the blood is bound to hemoglobin molecules, a complex that is formed in the capillaries of the lungs. When oxygen is needed in the tissues, the oxygen-hemoglobin complex breaks apart, and the tissues receive the oxygen.

New erythrocytes are constantly being produced in bone marrow, because they wear out from passing through the miles of capillaries in the body. Their life span is 100 to 120 days. Worn-out erythrocytes are removed from circulation by special cells in the bone marrow, spleen, and liver.

Leucocytes lack iron-red hemoglobin and are therefore called white blood cells. They have nuclei. As indicated in Table 7–4, some types of leucocytes engulf foreign particles such as bacteria, and others are important in forming antibodies and helping the body resist invasion by disease-causing bacteria and parasites. The white pus that forms at the site of an infected wound and in pimples and boils is largely an accumulation of a specific type of leucocyte, **neutrophils.**

The Lymphatic System

Another system of channels, the lymphatic vessels, forms part of the total circulatory system. When blood flows through capillaries, some of the fluid

TABLE 7–4
White Blood Cells (Leucocytes)

TYPE	NUMBER per Cubic Millimeter*	FUNCTION
Granular (Possessing Cytoplasmic Granules)		
Neutrophils	3,000–6,000	Involved in inflammation; cells have amoebic movement; infection or inflammation stimulates neutrophils to move out of blood vessels and to engulf the source of inflammation (for example, foreign bacteria); numbers increase during any of the "itises," such as tonsilitis and laryngitis; accumulation of dead neutrophils produces pus
Eosinophils	150–450	Weakly amoebic; associated with digestive system, respiratory system, and skin; become most numerous during allergic reactions and parasitic infections; may produce antihistamines to retard allergies
Basophils	50–75	Amoebic movement; engulfing of antigens in blood and tissue; function unknown but always associated with infection and inflammatory responses
Nongranular (Lacking Cytoplasmic Granules)		
Lymphocytes	1,000–3,000	Most frequent in lymph nodes, where many become incorporated into the lymph fluid; produce antibodies
Monocytes	150–700	Largest leucocytes; spherical when in blood, but change shape and have amoebic movement when in tissues; function believed to be the same as neutrophils; normal numbers increase during course of mononucleosis

*Normally, the body contains 5,000 to 9,000 white blood cells of all types per cubic millimeter of blood.

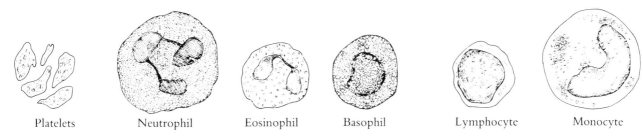

Platelets Neutrophil Eosinophil Basophil Lymphocyte Monocyte

portion leaks through the thin walls. This tissue fluid—blood minus its cells, platelets, and dissolved proteins—is called **lymph.** It bathes cells and eventually collects in tiny **lymph vessels,** such as those in the villi of the small intestine. These vessels interconnect to form a system that joins the circulatory system near the heart (see Figure 7–10).

Lymph vessels lack muscular walls, and the lymphatic system lacks a pump. Lymph moves because of pressure exerted on the vessels by contraction of body muscles.

Lymph vessels have a series of valves (much like those of veins) that direct the lymph fluid in only one direction. The smaller vessels lead to lymph nodes. Lymph vessels leading from lymph nodes join larger vessels, called lymph trunks. The trunks pass lymph to two major collecting vessels, the left thoracic duct and the right lymphatic duct. The right lymphatic duct collects lymph that circulates in the right arm and on the right side of the head, face, neck, and chest and carries it to the veins of the right arm. The left thoracic duct collects lymph from the remainder of the body and carries it to the veins of the left arm. Therefore, lymph is returned to the blood just before the arm veins merge with the superior vena cava.

Lymph nodes, which are also called lymph

CHAPTER 7 *Human Organ Systems: Digestive, Respiratory, and Circulatory*

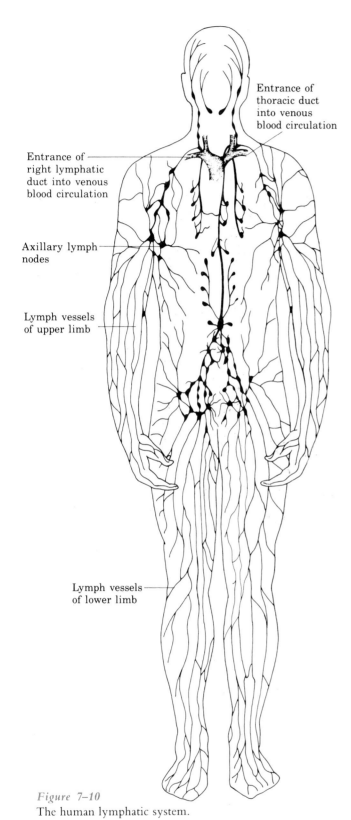

Entrance of
thoracic duct
into venous
blood circulation

Entrance of
right lymphatic
duct into venous
blood circulation

Axillary lymph
nodes

Lymph vessels
of upper limb

Lymph vessels
of lower limb

Figure 7–10
The human lymphatic system.

glands, have a complex structure. They are composed of an outer core of stored lymphocytes and lymphocytes in production and an inner network of lymphocytes arranged in strands or cords. Within the network are a variety of small spaces where lymph fluid and lymphocytes make contact. Lymphocytes can detect and destroy foreign matter in the lymph. This response is part of the immune system. A serious infection often causes lymph nodes to swell. The swelling is from an increased number of lymphocytes and an accumulation of fluid. The human body has many lymph nodes. (Figure 7–10 shows each node as a dark area.) The ones we are most familiar with are those that tend to hurt from time to time when we are ill: the cervical lymph nodes (those in the neck), the axillary nodes (those in the armpit), and the inguinal nodes (those in the groin).

THE IMMUNE SYSTEM

The system in the human body that attacks foreign agents entering the body is the immune system. The foreign agents include viruses, bacteria, parasites, pollen, proteins, and carbohydrates. The agents are classed together as **antigens.** The immune system produces specific **antibodies,** which attack the invading antigens. How the immune system does this is complicated. The explanation here is a simplified one.

The basis of the immune system is lymphocytes. There are two kinds of lymphocytes, **T-cells** and **B-cells.** During the embryo stage of development, a certain set of cells migrates from the bone marrow to the thymus gland, a small gland just under the upper end of the breastbone. These cells begin producing T-cells, and they continue to do so for the rest of the body's life. No one knows for sure where B-cells come from. The "B" indicates only that they do not come from the thymus gland. They do, however, come from lymphatic tissue. Recently, it has been suggested that they might come from bone marrow. T-cells and B-cells play different roles in the immune system.

Let's assume that a sore throat is developing. The antigen producing the infection could be a streptococcus bacterium□. Various white blood cells, known as phagocytes, can engulf antigens, much the way amoebas engulf food. This is probably the first cellular response in the immune reaction, and it is called nonspecific immunity. Many phagocytes may be on the front lines of battle, consuming the invading antigens. Eventually, some antigens may invade tissue fluids and get carried in the lymph to lymph nodes. The presence of these antigens stimulates the formation of T-cells and B-cells, which recognize and react to the specific antigens.

Each T-cell recognizes only a single specific antigen. Therefore, there is a different type of T-cell for each of the varieties of antigens to which the body has been exposed over the years. Each body contains thousands of different T-cells. The T-cells appear to live four or five years or longer. Immunologists have speculated that some T-cells may be twenty years old. When a T-cell is activated by the presence of an antigen, it divides many times and produces a whole population of similar cells, called a clone. The clone contains three kinds of T-cells: killers, helpers, and suppressors. Some members of the clone become memory cells. These cells live a long time and recognize the antigen at the next invasion (see Figure 7–11).

The killer T-cells directly attack and destroy antigens with potent chemicals. To do so, they leave the lymphatic tissue and move to the site of the antigen invasion. There they produce chemicals (such as interferon) that recruit more phagocytes and that make the phagocytes feed more aggressively on the antigens. They also produce chemicals that recruit other T-cells from the same clone to assist in the battle. Helper T-cells do most of their helping by stimulating B-cells to produce antibodies, which are also carried to the invasion site. Suppressor T-cells suppress the total immune reaction, keeping it from attacking the body's own cells.

The next line of support is the B-cells. Bodies have thousands of different B-cells because, just like the T-cells, they are antigen-specific. The membrane of each B-cell has been sensitized by previous encounters with the antigen. If a B-cell is not designated for a particular antigen, it dies quickly. However, new B-cells are manufactured all the time, so new antigens can always be identified. When B-cells are stimulated by the specific antigen, they multiply quickly and form a clone of plasma cells (see Figure 7–11). Each clone produces the same antibody at a rate of about two thousand antibody molecules per second. These

Figure 7–11
The proliferation of T-cells and B-cells in the presence of an antigen.

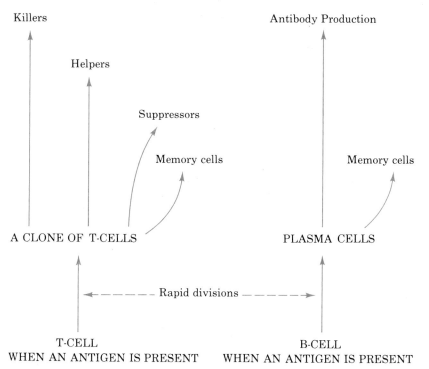

Killers

Helpers

Suppressors

Memory cells

A CLONE OF T-CELLS

Antibody Production

Memory cells

PLASMA CELLS

— — — — Rapid divisions — — — —

T-CELL
WHEN AN ANTIGEN IS PRESENT

B-CELL
WHEN AN ANTIGEN IS PRESENT

antibodies quickly fill the blood and join the T-cells at the front lines. Some of the B-cells do not produce antibodies. These cells remain as memory cells, and they function in the same way as memory T-cells.

Antibodies are proteins. An antigen is recognized by the chemical nature of its surface, which may bear specific sugars and proteins. This is the information that sensitizes and activates the original B-cell. A large bacterium may have many antigen sites. Thus the immune system may have many different antibodies attacking the same antigen. By a series of chemical reactions, the antibody produces several complement enzymes, which together eat into the antigen and thereby destroy it.

Immunity

Some of the B-cells producing specific antibodies are inherited. However, most of them were created in response to antigens to which the body was exposed during childhood diseases. This response is called **acquired** or **active immunity.** A person can have measles and chickenpox only once, because the body keeps their memory B-cells in circulation. Other diseases produce the same antibody responses, but the body forgets them over time. People can get several diseases more than once. Among such diseases are the flu, diphtheria, gonorrhea, and syphilis.

Some of the body's immunity is the result of immunization, a form of active immunity that comes from being injected with the weakened or killed disease organism. Smallpox and polio vaccines produce this sort of immunity. The immune system recognizes the injected substance as an antigen. Therefore, B-cells become sensitized to it and begin to produce antibodies. Often, a series of two or three injections, a week or more apart is necessary to build up complete immunity.

This situation demonstrates some important aspects of antibody production. The first time the body is exposed to an antigen, it produces antibodies. This is called the primary response. The body may develop some symptoms—fever or chills, for example. The next time the body is exposed to the antigen, it produces many more antibodies, and the person may never even know about the exposure to the disease. This is called the secondary response. Booster shots are often given to stimulate the secondary response acquired from active immunity. Probably they stimulate the memory of the B-cells.

Someone who comes in contact with a rare antigen (such as rabies or a food poison) whose consequences can include death is quickly inoculated with antibodies produced in the plasma of horses and cows. Although the person would produce antibodies, they might not be formed quickly enough. The transferred antibodies may be necessary for recovery. This type of immunity is called **passive immunity.**

The immune system is finely tuned to recognize and destroy all antigens. Thus it recognizes and tries to destroy transplanted tissues and organs from "foreign" donors. The T-cells quickly invade the foreign tissues. If the transplants are to be successful, the patients must be treated with drugs that suppress the immune reactions. T-cells also treat cancer cells as antigens. Therefore, those suffering from cancer may do so because of a failure of their immune system. Diseases are eradicated by the antibody production of B-cells. In rare cases, children are born without active T-cells or B-cells. They must live for years in their own antigen-free world, usually a special room or plastic bubble. Slowly, they are exposed first to one antigen, then to another, until their immune systems develop. Only then can they enter the real world. Without immunity to antigens, our lives would be extremely short.

SUMMARY

1. The digestive system is a tubular structure extending from mouth to anus. The major sections are the stomach and the small and large intestines.

2. Digestion of starch begins in the mouth by salivary amylase. Digestion of proteins begins in the stomach by the action of pepsin. Digestion is completed in the small intestine by a complex mixture of enzymes secreted by the pancreas and the intestine itself.

3. Digested nutrients are absorbed through the lining of the small intestine. The large intestine absorbs water and concentrates indigestible material into feces.

4. Good nutrition involves choosing among foods so that what is eaten is a well-balanced, nutritionally correct diet. The important groups of foods are carbohydrates, proteins, fats, vitamins, minerals, and fluids (water being the most important fluid).

5. Carbohydrates are the major source of energy for cells. Some proteins are the building blocks of cells and tissues. Others function as enzymes. Fats have many functions. They provide energy reserves, insulation, and shock absorption. They are also essential for healthy skin and hair and important to the structure of cells. Vitamins act as coenzymes for a multitude of metabolic pathways. Minerals become part of a large number of important organic molecules.

6. The respiratory system includes the nose, pharynx, larynx, trachea, bronchi, and lungs.

7. Respiration includes: breathing (external respiration) and internal respiration or cellular respiration.

8. Breathing, which consists of inhaling and exhaling, is controlled by a breathing center in the brain. The exchange of gases occurs in the alveoli of the lungs. Cellular respiration is the production of energy by cells.

9. The circulatory system consists of the heart and the following blood vessels: arteries, arterioles, capillaries, venules, and veins.

10. The heart consists of four chambers: two atria and two ventricles. The right atrium and right ventricle communicate via the tricuspid valve. The left atrium and left ventricle communicate via the bicuspid valve.

11. The heart has its own natural pacemaker. It consists of the sinoatrial (SA) node and the atrioventricular (AV) node. The electrical rhythm of the heart can be recorded. The recording is called an electrocardiogram (ECG).

12. The contraction of the heart muscles is called systole. The relaxation of the heart muscles is called diastole. The blood pressure is at its highest during systole and at its lowest during diastole.

13. The circulatory system consists of two major pathways, the systemic and the pulmonary. The coronary and hepatic portal pathways are divisions of the systemic circulation.

14. In the systemic circulation, blood passes from the heart through the large aorta. The first few branches from the aorta carry blood to the arms, neck, and head. Then the aorta bends downward to supply blood to the digestive system, other internal abdominal organs, and the trunk and legs.

15. The pulmonary circulation includes the passage of blood from the right ventricle through the pulmonary arteries to the lungs. There the blood is oxygenated and then returned to the left atrium through the pulmonary veins.

16. The coronary circulation includes two major coronary arteries that supply the cardiac muscles of the heart and then return the blood through the coronary sinus to the right atrium.

17. The hepatic portal circulation is a special circulation of oxygen-poor but nutrient-rich blood from the digestive system to the liver, where all nutrients and some wastes are filtered out by liver cells.

18. Blood is a complex liquid that contains red blood cells, white blood cells, platelets, dissolved gases, and a wide variety of plasma proteins.

19. The blood-clotting mechanism is a complex series of chemical reactions that ultimately form fibrous threads, called fibrin. The fibrin threads attach to tissues and platelets to seal wounds.

20. The lymph system is a circulatory system that carries lymph in lymph vessels from the tissues back to the systemic circulation. Lymph is tissue fluid.

21. Lymph nodes are glands that filter lymph during its return and detect and destroy any foreign matter.

22. The immune system consists of two types of lymphocytes, T-cells and B-cells. These cells attack foreign agents (antigens) that enter the body.

23. Antibodies are proteins manufactured by B-cells. They recognize and destroy specific antigens.

WILL A BODY SCANNER REPLACE YOUR DOCTOR?

Sometime in the next decade, when you go for a physical checkup, you may not see a physician with a stethoscope. If you see your physician at all, he or she will not even have a stethoscope. Your blood pressure won't be taken, and you won't have to leave a urine sample in a plastic cup.

Instead, you will be ushered into a small, cheerful room, where you will undress and put on a hospital gown. A technician will place you on a table and slowly move you into a large machine called a body scanner. Your physical exam will be short, painless, even pleasant. The only discomfort might be a claustrophobic feeling as you are moved through the scanner.

Later, your physician may call to tell you that if you don't change your diet, you will most likely have coronary artery disease in about fifteen years. Furthermore, your physician may be disturbed about some of the molecular changes in specific chemical reactions in a few epithelial cells in your lungs and may urge you to quit, or at least cut back on, your smoking. However, the physician may cheerfully add that there are no other precancerous cells in your entire body.

Is all this possible? In the next decade it will be, because of new breeds of body scanners. You have undoubtedly heard of body scanners. They made their first appearance in the 1970s. The early body scanners, called CAT (Computer Axial Tomography) scanners, use X-rays. While you slowly pass under the investigative eye of a CAT scanner, it makes thin-slice X-rays of your brain, your liver, or even your whole body. The images are fed to a computer, which reconstructs the organ, or your body, on a TV screen. CAT scanners are an amazing advancement over conventional X-ray diagnosis. The exposure to radiation is no more than that from a chest X-ray. The scanners can locate tumors, cancers, abnormal blood vessels, and so on. However, they provide no information on whether your organs are functioning properly. Your live body and the body of a corpse would produce the same kinds of images.

Newer on the body scanner scene are instruments that can both see inside organs and report in detail on the physical and chemical processes there. They can even watch complex chemical reactions as they occur. One of these scanners, the PET (Positron Emission Tomography) scanner, is showing physicians and researchers something they have never seen before—metabolism in action.

If you are examined by a PET scanner, you will first be injected with a small amount of a radioactive substance, for example, radioactive glucose. (Remember that all body cells use glucose for energy production and that the cells of the brain will accept no substitute for it.) As the radioactivity decays, positrons (a type of subatomic particle) are given off. The positrons are matter. When they meet antimatter, a painless microscopic nuclear explosion occurs. It emits radiation, which is recorded by the computer. The computer recreates the organ on the basis of its metabolism—in vivid color and in three dimensions. The colors of an organ indicate its different levels of activity. A PET scan of a heart will show normal, suffocating, damaged, and scarred muscles, all in different colors.

Because PET scanners are relatively new and expensive, they are used at only a limited number of facilities. Some hazards are associated with both CAT and PET scans. Excessive radiation, for example, may cause cancer or other problems.

NMR (Nuclear Magnetic Resonance) scanners are the scanners of the next decade. They do not use any type of dangerous radiation. Instead they work through magnets and radio waves. NMR scanners are capable of monitoring the activity of elements that have an odd number of protons and neutrons. This capability makes them electrically active and creates a tiny electromagnetic field at right angles to the field in the element. This causes the nuclei of the element to flip ninety degrees, producing a magnetic signal that can be picked up by a computer. The visual image produced by NMR scanners is the same as for other scanners. Only hydrogen is being watched at present, because it is the most common element in human bodies. However, new breeds of NMR scanners will detect phosphorous and other elements too.

NMR scanners have shown researchers astonishing things. They have shown, for example, that each time the heart beats, the brain throbs. This had been suspected but unseen before the use of the NMR scanner. With this scanner, the progression and cure of a disease can actually be watched as though it were a movie. If there is a new drug for arthritis or cancer, scientists can watch it at work.

Comparing the NMR scanners to the older CAT scanners is much like comparing an electron microscope to a light microscope. Improvements for the new body scanners are already being discussed. In the next decade, they will tell us more about our body functions than we ever learned before.

Body scanners will not replace physicians. Physicians will be in-volved in the interpretation of the body scans and will still decide about the treatment of the disorders discovered by the scans. In fact, physicians will be even more important than before because they will know so much more about each patient's body chemistry.

FOR DISCUSSION AND REVIEW

1. List the enzymes involved in the digestion of a piece of bread (assuming that the bread is made entirely of starch). Mention where each enzyme is produced.

2. Obtain a calorie counter. Over a period of three days, keep a diary of the foods you eat, their caloric content, and their relative amounts of carbohydrates, proteins, and fats. Then compute your average caloric intake per day and determine what percentages of your diet are carbohydrate, protein, and fat. If the percentages differ from the suggested percentages of 15, 30, and 55, respectively, explain how you can change your diet until you are eating properly.

3. Assume that you are going on a crash diet and that you want to consume only a thousand kilocalories per day. Compute the amounts of carbohydrates, proteins, and fats that you should eat for a nutritionally balanced diet at that level of intake.

4. Find any new diet craze. (Look in magazines, newspapers, and books). Read the details carefully, and determine if the diet is nutritionally balanced. Explain.

5. What are essential fatty acids and essential amino acids?

6. Describe the role of cholesterol, LDL, and HDL in the formation of plaque on the inside of artery walls.

7. List the amounts of vitamins and minerals you consume in one day. (Information is available on the labels of nearly every packaged product. The labels also give the recommended daily allowances.) Do the amounts add up to the RDA?

8. Describe all the passages through which your next breath of air will pass.

9. You can voluntarily breathe faster or slower. When you do so, what speeds and slows the breathing center in the brain? (Hint: Think of feedback from other parts of the brain.)

10. Describe the flow of blood through the heart. Include all the chambers and valves and the pulmonary circulation.

11. Describe the flow of blood from the heart to the digestive system (including specific vessels) and back to the heart. Include the hepatic portal system.

12. Explain the relationships between the arteries and the heart during normal and abnormal coronary circulation.

13. Describe the electrical events of a heart during a single heart cycle. Illustrate these events with a drawing of an electrocardiogram.

14. List the cell types found in blood, and explain their functions.

15. Pretend you have cut your finger. You cover the wound with a bandage. A few hours later, you remove the bandage and see that a clot has formed. Describe the events that occurred from the time you cut yourself until the clot formed.

16. Trace the passage of lymph from one of the inguinal (in the groin) lymph nodes back to the right atrium of the heart.

17. If you were exposed to antigens causing the common cold, what would all the activities of your immune system be?

SUGGESTED READING

BEHOLD MAN, by Lennart Nilsson. Little, Brown: 1978.

A remarkable color atlas of human anatomy and physiology, with many photographs of real organs and details of how they work.

A CHILD'S BODY: A PARENT'S MANUAL, by The Diagram Group. Bantam Books: 1977.

An illuminating look at what you should know about the anatomy and physiology of a growing child.

DISCOVERING THE HUMAN BODY, by Bernard Knight. Lippincott and Crowell: 1980.

An intriguing description of how the organs and organ systems were discovered and by whom. Contains anatomical detail not included in this chapter.

THE LIVING HEART, by Michael DeBakey and Antonio Gott. Charter Books: 1977.

Two renowned heart specialists describe how the cardiovascular system works, why it fails, and what you can do about it.

MAN'S BODY, by The Diagram Group. Bantam Books: 1976.

Male anatomy and physiology, from acne to cardiovascular problems.

WOMAN'S BODY, by The Diagram Group. Bantam Books: 1977.

Everything you wanted to know about female human anatomy and physiology.

8

HUMAN ORGAN SYSTEMS: EXCRETORY, NERVOUS, AND ENDOCRINE

THE CONCEPT OF HOMEOSTASIS

Homeostasis literally means "maintaining the same state." Homeostasis of an organism refers to the organism's **dynamic steady state**. The internal workings of the organism are maintained within the limits of that state (think of a set value with limits above and below that value). Thus they are dynamic.

Much of our everyday activity involves trying to keep our physical environment in homeostasis. The thermostat that controls the furnace is a good illustration of homeostasis. We set the thermostat for some comfortable temperature, say 68 degrees. When the temperature falls below that level, the burner unit in the furnace is ordered into operation by the thermostat. A second homeostatic mechanism determines when the blower, or fan, on the furnace is to operate. The fan does not come on when the burner unit is first ignited. If it did, cold air would circulate through the house. The fan comes on only after the fire chamber reaches a particular temperature. When the fan is activated, warm air circulates through the house. The house warms up to the preset level of 68 degrees. Then the burner unit is turned off. The fan continues to run for several minutes because the burner chamber is still hot and the circulated air can still be warmed. When the burner chamber temperature drops to a preset level, the fan is turned off.

PREVIEW QUESTIONS

After reading this chapter, you should be able to answer the following questions:

1. What is homeostasis? List some of the ways it is maintained.

2. How is urine formed, and what is it composed of?

3. What are the relationships of sensory and motor nerve cells in the nervous system?

4. What happens in your nervous system when you accidentally touch something very hot?

5. What are hormones, and how do they work?

What has happened in a single cycle of the furnace is that the home's temperature has been maintained in a dynamic steady state. Although the temperature was set for a precise 68 degrees, it probably fell to 67 degrees and rose to as much as 69 or 70 degrees in a single cycle. Some newer, electronic thermostats also control the air conditioning if the temperature in the home goes much above the preset level.

The internal state of living organisms is maintained by homeostatic mechanisms. In fact, every organism, including humans, has a large number of regulating, or homeostatic, mechanisms to keep the internal state in a dynamic steady state. Some of the regulated entities are the fluids in cells, the fluids that surround cells, blood sugar level, blood pressure, and the gases carried by blood cells.

The normal human body temperature of 37 degrees Celcius (98.6 degrees Fahrenheit) must be maintained because it is the temperature at which the biochemical activities of the cells occur at their optimal rates. However, the body's internal environment differs significantly from the external environment. On a cold winter day, the air temperature is many degrees lower than the body temperature. In midsummer, the air temperature may be several degrees higher than the body temperature.

Body heat is produced by cell activities. If this heat were not given off, the body would overheat dangerously. Most body heat is lost through skin. Blood carries the heat from the active cells through the small blood vessels in the skin, where the heat can easily dissipate. In addition, the sweat glands secrete perspiration, which carries away the heat as it evaporates. On a cold day, however, your body must conserve heat. Less blood circulates through the blood vessels in your skin, and if too much heat is lost, you begin to shiver. The increased activity of your muscle cells increases heat production.

Obviously, many different physiological activities are involved in maintaining a stable body temperature. All are coordinated and controlled by the central nervous system. Box 8–1 describes the problems that occur when organisms cannot regulate their body temperatures.

Under normal conditions, homeostasis does not prevent all changes within an organism. Nutrients entering the organism are processed to produce energy and to supply molecules for growth and cell maintenance. Waste products are produced and eliminated. Whole cells are destroyed and replaced. The organ-

BOX 8–1

ENDOTHERMS AND ECTOTHERMS

Endotherms are animals that maintain a constant body temperature. (They are often known as warm-blooded animals.) Only birds and mammals fit this category. All the other animals—reptiles, amphibians, fish, arthropods, worms, mollusks, and so on—are called ectotherms, or cold-blooded animals. Their body temperatures rise and fall with the temperature of the environment.

Even though ectotherms cannot maintain constant body temperatures, the members of each species function most efficiently within certain body temperature ranges. As environmental temperatures, and therefore body temperatures, fall below this range, the physiological processes of the animals slow and the animals become sluggish. At higher temperatures, the animals must seek shelter from the heat or they will suffer from overheating.

Marine animals, such as saltwater fish, have little difficulty with changing body temperatures. They are adapted to the water temperatures in which they live, and the temperatures fluctuate only a few degrees over a year's time. However, animals living in small bodies of fresh water and animals living on land have had to evolve special adaptations to deal with environmental temperatures that may fluctuate widely in a single day.

For example, the body temperatures of desert reptiles, such as lizards, drop at night. Before the reptiles can become active in the morning, their body temperatures must be elevated by heat from the sun. If the daytime environmental temperature becomes too high, the reptiles must seek shelter or assume postures that reduce the amount of body surface exposed to the sun. In temperate climates, many cold-blooded land and fresh water animals, such as frogs, are active only during the warmer parts of the year. In winter they seek shelter and are inactive. Birds and mammals are highly successful at colonizing many different environments. This is due in part to their ability to maintain a constant body temperature and therefore to be active over wide ranges of environmental temperatures.

ism, then, is in a dynamic steady state, constantly expending energy to maintain internal order while reacting to environmental changes.

In many cases, the homeostatic mechanisms fail

and the organism is placed in a state of stress. Illness is a case in point. All illness and disease symptoms are indications of the failure of dynamic steady state. These symptoms include chills, fever, nausea, vomiting, and pain. The medications prescribed by physicians are an attempt to reduce stress and return the internal condition to its homeostatic state.

Not all medications are cures, of course. Antihistamines taken for a cold, for example, simply mask the symptoms. The masking makes the body more comfortable for a short period of time. Although every organ system is active in maintaining homeostasis, the three systems discussed in this chapter—the excretory system, the nervous system, and the endocrine system—are the most heavily involved.

THE EXCRETORY SYSTEM

The circulatory system delivers oxygen and nutrients to body cells. It also removes toxic waste produced by the cells, transporting it to body sites where it can be eliminated. For example, carbon dioxide, a waste product of cellular respiration□, is transported to the lungs and exhaled.

Urea is one of the waste products transported in the blood. As cells function, amino acids are converted to other substances. During these conversions, the amino groups of the amino acids□ are often removed, forming the highly toxic substance ammonia. The ammonia is then converted to a much less toxic compound, urea. (Most people in our culture consume far more protein than their bodies need. However, the excess amino acids derived from this protein are not stored in the body. The liver converts them to substances that can be stored or used by the cells.) Although urea is much less poisonous to cells than ammonia is, it too would interfere with cell function if too much accumulated in the blood. Therefore, the kidneys, the primary organs of the human excretory system (see Figure 8–1), filter urea from the blood and concentrate it into urine.

Figure 8–1
The excretory system.

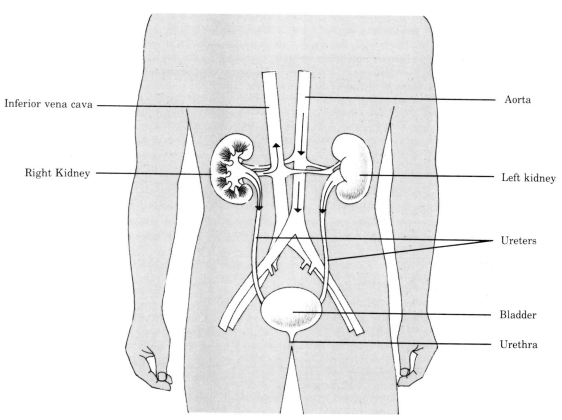

Inferior vena cava

Right Kidney

Aorta

Left kidney

Ureters

Bladder

Urethra

The elimination of urea is only part of the role kidneys play in maintaining homeostasis. During urine formation, the kidneys also regulate the levels of water and of a variety of other essential body chemicals in the blood. Thus the kidneys have three important functions: They eliminate urea and other wastes. They play an essential role in maintaining the proper level of body water. They help maintain critical levels of many constituents of blood and body fluids.

The two kidneys are small, bean-shaped organs located on the inner surface of the body wall near the small of the back. In adults, each kidney weighs between 125 and 150 grams and is about 12 centimeters long. Despite their small size, between 20 and 25 percent of the body's blood passes through the kilometers of kidney capillaries during any given period. Each kidney receives blood through a **renal artery** that branches directly from the aorta. Blood leaves each kidney through a **renal vein** that drains directly into one of the body's major veins, the **inferior vena cava.**

The urine produced by the kidneys flows through a duct called the **ureter** to the **bladder,** where it is stored. The bladder is an elastic muscular sac. When the muscles are contracted, urine is forced out of the body through the **urethra.** The urethra in women is only about three to four centimeters long. It opens to the outside just in front of the vagina. Because the urethra is so short, it is possible for bacteria to migrate up the urethra into the bladder. This is why bladder infections are common among women. The urethra in men is at least 15 centimeters long. It opens to the outside at the tip of the penis. When a man has an erection, the urethra is stretched several centimeters longer.

If a kidney were cut in half longitudinally, three internal areas would be exposed. The outermost area is a thin, dense tissue called the cortex. The innermost area is a much larger, less compact tissue called the medulla. The medulla connects to a cavity called the renal pelvis, where urine is collected before it drains into the ureter (see Figure 8–1).

Each kidney contains approximately 2 million functional units called **nephrons** (see Figure 8–2). Each nephron is divisible into several separate structures. One of these structures is the **Bowman's cap-**

Figure 8–2

A nephron and associated blood vessels. Blood pressure forces glomerular filtrate from the blood in the capillaries of the glomerulus through the wall of the Bowman's capsule and into the lumen of the nephron. As the filtrate passes along the nephron, much of the water and such substances as glucose, amino acids, and vitamins are reabsorbed into the blood passing through the capillary net enclosing the nephron. The remaining filtrate, urine, is a concentrated solution of salt, urea, and other excretory substances. It passes from the nephron into a collecting duct. Each nephron not only removes waste products from the blood but also helps regulate the body's water balance.

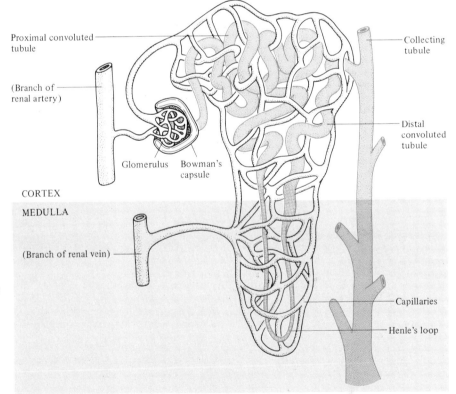

sule, a thin-walled, cup-shaped sac. The external layer of cells in this cup is dense. The inner layer of cells is specialized for transporting materials into the tubule that connects to the Bowman's capsule. Within Bowman's capsule, the circulatory and excretory systems come in close contact. The renal artery to each kidney branches and rebranches into 2 million tiny arterioles, one leading to each nephron. Each of the arterioles forms a capillary bed called a **glomerulus.** The glomerulus is almost completely surrounded by the Bowman's capsule. Together, these two structures are often called a **renal corpuscle.** An arteriole leaves the glomerulus and forms a tangled capillary network that encloses the tubules of the nephron. From here, blood flows through a system of veins, finally flowing out of the kidney in the renal vein.

A **proximal convoluted tubule** (*proximal* meaning "near") leads from Bowman's capsule. Eventually it straightens and forms an elongated, U-shaped tubule called **Henle's loop.** After this the tubule coils and becomes convoluted again. At this point, it is known as a **distal convoluted tubule.** The tubule then joins a **collecting tubule,** which eventually drains into the renal pelvis of the kidney.

Part of each nephron is in the cortex of the kidney, and part is in the medulla (see Figure 8–2). Henle's

CHAPTER 8   *Human Organ Systems: Excretory, Nervous, and Endocrine* loops and the collecting tubules are in the medulla. The other structures just mentioned are in the cortex.

Each nephron performs three important functions: pressure filtration, selective reabsorption, and tubular excretion. As blood flows into the capillary knot (the glomerulus), blood pressure forces about 20 percent of the fluid portion of the blood through the walls of the capillaries and into Bowman's capsule and the nephron tubule. This fluid, called **glomerular filtrate,** contains urea and many other chemicals, such as amino acids, vitamins, glucose, and salt. The blood cells and the molecules that are as large as proteins do not filter into Bowman's capsule. As glomerular filtrate passes along the proximal convoluted tubule toward the collecting duct, the cells in the wall of the nephron reabsorb water and essentially all of the amino acids, vitamins, and glucose.

This task is accomplished through active transport▫, which requires a considerable expenditure of energy. About 85 percent of the salt and water are reabsorbed. However, the water passes back into the blood by osmosis▫, not active transport. When the chemicals are actively transported through the wall of the proximal convoluted tubule, an osmotic gradient is created. Water diffuses from the nephron, where it is more abundant, into the body fluids and blood, where it is less abundant. More water is reabsorbed as the glomerular filtrate passes through the first half of Henle's loop. In the second half of the loop, salt is reabsorbed by means of an active sodium pump. This increases the concentration of salt in the fluids surrounding the collecting ducts, facilitating the reabsorption of more water by osmosis.

The final process, tubular excretion, occurs in the distal convoluted tubule. Large and small molecules are excreted from the nephron's blood supply into the fluid of the tubule. Large molecules that could not pass through Bowman's capsule are most of what is excreted. Penicillin and other drugs that interrupt homeostasis are some of the large molecules that are rapidly excreted into the distal convoluted tubule. Excesses of creatine (produced by the liver and important in the synthesis of ATP) are also expelled by tubular excretion. Many ions involved in maintaining blood pH, such as hydrogen and ammonia, are excreted into the distal convoluted tubule. Their addition to the urine makes the urine more acidic.

The cells forming the walls of a nephron play an active role in reabsorbing many chemicals. Other chemicals, such as urea, are reabsorbed to only a small degree. Thus they become concentrated in the urine as more water is returned to the body. Although the kidneys filter as many as 180 to 200 liters of fluid every day, only 1 to 2 liters of urine are formed. The amount of urine is controlled by a hormone▫ called the **antidiuretic hormone.**

How does this hormone work? If you drink a large amount of water, you produce a large volume of dilute urine. However, if you go without drinking for many hours, your kidneys produce small amounts of more concentrated urine. As your body dehydrates,

---

CONCEPT SUMMARY

EVENTS DURING URINE FORMATION

EVENT	RESULT	MECHANISM
Glomerular filtration	Water and substances of small molecular size, such as urea, glucose, and amino acids, are forced from the blood in the glomeruli through the walls of Bowman's capsules into the nephrons	Blood pressure
Reabsorption	Water and many other substances are returned to the blood, thereby concentrating the urine and regulating the water content of the body (occurs through the convoluted tubules and Henle's loop)	Sodium pumps, osmosis, and active transport
Tubular excretion	Molecules larger than those able to enter the nephron through Bowman's capsule are taken from the blood and passed into the urine through the distal convoluted tubules	Active transport

the salt content in your blood rises. (As blood loses water, dissolved substances in it become more concentrated.) Special sensors in your brain notice this change in salt concentration and stimulate the pituitary gland to secrete antidiuretic hormone. Carried to the kidneys in the blood, the hormone stimulates cells in the distal convoluted tubules and the collecting ducts to reabsorb more water, thereby conserving water and producing more concentrated urine. The reverse happens when there is a surplus of water. Less antidiuretic hormone is secreted, and more water is excreted with the urine. This finely tuned mechanism for achieving water balance is another example of homeostasis.

Urine is acid (pH around 5.0 to 6.0). Normally some shade of yellow, the color of urine is the result of a by-product of hemoglobin breakdown when red blood cells are destroyed. The smell of urine is regulated by many factors. Urea itself has a slight odor. When urine is not disposed of, the urea in it begins a chemical conversion to ammonia salts, which have a strong pungent odor. This is the odor found in many public restrooms. Certain ingested foods add their own odor to urine. Asparagus for example, adds methyl mercaptan, which gives the urine a distinctive odor.

When urine is tested, a great deal is learned about the body's general homeostasis. The composition of urine is a good indicator of the health of the kidneys and liver. It also provides clues to the general health of many other organs. Laboratories that test urine determine its specific gravity (concentration of solids), albumin/globulin (protein) content, glucose content (the presence of glucose may be caused by diabetes), and acidity. They also examine it under a microscope for bacteria, parasites, crystals, pus, and so on and test it for color, transparency, and odor. Box 8–2 describes the excretory systems of some nonhuman organisms.

The kidneys are not the only organs involved in excretion. The lungs and skin also play a role. Lungs excrete carbon dioxide and other gases. (The "alcohol breath" of overindulgence is produced not from alcohol in the mouth but from alcohol fumes expelled from the lungs when blood alcohol levels are high.) The sweat glands of the skin secrete small quantities of dilute urine. Bathing removes the fluid and the dissolved salts that are excreted. The skin is also the organ from which heat is lost from the body.

As animal cells carry on their basic biological functions, they create waste products that would harm cells if they were not eliminated. The process of eliminating these waste products is called excretion, and it is carried on by all living organisms. Excretion is often closely coupled with the processes of homeostasis and osmoregulation which regulates the amount of water and ionic concentrations in body tissues. This makes it impossible to discuss one of these processes without mentioning the others.

In the case of small aquatic animals and animals that have no cell or tissue far removed from circulating water (Cnidarians—jellyfish and their relatives—for example), excretion is accomplished by simple diffusion. Waste products such as carbon dioxide and ammonia diffuse from the tissues into the surrounding water, where they are diluted and carried away. If the organisms live in seawater, osmoregulation causes little or no problem, because the organism's body fluids are nearly isotonic (in the same concentration) with seawater (see Chapter 4). However, if the organism lives in fresh water, there are osmotic stresses, because the water is constantly moving from the environment into the body fluids of the organism, from which it has to be eliminated. In freshwater protozoans and even in some of the multicellular animals such as sponges, elimination is accomplished by pumping the water into contractile vacuoles in cells, which in turn eject the water outside the animals.

The flatworms (phylum Platyhelminthes) were probably the first animals to evolve structures, the protonephridia, to serve an excretory function. However, the protonephridia may have evolved first as an osmoregulatory device, later taking on an excretory role. Because flatworms are flat, no tissue is very far from either the aquatic environment or the water-filled digestive tract. Therefore, most flatworm excretion takes place by diffusion through the general body surface. The protonephridial system is a system of branching ducts that lead from the tissues to the outside of the organism through a nephridiopore. Internally, each duct is capped by an elaborate complex of two

Continued Next Page

EXCRETION AND OSMOREGULATION FOR ORGANISMS OTHER THAN HUMANS

connected cells that enclose a terminal cavity into which one or more flagella extend (see Figure A). Presumably, water and waste are actively drawn into the cavity and then pumped out through the duct system by the beating flagella. Ions may also be absorbed back out of these tubes.

Another type of nephridium, the metanephridium, evolved in mollusks and annelids. In its simplest form, the metanephridium is a funnel-shaped structure. The large mouth of the funnel, the nephrostome, opens into the animal's body cavity. The smaller end of the funnel, the nephridiopore, opens to the exterior (see Figure B). Wastes are collected from the fluid in the body cavity and transferred to the outside of the animal. Reabsorption of beneficial chemicals and water, if necessary, may take place through the wall of the metanephridium. Often, nephridia become elongated and highly convoluted, which gives them the appearance of much more complex structures. Part of

the duct may be expanded to form a bladder for waste storage. This is the case in earthworms and certain other annelids. Larger animals, such as modern mollusks, have become too large to rely on a pair of simple metanephridia. They have more elaborate, kidney-like structures that are efficient in waste removal, osmoregulation, and homeostasis.

Insects have evolved a unique solution to the problem of excretion. Attached to the anterior end of the rear portion of the digestive tract are a number of hollow tubules, called Malpighian tubules, that open into the intestine. These tubules extend from the digestive tract into the insect's body cavity, where they are bathed in blood. From the blood, they concentrate waste in the form of uric acid crystals. They conserve water and other substances by reabsorption, which contributes to the chemical homeostasis of the body fluids. The uric acid passes into the intestine and is eliminated with the feces. Carbon dioxide, the waste product of respiration, is eliminated through the tracheal respiratory system.

The chemical form of nitrogenous waste products eliminated by an animal is correlated with environment and with the animal's need to conserve water. With the exception of fish, most aquatic animals produce ammonia as a waste product. Although ammonia is highly toxic, it is easily and safely disposed of in an aquatic environment. Terrestrial animals with an easy

Figure A

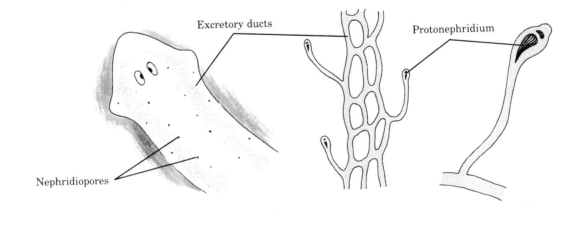

Excretory ducts

Protonephridium

Nephridiopores

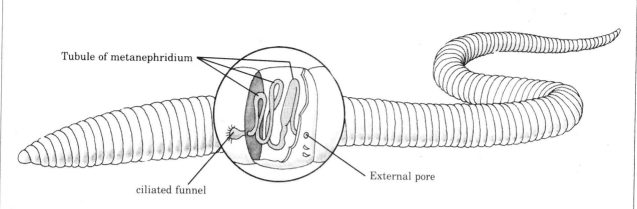

Tubule of metanephridium

ciliated funnel

External pore

Figure B

access to water usually convert ammonia to urea, a less toxic substance, which can be dissolved in urine. The elimination of urine causes a loss of water which must be replaced. However, animals that have little access to water and those that suffer high rates of water loss

conserve water as much as possible. These animals transform ammonia into uric acid, an even less toxic substance, which requires little, if any, water for disposal.

THE NERVOUS SYSTEM

Animals possess the ability to react to changes in their external environment. For example, an animal may flee if threatened by a predator. During flight, heartbeat and breathing increase, blood vessels to activated muscle systems dilate, and sensory information about the predator and its advance are rapidly processed. When flight ceases, the animal returns to rest and the changes are reversed. Two systems coordinate body functions: the nervous system and the endocrine system▯.

The nervous system receives information about the external environment from sensory organs such as the eyes, ears, and nose and from temperature and pressure receptors in the skin. **Internal receptors** provide information about pressure, touch, temperature, position, movement, and pain. (A cut finger signals external pain. An ulcer signals internal pain.) This information is coordinated and integrated at special sites in the nervous system, from which messages are sent to the appropriate structures—skeletal muscles, the

heart, and so on. The messages are sent through a network of nervous tissue extending throughout the body. The network, which is nearly as extensive as the circulatory system, is composed of the central nervous system and the peripheral nervous system. The **central nervous system** includes the brain and spinal cord. The **peripheral nervous system** includes all the nerves from sensory organs as well as those to muscles.

Central Nervous System

The coordination of all sensory information takes place in the central nervous system. The human **brain** (see Figure 8–3) is a complex structure and the center of body coordination. It is composed of three parts: the forebrain, the midbrain, and the hindbrain. The **forebrain** is the largest, and the greatest part of it is the **cerebrum**—the center for memory, thought, and speech. The cerebrum is divided into right and left hemispheres by a deep groove. The right hemisphere coordinates the left side of the body, and the

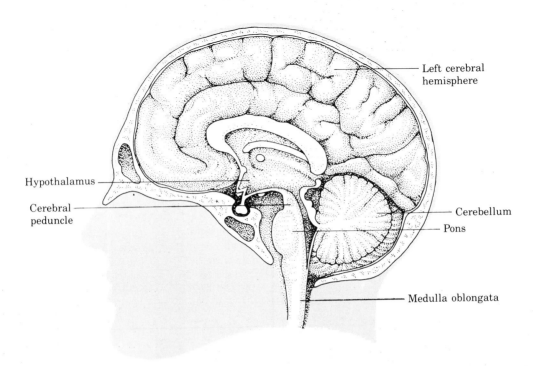

Left cerebral hemisphere

Hypothalamus

Cerebral peduncle

Cerebellum

Pons

Medulla oblongata

Figure 8–3

A midsagittal section through a human brain.

left hemisphere coordinates the right side. Different areas of the cerebrum are involved with different types of sensory stimuli. For example, certain areas are associated with sight, others with sound. (Figure 8–4 is a map of the left cerebral hemisphere. It illustrates the location of specific sensory areas.) Although the two cerebral hemispheres seem identical, they are different. The left side is associated with logical thought processes, such as those necessary for learning skills and ideas, and logic in general. The right side is associated with perceptual stimuli that are important in art, music, and other "intuitive" pursuits.

The **hypothalamus,** another part of the forebrain, coordinates body activities associated with homeostasis. It regulates body temperature and the activities of muscles and sweat glands. It is important in regulating the level of sugar in the blood, water balance, and activities of the kidneys. It also produces chemicals that affect the pituitary gland's role in regulating male and female sexuality.

The **midbrain** is the smallest of the three regions of the brain. It connects the forebrain and the hindbrain and functions mainly to transmit information between the two.

The **hindbrain** is composed of three subregions: the cerebellum, the medulla oblongata, and the pons.

The **cerebellum** is the second largest part of the brain. It is the coordinating center for muscular activities, including balance and posture. The **medulla oblongata** is the most posterior part of the brain. It connects the brain to the spinal cord. The medulla oblongata contains nerve bundles, called tracts, that transmit information from the spinal column to other parts of the brain. It also contains centers that regulate and coordinate specific body functions, such as rate of heartbeat and breathing. Finally, it regulates the diameter of blood vessels, thereby controlling the blood flow to specific parts of the body. Obviously, the medulla oblongata plays an important role in homeostasis.

The **pons** (meaning "a bridge") transmits information among the various regions of the hindbrain and the midbrain. The hindbrain and the midbrain form the **brain stem,** which controls and coordinates many of the unconscious functions of the body, including digestion, respiration, excretion, and circulation.

The **spinal cord** (see Figure 8–5) extends from the hindbrain to the pelvic region. It is protected by the vertebral, or spinal, column. The spinal cord is a soft white column of nerve tissue that contains peripheral nerve tracts. These tracts transmit sensory informa-

222

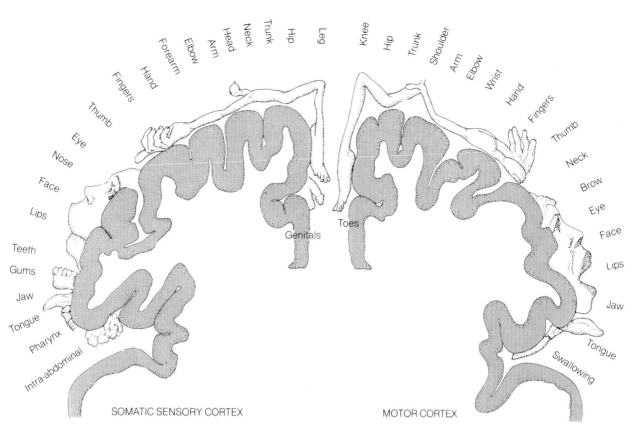

Labels on somatic sensory cortex (left side), top to bottom: Thumb, Fingers, Hand, Forearm, Elbow, Arm, Head, Neck, Trunk, Hip, Leg

Eye, Nose, Face, Lips, Teeth, Gums, Jaw, Tongue, Pharynx, Intra-abdominal

Genitals, Toes

Labels on motor cortex (right side): Knee, Hip, Trunk, Shoulder, Arm, Elbow, Wrist, Hand, Fingers, Thumb, Neck, Brow, Eye, Face, Lips, Jaw, Tongue, Swallowing

SOMATIC SENSORY CORTEX   MOTOR CORTEX

Figure 8–4

A posterior view of a section through the sensory area of the left side of the cerebrum, showing the location of specific sensory areas associated with various parts of the body. (Reprinted with permission from The Cerebral Cortex of Man. Copyright 1950 by Macmillan Publishing Co., Inc. Renewed 1978 by Theodore Rasmussen.)

tion from sensory receptors of the body up to the brain. The spinal cord also contains other, more central nerve tracts, which transmit information and messages from the brain to various effector organs of the body, such as muscles. They also transfer neural information from one region of the spinal cord to another.

The spinal cord is the center of reflex activities. If you touch a hot object, you pull your hand away quickly, before you are even conscious that you

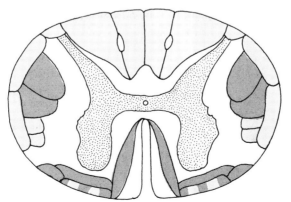

Figure 8–5

Cross section of the human spinal cord, showing the location of major spinal cord tracts that carry information to and from the brain. Ascending tracts (tan) carry sensory impulses to the brain. Descending tracts (grey) carry motor impulses down the spinal cord.

CHAPTER 8   *Human Organ Systems: Excretory, Nervous, and Endocrine*

touched something hot. The heat stimulus was processed in the spinal cord. The message to withdraw your hand was translated as a spinal reflex before the message was received in the pain centers of the brain. The brain and the spinal cord are hollow structures containing interconnecting fluid-filled chambers and channels. The fluid in these spaces is called cerebral spinal fluid.

Peripheral Nervous System

Twelve pairs of **cranial nerves** to and from the brain, thirty-one pairs of **spinal nerves** attached to the spinal cord, and all the branches of the spinal nerves (see Figure 8–6) comprise the **peripheral nervous system.** Some of the nerves, such as the optic and olfactory nerves, transmit sensory information from sense organs to the brain. Other nerves transmit motor instructions to muscle systems. The oculomotor nerve, for example, regulates muscles that control eye movements. Most of the peripheral nerves contain both sensory and motor pathways. They are called mixed nerves.

In many cases, people are able to consciously control their movements and responses. A person who decides to take a step can consciously cause the leg and thigh muscles to contract and so lift the leg. The system of consciously controlled peripheral nerves is called the **somatic system.** However, many activities in the body take place unconsciously. Among them are changes in the rate of heartbeat or in the rate of peristalsis.

The peripheral nerves involved in these unconscious activities comprise the **autonomic nervous system.** This system has two divisions: the sympathetic system and the parasympathetic system. In general, the **sympathetic system** readies the body for emergency situations. Nerves increase the rate of heartbeat, open blood vessels to skeletal muscles, open air passages, decrease digestive activities, and so on. In a sense the **parasympathetic system** is antagonistic to the sympathetic system, because it returns the body to a resting (nonemergency) state.

The nervous system is clearly a complex communication system. It receives information from both the external and the internal environments, and it transmits this information to coordination centers. From there, it transmits the information to peripheral nerve systems to bring about appropriate activities. This information is transmitted in the form of nerve impulses. The nervous systems of a variety of nonhuman organisms are described in Box 8–3.

Figure 8–6
Spinal nerves. Thirty-one pairs of spinal nerves attach to the spinal cord to form a portion of the peripheral nervous system.

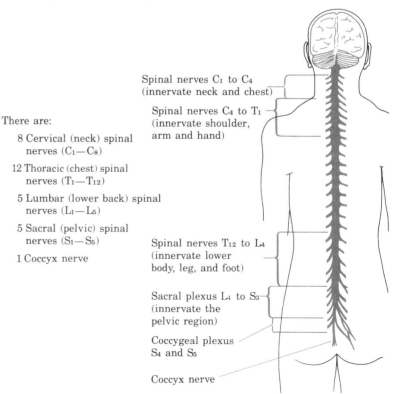

There are:

8 Cervical (neck) spinal nerves (C_1—C_8)

12 Thoracic (chest) spinal nerves (T_1—T_{12})

5 Lumbar (lower back) spinal nerves (L_1—L_5)

5 Sacral (pelvic) spinal nerves (S_1—S_5)

1 Coccyx nerve

Spinal nerves C_1 to C_4 (innervate neck and chest)

Spinal nerves C_4 to T_1 (innervate shoulder, arm and hand)

Spinal nerves T_{12} to L_4 (innervate lower body, leg, and foot)

Sacral plexus L_4 to S_3 (innervate the pelvic region)

Coccygeal plexus S_4 and S_5

Coccyx nerve

BOX 8–3

THE NERVOUS SYSTEMS OF ORGANISMS OTHER THAN HUMANS

Most animals have nervous systems that are simpler than and different from those of vertebrate animals. The nervous system enables animals to respond quickly and appropriately to changes in their environments. The simplest nervous systems are found in certain members of the phylum Cnidaria—namely the tiny freshwater polyps called hydras, which have a different net-like arrangement of nerve cells scattered among the bases of their epidermal cells (see Figure A). Each nerve cell synapses with several others. In many cases, impulses can pass across the synapse in either direction—in contrast to the one-way synapses of higher animals. Hydras have no main center, or brain, for nervous coordination. They are slow-moving and respond slowly to stimuli.

In more advanced animals, many neurons are concentrated into nerve cords. Some are concentrated at the anterior end of the animal in the region that is recognizable as a head. The concentration of neurons in nerve cords correlates with a more efficient transfer of nerve impulses. The concentration of sensory neurons at the head allows the organism to examine its environment as it moves forward. These simple nerve cord systems appear in the small aquatic flatworms. In the more primitive flatworms, three or four pairs of nerve cords extend posteriorly from the brain along the length of the body. These cords are joined at points along their lengths by interconnecting nerves (see Figure B). In the more advanced flatworms, all but two of these nerve cords (those in the lower, or ventral, body wall) have disappeared, leaving the brain and the two nerve cords as the largest structures in the nervous system.

In the segmented annelid worms, the brain is an important primary center of nervous coordination, and the nerve cords have ganglia at regular intervals along their lengths. The prominent brain is located in the upper, or dorsal, part of the head. The nerve cords are located in the lower, or ventral part of the body. The two parts of the nervous system are connected by nerves that encircle the digestive tract (see Figure C). This basic form of nervous system is also found in the arthropods and in many related phyla (see Figure D).

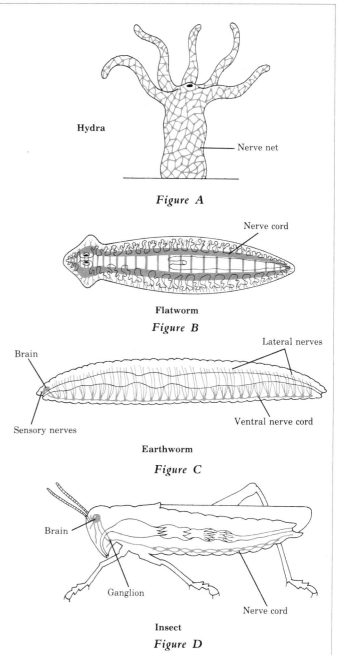

Hydra

Figure A

Flatworm

Figure B

Earthworm

Figure C

Insect

Figure D

The final change in the form of this type of nervous system is the fusion of the two ventral nerve cords into a single cord. Such systems are far different, structurally and organizationally, from the vertebrate system, which is primarily a dorsal hollow nervous system.

DRUGS AND THE NERVOUS SYSTEM

Hardly a day goes by that we do not hear or read about drug trafficking and drug addiction. Improper use of most drugs can cause serious health problems. Even the drugs that are legal to use or abuse, such as alcohol and tobacco, may cause serious disease. Drugs can be categorized into three general groups: depressants, stimulants, and hallucinogens. The charts here describe some of the more common of the dangerous drugs and explain how they affect the human nervous system.

Depressants.

Depressants reduce anxiety, tension, and mental activity. They act as sedatives and in general depress the central nervous system.

NAME	SLANG NAME	USERS PER MONTH IN U.S.*	LONG-TERM SYMPTOMS
Barbiturates	Phennies, candy, barbs, peanuts	1,060,000	Addiction, with both mental and physical dependence; withdrawal symptoms severe
Narcotics: Codeine Heroin Methadone Morphine	Schoolboy smack, horse dolly M, white stuff	522,000	Physical and mental addiction; loss of appetite with accompanying weight loss; loss of sexual drive; withdrawal symptoms severe
Ethyl alcohol	Booze	92,300,000	Probably none if used in moderation; otherwise, ulcers, liver and brain damage, and addiction; withdrawal symtoms possibly severe

*Source: National Institute of Drug Abuse.

Stimulants.

Stimulants excite the central and sympathetic nervous systems, increase mental activity and endurance, and decrease fatigue, appetite, and sensitivity to pain.

NAME	SLANG NAME	USERS PER MONTH IN U.S.*	LONG-TERM SYMPTOMS
Amphetamines	Pep pills, bennies, dexies	1,780,000	Psychological dependence that develops with increased tolerance to the drug; hallucinations; withdrawal symptoms including disorientation, severe depression, hunger, fatigue, and long periods of sleep
Cocaine	Coke, snow, gold dust	1,640,000	Similar to amphetamines; long-term snorting damaging to nasal tissues
Nicotine (tobacco)	Coffin nails	64,570,000	May lead to heart, lung, and respiratory diseases, including cancer; probably both psychological and physical dependence; withdrawal symptoms including anxiety, sleep disturbances, and general nervousness

*Source: National Institute of Drug Abuse. **(Continued Next Page)**

Hallucinogens.
Hallucinogens cause mood changes and perceptual distortions. Most are reportedly nonaddictive.

NAME	SLANG NAME	USERS PER MONTH IN U.S.*	LONG-TERM SYMPTOMS
Lysergic acid diethylamide (LSD)	Sugar, acid		Panic and increasingly psychotic symptoms
Mescaline	Mesc	1,140,000	Unknown
Psilocybin	Magic mushroom		Unknown
Marijuana (cannabis)	Tea, grass, pot	16,210,000	Opinions contradictory; reduced alertness and impairment of concentration and memory possible

The Neuron

Nerve tissue contains a variety of cells. Chief among them are the neurons, or nerve cells. Other cells support, protect, and insulate the neurons. Three categories of neurons are recognized: sensory neurons, interneurons, and motor neurons. Their differences are based on function and structure.

Sensory Neurons

Nerve impulses are transmitted to the central nervous system from sense organs such as the eye or ear, by **sensory neurons.** These impulses are also transmitted from other sensory endings, such as pressure and stretch receptors in the skin and muscles (see Figure 8–7). Sensory neurons usually stimulate interneurons.

Interneurons

The central nervous system is the only place where **interneurons** can be found. Here they are stimulated by sensory neurons and in turn stimulate either other interneurons or motor neurons. Interneurons provide billions of interconnections among different parts of the nervous system.

Motor Neurons

Nerve impulses are carried from the central nervous system to muscles, glands, and other response organs by motor neurons. A typical motor neuron (see Figure 8–7) consists of a cell body with a nucleus and other cellular organelles. On one end, it bears short, highly branched extensions, called **dendrites.** Extending from the opposite end of the nerve cell is a

long projection, called the **axon,** the end of which may also be branched. In large animals, the axons of some neurons are up to 2 meters long, extending from the spinal cord to the appendages. Many axons are enclosed by a chain of **Schwann cells** which form a protective and insulating **myelin sheath.**

Figure 8–7
Neurons. (a) A motor neuron. (b) A sensory neuron.

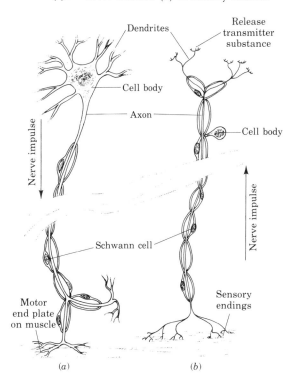

The neuron is the functional unit of the nervous system. When it is aroused by an appropriate stimulus, usually at the dendrites, electrochemical events move along the cell body and the axon. This wave is caused by the temporary halt of sodium pump, which maintains a specific distribution of charged ions on either side of the neuron's cell membrane. When this pump stops, sodium ions on the outside of the membrane rush into the neuron and alter the electrical charges on the membrane. The wave, called a depolarization because of the distribution of electrical charges on the membrane, represents a **nerve impulse.** It moves along the neuron to the end of the axon, where the nerve impulse stimulates another neuron, induces a muscle fiber to contract, or causes a gland to secrete. Immediately behind the wave, the pump restores the former charged state of the membrane.

Nerve impulses are transmitted from sensory structures into the central nervous system. In the central nervous system, they are transmitted via interneurons to various levels of the spinal cord or to coordination centers in the brain. Eventually, appropriate motor neurons are stimulated to produce responses somewhere in the body.

The place where one neuron comes close to and stimulates another neuron is called a **synapse** (see Figure 8–8). It is a small space between an axon of one neuron and the dendrite of another. When an impulse reaches the end of an axon, it induces tiny sacs in the axon, called synaptic vesicles, to release a transmitter chemical substance such as acetylcholine into the space between the two neurons. This substance quickly reaches the cell membrane of the second neuron and attaches itself to special chemicals in areas called receptor sites. When the amount of the transmitter chemical reaches a critical threshold level, it induces another wave of depolarization in the second neuron. The transmitter substance is quickly destroyed by enzymes.

Figure 8–8

Synapses. The point where one neuron comes near and stimulates another neuron is called a synapse. The insert shows a synapse as a small space between the axon of one neuron and the dendrite of the next. In the inset, two synaptic vesicles have just released a transmitter substance into the synapse.

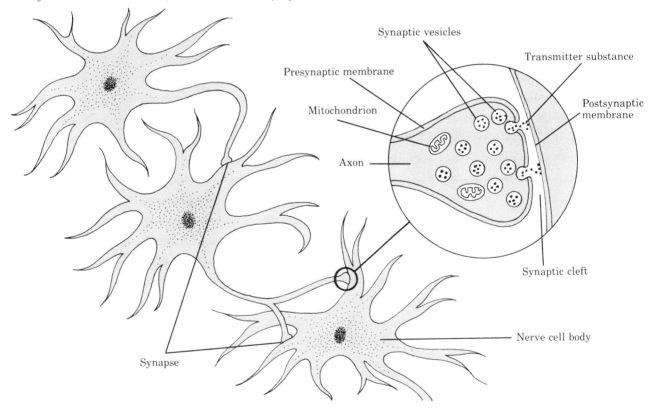

A nerve impulse is an all-or-nothing matter. It occurs when the threshold concentration is reached, and it continues to completion. After the impulse is completed, there is a short refractory period when the neuron is unable to transmit another impulse. This period lasts for only about 0.4 milliseconds.

The knee-jerk reflex provides a good illustration of the events that occur during neural transmission. It occurs when the tendon below the kneecap is tapped. In response to the tap, the leg quickly extends. What happens in terms of neural transmission? Special sensory endings called stretch receptors are present in the knee tendon. When the tendon is tapped, the stretch receptors originate a nerve impulse that travels along sensory neurons toward the spinal cord. Spinal nerves branch just before entering the spinal cord, and sensory neurons always follow the posterior (back), or dorsal, branch. The axons of these neurons extend into the interior of the spinal cord, where they form synapses directly with motor neurons. In this type of reflex, no interneurons are involved. When the impulses reach the synapses between the sensory neurons and the motor neurons, transmitter substances induce a nerve impulse in the motor neurons. This impulse travels along the motor neurons from the spinal cord out through the anterior (front), or ventral, branch of the spinal nerve. The cell bodies of the motor neurons are within the spinal cord, and the axons extend out from the spinal nerve to the leg muscles. When the muscles are stimulated to contract, the leg extends.

Neurons and nerves are not the same. A nerve is a bundle of neurons. If it contains only axons, as is the case with spinal nerves, it is smooth. However, some nerves also contain nerve cell bodies that are usually grouped together at some point along the nerve. They produce a swelling that is called a **ganglion.** There is a ganglion on the dorsal branch of the spinal nerve where the sensory neurons' cell bodies are located.

Neurons may branch and synapse with more than one other neuron. (Figure 8–9 illustrates some of the possible arrangements.) That way, messages can be sent simultaneously to different places in the nervous system via different nerves or nerve tracts. A number of body responses can occur because of a single sensory stimulus. The reverse is also true. Thousands of neurons may synapse with a single motor neuron.

The nervous system can also turn off or inhibit activities. **Excitatory neurons** induce an impulse to be

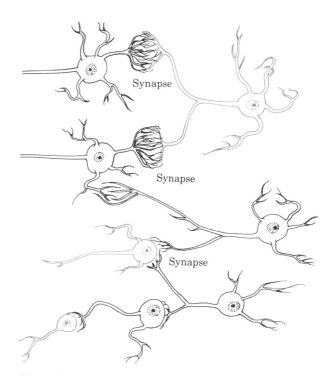

Figure 8–9

A neuron may synapse with more than one other neuron, producing the complex systems necessary for simultaneous transmission of information to many parts of the nervous system.

initiated in the motor neuron, but **inhibitory neurons** produce a transmitter substance that inhibits impulses in the motor neuron. A nerve impulse will occur in the motor neuron only when the excitatory stimuli it receives are greater than the inhibitory stimuli. This is the **principle of summation.** Obviously, then, the activities of the nervous system are very finely tuned.

The nervous system is complex. Neuroanatomists and neurophysiologists have traced many nerves and understand many of the functions regulated by these nerves. However, the total operation of the nervous system is not understood. Especially difficult to understand is the way the central nervous system coordinates the millions of nerve impulses that are fed into it during a simple activity such as walking across campus. The greatest mystery of all is understanding how the brain functions. (See Box 8–4.)

The Senses

Humans can detect many kinds of sensations in their environment. This ability is based on two kinds of

One of the excitatory neurotransmitter substances is the compound acetylcholine. This substance is released at many neuron junctions. It is also released at motor end plates, and it initiates the contraction of skeletal, cardiac, and smooth muscles. It is acetylcholine that depolarizes the dendrites of the next neuron or the sarcolemma of a muscle cell. In either case, once it is released, it is quickly deactivated by an enzyme. The deactivation allows the neuron or muscle cell to polarize again.

Another excitatory neurotransmitter substance is norepinephrine (NA). This substance is found at some motor end plates and in brain. It too is quickly deactivated, but in a slightly different manner. NA is rapidly pumped back into the axon, where enzymes deactivate it. It is then recycled during the next nerve impulse.

Today there is a great deal of interest in the chemistry of the brain because at least thirty different compounds are suspected of being neurotransmitters. Specific neurotransmitters are associated with certain areas of the brain. These substances not only excite or inhibit other neurons but are also thought to create the moods of happiness, sadness, apathy, and so on. When the normal functions of these neurotransmitters are either inhibited or enhanced by drugs, new moods, such as lengthy "highs" and "lows," are created. Many diseases of the nervous system and brain are now known to be the result of a lack of homeostasis among neurotransmitters.

One neurotransmitter that has been extensively studied is an amino acid called gamma aminobutyric acid (GABA), which a nerve inhibitor. Valium and many other common tranquilizers are known to enhance the effectiveness of GABA at synapses and to tranquilize (inhibit) large areas of the brain. Another neurotransmitter is glutamic acid, an amino acid that differs from GABA by only one chemical group. Other amino acids are also neurotransmitters. They include aspartic acid, which is excitatory, and glycine, which is inhibitory.

Another brain neurotransmitter is dopamine, which is probably keeping you awake right now as you read this page. It is also thought to be the excitatory neurotransmitter that stimulates dreaming. If you are sleepy, that is the effect of serotonin, which also is helpful in maintaining body temperature and is involved in sensory perception.

Another class of compounds, called peptides, is currently under study. A peptide is a chain of two to several amino acids. It appears that when neurons need to communicate rather than excite or inhibit, peptides do the job. It has been suggested that when peptides are released at a synapse, they may transmit complex information to a neuron. What they communicate is not known. However, the peptides have been implicated in regulating much of human behavior, including hunger and eating, the brain's blood pressure, and the sensation of pain. Some behavior is inherited. Maybe the peptides communicate the information on instincts from one neuron to another.

Neurotransmitter substances in the brain are not always inhibitory or always excitatory. They may be either, depending on where they are found in the brain.

The adage "you are what you eat" may be truer than we think. People commonly consume the precursors of the neurotransmitters mentioned here. Recently, relationships have been found between what people eat, the amount of neurotransmitter precursors found in their blood plasma, and the amount found in the brain, where they are transformed into neurotransmitters. These relationships are believed to dictate when we eat, what we eat, and how much we eat. A relationship may also exist between obesity and imbalances in a person's neurotransmitters. These imbalances may also explain why people on crash diets are often depressed.

senses, the general senses and the special senses. The **general senses** come from a multitude of microscopic sensory cells in the skin, joints, muscles, and so on. These cells provide many sensations, including touch, pain, heat, cold, and pressure. They also give the ability to detect the position of various body parts such as arms or toes. This is known as proprioception. The **special senses** include taste, smell, vision, and hearing and balance.

Taste

The sensation of taste is scientifically known as **gustation** (*gust* means "taste"). We are capable of tasting because we have taste buds on our tongues. Each taste bud has a pore that leads into a cluster of sensory cells. The substance to be tasted is first dissolved in saliva, which seeps into the pores of the taste buds. If the substance is salty, sweet, sour, or bitter, it stimulates a nerve fiber. Some taste buds are sensitive to only one or two tastes, such as sweetness and saltiness. Because of slight differences in the sensitivity of taste buds, we normally taste saltiness and sweetness best at the tip of the tongue, sourness on the sides of the tongue, and bitterness on the back of the tongue.

Most people believe that they can taste much more than saltiness, sweetness, sourness, and bitterness. The different tastes that people recall, however, are actually combinations of these four basic tastes. Also, much of what people think they taste is not tasted at all, but smelled. A TV commercial makes the statement "There is nothing like the good taste of coffee first thing in the morning." Wrong. Coffee is smelled, not tasted.

Smell

The things that are smelled are detected by the olfactory sense. Olfaction is possible because of numerous olfactory cells that line the inner surfaces of the nose. Each olfactory cell in the nasal epithelium has protruding from it a series of dendrites that are branches of olfactory nerve fibers. Therefore, the sensation of smelling is caused by direct stimulation of a nerve. Tasting, on the other hand, is first perceived by specialized cells, which eventually stimulate a nerve.

How olfactory cells distinguish among smells is not known. The current theory explains olfaction as a response to molecules in the gaseous state. Each olfactory cell is supposedly tuned to only one specific type of gas molecule. The specific molecule stimulates the cell to send an impulse.

The information from olfactory cells is relayed to the brain via the olfactory nerve. The sensation of smell is recorded in the most anterior portion of the forebrain. Experimentation has shown that humans rarely forget a smell.

Vision

The **eyes** are spherical structures approximately 2.5 centimeters in diameter. They are set in bony sockets of the skull called **orbits,** which are padded by connective tissue and fat. Extending from each eye to the walls of its orbit are six muscles that finely regulate eye movements. Extending from the back of each eye is an **optic nerve,** which eventually connects with the brain (see Figure 8–10). The outer wall of each eye is composed of three layers: the sclera, the choroid layer, and the retina, which is the innermost sensory layer.

The **sclera** is a tough layer of fibrous connective tissue that helps maintain the shape of the eye. At the front of the eye, it appears as the white. It is contin-

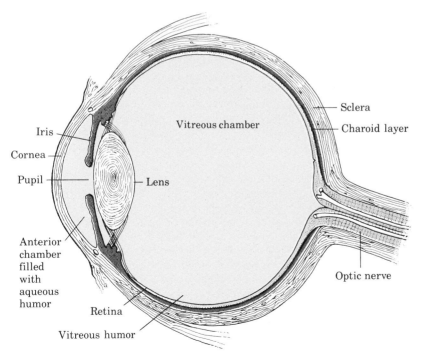

Figure 8–10
A sagittal section through a human eye.

Iris
Cornea
Pupil
Lens
Anterior chamber filled with aqueous humor
Retina
Vitreous humor
Vitreous chamber
Sclera
Charoid layer
Optic nerve

uous with the eye's transparent front covering, the **cornea.**

Inside the sclera is the second layer, the **choroid layer.** This layer contains blood vessels and is heavily pigmented to help darken the interior of the eye and prevent light reflection. Near the front of the eye, the choroid layer expands to form a ring of tissue that supports or gives rise to several important structures. One is a band of ligament and associated muscle tissue that supports and alters the shape of the transparent **lens.** Another is the **iris,** a circular disk of pigmented tissues that surrounds the central opening, the **pupil.** Muscles within the iris can increase or decrease the diameter of the pupil, allowing more or less light to enter the eye. The iris is suspended between the cornea and the lens.

The various structures of the eye divide the eyeball into three chambers (see Figure 8–10). The front anterior chamber is located behind the cornea and in front of the iris. The next chamber, the posterior chamber, is located behind the iris and in front of the lens. Both of these chambers are filled with a clear fluid, called aqueous humor. The pressure of the fluid creates part of the intraocular pressure that keeps the retina flat and in place so that clear images will be seen. The aqueous humor is constantly being replaced, and its amount can vary. In the elderly, more fluid may accumulate than is necessary, which increases the intraocular pressure. If this condition, which is called glaucoma, is left untreated, it can cause permanent damage to sight. Today, intraocular pressure is checked during all routine eye examinations.

The largest chamber of the eye, the vitreous chamber, is located behind the lens (see Figure 8–10). It is filled with a jelly-like substance called the vitreous humor. This substance is primarily responsible for the intraocular pressure that maintains the shape of the eyeball and keeps the retina smooth.

The **retina** lines the vitreous chamber, forming the third and final layer of the wall of this portion of the eye. It contains the **photoreceptors**—the cones and rods—as well as other neurons involved with vision. There are 6 to 8 million **cone** cells per eye, but they are concentrated mainly in one area of the retina, near the point where the optic nerve originates and directly behind the pupil and lens. Cones respond to bright light and are responsible for the ability to detect colors. **Rods** are much more numerous. There may be more than 100 million of them per eye. They

are sensitive to dim light and do not contribute to color vision. They are most abundant around the periphery of the retina.

The eye is similar to a camera in its ability to form visual images. Light rays reflected from objects pass into the eye through the cornea and on through the anterior chamber and the lens. As the light passes through these structures, it is focused on the retina. The iris regulates the amount of light entering the eye. The position and shape of the lens can be shifted slightly to allow the eye to focus on objects at varying distances. Tiny inverted images are focused on the retina. The varying intensities and wavelengths of the light energy stimulate combinations of cones and rods to initiate series of nerve impulses. These impulses are coordinated by other neurons in the retina. They are transmitted by the optic nerves to the optic centers of the brain for final interpretation and integration.

Part of the visual information from the right eye is relayed to the left side of the brain, and part of the visual information from the left eye is relayed to the right side of the brain. This arrangement gives us **stereoscopic (depth) vision.**

A look at the other students in the class will quickly show that vision is not always perfect. Many students wear glasses. The commonest visual problems are near- and farsightedness. If you are nearsighted, or myopic, you see nearby objects clearly, but distant objects are unclear. The problem is that the visual images you see at a distance are focused in front of the retina, because the natural lens is too thick or the eyeball is elongated. A concave glass or plastic lens worn either in a frame or as a contact lens corrects the problem.

If you are farsighted, you can see distant objects clearly, but nearby objects are unclear. The problem is the reverse of nearsightedness. Visual images of close objects are focused behind the retina (not literally, but images on the retina are not in focus). The eyeball may be too short or the lens too thin. Wearing a convex lens corrects the problem.

Astigmatism is also a common visual problem. It occurs when the surface of either the cornea or the lens is not perfectly regular, and it creates fuzzy images, particularly around the edges. The lens needed to correct this problem is one that is ground unevenly to correct for the unevenness of the cornea on the natural lens. It is not uncommon for people to be near- or farsighted and to have astigmatism.

Hearing and Balance

The **ears** are involved with two senses, hearing and balance. Ears are far more complex than the simple appendages attached to the sides of the head. Each consists of three sections: the outer ear, the middle ear, and the inner ear. All but a portion of the outer ear is embedded in cavities in the temporal bone of the skull (see Figure 8–11).

The **outer ear** consists of the **pinna,** the part of the ear you can see, and the **auditory canal,** which leads inward toward the middle ear. The inner portion of the auditory canal penetrates the temporal bone and ends at a thin, flexible sheet of connective tissue, the **eardrum,** which separates the outer and middle ear.

The **middle ear** occupies a small, air-filled cavity in the temporal bone. It is separated from the inner ear by a thin, bony partition. It is connected to the back part of the pharynx by the **eustachian tube,** a canal that equalizes air pressure between the middle ear and the outside environment. When you are in an airplane that is landing or taking off, you may experience the sensation of unequal air pressure. If you swallow or chew gum when your ears are "plugged," the eustachian tubes open and the pressure equilibrates.

The working parts of the middle ear are the three smallest bones. They are named after their shapes. The **malleus** (hammer) is attached to the upper portion of the eardrum at one end. It articulates with the **incus** (anvil) at the other. In turn, the anvil contacts the **stapes** (stirrup). The stapes abuts a thin membrane that covers the **oval window,** an opening in the thin bony partition that separates the middle and inner ears.

The **inner ear** is the most complex part of the ear. It consists of interconnected fluid-filled membranous structures located in three fluid-filled cavities in the temporal bone. These cavities are the vestibule, the semicircular canals, and the cochlea. The membranous structures in the **vestibule** and **semicircular canals** house sensory receptors associated with balance. The structures and receptors associated with hearing are in the cochlea.

The **cochlea** is shaped something like a snail shell. It spirals around a central column and tapers toward the apex. Inside the spiral, a thin shelf of bone divides part of the cavity into two parts. This shelf supports a membranous sensory structure called the **cochlear duct.** The auditory receptors are located on the floor of this duct. On either side of it are two other ducts. One of them begins at the oval window and parallels the cochlear duct up the spiral of the cochlea to its apex. At the apex, this duct melds into the second duct, which passes down the spiral along the opposite side of the chochlear duct. Near the base of the cochlea, the second duct ends at the **round window,** a membrane partition similar to the oval window. The receptors of hearing are cells with hairlike projections that extend to an overhanging membrane (see Figure 8–12).

How does this complicated structure work? Distur-

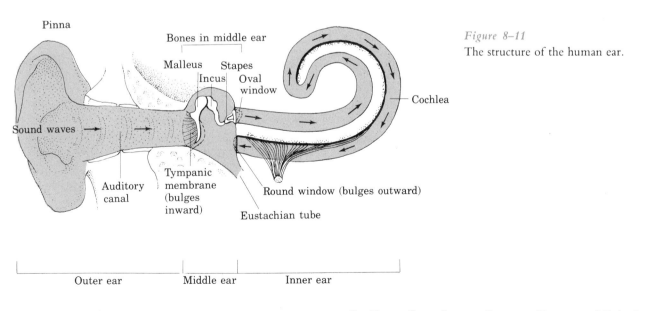

Pinna

Bones in middle ear

Malleus   Stapes
    Incus   Oval
            window

Cochlea

Sound waves

Auditory
canal

Tympanic
membrane
(bulges
inward)

Round window (bulges outward)

Eustachian tube

Outer ear   Middle ear   Inner ear

Figure 8–11
The structure of the human ear.

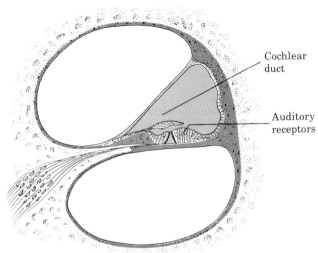

Figure 8–12

A section through the cochlea of the inner ear.

bances in the air—sound waves—are funneled by the pinna into the auditory canal, where they cause the eardrum to vibrate. The vibrations are transmitted through the three tiny bones of the middle ear. They cause the membrane of the oval window to vibrate, setting up wave motions in the fluid-filled ducts on either side of the cochlear duct. These waves induce the hairlike projections on the hearing receptors in the floor of the cochlear duct to stretch. This stretching initiates nerve impulses, which are transmitted to the brain.

Connected to the cochlea, but above it, are the semicircular canals (not shown in Figure 8–12). They give us our sense of balance. There are three canals, each projecting into a different plane. Like the cochlea, the canals are filled with fluid. A movement of the head moves the fluid and triggers tiny sensory projections on cells in the lining of the canals. The mechanism of balance is analagous to that of a carpenter's level. When the fluid is displaced (as shown by the bubble), various compensating movements of the body restore balance (the bubble returns to center.) If you spin around quickly several times and then suddenly stop, you feel dizzy. When you stop, the fluid in the semicircular canals is still displaced for several seconds. The upset in the general homeostasis of balance produces the sense of dizziness.

THE ENDOCRINE SYSTEM

The nervous system controls and regulates many body activities and functions. The endocrine system regulates and coordinates many of these activities too, as well as controlling growth, sexual maturation, and reproduction. The endocrine system consists of many widely scattered tissue masses called **endocrine glands** (see Figure 8–13). Sometimes endocrine tissues are incorporated within other organs. For example, the **pancreas** is a gland that secretes digestive enzymes into the small intestine. It also contains endocrine tissues called islets of Langerhans, which produce insulin. **Insulin** is a hormone that contributes to the control of blood sugar (homeostasis again).

Endocrine glands are different from exocrine glands (such as sweat glands), whose secretions drain from the glands through systems of ducts. Endocrine glands are ductless and secrete their products, **hormones,** directly into the bloodstream. The blood transports the hormones throughout the body.

Hormones have specific functions, as Table 8–1 indicates. They may stimulate or affect a body structure some distance from the endocrine gland that produces them. Often, several different hormones from different endocrine glands interact to induce changes in body functions. Some endocrine glands, especially the **pituitary,** secrete a variety of hormones, each of which may have a different function and may affect different organs. For this reason, the pituitary gland is called the "master gland." Malfunctions of endocrine glands and imbalances of hormones in the bloodstream cause a wide variety of problems, including goiters, diabetes, acromegaly (gigantism), and sterility.

It may seem that the endocrine and nervous systems have little in common other than their general roles in body coordination. However, the two systems are closely related. Special cells in the **hypothalamus,** a part of the forebrain, secrete chemicals called **releasing factors** into the capillary system that carries blood to the pituitary gland, which is located beneath the forebrain. The releasing factors induce the synthesis and release of some of the hormones of the pituitary. The hypothalamus secretes inhibiting factors, which inhibit release of pituitary hormones, in a form of negative feedback. Oxytocin and antidiuretic hormones are actually synthesized in the hypothalamus and then stored in the pituitary. A close physiological relationship exists between brain function and the pituitary. The hormones of the pituitary in turn regulate other endocrine tissues.

There is another relationship between the endocrine and the nervous systems. One of the hormones se-

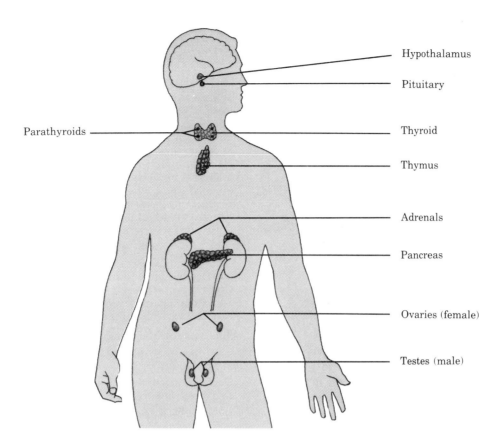

Figure 8–13
The major endocrine organs of humans.

creted by the adrenal glands is norepinephrine, which is also a neurotransmitter secreted by some of the neurons of the sympathetic portion of the autonomic nervous system. The two hormones have nearly identical effects on the body, causing increased heart rate and elevated blood pressure. The two systems interact and complement each other in maintaining properly regulated and controlled body functions, and they contribute equally to the maintenance of homeostasis. A close relationship between the nervous and endocrine systems also exists in other organisms (see Box 8–5).

Hormone Activity

Hormones have general effects throughout the body and specific effects on certain **target cells** of tissues. Many hormones are thought to generally affect the brain, particularly specific groups of neurons in the brain. Hormones may trigger certain groups of neurons to function. The result is an overall change in

behavior. Such behavioral changes are usually called **motivational states.** Much of sexual motivation, for example, is the result of brain chemistry. Hormones first affect brain chemistry. They reach target cells later.

Hormones stimulate cells in two different ways, depending on the chemistry of the hormones. Some hormones stimulate the metabolism of cells. Others stimulate genes.

The hormones of the adrenal cortex, ovaries, and testes are thought to function by stimulating genes in target cells. Table 8–1 shows that the hormones from these glands are involved primarily in stimulating tissues to grow.

Most hormones stimulate target cells by attaching to specific sites on the membranes of the cells. The hormone-membrane complex stimulates the activity of enzymes, which convert ATP□ to cyclic adenosine monophosphate (AMP). The AMP moves into the cytoplasm and directs the cell into some specific function. For example, insulin stimulates the cell mem-

Table 8–1
Human Hormones

ORIGIN	HORMONE	ACTION
Anterior pituitary gland	Follicle-stimulating hormone (FSH)	Stimulates production of eggs in females and sperm in males
	Luteinizing hormone (LH)	Stimulates hormone production in ovaries and testes; involved in stimulating ovulation
	Prolactin	Stimulates milk production in mammals; maintains parental instinct in all vertebrates
	Thyrotropin (TSH)	Stimulates production of thyroxine
	Adrenocorticotropin (ACTH)	Stimulates production of adrenal hormones in adrenal cortex
	Growth hormone	Regulates much of metabolism; stimulates bone growth
	Melanocyte-stimulating hormone (MSH)	Stimulates melanin production in skin
Posterior pituitary gland (manufactured in the hypothalamus of the brain)	Oxytocin	Stimulates milk release from breasts; stimulates uterine contractions during birth
	Antidiuretic hormone	Stimulates tubules of kidneys to reabsorb water; prevents excessive excretion of urine
Pineal gland of the brain	Melatonin	Inhibits ovarian function
Thyroid gland	Thyroxine	Stimulates general body metabolism; releases energy rapidly; maintains body temperature
	Calcitonin	Decreases amount of calcium in blood; inhibits bone resorption: involved with phosphorus metabolism
Parathyroid gland	Parathormone	Increases amount of calcium in blood; stimulates bone resorption
Adrenal gland cortex (periphery of gland)	Aldosterone	Stimulates kidney tubules to reabsorb sodium; increases blood pressure
	Cortisol	Stimulates available glucose by converting proteins to carbohydrates; reduces inflammatory responses; stimulates antibody production
Adrenal gland medulla (center of gland)	Epinephrine (adrenalin)	Stimulates glucose release by breakdown of muscle and liver glycogen; increases heart rate; shunts blood from skin to visceral organs; generally prepares body for "fight or flight"
	Norepinephrine	Similar to epinephrine but has strong effect on constriction of blood vessels; also a neurotransmitter
Pancreas	Insulin	Stimulates uptake of sugar by many tissues, thus decreasing blood sugar
	Glucagon	Stimulates breakdown of glycogen to blood sugar in muscle and liver
Ovary	Estrogen	Stimulates thickening of uterine walls during menstrual cycle; maintains secondary sexual characteristics of females
	Progesterone	Maintains uterine lining; maintains pregnancy
Testes	Testosterone	Maintains secondary sexual characteristics of males
Placenta	Chorionic gonadotropin	Maintains pregnancy
	Lactogen	Stimulates breast growth and milk production of mothers; mimics prolactin
	Relaxin (also some from ovary)	Relaxes pelvic ligaments, allowing widening of birth canal during labor

CHEMICAL COORDINATION IN NONVERTEBRATES

Endocrine systems are not unique to humans and other vertebrates. They are now known to occur in many other phyla. Probably more is known about the endocrine systems in arthropods than about the systems in any other nonvertebrates. Growth, reproduction, metamorphosis, molting, color change, and many other physiological processes are regulated in arthropods by chemical control systems.

One of the more fascinating systems is the one that controls molting, growth, and metamorphosis in insects. Like other arthropods, insects have hard exoskeletons. Periodically, the insects molt and replace these exoskeletons with new, larger ones to accommodate growth. Many insects pass through a number of larval stages, each separated by a molt, before metamorphosing to the adult form. In the case of moths, the larval stage creatures are called caterpillars. The final larval molt results in a pupal stage. The insect appears to be quiescent at this stage, but actually the larval tissues are being rebuilt into adult types and forms. When the pupa molts, the adult emerges. The adult does not molt.

All of these changes are regulated by two hormones. Each is produced in a separate part of the insect's brain. The first hormone stimulates an endocrine gland in the insect's thorax to produce the hormone ecdysone, which stimulates growth and molting. The second hormone determines what will emerge from the molt. It is called juvenile hormone because it inhibits metamorphosis and maintains larval structures. As long as a critical level of juvenile hormone is present in the blood at the time of molting, the molt will result in another larval stage. When the juvenile hormone level drops below this critical level, the molt results in a pupa. When there is no juvenile hormone, the pupa gives rise to the adult. This has been demonstrated experimentally by the injection of additional juvenile hormone into moth larvae that are ready to begin metamorphosis. The injected juvenile hormone produces larval stages beyond the number in the normal life cycle.

branes of liver, muscle, and diaphragm cells. The resulting cyclic AMP causes these cells to transport glucose into themselves. There it is converted into glycogen. Epinephrine, from the adrenal cortex, does the reverse. It stimulates these cells to split glycogen back into glucose molecules and to transport them into the blood. The blood-glucose level rises, making more glucose available for the "fight or flight" reactions that follow (see Table 8–1).

Prostaglandins

Considered hormones even though they are not produced in glands, the **prostaglandins** are lipids□. They are composed of the fatty acids that are produced by nearly every cell membrane in the body. The first prostaglandins were identified from the secretions of the male prostate gland, for which they were named. The prostaglandins in semen ejaculated into the female vagina are thought to stimulate contractions in the uterus. These contractions may help pass the sperm into the Fallopian tubes, or oviducts. Prostaglandins in human fetuses are produced by the amniotic membrane, which stimulates the uterine contractions that begin labor.

The prostaglandins are involved in a variety of activities, ranging from keeping the breathing passages open to regulating stomach secretions and aiding in the clotting of blood. They are nearly always produced when tissues are inflamed. Recently, it has been found that aspirin inhibits prostaglandin production. This explains in part why aspirin is such a powerful drug in the treatment of inflammatory diseases such as arthritis.

The prostaglandins also appear to stimulate cyclic AMP activity in cells. Because they are produced in membranes, they are on the front line in stimulating enzyme activities that lead to cyclic AMP. Again, their messages are for specific tasks to be performed in specific cells.

Recently, considerable interest has been generated in the prostaglandins. Many researchers believe that when prostaglandins are fully understood, they may turn out to be the keys to curing and treating a wide variety of diseases.

SUMMARY

1. Homeostasis refers to the maintenance of a steady, stable internal environment.

2. The excretory system consists of the kidneys, the ureters, the bladder, and the urethra.

3. The functional unit of the kidney is the nephron. Each nephron is associated with a capillary knob, the glomerulus, inside Bowman's capsule. The remainder of the nephron is a continuous tubule, composed of three parts, the proximal convoluted tubule, Henle's loop, and the distal convoluted tubule.

4. Bowman's capsule and the glomerulus are involved in pressure filtration. The rest of the nephron is involved in selective reabsorption and tubular excretion.

5. The nervous system is composed of the central nervous system and the peripheral nervous system. The central nervous system includes the brain and spinal cord. The peripheral nervous system includes the cranial and spinal nerves and the autonomic nervous system.

6. Neurons are the functional units of the nervous system. There are three types: sensory neurons, interneurons, and motor neurons.

7. A nerve impulse is an electric wave of depolarization that moves down a neuron because of a shift in the location of various ions inside and outside the neuron.

8. The input to the nervous system includes the general senses. These senses include such special senses as taste, smell, vision, and hearing and balance.

9. The endocrine system consists of many ductless glands that produce hormones.

10. Hormones affect target cells by stimulating either the metabolism or the genes of these cells.

11. The prostaglandins are considered hormones even though they are produced in the cell membranes of nearly every body cell. They are involved in a wide variety of activities, from inflammation to stomach secretions.

THE NEXT DECADE

YOUR AGING AND ACHING BODY

One thing will happen for sure in the next decade. You will age. The exact biological cause of aging remains unknown, but there are several suggestions about what is happening. Many authorities believe that the various cells and tissues of the body have a finite number of times that they can replicate themselves. When the magic number is reached, natural repair can no longer occur. Organs thus wear out, and people die. For example, when a woman reaches menopause, the cells of her ovaries die. Another suggestion is that cells are genetically programmed by their DNA to slow down their activities or to stop functioning altogether. Aged cells may self-destruct. Like tiny time capsules, they may command themselves to give up life. Yet another suggestion is that aging is a slowing down of the immune system, which makes people vulnerable to all kinds of diseases, including arthritis, cancer, and cardiovascular disease.

The cause of aging may not be known, but what happens to the organs and organ systems as they age has been well studied. Let's assume that you are twenty years old. What will happen to you five decades from now?

SKIN

One of the commonest measures we use for determining age involves the skin. In the next decade you will add some worry lines to your forehead. In the following decade, plan on crow's feet at the eyes. In the fourth decade, the skin of your body will

sag and bag. This will be most noticeable on the face. The next decade will bring more and more wrinkles. The skin will become coarse and perhaps red.

Many of the skin changes are caused by overexposure to ultraviolet light (the sun) and by exposure to harsh chemicals, such as makeup and detergents. Even now, the skin on your arms and face appears older than your other skin because of constant exposure to the sun. If you are a sun worshipper, then the skin of your torso and legs may also have aged considerably by now. The youngest skin on your body is always that of your buttocks and of other areas that have the least exposure to the environment.

HAIR

Decade by decade, your hair will become thinner. Each hair on your body will become thinner with age. The average hair shaft at age twenty is about a hundred microns in diameter. At age seventy it will be less than eighty microns. Hair turns gray because the darker pigments are no longer produced. A large amount of hair falls out with age. Men may go bald. They will also develop a few patches of new hair as they age—but in the nose and ear canals only.

STRENGTH

If you tested the strength of your hand grip every decade, you could expect it to decline by about five pounds per decade.

WAISTLINE

Data about waistlines are for men only, because women are more conscious of their waistlines and diet more frequently. Men can anticipate adding one inch per decade to their waistline. The addition is stored fat. Fat will also accumulate somewhat more slowly in the shoulders and chest, but men can count on an extra inch per decade.

WEIGHT

You can expect an increase in weight of about ten pounds per decade for the next two decades. After that, your weight will stay constant for a couple of decades. In the last decades, you will lose weight. Today, about 15 percent of your total body weight is fat. That amount will double during your heavy years.

EYESIGHT

The lenses of your eyes are already beginning to harden. In the next few decades, they will become so hard that the eye muscles will be unable to change their shape easily. As a consequence, you will not see close objects clearly unless you wear corrective lenses. Less light will reach the retina with age, and you will require brighter light when reading. You will also have difficulty seeing anything in the dark.

HEARING

If you are not a fan of loud rock (which has been shown to damage hearing), you now hear sounds as high as 20,000 hertz. Your ability to hear higher frequencies will decline by more than 2,500 hertz per decade. By age seventy you will be unable to discern sounds above 6,000 hertz. Normal conversation falls within the range of 4,000 to 6,000 hertz.

TEETH

You probably have lost at least one tooth already. Expect to lose about two more per decade in the future. New dental techniques may be used to restore old teeth.

HEART AND BLOOD

Your resting heart rate will not change. However, the heart weakens with age and pumps less and less blood. The amount of blood your heart can pump will decrease 6 percent per decade. Now, at age twenty, your pulse rate may climb to 200 when you are exercising. It will return to normal in about two or three minutes. (This is a good test of coronary homeostasis.) Expect your pulse rate during exercise to decline about ten beats per decade. Also expect the time required for the heart rate to return to normal to be longer and longer. The cholesterol content of your blood plasma will increase 4 percent per decade.

KIDNEYS

Expect a decrease of 5 percent per decade in the kidneys' filtering ability.

INTELLIGENCE

You will become forgetful, and your recall of words and numbers in a sequence will decline because of the death of many neurons in your brain. So many neurons die that your brain will actually shrink. If you took an IQ test every decade, you could anticipate your score dropping about six points each time.

SEX

Less and less is about the best thing that can be said for sex as the decades go by. The average rate of sex for your age group now is about twice per week (either solo or with a partner). That rate may increase by one-half per week in the next decade. From then on, it is all downhill, until the frequency of sex is about once every third week. It is not that you won't want sex. It is that you will lack the stamina for it. Exercise and good diet can help, however.

There are, of course, many exceptions to this long list of damages that will occur to your body in future decades. The best exception may be the psychology of thinking young, which can mask the signs of an aging body.

FOR DISCUSSION AND REVIEW

1. Explain the formation of urine. Begin with the blood flowing to the kidneys in a renal artery. Mention also the transport of urine from the renal pelvis of the kidneys to the bladder.

2. What can the examination of a sample of your urine tell you about your homeostasis?

3. List the major subdivisions of the brain and a function for each one.

4. Distinguish among the following terms: peripheral nervous system, central nervous system, autonomic nervous system, sympathetic nervous system, and parasympathetic nervous system. Give the major function of each.

5. Explain how a nerve impulse passes from one neuron to another. Include the following terms in your description: depolarization, the sodium pump, axon, dendrite, synapse, and neurotransmitter.

6. You are just beginning to eat a bowl of Rice Crispies and milk, sprinkled with sugar. When you pour the milk over the cereal, you hear the familiar "snap, crackle, and pop." As you eat the first spoonful of cereal, you taste the sweetness. Describe how you detected the sound of the cereal, the shape of the spoon, the smell of the milk, and the taste of the sugar.

7. What are the relationships between the hypothalamus and the pituitary? What hormones are produced in the hypothalamus? In the pituitary?

8. What are prostaglandins? Why are they important?

SUGGESTED READINGS

The Suggested Readings for chapters 6 and 7 contain relevant material for the topics discussed in this chapter. In addition, the following will be helpful.

THE HUMAN BODY, by a variety of authors and the editors of U.S. News Books. First published in 1981, new volumes appear every few months. The titles relevant to the topics discussed in this chapter are THE HEART, THE BLOOD, THE EYE, and THE BRAIN.
These beautifully illustrated books are easy to read and cover a wide range of topics, including medical history, anatomy, physiology, and new treatments of diseases.

9
PLANT STRUCTURE AND FUNCTIONS

Plants are living organisms. Like animals, they have tissues, organs, and systems. However, their systems are somewhat different from those of animals. For example, plants do not have digestive systems, although they do process nutrients. Plants are stationary, so they have no need for a muscular system. They also have no nervous system. On the other hand, plants do have structures that support and protect their more delicate tissues much the way that the integumentary and skeletal systems protect animals. Plants respire in a process much like that of animals. They also have conductive tissue for transporting water and nutrients. Finally, like animals, plants have hormones.

This chapter details the major plant systems and their functions, using the common garden bean as an illustration (see Figure 9–1). However, the generalities can be extended to any common plant.

Plant structures can be classified as vegetative or reproductive. The vegetative structures include roots, stems, and leaves, the major organs of a plant. The reproductive structures, flowers and cones, are major organs involved with sexual reproduction.

PLANT TISSUES

The bulk of any multicellular organism, plant or animal, is composed of tissues. Plant tissues are divided into two main categories, meristematic and permanent.

205

94

PREVIEW QUESTIONS

After reading this chapter, you should be able to answer the following questions:

1. Of what types of tissues are plants composed?

2. List the functions of the various organs of a plant.

3. Which factors influence the amount of water a plant needs?

4. How are water and nutrients transported through a plant?

5. What hormones do plants have?

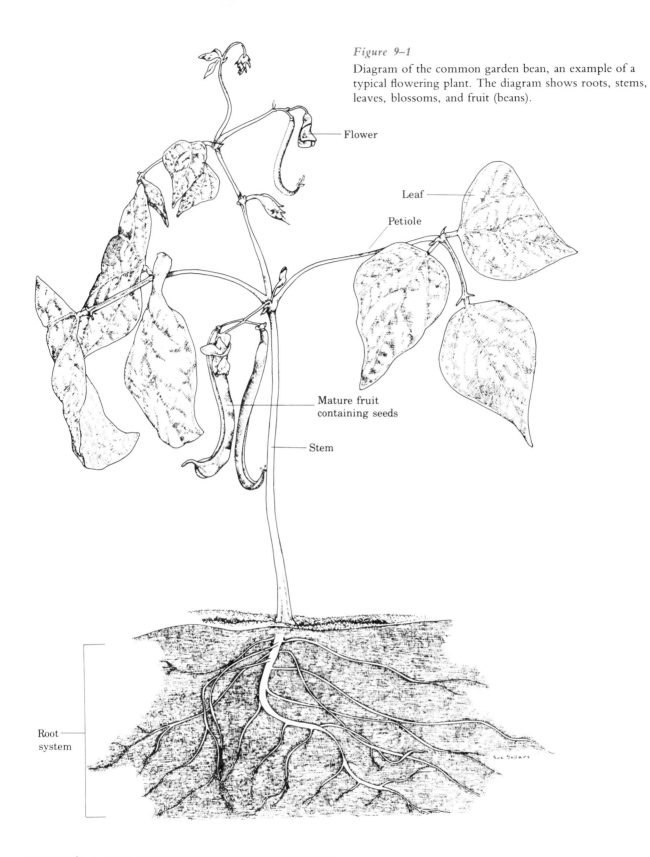

Figure 9–1

Diagram of the common garden bean, an example of a typical flowering plant. The diagram shows roots, stems, leaves, blossoms, and fruit (beans).

Flower

Leaf

Petiole

Mature fruit containing seeds

Stem

Root system

Meristematic Tissue

Tissue that is capable of dividing or reproducing throughout the life of the plant is **meristematic tissue**. It can be found in many places in a typical plant—at the tips of growing roots, at the ends of stems (called the apical areas), and at the bases of leaves, for example.

You have probably pulled a blade of grass and sucked on the soft, sweet tissues at the end of the blade. These tissues are a permanent band of meristematic tissue, which is found at the bases of monocot◻ leaves. It is the meristematic tissue that causes mown grass to grow again. The diameter of stems and tree trunks increases because of a circular band of lateral meristematic tissue, called **cambium.**

Meristematic tissues give plants an advantage that most animals don't have: Damaged plants can grow new buds that may become branches or leaves. Only a few animals are capable of similar replacement. For example, starfish can bud new arms, crustaceans can grow new legs, and some amphibians can bud new tails. In addition, plants continue to grow as long as they live; most animals do not.

Permanent Tissues

Unlike the cells of meristematic tissue, the cells of various permanent tissues do not divide. They make up the structures that are essential to the plant's existence.

Epidermis

The outer surfaces of all plant parts—including roots, stems, leaves, and flowers—are covered by epidermal cells. These cells serve the same function as the epidermal tissue in animals—protection of underlying tissues. The plant epidermis is composed of cells that secrete a waxy material called cutin onto the cell walls. The cutin waterproofs this layer of cells, much as keratin protects human skin cells. Cutin also prevents the loss of water from internal tissues. One of the most important adaptations of plants is water conservation. Plants wilt and die quickly when too much water is lost.

The epidermis of plants is permeated by openings, or pores. Leaves, for example, always have openings for the diffusion of atmospheric gases. These openings are called **stomata.** Each stoma (singular) is sur-rounded by two kidney-shaped cells, known as **guard cells.** When the guard cells swell with water, the stoma opens. When the guard cells lose water and shrink, they collapse against one another and the stoma closes. In this way, the stoma and guard cells allow gases to pass into and out of the internal tissues of the leaves. Stomata usually appear on the lower surfaces of leaves. Aquatic plants, such as water lilies, whose leaves float on the surface of water, have stomata only on the upper surface.

Woody stems and small, young trees have small slits or small warty porous growths on them. These openings into the internal tissues are called lenticels. They are particularly noticeable on the trunks of young fruit trees and on the new stems of most trees. Lenticels have the same function as stomata, but they remain open all the time.

Parenchyma

The cells of **parenchyma tissue** are usually spherical or box-shaped. They have thin cell walls, so their shapes are often distorted by the pressure of surrounding cells. Leaves are composed almost entirely of parenchyma tissue. The cells of leaf parenchyma are large. They have large vacuoles◻, and they contain **chloroplasts◻.** Leaf parenchyma is the main site of photosynthesis.

Parenchyma cells do occur elsewhere in plants—for example, in stems or roots—but such cells commonly lack chloroplasts. They are usually storage cells for water and nutrients. The delicious fleshy fruits are composed largely of parenchyma tissue. The different tastes of fruits and other plant parts come from the stored nutrients in parenchyma cells. Watermelon parenchyma cells contain large amounts of water and sugars. Potato parenchyma cells have less water and contain starch.

Sclerenchyma

The cells of **sclerenchyma tissue** have thick secondary cell walls —cell walls secreted inside the primary cell wall. After a sclerenchyma cell manufactures the secondary cell wall, it dies, leaving behind the hard and permanent secondary cell wall (see Figure 9–2). Sclerenchyma cells facilitate the main function of sclerenchyma, which is protection and support of the plant.

Some of the material deposited in the secondary cell walls of sclerenchyma cells is extremely hard.

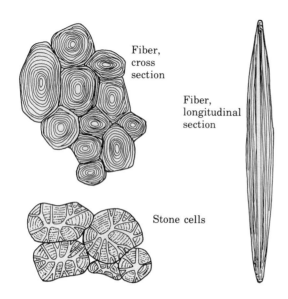

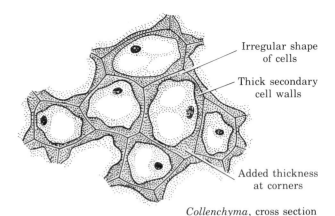

Irregular shape
of cells

Thick secondary
cell walls

Added thickness
at corners

Collenchyma, cross section

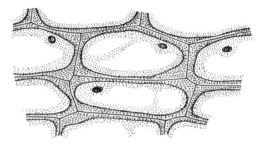

Collenchyma, longitudinal section

Figure 9–2

Typical cells of sclerenchyma. After these cells manufacture a thick secondary cell wall, they die, leaving behind hard cell walls that become an important form of support for the plant. Two kinds of sclerenchyma cells are shown—fiber cells and stone cells. A group of fiber cells is shown in cross section (left). A single fiber is shown as it would appear from the side (right). Stone cells give pears their grittiness.

Cells found in the shells of nuts are an example. In some cases, the deposits are softer. The grittiness of some varieties of pears is caused by stone cells, which are sclerenchyma cells dispersed in the softer parenchyma tissue. Some sclerenchyma cells are much longer and more flexible, and the space the cytoplasm previously occupied is completely obliterated by them. These long, tapering cells are called fibers. Hemp, flax, and other plants are economically important for their fibers.

Collenchyma

Some irregularly shaped cells have thickened primary cell walls. These cells are part of **collenchyma tissue** (see Figure 9–3). The cell walls are usually thickest at the corners, where they join the walls of other collenchyma cells. Although reasonably rigid, the cell walls are more flexible than the cell walls of sclerenchyma. The semi-rigid, but flexible, cell walls of collenchyma provide excellent support for all the growing parts of a plant. This tissue is particularly abundant in leaves and young stems. A bean seedling stands erect because it is supported by collenchyma tissue.

Figure 9–3

Typical cells of collenchyma tissue. Collenchyma cells are seen in both cross section and longitudinal section. The important characteristics of these cells are obvious: irregular shapes and thickened secondary cell walls, which are usually thickest at the corners. Collenchyma cells are important support tissues of plants, but they are more flexible than sclerenchyma cells.

Cork

Everyone is familiar with cork. It has been used to stopper everything from wine to shaving lotion. These days, though, a number of substitutes (usually rubber and plastic) often replace it. The cork used to stopper bottles comes from a special tree, the cork oak. This tree grows only along the Mediterranean Sea. It produces an unusually thick cork layer.

The outer thick layer of tree bark is cork tissue. Cork is composed of cells with thick cell walls. The cell walls are composed of suberin, a waxy material that gives cork its sponginess and ability to shed water. Cork is formed by the layer of meristematic cells called the cork cambium (see Figure 9–4). As the cork cells mature, the cellulose cell walls become impregnated with suberin. The cell dies, leaving only the suberin-impregnated cellulose walls.

HEMP

Marijuana has many names. Among them are hemp, pot, cannabis, hashish, and grass. Technically, though, marijuana is *Cannabis sativa*. The plant has the reputation of "growing like a weed." In fact, it has been weeded out of many gardens by unsuspecting gardeners. Hemp leaves are notched or toothed and occur in groups of five to eleven. The plants can grow as tall as 15 feet, but usually they are only a few feet tall.

Glandular hairs on hemp flowers (male and female flowers are borne on separate plants) produce a resin that contains the intoxicant compound tetrahydrocannabinol (THC). Hashish, which is far more potent than marijuana, is composed mostly of resin and accompanying flower parts. Marijuana usually consists of dried leaves and flowers crushed together.

Hemp has a long history. The Chinese wrote about it centuries ago. The Scythians (from an ancient kingdom that is now part of the Soviet Union) were probably the first people known to smoke marijuana. Around 600 BC, the Greeks reported that the Scythians burned the seeds of the plant and inhaled the smoke. Marco Polo wrote of a Moslem who prepared his men to murder Christian crusaders by having them drink a potion containing hashish. These murderers became known as *hashshashins,* and it is thought that the word *assassin* may be derived from the source.

In the United States, hemp was first planted in large quantities in Jamestown, Virginia, in 1611—for making rope. It wasn't until the early 1900s that it was smoked, a custom introduced by the Mexican people. In 1937, the Federal Marijuana Act was passed by Congress. It made marijuana illegal except for approved industrial and medical purposes.

The hemp plant has more uses than most other plants. It produces fibers that can be used to make twine, rope, cloth, and sacks. Its seeds have been used for lamp oil, paints, varnishes, soaps, and birdseed. Marijuana can be eaten, and it is frequently smoked.

Marijuana is not commonly used in medicine today, but it has been used as an analgesic, much as aspirin is used today. It was prescribed to treat ulcers, asthma, rheumatism, childbirth pain, and migraine headache; to relax muscles and prevent convulsions, menstrual cramps, and epilepsy; and to induce sleep, boost appetite, and relieve senility.

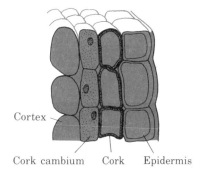

Cortex — Cork cambium — Cork — Epidermis

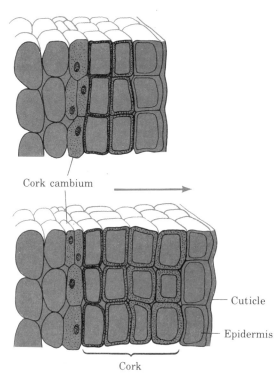

Cork cambium

Cuticle

Epidermis

Cork

Figure 9–4

The formation of cork. Cork is found between the epidermis and the other internal tissues of a plant. The cork layer is formed by a layer of meristematic cells called the cork cambium. As the layer of cork increases in size, the outer layer of epidermal cells is pushed farther from nutrients and water. The epidermis eventually dies. The cork layer gives a plant much better protection than did the thin epidermis. Most plants form a cork layer during their first year.

The cork layer grows between the epidermis and the other internal plant tissues. It pushes the epidermis away from water and nutrient sources, causing the epidermis to eventually die and scale away. Cork then takes over the function of protecting the plant.

Most stems begin the development of a cork layer during their first year.

Cork gives plants much better protection than the thin epidermis can and nearly all plants form a layer of cork on the outside of their stems. The thick bark of many trees not only prevents water loss, but also insulates the trees against environmental extremes. It takes an unusually severe winter to kill trees, although even an average winter can kill many small plants and shrubs—because they have thinner layers of cork. After forest fires, many of the larger trees survive despite bad burns because their thick layers of cork protected them from the intense heat.

Some trees have smooth, thin bark. Others have rough, deeply fissured bark with coarse pieces peeling away from it. The trees with smooth bark have a cork cambium that lives until the tree itself dies. The continuous growth of the cork adds to the diameter of the tree, and new layers of cork are stretched smooth. Trees with flaky bark have a discontinuous cork cambium. As new cambium is formed, phloem produced by the vascular cambium becomes flattened and pushed to the outside, where it flakes off and becomes part of the bark.

Many vegetables are covered by cork. New potatoes, for example, have a thin layer of cork that is probably scrubbed away when the potatoes are washed. If potatoes are stored for later use, the cork layer continues to grow thicker. Eventually it forms the familiar peel. The continued growth of the cork layer is why potatoes store so well.

Xylem

The permanent tissues known as xylem and phloem are the main conductive tissues of plants. The conduction of water and minerals from the root to the stem is a function of **xylem** (see Figure 9–5). Most of this transport occurs through specialized cells called **tracheids.** These cells have tapering ends, and their secondary cellulose walls are impregnated with lignin. When tracheids are mature, the cell inside dies, leaving behind the tapering cell walls it formed. Tracheids are arranged end-to-end in plant tissues and rapidly conduct water and minerals.

Vessel elements are another form of xylem. Like tracheids, they are formed by cells that die when they are mature. The central opening of vessel elements is much larger than that of tracheids. Vessel elements have thin cell walls of lignin and may have thicken-

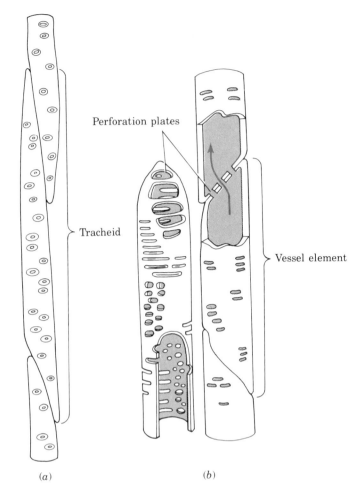

(a)   (b)

Figure 9–5

Some xylem elements (a) Tracheids, the most common of the xylem elements that conduct water and minerals in plants, have tapering ends. When tracheids are mature, the cell inside them dies. Tracheids are arranged end-on-end. They rapidly conduct water and minerals. (b) Vessel elements are also xylem elements, but they have much larger internal spaces. They too are formed by cells that die when they are mature. Vessel elements have thin cell walls, and many have spirals and other thickenings that strengthen them. Vessel elements are stacked end-on-end. The cell walls at the ends dissolve, creating continuous tubes. Vessel elements communicate by perforation plates.

ings and spirals that add strength to the walls. They are arranged end-on-end, but their ends dissolve to form tubes that look like stacked sewer tiles.

Wood is composed of xylem. Softwood contains only tracheids. Hardwood is composed of both tracheids and vessel elements.

Phloem

Another conductive tissue in plants is **phloem.** It is composed of several elements. The most common are the sieve elements. When these elements are stacked end-to-end, they form **sieve tubes.** As the sieve tubes develop, the primary cell wall becomes thin in areas and is perforated with pores (resembling a sieve). Sieve tubes have no secondary cell walls. When the element is mature, the nucleus disintegrates and the cytoplasm remains. Small projections of the cytoplasm of one sieve tube cell reach through the small perforations to the next sieve tube. The sieves at the end of each cell are called sieve plates.

Each sieve tube is usually accompanied by a companion cell, a cell that retains its nucleus and cytoplasm for life (see Figure 9–6). Companion cells probably contribute nuclear information to the cytoplasm of sieve tube cells.

These conductive elements transport food material from leaves to stems and roots. They also transport food material upward in the plant. Phloem is often associated with parenchymal cells, which act as storage sites. It may contain sclerenchymal cells with fibers, called **phloem fibers,** which provide support.

Phloem is constantly being replaced with secondary growth. Its cytoplasm does not survive for long, and the pressures of surrounding tissues crush the sieve tubes. The life span of phloem vessels is about one growing season.

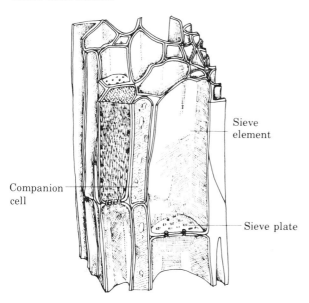

Figure 9–6
Some phloem elements. Sieve elements, the more common elements in phloem, are stacked end-on-end to form sieve tubes. The perforated plates at the end of each element are called sieve plates. The cells with discrete nuclei alongside the sieve elements are companion cells. These cells are believed to contribute nuclear information to the cytoplasm of sieve elements, which lack their own nuclei when mature.

Companion cell

Sieve element

Sieve plate

CONCEPT SUMMARY
FUNCTIONS OF PLANT TISSUES

TISSUE TYPE	FUNCTION
Meristematic	Plant growth and regeneration
Epidermis	Protection of underlying tissues; production of cutin for waterproofing
Parenchyma	Primary site of photosynthesis in leaves; storage cells for water and nutrients in other plant structures where chloroplasts are lacking
Sclerenchyma	Protection and support
Collenchyma	Support, especially for growing parts of plants
Cork	Protection
Xylem	Conduction of water and minerals from root systems to other parts of plants
Phloem	Conduction throughout plants of nutrients that are formed by or that follow photosynthesis

Celery

The next time you eat a piece of celery, you can easily review plant tissues. As you bite down, you will first penetrate the epidermis. The celery's watery texture is due to the water content of parenchyma cells. The crispness is a result of water pressure in parenchyma and collenchyma cells. Limp celery can become crisp again by being submerged in water, which restores **turgor.** The juices you taste are the fluids of parenchyma. The annoying fibers that get in your teeth are vascular bundles composed of xylem and phloem. A celery stalk is a rapidly growing part of the plant, and its obvious flexibility comes from collenchyma tissue.

PLANT ORGANS

Leaves

The principal photosynthetic organs of plants are their leaves, which come in a variety of sizes and shapes. They range from the gigantic, broad leaves of the elephant plant to the fine, sharp needles of pine trees. The leaves of the bean plant are typical. They are connected to the main stem by a slender stalk, the **petiole. Vascular bundles,** a network of xylem and phloem, run from the main stem into the petiole and then into the leaf. They are visible as the leaf veins that divide many times so that no leaf cell is far removed from conducting tissue. This is analogous to

Figure 9–7

The major tissues and cells of a leaf in cross section. Both the palisade and spongy cell layers can be seen between the upper and lower epidermal layers. The openings on the lower epidermis are the stomata, the sites of gas exchange between the environment and the internal tissues. The fine projections nearby are hairs. Alongside the diagram is a photograph of the epidermis of a leaf (courtesy of the Carolina Biological Supply Company). The stomata are distinct and open. The epidermal cells are irregularly shaped. The dark spots are the nuclei of epidermal cells and guard cells.

the way capillaries lie close to every cell in the human body.

Figure 9–7 illustrates the internal structure of a leaf. As noted earlier in the chapter, the outer covering is a layer of epidermal cells that secrete a cuticle, a thin, waxy film. Many of the epidermal cells have small projections called hairs. Some of the hairs are short and fine; others are long and stout. Many are defense

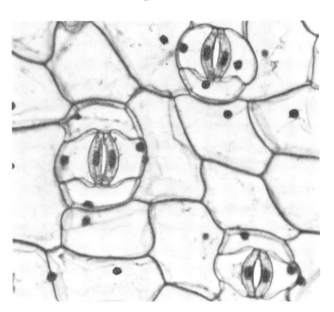

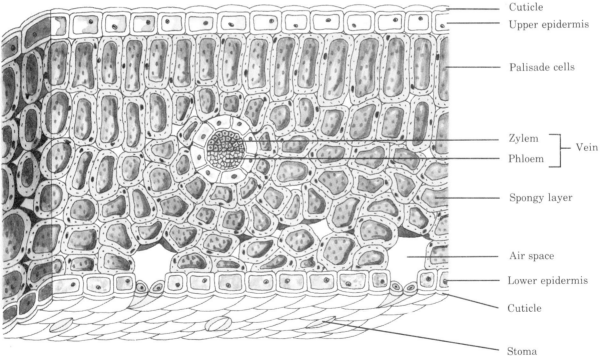

Cuticle
Upper epidermis

Palisade cells

Zylem
Phloem — Vein

Spongy layer

Air space

Lower epidermis

Cuticle

Stoma

mechanisms intended to discourage animals from eating the leaf. The epidermal hairs on nettles can penetrate human skin, break off, and inject a poison that produces a burning sensation.

As noted, the pores, or openings, into the internal tissues of leaves are **stomata.** These and the accompanying **guard cells** are the sites of gas exchange between the environment and the internal tissues.

The internal tissues are composed almost entirely of parenchyma cells with **chloroplasts.** Here is where nearly all photosynthesis□ occurs. The parenchyma cells are arranged in two layers. The upper layer, the **palisade layer,** is in neat, dense, vertical stacks. Its name comes from its resemblance to a cliff. The lower layer, the **spongy layer,** is composed of cells in a more random, loose arrangement.

Some plants are called evergreens because they always have green leaves. An evergreen does lose and grow new leaves, but not all at once. The trees that lose their leaves each fall are **deciduous.** Before a leaf falls, a special layer of cells forms at the base of the petiole, where it is attached to the stem. This layer has soft flexible cell walls that allow the petiole to bend and a layer of cork to form between the petiole and stem. Wind or rain causes the softened tissue region to break, and the leaf falls. The layer of cork seals the wound on the stem.

The most spectacular thing about deciduous trees is that the leaves often change color prior to falling. Color changes are induced by cool, moist weather and short days. When the temperature is near freezing, chlorophyll is no longer produced. The greenness fades from the leaves as the amount of chlorophyll is gradually reduced. The lack of chlorophyll allows other leaf pigments, which were previously masked, to become visible. These pigments may be gold, yellow, or red (cool temperatures stimulate the production of red pigments). The result is a beautiful spectacle. When the leaf is about to fall, its lifeline to the stem is interrupted and the leaf dies. After death, the leaf turns brown because of tannic acid.

Stems

There are two basic kinds of stems—herbaceous and woody. The common bean plant has an herbaceous stem. Such stems tend to grow little in diameter and have no strong woody tissue. The plants that propagate these stems are commonly short-lived. Some are known as **annual plants,** because they tend to last

only one year. Others are known as **biennial plants,** because they live a second year. (They flower in the second year.) Still others are called herbaceous perennials, because they survive numerous years, producing new growth each year. Herbaceous stems are often green. Their cells contain chlorophyll and are involved in photosynthesis. Such stems have a thin epidermis, with stomata present in most of it. Herbaceous stems are composed of both parenchyma and sclerenchyma cells, which add support.

Woody stems of plants are quite different from herbaceous stems (see Figure 9–8). Woody stems are part of **perennial plants** (those living more than two years). The outer surface of even the youngest of these stems is covered by cork. Lenticels, or air spaces, are numerous. Many of the peripheral cells remain active in photosynthesis. As the layer of cork increases, however, this function wanes, because light cannot reach the tissues. As a woody stem matures, a ring of cambium forms. On the outside, the cambium forms phloem. On the inside, it forms a layer of xylem. The rapid growth of the xylem (actually wood composed of lignified cellulose cell walls—the cells that transport water) increases the diameter of the woody stem. As a result, the delicate phloem cells are crushed. New phloem is then produced by the cambium during the next growing season. The cambium also produces a wide zone of xylem during the next growing season.

The layers of xylem in woody stems form rings. Each ring represents a growing season. People often count the rings to determine the approximate age of the plant (see Figure 9–9). Usually, wet springs produce a xylem layer that has wide cells with thin cell walls. In drier seasons, the cells are smaller and have thicker cell walls. The spring layer of xylem is easily compressed. Therefore, it appears as a thinner and

Figure 9–8

A woody stem as it would appear in winter. (Courtesy of Gary Fitzhugh.)

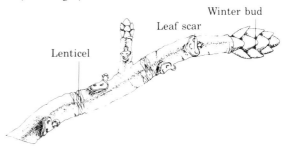

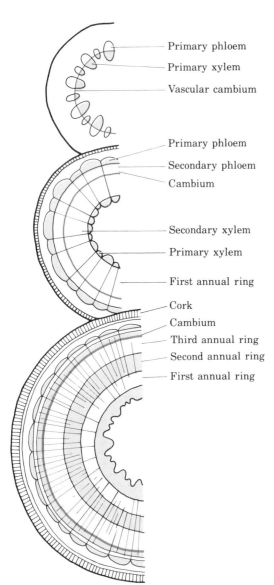

Primary phloem
Primary xylem
Vascular cambium

Primary phloem
Secondary phloem
Cambium

Secondary xylem
Primary xylem

First annual ring

Cork
Cambium
Third annual ring
Second annual ring
First annual ring

Figure 9–9
The growth of a woody stem. In the first year *(top)*, the meristematic layer—the cambium—manufactures phloem to the outside and xylem to the inside. During the second year *(middle)*, the primary xylem (the first formed) is surrounded as the cambium produces a second xylem layer. Growth of the stem crushes the more delicate phloem layer. Although the phloem is constantly replaced by the cambium, no distinct layers form. In the third year *(bottom)*, the cambium forms a third layer of xylem.

denser layer. The rings of a woody stem do not always indicate age accurately. The environment—conditions that accelerate growth, moisture, and nutrients—influences the addition of new layers.

The plant's first ring of xylem is always the ring closest to the center. The later rings are deposited around it. The cambium is pushed outward with this growth and constantly produces new phloem vessels on the outside. The cork cambium produces more layers of cork. This is how the tree increases in diameter.

Roots

Plants have extensive **roots.** The total surface area of the root system almost always exceeds the aboveground surface area of the plant. The roots anchor the plant, give it a source of water and minerals, and store food.

The tip of each root is composed of a band of meristematic tissue (see Figure 9–10). Scientists often use root tips to study the process of cell division (mitosis)□. The actively dividing cells of the root tip create layers that push the root downward. This proliferation of cells within the root tip causes the root to grow and elongate. The root tip is covered with a root cap, a layer of cells that prevent damage to the root tip as it is pushed through the soil.

The area of cells just above the root tip is the zone of elongation. In this area, cells mainly enlarge or grow. Cells above this area have already grown, and they begin to mature. These cells form the zone of maturation. Here the elongated cells differentiate into specific root tissues.

The epidermal cells around the zone of maturation usually have small tubular extensions called root hairs. When plants are dug or pulled from the ground, the hairs are often stripped off. However, they are reasonably conspicuous on such young seedlings as those of radishes or beans. The root hairs absorb water and minerals from the soil. Thus the zone of maturation, with its root hairs, is one of the most important structures for water uptake. Root hairs are delicate. They are constantly broken away and replaced by new epidermal cells.

The main tissues of the root, from the outside inward, are the epidermis, the cortex, the endodermis, the pericycle, and, in the center, the vascular bundle (see Figure 9–11). The function of the **epidermis** is protection. The **cortex,** the widest layer of the root, is composed of parenchymal tissue. Water and minerals absorbed by the epidermis pass through this layer. This tissue is also the storage tissue of roots that store starch. The **endodermis,** between the cor-

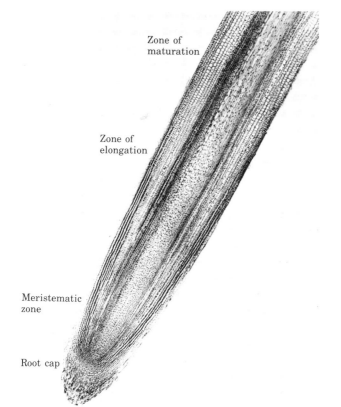

Zone of
maturation

Zone of
elongation

Meristematic
zone

Root cap

Figure 9–10
The root tip. The newly germinated mustard seed (above) has an obvious growing root tip with many root hairs. The growing root tip (left) is shown here in a longitudinal section as viewed under a microscope. The root cap consists of several layers of protective cells over the end of the root tip. These cells protect the more delicate meristematic layer, where cells actively divide. After division, cells grow and enlarge in the zone of elongation. Farther up, in the zone of maturation, the cells begin to differentiate into specific root tissues.

Figure 9–11
Cross section of a root, showing the arrangement and types of tissues.

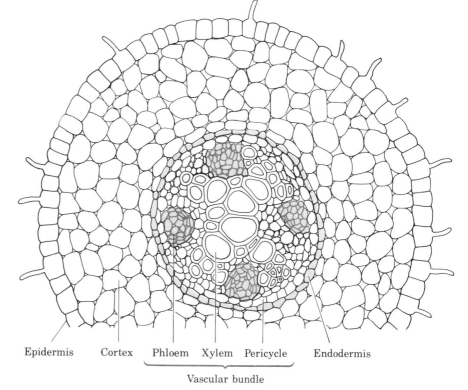

Epidermis   Cortex   Phloem   Xylem   Pericycle   Endodermis

Vascular bundle

tex and the pericycle, is a single layer of cells. The cell walls of endodermal cells have waxy stripes. Endodermal cells ensure that only substances that can penetrate the plasma membrane will enter the vascular bundle. Between the endodermis and the vascular bundle is a thin layer of cells called the **pericycle.** This layer is meristematic tissue in which new lateral roots develop. Xylem and phloem vessels are distributed in the **vascular bundle** in the center of the root. The xylem is shaped like a plus sign, with vessels arranged in radiating arms from the center. The phloem lies among the radiating arms of the xylem.

Many stout, robust roots are used as food. These roots, called tap roots, include carrots, parsnips, and beets. One of the most economically important root crops is sugar beets, from which some table sugar is refined. For further information about edible plant parts, see Box 9–1.

PLANT SYSTEMS

Photosynthesis and Respiration

The complex biochemical process by which plants assimilate water and carbon dioxide to produce sugar is |57| **photosynthesis**. The single most important process in the balance of all ecosystems, photosynthesis requires light and chlorophyll, and produces oxygen as a by-product. Most of the organs and systems of plants exist to support this process.

Like animals, plants must respire. The majority |63| have **aerobic respiration**: The cells use oxygen and sugar and release energy and carbon dioxide—twenty-four hours a day. In the 1940s and 1950s, hospitals removed plants from patients' rooms at night so the plants would not suffocate the patients by using up all the oxygen. Aerobic respiration by plants is no different from aerobic respiration by animals. Table 9–1 clarifies the differences between photosynthesis and respiration in plants.

Transpiration

The loss of water from plants, usually from the leaves, is called transpiration. It occurs constantly but is most marked during the day and accelerates on hot days. Some plants transpire 99 percent of the water taken up by the roots. A single corn plant, for example, may transpire as much as 2 quarts of water

Table 9–1

Photosynthesis and Plant Respiration

	PHOTOSYNTHESIS	PLANT RESPIRATION
Where	In plant cells containing chloroplasts	In all live plant cells
When	In the presence of sunlight	Twenty-four hours a day, regardless of sunlight
Requirements	Water, carbon dioxide, light energy, chlorophyll	Oxygen, simple sugars (from photosynthesis)
Products	Simple sugars (later converted to starch), oxygen, ATP molecules, water	ATP molecules, carbon dioxide, and water

per day. An acre of corn plants may give off the equivalent of 300,000 gallons of water per day. Plant transpiration is an important aspect of water recycling in all ecosystems.

Plants lose most of the water through the stomata in their leaves. Many common plants have as many as 250,000 stomata per leaf, so it is no wonder that they lose so much water. Although such a great water loss may seem risky for the plant, it is essential, because the stomata must open in order for the carbon dioxide necessary for photosynthesis to enter the leaves efficiently. Transpiration accomplishes another vital function too. It is the force that draws water from the soil up through the vascular system of the plant, bringing with it nutrients such as nitrates and phosphates.

As Figure 9–12 shows, the stomata of a typical plant are usually open during the day, although some close during the late afternoon. During daylight, the guard cells surrounding the stomata take up enough carbon dioxide from the atmosphere to reduce the acidity of plant fluids. This reduction in acidity stimulates chloroplasts in the guard cells to convert stored starch to simple sugars. The sugars become concentrated in the plant fluid. Their concentration signals the guard cells to absorb water, and the stomata open. At night the reverse occurs. The acidity of the cell fluid decreases, starch is stored, water is lost, and the stomata close.

If you are a plant fancier, you know that certain

BOX 9-1

EDIBLE PLANT PARTS

After reading this box, you will be able to amaze your friends and the produce staff at your local supermarket by displaying your knowledge of plant anatomy as you select your purchases. The following shopping list has been arranged according to the basic anatomy of flowering plants—fruits, flowers, seeds, leaves, petioles, stems, and roots.

FRUITS

The edible plant part developed from the ovary of a flower is known as the fruit. All the other edible parts are known as vegetables. Tomatoes, cucumbers, squashes, and pea pods, are actually fruits—not vegetables. Their development is similar to the development of the more commonly accepted fruits, such as apples, oranges, pears, and watermelons. A good rule of thumb: If the edible plant part contains seeds, then it is a fruit. Seedless varieties of fruits may confuse this rule. However, most of them merely have fewer seeds. Only rarely are they totally seedless.

FLOWERS

Are flowers themselves regularly eaten? Yes. It is a custom in the Midwest to batter and fry squash and pumpkin blossoms. They taste something like mushrooms. Broccoli, cauliflower, and artichokes are actually flower heads. The part of broccoli that is called the stem is actually the flower stalk. The tastiest portion of the artichoke is the heart—the receptacle on which the flower sits. The flower, or choke, is inedible. To eat artichokes, you must remove the surrounding bracts (whose softer portions are edible) to unveil the delicate heart. Other flowers are used in making some herbal teas. (What is usually called tea is made from fermented dried leaves.) Some spices, such as cloves, are dried flowers.

SEEDS

What about seeds? We avoid some seeds, but we relish others. Beans, peas, corn, and wheat, all seeds, are the most nutritious parts of the plants that produce them. In fact, seeds are the most important source of food in the world. Without them, most of the world's people could never be fed. Cashews, walnuts, peanuts, and other common nuts are seeds. So are many of the commoner spices, such as nutmeg, caraway, fennel seed, dill seed, mustard seed, celery seed, and poppy seed.

LEAVES

We eat many leaves, calling them leaf vegetables. Lettuce, spinach, cabbage, mustard greens, turnip greens, and collards are all leaf vegetables. And what gourmet could cook without onions, shallots, or garlic? These so-called bulbs are actually underground stems with overlapping leaves. Many spices and seasonings are also leaves. Among them are sage, oregano, bay leaf, basil, and parsley.

PETIOLES

When we eat spinach, we often eat the petioles attached to the leaves. Other petioles are celery (most of the leaf is usually removed) and rhubarb (the leaf is almost always missing, but if attached, beware—the leaves are poisonous).

STEMS

Do we eat stems? Not usually, but two that we commonly consume are asparagus and bamboo shoots. The scales along the asparagus spears are actually leaves. Therefore, we eat both stems and leaves. Bamboo shoots, usually purchased in cans, are a succulent addition to salads and to Chinese-style dishes.

ROOTS

Some common roots that we eat are carrots, radishes, parsnips, beets, and sweet potatoes. White and red potatoes are not roots. They are underground stems known as tubers. (Anatomically, they have the structure of stems, not roots.) Other underground stems, called rhizomes, play a prominent role in the spice section of the supermarket. They are ginger, turmeric (used in curries), and arrowroot (a thickener often used as a substitute for cornstarch). In the refrigerated section of the supermarket, we find ground horseradish root. In the pudding-mix section, we find tapioca, the root of the cassava plant.

Of the many flowering plants, only a small fraction are cultivated for food. Among those used for food, many are closely related. Some of the closest relatives are in the mustard family. It is assumed that the wild flowering ancestor of many edible plants in this family was colewort *(Brassica oleracea).* Artificial selection of this species by humans created three varieties (all of the same species): kale, cabbage, and cauliflower and broccoli (cauliflower and broccoli are even the same variety). Another line of coleworts is turnips, rutabagas, and kohlrabi.

The next time you are in a supermarket, notice how many of the foods you buy are plant products. With the exception of foods in the meat and dairy cases much of what you buy is plant matter. The cereals or grains are the mainstay of our diet.

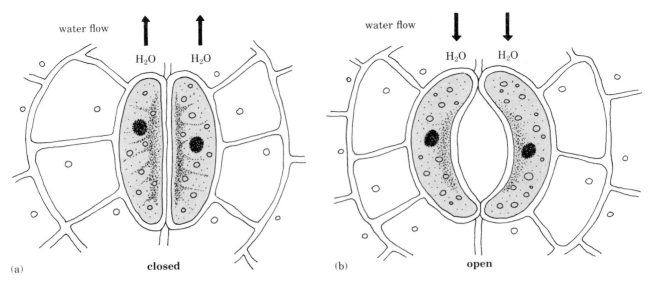

Figure 9–12

Open and closed stomata. (a) When water is lost from guard cells, they decrease in volume and collapse against one another, causing the stoma to close. (b) When the guard cells absorb water, they swell, pulling or arching apart and causing the stoma to open. Most plants have their stomata open during the day and closed at night.

plants grow better when they are misted with water. Misting increases the humidity, thereby reducing water loss. The stomata stay open, photosynthesis operates at full throttle, and the plant flourishes. But without misting the reverse may occur. In dry air, the stomata close and photosynthesis and respiration decline, perhaps to the point where the plant dies. Many tropical houseplants do not thrive in air-conditioned homes because air conditioning removes humidity. The plants that require misting are usually tropical plants that need a humid environment. Most of the desirable houseplants are of this type. Some tropical plants, such as the pineapple, open their stomata at night and close them during the day, thereby reducing water loss. Many ornamental plants do not require misting because they have thick epidermal layers or cork, and water loss is minimal.

Temperature also affects the rate of transpiration. Hot, dry air speeds transpiration. (People respond the same way. On hot days they perspire profusely, and great amounts of water are lost from their bodies. They drink all kinds of beverages to make it up.) Houseplants generally need more water in the winter, when the furnace is on, because of the dry heat.

Many other factors affect transpiration and, ultimately, photosynthesis and respiration. For example, as a plant absorbs sunlight, it also absorbs heat energy. Usually there is more than enough light energy for photosynthesis. The extra is absorbed as heat energy. The heat warms the plant leaves, which become hotter than the surrounding air. This increases transpiration. People experience a similar effect when they sunbathe. They want enough sunlight for melanin synthesis, but much of the sunlight is absorbed as heat. Thus people get hot and sticky, perspire, and lose body fluids.

Wind also affects transpiration. In tanning, wind helps cool the body by evaporating sweat. The same is true of plants. However, the reverse can also occur. Wind can reduce humidity and increase water loss.

The amount of water in the soil directly affects transpiration in plants. Water loss from plants is greatest during the day. Most plants recover from any temporary water deficiency at night if the soil is moist. Do you now wonder why lawns and flower beds should be watered late in the evening? The answer is obvious. During the night, the rate of evaporation of water in the soil is decreased. More water is retained in the soil for plant uptake, which helps make up for daytime transpiration. If you water during the hottest part of the day, when sunlight is plentiful, much of the water will be lost from the soil and will not be of use to the plant in supporting photosynthesis.

Transpiration is a complicated process. Basically, however, it is water loss that is governed by sunlight, humidity, wind, and available water. It affects the rate of photosynthesis and the rate of respiration.

Digestion

Some of the sugar produced during photosynthesis is converted to starch for storage. When plant cells become depleted of simple sugar, as they do when the plant is kept in the dark and is unable to carry on photosynthesis, the starch is converted into the simple sugars that are used in respiration. Plant digestion is very much like animal digestion, with one major difference. Most animals have digestive systems—series of open passages into which cells secrete enzymes that break food into smaller and smaller molecules. This form of digestion is called extracellular (occurring outside the cells). Digestion in plants is intracellular. It occurs within individual cells.

The digestion of stored starch is one of the more common processes in plant cells. Parenchyma cells often have large stores of starch, and the intracellular breakdown of starch takes place in these cells. Like animal cells, plant cells secrete the enzyme amylase. This enzyme breaks starch down into maltose, a double sugar (or disaccharide). The maltose is further converted into two molecules of glucose. Sucrose, the form of sugar that plants translocate through the phloem of their vascular systems, is converted into glucose and fructose. Plant cells also have another unique enzyme for starch digestion: starch phosphorylase. This enzyme breaks starch down into glucose phosphate, which is then converted into ATP molecules[59]. These molecules have high energy phosphate bonds. What makes starch phosphorylase unique is that, within certain pH levels[20], it can also convert simple sugars back to starch.

You are probably familiar with intracellular starch formation. Have you heard that corn on the cob is best eaten the same day that it is picked? Corn growing in sunlight produces large quantities of simple sugars, many of which are stored in the kernels. Therefore, when picked and eaten fresh, the corn is very sweet. But if the corn is stored for a day or two, the simple sugars are converted into starch molecules, and the corn tastes starchy. Varieties of corn that taste much sweeter than conventional varieties are now available. Their sugar is converted to starch very slowly. Such corn retains its sweet taste for a week or longer.

The reverse reaction occurs when potatoes are stored too long under cool conditions. For potatoes, we prefer the starch flavor, not sweetness. When potatoes are stored too long, the intracellular digestion of starch causes them to become soft and watery and sweet when cooked. The process can be reversed if the potatoes are left at room temperature for a short time before cooking. Then the sugars are converted back into starch.

Many root vegetables—for example, carrots, beets, and parsnips—are best left in the ground until eaten. In the cool earth, their sugars are slowly converted to starch. But in the warmth of the produce department of a supermarket, the conversion occurs quickly, and the vegetables soon lose their palatability. Root cellars were popular at one time because their cool temperatures delayed the conversion of sugar to starch. Refrigerators do the same job today.

Plants also contain fats, or lipids[26]. Many seeds have large amounts of oils (liquid fats). Corn kernels and soybean seeds are economically important for this reason. Corn oil is converted into many products, from cooking oil and margarine to plastics. Many nuts, such as cashews and walnuts, are also known for their oil content. Plant cells use lipase to break these fats down into glycerol and fatty acids in the same way they are broken down by animals. No one knows how the plant uses the glycerol and fatty acids, although they apparently are not transported within the plant. Current theory suggests that they are used within the cell that digests them.

Plants also store proteins[22], which are broken down by enzymes similar to those found in animals. The protein-digesting enzyme complexes are commonly referred to as proteinases. Their products are various amino acids. Several plants are known for their powerful protein-digesting enzymes. Boxes of gelatin contain the warning "Avoid mixing with fresh pineapple". A protein-digesting enzyme known as bromelin is concentrated in pineapple juice. Gelatin is almost entirely protein. Mix the two together, and the result is gelatin soup with floating chunks of pineapple. The bromelin digests the protein molecules of the gelatin.

It is becoming more and more common for homemakers and fast food restaurants to use cheaper cuts of meat treated with meat tenderizer to make them more tender. Some meat tenderizers contain papain, a powerful protein-digesting enzyme that comes from unripe papayas. Papain is used in both medicine and industry.

In summary, we can say that plant digestion is similar to animal digestion. However, the process is intracellular in plants and extracellular in most animals.

Translocation

The movement of water, minerals, and food materials within a plant is **translocation.** This movement is analogous to circulation in animals, although its details are much different.

Trimming trees or shrubs during the growing season can cause the newly cut ends to "bleed." The liquid is plant sap (water, minerals, and food materials). The bleeding is a visible sign of translocation. A seriously wounded plant may die. To prevent this, people seal the exposed xylem and phloem with various sprays—plant first aid. Better yet, trim the trees and shrubs before the growing season, when the plants are dormant and the sap is not being translocated. Water transport in plants is part of translocation. It occurs in the xylem. In trees, most of the material called wood is xylem—lignified cellulose cell walls that hold water.

The food materials a plant needs for respiration move up and down the plant in sieve tubes in the phloem. Most of the carbohydrates in phloem are in the form of sucrose (a simple sugar) mixed with water. Simple sugars are often converted to starches in underground tubers and roots. Later, the starches may be converted back to simple sugars and again circulated in the plant sap.

If a continuous ring of bark is stripped from a tree (called girdling), its phloem ring is destroyed and the translocation of nutrient materials is interrupted. A bulge in the bark above the wound may become obvious because there is a pronounced accumulation of carbohydrates above the girdle. This phenomenon demonstrates that most of the materials in the phloem flow downward during the growing season. If girdling occurs during the summer, the tree usually survives the season but shows no increase in the bulk of roots. Eventually, the tree dies because its roots starve to death. Deer and squirrels often girdle trees as they seek water and dissolved sugars, particularly during periods of drought.

Many kinds of plants, particularly deciduous trees, become dormant in the autumn, after all the leaves have fallen. Without the leaves, there are no organs of photosynthesis and no new production of nutrients. However, the root system is filled with starch reservoirs from the nearly six months of active photosynthesis before the leaves dropped. During dormancy, there is little translocation of plant fluids. In temperate regions, much of the water needed for translocation is locked away in the form of ice.

In the spring, melted ice and rain are absorbed by the tree, and translocation begins again. Many of the stored starches in the roots are broken down into sugars and are transported upward for the growth of new stems and leaves. The amount of sap moving upward in early spring varies from tree to tree. Sap from the sugar maple is collected and concentrated, by boiling and evaporation, into maple sugar. A small spigot is driven into the phloem of the bark, and the sap drips into a bucket. Although considerable amounts of sap are collected from each maple tree, the trees are not harmed.

Translocation is complicated. An understanding of the mechanisms responsible for the movement of materials in plants requires a knowledge of physics, particularly fluid dynamics, a topic that goes beyond the scope of this book. Plants do not have pumps, or hearts, to move fluids. Therefore, the fluids are moved by the processes of diffusion, osmosis, and active transport.

Hormones

Like animals, plants have **hormones.** Plant hormones are usually produced by the most rapidly growing parts, including the root tips and stems (see Figure 9–13). Most of the known plant hormones are involved with promoting growth and flowering. Table 9–2 lists some of the better-understood plant hormones and their functions. Like animal hormones, plant hormones are intracellular products. They are transported by circulating sap to various parts of the plant. Hormones produced in one part of the plant may affect a distant part of the plant. Animal hormones function much the same way.

Various combinations of plant hormones stimulate or inhibit the movement of plant parts toward light, the opening of buds and leaves, the production of flowers, the healing of wounds, and rapid growth of the plant.

Reproduction

Plants reproduce sexually and asexually. The details are given in Chapter 10.

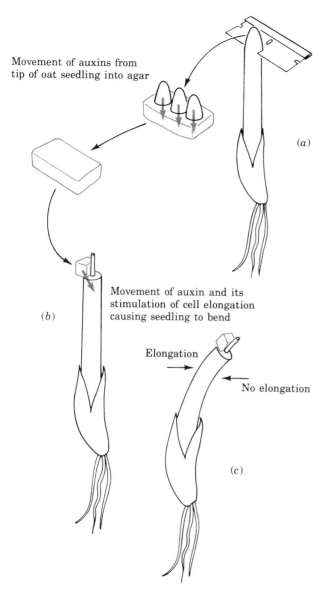

Movement of auxins from tip of oat seedling into agar

(a)

Movement of auxin and its stimulation of cell elongation causing seedling to bend

(b)

Elongation →   ← No elongation

(c)

Figure 9–13
The behavior of an oat seedling. (a) The top of an oat seedling is removed, and the cut area is covered with a small block of agar (gelatin). The block absorbs the plant hormones known as auxins. (b) A small block of the agar with the auxins is placed to one side of a similar oat seedling with the tip removed. (c) The auxins from the agar block diffuse into the cells on that side, causing them to enlarge and elongate. As they do, the growing tip bends in the opposite direction. The response takes less than an hour. The reverse could be demonstrated if the block of agar were laced on the opposite side of the seedling. Light influences plants in the same way: More auxins are produced by cells on the dark side than by cells on the light side. As a result, the stem, leaf, or entire plant bends toward the light side.

Table 9–2
Plant Hormones

HORMONE*	FUNCTION
Florigen (a hypothetical hormone not yet isolated)	Promotes flowering
Auxin	Promotes growth by causing cells to enlarge; most pronounced in regions of elongation, stems, roots, and leaves; causes vertical growth, the dominant direction of growth; involved in fall of leaves and fruits and in formation of tissues over wounds
Gibberellins	Similar to auxin in promoting growth but may be restricted to elongation of stems only; stimulates seed germination; involved in flowering
Cytokinin	Stimulates cell division; associated with meristematic tissues, embryos, and fruits

*Each hormone is actually a complex of several interacting hormones.

SUMMARY

1. The anatomical structures of seed plants can be divided into two groups, vegetative and reproductive. The vegetative organs include the root, stem, and leaves. The sexual reproductive organs are flowers or cones.

2. The tissues of plants are divided into two groups, meristematic and permanent. Meristematic tissue is capable of reproducing throughout the life of the plant. It is found at the tips of growing roots, at the ends of stems, and at the bases of leaves.

3. The permanent plant tissues include the following. The epidermis covers the outer surface of the plant tissues and organs. Parenchymal tissue composes the bulk of organs such as leaves and becomes the major tissue of photosynthesis. Sclerenchyma tissue protects and supports the plant. Collenchyma tissue supports the growing parts of the plant. Cork protects the plant, prevents water loss, and provides insulation. Xylem and phloem conduct water, minerals, and nutrients.

4. Leaves are the principal photosynthetic organs of plants. The openings into their internal tissues, called stomata, are the sites of gas exchange.

5. Most annual plants have herbaceous stems that grow little in diameter. Many perennial plants have woody stems, which constantly increase in diameter. As they increase, layers of xylem form rings, each ring representing a growing season.

6. The root systems of most plants are as extensive as the above-ground parts of the plants. The roots provide plants with water and minerals and are often the site of food storage.

7. Photosynthesis is the process by which plants convert light energy to chemical energy that can be held in the bonds of sugar molecules. Like animals, plants respire, a process whereby chemical energy is released for growth and reproduction.

8. Transpiration is the loss of water from plants. Most of the loss occurs through the stomata. Many factors influence the transpiration rate. Among them are humidity, temperature, and amount of water in the soil.

9. Plant digestion is similar to animal digestion. However, because plants lack a digestive system, plant digestion is intracellular (within cells).

10. Translocation is the movement of water, minerals, and food materials within a plant. It is analogous to circulation in animals.

11. Like animals, plants have circulating hormones. Most plant hormones are produced by the most rapidly growing tissues and are involved with growth and flowering.

THE NEXT DECADE

FUTURE PRODUCTS FROM PLANTS

Plants currently produce many useful and economically important products. Scientists are searching, however, for plants that can produce substitute or new products for human use. They are interested in new sources of oil, plant saps for rubber, and even naturally occuring insecticides found in plant tissues.

What started the search was the recent shortage of fossil fuels. The most favorable prediction is that, with conservation, we may have sufficient supplies for another century or two. The least favorable prediction is that we could run out of our supplies in the next decade. A large number of products are made from fossil fuels. They include gasoline, synthetic rubber, plastics, fertilizers, and insecticides.

The search has taken scientists to many of the world's deserts. One plant of special interest is a desert shrub called guayule (pronounced whe-you-lay). The shrub produces a rubber that is essentially the same as that produced by the famous rubber tree of Asia. Guayule was an important commercial source of rubber earlier in this century. However, because imported rubber was easily obtained, guayule propagation stopped. During World War II, the Japanese cut off U.S. supplies of rubber from Asia, so the United States once again began cultivating guayule. After the war, artificial rubber, made from petroleum, replaced natural rubber. At that point, all interest in guayule waned.

Now scientists and the federal government are encouraging gua-

yule farming. Guayule is easily grown in the United States and can provide all the natural rubber we need. The government has authorized a multimillion-dollar project for cultivation of guayule, and the military is testing the sap as a substitute for rubber from rubber trees. Some commercial manufacturers have already made sample truck and tractor tires from guayule rubber.

Guayule sounds like a sure thing, except for one problem: The plant has unusual and unpredictable breeding habits. These habits must be studied in more detail by biologists so cultivation will be successful. One odd habit is that both diploid and tetraploid guayule plants exist. Each type is capable of both sexual and asexual reproduction. Frequently, diploid plants cross with tetraploid plants, producing triploid plants, which can reproduce only asexually. The life cycle of guayule is thus complicated and unpredictable. Plant breeders, however, are seeking new hybrids with desirable characteristics—hybrids that can reproduce only asexually so that all offspring will possess the same desirable characteristics as the parent plants.

Other noncultivated plants that produce oil are being considered for extensive cultivation. Several species of desert gourds, for example, produce an oil similar to the oil of soybeans and cottonseeds. The gourd oil is low in saturated fatty acids and contains large amounts of linoleic acid, an essential fatty acid. The gourd oil could become a key ingre-

dient of supermarket vegetable oils, margarines, and so on.

Another desert plant is jojoba (ho-ho-bah). This plant produces a seed oil that is essentially the same as sperm whale oil. Scientists are planning to use it in many products, including lubricants, floor and auto waxes, waxed paper for food containers, and even slow-burning candles. Currently, jojoba oil is being used in skin and facial creams and shampoos.

The desert gopher plant produces an oil that is considered a good substitute for crude oil. Some test plots have yielded more than 25 barrels of oil per acre.

Many plants, particular desert species, produce their own insecticides. The bug sprays that people buy during the summer are all made from petroleum products. The insecticides made by some plant tissues repel insects rather than killing them. These natural insecticides keep away insects that might otherwise eat the plants. Other plants manufacture as many as 10 different compounds that prevent insects from producing the hormone that is necessary for insects to develop into fertile adults. The plants actually keep insects from reproducing. These compounds and similar ones are being studied. One day, they may be used on a wide scale as a new form of biological control of insect pests. Other plant compounds have been shown to prevent the growth of cancer cells. In the next decade, they may be used in the treatment and cure of various cancers.

Many more plant uses will be found during the next decade. There is, however, one economic problem. Land for the cultivation of potentially important crops should not be taken from land already in use for food and fiber crops. One reason that desert plants are being studied is that most deserts are not currently used for crops.

Deserts are not easily managed; however, botanists are considering tissue culture for desert crops. Many plant tissues and cells are currently being cultured in huge vats, some of which have a capacity of 20,000 liters. Botanists have discovered that certain tissue cultures produce more of the desired compounds than do mature cultivated plants. For example, nearly any type of cell from a coffee tree will produce caffeine when cultured. The amount of caffeine produced by tissue cultures is 1.5 percent of the dry weight of the cultured tissue. This amount is significantly greater than the amount produced by the coffee tree. Although we have little need for additional caffeine, this example does show the potential of plant tissue cultures. Other substances being produced by tissue culture include the ingredients in a popular laxative, a pigment, and terpenes (a substance used in the manufacture of such products as paint).

In the next decade, we probably won't see fields of cultivated guayule or jojoba, because their components will be produced by tissue culture techniques. Tissue cultures will provide an ever-increasing variety of new products that will benefit humanity. Many of the cells in vats will be clones (genetically identical). They will be "genetically tuned" by scientific gene splicers to produce what humans need.

FOR DISCUSSION

1. What are the differences between meristematic and permanent tissues?

2. List the permanent tissues of plants, and give the function of each.

3. Define the following terms: cutin, primary and secondary cell walls, stomata, cork, xylem, phloem, vessel elements, sieve tubes and plates, and companion cells.

4. What are the primary functions of leaves, stems, and roots?

5. Draw a section of a leaf, and label the external and internal tissues. Give the function of all these tissues.

6. Compare and contrast the internal tissues of stems and roots.

7. What are the differences between herbaceous and woody stems?

8. Define *transpiration*. What are the factors that influence the rate of transpiration?

9. Compare and contrast digestion in plants and animals.

10. What is translocation? What vessel elements are involved in plant translocation?

11. Compare and contrast the endocrine control of plants and animals.

SUGGESTED READINGS

BOTANY, 5th ed., by Carl L. Wilson, Walter E. Loomis, and Taylor A. Steeves. Holt, Rinehart and Winston: 1971.
One of the classic botany books and still one of the best. Tremendous detail and clear explanations.
GROWING PLANTS INDOORS, by J. Lee Taylor. Burgess: 1977.
A good primer to basic plant anatomy and function, including how to grow your own plants indoors.
INTRODUCTORY PLANT BIOLOGY, by Kingsley R. Stern. William C. Brown: 1979.
An easy-to-read introduction to botany.
THE PLANTS, by Frits W. Went and the editors of *Life*. Time: 1963.
Good coverage of plant anatomy and types of plants. Well worth consulting for the illustrations.

SECTION FOUR

REPRODUCTION AND DEVELOPMENT

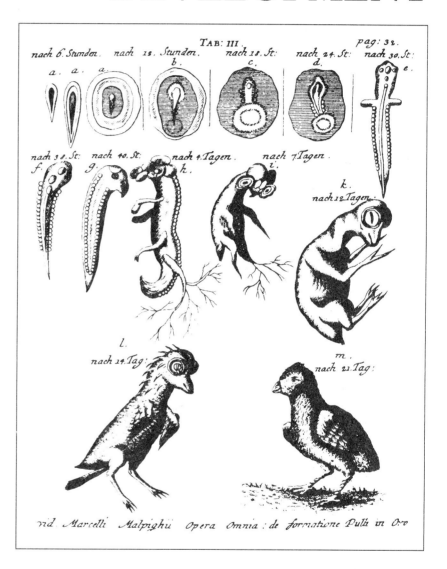

In 1672 Marcello Malpighi made these drawings of the development of a chick embryo.

10

ANIMAL AND PLANT REPRODUCTION

PREVIEW QUESTIONS

After reading this chapter, you should be able to answer the following questions:

1. What kinds of organisms can reproduce without sex (asexually)? How is asexual reproduction accomplished?

2. What are the basic differences between asexual and sexual reproduction?

3. Sexual reproduction of animals occurs in a variety of ways. Describe several of these ways.

4. How do mosses, ferns, and seed plants reproduce sexually?

A unique characteristic of living organisms is their ability to reproduce. Most people would probably define reproduction in terms of sex, external genitals, and copulation. These terms describe certain aspects of reproduction for humans and many other organisms, but not for all organisms. In fact, some organisms reproduce without sex. Asexual reproduction, (*a* means "no" or "without") as sexless reproduction is called, is an effective way to create offspring. The offspring produced by this method are carbon copies of their parents, although they begin life as smaller individuals. Sexual reproduction, on the other hand, creates genetic variability.

Asexual reproduction is relatively common, but it threatens survival because it guarantees uniformity. Offspring that differ genetically from either parent may be more likely to be able to adapt to changing environments, which helps ensure survival of the species. Although sexual reproduction is more common than asexual reproduction, some organisms can reproduce both ways.

ASEXUAL REPRODUCTION

Organisms that reproduce asexually do so by the process of simple cell division▫. A parent cell becomes two daughter cells that are identical to the parent in genetic makeup.

Asexual Reproduction of Unicellular Organisms

Most unicellular (single-celled) organisms are capable of reproducing asexually. Each organism is a free-

77

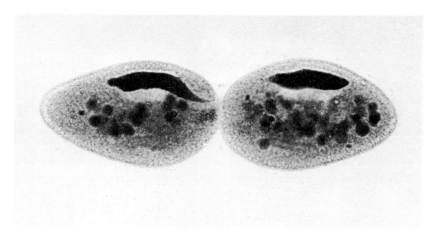

Figure 10–1
Fission in *Paramecia*. This *paramecium* has nearly completed a cycle of asexual reproduction known as fission. An individual is divided in half by mitosis, creating two new individuals that are genetically identical. (Courtesy Carolina Biological Supply Company.)

living cell. When it undergoes mitosis, it produces two identical free-living cells. Generally, the process of mitosis in a unicellular organism is referred to as fission (meaning "splitting"). It is a mode of asexual reproduction (see Figure 10–1).

Many unicellular organisms undergo fission. Among them are the common paramecium, other protozoans, bacteria, and many algae.

Asexual Reproduction of Multicellular Organisms

Many multicellular organisms are capable of asexual reproduction. The method is the same as for unicellular organisms. However, it involves the synchronous division of many cells (instead of only one) to produce an outgrowth that is commonly referred to as a bud.

One of the best examples of **budding** can be seen in asexual reproduction of a freshwater hydra, a common aquatic organism that is particularly abundant in ponds and streams during the spring. Hydra are small and delicate, usually less than a centimeter in length. They are often visible in great numbers on rocks and pilings, where they look like white fuzz. They may be equally common in home aquariums. Under a microscope, they resemble miniature sea anemones. They possess a foot that attaches to the substratum (the base the organism sits on) and a column crowned by a row of tentacles that carry food into their mouths. Close examination of the column often reveals small outgrowths, which are the beginnings of asexual buds (see Figure 10–2).

As the buds grow, they too develop a column and

Figure 10–2
Hydra asexual reproduction by budding. The bud from the column of an adult hydra develops into a much smaller individual that is a faithful copy of the adult. This copy will soon detach itself from the parent and attach itself to the substratum, where it can bud new individuals. (Courtesy Carolina Biological Supply Company.)

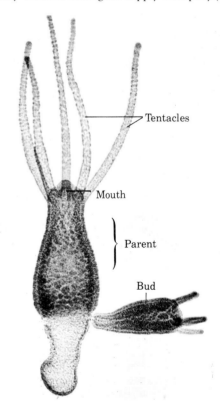

Tentacles

Mouth

Parent

Bud

a crown of tentacles with a centrally located mouth. The mature bud is an exact copy of the parent but only a fraction of the parent's size. When the duplication is completed, the bud detaches itself from the parent and is swept into the environment. There it attaches to the substratum, grows to adult size, and becomes capable of producing its own asexual buds.

Small cross-eyed flatworms (about one to two centimeters long), known as planarians, are also capable of asexual reproduction. Commonly found in ponds and streams, planarians are frequently seen when rocks are overturned. Under certain environmental conditions, a single planarian divides in half by a transverse constriction halfway along its body. By budding (in this case better called regeneration), each half is able to create the missing components of the adult. Each fragment thus regenerates an entire worm.

The asexual reproductive powers of planarians can be demonstrated in the laboratory. If a planarian is bisected with a blade, it does not die. Instead, the anterior (head) end heals and in time completely regenerates the missing posterior (tail) portion. What is even more miraculous is that the tail end also heals and in time regenerates the missing head end, including the eyes and the brain (see Fig. 10–3).

Many other organisms are also capable of regenerating lost parts. For example, suppose that one of the arms of the common starfish is removed. As long as the arm is still attached to a small portion of the central disk, it will asexually regenerate all the rest of the starfish (see Fig. 10–4). The parent starfish will also regenerate the missing arm.

Asexual Reproduction of Plants

Plants, too, are capable of reproducing asexually. In their case, however, asexual reproduction is commonly referred to as **vegetative reproduction.** If you raise houseplants, you are probably aware of vegetative reproduction. When you began your collection, you most likely bought adult plants from a nursery. You probably did not try to grow the plants from seed, because the process is difficult. (The formation and dispersal of seeds is a method of sexual reproduction.) Later, you may have traded "starts" with other enthusiasts to increase your collection. These starts are cuttings, or stems, taken from parent plants. When nurtured in water or damp soil, the stems develop roots and asexually create other plants.

Obviously, asexual reproduction of plants has important economic consequences.

One of the best examples of vegetative reproduction can be seen in the potato. A specialized organ, known as the tuber, stores starch. Figure 10–5 shows the whole potato plant, a green, lush bush that matures late in the summer and often has small yellow blossoms. When the leaves wither, the gardener knows that potatoes are fully formed and can be dug. To propagate potatoes, the gardener cuts each tuber into a number of pieces. Each piece must contain a bud, often called an "eye," which will become the source of the new plant. When planted, each potato fragment will asexually re-create the entire parent plant.

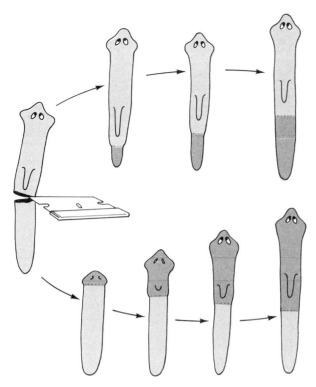

Figure 10–3
Regeneration of a planarian. A planarian is cut in half with a razor blade. It does not die. In time, its head end generates a new tail, and its tail end generates a new head. Regeneration is a mode of asexual reproduction for planarians and other species.

Figure 10–4
A starfish regenerating two arms lost through injury (notice the two very short arms). It is quite possible that the missing portion of the starfish, the part with the two lost arms, is regenerating three arms. (Runk/Schoenberg; Grant Heilman Photography.)

The common strawberry plant provides another example of vegetative reproduction. Strawberries are commonly propagated not by seeds but by "runners"—horizontal stems known as **stolons.** As the parent plant grows, the stolons extend in all directions. At regular intervals, they penetrate the soil. At each of these points, a root system develops, and a perfect copy of the parent plant is asexually created. The stolon and its new underground root system are commonly used to propagate the strawberry plant.

Limitations of Asexual Reproduction

Obviously, many organisms can create new individuals simply by reproducing themselves asexually. However, a species that reproduced exclusively by asexual methods could be heading toward extinction. The offspring resulting from asexual reproduction have the same **genotypes** (genetic characteristics) as

their parents—and therefore little genetic variation for **natural selection** to act on.

Sooner or later, a generation of asexually reproduced offspring might find itself in a changed environment. Unequipped genetically to cope with the change, it might perish, and the species would become extinct. Most of the organisms that frequently reproduce asexually do so when environmental conditions are at their best. They resort to sexual reproduction when the environment is at its worst.

Economic Importance of Asexual Reproduction

Agriculture has often capitalized on the asexual reproduction abilities of some plants. For example, strawberries and potatoes are usually propagated asexually. Thus the genotype is preserved, and the edible parts are consistently good from generation to generation. If strawberries were propagated sexually (by seeds),

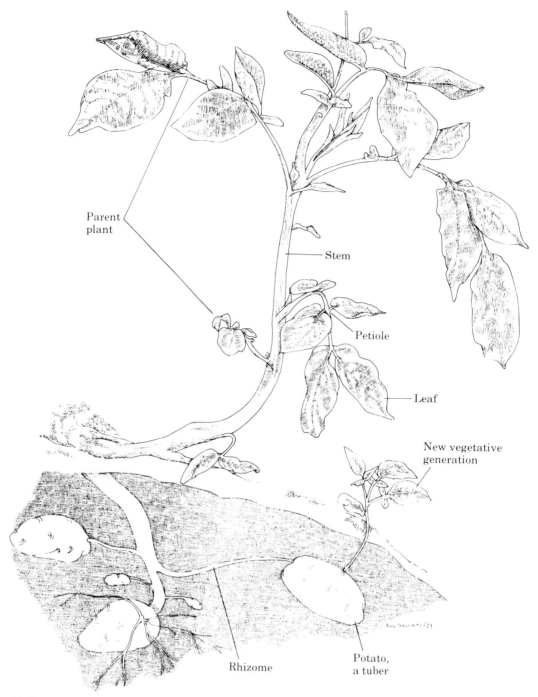

Parent plant

Stem

Petiole

Leaf

New vegetative generation

Rhizome

Potato, a tuber

Figure 10–5

The anatomy of a potato plant. Humans propagate potatoes by asexual methods. These methods preserve the genetic makeup of each potato variety and thereby help guarantee the quality of the potato crop year after year. The potato itself is a tuber, a modified stem. When the potato matures, it can be dug up and eaten. Some of each crop is cut into small pieces and saved as "seed" potatoes. Each piece planted must possess an "eye". When planted, each eye-containing fragment will grow an entire new plant, producing stem, leaves, and blossoms as shown. More importantly, each new plant will produce many new tubers, or potatoes.

they would produce a certain number of plants of the original parental type—the small wild strawberries that have almost no economic importance.

Suppose that while you are walking in the woods on your property, you come upon a small apple tree that has the most perfect apples you have ever encountered. They are beautifully red, large, crisp, and sweet. You have never tasted anything like them. While eating an apple, you contemplate the fortune you could make if you could market this variety. But there is only one tree, and you can't get rich selling from only one tree. Could you get more of the same type of tree by planting the seeds from your apple? No. Planting the seeds would be a form of sexual reproduction. Therefore, only a few of the offspring would have the fruit of the parent tree. The solution to your problem is grafting.

Grafting is a human-created method of vetetative reproduction. A branch from one tree (one genotype), called the **scion**, is grafted to the stem or trunk of another tree (a different genotype), called the **stock** (see Figure 10–6). An incision is made in the stock so that the tissues of the scion and stock can eventually

Figure 10–6
The process of grafting. A branch from one tree, called the scion, is grafted to the stem or trunk of another tree, called the stock. The graft does not change the flowering or fruiting ability of either scion or stock. After grafting, trees bear the fruits and flowers of both scion and stock.

Scion

Stock

fuse together, completing the graft. The graft does not alter the flowering or fruiting ability of the scion or the stock.

Nearly all apple trees are propagated by grafting, as are many common varieties of fruits and nuts. In fact, several different varieties of apples can come from a single tree. All Red Delicious apples look and taste the same, crop after crop, whenever and wherever you buy them. The genotype has been preserved by grafting.

The citrus industry depends on grafting. Many orange trees in Florida have several varieties of oranges on the same tree. It is common to graft sweet varieties on a particular species of sour orange that has a disease-resistant root system. The sour orange stock can still produce sour oranges, but they are not picked or sold. Similarly, nearly all of the English walnuts grown in the western United States are grafted on black walnut stocks, because the black walnut root system is resistant to many diseases that plague the English walnut.

SEXUAL REPRODUCTION IN ANIMALS

Organisms that reproduce sexually create genetic variability—new gene combinations in their offspring. Thus each organism produced is genetically different from even its closest relative. Genetic variability is tested by the natural environment. If an organism's mix of genes allows it to succeed to reproductive age, it will pass its genetic advantage to its offspring. If the organism fails to reproduce, that genetic combination ceases to exist.

All sexually reproducing animals have the following characteristics:

● All produce **gametes**, or sex cells, by the process of **meiosis**▫.
● All the gametes are **haploid**—having half the parent's chromosomes.
● The gametes produced by a species unite, or fuse, in the process known as **fertilization**.
● The single **diploid** cell resulting from fertilization is known as a **zygote**.

New genetic combinations inherited by the offspring of sexually reproducing parents are a result of these characteristics (which are detailed in Chapter 5).

A great amount of genetic variation is created during meiosis□ when crossing over and random assortment of chromosomes□ occur. More genetic variation comes about by the chance effect of gamete fusion, which brings together genetic material from two parents and thus creates a new genotype. The resulting zygote possesses different gene combinations, depending on which gametes have fused. For example, the average human male releases approximately 160 million sperm cells in a single ejaculation, even though only one cell is needed to fertilize an ovum. Which sperm cell fertilizes the ovum is entirely a matter of chance. It is likely that the sperm cells differ genetically from one another because of crossing over and random assortment of genes. Thus the chance of predicting the genetic contribution of a sperm cell in a human zygote is essentially 1 in 160 million, assuming that each sperm cell is a unique type.

Sexual reproduction, then, is a mechanism that allows for the creation of new, unique gene combinations. The mechanism can be expressed as a cycle. Figure 10–7, for example, shows a diploid parent creating haploid gametes that unite (fertilization). They form a zygote that develops into a diploid organism

Figure 10–7
The life cycle of any sexually reproducing animal. Diploid (2n) male and female parents produce haploid (n) gametes by the process of meiosis. Eventually, the gametes fuse during fertilization. The resulting diploid cell (2n) is known as a zygote. Further cell division produces an embryo that eventually grows to become an adult male or female.

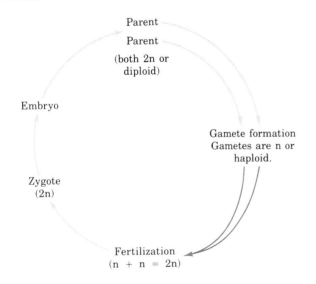

Parent
Parent
(both 2n or diploid)

Embryo

Gamete formation
Gametes are n or haploid.

Zygote
(2n)

Fertilization
(n + n = 2n)

capable of repeating the cycle. The life cycles of all sexually reproducing animals are covered by the generalized cycle shown in the figure.

There are many ways to catalog animals according to how they reproduce sexually. Among them is classification according to the adaptations that have evolved to ensure fertilization of gametes. These adaptations are often unique and bizarre.

External Fertilization

When the gametes of an organism fuse outside or away from either parent, the organism is said to reproduce by external fertilization. Such organisms broadcast large numbers (millions or billions) of gametes into the environment. The numbers involved make it probable that many gametes will fuse and create zygotes and, later, new individuals. The broadcasting animals are all aquatic organisms, mainly marine invertebrates and fish. Despite its simplicity, this method of reproduction is an expensive way to perpetuate the species. The organisms must devote an enormous amount of energy to producing the large numbers of gametes. Most of their other activities therefore receive lower priorities.

An interesting broadcaster is the palolo worm, found in the South Pacific and best known around the Samoan islands. Palolo worms are small annelids that are related to the common earthworm. They burrow into coral reefs. The posterior segments of both males and females become specialized gamete-producing organs. The males produce millions of sperm cells, and the females produce thousands of eggs. When the moon is right, the gamete-producing segments break loose from the adult worms and swim to the surface, where they congregate in enormous numbers, known as swarms.

In the South Pacific, swarms occur only during October and November at dawn on the morning before and the morning after the third quarter of the moon. At these times, the surface waters may be so thick with sexual segments that it becomes difficult to row a boat. Soon the sexual segments burst, liberating their stores of gametes, and fertilization occurs. Many of the zygotes do not survive, but those that do develop into free-swimming **larvae** that eventually settle on coral reefs and there become adult worms. The local islanders consider the sexual segments a delicacy, much like caviar.

INTERNAL OR EXTERNAL FERTILIZATION?

Some tropical fishes are mouth brooders. The male or female of these species (usually the female) carries the eggs in the mouth until the eggs become free-swimming juveniles. This type of parental care lasts from thirteen to forty-three days, depending on the species. The advantages are obvious.

When a female approaches the breeding territory of a male mouth brooder, a brief courtship follows. At the end, the female releases eggs, and the male then releases sperm. The female reflexively scoops the eggs into her mouth. The reflex in some species is so fast that the males do not have enough time to fertilize the eggs before the female scoops them up.

But fertilization is ensured by a unique reproductive strategy. On the anal fins of the males are one or more colored spots referred to as egg dummies. In most species, the dummies are nearly the same size and color as the eggs produced by the female. While the female picks up the unfertilized eggs, the male displays his egg dummies. He tips to one side, extends the anal fin, and quivers his body, giving motion to the egg dummies. The female snaps at the egg dummies, and the male releases sperm that run down the anal fin into the mouth of the female, thereby ensuring fertilization of the real eggs.

External Fertilization with Amplexus

The form of reproduction known as external fertilization with amplexus involves physical contact between males and females (*amplexus* means "embrance"). However, the males lack an intromittent organ such as a penis. This kind of reproduction is typical of frogs and toads.

During the mating season, the males give vocal calls that attract the females to their breeding sites, usually ponds or swamps. When a female approaches, the male mounts her. His forearms clasp her in an embrace. In this position, the cloacae of the male and female are close together. (*Cloaca* literally means "sewer"; the great sewage system in ancient Rome was called the *cloaca maximus*.) In animals, a cloaca is a passageway for excretion, passage of gametes, and waste elimination from the digestive tract. The fe-

male releases her eggs in gelatinous strings. As they pass from her body, they are immediately fertilized by the male. During the reproductive season, male frogs are highly motivated sexually. They will therefore attempt to amplex anything in their vicinity, including other male frogs and an occasional fish.

Many fish also rely on external fertilization with amplexus. The Siamese fighting fish and its close relatives are an example. Male and female Siamese fighting fish (*Betta splendens*) engage in a beautiful and lengthy courtship that involves colorful displays by both participants. Courtship culminates in amplexus, with the male clasping his body around the female to bring his urogenital pore close to hers. The male's body then goes into convulsive spasms, which initiate the spawning reflex in the female. A few eggs fall from the cloaca of the female and are immediately fertilized by the male. As the fertilized eggs continue to fall toward the substratum, the male releases his grip on the female and quickly seizes the eggs in his mouth. He then spits them into a nest of bubbles he has previously built. Once the zygotes are resting in their individual bubbles, courtship resumes, and the ceremony is repeated again and again.

Internal Fertilization without Copulation

Fertilization that occurs in the urogenital or reproductive tract of the female is known as internal fertilization. Some species achieve it even though the male and female do not copulate. (Copulation is sexual union during which the male inserts his penis or a similar organ into the body of a female.) Bizarre as this mode of reproduction may seem, it has many practitioners—for example, the newt, a small aquatic amphibian. Newt reproduction is preceded by a lengthy courtship. The male poses in front of the female, holding his broad, elaborately frilled tail at an angle and then violently whips it about. The most dramatic part of the courtship display is when the male folds his tail alongside his body and lashes it backward, directing a stream of water toward the female. At the instant the tail is lashed, stimulating secretions (**pheromones**) are released from the male's cloaca and transmitted to the female.

The female is at first hesitant. Then she begins to move toward the male. The male turns, repeats his display, and creeps forward as the female follows. After creeping a few inches, he stops and wiggles his tail. The female moves forward until her snout touches the male's tail. At this instant, the male releases a **spermatophore**—a gelatinous capsule of sperm on a short stalk—from his cloaca. At the next advance of the male, the female straddles the freshly deposited spermatophore and draws it into her cloaca. The action of the male guides the female. He must move only one body length at a time so she can position herself correctly. The result is internal fertilization without actual copulation.

This mode of reproduction is also used by certain insects and spiders. In some instances, spermatophores are transferred from male to female during courtship. In others, the males may leave spermatophores at various locations on the chance that females will find them.

Internal Fertilization with Copulation

The most familiar mode of reproduction is internal fertilization with copulation—the mode used by humans. Although we usually think of sexual reproduction in terms of male and female genitals, particularly the penis, many animals copulate without a penis. They have evolved some other type of intromittent organ.

The common guppy is an example. The male's anal fin is an intromittent organ known as a gonopodium. It is fashioned as a trough or tube. A courting male attempts to get as close to the female as possible, approaching from the rear. When the male is highly motivated sexually, he thrusts his gonopodium forward so that the base of the fin trough is positioned below his urogenital pore. He then attempts to insert the tip of the gonopodium into the urogenital pore of the female. When insertion is achieved, the male releases sperm, which are transferred to the female.

Octopuses also copulate without a penis. Male and female octopuses engage in an elaborate courtship involving synchronous movements of their sixteen arms as well as subtle and dramatic color changes. The courtship sequence ends when the male copulates with the female with one of his eight arms—a special arm known as a hectocotylus. The male inserts the hectocotylus into his mantle cavity and withdraws a spermatophore. He then uses the hectocotylus to insert the spermatophore into the mantle of the female.

Courtship displays of birds are probably more elaborate than those of any other animal group. The lengthy, often ballet-like sequences end in copulation, even though most birds lack an intromittent organ. Copulation is similar from species to species. The male's display entices the female to assume the copu-

latory posture. In this posture, the female sits with her breast touching the ground, wings spread to the side, and head often tilted downward. The male hops on her back and tries to position his cloaca against hers. Because positioning is awkward for the male, he may grasp the female's head feathers with his beak or extend his wings as side supports. His efforts cause the female to arch her back and extrude her cloaca. The male does likewise until their cloacae touch. At this point, the sperm are transferred to the female. This act of reproduction is often referred to as treading, because the male bird literally walks onto the female's back.

Most male birds lack a penis. In some species, however (ducks, ostriches, and flamingos are examples), a small curved penis housed inside the male's cloaca is common.

Hermaphroditism

In Greek mythology, Hermaphroditus was the son of the god Hermes, the messenger of all the gods, and Aphrodite, the goddess of love. Hermaphroditus loved a nymph in a fountain and wanted only to be with her always. The gods granted his wish by uniting them in one body. In the animal kindgom, **hermaphrodites** are animals that have both male and female gonads and can produce both sperm and ova.

The common earthworm is a hermaphrodite. On warm, moist spring evenings, earthworms can be seen in the act of copulation. Two worms lie head to tail, held together by a mucous tube. In this posture each worm transfers sperm to the other. After copulation, the worms separate, and each lays eggs. When the ova are ready to be extruded, the worm secretes a cocoon from the collar region, or clitellum. The cocoon receives ova and sperm as it slides forward and off the anterior end of the worm. The eggs are fertilized inside the cocoon.

The mating of hermaphrodites is referred to as cross-fertilization. However, tapeworms, which are hermaphrodites, are also capable of self-fertilization. Hermaphrodites mate with other hermaphrodites to increase genetic variation in their offspring. The number of new genetic combinations among zygotes is drastically less in the case of self-fertilization. Nevertheless, self-fertilization is common among parasites, whose life-styles make it difficult for them to find mates. For example, an adult tapeworm may live alone in its host, with no opportunity to mate with another worm. Thus it is better for the tapeworm to reproduce by self-fertilization than not to reproduce at all.

There are many other hermaphroditic animals. Among them are the majority of land snails and garden slugs.

Parthenogenesis

Many organisms can reproduce without any sexual union. Their eggs develop without the stimulation of sperm penetration, a condition known as **parthenogenesis** (literally "a virgin beginning"). Male honeybees, for example, are the product of parthenogenesis. Each colony of honeybees (sometimes more than 50,000 individuals) is under the control of a single queen. A newly emerged queen who is destined to reign over a colony begins her life by challenging and killing any would-be contenders. When she reigns supreme, the new queen leaves the hive.

On her nuptial flight, the queen is inseminated in midair by one or several males, called drones. The drones explode their genitals into the queen, forcing sperm into her body. Then they die. The queen participates in three to five nuptial flights a day for as long as a week. During these flights, she stores sperm sufficient for fertilizing eggs for the rest of her life, which may be several years. She never has to mate again.

When mating is over, the queen settles down to the routine business of laying approximately 1,500 eggs per day. She uses her forelegs to determine the size and depth of each cell in the honeycomb in which she lays an egg. In the smaller cells, she lays fertilized eggs. These eggs develop into workers, which are all sterile females. They have the normal species-specific diploid number of chromosomes. In the larger cells, the queen lays unfertilized eggs. These eggs also develop, but they all become males or drones, fertile but possessing only the haploid chromosome number.

Parthenogenesis occurs in many groups of animals, including aphids. (That is why aphids can devastate a garden in a day or two.) Crustaceans, rotifers, nematodes, mollusks, fish, and reptiles also may reproduce through parthenogenesis. One species of lizard in Texas has no known males, so females must reproduce through parthenogenesis.

Alternating Sexes

Some organisms are capable of changing from male to female and, often, back again. Many mollusks are known to change their sex, but perhaps the most remarkable example of sex reversals is that of the fish known as cleaning wrasses (Labroides dimidiatus), which inhabit coral reefs. A typical social complex of cleaners consists of a male and several females that constitute his harem. Members of the harem have their own small territories near the territory of the male. The females, which are egg scatterers, are both aggressively threatened by and courted by the single male.

All members of the social group engage in cleaning. When larger fish swim into the community, both male and female wrasses search their bodies carefully for infections and parasites. Some of the fish receiving the service are so large that the cleaners can swim into their gill chambers and mouths. The relationship is known as **symbiosis**▫. The cleaner fish benefits by receiving morsels of food, and the cleaned fish benefits by the removal of disease-causing organisms.

Each social group of cleaner wrasses contains only a single male. Until recently, no one knew where new males came from. It is now known, however, that if the lone male is removed from the group, the most dominant female will assume his place. Within hours, she will begin acting like a male. In two days she (now he) will be capable of mating with other females of the harem. Close examination has revealed that a patch of tissue in the center of each female's ovary can enlarge and produce sperm when the female changes sex.

SEXUAL REPRODUCTION IN HIGHER PLANTS

The descriptions of sexual reproduction in animals need only minor modification to make them applicable to plants. One difference is that higher plants display an **alternation of generations.** That is, a sexually reproducing generation alternates with an asexually reproducing generation. Figure 10–8 illustrates a typical cycle for the alternation of generations. The two generations in the life cycle are the **gametophyte** (gamete-producing, or sexual) **generation** and the **sporophyte** (spore-producing, or asexual) **generation.** One generation is usually more conspicuous than the other. In general, the higher the plant is on the evolutionary scale, the more dominant the sporophyte generation.

The sporophyte generation of a plant produces spores by **meiosis**▫. Spores are resistant bodies that

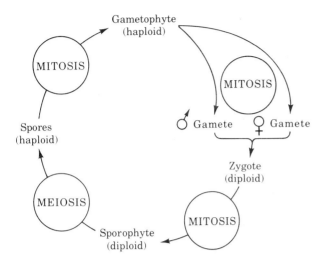

Figure 10–8

Alternation of generations. The alternation of a sexual generation with an asexual generation is typical of the life cycle of all higher plants. In the sexual phase, the gametophyte (haploid, or *n*) produces gametes (eggs and sperm) by mitosis. The gametes eventually fuse, creating a diploid zygote (2*n*). This marks the beginning of the asexual generation of the life cycle—the sporophyte generation. The sporophyte produces spores *(n)* by meiosis. When the spores are liberated, they eventually germinate into haploid gametophytes, completing the alternation of generations.

can withstand harsh environmental conditions, such as cold and drought. They germinate and develop into a haploid generation, the gametophyte generation. The gametophyte generation produces gametes (often both male and female) by mitosis□. (This generation is already haploid. Because gametes are always haploid, another meiotic reduction is not necessary.) The gametes fuse in fertilization, and the diploid zygote develops into the next diploid sporophyte generation, completing the alternation of the two generations. In summary, the sporophyte generation forms the gametophyte generation, and the gametophyte generation forms the next sporophyte generation.

Mosses, ferns, and seed plants, which constitute the subkingdom Embryophyta, all retain the developing **embryo** in the female parent. Embryophytes also have other characteristics in common. They create their own nourishment by photosynthesis□ and thus are called **autotrophic** ("self-feeding"). They are also mostly terrestrial and macroscopic.

The Mosses

People often overlook mosses, even though they are common. Mosses are often less than one centimeter in height. They generally occur in masses composed of hundreds of individual plants, as in the green mats on the floors of forests, the green growths on the sides of trees, and the green fuzz on sidewalks, between bricks, and on buildings.

Close inspection of a moss plant reveals two distinct parts. The short, stubby, leafy portion is the gametophyte generation. It is autotrophic and **perennial** (lasting for more than two years). If environmental conditions are favorable (damp and moist) at the time of examination, a leafless stalk can be seen growing from the gametophyte. This stalk terminates in a cap, or capsule. The stalk and capsule are the sporophyte generation.

The typical moss life cycle (see Figure 10–9) begins when the gametophyte generation produces gametes by mitosis. Most mosses are **dioecious** (*di* means "two," and *ecious* means "house"; hence "two houses"). That is, male and female sex organs are on separate plants. Therefore, some leafy gametophytes are male and others are female.

Before the sexual generation of a moss can reproduce, the male gametophyte needs to be nearly submerged in water. The requirement of water for reproduction has earned the mosses the nickname "amphibians of plants." Water causes the cap of cells on each male gametophyte to burst, releasing the sperm cells. These cells were produced in the male sex organs, the **antheridia.** The sperm swim in the watery environment until they come in contact with a female gametophyte. The female sex organs, called **archegonia**, are shaped like miniature flasks with swollen bases and elongated necks. When a sperm cell reaches an egg within the archegonium, they fuse (fertilization), and a diploid zygote is formed.

The sporophyte (asexual) generation begins to develop immediately. The first few cleavages of the embryo are restricted to the inside of the archegonium. Shortly thereafter, though, the leafless stalk of the sporophyte grows upward from its gametophyte parent. (Sporophytes are found only on female gametophytes.) As the stalk pushes through the gametophyte, it often takes with it the gametophyte cap of cells, called the calyptra. The sporophyte is nourished by the female gametophyte.

When the sporophyte is nearly mature, the uppermost end forms a capsule in which **spore mother**

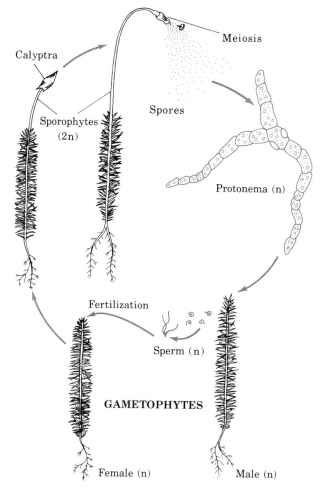

Calyptra

Meiosis

Sporophytes
(2n)

Spores

Protonema (n)

Fertilization

Sperm (n)

GAMETOPHYTES

Female (n)   Male (n)

Figure 10–9

The life cycle of a moss (alternation of generations). The more obvious and longer-lasting generation of the moss is the gametophyte (haploid, or *n*). Mosses have separate male and female gametophytes. Sperm produced by male gametophytes swim to female gametophytes, where eggs within the female sex organs—the archegonia–are fertilized. The resulting zygote (diploid, or *2n*) grows as a stalk from the archegonium and eventually forms a terminal capsule. Within the capsule, spores are produced by meiosis. Liberated spores (haploid, or *n*) eventually germinate. They first develop into an inconspicuous protenema, which eventually grows into a mature gametophyte, completing the alternation of generations.

cells give rise to spores by meiosis. The capsule opens, releasing its store of spores, which are dispersed by the wind. If the environment is suitable, the spores may immediately begin to form another gametophyte generation. If conditions are unfavora-

ble, the spores can withstand environmental extremes until the conditions improve. Each spore germinates into a small, green, ribbon-like growth, the protonema. The protonema develops several swellings. Each gives rise to a new gametophyte, thus completing the life cycle.

The gametophyte of a typical moss is much more conspicuous than the sporophyte is. The gametophyte is also considered to be the more dominant and longer lasting generation.

The Ferns

Among the first house plants people tend to buy are ferns. If you have ferns, you know how beautiful they are and how easy they are to care for. They also come in many varieties, ranging in size from small to gigantic. Some tropical species are called tree ferns because they look like palm trees and are quite large.

Ferns are usually propagated for plant collections asexually, from "starts." Early admirers of these plants had no idea how they reproduced and thought they had a magical power that was released only in the absence of human observation. There was even a short-lived belief that the discoverer of the secret of fern reproduction would become endowed with the power to be as invisible as was the reproduction of ferns.

The large green and leafy fern plant that we are familiar with is the sporophyte generation—the dominant and asexually reproducing one. (The fern's life cycle is illustrated in Figure 10–10.) The lacy leaves are called **fronds**. Inspection of the underside of a frond may reveal numerous brown spots, called fruit spots or **sori**. More than one irate fern purchaser has complained to the shop proprietor about newly bought ferns that had a disease characterized by spots or an infestation of brown scale. (It was a long time before early observers found out that the brown spots are the sites of spore production.)

Microscopic examination reveals the nature of the spots. Each sorus is an assembly of many spore-producing organs, known as **sporangia**. Each sporangium, or spore capsule, has a stalk and a terminal capsule in which meiosis creates spores. Sufficient moisture around the fronds causes the stalk of the sporangium to recoil and to snap the capsule forward as it opens. Spores are catapulted considerable distances (the action being much like the snap of a whip).

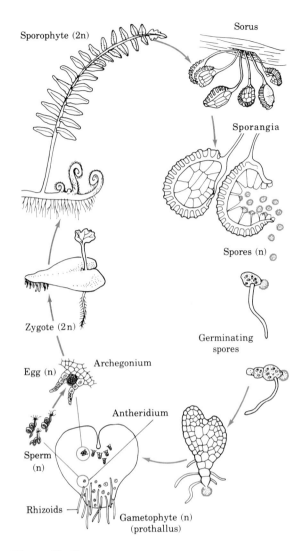

Sporophyte (2n)

Sorus

Sporangia

Spores (n)

Germinating spores

Zygote (2n)

Egg (n)

Archegonium

Antheridium

Sperm (n)

Rhizoids

Gametophyte (n) (prothallus)

Figure 10–10
The life cycle of a fern (alternation of generations). The most obvious generation is the sporophyte (diploid, or 2*n*). The gametophyte generation is inconspicuous. It consists of a heart-shaped structure called the prothallus. Sperm produced by prothalli eventually make their way to eggs held within the archegonia. The resulting zygote (diploid, or 2*n*) develops into the sporophyte generation, which grows from the prothallus. As sporophytes become obvious, they often look like the bowed head on a fiddle and thus are commonly called fiddleheads. As growth continues, the fiddleheads open and the stems and fronds of the sporophyte become more obvious. The spore-producing organs, located on the fronds, are called sori. Each sorus is composed of many sporangia. Each sporangia produces numerous spores by meiosis. When spores *(n)* are liberated, they eventually germinate and develop into new gametophytes, completing the alternation of generations.

The spores produced by the sporophyte are resistant to unfavorable environmental conditions. But when conditions are favorable, they germinate and form very small heart-shaped gametophytes called **prothalli**. Because each prothallus is usually no more than about an eighth of an inch across, it escapes notice by all but the most careful observers. (You might begin looking at the soil surface of fern pots for prothalli. They should be common around the base of indoor ferns that are kept moist.) The prothallus obtains nourishment by photosynthesis□. The water it needs is supplied by short, rootlike structures on the lower surface, called **rhizoids**. Male and female sex organs—antheridia and archegonia—develop on the under surface of the same prothallus, among the rhizoids. Fern prothalli are **monoecious** (one "house" for both sexes).

At maturity, the fern's antheridia produce swimming sperm cells. With sufficient moisture, the sperm swim to the archegonia and down the necks of these flasklike structures. There a sperm cell fertilizes the mature egg cell, forming a diploid zygote. Early cell division of the zygote, which begins the sporophyte generation, produces an embryo that is retained within the archegonium. As the embryo grows, it develops roots, stems, and fronds.

The young sporophyte is coiled as it emerges from the soil layer in which it developed. Before the fronds unfold, the sporophyte resembles the bowed neck of a violin. For that reason, early-emerging fern sporophytes are often called fiddleheads. In the spring, some New Englanders collect fiddleheads of certain species, cook them like spinach or any other greens, and eat them.

As the fronds unfold, the expanding surface area is exposed to the sun, thereby increasing the rate of photosynthesis. The fronds grow rapidly, and before long sori develop on their undersides, completing the life cycle.

As we have seen, although embryos and spores are involved, there are no seeds in the life cycle of mosses and ferns.

Seed Plants

The conifers and the flowering plants, or angiosperms, are seed plants. They have evolved adaptations similar to those developed by terrestrial vertebrates. They too are terrestrial. Unlike mosses and ferns, they do not depend on water for reproduction.

Their independence came about with the evolution of **seeds**. The male gametes of most seed plants are found within **pollen grains** (see Box 10–1).

The sporophyte generation is the dominant generation of all seed plants; the gametophyte generation is microscopic. The sporophyte typically produces two kinds of spores, megaspores and microspores. **Megaspores** develop into female gametophytes. **Microspores** develop into male gametophytes. Both gametophytes are held within the tissues of the sporophyte—the female permanently, the male temporarily. Male gametophytes, the pollen grains, are dispersed by wind or insects. When they reach female gametophytes, fertilization occurs. It forms the diploid zygote, the beginning of the new sporophyte generation contained in the seed. New sporophytes are germinated from seeds that contain the embryo, nutritive tissue, and a seed coat. This process will be covered in more detail as we study the life cycles of the two major groups of seed plants, the angiosperms and gymnosperms.

Angiosperms are more familiar than gymnosperms because they flower. Some plants (for example, sunflowers and roses) have conspicuous flowers. Others (for example, grasses) may have nearly invisible flowers. Gymnosperms are nearly as common as angiosperms. They produce cones rather than flowers. Some (for example, pines and redwoods) produce cones that are extremely large and conspicuous. Others (for example, many of the ornamental shrubs) produce less conspicuous cones.

Angiosperms

Approximately a quarter million species of flowering plants, or angiosperms, exist. One of the largest groups of flowering plants is the composites, which includes sunflowers, dandelions, chrysanthemums, and daisies. The conspicuous "flowers" of these plants are not individual flowers at all but a group of many smaller flowers.

The reproductive organs of flowering plants are part of the flowers. Flowers can be **pistillate** (having only female sex organs), **staminate** (having only male sex organs), or **perfect** (bearing both male and female sex organs). Those that have either male or female sex organs, instead of both, are **imperfect flowers**.

Plants with both pistillate and staminate flowers and plants with perfect flowers, are monoecious (one "house" for both sexes). Begonias, corn, pumpkins,

---

BOX 10–1
HAY FEVER

Approximately one out of every fifteen people who read this book suffers from an allergy to pollen grains called hay fever. Sufferers are most bothered in late spring and in summer, when flowering plants generally reproduce, releasing large amounts of pollen. People afflicted with pollen allergies usually have no idea which pollens are the most bothersome.

Many common trees—maples, walnuts, and oaks, for example—produce tremendous amounts of pollen early in the spring, causing an allergic reaction in some people. Other people are more bothered by the midsummer pollen production of many species of grasses. Yet other hay fever victims suffer from the pollen produced by the common ragweed plant. Ragweed flourishes throughout most of the United States, along roadways, in vacant lots, and almost anywhere that natural vegetation has recently been disturbed. Like most weeds, ragweed quickly invades disturbed areas.

Hay fever, called rhinitis ("nose inflammation"), is an inflammation of the linings of the nose and upper respiratory system, typically including the sinuses. The allergic person's mucous membranes respond to the invasion of pollen proteins (antigens) by producing histamine. Histamine causes capillaries to dilate and leak plasma and causes smooth muscles to contract. As a result, more blood and plasma are brought to the annoyed membranes. The influx of new fluids and the contraction of the smooth muscles of these membranes cause fluid discharge—runny noses, watering eyes, and stuffiness. Some individuals suffer more severe reactions. The smooth muscles in the bronchioles of their lungs may contract and restrict breathing.

Many hay fever victims try to avoid geographical regions where pollen is common. The most frequent defense, however, is to take antihistamines. These drugs inhibit the production of histamines by the affected cells, thereby giving relief.

---

cucumbers, and squash are monoecious plants. Plants that are either male or female (having only staminate or pistillate flowers) are dioecious. Maples, ashes, spinach, and asparagus are dioecious plants.

Examination of a perfect flower will reveal the four kinds of floral organs diagrammed in Figure 10–11: sepals, petals, pistils (female sex organs), and stamens

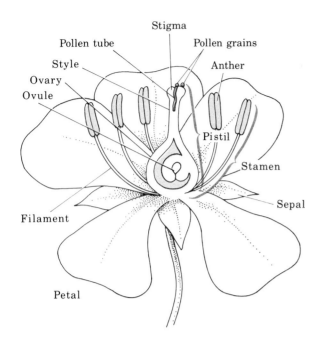

Stigma

Pollen grains

Pollen tube

Anther

Style

Ovary

Ovule

Pistil

Stamen

Filament

Sepal

Petal

Figure 10–11

A typical flower. A complete flower possesses four organs: sepals, petals, pistils, and stamens. An incomplete flower may lack one or more of these organs. A perfect flower has both pistils and stamens. An imperfect flower has either pistils or stamens. A stamen is composed of an anther (pollen producer) and filament. A pistil is composed of a stigma (a sticky landing platform for pollen), a style (neck), and an ovary that contains one or more ovules.

(male sex organs). **Complete flowers** have all four types of organs. **Incomplete flowers** lack one or more of them.

The **sepals** are leaflike and usually green, although they can be a variety of colors. They are lateral organs that cover the flower in the bud stage and often are more conspicuous on buds than on mature flowers.

Petals, the floral organs for which flowers are most appreciated, can be any color of the rainbow. Their colors encourage the visits from insects or birds that assist in transferring pollen from flower to flower in the process of **pollination**. Insects do not indiscriminately visit flowers of any color. Many insects see only particular colors and therefore pollinate only particular flowers. For example, honeybees cannot see the color red and therefore rarely pollinate red flowers. Furthermore, they are attracted mainly to the flowers that have the most numerous petals and that produce the most exaggerated visual profiles. Flowers

that are red or that have only a few petals must depend on other pollinators, such as hummingbirds.

The reproductive organs of flowers include the male organs, called **stamens**, and the female organs, called **pistils**. A stamen is composed of a slender **filament** topped with an **anther**, which produces **pollen**. Pistils have an enlarged base, the **ovary**, that contains one or more **ovules**. Extending from the ovary is an elongated **style**, which has a sticky top called the **stigma** (see Figure 10–12).

Reproduction in flowering plants is an example of alternation of generations. The phases of the life cycle are not as obvious, however, as they were in mosses and ferns. (Figure 10–12 illustrates alternation of generations in flowering plants.)

Figure 10–12

The life cycle of a flowering plant (angiosperm). The most conspicuous generation of flowering plants is the sporophyte generation ($2n$). This generation produces two kinds of spores (n), microspores and megaspores. The anthers of stamens contain many pollen mother cells ($2n$), which divide by meiosis to form microspores. Microspores are transformed by mitosis into the male gametophytes (n) or pollen grains. Each microspore has two nuclei—a tube nucleus and a generative nucleus. Ovules within the ovary of a pistil contain megaspore mother cells ($2n$) that also divide by meiosis, producing four megaspores (n). Three of them disintegrate. The fourth forms the embryo sac, or megagametophyte (n). During the development of the megagametophyte, several cell divisions occur. These divisions ultimately result in seven cells and eight nuclei. Three of the cells, including the egg cell, are located at the lower end of the ovule, near the micropyle. Three cells are at the opposite end. Midway between them reside two polar nuclei within a single cell. Pollen grains (microgametophytes) alight on the stigmas, and each grows a pollen tube down the style. The tube nucleus of each pollen grain is at the tip of the pollen tube as it grows downward. The generative nucleus follows but in its journey, it divides, forming two sperm nuclei. When the pollen tube reaches the ovule, one sperm nucleus unites with the egg (fertilization) to create a $2n$ zygote. The other sperm nucleus unites with the two polar nuclei (also fertilization) to create a $3n$ cell. (Fertilization among angiosperms is referred to as double fertilization.) The zygote divides many times and becomes an embryo. The endosperm nucleus ($3n$) also divides many times to form nutritive tissue called the endosperm. The ovule forms a tough outer layer. All three become the seed. The germination of the seed completes the alternation of generations.

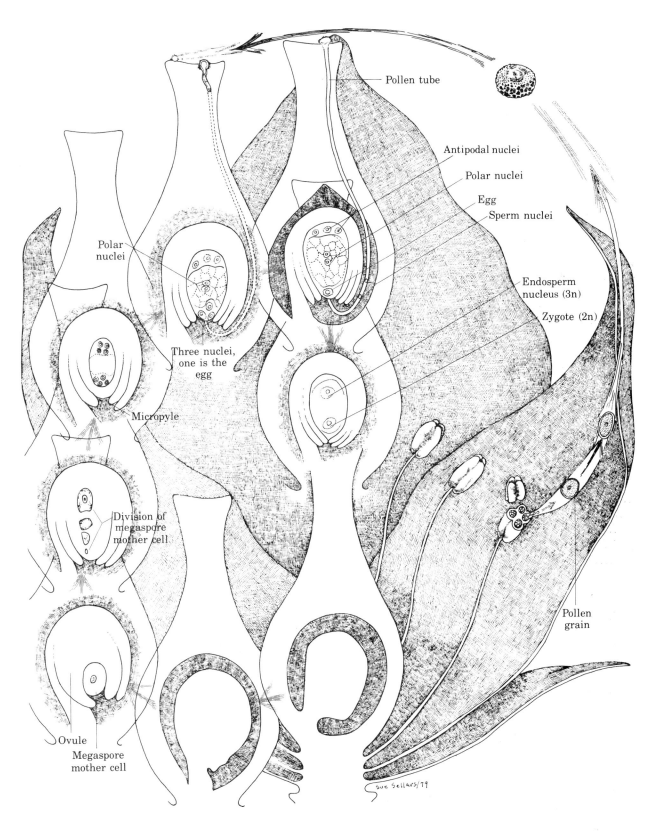

Pollen tube

Antipodal nuclei

Polar nuclei

Egg

Sperm nuclei

Polar
nuclei

Endosperm
nucleus (3n)

Zygote (2n)

Three nuclei,
one is the
egg

Micropyle

Division of
megaspore
mother cell

Pollen
grain

Ovule

Megaspore
mother cell

Sue Sellars/79

Within the pistil are one or more ovules. These structures are part of the sporophyte generation. Within each ovule, a megaspore mother cell divides by meiosis to produce four haploid megaspores (in plants meiosis produces spores). Three of the megaspores disintegrate. One becomes the embryo sac, or megagametophyte, the first cell of the gametophyte generation from which the new plant embryo will form (spores in higher plants develop into gametophytes).

The megagametophyte undergoes several complicated mitotic divisions. The result of these divisions is seven cells with a total of eight nuclei. Two of the cells are of particular interest. One is the egg cell. The other is a cell with two nuclei, called polar nuclei. (This cell accounts for the extra nucleus.) Three of the seven cells, including the egg cell, are located near the micropyle (an opening in the covering of the embryo sac). Three other cells are at the opposite end. In the center are the two polar nuclei within a single cell. After fertilization, all the cells except for the egg cell and the cell containing the polar nuclei disintegrate (see Figure 10–12).

Other cellular events occur in the anthers, the male reproductive organs. The anthers are actually pollen sacs, or microsporangia. Each pollen sac is composed of many pollen mother cells that divide meiotically to form four haploid microspores. Each microspore divides by mitosis to form two nuclei, a tube nucleus and a generative nucleus. When a microspore develops its tough outer coat, it becomes a **pollen grain**, the gametophyte generation (see Figure 10–12).

These cellular events occur before the flower blooms. When the plant is ready to reproduce, the sepals peel away from the bud, and the petals of the flower open, exposing the reproductive organs. The pollen is carried by wind, insects, birds, and even humans to the pistils of the same plant (self-fertilization) or to the pistils of other plants (cross-fertilization). When a pollen grain settles on the pistil, it is caught on the sticky stigma at the top of the style. Pollen grains germinate immediately. Each begins to grow a pollen tube down the length of the style. Eventually, the tube reaches into the female gametophyte to the mature egg cell (see Figure 10–12).

The tube nucleus of each pollen grain remains at the tip of the pollen tube as it grows downward. The generative nucleus follows behind and divides en route, forming two sperm nuclei. The pollen tube eventually reaches the micropyle of a megagametophyte. Immediately, one sperm nucleus unites with the egg cell, forming a diploid zygote, the beginning of the new sporophyte generation. The other sperm nucleus moves toward the two polar nuclei. The three nuclei fuse, forming a triploid ($3n$) nucleus called the endosperm nucleus. Both events—fertilization and the fusing of the three nuclei—occur simultaneously. Collectively they are referred to as **double fertilization**.

After fertilization, the zygote divides many times to form the multicellular embryo of the sporophyte generation. The endosperm nucleus likewise divides many times, forming a mass of nutritive tissue, the **endosperm**. The ovule forms a tough outer layer, the seed coat. The final outcome is a seed that consists of an embryo, nourishing tissue, and seed coat.

As the fruit forms, the walls of the ovary (which contains the ovules) are transformed. The fruit may have one or many seeds (compare a peach and an apple). The number of seeds represents the number of ovules originally held by the ovary (see Box 10–2).

There are three main kinds of fruit: dry, fleshy, and hard (nuts). The walls of the ovary develop in a different way for each kind of fruit, and each represents a different adaptation. For example, the walls of the ovary of a dry fruit are lightweight, often thin, and sometimes transparent. The "winged" dry fruit of a maple tree is adapted for dispersal by the wind. Pea pods and peanuts are also dry fruits.

The best known fruits are the fleshy fruits such as apples, peaches, pears, plums, watermelons, and grapes. The walls of the ovary enlarge and thicken to produce the fleshiness of the fruit. Fleshy fruits are important human food and are essential to the agricultural economy of most nations. They often have appealing flavors. Most are sweet, and many animals eat them. Often, only the fleshy part of the fruit is eaten. But the seeds may also be swallowed and may pass undigested in the animal's feces at some other time in some other place, an effective means of species dispersal. This is the major adaptation of fruits to their parent plants: The fruits ensure propagation of the species. Humans often avoid eating the seeds of many fruits and thereby limit the dispersal of the flowering plants that produced them. However, people do commonly eat tomato, berry, and grape seeds.

The hard fruit, or nut, is a specialized kind of fruit in which the walls of the ovary have become extremely hard. The seed is thus protected from many

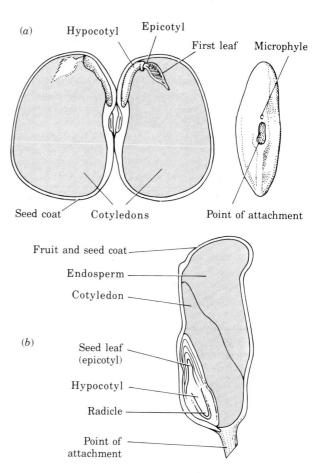

Figure 10–13

Comparison of bean and corn seeds. (a) The bean seed possesses two cotyledons. (b) The corn seed has only a single cotyledon.

BOX 10–2

FRUIT OR VEGETABLE

Few people have difficulty deciding which type of produce to buy, but they may not know which are fruits and which are vegetables. Technically, a fruit develops from the ovary of a flower after pollination and contains the seeds. Therefore, apples, oranges, pears, peaches, watermelons, and grapes are true fruits.

But many foods that we call vegetables, such as green beans, peas, tomatoes, squash, and cucumbers, are actually fruits. They develop from the ovaries of flowers. A vegetable is any other edible plant part besides the fruit. Vegetables include potatoes (tubers), onions (bulbs), carrots (roots), asparagus (stems), and cabbage (leaves). Despite the error of calling some fruits vegetables, there seems to be no real confusion. Early in childhood everyone learns the same list of vegetables. Thus the error is perpetuated generation after generation. Just to avoid any future confusion, though, the Department of Agriculture recently ruled that for all commercial purposes, tomatoes will be known as vegetables—even though they are fruits.

elements in the environment. The shells of nuts that are buried or exposed to the weather finally weaken and rot, releasing the seed to germinate into a new sporophyte. Many animals, among them squirrels, specialize in a diet of nuts. After squirrels eat their fill, they bury the rest of their cache. Some shells then deteriorate and release the seeds to germinate.

Whatever the fate of the seed, its germination completes the life cycle of the flowering plant. The embryo within a seed develops one or two seed leaves, or **cotyledons**. The number of cotyledons is the basis for separating flowering plants into two major divisions, the dicotyledons (dicots) and the monocotyledons (monocots). Compare the bean seed and corn kernel in Figure 10–13. One has two cotyledons, the other only one. Monocots and dicots have other distinguishing characteristics too. The leaf veins of monocots are parallel to one another, whereas those of dicots run in many directions and are rarely parallel (compare the leaves of an iris and a rose). Monocots have floral parts in groups of three or multiples of three, dicots in groups of four or five or multiples of four or five. Some common monocots are lilies,

palms, grasses, corn, bananas, and orchids. Some common dicots are poppies, violets, geraniums, nettles, olives, most nut trees, snapdragons, and dandelions.

Figure 10–13 shows a bean seed that has been opened to expose both cotyledons. The small embryo is attached to its cotyledons at a point called the axis, which is a reference point for tissues that occur above and below the cotyledons. The tissue of the embryo above the axis is called the **epicotyl** (*epi* means "above"; hence "above the cotyledon"). It is composed of miniature folded leaves. (Sometimes the epicotyl is referred to as embryonic leaves). The tissue below the axis is the **hypocotyl** (*hypo* means "below"). It is thicker than the epicotyl and terminates in a blunt end, the **radicle**.

As the seed germinates, the hypocotyl elongates, pushing the epicotyl to the surface. There the epicotyl gives rise to the stem and leaves of the young sporophyte. The radicle develops into the primary root and later the entire root system (see Figure 10–14). The cotyledons of many plants are lost shortly after the seed germinates. In other plants they become true leaves. In either event, the cotyledons absorb, digest, and store food from the endosperm, nourishing the young plant until it is large enough and has sufficient quantities of chlorophyll to nourish itself. For example, a young bean seedling just emerging from the soil has conspicuous cotyledons. As the young plant absorbs the nourishment from the cotyledons, they wither and fall off. Photosynthesis in the newly formed leaves nourishes further growth of the plant.

The seed is indisputably the most important food source for humans. Wheat, oats, rice, and corn are all seeds that nourish millions of people every day. The nutritive part of these seeds is the endosperm. (Monocot seeds have a large endosperm. The endosperm of dicots is often absorbed by the embryo before the embryo is released from the parent.) The oil in the endosperm of many seeds, including corn, is used to make margarines, cooking oils, salad oils, and salad dressings.

Gymnosperms

Everyone knows about gymnosperms. The most common are the conifers, or cone-bearing trees. Most are evergreen shrubs and trees that have leaves shaped like needles or scales. Conifers include small ornamental shrubs, the cedars and yews, and the world's largest trees, the giant redwoods. Various species of gymnosperms, including firs, pines, cedars, cypresses, and spruces, make up large forests. From these trees come hundreds of products that are used daily. Houses and other buildings are often constructed of and furnished with conifer products. In fact, conifers provide more than 85 percent of the lumber and wood products used in building. They are also used in the manufacture of such products as paper, cellophane, rayon, and film. In one week, a daily newspaper of a large U.S. city may use the paper made from 300 to 400 acres of conifers.

The life cycle of gymnosperms is similar to that of angiosperms, with minor variations (see Figure 10–15). The dominant generation, the sporophyte generation, is the pine or fir tree. On each sporophyte are two types of cones, staminate and ovulate. The staminate cones are the male cones, and they produce pollen. They are usually small and often go unnoticed or are thought to be the immature stage of the larger and more conspicuous ovulate cones. The ovulate cones are the female cones. They are large and frequently are collected for decorative purposes. Cones of both types are composed of scales. The events leading to the formation of the male and female gametophyte generation occur on the scales.

The cell divisions that produce gymnosperm gametes are similar to the divisions in flowering plants. When a pollen grain reaches an ovulate cone, it penetrates the ovule with its pollen tube, which has germinated from the pollen grain. The tube may take a year or more to grow the distance needed to reach the egg cell. When it finally reaches the egg, it bursts open, and two nuclei rush into the female gametophyte. One fertilizes the egg, and the other disintegrates. Fertilization of the egg produces a diploid zygote.

What is described here is one of the major differences between angiosperms and gymnosperms. Angiosperms have double fertilization. Gymnosperms do not. It is not necessary for gymnosperms because the female gametophyte provides the nutritive tissue for the embryo. In angiosperms, the nutritive tissue (endosperm) is stimulated to grow through fertilization by one of the sperm nuclei.

Shortly after fertilization in gymnosperms, an embryo begins to grow. Tissues of the female gametophyte form the nutritive tissues for embryonic growth. The sporophyte tissues (ovule) produce a seed coat. When the seeds are mature, they drop from the cone, or the entire cone may be shed. When conditions are favorable, the seeds germinate into new sporophytes, completing the life cycle.

Figure 10–14

Seed germination. When seeds germinate, the growth of the hypocotyl pushes the epicotyl and cotyledons above the ground. The epicotyl gives rise to the stem and leaves. The hypocotyl develops the root system. Cotyledons wither and fall when leaves begin photosynthesis. When the plant matures, it blossoms. The flowers hold the gametophytes. Pollen is transferred to pistils. After fertilization, fruits form. They contain seeds that can germinate new sporophytes.

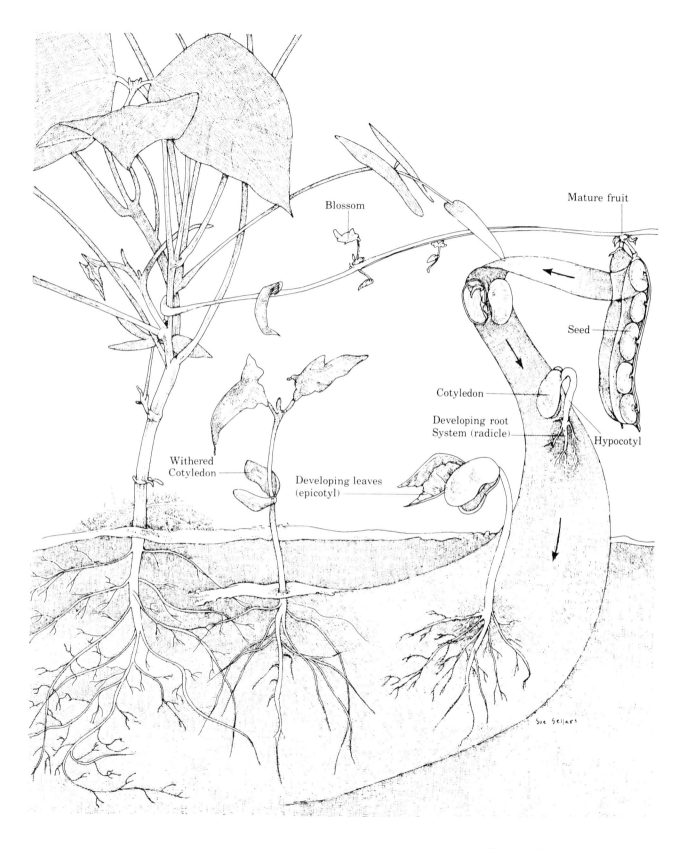

Blossom

Mature fruit

Seed

Cotyledon

Developing root
System (radicle)

Hypocotyl

Withered
Cotyledon

Developing leaves
(epicotyl)

Sue Sellars

COMPARISON OF SPOROPHYTE AND GAMETOPHYTE GENERATIONS OF MOSSES, FERNS, AND SEED PLANTS

PLANT GROUP	SPOROPHYTE (DIPLOID) GENERATION, PRODUCING SPORES BY MEIOSIS	GAMETOPHYTE (HAPLOID) GENERATION, PRODUCING GAMETES BY MITOSIS
Mosses	Leafless stalk bearing terminal capsule in which spores are produced; grows from top of female gametophyte plants	Conspicuous short, green, leafy plants that grow in mats; separate male and female plants
Ferns	Conspicuous green, leafy plants that produce spores in sporangia that commonly are clustered in brown spots (called sori) on undersides of fronds	Inconspicuous small, heart-shaped plants that bear both male (antheridia) and female (archegonia) sex organs on the underside of the same plant
Seed plants	Large, conspicuous dominant generation, including trees, flowers, and evergreens; produce two types of spores: microspores (which develop into male gametophytes) and macrospores (which develop into female gametophytes)	Microscopic and inconspicuous; male gametophytes are the pollen grains; female gametophytes develop within sporophyte ovules and are retained there

SEXUAL REPRODUCTION IN SINGLE-CELLED ORGANISMS

It was assumed for a long time that single-celled prokaryotes, such as bacteria□, reproduced only asexually, by fission. (**Prokaryotic cells** lack a nuclei, distinct chromosomes, and most cellular organelles.) However, almost four decades ago, it was discovered that bacteria do have sex. In fact, microbiologists now refer to different mating strains. Other terms they use are positive and negative strains and donor and recipient strains.

During mating, male and female bacteria are united by a cytoplasmic bridge. This temporary union is called **conjugation**. Male bacteria may conjugate with more than one female at a time. During conjugation, the male bacterium replicates its single circular strand of DNA. One of the two strands then uncoils and is transferred to the female.

After the two bacteria separate, the female is temporarily $2n$. That is, the number of chromosomes usually present in the female has been doubled by the addition of the male's chromosome. The $2n$ female incorporates part of the male's DNA strand into the DNA already present, in a process similar to crossing over. The pieces of the female's DNA strand that have been replaced by male DNA are eventually expelled from the cell. Then the female bacterium becomes haploid again—although with new genetic combinations that will be transmitted to daughter cells during fission.

Prior to learning that bacteria have private sex lives, scientists were aware that many single-celled eukaryotes mate sexually during conjugation. (**Eukaryotic cells** have nuclei, chromosomes, and cytoplasmic organelles.) An example is the paramecium, the organism so commonly observed in biology laboratories. Paramecia have different mating types, with as many as eight "sexes" in some instances. Opposite mating types engage in courtship. In sexual reproduction, two paramecia press their oral surfaces together to conjugate (see Figure 10–16).

Typically, each paramecium has two nuclei—a large macronucleus and a smaller micronucleus. During conjugation, the macronuclei of both paramecia disintegrate. The micronucleus of each undergoes meiosis to form four haploid nuclei. Three disintegrate, and the remaining nucleus divides mitotically□ to form two. One of the two nuclei becomes motile (capable of moving). The other remains stationary. The motile nucleus crosses into and fuses with the stationary nucleus of the partner (fertilization), forming a diploid zygote.

In effect, the nuclei act as gametes. Fertilization is double because it occurs simultaneously in each partner. The resulting two zygotes are genetically identical. After the exchange of nuclei, the paramecia separate, but they now contain new genetic combinations not present prior to conjugation. Each is capable of continued rapid asexual reproduction, creating many new individuals.

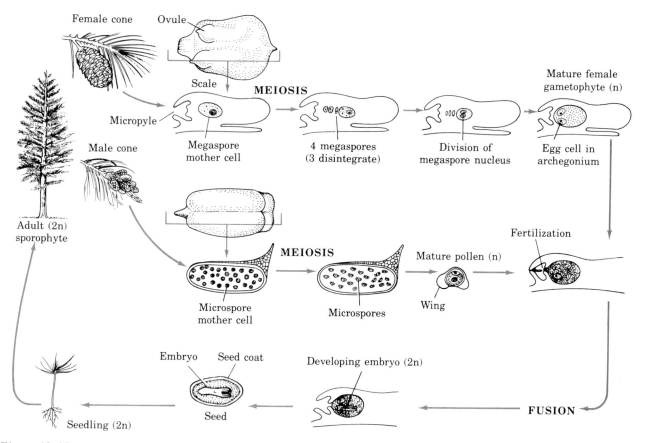

Figure 10–15

The life cycle of cone-bearing plants (gymnosperms). The conspicuous sporophyte generation of gymnosperms is the cone-bearing tree itself, such as a pine or fir tree. The sporophytes produces two kinds of spores, microspores and megaspores. Microspores are formed by meiosis in specialized structures on the lower surfaces of the scales that make up a male cone. The megaspores are formed by meiosis in the large and more conspicuous female cones. Ovules on the upper surfaces of cone scales contain one megaspore mother cell (2*n*) that divides by meiosis to form four cells *(n)*. Three disintegrate, and only one becomes the megaspore. The single megaspore divides by mitosis many times, forming a gametophyte with many distinct nuclei but indistinct cell walls. Later, several archegonia develop from the multinucleated mass, each containing one egg. Usually, only one egg in each ovule is fertilized. Microspore mother cells in male cones divide by meiosis from four microspores *(n)*. Each microspore has a thick cell wall and develops wings. The entire structure represents the gametophyte generation. The nucleus divides further by mitosis. Finally, a mature pollen grain is formed. When pollen grains are liberated, they are dispersed by the wind. Eventually, some reach female cones, where they lodge between the scales. The pollen grain germinates and a pollen tube grows toward the ovule. In most instances, fertilization does not occur until the next growing season, even though pollen tubes have begun to grow. When the pollen tube reaches the female gametophyte, its sperm nucleus fertilizes the egg cell, and a diploid (2*n*) zygote is formed. Eventually, a seed develops. The germination of the seed completes the alternation of generations.

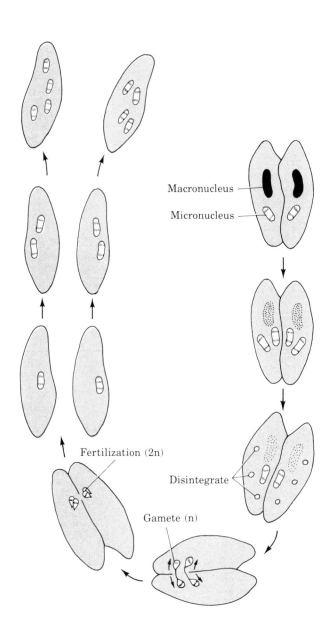

Macronucleus

Micronucleus

Fertilization (2n)

Disintegrate

Gamete (n)

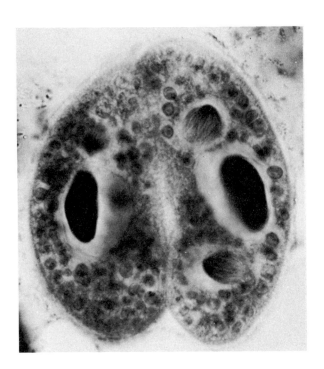

Figure 10–16

Conjugation in paramecia. Opposite mating types of paramecia are attracted and press their oral surfaces together in conjugation. The larger nucleus of each—the macronucleus (dark)—disintegrates. The smaller nucleus—the micronucleus (light)—undergoes meiosis. Eventually, the two micronuclei form four haploid *(n)* nuclei. Three of the four micronuclei in each paramecium disintegrate. The remaining nucleus divides once more by mitosis to form two nuclei. One of the two nuclei becomes motile. The other remains stationary. The motile nucleus of each paramecium crosses over into the other and fuses with the stationary nucleus (fertilization), which forms a diploid zygote (2*n*). Later, the paramecia separate. Each now contains new genetic combinations. The exconjugates can create more paramecia by fission. The photograph shows two paramecia in the process of conjugation. (Courtesy Carolina Biological Supply Company.) Each had a distinct macronucleus *(dark)*. One has two distinct micronuclei *(light)*. The other has only one visible micronucleus.

SUMMARY

1. Many organisms reproduce asexually by simple cell division.

2. Fission is asexual reproduction by single-celled organisms. Budding is asexual reproduction by multicellular animals. Vegetative reproduction is asexual reproduction by plants.

3. Plants and animals that reproduce asexually produce offspring with the same genotypes as the parent. There is no genetic variation for natural selection to act on.

4. Asexual reproduction has important economic significance because desirable characteristics of edible plants can be preserved.

5. Sexual reproduction creates offspring that are ge-

netically different from the parents. Sexually reproducing organisms produce haploid gametes by meiosis. Then male gametes fertilize female gametes, which begins a new diploid generation.

6. Genetic variability is created during meiosis by the chromosomal process of crossing over and random assortment. Fertilization is another random event that increases genetic variability among offspring.

7. Multicellular animals exchange gametes (sexual reproduction) in a variety of ways. Some use external fertilization without amplexus (broadcasters) or with amplexus (frogs and toads). Others use internal fertilization without copulation (newts) or with copulation (humans).

8. Hermaphroditic animals (such as earthworms are male and female at the same time. However, generally they cross-fertilize one another, which increases genetic variability. Some eggs begin development without fertilization, which is known as parthenogenesis. Some organisms (cleaning wrasses, for example) are capable of changing their sex according to season or social stimuli.

9. Sexual reproduction of plants is essentially the same as for animals—except that plants demonstrate alternation of generations, an alternation of asexual and sexual phases.

10. The gamete-producing generation of plants is the haploid gametophyte, or sexual, phase. The sporophyte generation is the diploid asexual phase, which produces spores by meiosis. Germination of spores creates another haploid gametophyte stage. Among mosses, the gametophyte generation is the more conspicuous. Among ferns, the sporophyte generation is dominant.

11. Flowering plants (angiosperms) are the sporophyte stage. Gametophytes of flowering plants are microscopic and are housed in the flowers. Flowers contain the male reproductive organs (stamens) and female reproductive organs (pistils). These organs are often surrounded by petals and sepals.

12. The anthers of the stamens produce pollen grains, which are male gametophytes. The ovary of each pistil produces embryo sacs (megagametophytes), which are female gametophytes.

13. Double fertilization occurs in flowering plants. When a pollen grain reaches a pistil, its generative nucleus forms two sperm nuclei. On reaching the female gametophyte, one sperm nucleus fertilizes the egg cell, and the other fertilizes the cell with two polar nuclei. The latter cell eventually becomes the endosperm, which will nourish the developing embryo.

14. The embryo of a flowering plant is within a seed. It consists of an epicotyl, which will give rise to stems and leaves, and a hypocotyl whose tip, the radicle, will become the root system. Early development of the embryo is supported by the nutritive tissue within the seed, the endosperm.

15. The walls of the flowering plant's ovary become the fruit that contains the seeds. The three main types of fruits are dry, fleshy, and hard.

16. The life cycle of gymnosperms (cone-bearing plants) is similar to that of angiosperms. Coniferous trees represent the diploid sporophyte stage. These trees have staminate cones (male) and ovulate cones (female) that house the sex organs. Staminate cones produce pollen. When the pollen grains reach female cones, they generate pollen tubes that eventually reach the female gametophyte and the egg cell. Then the sperm nucleus fertilizes the egg. Gymnosperm embryos are housed in the scales of ovulate cones. When they are shed, they germinate into new sporophytes.

17. Unicellular organisms such as bacteria and paramecia may also reproduce sexually by conjugation.

THE NEXT DECADE

CLONING

In many science fiction accounts, humans and other organisms reproduce asexually by splitting into two or more identical organisms. Each time the futuristic organism splits, the offspring become exact carbon copies. The results are thousands and thousands of genetically identical individuals collectively called **clones**.

Natural clones do exist. When a single paramecium undergoes fission, the daughter individuals are all clones. But paramecia also engage in conjugation, which provides genetic variation. The next fission cycle creates identical twins or quadruplets, but they are not identical to those created before conjugation.

If you were to look down on a bed of certain sea anemones (see the photo), you would see something that looks like a jigsaw puzzle. Each piece of the puzzle consists of a number of genetically identical individuals—clones. The "lines" between the various clones are spaces devoid of sea anemones. They are created when individuals of different clones come in contact. Members of competing clones sting one another

to death. When their bodies wash away, the space is created.

The next time you see a vineyard or an orchard of Delicious or McIntosh apples, you are seeing clones. However, these clones are not natural. They exist only through selective grafting by humans.

Other unnatural clones have also been created. J. B. Gurdon, an embryologist at Oxford University in England, reported in the 1960s the results of some rather bizarre experiments he had conducted with South American clawed frogs. He removed ova from female frogs and destroyed their haploid nuclei by radiation. He then removed diploid nuclei from the healthy skin cells of a single tadpole and inserted a skin nucleus into each of the ova. Because all of the donor nuclei came from the same tadpole, all of the ova contained the same set of genetic information. Many of the ova developed normally, producing adult frogs that were all genetically identical. Gurdon has created a clone, a lineage of individuals that all have the same genetic makeup.

Human identical twins are clones, but can humans be cloned like Gurdon's frogs? Let your scientific imagination predict how cloning might have human application. If similar techniques could be developed using human ova and donor nuclei from a single individual, then, theoretically, clones of humans could be created. That, at least, has been the basis for much widespread speculation.

A book on this subject, *In His Im-*

age: The Cloning of Man, by David Rorvik, became a bestseller in 1978. The book reported that a human being had been created by cloning. The author claimed that at the time of publication the infant was fourteen months old and the identical genetical copy of its male donor. The donor was supposedly a millionaire who paid for "his own image." The book triggered considerable debate, with scientists energetically refuting its claim. Eventually, the book was relegated to its rightful place among other scientific hoaxes. Most scientists agree that the technical problems of cloning humans are immense. Human clones remain a long way in the future and may, in fact, never become a reality. However, the idea continues to intrigue people. It is certainly a popular topic of science fiction writers.

Human-made clones of other animals do exist, and they already play an important role in various aspects of scientific research. Using Gurdon's basic techniques, Robert McKinnell of the University of Minnesota has created hundreds of genetically identical frogs. McKinnell and his colleagues contend that such frog clones will act as unprecedented controls for cancer therapy. Any changes in frogs with cancer must be the effect of therapy, not of genetic influences. Furthermore, one of the McKinnell frogs was cloned by replacing the normal nucleus of an ova with the nucleus of a cancerous cell. The ova did not become cancerous but instead developed into a normal swimming tad-

(a)

(b)

Many plants reproduce both sexually and asexually. The strawberry plant reproduces sexually by its floral structures **(a)** and the development of fruit and asexually by growing runners **(b)** which are aerial roots called stolons. From the ends of stolons new small strawberry plants develop.

(c)

Pine trees which are gymnosperms represent the dominant sporophyte generation. The spore-producing structures on pines and other evergreens are the cones. Shown above are the male **(a)** and the female **(b)** cones of a Jack Pine tree. To the left **(c)** male, or staminate, cones are releasing pollen, the male gametophytes, which developed from microspores produced in the male cone. The female or ovulate cone bears the female gametophytes on the upper side of its scales. These developed from megaspores.

(a)

(b)

Mosses and other higher plants show alternation of generations. Here we see the green leafy female gametophyte of a moss plant **(a)** with the sporophyte generation growing out of it as a stalk and capsule from which spores are being released **(b).** The gametophytes are haploid and produce gametes by mitosis; the sporophytes are diploid and produce spores by meiosis.

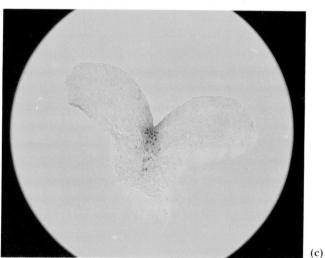

(c)

Ferns **(c)** also display alternation of generations. Here we can see the brown spore producing sori in the underside of a fern frond which represents the diploid sporophyte generation. The prothallus **(d)** is the haploid gametophyte that produces male and female gametes in reproduction structures in its underside.

(d)

Sexual reproduction on an apple tree. **(a)** After pollen has been carried to apple blossoms and fertilization has been achieved, the petals fall from the flower. The parts remaining are only the sepals and stamens. **(b)** Below the remaining flower parts the ovary of the apple begins to expand as the fruit begins to develop. **(c)** When the fruit is mature, the sepals and stamens are still present but not as conspicuous. The seeds, when germinated, will produce new trees also capable of sexual reproduction.

(c)

pole. This single demonstration has changed many ideas about cancer.

Scientists are applying the principles of cloning to other problems too. Cloning may have real value in the propagation of desirable plant species. Today, scientists are taking cells from the immature stems of grape plants and nurturing them in a special growth medium. Each cell divides again and again and eventually forms a mass of cells. The mass gradually differentiates into a grape seedling. All the new grape plants produced in this way have genotypes identical to that of the original plant. Plant cloning can maintain desirable genotypes indefinitely. The process is also economical. It is estimated that 10,000 seedlings per month can be produced by cloning.

Cloning in horticulture has become a multimillion-dollar business during the past 10 years. Most agricultural crops in Europe, including strawberries and potatoes, are cloned to eliminate disease. Orchids are routinely cloned, as are poinsettias, some daisies, and most tropical leafy houseplants. The common Boston fern was propagated (cloned) from a single specimen that appeared in Boston more than a hundred years ago. The millions of Boston ferns now sold are all descendants of the same parent.

FOR DISCUSSION AND REVIEW

1. What is the difference between asexual and sexual reproduction?

2. Why is vegetative reproduction economically important? How are crops propagated asexually? What are the advantages of asexual propagation of crops?

3. Sexual reproduction involves the creation of genetic variation. How is this accomplished? What are the major sources of variation?

4. Describe conjugation, giving details of the process for a specific organism.

5. Some organisms reproduce by external fertilization without amplexus. Where would you expect to find such organisms in nature? Explain.

6. Give examples beyond those in the chapter of organisms that reproduce by external fertilization without amplexus. By external fertilization with amplexus. By internal fertilization without copulation. By internal fertilization with copulation.

7. It is not uncommon in the early spring to see garden slugs or earthworms copulating in yards and gardens. But both of these types of organisms are hermaphroditic, so why do they bother to copulate?

8. Give the general life cycle of the honeybee, emphasizing when and how parthenogenesis plays a role.

9. What is the adaptive advantage for an animal capable of changing its sex?

10. In terms of life cycle, what is meant by the phrase *alternation of generations*? What organisms display this mode of reproduction?

11. Describe the sporophyte generation for a moss, a fern, a flowering plant, and a cone-bearing plant. What are the major differences in this generation among these plants?

12. Describe the gametophyte generation for moss, a fern, a flowering plant, and a cone-bearing plant. What are the major differences in this generation among these plants?

13. Describe the organs of a typical flower. What are their functions?

14. Explain this statement: A complete flower is always perfect, but an incomplete flower may not always be imperfect.

15. Compare and contrast the seeds of an angiosperm and a gymnosperm.

SUGGESTED READINGS

THE GREEN WORLD: AN INTRODUCTION TO PLANTS AND PEOPLE, by Richard M. Klein. Harper & Row: 1979.
A remarkable and detailed account of the interrelationships of humans and plants. Interesting reading on a wide range of topics, from plants in religion to plants in liquor.

PLANT LIFE CYCLES, by Thomas R. Mertens and Forrest F. Stevenson. Wiley: 1975.
A self-teaching guide with numerous illustrations and simplified text. If you think plant life cycles are confusing, this volume is a guide to understanding them.

THE SEXUAL CONNECTION: MATING THE WILD WAY, by John Sparks. David and Charles: 1977.
An entertaining look at reproduction of animals—with humor and cartoons. A joy to read and an excellent factual account of "unusual sex."

STALKING THE WILD ASPARAGUS, by Euell Gibbons. McKay: 1970.
Interesting and fun for the naturalist, hiker, or average nature-trail walker. If you get hooked, there are two more Gibbons books.

11

HUMAN SEXUALITY AND REPRODUCTION

PREVIEW QUESTIONS

After reading this chapter, you should be able to answer the following questions:

1. What are the anatomical parts of the complete male and female reproductive system? What is the function of each part?

2. What hormones are involved with male sexuality? What is the function of each hormone?

3. What events occur during a menstrual cycle? Which hormones regulate each event?

4. Describe as many forms of birth control as you can.

5. What are the symptoms of gonorrhea, syphilis, and genital herpes?

Our knowledge of the processes of human reproduction has increased tremendously in recent decades. Much of this knowledge is a direct result of the research of William Masters and Virginia Johnson, who have studied human sexual response in a clinical setting. They have measured the physiological and emotional responses of more than two thousand couples performing the sex act, or coitus. In other laboratories, endocrinologists, neurophysiologists, neuroanatomists, fertility experts, and other specialists have achieved further understanding of human reproduction. All this new information has drastically changed many assumptions about sex and human sexuality.

One of the most significant recent changes is a new openness about sex and human reproduction. Students must be given some credit for this change. They helped force much of the adult population into being more open about sex as a part of everyday life and a natural biological process.

This chapter is intended as a summary of the facts about human sex and reproduction. Dozens of more sophisticated treatments are available for those who want more information.

THE MALE REPRODUCTIVE SYSTEM

The organs of the male reproductive system are diagrammed in Figure 11–1. The figure can be referred to throughout this section.

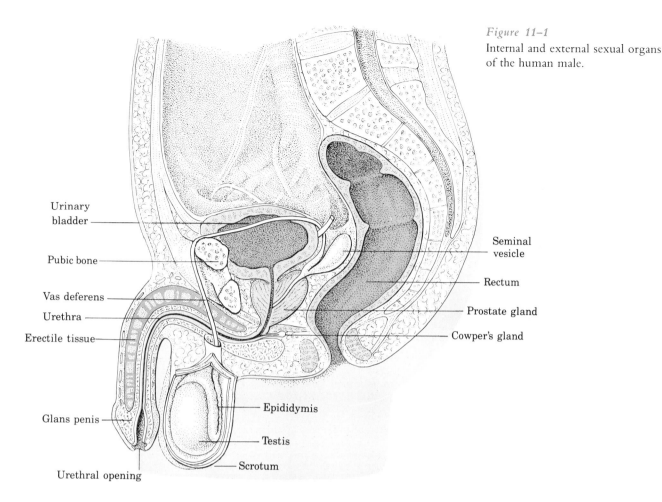

Figure 11–1

Internal and external sexual organs of the human male.

Urinary bladder

Pubic bone

Vas deferens

Urethra

Erectile tissue

Glans penis

Urethral opening

Epididymis

Testis

Scrotum

Seminal vesicle

Rectum

Prostate gland

Cowper's gland

External Anatomy

The most conspicuous elements of the male's external sexual anatomy are the penis and the scrotum.

The Penis

Human males have an intromittent organ called a **penis**. The average human penis in the flaccid, or soft, stage is ten to fifteen centimeters in length. The penis is composed of a long shaft and a rounded head, the glans penis. The glans, the most sensitive area of the penis, is endowed with many sensory receptors. The base of the glans forms a ridge, called the corona that separates the glans from the shaft. The corona is the most sensitive region of the glans. If a male has not been circumcised, the glans is normally covered by a layer of skin called the foreskin or prepuce. The prepuce is often removed by circumcision, a minor surgical procedure that some think improves genital hygiene.

The prepuce contains preputial glands. These glands secrete a curdlike substance. When that substance is mixed with bacteria and dead cells, it is called smegma. The accumulation of smegma between the glans and the prepuce can create a strong odor. In sufficient quantities, it can also cause the foreskin and glans to grow painfully together. Smegma is rich in protein and accumulates in dark, moist, warm places. It therefore can provide the perfect environment for disease-causing organisms. For this reason, circumcision has been a standard procedure in U.S. hospitals since the late 1930s. Today, about 90 percent of the male babies born in the United States are circumcised. However, good genital hygiene can eliminate the disadvantages of having a prepuce. Uncircumcised men probably suffer no disadvantages if the glans is frequently bathed.

During sexual arousal, the penis increases in length and diameter and becomes firm. This is called an erection. The erect penis of the average male is 14.0 to 16.5 centimeters in length and about 3.8 centimeters in diameter. Although the flaccid penis varies greatly in size and shape from man to man, there is much less variation in the size of the erect penis. If there is any correlation between the size of the flaccid and erect penis, it is an inverse one. A long flaccid penis increases very little in size when erect, but a small flaccid penis may double in size.

The ability of the penis of humans and other animals to drastically change its anatomical state is a direct function of the erectile tissue that composes most of it. The penis has three columns of erectile tissue—one bundle on the underside that houses the **urethra** (the urinary duct) and two bundles above and somewhat to the side (see Figure 11–2).

Upon sexual arousal, the arteries that supply the penis dilate and deliver more blood into the spongy erectile tissue than the veins (which contract somewhat) can carry away. As a result, blood is trapped. As the blood accumulates, the penis becomes longer, thicker, and more rigid. The reverse response causes the penis to once again become flaccid. Arteries regain their normal size, and the blood they brought to the penis is quickly carried away by the relaxed veins.

The Scrotum

The second prominent feature of the human male's external reproductive anatomy, the scrotum, or scrotal sac, is composed of a thin layer of skin and underlying layers of muscle. The sac is sparsely covered with pubic hair. It contains the **testicles**, or **testes**, which are the sites of sperm production and of male sex hormones. (*Testis* is Latin for "witness." During Roman times it was customary for those testifying under oath to cover the scrotal area with the hand.) Examination of the scrotum by touch will reveal the oblong, egg-shaped testes and the prominent ridge, the **epididymis**, along the back of each testis. Further examination may also reveal the sperm-carrying duct ascending from each testis, the **vas deferens**.

The scrotum and testes respond to sexual arousal, although their response is less noticeable than the response of the penis. Physiologically, the response of the testes is similar to an erection. It too involves the dilatation of arteries, so that more blood is delivered than can be carried away by nearby veins. Similar vasocongestion causes the wall of the scrotum to double in thickness. The scrotum pulls upward and presses tightly against the abdominal wall. This response decreases the total volumn of the scrotal sac, which makes it necessary for one testis to sit on top of the other. For some unknown reason, the left testis rises above the right one in most men.

Internal Anatomy

The testes are composed of many separate compartments, called lobules (see Figure 11–3). Each lobule is separated from the next by a curtain of tissue, and each contains from one to four coiled **seminiferous tubules**. The tubules join into an efferent duct (*efferent* means "going out"). The ducts from all the lobules of a testis eventually join in the epididymis, the organ of sperm storage, which is a tightly coiled spiral duct.

The tail of each epididymis becomes the vas deferens, which ascends from the epididymis through the scrotum and the muscles of the lower abdominal wall and enters the abdominal cavity through the inguinal canal (a hole in the abdominal muscles for passage of the vas deferens).

Some men suffer hernias, ruptures in the abdominal muscles that often allow nearby internal organs to protrude. Most hernias in men are inguinal, which means that the inguinal canal has probably been ripped or ruptured. Sometimes the opening becomes large enough to accommodate not only the vas deferens but also portions of the intestine. In severe cases a loop of the intestine lies in the scrotal sac, obviously a painful malady. However, a routine surgical procedure can easily repair the rupture.

Figure 11–2
Cross section of the human penis, showing the three bundles of erectile tissue. The urethra passes through the lower bundle.

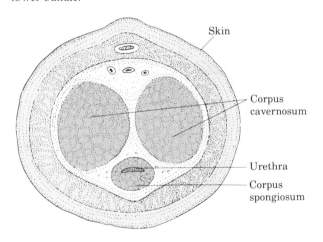

Skin

Corpus cavernosum

Urethra

Corpus spongiosum

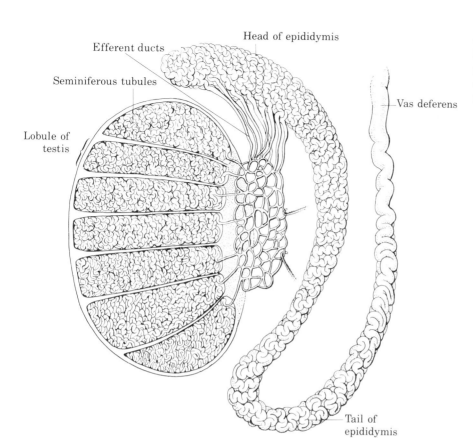

Efferent ducts

Head of epididymis

Seminiferous tubules

Lobule of
testis

Vas deferens

Tail of
epididymis

Figure 11–3

Internal structures of the human
male testis, illustrating the complex
of ducts involved in sperm
manufacture and transport.

Within the abdominal cavity, each vas deferens loops over the urinary bladder and becomes an **ejaculatory duct**. Behind the bladder, the paired seminal vesicles communicate with each ejaculatory duct. The ejaculatory ducts unite with the urinary duct, the urethra. The two types of ducts fuse within the walnut-size **prostate gland**, which secretes its products into the urethra at the time of ejaculation.

If the prostate becomes inflamed, sexual function and urination may be impaired. A diseased prostate may also become cancerous. (Men over the age of thirty should request a prostate examination as part of their annual physical. The physician will palpate the prostate through the anus to determine its physiological condition.) At the base of the prostate, the paired Cowper's glands, or bulbourethral glands, join the urethra. The urethra continues into the body of the penis and opens to the outside through the urethral meatus at the tip of the penis.

Sperm Production, Semen, and Ejaculation

The production of sperm, known as **spermatogenesis**, takes place within the cells lining the seminiferous

tubules of the testes (see Figure 11–4). The outermost cells of the seminiferous tubules produce primary spermatocytes. These spermatocytes undergo two successive meiotic divisions□, usually requiring fifty-three days. Each primary spermatocyte forms four spermatids, which later develop into mature sperm cells. Because of the long time it takes to produce sperm, one may wonder how a male even ejaculates millions of sperm cells at once. This is possible because the seminiferous tubules are long (approximately 120 centimeters). A single millimeter of tubule produces sufficient sperm every fifty-three days to ensure fertilization.

As the spermatids are transformed into sperm cells, they are moved along the seminiferous tubules toward the epididymis, where they will be stored. The movement is facilitated by the secretion of fluids, the microcontractions of the walls of the tubules, and the movement of hairlike cilia on the walls of the efferent ducts. The entire trip for a sperm cell, from its origin to its storage point, takes eight to fourteen days. Therefore, the seminiferous tubules are like microscopic conveyer belts carrying a nearly endless supply of sperm. An average male produces a minimum of

82

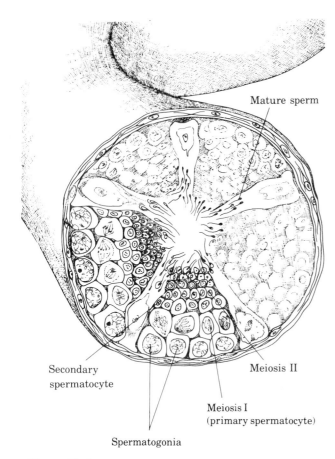

Mature sperm

Secondary
spermatocyte

Meiosis II

Meiosis I
(primary spermatocyte)

Spermatogonia

Figure 11–4

Cross section of a seminiferous tubule, showing cells involved in various stages of sperm production. The primary spermatocytes, which are peripheral cells, undergo meiosis I and become secondary spermatocytes. Some of the nuclear events in meiosis I are illustrated. Secondary spermatocytes undergo meiosis II to become spermatids. Spermatids become sperm cells after cellular transformations. The cellular events leading to the formation of four sperm cells from a single primary spermatocyte usually require fifty-three days. The sperm cells are sloughed off into the center of the seminiferous tubule and eventually are carried to the epididymis by microcontractions of the tubule wall.

100 million mature sperm cells every day and as many as 12 trillion in a lifetime.

Sperm are stored in the epididymis, where they undergo further maturation. The epididymis consists of eighty centimeters of coiled tubules. Once the sperm are mature (about seventy-four days from spermatogonia to maturity) and capable of fertilization, they

can remain viable in the epididymis for six weeks. Those that are never ejaculated are absorbed by the tissues of the epididymus.

Sperm are viable only if they are produced and stored at about three degrees below normal body temperature. The scrotum then acts as a thermostat, maintaining the temperature of the testes near the necessary 95.6 degrees Fahrenheit. When cold, the muscles of the walls of the scrotum contract, pulling the scrotum closer to the abdominal cavity and thereby warming the testes. When warm, the muscles relax, allowing the testes to cool as they move away from the body. Every man knows that on a cold day the scrotum is tightly drawn against the body and that after a hot shower it hangs lower between the legs, with the testes at the bottom. The newspaper column "Dear Abby" has advised more than one couple unsuccessful at having children that the male should wear boxer shorts rather than tight-fitting shorts. Tight-fitting shorts force the scrotum against the body wall, where it absorbs body heat that may kill sperm and cause the male to be temporarily sterile. Although the advice has not been scientifically tested, it may well work.

At the time of **ejaculation**, masses of sperm cells are mobilized by rhythmic contractions of the epididymides and vasa deferentia. When the sperm reach the seminal vesicle near the bladder, these glands secrete a thick discharge. The volume varies, but normally this secretion makes up about 70 per cent of the male ejaculate. The secretion stimulates the rapid swimming motions of the sperm, gives the sperm a high concentration of sugar (fructose) from which they can get energy for their swimming movements, and provides protective compounds that maintain the integrity of the sperm. Almost simultaneously, the **prostate gland** contracts, releasing a watery whitish fluid. It is probable that sperm can swim in the discharge of the seminal vesicles only after it is diluted with prostatic secretion. This secretion contains many compounds, including a number of enzymes, vitamin C, cholesterol, and minerals. The compounds are important for the metabolism of sperm.

The urethra is a double-purpose tube, conducting both urine and sperm. Nearly a liter of urine per day passes from the bladder of the average man to the outside of the body. Urine is acidic and toxic to cells of the body. After it passes through the urethra, the urethra too becomes toxic and acidic, especially for sperm cells. Without some preparation, most sperm being ejaculated would die in the urethra.

Prior to ejaculation, the Cowper's glands at the base of the prostate discharge a few droplets of colorless sticky fluid. The secretions bathe the urethra, buffering the otherwise hostile environment. It is common for secretions from the Cowper's glands to pass through the urethral meatus at the tip of the penis before ejaculation. The secretion, which can be copious, also acts as a lubricant during coitus.

Just prior to ejaculation, an additional surge of blood to the penis causes it to become distinctly firmer and more erect. The penis often bows upward at this time, and the increased blood supply causes the glans to become a deeper red or even bluish in color. The urethra visibly widens, increasing about twice in total diameter. Orgasm is imminent at this time.

Orgasm has two distinct phases. The first consists of rhythmic contractions of the epididymides, vasa differentia, seminal vesicles, and prostate gland. A muscle surrounding the prostatic part of the urethra holds the semen back for two or three seconds. This telltale signal lets the man know that he is about to ejaculate—whether he wants to or not. Some men ejaculate too quickly, a problem known as premature ejaculation. Experiments by Masters and Johnson have shown that some men can avoid premature ejaculation by practicing a special technique. When these men experience the first phase of orgasm, they are to hold the glans penis tightly between forefinger and thumb. This seems to inhibit the second phase. With practice, the time to ejaculation is lengthened and the problem of premature ejaculation is solved.

The second phase of orgasm consists of powerful contractions of the urethra, which propel semen from the urethral meatus. At the same time, the muscles at the base of the penis and the anal sphincter (a muscular ring) contract. There are three to four major contractions spaced six-tenths of a second apart, followed by at least as many irregularly spaced weaker contractions.

The ejaculate, called **semen**, is the combined secretions of the seminal vesicles, prostate gland, and Cowper's glands mixed with the masses of sperm delivered from the epididymides (see Figure 11–5). The average man ejaculates some four milliliters of semen (about a tablespoon) per orgasm. The consistency of semen varies. Men who have ejaculated recently (and frequently) often have semen with a watery consis-

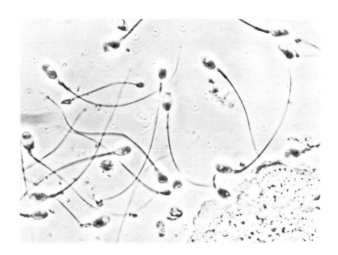

Figure 11–5

Human sperm. The headpieces and tailpieces (flagella) of the sperm are obvious. The average human male ejaculates about 360 million sperm cells at a time. (Photo by Peter Arnold, Inc.)

tency, whereas men who have not ejaculated for a few days have thick, viscous semen.

The average sperm count for normal, fertile men is about 40 million sperm cells per milliliter of semen. However, some men may produce a billion or more sperm cells per milliliter. Men whose sperm counts are 10 million or less are considered to be clinically sterile. Men are also considered to be sterile if 25 per cent or more of their sperm cells are abnormally formed—having two heads, three heads, fused tails, and so on. The number of sperm per ejaculation decreases as the frequency of ejaculation increases and as men age. It is nearly impossibile for a man to deplete his store of sperm simply by ejaculating. It would take two or three ejaculations each day for ten days before sperm counts would reach the infertile level. Any abstinence would then return the counts to normal, fertile levels.

Hormonal Control of Male Sexuality

At puberty, the **pituitary gland** produces sufficient quantities of two gonadoptrophic hormones to begin male functioning. The pituitary is activated by releasing factors produced in the **hypothalamus** of the brain (see Figure 11–6). **Follicle-stimulating hormone (FSH)**, one of these pituitary secretions, has a direct effect on sperm production. It does not seem

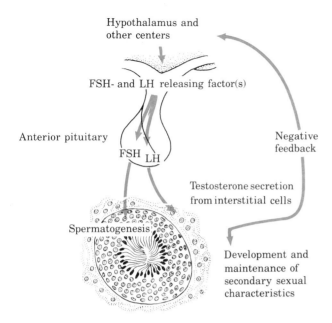

Figure 11–6

Hormonal control of human male sexuality and sperm production. Control involves interplay among the hypothalamus, pituitary gland, and testes.

to be required to initiate spermatogenesis, but it is necessary for the last meiotic division and the maturation of sperm cells. The second pituitary hormone, **luteinizing hormone (LH)**, has a direct effect on cells in the connective tissue between tubules, known as interstitial cells. These cells produce the male hormone **testosterone**, which is necessary for the reduction division of meiosis▫. Because of the action of LH in the male, LH is often called interstitial cell stimulating hormone (ICSH).

As the amount of testosterone in a boy's body increases, the characteristics of maturity are manifested. They include rapid growth of the penis, scrotum, and testes; distribution of body hair, particularly on the face (the beard), chest, and abdomen and in the pubic area; thickening of the skin; enlargement of the larynx and subsequent lowering of the voice; increased activity of the oil glands of the skin (which often become infected and produce acne in the young postpubescent male); and a dramatic increase in protein metabolism (the appetite of the pubescent and young postpubescent male is well known). The increase in protein metabolism gives men larger muscles and longer bones than women have.

There is no direct evidence that men experience any type of monthly hormone cycle, as women do. Men

also do not have a characteristic cessation of reproductive ability, which women experience during menopause▫. (However, a man's ability to achieve erection and orgasm diminishes steadily after age twenty.) The end of sperm production, and thus of reproductive life, is called the climacteric, but many men never experience it. They actively produce sperm until their death. Some elderly men are the fathers of young children. Sperm production is thought to cease because the seminiferous tubules and accompanying interstitial cells degenerate. As a result, testosterone production also falls off. Men who develop psychological difficulties during the climacteric are often given testosterone to alleviate their problems. The climacteric need not interfere with sexual desire or sexual activity.

THE FEMALE REPRODUCTIVE SYSTEM

External Anatomy

The female genitals are not as conspicuous externally as those of the male (see Figure 11–7). The **mons veneris**, a fatty pad of tissue that is densely covered with pubic hair, marks the upper border of the external genitals—the **vulva**—which include the labia majora, labia minora, and clitoris.

The **labia majora**, or major lips, are actually folds of fatty tissue that extend from the mons downward

Figure 11–7

External genitals of the human female.

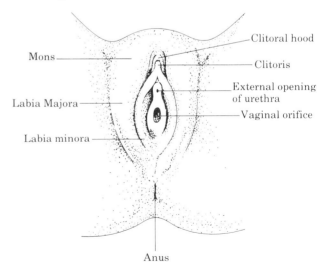

Many animals, particularly mammals and insects, produce a number of chemical scents called pheromones that are involved in the game of finding a mate.

One group of compounds deserves special attention here. The compounds, called copulins, are secreted in the vaginal fluids of female mammals. Their name comes from the copulating action they stimulate in males. Copulins are a number of compounds that together produce their effect. Some scientists have estimated that there are more than two dozen compounds in copulins. Female primates (for example, chimpanzees) produce copulins only a few days each month, just prior to ovulation. The copulins alert males that the females are in heat (estrus) and capable of mating. The concentration is believed to increase as ovulation nears. It is common for male primates to touch the genitals of females in heat and then smell their fingers. Males evidently can detect changes in pheromone production. If copulins collected from a female in heat are rubbed on the genitals of a nonestrus female, the males become just as interested in the nonestrus female as they are in the estrus female. This behavior shows that the copulins, not the physical appearance of the females, are the stimulus.

Humans are also known to produce copulins. One study examined the chemical secretions absorbed into tampons worn by a group of women throughout their menstrual cycles. Some of the women produced all the compounds that comprise the copulins produced by nonhuman female primates, and 97.5 percent of them produced some of the compounds common to copulins. The study also showed that the amount of copulins peaked near the time of ovulation.

In a follow-up study, sixty-two married women were asked to rub one of four different perfumes on their chests each night before retiring. Unknown to the couples in the study, one of the perfumes contained copulins. Each morning, husbands and wives filled out questionnaires about their sexual activities the night before. Twelve of the couples showed increased sexual activity when the perfume containing couplins was used. The others couples did not show any correlation between perfume and sexual activity. Although the study is not conclusive, it does suggest that copulins may regulate the sexual activities of some humans.

Men also have scents that may attract women. A manufacturer of men's colognes and after-shaves released a new fragrance that was supposed to be liberally dosed with human pheromones. The magic ingredient was at first kept secret, but a television show informed the public that it was human male sweat. (We are not aware of any scientific studies that have found sexual pheromones in male sweat.) Another report said that the magic ingredient was not sweat but male hog saliva. Adult male hogs produce a potent sexual pheromone in their testes that is released in their saliva and attracts crowds of female suitors. Male students, however, reported that the cologne did not improve their sex lives. Local stores soon began selling it at a 50 per cent reduction in price because its popularity was waning.

between the legs. Their outer borders are distinct and often pigmented. Like the mons, they possess pubic hair. Their inner borders lack pubic hair, are smooth, and consist of moist mucous membranes. In the young adult woman who has not given birth (nulliparous), the libia majora meet, occluding the other structures of the vulva. The labia majora of women who have given birth (multiparous) often lack muscle tone and flare apart, exposing the labia minora.

The **labia minora**, or minor lips, lie within the folds of the labia majora. They do not have fatty deposits and therefore are thinner, more veil-like mucous membranes. Where the labia minora come together, they form a hood that covers the sensitive clitoris.

The **clitoris** is small, usually less than an inch in length and a quarter inch in diameter. It has a shaft portion and a terminal glans. The clitoris is composed of erectile tissue that is the same as the tissue of the penis. The prepuce of the clitoris and adjoining surfaces house glands that produce the curdlike smegma, presumably the same substance produced by the prepuce in males. Accumulation of female smegma also produces a strong odor (probably a metabolic by-product of bacteria that digest the secretion) and provides a moist, dark, warm home for disease-causing organisms. If the accumulation is not regularly removed by good genital hygiene, the clitoral hood and clitoris may stick together, causing severe pain—particularly during sexual arousal. The problem can be solved by a gynecologist.

Sexual arousal causes obvious changes in position and size of the structures of the vulva. Vasocongestion of these spongy tissues engorges them with blood. The labia majora increase in size and flare to the sides, exposing the labia minora and vaginal opening in nulliparous women. In multiparous women, the labia majora more commonly flare only slightly but hang in a pendulous fashion. The labia minora also become engorged with blood during arousal, often increasing to two or three times their unstimulated size. In fact they may protrude between the labia majora. The clitoris may also grow to twice its usual size, with the glans protruding from the clitoral hood. The labia minora and clitoris are very sensitive. Both are supplied with many nerve endings from which most of the pleasure of sexual stimulation is derived.

When sexual arousal is sustained and orgasm is imminent, the clitoris withdraws into its hood, becoming flatter and often smaller than in its unstimulated state. During this stage, intense vasocongestion of the vulva causes the labia minora to change from pink to red or burgundy. Some authorities believe this is a vestige of the skin reaction during estrus, the period of sexual receptivity in all other primates. (See Box 11–1.)

Internal Anatomy

Between the labia minora is the vestibule (meaning "opening"). It contains the openings of both the vagina and the urethra. The vestibule is the gateway to the reproductive and excretory structures (diagrammed in Figure 11–8). The urethral opening lies between the clitoris and vagina and is hardly discernible. The urethra conveys urine from the bladder to the outside, its only function in the woman.

The opening of the vagina in young women may be partly occluded by the **hymen** (Figure 11–9), a thin membrane that originates during sexual differentiation in fetal development. The hymen rarely

Figure 11–8
The internal structures of the human female's reproductive system.

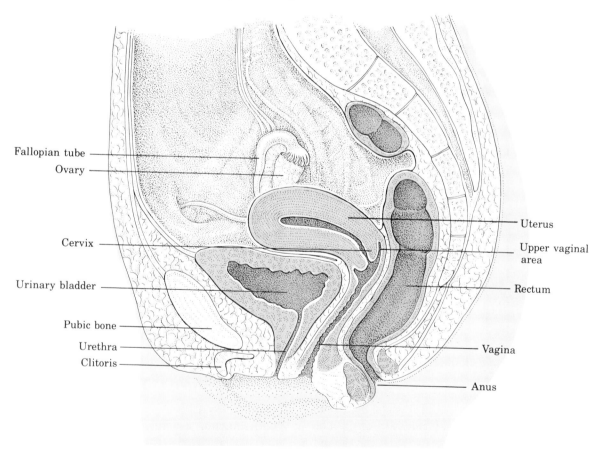

Most female mammals have estrus cycles, or periods of "heat"—physiological times when the females are receptive to males. Estrus cycles are almost always seasonal, occurring once or twice to several times per year. For example, the domestic dog generally has two estrus periods per year, about six months apart. Each lasts about three weeks. Ovulation occurs during the second week. This is when the female is most likely to conceive.

Animals advertise their estrus periods by anatomical releasers and pheromones that attract potential mates. It is common, for example, to see female monkeys in the zoo with swollen, bright red genitals that extend from the body. This is a signal to the male of the species that the female is ready to mate. Often, vaginal discharges are also obvious. It is common for male monkeys to taste or smell the discharges, presumably to receive information about the state of estrus.

Female dogs also have swollen genitals and voluminous discharges from the vagina during estrus. The discharges contain a potent pheromone that attracts male dogs from long distances. Experiments have shown that male dogs are more attracted to the odor of the vaginal discharge than to the sight of the estrus female.

If an estrus mammal does not conceive, the thickened uterine lining is not sloughed off (as in human menstruation). Instead it is absorbed.

covers the entire opening of the vagina. Often, it is shaped like a half moon, covering only part of the vaginal opening. If the hymen is a complete covering, it may have one or more holes in it, called fenestra ("windows"). Some women retain the hymen until their first sexual experience, when penetration by the erect penis ruptures it. This event may be accompanied by some pain and blood loss. However, many women lose the hymen earlier in life, often without noticing it, while bicycling, swimming, or pursuing some athletic endeavor. Still other women have such a thick, elastic hymen that it is never lost, even through they may bear several children. The hymen rarely interferes with normal hygiene, tampon use, or sexual relations. In cases where it does interfere, it can be painlessly removed by a gynecologist.

There is probably more folklore associated with the hymen than with any other reproductive structure. In almost all societies, it is associated with virginity. Premarital rites in some primitive cultures included removal of the hymen—an obvious demonstration of its presence and therefore the bride's virginity. Certain modern societies still demand a token of the presence of a hymen, maybe just a spot of blood on a sheet taken from the bed of the newlyweds. However, there is little correlation between virginity and the presence of a hymen.

On either side of the vaginal opening, near the hymen, are the Bartholin's glands. Scientists used to believe that these glands produced the lubricating fluid that was discharged during sexual arousal. However,

Figure 11–9

Various shapes of the human female's hymen, a membrane that partly covers the entrance to the vagina. The hymen never completely covers the vaginal opening. Because of its shape, the vaginal opening in a female with an intact hymen may be (a) circular, (b) a series of small openings, or fenestra, (c) two or more larger openings, or (d) a slit. The presence of the hymen does not guarantee virginity. Many women lose or break the hymen without noticing it, while running, biking, or engaging in some other activity. Others have an elastic and thick hymen that it remains even after childbirth.

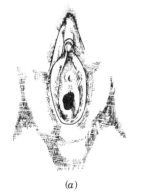   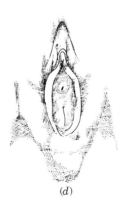

(a)　　　　　　　(b)　　　　　　　(c)　　　　　　　(d)

CHAPTER 11　*Human Sexuality and Reproduction*

research by Masters and Johnson has shown that these glands produce only a drop or two of fluid during sexual arousal, not enough to account for the copious fluid produced during lubrication. The function of Bartholin's glands remains a mystery.

The **vagina**, a muscular tube that serves as both copulatory organ and birth canal, is a major structure in the female anatomy. In the unstimulated state, it is no more than 7.6 to 10.2 centimeters in length, and its ridged walls are collapsed. It is very flexible, however, and can expand considerably during coitus or childbirth. This expansion results in part from a flattening of the folds. The portion of the vagina near the vestibule has nerve endings sensitive to touch and pressure, but the inner portion has only a few sparsely located nerve endings. At its upper end, the vagina attaches to the uterus.

The lower end of the uterus, where the opening between the vagina and the uterus appears, is called the **cervix**. The wall of the cervix contains a ring of muscle and cervical glands. The second most common form of cancer among women is cervical cancer (breast cancer is the most common). The best "cure" for cervical cancer is early detection. Women, particularly middle-aged women, should request annual Pap smears (named after George Papanicolaou) from their physicians. The Pap smear involves the removal of a few cells from the cervix. These cells are examined under a microscope to determine whether any of them are cancerous.

The **uterus** (womb) in a woman who is not pregnant and not sexually aroused is quite small. It is approximately the size, shape, and texture of a large prune (see Figure 11–10). Its walls are thick and muscular. The walls touch, and the outer muscle layers are thrown into folds or wrinkles. The uterus lies between the bladder and the rectum. In the unaroused woman, it rests almost at a right angle to the vagina (see Figure 11–8). When a woman becomes pregnant, the uterus stretches to accommodate the developing fetus. At the time of birth, its muscles contract powerfully to propel the infant into the world. The Greeks referred to the uterus as the *hystera* and considered it the site of female emotionality (hence the word *hysteria*). Today, *hysterectomy* means the removal of the uterus.

The two **Fallopian tubes**, or uterine tubes (named after the seventeenth century Italian anatomist Gabriello Fallopio), ascend from the upper corners of the uterus (see Figure 11–10). They are sometimes called oviducts ("egg tubes") because they convey eggs to the uterus. Each Fallopian tube terminates in a funnel-shaped opening. The two **ovaries**, the almond-sized sites of egg production, lie near each opening but are not directly connected. Instead, the ovaries are held in the lower abdominal cavity by ligaments.

Sexual arousal causes a number of internal changes in addition to those described for the vulva. The most immediate response is dilatation of the vagina—expansion of the vagina in length and diameter. These

Figure 11–10

The anatomical relationships of the vagina, cervix, uterus, Fallopian tubes, and ovaries. The ovaries are secured in the abdominal cavity by ligaments. The right ovary in the diagram has been opened to illustrate the progressive stages in the development of a follicle. A follicle has just ruptured, and the mature egg cell is swept into the funneled end of the Fallopian tube, the fimbria, by the action of cilia.

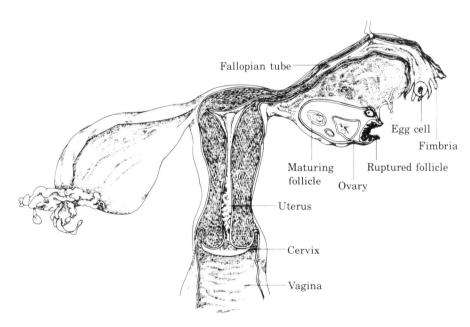

changes are caused by vasocongestion of the vaginal walls.

At the same time, copious amounts of mucous fluid gush through the cell membranes of the vagina, causing the lubrication reponse, often called "sweating." The vagina can become lubricated in as little as fifteen seconds after the woman sees, hears, or smells a sexually arousing stimulus. Lubrication is brought about by the same mechanisms that create erections in men, and it is thought to be the equivalent response.

If sexual stimulation is sustained and vasocongestion becomes more intense, the top third of the vagina expands like a balloon, a response called tenting. This response produces a potential reservoir for semen, called the seminal pool. The uterus begins to contract rhythmically and to tip upward. (It is thought that the sensations a woman feels during sexual arousal are the contractions of the uterus.) These contractions push the cervix, at the lower end of the uterus, into the semen in the seminal pool. A woman's orgasm creates a vacuum in the uterus that aids the sperm in moving into it. Orgasm is not necessary, however, for sperm to enter the uterus.

Female orgasm occurs in two distinct phases. The sensation that women usually associate with orgasms is only the first phase. It involves rhythmic contractions of the orgasmic platform (the labia minora and the muscles of the lower third of the vagina). There are three to four powerful, major contractions, spaced at intervals of eight-tenths of a second, followed by at least as many weaker, more irregular contractions. Immediately after the major contractions, the second phase of orgasm begins. It consists of weak, irregular contractions of the uterine muscles and the anal sphincter. Many women, however, do not experience orgasm.

Egg Production and Ovulation

Men may continue to produce sperm for their entire adult lives, but women do not produce eggs forever. The average woman is born with some 200,000 egg cells in each ovary. About every twenty-eight days, one egg cell matures into an **ovum**, for a total of thirteen ova per year (see Box 11–2). The average woman remains fertile for about thirty years, so a total of some 400 ova reach maturity. The remaining egg cells degenerate, most before puberty, and are absorbed by other cells in the ovaries. Therefore, a woman has a predetermined number of egg cells that

---

BOX 11–2
TWENTY-EIGHT DAYS

The approximate twenty-eight-day duration of the menstrual cycle is significant. Twenty-eight days is a perfect lunar month. From earliest times, romance and courtship have been associated with the moon.

There is no direct evidence of a link between the moon and reproduction. However, there is considerable evidence that much animal behavior is triggered by environmental change, such as amount of light and changing seasons.

Experiments that eliminate the effects of night and day have shown that some animals, when kept under constant illumination, remain constantly in estrus. Kept in the dark, they never come into estrus. Other animals (many fish and birds, for example) need a critical amount of light to trigger reproduction. With too little or too much light, they do not become reproductively active.

Other research has demonstrated that generally the organ sensitive to light is not the eye but rather a portion of the brain known as the pineal gland. This gland stimulates the hypothalamus to produce hypothalamic stimulating factors that are the precursors of pituitary hormones. The pituitary hormones inititate sexual cycles.

This information suggests that the human animal is not immune to environmental change. Perhaps early in our evolutionary history it was a significant advantage for menstrual cycles to be controlled by lunar months. If this kind of trigger did exist, most females would have ovulated near the same time and would have had children near the same time. Coincident births would have stimulated cooperation among mothers raising children and made it easier to adopt and nurse a child if its genetic mother died.

Perhaps other environmental events restricted breeding to those times of year when there would be the best likelihood of successfully carrying and delivering the fetus. This kind of timing would have favored human survival.

Today, humans no longer seem to respond to obvious environmental events. Studies of Eskimos, who live in the Arctic in near total darkness during the winter and in continual light during the summer, have shown no differences in reproductive physiology during these contrasting periods. However, other studies have shown that blind girls begin to menstruate much earlier than sighted girls do.

can mature. The events involved in the maturation of an ovum and its release from the ovary make up the menstrual cycle.

Puberty

Females begin puberty between the ages of ten and twelve. At this time, the pituitary gland begins producing follicle-stimulating hormone. The FSH induces the development of the ovaries, which in turn produce the hormone estrogen. This hormone is responsible for the development of the secondary female sexual characteristics, including a change in voice and the development of external genitalia, breasts, body hair, pubic hair, and the feminine shape. This shape means a widening of the pelvis and deposits of fat in the thighs, buttocks and face.

Menstrual Cycle

The cycle during which one or more ova mature and are released is known as the menstrual cycle. It is regulated by several hormones, some of which are secreted by the pituitary gland. The pituitary hormones are the same hormones that control male sexuality. The pituitary gland is stimulated by releasing factors produced in the hypothalamus (see Figure 11–11). The two most important hormones released by the pituitary gland are the follicle-stimulating hormone (FSH) and the luteinizing hormone (LH), both of which play a prominent role in the menstrual cycle. The cycle is divided into three phases: proliferation, secretion, and menstruation. (See Figure 11–12 for graphs showing how hormones affect the menstrual cycle.)

PROLIFERATION PHASE   The first half of the menstrual cycle is the proliferation phase. During this phase, the pituitary gland actively secretes FSH. The FSH directly stimulates one or more of the immature egg cells to grow. The egg cells are stored in the ovaries in follicles, structures that account for FSH's full name—follicle-stimulating hormone. Specialized cells within the follicle begin the production of another hormone, estrogen, under the influence of a steady flow of LH from the pituitary. As the egg cells continue to mature and the follicles begin to swell, more and more estrogen is produced. The rising amount of estrogen in the blood signals the hypothalamus to produce inhibiting substances that reduce the output of FSH. This is an example of biological feedback, or, more precisely, negative feedback. That is, one physiological event has a negative or inhibitory effect on another event. The inhibition of FSH stops the growth of any more egg cells at this time.

The importance of estrogen during the proliferation phase concerns its effect on the lining of the uterus. The inner layer of the smooth muscle of the uterus has a lining, called the **endometrium,** that contains a rich supply of blood vessels. Estrogen stimulates the endometrium to thicken through an increase of general tissue, blood vessels, and endometrial glands. These glands grow taller than the general tissues that make up the lining and give the surface a wavy appearance. The general increase in thickness of

Figure 11–11

Hormonal control of human female sexuality and ovulation. Control of the menstrual cycle involves interplay (feedback) among the hypothalamus, pituitary gland, and ovaries.

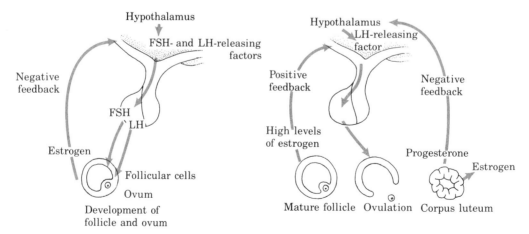

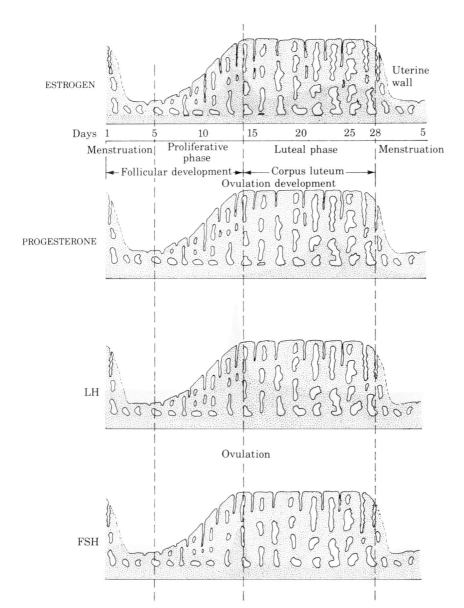

ESTROGEN

Uterine wall

Days 1   5   10   15   20   25   28   5

Menstruation | Proliferative phase | Luteal phase | Menstruation

← Follicular development → ← Corpus luteum →
Ovulation   development

PROGESTERONE

LH

Ovulation

FSH

Figure 11–12
Hormonal control of the menstrual cycle. The graphs show the cyclic levels of various hormones and their effects on a typical menstrual cycle. During menstruation, the pituitary gland produces large amounts of FSH (follicle-stimulating hormone), which stimulates an ovarian follicle to begin development. As the follicle grows, its cells begin the production of estrogen (top graph). Estrogen stimulates the repair of the uterine lining and causes it to proliferate. Because of negative feedback, the level of FSH begins to drop as estrogen increases (compare top and bottom graphs). About Day 14, estrogen levels reach their highest level, and by positive feedback cause the pituitary to produce a surge of both FSH and LH (luteinizing hormone), which jointly stimulate ovulation. After ovulation, a corpus luteum forms in the ovary and actively secretes both estrogen and progesterone. These two hormones cause the uterine lining to fully develop. During the luteal phase, there is a slight decrease in both FSH and LH. The corpus luteum begins to degenerate about fourteen days after its formation, and both estrogen and progesterone decline. The uterine lining breaks down, and menstruation begins.

the endometrium is called proliferation, which lends its name to this phase.

The proliferation phase lasts for about nine to thirteen days. At the end of this time, when the estrogen level has reached its peak, the hypothalamus signals the pituitary to release large amounts of FSH and LH for a short time. The sudden surge of FSH and LH occurs at about the fourteenth day of a twenty-eight-day cycle. Their combined effect on the now-bulging follicle causes it to erupt, releasing the mature ovum (see Figure 11–13). This event is called **ovulation**.

At first, the ovum probably floats weightlessly in the abdominal fluids. However, the funnel end of one of the Fallopian tubes is positioned nearby. Hairlike cilia beat inward toward the uterus, creating small currents that draw the ovum into the Fallopian tube (see Figure 11–10). Recent research has suggested that, in some women, fingerlike projections of the funnel, called fimbriae, gently stroke the ovary at the time of ovulation, assisting ovulation and scooping up the ovum. Some women can tell the exact moment of ovulation, perhaps from the sensation of the

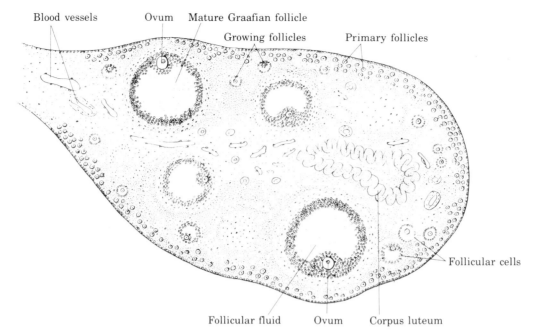

Blood vessels   Ovum   Mature Graafian follicle

Growing follicles   Primary follicles

Follicular cells

Follicular fluid   Ovum   Corpus luteum

Figure 11–13

Cross section of the ovary and its contents, showing the development of a follicle, the ovulation of an egg, and the formation of the corpus luteum. The cells of the walls of a developing follicle produce the hormone estrogen. After ovulation, the follicle collapses, and the cells form a yellowish mass, the corpus luteum. This mass becomes a temporary organ for hormone production. It produces estrogen and progesterone, hormones that influence the final thickening of the walls of the uterus after ovulation.

contracting fimbriae. The ovaries take turns producing mature ova, but the alternation is irregular and unpredictable. On occasion, both ovaries may release an ovum. The result could be fraternal twins.

SECRETORY PHASE   While the ovum makes its six- or seven-day journey toward the uterus, hormones stimulate the endometrium to prepare for the arrival of an ovum that may have been fertilized. This period of time is the secretory phase. Immediately after ovulation, the follicle collapses and its cells form a yellow mass, the **corpus luteum**, which becomes a temporary site for production of other hormones (see Figure 11–13). LH affects the corpus luteum, causing continued production of estrogen and initiating production of a second ovarian hormone, **progesterone**.

Estrogen and progestrone have a dual effect. They stimulate the endometrium to thicken even more through the enlargement of cells produced during the proliferation phase and through increased production of tissue fluids. Blood vessels permeate the entire endometrium.

If fertilization takes place, the corpus lateum must continue to produce enough estrogen and progesterone to maintain the endometrium. Pituitary LH stimulates the formation of the corpus luteum. However, high levels of progesterone produced by the corpus luteum inhibit further production of LH. So what maintains the corpus luteum during pregnancy? About a week and a half after the ovum and sperm unite, the young embryo is implanted in the uterine wall. Certain cells of this embryo begin to produce the hormone **chorionic gonadotropin** (CG), which affects the corpus luteum much as LH does. CG levels become quite high during pregnancy. By the end of the first month, so much CG is produced that the mother secretes some in her urine. The detection of CG in urine is the basis of all pregnancy tests, including the home pregnancy test kits available at any drugstore. By the end of the third month, CG levels drop and the corpus luteum begins to degenerate. However, by this time the placenta produces enough progesterone and estrogen to maintain the uterine lining.

If the ovum is not fertilized, the hypothalamus reacts to the high levels of progesterone being produced by the corpus luteum and signals the pituitary to diminish its output of FSH and LH. As LH diminishes, so do the corpus luteum and the amounts of estrogen and progesterone. The corpus luteum ceases to function about the twenty-eighth day of the cycle, fourteen days after its formation. The rapid decline in the levels of estrogen and progesterone causes the endometrium to break down and slough away. The resulting menstrual fluid is discharged through the vagina.

MENSTRUAL PHASE   Menstruation, or the passage of menstrual fluid, marks the menstrual phase, which lasts for one to five days on the average. The amount of menstrual fluid varies from woman to woman. It averages from seven to nine teaspoons. Most of the fluid is tissue fluid from the endometrial lining. However, as the lining breaks down, many of the arteries spasmodically contract and reopen. Without a steady supply of oxygen, the tissues begin to die and leak blood. Each reopening of the arteries causes minihemorrhages. The actual amount of blood lost is generally very little (probably less than a tablespoon), and it is old, stagnant blood left behind in the dead tissues, so the body scarcely notices the loss. However, some women lose considerable amounts of blood and must replenish lost minerals (usually iron) with supplements.

At menstruation, all female hormones are at their lowest levels. In response, the hypothalamus produces releasing factors that stimulate the pituitary to once again release FSH and LH. Secretion of FSH stimulates other follicles to begin the maturation process, and the estrogen produced under the stimulating effect of LH begins another proliferation phase.

Just prior to actual menstruation, when hormone levels are falling, many women experience a physiological state called premenstrual tension (PMT). Hormone levels at this time are so low that a number of physiological and emotional changes occur.

Most women experience some form of depression at this time. In fact, many women regard that feeling as a signal for the impending beginning of the next menstrual phase. The lack of female hormones also causes water retention, which is often manifested as swollen ankles and feet and even puffy faces. Water retention also accounts for an increase in breast size and firmness during menstruation. Some women gain as much as five pounds in retained water but lose the weight once hormone levels are restored. Women with long-term histories of skin disorders (blemishes or acne) often report outbreaks that coincide with their menstrual periods.

There is some evidence that low hormone levels may cause serious problems for some women. For instance, a recent survey has shown that 50 percent of the women who commit serious crimes—murder and robbery, for example—do so while menstruating. The same survey has found that 90 percent of the women who commit misdemeanors—such as running stop signs, causing fender-bender accidents, and violating parking rules—do so while menstruating.

Low hormone levels and the physiological and emotional conditions that result from them are usually no cause for concern. Some women experience few, if any, changes in their normal functioning. And many of those who do experience depression, tension, or physical discomfort have learned to compensate—for example, by consuming less salt to reduce water retention, by planning critical events for a less stressful time, or by getting extra rest and a good diet. Indeed, most women can expect to function effectively and efficiently throughout their cycles.

Although the average menstrual cycle—including the proliferation, secretory, and menstrual phases—lasts twenty-eight days, the cycles can be irregular and unpredictable. Research has shown that some women have normal menstrual cycles as short as sixteen days and others have cycles as long as ninety days. In addition, menstrual cycles are often quite irregular in women under twenty. Emotional states also have an effect on the length of the cycle. Worry, stress, and anxiety can make it shorter or longer than usual. If you are a female, the trauma of getting ready for your next biology exam may have an effect on the timing of your current cycle.

Menopause

On the average, women stop ovulating and menstruating, a phenomenon called **menopause**, in their mid to late forties. Some women have irregular cycles for months or years prior to menopause. Others simply stop menstruating abruptly. One theory is that menopause is a result of changes in the pituitary gland and the nearby hypothalamus. Another theory suggests that menopause may begin when no follicles are left in the ovaries.

Without developing follicles, there is a reduced supply of estrogen and progesterone. Women may therefore suffer temporary depression, hot flashes, and other physiological and psychological problems at menopause. While they last, most of these symptoms can be relieved by taking hormones prescribed by a physician. Unfortunately, estrogen-replacement therapy is suspected of encouraging certain types of cancer. On the plus side, menopause need not necessarily affect sexual desire or sexual activity. It simply marks the beginning of a period in the life cycle when a woman no longer needs to worry about birth control.

BIRTH CONTROL

Human reproduction is elaborate and complex, but the reproductive machinery is perfectly adapted to

---

CONCEPT SUMMARY
EVENTS IN THE MENSTRUAL CYCLE

Proliferation phase (nine to thirteen days)

1. Pituitary secretes follicle-stimulating hormone (FSH).

2. FSH stimulates one or more follicles and eggs to mature.

3. Follicle cells secrete estrogen.

4. Estrogen stimulates endometrium to proliferate (thicken).

5. Near end of phase, pituitary releases a surge of FSH and luteinizing hormone (LH), which stimulate ovulation.

Secretory phase (next six to nine days)

1 Ovum moves down Fallopian tube to uterus.

2. LH stimulates development of corpus luteum at site of ruptured follicle.

3. Corpus luteum secretes estrogen and progesterone, which stimulate endometrium to continue to thicken.

4. If ovum is not fertilized, pituitary dramatically decreases output of FSH and LH and corpus luteum ceases to function.

5. Rapid decline in levels of estrogen and progesterone induce breakdown of endometrium.

Menstrual phase (one to five days)

1. Menstrual flow occurs because of breakdown of endometrium.

---

making sure that individuals produce as many offspring as possible. Of course, if humans produced as many offspring as possible, the population of the world would be many times what it is today. From earliest recorded history, most humans have felt the need to control the number of births.

Birth control is any attempt, by any method, to control the number of offspring produced. One of its more common forms is contraception—methods that prevent or decrease the frequency of fertilization of egg cells by sperm. Contraceptive methods can be natural (including the rhythm or ovulatory method) or artificial (including condoms and diaphragms).

Mechanical Methods of Birth Control

Condoms, cervical caps, diaphragms, and intrauterine devices (IUDs) are all mechanical birth control devices. The first three prevent sperm from entering the uterus.

Condoms are rubber or plastic sheaths that are worn over the penis. They are the most common form of birth control because of their wide availability, low cost, and reliability. The failure rate for condoms is between 3 and 15 percent. As condoms age, the chances of tearing and perforation increase. Indeed, it is likely that the higher failure rates are associated with using old or damaged condoms. Condoms can be tested by blowing them up like balloons. If they hold air, they will hold semen. Perhaps the most serious disadvantage of condoms is psychological. Many people complain that sex is ruined for them if they have to take the time to fuss with a condom. Women and men also complain that condoms dull the sensations of coitus. However, many new brands are extremely thin and therefore interfere little with sensation. One great advantage of condoms is the protection they provide against venereal disease. No other form of contraception provides this kind of protection. Therefore, the use of condoms is highly recommended for coitus with strangers.

Cervical caps are rubber or plastic nipples that fit snugly over and around the cervix. They are manufactured in sizes ranging from twenty-two to thirty-one millimeters in diameter. They fit tightly, creating a partial vacuum that helps them stay in place. Size is estimated by a physician. The fit is refined by trial and error, with the user suggesting larger or smaller caps.

Cervical caps have one big advantage. They can be

worn for days at a time. It is recommended, however, that they be removed every two or three days, cleaned thoroughly, and reinserted with a new supply of spermicidal jelly. Cervical caps have a failure rate of about 8 percent. They fail because they are improperly placed or become dislodged. They are more popular in Europe than in the United States.

Diaphragms are commonly used by women in the United States. They are considered the female counterpart of condoms. However, they are not particularly popular in other countries. Like cervical caps, diaphragms are manufactured in a variety of sizes, ranging from forty-five to a hundred millimeters in diameter. Rather than being nipple shaped, diaphragms are shaped like shallow cups. A spring built into the rim retains the shape. Insertion is difficult at first. With experience, however, a diaphragm can be easily inserted. (Instruction by a qualified physician or technician is important.) When properly positioned, the diaphragm fits between the back wall of the vagina and the upper edge of the pubic bone, completely covering the cervix (see Figure 11–14).

The main advantage of the diaphragm is the low health risk to the user (in comparison to IUDs or chemical methods of contraception). A disadvantage is that, like the condom, the diaphragm may require some interruption of the sex act for insertion. However, the main disadvantage is its failure rate, which may be as high as 20 per cent. The low effectiveness of diaphragms is due to improper insertion, premature insertion (insertion longer than two hours before coitus), premature removal (before eight hours after coitus), and jolting out of position during coitus. Use of a spermicidal cream or foam with the diaphragm increases its effectiveness dramatically.

Intrauterine devices are small plastic objects that are placed in the uterus. They are made in a variety of shapes, including coils, loops, spirals, rings, bows, and shields. Before insertion, the IUD is collapsed. It is inserted through a tube with a diameter small enough that the tube can enter the cervix without the cervix being dilated. Once the tube is inserted, a plunger forces the IUD out of the tube and into the uterus, where it expands into its full shape. Insertion must be done by a trained technician, but anesthetics usually are not required. The UID must touch the lining of the uterus to be effective. The more points of contact, the better the effect—a fact that stimulated the production of the many shapes now available. Most IUDs have an attached plastic thread, or tail,

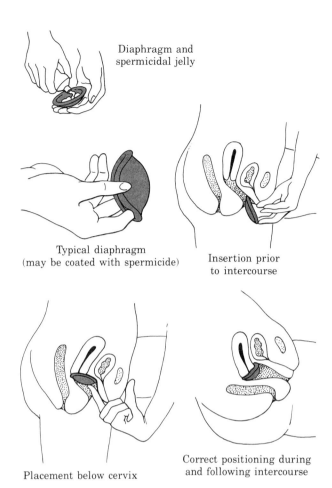

Diaphragm and spermicidal jelly

Typical diaphragm (may be coated with spermicide)

Insertion prior to intercourse

Placement below cervix

Correct positioning during and following intercourse

Figure 11–14
Diaphragm insertion. Diaphragms come in various sizes and must be fitted by a physician. They are most effective when used with a spermicidal jelly or cream. For maximum effect as a birth control device, a diaphragm should be inserted no more than two hours before coitus and should be left in place for at least eight hours after coitus.

that hangs from the cervix, facilitating the removal of the IUD. The way in which IUDs prevent pregnancy is unknown. They do not interfere with ovulation or menstruation, and reproductive function appears normal. The most widely accepted hypothesis is that the presence of such a device in the uterus causes a mild inflammation of the uterine lining, which prevents successful implantation of the embryo. Why implantation is prevented is not known. Failure rates for IUDs are less than 2 percent.

Surgical Methods of Birth Control

Sterilization and abortion are two surgical methods of birth control.

Male Sterilization

The accepted method for sterilizing males is called vasectomy. It is a surgical procedure in which each vas deferens is cut just above the epididymis. (See Figure 11–15.) Small incisions are made high on each side of the scrotal sac. The vasa deferentia are pulled through the incisions, cut, folded, and finally sutured at each end. Usually, only one or two stitches are required to close the incisions. Sperm may remain in the upper regions of the vasa deferentia for several weeks and may be passed in sufficient quantities for fertilization. Therefore, men are advised to use some other method of contraception (such as condoms) for about six weeks after a vasectomy or until a sperm count indicates complete sterility. A man's sexual activity is unchanged by vasectomy. He is still capable of orgasm, but he ejaculates semen free of sperm. (Most of the fluid of semen is produced in the seminal vesicles.)

Female Sterilization

Methods for female sterilization are similar to those for male sterilization. They involve tying (tubal ligation) and cutting the Fallopian tubes. Procedures vary. In one method, each tube is tied only once, but in another each tube is tied twice and then cut between the ties (see Figure 11–15). Both procedures require a short hospital stay and a local or general anesthetic. The tubes are often tied immediately after delivery of a child, because they are larger than usual and easier to handle.

If done properly, both male and female sterilization are almost 100 percent effective as birth control measures. However, if the people later change their minds, they may find it difficult or impossible to have the cut ducts or tubes rejoined. Recently, though, a method of inserting removable plastic plugs in Fallopian tubes was developed. It may become a widely accepted and reversible method of female sterilization. Vasectomies may also be reversed by surgically reconnecting the ducts, but the success rate is not high.

Abortion

Abortion became legal in the United States in 1974, but it remains a topic of debate. Most of the arguments for or against it are based on personal philosophies and religious beliefs. Despite the debates, though, abortion is practiced throughout the world, legally and illegally, to control the number of offspring.

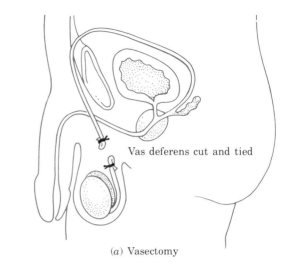

(a) Vasectomy

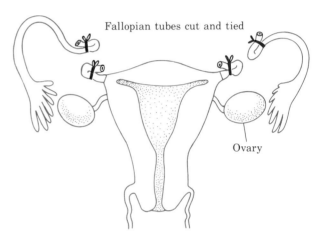

Figure 11–15
Male and female methods of sterilization. (a) A vasectomy involves tying and then cutting the right and left vasa deferentia (only the left is shown). Sperm transfer to the outside during ejaculation is then impossible. (b) Tubal ligation, a similar procedure, involves tying and then cutting the Fallopian tubes. Passage of eggs from ovary to uterus is then impossible. Some methods of male and female sterilization are reversible, but sterilization is most often a permanent change.

Most legal abortions in the United States are done for one of two reasons. Either the pregnant woman is unwed and does not want a child or her life is jeopardized by carrying a child. Abortions supervised by certified clinics or hospitals are considered safe. The surgical procedure most used today is called curettage. It means "to cleanse." More technically, it is a scraping of the uterine lining. The most common procedure is dilation and curettage (D and C). First the cervix is dilated. Then the uterus is scraped to remove the fetus. Suction curettage, which is considered an improvement over dilation and curettage, removes the fetus with a vacuum-like device rather than by scraping. Both methods are used during the first trimester of pregnancy□. (Pregnancy is commonly divided into three trimesters of about three months each.) Suction is usually used if the woman is less than twelve weeks pregnant. D and C is used if she is twelve to fourteen weeks pregnant.

Abortions can be induced during the second trimester with different methods. A saline (salt) solution may be injected into the uterus to induce labor and delivery. A surgical incision into the uterus may be made to remove the fetus. (This is termed a hysterotomy). Prostaglandins□, hormone-like substances that induce labor and premature birth, may be injected into the veins. This last method is the most common one used today. However, the decision to use a particular method is made on an individual basis. The effectiveness of abortion is 100 percent.

Chemical and Hormonal Methods of Birth Control

Spermicides, douches, and birth control pills constitute the chemical and hormonal methods of birth control. The introduction of substances into the vagina to kill sperm is an age-old technique. Many agents can kill sperm, but those used must be mild, so they will not harm the membranes of the vagina. Common spermicides are lactic acid, citric acid, boric acid, potassium permanganate, and zinc sulfate.

Most spermicides are mixed with foams or jellies that adhere to the vagina and cervix to simplify their application and improve their effectiveness. When foams are used alone their effectiveness is only about 75 percent. The effectiveness of creams and jellies is even less, about 65 percent. The effectiveness of each increases if it is used with a diaphragm.

Douching the vagina with a jet of warm water or a spermicide to wash away semen after coitus is probably a waste of time. Generally, the sperm have already passed through the cervix and are on their way to the Fallopian tubes before the douche is begun.

There are many birth control pills on the market today. The number of different kinds is estimated to be more than forty. They are all pretty much the same, differing mainly in the amount of synthetic hormone they contain. Physicians determine how much synthetic hormone each woman will need and prescribe a specific pill. (By law, the pill must be prescribed by a physician.) A woman who begins using the pill may find it necessary to change the dosage two or three times until the correct one is found.

The most common type of pill is the so-called "combined pill." It contains a combination of synthetic progestins (acting like progesterone) and estrogen. After the last pill of each series is taken, no more pills are taken for seven days. Menstruation occurs during this time. Placebos (sugar pills without hormones) are often included in pill kits, one for each of the seven days of menstruation, so pills can be taken every day. Combined pills inhibit ovulation by inhibiting the normal release of FSH and LH from the pituitary. They mimic the hormones produced by the corpus luteum, causing the uterine walls to thicken, as during a normal menstrual cycle, and suppressing the release of FSH and LH. Withdrawal of the pills for a short time results in menstruation.

All birth control pills, regardless of type, have some side effects. Among them are nausea, breast tenderness, water retention, weight gain, and breakthrough bleeding (slight blood loss between menstrual periods).

Pills are often considered 100 percent effective, but they do fail in some cases. Clinical studies have shown pills to be about 99.5 percent effective.

Natural Methods of Birth Control

Birth control methods that depend on the natural rhythms of a woman's body, rather than on chemical, surgical, or mechanical devices, are known as natural methods of birth control. In practicing natural birth control, a woman determines when she ovulates and abstains from sexual relations during this period. She engages in intercourse only when she cannot get pregnant.

Three techniques help women determine when they ovulate: the rhythm, or calendar, method; the basal body temperature method; and the mucus method, commonly known as the ovulatory or Billings method. The effectiveness of the first two methods, when used alone, rarely exceeds 85 percent. The effectiveness of the third method is much better. Although reports vary, it seems to be at least 95 percent. When all three are used together, the effectiveness is 98.5 percent. There are two types of failures with these methods. First are user failures—pregnancies that result from the user having sex at a time not recommended by the method. Second are method failures—pregnancies that result from the user failing to use the method correctly, presumably because of a lack of knowledge.

Rhythm, or Calendar, Method

The basic idea of the rhythm method is simple. It is based on the rhythm of the menstrual cycle, particularly the rhythm of the safe and unsafe days for coitus. The average menstrual cycle lasts twenty-eight days. Ideally, menstruation lasts from Day 1 through Day 5, ovulation occurs about Day 14, and the cycle ends about Day 28. Because the egg is viable for about twenty-four hours and sperm are capable of fertilizing it for forty-eight hours, conception can occur during a period of about three days—in a typical cycle, Days 14, 15, and 16. Abstinence should begin two days earlier, just in case viable sperm remain in the Fallopian tubes until the time of ovulation. It should be extended an extra day, just in case ovulation occurs on Day 16 and the egg lives until Day 17.

In summary, the unsafe days for coitus are, statistically, Days 10 through 18 of a typical menstrual cycle (see Figure 11–16). All other days are considered safe.

Basal Body Temperature Method (BBT)

A woman's basal body temperature varies predictably during a typical menstrual period. The temperature changes, which can be charted, are another index to the approximate time of ovulation. A woman's average body temperature fluctuates above and below 98 degrees Fahrenheit. Just after menstruation, however, the body temperature drops as low as 97.3 degrees. This lower temperature continues until near the time of ovulation. At that point, the temperature dramatically increases in a day or two, until it reaches the norm of 98.6 degrees. The temperature remains relatively normal until the time of menstruation, when it drops again.

The body temperature of the average woman therefore has two discrete phases, a lower preovulatory phase and a higher postovulatory phase. In order to chart her temperature changes, the woman must use a special thermometer that can be read in tenths of degrees. For maximum effectiveness, she must take her temperature orally every day before getting out of bed in the morning. Temperature records can become somewhat complicated if the woman comes down with any illness that causes a slight fever.

Mucus (Ovulatory) Method

Commonly called the Billings method because it was first described (in the late 1960s) by physicians John and Evelyn Billings of Australia, the mucus or ovulatory, method has achieved considerable popularity. It is based on the chemical and physical changes of cervical mucus during each ovulatory period. The method must be taught by someone trained in its use. Using a cleansed finger, the woman collects a sample of mucus from the cervix or vulva daily—usually before and after urination.

Figure 11–17 illustrates the typical changes of mucus during a menstrual cycle of four distinct phases: menstruation; early dry days, when no mucus is produced; wet days, when mucus is produced; and late dry days, when no mucus is produced. As the chemical and physical properties of the mucus change, they indicate the most fertile time of the cycle. Menstruation, of course, is detectable because of the bleeding. Early dry days are detectable because mucus is not produced.

The first mucus produced after menstruation is infertile mucus. In appearance, it is thick, tacky, cloudy, creamy or lumpy, and inelastic. It is often called flu mucus because it is similar in appearance to the nasal mucus produced during a cold or the flu.

After a day or two, the mucus changes dramatically in appearance, becoming clear, sticky, and elastic. Now called fertile mucus, it is similar in appearance to the white portion of a raw egg. The fertile mucus can sometimes be stretched several inches without breaking. Some women produce larger amounts of fertile mucus than infertile mucus.

After a few days, the mucus suddenly becomes cloudy again. The first day of the newly cloudy mucus indicates that the day of ovulation, the so-called "peak day," was the previous day, when the mucus

THE THEORY OF RHYTHM

28 day cycle

1 Menstruation begins	2	3	4	5	6	7
8	9	10	11	12	13	14
		Intercourse on these days leaves live sperm to fertilize egg		Ripe egg may be released on any of these days		
15	16	17	18	19	20	21
Ripe egg may also be released on these days		Egg may still be present				
22	23	24	25	26	27	28
1 Menstruation begins again						

Black numbers—"Safe days"

Colored numbers—"Unsafe days"

Grey numbers—"Safe but possible psychological barriers"

Figure 11–16

A calendar for the rhythm method of birth control. This calendar is appropriate only for women with an average twenty-eight-day menstrual cycle. Because the cycle for a specific woman may be longer or shorter, a special calendar must be constructed for each woman. Most couples abstain from coitus during both "unsafe" and "safe but possible psychological barriers" days, a period which may total two weeks.

was still clear. Thus the peak day can be determined only in retrospect. A fertile egg will be present on the cloudy days after the peak, so this is obviously a time for abstinence. Eventually, mucus production stops, and the dry days before menstruation begin.

Not all women show this pattern, so mucus characteristics must be charted daily. (The available charts cover an entire year on one page.) Some women begin mucus production on the last days of menstruation and experience no early dry days. Various illnesses, going off the pill, stress, and medication may cause changes in the ovulatory cycle. However, once the woman is able to correctly identify infertile and fertile mucus, the ovulatory cycle can be clearly charted by daily inspection and interpretation of the

mucus. Thus any abnormal fluctuations are simply part of the record.

Women practicing the ovulatory method are generally advised that intercourse is safe during early and late dry days and days of infertile mucus. However, abstinence must be practiced on all fertile mucus days.

Coitus Interruptus

A natural method of contraception that is based on chance rather than on biological rhythms is coitus interruptus (withdrawal). The Latin name means "interrupting coitus." The method involves stopping intercourse just before the male ejaculates so that semen is deposited outside the vagina.

Figure 11–17

Typical changes in mucus during a menstrual cycle. The outer ring indicates the types of mucus found on various days of the menstrual cycle. The middle ring indicates the days of the menstrual cycle and the inner ring indicates the days that are safe and unsafe to have intercourse.

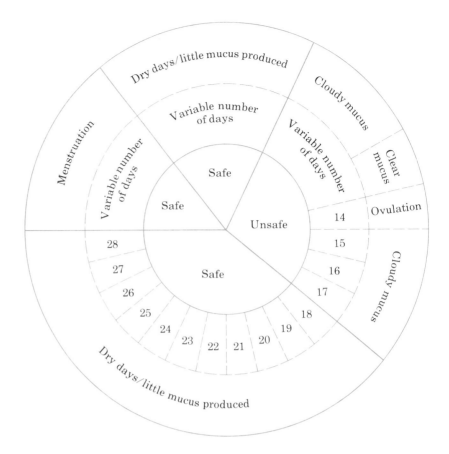

Probably less than 4 percent of U.S. couples routinely use withdrawal as a method of birth control. It is difficult to evaluate the effectiveness of this method. Most authorities simply state that it is not effective. Actually, it is rarely more than 50 percent effective.

Abstinence

Literally meaning "to refrain from indulgence," abstinence is not considered a natural form of birth control. Many couples practice abstinence successfully at specific times. If faithfully practiced, it is the most effective method of all.

SEXUALLY TRANSMITTED DISEASES

The health and well-being of tens of millions of humans are threatened by sexually transmitted diseases. These diseases, commonly known as venereal diseases, are transmitted by sexual intercourse. Some

have reached epidemic proportions. Recent statistics show that the three most common communicable diseases in the United States are the common cold, flu, and gonorrhea, in that order. Other common sexually transmitted diseases are syphilis and genital herpes.

Gonorrhea

Late in the last century, the cause of gonorrhea was determined to be the bacterium□ *Neisseria gonorrhoeae* (see Figure 11–18).

So many people in the United States have gonorrhea currently (perhaps as many as 3 to 4 million, mostly between the ages of fifteen and twenty-nine) that it is considered an epidemic. Why is the disease so common? Many explanations have been offered. Some claim that sexual permissiveness is a factor, but it is certainly not the primary cause. There has been a tremendous increase in the number of women who use birth control pills. As a result, men are not using condoms as often as they used to. Condoms prevent

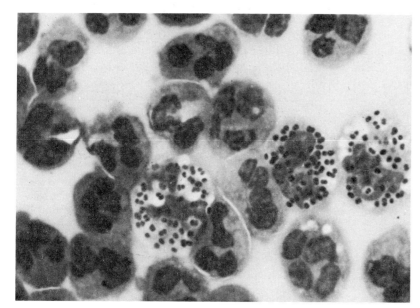

Figure 11–18.
White blood cells that have engulfed the bacterium that causes gonorrhea. (Courtesy of Carolina Biological Supply Company.)

the transfer of gonorrhea. Their lack of current use is probably the main reason for the increased incidence of the disease.

In men, the gonorrhea bacterium invades the urethra. Within five to fourteen days after infection, pus and even blood may be discharged from the penis. (The photograph in Figure 11–18 is of pus from an infected man. The white blood cells in the pus have engulfed the gonorrhea bacteria, which appear here as dark dots.) Urination can become painful. If not treated, the infection may spread up the urethra and eventually into the prostate gland, seminal vesicles, epididymis, and even the testes. Scar tissue left in these organs by the infection often cause sterility.

About 80 percent of the women who contract gonorrhea have no symptoms during the first thirty days. The bacterium invades the cervix first, causing a cervical inflammation. If untreated, the infection may spread outward to the Bartholin's glands, urethra, and anus. If the urethra becomes involved, urination can be painful. More serious complications arise when the infection spreads through the uterus and into the Fallopian tubes. The tubes swell, causing pelvic pain—now often called pelvic inflammatory disease. If serious scarring results, the tubes become blocked and the woman becomes sterile.

Because women do not experience symptoms as early as men do (some never show symptoms), it is important that men who have contracted gonorrhea inform their female sex partners. Many women become carriers, spreading the disease to others without ever knowing that they have it.

Gonorrhea can also cause infections in the anus, mouth, throat, and eyes. Infection of the eyes is most common among newborns who come in contact with the infected cervix of their mothers during birth. The eyes of most newborns are therefore washed with silver nitrate as a preventive measure.

The best treatment for gonorrhea is antibiotics, but the bacteria are becoming resistant to many of them. If you think you have gonorrhea, consult a physician immediately and inform your sex partner of the possibility of having the disease. Never try to self-diagnose venereal diseases. Physicians must use laboratory tests to determine their presence.

Syphilis

In 1905, the cause of syphilis was determined to be *Treponema pallidum,* an elongated, spiral-shaped bacterium□. Syphilis is not as common as gonorrhea. (There may be only a hundred thousand infected people in the United States.)

People who contract syphilis show few symptoms. The bacterium creates painless sores (called chancres) where it enters the body. These sores are usually found on external genitals—the penis and the lips of the vulva. They signal the primary stage of infection, and they eventually heal and disappear.

Several months later, the afflicted person may enter the second stage of the disease. By then the bacteria

 50

are circulating in the blood, causing bumps or a rash on the palms of the hands, the soles of the feet, the groin, and other areas of the body. The rash does not usually itch, but the bumps often leak pus, which contains many of the disease-causing bacteria. At this stage, the person is highly infective and may also have a sore throat, headache, fever, and general aches and pains. Because these symptoms are common to many disorders, syphilis is rarely suspected.

After a few weeks or months in the second stage, the infected person enters the latent (hiding) stage, when all symptoms disappear. Half of the people in this stage remain latent for the rest of their lives. The other half experience the third stage of the disease (years later), which involves serious deterioration of the circulatory, skeletal, and nervous systems. Most of these complications are fatal.

The diagnosing of syphilis can be difficult, but several blood tests are commonly used. A premarital blood test is required in most states to determine if either individual has syphilis. If the disease is discovered in the primary stage, antibiotic therapy can be used to stop it.

Herpes

One of the sexually transmitted diseases caused by a virus□ is herpes. Common cold sores, which are caused by herpes simplex 1, are called oral herpes. The painful sores on or in the penis, vagina, and anus, which are caused by herpes simplex 2, are called genital herpes. However, it is possible to have genital herpes in other areas of the body, including the mouth.

Most health officials consider genital herpes to be at epidemic levels. The virus is passed from individual to individual by direct contact with the sores. The virus can enter the body through any moist body structure, including the mouth, eyes, genitals, anus, and even skin pores. A person who contracts herpes may show symptoms within a few days or not for months or even a year. The first symptoms are similar to those of the flu: muscle aches, slight fever, listlessness, and swollen glands.

People do not really know if they have herpes until the sores appear. The sores, which are usually preceded by itching in and around the genitals, are very painful. They turn into large red blisters that eventually erupt. People are contagious only when they have the blisters and for about a day before the blisters appear. The sores are similar to cold sores, but

they are usually larger and more painful. Some people may have dozens of these painful blisters at a time and be in so much pain that they require hospitalization and pain killers until the sores heal. Moreover, those infected will always have the virus in their bodies. The virus is thought to remain inactive in and around nerve cells. If the carrier contracts another disease, becomes generally rundown, or suffers emotional stress, the herpes virus is often activated. New sores then erupt, and the painful cycle of blisters and healing begins again.

The disease must be diagnosed by a physician. The diagnosis usually involves a blood and urine analysis and a sample of the pus from one of the blisters. The doctor may prescribe a variety of drugs to reduce the discomfort and pain, but that is all that can be done. As yet, there is no known cure for herpes—which is why there is so much concern about its spread and why it has become an epidemic. What complicates the problem is that a single individual can become contagious again and again, infecting other people who then can also spread the disease for the rest of their lives.

Herpes is not just a matter of painful blisters. The disease is dangerous and can have many long-term complications. In both men and women, herpes can cause blindness. There is also a strong link between women who have suffered from herpes and those who develop cervical cancer. For this reason, women who have contracted herpes are advised to have Pap smears twice a year. Other serious complications arise when a herpes victim becomes pregnant. The disease can cause miscarriages, and it can be transmitted to unborn children, causing brain damage and other birth defects as well as death. An unborn child may be spared the disease in the womb only to contract it at birth from sores in the birth canal. Because of these problems some doctors these days tend to do caesareans on women with herpes to avoid the problem of the child contracting it at birth.

AIDS (Acquired Immune Deficiency Syndrome)

AIDS, a commonly fatal disease, has been on the increase in the United States since 1980. So far 1600 (September 1983) people have been afflicted and about 165 new cases are reported each month. High-risk groups include gay (homosexual) men, heroin addicts, Haitians, and hemophiliacs. The disease may be transmitted via blood transfusions, the use of dirty hypodermic needles, and through sexual contact. The disease, which results in a breakdown of a person's immune system, may be caused by a viral infection. However, the cause and mode of transmission are not yet certain. Recently the disease has been reported in women who have had sexual relations with male AIDS victims and perhaps in infants who have received blood transfusions.

SUMMARY

1. The penis consists of the shaft and the glans penis. The foreskin (prepuce), which normally covers the glans of the uncircumsized male, contains preputial glands that produce smegma. The other major external structure of the male reproductive system is the scrotum, which houses the testes.

2. During sexual arousal, more blood is delivered to erectile tissue in the penis than can be drained away by veins, a response called vasocongestion. The result is an erection. Arousal also causes vasocongestion of the scrotum and testes.

3. Testes contain seminiferous tubules, which are the sites of sperm production. The tubules join a prominent ridge of tissue on each testis, the epididymis, where sperm are stored. A vas deferens ascends from each epididymis and eventually reaches the abdominal cavity.

4. In the abdominal cavity, near the bladder, the vasa deferentia become ejaculatory ducts. Seminal vesicles join the ducts, which in turn unite with the urethra within the prostate gland.

5. Ejaculation consists of the forceful movement of stored sperm cells from the epididymides through the vasa deferentia. Seminal vesicles secrete a fluid that stimulates sperm mobility. A secretion of the prostate gland provides additional fluid for the semen. Orgasms propels semen (sperm and fluid) through the urethra by rhythmic contractions of the urethra and of muscles at the base of the penis.

6. Male sexuality is controlled by follicle-stimulating hormone (FSH), which stimulates sperm production, and luteinizing hormone (LH), which promotes the production of testosterone. Testosterone affects sperm production and the male sex drive.

7. The external genitals of the woman—the vulva—consist of the labia majora, labia minora and clitoris.

The labia majora and labia minora border the vestibule, where the urethra and vagina open.

8. Sexual arousal promotes vasocongestion of the vulva. The labia majora and labia minora enlarge, and the clitoris lengthens and widens.

9. The entrance to the vagina is often partly blocked by the hymen. The absence of the hymen does not necessarily signify virginity.

10. The vagina is a short, muscular tube. When unaroused, its walls touch. The cervix, a protuberance into the upper end of the vagina, is the gateway to the uterus. The uterus, or womb, is generally small. Its walls fold and touch.

11. During sexual arousal, vasocongestion causes the walls of the vagina to expand and become lubricated and causes the uterus to tip upward.

12. Female orgasm consists of strong rhythmic contractions of the orgasmic platform (the labia minora and the muscles of the lower third of the vaginal walls) and gentler contractions of the uterus and anus.

13. During coitus, tenting of the vagina may occur, forming a seminal pool that receives ejaculated semen. Contractions of the uterus push the cervix into the semen, and the woman's orgasm creates a vacuum in the uterus that helps draw sperm into the uterus.

14. At one end, the Fallopian tubes join the uterus. At the other, they open as a ciliated funnel into the abdominal cavity, near the ovaries but not in direct contact with them. The Fallopian tubes convey mature ova from the ovaries to the uterus. The average woman produces viable ova from about the age of thirteen until well into the forties. Cessation of the menstrual cycle is called menopause.

15. The menstrual cycle can be divided into three phases. The proliferation phase begins with the development of a follicle under the stimulation of pituitary FSH. Cells within the follicle secrete estrogen under the influence of pituitary LH. This stage ends when the amounts of FSH and LH peak, inducing ovulation.

16. The secretory phase begins at ovulation. The follicle (now minus its egg) collapses and forms the corpus luteum, which produces estrogen and progesterone. These hormones cause the cells of the uterine wall to increase in size, preparing them for implantation of an embryo.

17. If fertilization does not occur, the amount of LH diminishes. As a result, the corpus luteum disintegrates and the production of estrogen and progesterone declines. The uterine lining breaks down, and the menstrual phase (the third and final phase) commences.

18. As hormone levels drop near the end of the menstrual cycle, many women experience premenstrual tension—depression and other psychological problems.

19. Birth control is any attempt, by any method, to control the number of offspring. Contraception is a form of birth control that prevents or decreases the frequency of fertilization of egg cells by sperm.

20. Contraceptive methods can be natural or artificial.

21. Artificial methods of birth control include mechanical devices such as condoms, cervical caps, diaphragms, and intrauterine devices.

22. Surgical methods of birth control include male and female sterilization and abortion.

23. Chemical and hormonal methods of birth control include spermicides, douches, and birth control pills.

24. Natural methods of birth control include the rhythm method, the basal body temperature method, and the ovulatory method.

25. Gonorrhea is the most common sexually transmitted disease and is caused by a bacterium. It ranks as the third most common communicable disease in the United States.

26. Syphilis, which is also caused by a bacterium, is less common than gonorrhea. The complications it can cause if it is left untreated, however, can be extremely serious. It may even result in death.

27. Herpes is a new and widespread sexually transmitted disease. It is caused by the virus herpes simplex 2. There is no cure for herpes. Once you contract the virus, you will always have it in your body.

ALTERNATE ROUTES TO REPRODUCTION

No one is predicting that the human reproductive anatomy will change in the next decade or that human sexual habits and appetites will be altered appreciably. However, what seems likely is the widespread acceptance of methods of reproduction that exclude the time-honored but unreliable act of sexual intercourse. Nearly 4 million couples in the United States want children but cannot conceive them. Many of these couples hope that in the next decade they will be candidates for some of the alternate routes to reproduction—sperm banks, artificial insemination, surrogate motherhood, and test-tube gestation of babies, for example.

SPERM BANKS

More common today than ever before, sperm banks will become even more common in the next decade. The most frequent visitors to sperm banks are men who are contemplating vasectomy. They wonder if they might marry again and want a second family. The new microsurgical vasectomies are often reversible, but they come with no guarantees. Therefore, many men go to sperm banks before the vasectomy to deposit their semen for possible future need.

The semen is frozen at −385 degrees Fahrenheit. It retains more than half its ability to fertilize an egg if it is thawed within five years. If the donor remarries and wants to father more children, the sperm can be thawed and the new female partner can be artificially inseminated. This is the major role of sperm banks at present.

Some people are advocating that men of great talent deposit semen samples for possible future use. These people believe that the genius of today should be preserved for use tomorrow. Sperm banks have asked many such men for semen samples. Among those asked are Nobel Prize winners, movie stars, and industrial executives. (One sperm bank in California specializes in semen from "very special donors.") This idea borders on what was once called eugenics—the control of genetic variation among humans. Many people worry that if this kind of selection becomes widespread, it could decrease human variation and reduce the ability to evolve in the future.

Sperm banks will probably thrive in the next decade. However, their major customers will most likely continue to be ordinary men who are hoping for immortality.

ARTIFICIAL INSEMINATION

One solution to the problem of childlessness is artificial insemination. In the next decade, more and more couples will probably select this alternative. Most of them will do so because the man is sterile.

Artificial insemination of animals has been practiced for a long time. The first instance is thought to date back to the fourteenth century, when a horse was artificially inseminated. Today, the practice is common among breeders of all kinds of livestock. More than 10 million cows and 150 million turkeys are bred in this manner every year.

The number of human females who are artificially inseminated is estimated at more than twenty thousand per year. The procedure is simple. First, the couple select the semen. Most sperm banks record some data about donors—eye and hair color, height, and so on. Once the selection is made, about one milliliter of thawed sperm is placed into the woman's vagina with a syringe. The entrance to the vagina is plugged temporarily so none of the sperm are lost. The time of insemination is critical. It must occur near the time of ovulation. Many women inseminate themselves at home after selecting the donor sperm. The procedure may be repeated every three or four months until pregnancy occurs.

One recent concern about artificial insemination is the background of the donors, who are generally anonymous. Many are physicians, or medical students who are paid $30 or more per donation. For convenience, many sperm banks use the same donors again and again. One clinic reported that a single donor fathered fifty babies. What if the donors harbor genetic diseases? Their offspring will carry the burden.

Most sperm banks do not screen donors to determine whether they carry genetic disorders. Furthermore, many do not keep records of donors or the children fathered by them. This sort of laxity is under attack by people who argue that sperm donors should be meticulously screened and that records about their physical characteristics (height, skin color, IQ, and so on) and genetic background should be kept. Such records are kept for livestock and pets. Surely records for humans are even more important. Perhaps in the next decade donors will be more completely screened, and detailed records will be available to recipients and their families, especially to the children resulting from artificial insemination.

SURROGATE MOTHERHOOD

Another alternate route to having a family is hiring a surrogate mother. This route is used when the woman who wants a child is fertile but has obstructed Fallopian tubes or when she is sterile. The male partner donates semen, and an unrelated woman is artificially inseminated (although in one case the surrogate mother was the woman's sister). The couple wanting the child usually pay the surrogate mother a fee for her nine-month service. They also pay for prenatal care and delivery of the child.

Most surrogate mothers are married and have families of their own. They do not want more children themselves, but they are willing to offer their reproductive abilities to others. Some surrogate mothers are unmarried, a situation that has raised moral and philosophical questions. In some cases, the surrogate mother is or becomes a friend of the couple desiring the child. In other cases, she may choose to remain anonymous.

Some problems occur with surrogate motherhood. At the end of the pregnancy, some women refuse to give up the child. This situation has resulted in lawsuits. Certain states are examining whether surrogate mothers should be paid for their services. Although some women volunteer themselves to clinics and physicians, many receive thousands

of dollars. In the next decade, surrogate mothers will probably be screened by national or regional groups of medical experts.

There are, of course, risks involved when one's future child is nurtured in the womb of a stranger. Many genetic disorders, including Down's syndrome, could be inherited from the surrogate mother. Perhaps the woman will smoke heavily during pregnancy, and this will affect fetal development. Or maybe she will turn out to have a history of venereal disease or alcoholism.

TEST-TUBE BABIES

It has been common for science fiction writers to imagine human fetuses developing in giant seethrough artificial wombs. This will probably not happen in the next decade. What will become widespread, however, is the fertilization of human ova in petri dishes—an effort to create viable embyros that will then be transplanted to wombs.

Some women are fertile but unable to conceive because of obstructions in the Fallopian tubes. They regularly ovulate, but fertilization is impossible. Such women (and their husbands) now have some hope of bearing children who are genetically theirs.

The first test-tube baby was born to Lesley and Gilbert Brown on July 26, 1978, in Oldham, England. The baby girl was full term and weighed five pounds, 12 ounces at birth. Like many other women, Mrs. Brown had obstructed Fallopian tubes. Dr. Patrick Steptoe and Dr. Robert Edwards, both of England, opened her abdomen near the time they calculated she would be ovulating. (She had been given a hormone injection before surgery to encourage ovulation.) The doctors located her ovary with a laparoscope and then used a fine glass tube to remove mature ova from the follicles. The egg cells

were placed in a culture dish of nutrient broth. A few hours later Mr. Brown's semen was added. Fertilization occurred and two days later a young embryo had advanced to the thirty-two cell stage. It was then placed in Mrs. Brown's uterus, where it developed.

Later in 1978 and again in 1979 (also in England), two more testtube babies entered the world. Shortly afterwards, a test-tube baby was born in Australia.

In 1974, the United States government banned the use of federal funds for research on human eggs. In 1979, this decision was reversed, partly because of the success of testtube babies elsewhere in the world. The new guidelines proposed by the government were as follows: (1) no embryo was to be left outside the mother for more than fourteen days, (2) egg and sperm donors for embryo transfers were to be married, and (3) federally funded research would not involve embryo transfers from one woman to another.

The new guidelines opened the door for possible test-tube babies in the United States. On December 28, 1981, Elizabeth Jordan Carr was born. She was the first test-tube baby born in the United States. By the middle of 1982, three more testtube babies had been born in the United States. One of them was the first test-tube baby to be delivered by natural childbirth. (The others were all delivered by Caesarean section.) Recently, the first test-tube twins were born. The number of test-tube babies throughout the world is now nearly two dozen. Mrs. Brown in England has given birth to her second test-tube baby.

Test-tube babies seem like miracles of science, but they are giving the scientific community a lot to think about. What are the risks involved? Many believe that the manipulations involved increase the risk of deformities in the newborns. What happens if the woman supplies more than one egg? Can the extras be given to other women? So far, all the egg and sperm donors have been husbands and wives (by law in the United States). What will be the implications if and when anonymous donors become involved? Will Mrs. X be able to use the uterus of Mrs. Y to incubate her fertilized egg? How will these decisions be made? Who will make them?

There have been considerably more advances in embryo transplants for livestock than for humans. For example, cows are given fertility drugs so they will produce many more eggs than usual. When they conceive, the multitude of tiny embryos are flushed out with a saline solution. The embryos are then transferred to other cows, where they develop to full term. The procedure is safe. The advantage is that a specific cow and bull can produce many more offspring than they normally would in their lifetimes. Cows usually give birth only once a year and for only a few years. The technology of embryo transplants has made it possible for one cow to produce as many as eighty-nine calves in one year.

Embryos from livestock donors are also being shipped around the world in special fluids. Recently, a horse embryo was flown from one part of the world to another in the uterus of a live rabbit. Fertilized eggs of cows can now be frozen up to a year. When thawed, they can develop into normal calves. Many concur that the endangered species of the world could be saved by using embryo transplants into the females of other species. Will these advancements in embryo transplantation be available to humans? Some of them undoubtedly will.

It has already been suggested that surrogate mothers may become obsolete if embryo transplants become more widespread. A fertile woman with blocked tubes could bear her own child. In addition, if embryo transplants beyond those for married couples are ever allowed, then a completely sterile woman could gestate another woman's egg that was fertilized by the latter's husband. Although this would require some hormonal preparation and maintenance for the sterile woman, it is within the realm of biological possibility in the next decade. Obviously, it would be a much more fulfilling experience for a woman to receive an embryo transplant so that she actually becomes pregnant and delivers a child. It is likely that in the next decade the rules of embryo transplants for humans will be extensively modified and that the improved techniques of embryo transplantation will make many couples happier.

FOR DISCUSSION AND REVIEW

1. What constitutes the external genitalia of males?

2. What are the various glandular secretions that make up semen? Where does each originate? What is the function of each?

3. What are the events that constitute male orgasm? Which reproductive structures are involved? What is the function of male orgasm?

4. What constitutes the external genitalia of females?

5. Distinguish between the proliferation and secretory phases of the menstrual cycle. Outline the hormonal cycles that trigger each phase.

6. What are the events that constitute female orgasm? Which reproductive structures are involved? What is the function of female orgasm?

7. Which hormones are involved in the maturation of an egg cell? In the maturation of a sperm cell?

8. Distinguish between menopause and the climacteric.

9. Describe negative feedback in relation to the menstrual cycle.

10. How are the physiological events of sexual arousal similar in males and in females? How are they different?

SUGGESTED READINGS

HUMAN SEXUAL RESPONSE, by William Masters and Virginia Johnson. Little, Brown: 1966.
The original report of the first detailed studies of human sexuality. Much of the information in this chapter was adapted from this work. For those who get hooked, they have written two other books as well.

HUMAN SEXUALITY: ESSENTIALS, by Bryan Strong, Sam Wilson, Leah Miller Clarke, and Thomas Johns. West: 1978.
A collection of original articles and chapters from other sources covering diverse topics. Good articles on all aspects of sexuality discussed in this chapter as well as sections reprinted from Masters and Johnson.

THE NATURE OF HUMAN SEXUALITY, by A. M. Winchester. Merrill: 1973.
A handbook for the beginner and a book for every library. An excellent combination of history and current knowledge.

REPRODUCTION, by Karen Jensen and the Editors of U.S. News Books. One of the volumes from the series THE HUMAN BODY. U.S. News Books: 1982.
A well-illustrated and easy-to-read account of all aspects of human reproduction, including reproductive anatomy and physiology, birth control, fetal development, and much more. The historical notes about past misconceptions about sex and sexual taboos are particularly interesting.

UNDERSTANDING HUMAN SEXUALITY, by Janet Hyde. McGraw-Hill: 1979.
A very readable account of many topics related to human sexuality, such as sexual anatomy, the menstrual cycle, human sexual response, and coitus.

(a)

(b)

One of the unique and fundamental characteristics of living organisms is their ability to reproduce themselves. In the Animal Kingdom a large variety of animals show parental care of the eggs and/or the young. Parental care is well known for mammals such as this troop of baboons **(a)** which includes males, females, and young of various ages. But animals as diverse as certain arthropods (for example, scorpions), molluscs such as octopuses, and many fish such as the cichlid fish **(b)** also display these types of behavior.

(a)

Frogs, toads and similar creatures reproduce by external fertilization, but the male mounts the female in a posture called amplexus. These frogs are shown in amplexus **(a);** when the male mounts the female, the grip of his forearms around her stimulates the release of eggs. The eggs pass from the female's cloaca on gelatinous strands. As they are extruded, the male releases sperm to fertilize them. A large cluster of eggs can be seen in front of the amplexing pair; the eggs have already been fertilized by the male. Shortly after being fertilized, each zygote begins to divide in a series of cleavage stages. Shown here are the first cleavage plane producing a two-celled embryo **(b)** and the second cleavage plane resulting in a four-celled embryo **(c). (d)** The third cleavage plane is the development of a frog embryo produces an eight-celled stage. **(e)** Continued division of the embryo produces 16- and 32-cell stages. **(f)** The embryonic stage, known as a blastula, is composed of many cells and is a hollow ball of cells. The events of cleavage occur quickly, and the cells do not increase in volume. Therefore, the blastula of a frog embryo is not much larger than the original fertilized egg. The blastula stage sets the stage for subsequent stages in development and differentiation (photos **b-f** courtesy Carolina Biological Supply Company).

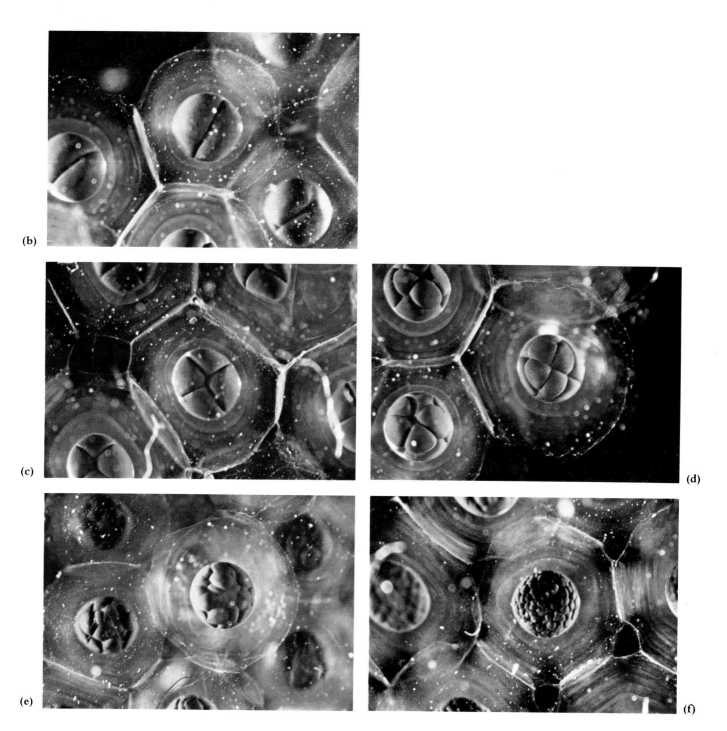

(a)

(b)

(a) The large fish, called a sweet lips, is being cleaned by two smaller fish known as cleaning wrasses. The wrasses remove parasites and other encrustations from their hosts. Socially, cleaning wrasses live in harems consisting of one male and several females. If the male is lost, the highest ranking female becomes a male. Therefore, all males are first females. (b) Earthworms are hermaphrodites and are functionally male and female at the same time. Earthworms copulate to cross fertilize one another, creating new genetic variations. (c) A male and female mosquito fish; the male is the smaller. The obvious fin on the underside of the male is modified into an intromittant organ with which he copulates with females.

(c)

 (a)

(a) In service of the queen. Many worker bees (sterile females) crowd around and attend the queen who is shown in the middle with an "8" attached to her thorax. The queen mates only during one period in her life and from then on retains live sperm in her body for the rest of her four-year existence. The workers attend the queen by both feeding and grooming her. (b) The queen honeybee lays only one egg in a cell. If the queen fertilizes an egg, it will develop into a worker, a sterile female (2n). If the queen does not fertilize an egg it develops parthenogenetically into a drone, a fertile male (n). During her life the queen is nothing more than an egg-laying machine; during the warm months of the year the queen may lay as many as 1,500 eggs per day. (c) Workers is the caste within a honeybee colony that searches for food. Here a worker is visiting a flower to collect both nectar and pollen. This worker has at least one pollen basket filled with yellow pollen. When this worker returns to the hive, she will "dance" the location of the food source she visited so that other workers can help in the harvest.

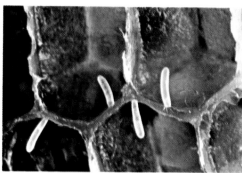

(b)

(c)

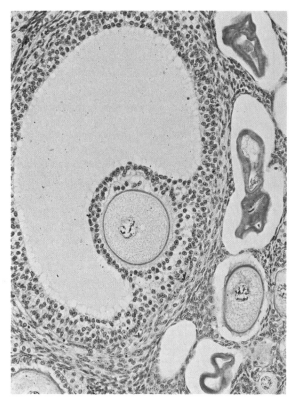

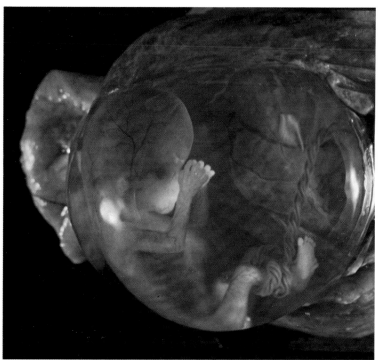

(a) A developing human egg cell within the ovary. The largest structure in the photo (mostly open space) is known as a follicle. Lining the walls of the follicle are the cells that secrete estrogen and progesterone as the follicle develops. At one side of the follicle, a larger structure bulges inward; this is the egg cell surrounded by follicular cells. This egg is nearly mature and will be ovulated in a few days. Two developing human fetuses are shown in **(b)** and **(c).** Fetus (b) is about 12 weeks old; fetus (c) is only 8 weeks old. The mother harboring either fetus has probably just learned with certainty that she is pregnant. The internal organs of both fetuses are fully formed, and many organs and systems are functional. Both fetuses clearly show their human characteristics. Both are suspended in amniotic fluid within the amnion and are connected to the placenta by the umbilical cord. Notice that the umbilical cord of fetus (b) is looped around its legs. This will not interfere with the function of the umbilicus—it probably became looped as the fetus moved about and will be kicked loose again. Fetus (b) has ended its first trimester—notice the additional refinement in the formation of arms, legs, fingers, and toes. The muscles of its neck are sufficiently strong to allow it to move its head up and down.

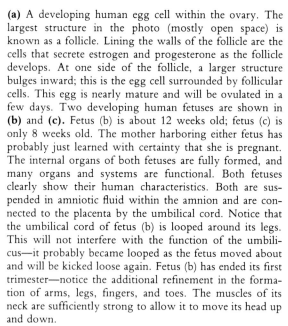

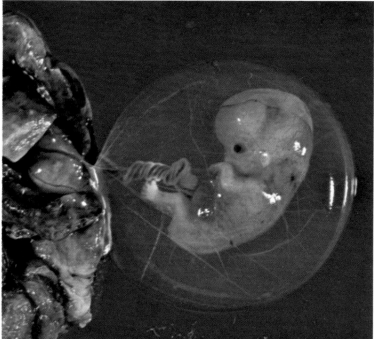

(c)

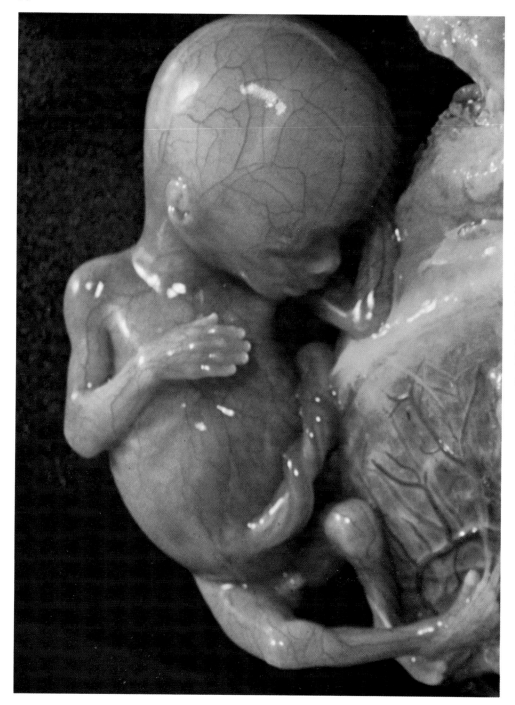

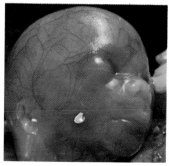

This human fetus is approximately 16 weeks old. At this stage in development a fetus has fully formed eyelids and is capable of opening and closing them. The mouth can open and close as the fetus drinks amniotic fluid. The fetus at this time has regular cycles of sleeping and wakefulness and may suck its thumb while sleeping. When awake it will become quite active, and its movements can be felt by its mother. The umbilical cord in **(a)** is clearly visible.

(a)

(a) When reproducing, male and female Siamese fighting fish clasp their bodies together in an embrace (amplexus). Eggs are fertilized externally. Notice the eggs falling from the female; all have been fertilized by the male as they passed from the female. Many organisms broadcast their gametes, a form of reproduction by external fertilization; there is no contact between the adults that produce the gametes. **(b)** The sponge is broadcasting gametes. **(c)** The pair of penguins in the middle of the photograph are in the act of copulation. Birds have internal fertilization, and they copulate. However, most male birds, with the exception of ducks and geese, lack a penis; females are inseminated by males pressing their cloacae to that of females and then ejecting sperm. **(d)** Grasshoppers copulate and thus have internal fertilization. Males possess an intromittent organ (photo by Richard S. Funk).

(b)

(c)

(d)

12

THE DEVELOPMENT OF VERTEBRATES

In the last century, the German biologist Ernst Haeckel proposed that developing embryos retrace the evolutionary history of their species. For example, say that a reptile represents a third major stage in the evolutionary development of vertebrates and that a bird represents a fourth, more advanced stage. Haeckel proposed that an embryonic bird would retrace all three earlier stages and then proceed to the fourth stage. Likewise, human embryos would pass through all the stages representing less sophisticated creatures and would add to their development a fifth or sixth stage. This idea, called the principle of recapitulation and later the law of biogenesis, was popular until well into the twentieth century. Textbooks commonly expressed the principle as "ontogeny recapitulates phylogeny" or as "the development of an individual (**ontogeny**) retraces the evolutionary descent of the entire species (**phylogeny**)." The concept is an oversimplification, however, and today is only historically significant. The important point about recapitulation is that all vertebrate embryos pass through similar developmental phases, which suggests a strong evolutionary relationship.

VERTEBRATE EMBRYO DEVELOPMENT

The detailed knowledge we have about vertebrate development (**embryology**) has been derived from studies of embryos of only a few species, such as frogs and chickens, in laboratories. The major stages

PREVIEW QUESTIONS

After reading this chapter, you should be able to answer the following questions:

1. What is accomplished during each of these processes—fertilization, cleavage, gastrulation, neurulation, and morphogenesis?

2. Name the extraembryonic membranes associated with a human fetus, and give the function of each.

3. What events occur during the first, second and third trimesters of human development?

4. List the differences between fraternal and identical twins.

5. Describe some of the factors that may cause developmental anomalies.

of development described in these studies have been generalized to vertebrates as a group. Figure 12–1 shows how a vertebrate embryo begins its life.

Fertilization

Vertebrates employ various strategies to ensure fertilization, the union of sperm and egg cells. A **sperm cell** generally has a long, whiplike tail, which is actually a flagellum; a head portion containing the genetic information for the next generation; and a head cap, the acrosome. The shape of the head portion varies from species to species. It may, for example, be rounded, pointed, blunted, humped, or doubled. Specialists can look at the shape of a sperm head and identify the species that produced it.

Most **ova,** or eggs, have some form of jelly coat, or layer of cells, adhering to the cell membrane. A sperm must penetrate this layer in order to fertilize the ovum. Here is where the acrosome plays its role. It contains enzymes that allow the sperm to digest its way through the barrier. Eventually, it penetrates the barrier and reaches the cell membrane of the ovum.

Figure 12–1
Early embryonic stages in the development of a typical vertebrate. Fertilization of an egg by a sperm stimulates the egg to erect a physical barrier, the fertilization membrane, to prevent penetration by other sperm. The zygote divides (cleaves) many times to become a hollow ball of many cells, the blastula. The next phase, gastrulation, forms the primary germ layers. During neurulation, the nervous system begins development.

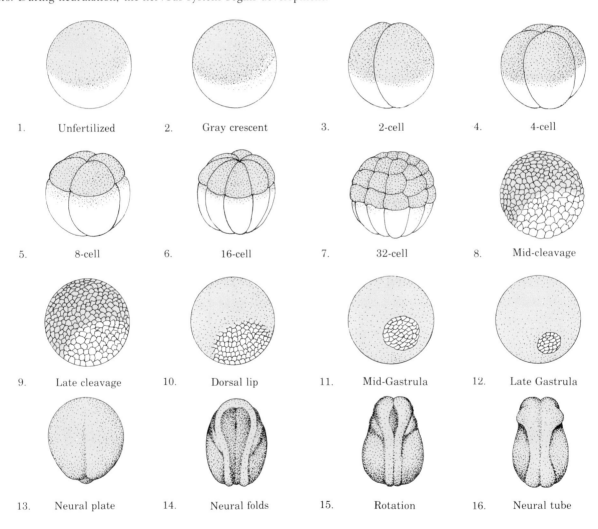

1.   Unfertilized
2.   Gray crescent
3.   2-cell
4.   4-cell
5.   8-cell
6.   16-cell
7.   32-cell
8.   Mid-cleavage
9.   Late cleavage
10.   Dorsal lip
11.   Mid-Gastrula
12.   Late Gastrula
13.   Neural plate
14.   Neural folds
15.   Rotation
16.   Neural tube

Immediately after the sperm penetrates the cell membrane, a fertilization membrane lifted from the cell membrane creates an effective physical barrier to the entrance of other sperm (see Figure 12–1). The ovum also releases chemical substances that inhibit the activity of all sperm in the vicinity of the newly penetrated egg. The motionless sperm die and begin to disintegrate.

The single sperm that penetrated the egg loses its tail. The head portion becomes rounded and migrates with its genetic load toward the egg's nucleus. The sperm and egg nucleus are both haploid, and their fusion restores the diploid chromosome number characteristic of the species. The fusion of the two nuclei is **fertilization**, and the resulting diploid cell is called the **zygote**. (See Box 12–1).

Cleavage

Shortly after fertilization, the zygote divides by mitosis□ to form the two-celled stage. It divides again to create four cells, again to create eight cells, again to create sixteen cells, and again and again—until the embryo is a mass of at least sixty-four cells. These **cleavages** occur quickly, and the cells do not increase in volume. The mass of cells is seldom much larger than the ovum was before fertilization. In fact, the volume of cytoplasm to nuclei decreases during cleavage, which facilitates the later process of cellular differentiation.

During these early divisions, cells often become segregated by type. The more inert yolk cells form in the lower half of the cell mass, the vegetal pole. Cells without yolk occupy the upper half of the cell mass, the animal pole. As the cells continue to divide, a cavity called the **blastocoel** forms within them. The cell mass is now a hollow ball of cells, called the **blastula**. The blastula has cells arranged so that the next stage, gastrulation, can take place.

Gastrulation

Cells in the blastula move in two directions. Some grow downward from the animal pole until they have completely engulfed the cells of the vegetal pole. Others grow rapidly from the animal pole into the blastocoel through an equatorial opening, the **blastopore**, in a process called **involution.** As the cells move inside, the blastocoel becomes smaller and smaller. The rapid growth of cells into the blastocoel,

<table><tr><td>

BOX 12–1

OVISTS AND SPERMISTS

During the seventeenth century, biologists debated about the contributions of the egg and the sperm to the developing embryo. One school of thought was led by William Harvey, who gave us our first understanding of the circulatory system. Harvey and his supporters were labeled "ovists," because they believed that the new human was stored in the egg in miniature form and that sperm penetration caused it to grow to adult size.

The many opponents to the ovist point of view were collectively known as the "spermists." They insisted that the preformed individual actually rode within the sperm and was nurtured by the constituents of the egg upon fertilization. In fact, one scientist drew a sperm cell with a small man, a homunculus, in its head.

Both points of view support the notion that the human embryo is preformed and simply grows to adult size. It wasn't until the end of the eighteenth century that the notion of preformation lost popularity and the theory of epigenesis was accepted. This theory states that the egg and sperm supply the basic constituents from which an embryo is progressively differentiated.

</td></tr></table>

however, creates a new cavity, the **archenteron** (primitive gut). These are the major events of **gastrulation**. The result is a **gastrula**, a two-cell-layer stage.

By way of analogy, the blastula can be thought of as a blown up balloon. The cavity within the balloon represents the blastocoel, the balloon itself the cells produced during cleavage. You can simulate involution by pushing your thumb into the skin of the balloon near its equator. The subsequent stretching of the balloon's skin demonstrates how cells grow inward as well as downward and from side to side. As you push your thumb farther in, the original cavity of the balloon (the blastocoel) becomes smaller. At the same time, a new cavity forms around your thumb (the **archenteron**). The process of gastrulation continues until the original cavity is completely obliterated and replaced by the new cavity. This can be simulated by using several fingers or your entire fist to press against one side of the balloon.

At the end of gastrulation, the cells of the archenteron press against the cells of the animal pole and

255

begin to develop into the **primary germ layers.** These layers differentiate into the organs and organ systems of the new individual. The layer of cells lining the archenteron develops into the first germ layer, the **endoderm**. It will give rise to the digestive system and associated structures (such as the liver and pancreas) and the respiratory system. The outer layer of cells will develop into the second germ layer, the **ectoderm**. It will give rise to the new covering—skin, scales, feathers, or hair, depending on the species—and to the nervous system. A layer of cells that becomes sandwiched between the ectoderm and endoderm is the third germ layer, the **mesoderm**. It gives rise to the remaining major organ systems, including the skeletal, muscular, excretory, circulatory, and reproductive systems.

CONCEPT SUMMARY
PRIMARY GERM LAYERS AND THEIR FATES

GERM LAYER	FATE
Endoderm	Forms epithelial lining of digestive tract and associated structures, such as pancreas, liver, and respiratory system
Ectoderm	Forms epidermis of skin and structures (such as hair, feathers, and scales) derived from epidermis, as well as nervous system
Mesoderm	Forms all other systems, including skeletal, muscular, excretory, circulatory, and reproductive

---

Neurulation

When does the rather simple ectoderm become the complicated nervous system? During the next embryonic process, **neurulation**.

Shortly after gastrulation, the ectoderm begins to grow inward along the upper midline of the embryo and directly above the developing digestive system, forming a trough, or groove. As the trough becomes deeper, its edges grow rapidly, finally meeting. The result is a tube of ectoderm, the **neural tube**, that detaches from the outside layer of ectoderm. The neural tube later differentiates into the brain, spinal cord, and nerves.

Again, the balloon simulation can help explain the process of neurulation. This time, the blown-up balloon represents the gastrula, with its definitive germ layers. Press the edge of your hand into the side of the balloon, holding your hand parallel to the poles of the balloon. The depression you make represents the neural groove. As you continue to push inward, your entire hand becomes engulfed in the balloon, and the edges of the original groove eventually meet. The tube that has formed inside represents the neural tube. It is formed from the outer skin of the balloon, the ectoderm. (See Box 12–2.)

Morphogenesis

From neurulation onward, embryonic development is different for different vertebrate species. **Morphogenesis** involves the further differentiation of the primary germ layers and the formation of organs and organ systems—**organogenesis**. The embryo grows rapidly at this time, with an obvious polarization into head and tail ends. Arm, wing or leg buds, and the circulatory and nervous systems all become increasingly obvious. Tissues stretch, fold, expand, and shrink as the embryo grows to its final form.

TERRESTRIAL VERTEBRATE DEVELOPMENT

The embryos of many organisms develop in an aquatic medium. Among vertebrates, such organisms include fish and amphibians. The watery habitat has many properties that make it ideal for embryonic development:

— There are no sudden changes in temperature (which might kill a developing embryo), because water warms and cools slowly.
— Most bodies of water are rich in nutrients and have sufficient amounts of dissolved oxygen to support embryonic life. Currents continually replace the nutrients and oxygen and carry away the wastes produced by embryos.
— Water is a buoyant medium, allowing embryos to float almost weightlessly and giving them freedom of movement.
— Water is a good shock absorber, softening any mechanical disturbances that might affect embryos.

BOX 12–2

CELLULAR DIFFERENTIATION AND INDUCERS

It is a mystery how, during embryonic development, certain groups of cells become differentiated into specialized tissues and organs. It is assumed that the physical and chemical environment in which the cells exist gives them "messages" about what their fate will be in the new organism. Each cell created during embryonic growth has an exact copy of the genetic material originally present in the zygote.

However, because each cell has the same set of genetic messages, some of the messages must be turned on briefly and then turned off once the cell becomes differentiated. The most popular theory of cellular differentiation is that the genetic messages being translated at a certain time create chemical substances that affect genetic expression in neighboring cells and tissues. These substances are called inducers. Some recent work suggests that inducers may be molecules of messenger-RNA (see Chapter 15).

One of the most common experiments in studying inducers is the so-called transplant experiment. Aggregates of embryonic cells from laboratory animals are transplanted from one site in an embryo to another or from one embryo to another. Tissues then develop in an environment that is different from the normal one. Thus researchers can evaluate whether genetic or environmental factors determine differentiation.

For example, the vertebrate eye first appears as an optic cup differentiated from the developing nervous system. Where the cup comes in contact with ectoderm on the outside of the embryo, the ectoderm differentiates into the lens of the eye. But when the optic cup is removed from the embryo, the layer of ectoderm left behind does not differentiate into a lens. Clearly, then, the cup induces the development of the lens. In other experiments, the optic cup is surgically removed and placed in an unusual site, such as the flank or thigh of a frog embryo. Wherever the optic cup is transplanted, the overlying ectoderm forms the lens. Under normal circumstances, it would most likely form a layer of skin.

The first terrestrial vertebrates were confronted with a serious problem of adaptation. Their embryos could never develop in the harsh terrestrial environment without some mechanism for protection and as-

sistance. The problem was solved in more than one way. Reptiles and birds evolved **oviparity**, which means that the ova expelled from the body are encased in shells, and the embryos develop outside the parent. An egg and its shell protect a developing embryo. The yolk contains the nutrients that support development. For the most part, mammals and some reptiles evolved as **viviparous** creatures. Their young are born alive after embryonic development within the female parent. Seclusion of the embryo within its mother provides both protection and nourishment, allowing the developing young to grow to a larger size.

The eggs of both reptiles and birds have a large amount of yolk. As a result, the stages of cleavage and formation of the blastula and gastrula are different from those previously described. The yolk does not divide at all. The cytoplasmic division occurs in a small restricted area of tissue that contains no yolk. Cleavage produces a flattened mass of cells above the yolk, called the blastodisc. It is here that the events of gastrulation occur.

The eggs of mammals contain little yolk, and the early phases of embryonic development are similar to those described earlier. However, most mammalian embryos are nourished by the placenta. That connection develops first, while some of the events of gastrulation are delayed.

In either case, the embryos of all terrestrial vertebrates have evolved a system of membranes that envelop their embryos and contain fluids which in many ways mimic the aquatic habitat of their ancestors. These membranes are called the **extraembryonic membranes**. *Extra* refers to the fact that the membranes are not part of the embryo, even though they are formed by it. The chorion, amnion, yolk sac, and allantois are the extraembryonic membranes.

The development of a chick embryo illustrates the development and function of extraembrhonic membranes (see Figure 12–2). Early in its development, a double fold of tissue grows from the embryo. In time, it completely wraps the embryo. The outer layer of the fold is the **chorion**, and the inner layer is the **amnion**. The chorion in part allows the embryo to exchange gases through the porous egg shell.

The amniotic fluid that accumulates within the amnion is similar to blood plasma and has many of the properties of sea water, particularly its ion concentrations and kinds of elements. The embryo floats nearly weightlessly in the amniotic fluid, which acts as a

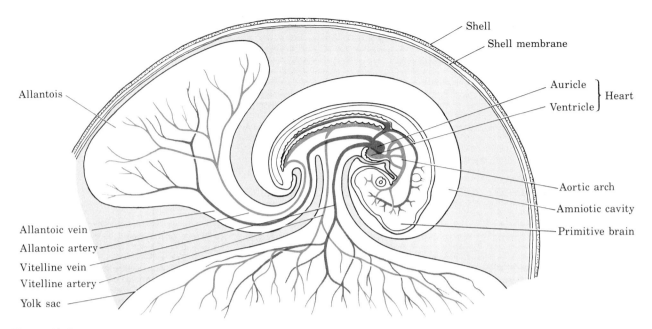

Allantois

Shell
Shell membrane

Auricle ⎫
Ventricle ⎭ Heart

Aortic arch
Amniotic cavity
Primitive brain

Allantoic vein
Allantoic artery
Vitelline vein
Vitelline artery
Yolk sac

Figure 12–2

The extraembryonic membranes of a chick embryo. The young chick embryo is suspended among membranes that provide it with a miniature aquarium during its development. The extraembryonic membranes form the chorion, amniotic sac, allantois, and yolk sac.

shock absorber and allows the embryo freedom of movement. The embryo develops, then, as if in a double sac. Embryologist Bradley M. Patten has said that the chorion and amnion create a miniature aquarium in which the embryos of terrestrial vertebrates recapitulate their aquatic ancestry.

The midgut region of the terrestrial embryo develops another membrane, which grows downward engulfing the yolk and becoming the **yolk sac**. The development of the yolk sac is closely associated with the development of a rich supply of blood vessels to the yolk. From the yolk, nutrients are mobilized and transported to the embryo.

Finally, the hindgut region develops a membranous sac that becomes the **allantois**. Associated with this sac are the allantoic arteries and veins, which connect the allantois with the embryonic circulation. In oviparious vertebrates, the allantois presses against the chorion. Together they become the organ of gas exchange. The allantois also serves as a reservoir for waste products. As the embryo grows, so does the allantois.

The development of human extraembryonic membranes is similar to that of other terrestrial vertebrates.

HUMAN EMBRYONIC DEVELOPMENT

Human copulation is commonly referred to as intercourse, although technically it is called coitus. During coitus, sperm are deposited into the vagina, an event that makes fertilization possible.

Fertilization

Millions of sperm are released into the vagina by ejaculation but only a few thousand are likely to make the trip to the uppermost regions of the Fallopian tubes. Sperm can cross this great distance in a short time. In one investigation, they reached the Fallopian tubes in twenty-eight minutes. The time is even shorter for the sperm of cows and sheep. Even dead sperm are quickly moved into the Fallopian tubes, which suggests that contractions of the uterus and Fallopian tubes assist in sperm movement.

Even though sperm arrive at the point of potential fertilization quickly, they are unable to fertilize immediately. They must first go through an aging phase that enables them to penetrate the ovum. There is some evidence that a single sperm cannot digest the protective materials around the ovum. Such digestion seems to require the cooperation of many sperm and the digestive enzymes produced in their acrosomes.

The life span of sperm is limited. They can swim for three or four days but probably lose their power to fertilize within forty-eight hours. The ovum is capable of being fertilized for only about twenty-four hours after ovulation. If the timing is right, fertilization occurs when the egg is in the upper portion of the Fallopian tube. It can occur later, but then the embryo may be too young when it arrives at the uterus to implant itself on the uterine wall. As a result, it will die.

Cleavage

About thirty hours after fertilization, the newly formed zygote divides into two cells in the upper portion of the Fallopian tube. This is the first cleavage. The next division occurs within forty hours after fertilization. The third occurs about three days after fertilization. During these early cleavages, the young

embryo is slowly moving down the Fallopian tube toward the uterus, at the rate of about two centimeters per day.

At the end of four days, the embryo reaches the uterus. It has thirty-two cells, but the boundaries between the cells are indistinct. This mass of cells is known as a **morula**.

At the next cleavage, which produces an embryo with about sixty-four cells, a cavity forms within the cell mass. The cavity is the blastocoel, and the mass of cells is the **blastocyst** (blastula stage). On one side of the blastocyst is a mass of cells, the **inner cell mass.** A thin layer of cells, the **trophoblast**, encircles the blastocoel and the inner cell mass (see Figure 12–3).

Gastrulation

The first change in the inner cell mass is its differentiation into two types of cells about eight days after fertilization. The outer layer, facing the blastocoel, becomes the endoderm, and the cell layer below becomes the ectoderm (see Figure 12–4). These two pri-

Figure 12–3

A summary of the early events during the embryonic development of a human. The egg is fertilized in the upper third of the Fallopian tube. Digestive enzymes secreted by the head of the sperm digest the egg's gelatinous coat. After fertilization, the developing embryo moves about an inch a day en route to the uterus. By the fourth day after fertilization, the embryo is a mass of about thirty-two cells (called a morula) and has reached the uterus. The embryo implants itself on the uterine wall as a blastocyst on about the tenth or eleventh day.

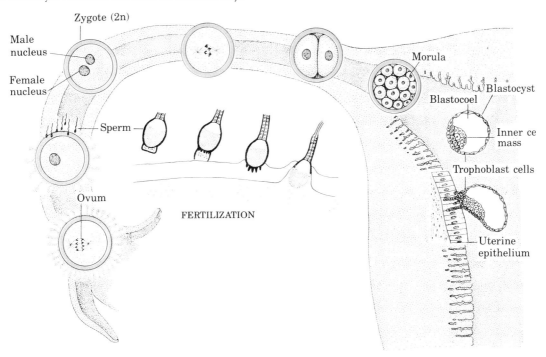

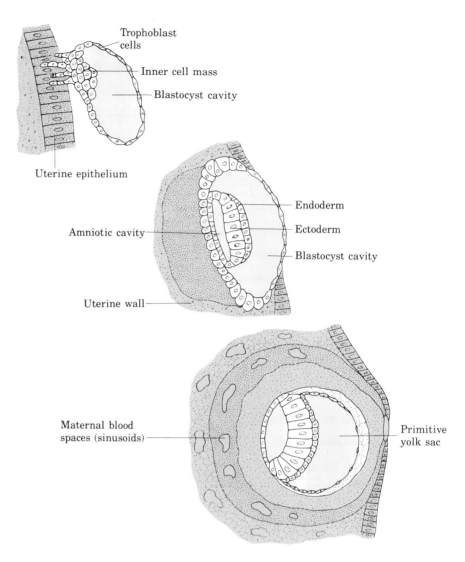

Figure 12-4
The stages of implantation of a
human embryo on the uterine
wall and the formation of the
primary germ layers. The
trophoblast cells of the blastocyst
begin to penetrate the uterine
lining (top). As implantation
proceeds, the inner cell mass
differentiates into the ectoderm,
which becomes the embryonic
disk (center). When implantation
is completed (bottom), the
maternal tissue and embryonic
trophoblastic layer form the
placenta.

Trophoblast
cells

Inner cell mass

Blastocyst cavity

Uterine epithelium

Endoderm

Ectoderm

Blastocyst cavity

Amniotic cavity

Uterine wall

Maternal blood
spaces (sinusoids)

Primitive
yolk sac

mary germ layers flatten together, forming the embryonic disk from which the fetus will develop.

The separation of the ectoderm from the trophoblast creates a cavity, the amnionic cavity, which is
soon lined with ectoderm. At that point the amnion
pulls away from the trophoblast. Simultaneously, the
endoderm grows to line the blastocoel. When the lining is completed, the blastocoel pulls away from the
trophoblast and becomes the primitive yolk sac. The
entire process is completed by the ninth or tenth day
after fertilization.

The third germ layer, the mesoderm, does not differentiate until a week after the embryo is implanted
on the uterine wall. The mesoderm differentiates between the ectoderm and the endoderm of the embryonic disk.

The primary germ layers give rise to the major organ systems of the new individual. The ectoderm
forms skin, hair, nails, skin glands, nasal and mouth
cavities, tooth enamel, central and peripheral nervous
systems, and linings of all the sense organs. The endoderm gives rise to part of the pituitary gland, the
respiratory system, the lining of the digestive system,
and the liver and pancreas. The mesoderm forms the
muscles, bone, blood vessels, kidneys, gonads, and
part of the adrenal glands.

Implantation

On about the tenth or eleventh day after fertilization,
the blastocyst begins to affix itself to the uterine wall.
From the time the morula arrived in the uterus until

this time of **implantation**, the embryo has been floating freely in the fluids of the uterus. But now it must become securely implanted and signal its presence to the body. It has to work fast, though, because menstruation□ is due to begin. Quite often, a new embryo does not become implanted soon enough and is swept away in the menstrual tide. The factors that prevent it from attaching in time are unknown.

Implantation is accomplished by further differentiation of the trophoblast layer of cells into fingers, or villi, that erode the **endometrium** lining the uterus and thus establish firm contact with the uterine wall. These new cells are thought to begin immediate production of the hormone **chorionic gonadrotropin** (CG), which maintains the corpus luteum in the ovary and thereby prevents the next menstruation.

Neurulation

Shortly after formation of the mesoderm, the ectoderm along the dorsal (back) midline of the embryo begins to thicken and form a groove. The two sides of the groove close, forming portions of the neural tube, by the end of the fourth week. The forward end of the neural tube grows rapidly, ballooning in the back and front as well as to the side. This enlargement is the first indication of the development of the brain. Growth continues so rapidly that the head end of the embryo begins to bend downward (the beginning of the fetal posture). The rest of the neural groove closes later, forming the spinal cord.

Placenta

At first the human embryo is nourished by the products of the breakdown of the endometrium during implantation. Later, the **placenta** becomes the major life-support system, and it remains so until birth. When completely formed, the human placenta is disk-shaped and about 15 to 23 centimeters across and 2.5 centimeters thick. It may weigh half a kilogram. The placenta allows the woman to carry out digestive, excretory, and respiratory functions for the embryo. It begins to form where the trophoblast invades the endometrium. One of the extraembryonic membranes, the chorion participates in its formation.

In the early embryo, the trophoblast fused with a layer of mesoderm to form the chorion (a membrane composed of a layer of ectoderm and layer of mesoderm). The chorion now completely encircles the developing embryo, and the fingerlike **villi** grow outward from it (see Figure 12–5). Villi on the side that implants become large and branch many times. They form treelike structures that penetrate deeply into the endometrial lining of the uterus. The villi on the rest of the chorion degenerate, and the chorionic surface becomes smooth.

The first blood vessels appear during the third week, but they belong to the chorionic villi, not the embryo. The embryo's blood vessels begin developing a few days later. When the embryo's heart becomes functional, two umbilical arteries carry a rich blood supply to the placenta. When the blood reaches the capillary beds of the villi, fetal wastes are forced across the capillary walls into the maternal bloodstream, which generously bathes the maternal side of the villi. Nutrients, oxygen, and other materials from the maternal blood reach the fetus in the reverse process. Blood returns to the fetal heart by way of a single umbilical vein and is recirculated. Fetal and maternal blood do not normally mix.

Extraembryonic Membranes

The chorion, amnion, yolk sac, and allantois—the extraembryonic membranes—are the same in the human embryo as in all other terrestrial vertebrate embryos□. The chorion and its relationship to the placenta have just been described. The amnion forms during gastrulation. With subsequent growth, it completely engulfs the developing embryo. However, it is separated from the embryo by a large, fluid-filled amniotic cavity (see Figure 12–5). The amniotic fluid gives the developing embryo the advantages of an aquatic environment. In fact, the composition of human amniotic fluid, in terms of salts and ions, is remarkably similar to that of seawater.

Although human egg cells possess virtually no yolk, the human embryo still develops a yolk sac. (Vertebrate embryos all develop similarly, even though the result may be a structure of little use to the embryo.) The yolk sac produces the first blood cells, a function quickly taken over by the liver and spleen. After that, the yolk sac is nonfunctional.

As is the case with other vertebrate embryos, the human embryo develops an outgrowth of the hindgut, the allantois. In the human embryo, the allantois is quite small and is considered vestigial, because the placenta takes its place as the organ of respiration and excretion. The allantoic vessels, however, become the

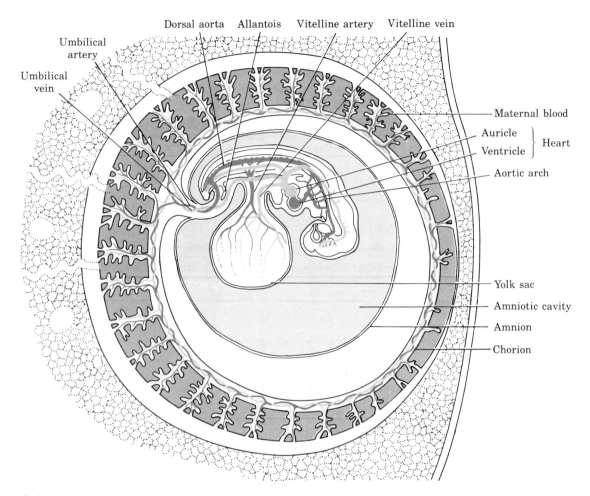

Labels on figure:
Umbilical artery
Umbilical vein
Dorsal aorta Allantois Vitelline artery Vitelline vein
Maternal blood
Auricle } Heart
Ventricle }
Aortic arch
Yolk sac
Amniotic cavity
Amnion
Chorion

Figure 12–5

The human embryo shortly after implantation. Like all other embryos of terrestrial organisms, the human embryo is suspended among extraembryonic membranes. The highly branched tissue layer surrounding it is the placenta, whose inward layer was derived from the chorion. The embryo floats weightlessly in amniotic fluid within the amnion. The small yolk sac is the first site of blood cell formation, but this function is quickly taken over by the liver and spleen. The allantois is vestigial, but its vessels become the arteries and vein that are housed in the umbilical cord.

umbilical arteries and vein and the main component of the **umbilical cord**. Thus they play an important role in blood circulation to the placenta.

The events leading up to this point are crucial and carefully timed. The calendar in Figure 12–6 helps illustrate how the embryo accomplishes so much in such a short time.

SEQUENCE OF DEVELOPMENT

A human embryo remains within a woman's body for about nine months (266 days ± 2 weeks). This

period is typically divided into three **trimesters** of three months each.

First Trimester

The events in human embryological development that have been described all occurred within the first four weeks after fertilization. During the first month, an embryo has been implanted on the uterine wall. By the time it is four weeks old, the embryo is about the size of a pea. It has recognizable head and tail ends and a stout tubular heart that vigorously pumps blood to the life-supporting placenta. All of this has

SUN	MON	TUE	WED	THU	FRI	SAT
1	2	3	4	5	6	7
			MENSTRUAL PERIOD			
8	9	10	11	12	13	14
Ovulation	Fertilization	Cleavage	18	Morula	Blastocyst	21
	Implantation begins	24 Embryo one week old	Two germ layers	26	Implantation completed	28
29	30	31				

probably occurred before the woman is aware that she might be pregnant.

During the second month, the embryo demonstrates why this trimester is also known as the period of organogenesis. The head and brain grow very rapidly, so the embryo seems to be all head with a small torso. The ears develop, and their external canals open to the outside. The eyes become fully formed but remain lidless. The small torso develops paddle-like arm and leg buds and a rudimentary tail.

By the end of the second month, arms and legs have formed from the buds, and they have distinct fingers and toes. The tail bud has been reabsorbed. The heart has developed all of its chambers and valves. Distinct sculpturing has continued around the head, so that now the jaw is formed and the embryo has a distinct face, with mouth, nose, ears, and eyes. At this time, the embryo is an inch long and weighs about a tenth of an ounce. Morphogenesis has created unmistakably human features. From this point forward, the embryo is called a **fetus**—a name it will keep until birth.

By the end of the third month, all of the organ systems have been formed and are clearly recogniz-able and functional. For example, the bones of the fetus were first laid down as cartilage, but now bone deposition begins at the center of each long bone. The fetus now has a completely functional muscular system and begins to move. Arms and legs thrash about, the head moves up and down, and the mouth opens and closes, swallowing and expelling amniotic fluid. The three-month-old fetus even smiles and grimaces. Although the fetus is a full 10 centimeters long and weighs 28 grams, its movements go completely undetected by its mother. By this time, however, she probably has reason to believe she is pregnant. She may have swollen and tender breasts. She may feel nauseated from time to time, particularly in the morning. She probably urinates more frequently than usual.

Second Trimester

The fourth through sixth months are known as the period of growth. The torso grows more rapidly than the head region and eventually catches up to it. The fetus and its accompanying extraembryonic membranes also grow, to the point that the chorion occupies the entire uterine cavity. Soon the uterus must stretch to accommodate the growing fetus, and the pregnancy begins to "show" for the first time. By the end of the second trimester, the pregnancy is obvious.

The rapid growth of the trunk during this period gives the fetus a straighter posture. With increased muscle strength, the fetus can also hold its head up. It begins to thrash about, often kicking and bumping violently. These movements are often forceful enough to be observed from the outside. Many an expectant mother has been embarrassed in public by uncontrollable gyrations of her abdomen. The fetus now sleeps from time to time, and its activities become more periodic. During sleep, the fetus may suck its thumb or fingers.

At the end of the second trimester, the organ systems of the fetus are fully formed and can be completely functional, although there is no food to eat or air to breathe. Now the bone marrow begins production of blood cells, a function previously handled by the liver and spleen. The circulatory system carries nitrogenous wastes to the placenta by powerful strokes of the well-developed heart. The heartbeat can easily be heard with a stethoscope.

One of the most significant processes that occurs during the second trimester is the continual development of the external genitalia (see Figure 12–7). By the end of the first trimester, the external rear opening of the fetus (actually a cloaca) is divided by a thin bank of tissue. The back portion becomes the anus and the front portion becomes the urogenital sinus. The urogenital sinus is flanked by folds of tissue, the genital folds. This is the so-called **indifferent sex stage,** because the sex of the fetus cannot be determined by observation even though it was genetically programmed at the time of fertilization. (The sex depends on whether the sperm cell that fertilized the ovum was an X or Y cell.) After about thirteen or fourteen weeks of fetal development (very early in the second trimester), the sex can be determined externally.

The urogenital sinus and accompanying genital folds are the rudiments of external genitalia. This area forms male genitalia if certain substances produced in the developing testes induce it to. (These substances have not been identified, but testosterone is not one of them.) Otherwise, female genitalia form.

The front portion of the urogenital sinus has a rounded mass, the genital tubercle. In males, the inducer substances from the testes cause rapid widening of this area and rapid lengthening of the urogenital sinus. The lips of the sinus eventually fold over and fuse to create an internal canal, the urethra. The genital tubercle becomes the glans penis. The tubular, fused urogenital sinus becomes the shaft of the penis. Simultaneously, the genital folds grow backward and fuse to form the scrotal sac. (The human male genitals are described in greater detail in Chapter 11.) Fusion of the urogenital sinus and the genital folds leaves an external scar, a line suggesting the fusion of two halves, that every male carries for life. It can easily be traced from under the glans penis down the shaft and across the scrotal sac.

The testes complete their development within the abdominal wall of the fetus. They descend through the inguinal canals into the scrotum before the child is born. For unknown reasons, this descent sometimes does not occur. The condition is known as cryptorchism (*crypto* meaning "hidden" and *orchis* meaning "testis"; hence "hidden testis"). Because the temperature in the abdomen is higher than the temperature in the scrotum, a man with this condition cannot manufacture viable sperm. A simple surgical procedure can be used to assist the testes in their descent and place them in the scrotal sac.

In females, the lack of inducer substances causes little enlargement of the genital tubercle but more pronounced enlargement of the walls of the urogenital sinus and the genital folds. As a result, the genital tubercle remains small, becoming the clitoris. The walls of the urogenital sinus do not fuse but instead enlarge and elongate to form the cavity of the vestibule and the labia minora. The genital folds undergo tremendous enlargement but never fuse. As large, fleshy folds of skin and blood vessels, they become the labia majora. The ovaries develop and remain in the abdominal cavity. (The human female reproductive system is described in more detail in Chapter 11).

At the end of the second trimester, the fetus has grown to a length of 30 to 36 centimeters and a weight of 0.7 to 0.9 kilograms. Its organs are all in place. The developmental tasks that remain are primarily refinements of the major systems.

Third Trimester

The last three months of pregnancy involves many subtle but important physiological changes. The best

UNDIFFERENTIATED

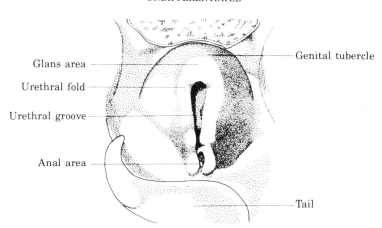

Glans area

Urethral fold

Urethral groove

Anal area

Genital tubercle

Tail

Figure 12–7
The development of the human male and female external genitalia from the tissues of the indifferent sex stage.

MALE   EMBRYO   FEMALE

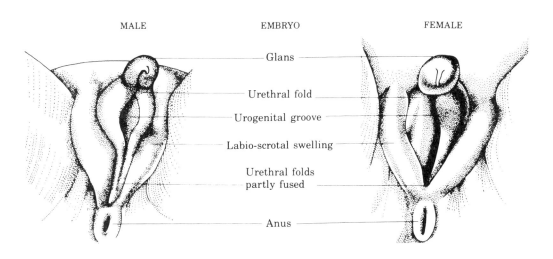

Glans

Urethral fold

Urogenital groove

Labio-scrotal swelling

Urethral folds partly fused

Anus

FULLY DEVELOPED

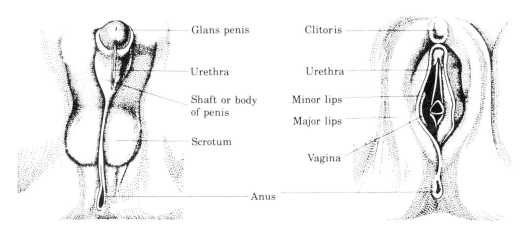

Glans penis

Urethra

Shaft or body of penis

Scrotum

Clitoris

Urethra

Minor lips

Major lips

Vagina

Anus

understood ones are those concerned with the refinement of the nervous system.

The nervous system forms during the first trimester, but a large number of connections within the nervous system are believed to be established during the last trimester. At this time, it becomes possible for sensations from the skin and other parts of the body to be relayed to the appropriate regions of the **cerebral cortex**. The brain develops important relays, which send impulses to specific muscles and organs. Intricate connections develop in the spinal cord. The cerebral cortex gains many more convolutions than it had previously.

More than likely, most of the synaptic networks that develop prior to birth are associated with basic reflexes and human instincts. It is well established that these networks continue to develop (and the number of nerve cells continue to increase) most dramatically from birth until about age four. They probably develop through experience.

The brain also undergoes sexual differentiation during the third trimester. If the fetus is male, there is a masculinization of the **hypothalamus**. The **pituitary gland** is not influenced by the sex of the fetus, although its output of gonadotropic hormones is steady in males and cyclic in females. That is because the hypothalamus produces gonadotropic-releasing factors that affect the activity of the pituitary. Therefore,

the hypothalamus must be programmed during the third trimester as male or female. If the fetus is male, its developing testes produce male hormones (**androgens**), which cause masculinization of the hypothalamus. If androgens are not present, then the hypothalamus becomes the cyclic hypothalamus of the female. Female hormones (**estrogens**) do not affect the hypothalamus, because they are produced by the placenta□ accompanying both male and female fetuses. Experiments with laboratory animals have shown that female fetuses exposed to androgens may have permanent masculinization of the hypothalamus. The hypothalamus becomes programmed for steady, noncyclic activity, causing the female laboratory animals to be in estrus□ continually.

The third trimester is often referred to as the "period of intelligence," because intellectual abilities are thought to be determined at this time. The fetus is now quite large and continues to make more and more demands on its mother's physiology and diet. Women who do not maintain a proper diet at this time give birth to malnourished babies. Protein deficiency has its most profound effects during the third trimester. A lack of protein in the mother's diet may give the baby a low IQ.

The way a zygote becomes a fully functioning fetus in only nine months is little short of a miracle. For a review of the developmental steps through which a fetus passes, see Figure 12–8.

CHILDBIRTH

During the ninth month of fetal development, the placenta begins to break down, and areas of it become nonfunctional. This slow erosion of the fetal lifeline to the mother is preparation for birth, which occurs about nine months after fertilization.

The birth process has three phases. The first phase, labor, lasts about ten to sixteen hours for women who are having their first child and often half that time for subsequent births. The first contractions of the uterus are usually fifteen to twenty minutes apart and last only a few seconds. However, as labor progresses the contractions become more frequent, last considerably longer, and become more intense. At the onset of the next phase, contractions may only be minutes apart and last for a whole minute each. During labor, the cervical opening slowly dilates from a

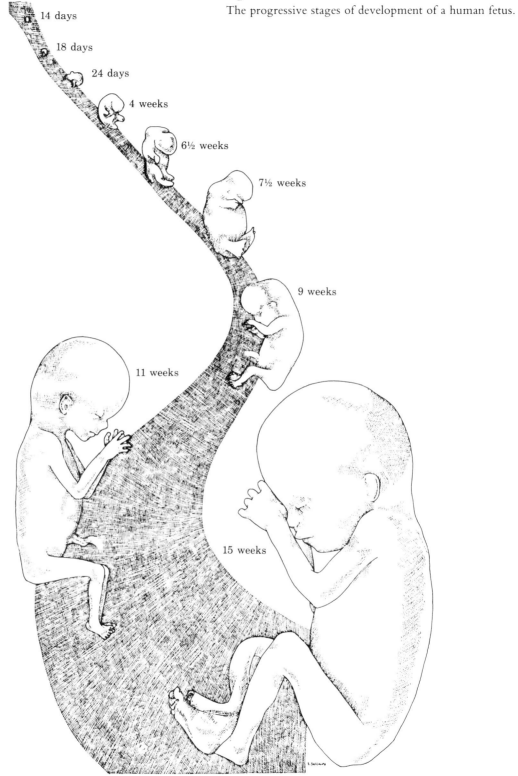

Figure 12–8
The progressive stages of development of a human fetus.

14 days

18 days

24 days

4 weeks

6½ weeks

7½ weeks

9 weeks

11 weeks

15 weeks

diameter of ½ or 1 centimeter to a diameter of 10 centimeters or more.

The second phase, delivery, begins when the head of the fetus pushes through the cervix and into the vagina. Delivery is the most difficult part for the mother, because she is asked to "bear down," using her abdominal muscles to force the head and shoulders of the fetus through the cervical opening. When they have cleared the cervix, the rest of the fetus rushes from the uterus.

After the fetus has left its mother's body entirely, the umbilical cord is cut. Mild contractions of the uterus continue until the placenta, or afterbirth, is expelled. This final phase of childbirth may require only a few minutes or as much as half an hour (see Figure 12–9).

Figure 12–9

The stages of birth of a human infant: (a) before labor but near the end of pregnancy; (b) at the beginning of labor, which is marked by contractions of the uterus and dilatation of the cervix; (c) at the "bear-down" stage, late in labor, when dilatation of the cervix is maximal and the infant's head and body slowly turn; (d) at the delivery phase.

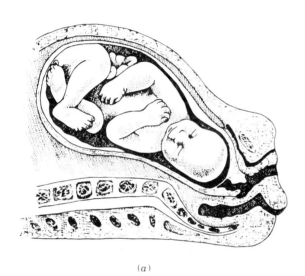

(a)

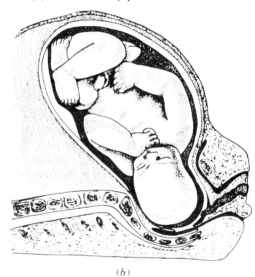

(b)

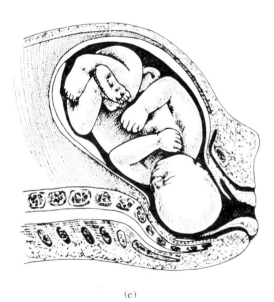

(c)

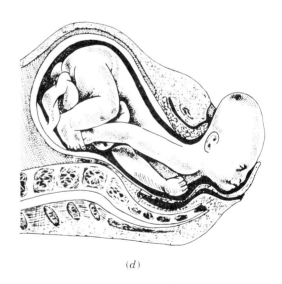

(d)

METHODS OF CHILDBIRTH

The Lamaze method of childbirth is frequently discussed but usually understood only by those who have had the special classes. Fernand Lamaze, a French obstetrician, created the technique. It was popular in Europe before its introduction into the United States in 1959. Such labels as "natural childbirth" do not accurately express what this method involves. Although the Lamaze method is certainly more "natural" than the usual delivery procedures, it does not preclude the use of anesthetics. Each woman may ask for any technological assistance that is available.

The method mainly involves relaxation and controlled breathing. The pregnant woman attends classes where she learns to relax most of the muscles of her body while tightening specific muscles or groups of muscles. In short, she learns muscle control, particularly of the abdomen, groin, and legs. During birth, she helps expel the fetus by contracting specific muscles. The breathing exercises help her cope with the labor contractions.

There are two other important aspects of the Lamaze method. The first aspect is that the woman must have a coach, usually her husband, who also attends the classes. During the birth, the coach times contractions, directs breathing, and tells the woman when "pushing down" is most desirable. The coach also gives the woman moral support. The Lamaze method allows the father to participate in the birth of his child. The second aspect is that both the woman and her coach learn a great deal about pregnancy, fetal development, and childbirth. The instructional sessions expand awareness of what human reproduction is all about—an awareness that few couples have.

The Lamaze method is often combined with the Leboyer method, which is based on the idea that childbirth should be less violent. Frederick Leboyer, a French obstetrician, argues that childbirth should take place in a homelike atmosphere, with other family members (even other children) present, with subdued lighting, pleasant surroundings and music. These elements contrast sharply with the brightly lit, sterile atmosphere of standard delivery rooms. Leboyer suggests that more normal surroundings may be less stressful for both mother and infant. Various studies on laboratory animals support this contention.

Many pregnant women prefer childbirth at home. If this is planned, emergency services should be nearby in case of complications. As an alternative, many hospitals are now equipped with family rooms for childbirth. These rooms are more like bedrooms at home than like the usual delivery rooms. Family members of the woman's choice can be present, or she can deliver unattended or attended by a coach. If complications develop, she can quickly be seen by the medical staff.

Another so-called innovation in childbirth is the development of the birthing chair or stool. The woman sits upright on it, rather than lying horizontally on a delivery table. Squatting during labor has been common in most cultures of the past. It has probably been practiced by women since earliest times. The general posture was adopted centuries ago by Europeans, but the woman sat in a chair instead of squatting. This was the invention of the birthing stool.

Until the seventeenth century, only women (known as midwives) assisted other women in childbirth. Even physicians were not allowed to witness births. One physician disguised himself as a female midwife. When he was found out, he was burned at the stake.

King Louis XIV of France might have been the first modern man to witness the birth of a human child. He did not want his court to know that his mistress was pregnant. (The clothing styles of the time easily covered up the fact.) To keep the secret, when she went into labor, the king asked his own physician, a man, to attend her. He also instructed that she lie in a bed so he could see the child born. Somehow, lying prone on a bed caught on and became the acceptable posture for childbirth. Also, for the first time in history, men became midwives, presumably because of the male physician who delivered the king's son. The prone posture for childbirth has continued into modern times because it gives the midwife or obstetrician a better view of what is going on. It is an uncomfortable and unnatural posture for the pregnant women, however.

The recently revived birthing chair is a more modern model than that used a few centuries ago. It is made of plastic or steel and has electrical controls for elevation, tilting, and so on. The birthing chair shows definite advantages over the flat table. It makes it easier for the woman to push down and gives her the assistance of gravity. The labor is usually easier and shorter. Birthing chairs are becoming common in clinics and hospitals.

TWINNING

In the normal course of reproductive events, more than one fetus may develop, a phenomenon known as twinning.

Identical twins occur when a single embryo splits during its development, creating two embryos. The splitting occurs in the embryonic disk, probably before the fourteenth day after fertilization. The paired fetuses share the same set of extraembryonic membranes and a single amniotic cavity. They are two individuals of the same sex and genotype⬚. Because the twins are produced from a single zygote, they are referred to as **monozygotic twins**.

Fraternal twins develop from two eggs that were ovulated near or at the same time. Each zygote passes separately from the Fallopian tubes to the uterus. Each embryo is implanted separately, develops its own extraembryonic membranes, is nourished by its own placenta, and floats in its own amniotic cavity. Fraternal twins can be either the same sex or different sexes. They always have different genotypes. The genetic relationship between fraternal twins is no greater than the relationship between any siblings. Because these twins develop from separate zygotes, they are often called **dizygotic twins**. (Figure 12–10 compares fraternal and identical twins.)

The rate of monozygotic twinning in the United States is about 3.4 sets of twins per 1,000 births. The rate is almost the same all over the world and is due simply to chance, not inheritance.

Dizygotic twinning rates are different from country to country. In the United States, the rate is currently 8.1 sets of twins per 1,000 births. The rate may be much higher than this in some families and much lower in others. What this suggests is that dizygotic twinning is a familial trait—it runs in families. Therefore, it must have a hereditary component. It occurs more frequently among women whose fathers or mothers have a family history of dizygotic twinning. The ability to multiovulate is a simple autosomal recessive trait⬚. It is expressed only in women because only they can ovulate. The trait, then, is not sex-linked but sex-limited⬚. Future data on dizygotic twinning may become quite confused, because fertility drugs enhance a woman's ability to multiovulate.

If more than two fetuses develop—triplets or quadruplets, for example—various combinations of dizygotic and monozygotic relationships can occur. In the case of triplets, the combination can be monozygotic

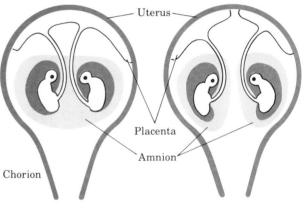

Figure 12–10

Identical and fraternal twins. The two adult women are identical twins. They began as a single embryo that split into two embryos near the time of implantation. As they developed, they shared the same extraembryonic membranes. The woman on the left gave birth to the children pictured, who are fraternal twins. They developed from separate fertilized eggs, and each had separate extraembryonic membranes. They are no more closely related than any other brother and sister. (Photo by authors.)

SIAMESE TWINS

The most famous conjoined twins were Chang and Eng Bunker, who were born in Siam (now Thailand) in 1811 (see photo). Their worldwide reputation originated the term *Siamese twins*. Chang and Eng were linked at the chest by a short, thick, rope-like tissue mass. When they were young, one twin walked forward while the other walked backward. But as they grew older, the tissues surrounding their point of fusion stretched until they were able to walk side by side.

The twins were discovered by a talent scout for the Barnum and Bailey Circus when they were young boys. The most popular attraction of the circus for thirty-five years, they were seen by thousands of people around the world. They retired wealthy and bought a farm in White Plains, North Carolina. Subsequently, each married. One fathered ten children, the other eleven. They died on their farm in 1874 at the age of sixty-three. Chang died of partial paralysis and pneumonia. Eng died about ten minutes later from what was described as shock.

During their lifetime, medical authorities constantly debated whether they could be separated and survive. They themselves often discussed whether they should allow doctors to separate them. An autopsy after their death showed that they could not have survived without each other. They were buried still attached.

Another famous pair of conjoined twins were the Blazek sisters, born in the 1870s in Czechoslovakia. They were joined at the hip and shared some bony elements of the hip. They had a common vulva, but separate vaginas and uteri, and despite their joinings were able to walk alongside each other. The Blazek sisters became associated with high society. They were proud of their condition and were often seen in public dressed in the best finery.

Once, one of the sisters complained of abdominal pains. When examined by a physician, she was

(Photo courtesy of The Bettmann Archive.)

discovered to be pregnant. The sisters always denied "playing around." Soon after the diagnosis, one of them married. (Presumably, the groom was the father of her unborn child.)

The sisters wanted no more than to retreat to the country and live with their new husband. Unfortunately, they were constantly harassed by legal authorities. First the groom was charged with bigamy, but the charge was dropped. Then the law reasoned that, if not a bigamist, he must be an adulterer, because it would be easy for him to penetrate the wrong vagina. Peace came only when the sisters died, within minutes of each other, at the age of forty-three.

(all three fetuses from a single zygote), dizygotic (one set of monozygotic twins and one monozygotic individual), or trizygotic (all three from separate zygotes). The rates in the United States are, respectively, 21, 58, and 34 per million births.

On rare occasions, the space available for the de-velopment of monozygotic twins becomes limited and the fetuses occupying a single amnion become fused. The result is conjoined, or Siamese, twins. Modern surgical techniques have made it possible to separate many Siamese twins, as long as common tissues do not unite too many vital organ systems.

CHAPTER 12 *The Development of Vertebrates*

DEVELOPMENTAL ANOMALIES

The development of a new human being is obviously a complicated and delicate process. It is no wonder that occasionally something goes wrong and the newborn is in some way deformed. Most babies, however, are born normal. (See Box 12–3.)

Teratology

Various environmental agents either adversely affect fetal development directly or adversely affect the woman's body (physiology), thereby harming the fetus. Agents known to do this are referred to as teratogens (*terato* meaning "monster"). Teratology is a rather new science, but already it has provided a massive list of environmental agents suspected of causing developmental anomalies.

Teratology was responsible for proving the effect of the tranquilizer thalidomide, taken by thousands of pregnant German women in the early 1960s. Thalidomide inhibits the development of fetal arm and leg buds during the first trimester. Many children were born armless, legless, or both (see Figure 12–11) before the cause of their deformation was established. Some women chose therapeutic abortions over deformed children. However, many of the children born to women who took thalidomide are now normal adults (some are unusually talented) except for their lack of limbs. This incident put all sleeping aids and tranquilizers on the suspect list, and pregnant women are warned against taking them.

Narcotics are also known teratogens. Heroin often causes fetal convulsions and miscarriage. Morphine and other hard drugs can cause fetal addiction.

The pregnant woman may have or may come in contact with diseases that cause fetal anomalies. For example, if a prospective mother comes down with German measles (rubella) during the first trimester, the fetus may become blind, deaf, or both. But if she contracts German measles in the last trimester, the fetus is seldom harmed. A prospective mother may have natural or acquired immunity to German measles. If so, her antibodies give the fetus protection that lasts until a few months after birth. The campaign to inoculate preschool children against diseases like German measles is not only for their protection but also for the protection of their unborn brothers and sisters.

What are the chances of having a handicapped or deformed child? It is generally accepted that about one of every fourteen full-term pregnancies will result in a child with a minor or major birth defect. In the United States, one family in every ten will attempt to raise such a child. Not all birth defects are obvious at birth. Many chemical disorders and diseases, such as muscular dystrophy, don't develop until the child is a few months or even a few years old. (See Chapter 16 for specific details of some human genetic disorders.)

Can birth defects be corrected? Many birth defects can be corrected—but not all of them. Chemical disorders, such as phenylketonuria, can be corrected once diagnosed. Circulatory, digestive, excretory, and neural defects can often be corrected surgically. For example, the surgical correction of heart defects that limit the oxygen in the baby's blood (blue babies) has become almost routine today, as have corrections of cleft lip, clubfoot, open spine, and even water on the brain. There is no cure, however, for most disorders that cause mental retardation.

Can birth defects be prevented? Yes. Many birth defects could be prevented if all girls were vaccinated against German measles. If transmitted to the fetus the virus can cause blindness, deformities, and mental retardation. During 1964–1965, there was an epidemic of measles, and that single factor increased the number of birth defects in subsequent months by over 50,000.

A second method of prevention is for the pregnant woman to be careful of her health and environment. She should not smoke, consume alcohol, take drugs (except those prescribed), or expose herself to radiation. A woman can seldom be too careful or live too conservatively during her pregnancy. However, the list of environmental factors known to affect fetal development would fill several pages of this book. The best bet is a good diet, good health, and the services of a well-qualified obstetrician.

Unfortunately, the fetus is most susceptible to serious damage during the first trimester, often before the woman knows she is pregnant. The new home pregnancy test kits can help women determine if they are pregnant within six to ten days after fertilization, and they are 95 percent accurate. The newly pregnant woman can then consult her obstetrician for advice about diet and a generally safe life-style.

Can a normal fetus be harmed during childbirth? Yes. Many of the obstetrical emergencies that occur in the delivery room can affect later development of the child. One routine procedure being carefully scrutinized is the use of anesthetics during childbirth, whether inhaled or injected. Investigation has shown that inhalants are particularly harmful to the nervous system and affect the development of motor coordination in young children. Exposed children may be shorter and have lower blood pressure and a slower heart rate. Analgesic drugs injected during childbirth to lessen pain are also suspect, but the results of their use are thought to be less serious.

Should the pregnant woman undergo amniocentesis (taking amniotic fluid from the uterus) to determine any potential problems? Generally no. Although amniocentesis is becoming more and more of a regular service, it should be undertaken only if there is a family history of some genetic disorder or if the woman is older and there is an increased chance of having a child with Down's syndrome. Even then, it should be done only after complete genetic counseling. The discovery of serious complications by means of amniocentesis is not primarily of benefit to the fetus. Instead, it helps the parents decide whether to abort the fetus or to raise a child with birth defects.

It is hoped that solutions to many of the problems of pregnancy and childbirth will soon be found. Most of the solutions will come from experiments with rats, rabbits, mice, monkeys, and other laboratory animals. This research is usually basic and does not appear to have immediate human application. However, the better we understand fetal development and birth in other animals, the better we will understand these processes in humans. Thus basic research plays a key role in science.

Figure 12–11

This child suffers from a birth defect that can now be prevented. His mother took the tranquilizer thalidomide while she was pregnant. It is now known that thalidomide is a teratogen that interferes with normal limb development during the first trimester of pregnancy. Before this fact was known, thousands of children were born with similar deformities. (Photo by Leonard McConkell/Life Magazine.)

Many other infections are known to have adverse effects on the fetus. Among them are mumps, syphilis, polio, and even the common flu. These viruses and bacteria are small enough to cross the placental barrier between maternal and fetal circulatory systems.

Cigarette smoking has been under attack for many years and for many reasons. Some serious diseases, including lung cancer and heart disease, are influenced by smoking. Cigarette smoking by the pregnant woman can also affect her developing fetus. It is believed that smoking diminishes the amount of oxygen available to the fetus and increases the amount of poisonous carbon monoxide in the fetal blood. Although the results are variable, the following effects may occur as a result of heavy smoking by the pregnant

woman: diminished fetal weight, cleft lip, intestinal abnormalities, respiratory difficulties (including sudden infant death syndrome—crib death), and a greater than normal chance of mortality.

Irradiation in any form is considered potentially dangerous to a developing fetus. It can cause chromosomal anomalies that in turn can cause malformations and physiological deficiencies. One potentially hazardous source of irradiation is X-rays. The microwaves produced by color television sets and by microwave ovens are also under suspicion.

During the last trimester, alcohol consumption by the mother can be particularly harmful to the fetus. For some unknown reason, the fetal brain and liver absorb large amounts of alcohol, which can later have drastic effects on both behavior and metabolism. Heavy drinking by pregnant women may also cause fetal heart defects, defective joints, a reduction in head size, and depressed IQs. This is called fetal alcohol syndrome and is now known to be the cause of mental retardation in many children. It is estimated that at least one child of every thousand born suffers from the results of fetal alcohol syndrome. Alcohol is probably the major teratogen.

Even drugs that are thought of as helpful or that are necessary to combat certain infections may harm the fetus. When a pregnant woman takes the commonly used antibiotic tetracycline, she may suffer serious, even fatal, liver disease. At the same time, fetal teeth and bones may be damaged. Infants who have been poisoned by tetracycline are often born stunted in growth and with deformed limbs. Their teeth, when they come in, are often discolored.

The list of teratogens increases daily. The range is from simple aspirin to household cleaning agents to chemicals in the workplace. Many over-the-counter remedies now bear labels warning pregnant women to consult a physician before taking them. Such women should read all directions and precautions on any substance that she regularly ingests or uses.

Teratogens can also be hidden in unsuspected sources. For example, potatoes that become blighted by a fungal infection are often sprayed with a fungicide. If a pregnant woman eats the potatoes and their chemical contaminant during the first month of fetal development, the fetus may suffer an opening in the vertebral column that exposes the spinal cord (spina bifida), which can be fatal. The same substance can also restrict the development of the top of the head and brain of the fetus.

Hormonal Anomalies

Many defects that arise during fetal development are the result of hormone malfunctions. Perhaps the most noteworthy are those that occur during differentiation of the external genitalia. Some of these defects are genetic in nature. Others are caused by the hormonal environment in which the fetus develops.

In some cases, the fetus is genetically male but during the indifferent sex stage does not react to the hormones that normally cause differentiation of male genitalia. Some authorities believe that the fetus actually manufactures substances that inhibit the hormones. In all cases, the genetic male develops complete or partial female external genitalia.

When the genitals are entirely female, the syndrome may never be detected—except that the individual is sterile. People with this syndrome are called female pseudohermaphrodites, because they look like women but are genetically men. (There are no known true hermaphrodites—people who are functional males and females at the same time—among humans.) In some cases, the female pseudohermaphrodite may be detected by swelling testes in the labia majora that are attempting to descend into the nonexistent scrotum. More commonly, the clitoris undergoes slight enlargement, appearing as a smaller than average penis. The crude lips of the vulva below often look like an opened scrotum.

Genitalia of this kind often mean that the attending physician will indicate "male" on the birth certificate. Unfortunately, many of these individuals never receive treatment. They are raised as boys, even though they develop breasts and have widened hips and pelvis.

There is no direct way to treat female pseudohermaphrodites, because they remain insensitive to androgens for life. At best, cosmetic surgery can correct and reshape the genitalia to make them look more like those of a normal woman.

As an example, consider the following case history of a female pseudohermaphrodite. An attractive young teenage girl, winner of a beauty contest, complained of abdominal pain. Her physician diagnosed her complaint as possible ovarian cysts. Surgery demonstrated that the cysts were fully formed testes in her abdomen.

The reverse condition, male pseudohermaphrodism, may result from the adrenogenital syndrome. In these cases the adrenals (hormone-producing glands near the kidneys) of genetic females produce male hormones during fetal development. (All females produce some male hormones in the adrenal glands, but the amounts are usually very small.) The male hormones cause the structures at the indifferent sex stage to differentiate into male external genitalia, even though the fetus develops a uterus and ovaries internally.

The most obvious sign of male pseudohermaphrodism is an enlarged clitoris. In some cases the clitoris may be large enough to appear as a normal penis, and the labia majora develop into a scrotum without testes.

Male pseudohermaphrodites often have such complete male genitalia that they are never detected—but they are sterile. These individuals are reared as males. When the transformation is only partial, surgery can transform the genitals to female again. Hormone treatment restores the female physiology, and many of these individuals marry and have children.

Pseudohermaphrodism was unintentionally induced during the 1950s. Artificial progesterones were used to help pregnant women avoid miscarriages. However, the hormones caused masculinization of the genitalia of female fetuses. Surgery and hormone replacements were used to treat these children. The results were similar to those achieved when the same treatments were used for the adrenogenital syndrome.

SUMMARY

1. The law of biogenesis states that all vertebrate embryos pass through similar developmental phases.

2. Fertilization is the union of egg and sperm nuclei. The resulting cell is a zygote.

3. The early cellular divisions that occur after fertilization are called cleavages. After several cleavages, a hollow ball called the blastula results.

4. Gastrulation is the rapid growth of cells into the cavity of the blastula. The resulting embryonic stage is the gastrula. By the end of gastrulation, the three primary germ layers—ectoderm, mesoderm, and endoderm—have formed.

5. Neurulation is the process by which a tube of ectoderm that will become the nervous system forms in the embryo.

6. Terrestrial vertebrates form a series of extraembryonic membranes that allow them to develop in essentially the same way aquatic organisms develop—in a fluid medium. These membranes are the chorion, amnion, yolk sac, and allantois.

7. Fertilization of human egg cells occurs in the upper third of the Fallopian tubes. Sperm may reach this region less than thirty minutes after they have been ejaculated.

8. The fertilized human egg cell (zygote) begins cleavage as it slowly moves down the Fallopian tube toward the uterus. The embryo reaches the uterus on the fourth day after fertilization.

9. The human embryo passes through the morula, blastocyst, and early gastrula stages before being implanted in the uterus. The embryo has differentiated endoderm and ectoderm before implantation. The mesoderm will not differentiate until about a week after implantation.

10. Implantation occurs about the tenth or eleventh day after fertilization. The cells that begin the development of the placenta produce chorionic gonadotropin (CG), which prevents the onset of menstruation that would wash away the embryo.

11. The nervous system begins to form after implantation. A complete neural tube is formed by the end of the fourth week.

12. The placenta is the major life-support system of the human embryo until birth. The embryo is attached by the umbilical cord.

13. The human embryo develops extraembryonic membranes, as do all terrestrial embryos. The chorion forms the placenta. The amnion completely encircles the embryo and fills with amniotic fluid. The allantois produces the umbilical arteries and vein. The yolk sac produces the first red blood cells, an activity soon taken over by the liver and spleen.

14. Human gestation is divided into three trimesters. The first trimester is the period of organ formation, or organogenesis. The second trimester is the period of rapid growth. The third trimester is the period of refinement of the nervous system.

15. The external genitals of the human fetus start to develop during the first trimester and continue to develop in the second trimester. Whether male or female, the fetus passes through an indifferent sex stage. Male and female external genitalia are homologous (arising from the same embryonic tissues).

16. Childbirth has three distinct phases: labor, delivery, and afterbirth.

17. Identical (monozygotic) twins develop from a single fertilized egg. Fraternal (dizygotic) twins develop from separate fertilized eggs and are therefore no more closely related than other children in a family. Conjoined, or Siamese, twins develop when monozygotic twins fuse together during development.

18. Birth defects are caused by genetic, hormonal, and physiological agents. Environmental agents causing birth defects are called teratogens. They include many drugs, other substances, and diseases Other birth defects result from mistakes during development, usually due to hormonal imbalances.

THE NEXT DECADE

FETAL THERAPY

As indicated in this chapter, many inborn errors influence fetal development. These errors can result in the death of the fetus or in serious lifelong handicaps. Many errors in human development are inherited. Others, such as fetal alcohol syndrome, are environmentally induced. In past decades, the pregnant woman had no way of knowing that her fetus was abnormal. However, during the last decade, it has become possible to monitor the development of the fetus by amniocentesis, X-rays, and ultrasound. These new techniques determine everything from vitamin deficiencies to serious anatomical deformities such as spina bifida.

When life-threatening defects are detected, the parents are given the option of abortion. Until recently, parents had only two choices: raise a deformed or mentally retarded child or abort it. In the next decade, some parents will have a third choice: fetal therapy to correct the defect so the child will be normal. It is estimated that about 3 percent of all children in the United States are born with birth defects. Fetal therapy should drastically reduce this percentage.

The idea of such therapy is not new. Many obstetricians and pediatricians have wanted to treat children before birth. The first fetus to be successfully treated while still in its mother's womb had an inherited form of vitamin B_{12} deficiency. Once the defect was detected, massive injections of vitamin B_{12} were given to the mother. The vitamin passed through the placenta and cured the deficiency in the fetus. The next incident was similar: an inherited deficiency in biotin (another vitamin) metabolism. Again, the mother was given the vitamin, which reached the fetus via the placenta.

In the next decade, many chemical errors, ranging from enzyme to nutrient to hormone deficiencies, will probably be corrected in this fashion. Drugs as well as vitamins may be administered to treat fetuses suffering from infections and other diseases. If the need is for compounds that do not cross the placental barrier, then injection could be made through the mother's abdominal wall into the amniotic fluid. The fetus could then absorb or even swallow the substances.

The first surgical procedure done on a fetus involved the removal of a block in the urethra. This condition is usually fatal shortly after birth because the blockage is not normally detected in newborns. This particular fetus was a male twin. Surgeons inserted a small tube (catheter) through the abdomen of the mother and into the opening of the blocked urethra at the tip of the penis. They then pushed the catheter to the bladder, opening the blockage. The urine drained through the catheter into the amniotic fluid until the birth.

Recently, a fetus was moved part way out of its mother's womb so surgeons could correct a kidney problem. After the therapy, the fetus was placed back in the womb.

The fetal therapies mentioned here are pioneering efforts to correct problems of development before it is too late. Many experiments are being conducted with animals to determine the limits of surgical manipulation of fetuses. For example, experiments demonstrated that it was possible to insert a valve in the brain of a monkey fetus while the fetus was still in the mother's womb. This was first done on a human fetus in 1981. The fetus suffered from hydrocephalus (water on the brain). In the United States, nearly five thousand human infants are born each year with hydrocephalus. Cavities within the brain accumulate fluid. As this fluid increases, the brain often swells to enormous proportions. Normal development of brain cells is inhibited, and the child either dies or is mentally retarded. This condition can also cause blindness, paralysis, and many other disorders of the nervous system.

We will hear much more about fetal therapy in the next decade. Surgeons are already proposing to correct fetal hernias of the diaphragm. If such hernias are left untreated, the abdominal organs compress the lungs and the child dies. Ruptures in fetal membranes may also be surgically corrected. If they are left unrepaired, they can strangle the fetus or cause its premature expulsion from the uterus. Many other repairs, including those to a variety of other hernias, perforations, and malformed organs, will be possible in the coming decade.

Yet, before fetal therapy can become a common practice, many ethical and philosophical issues will have to be settled. One of the most important is that of the risk of fetal therapy to the mother. There are also risks for the fetus. Some have proposed that it might be easier to induce premature birth in the third trimester and surgically correct fetal defects outside the womb. However, this would bring greater risks for the fetus in both premature birth and therapy.

Another problem is that of telling parents that the fetus might be treatable while allowing abortion as an alternative. How will these decisions be made? Who will make them? Some specialists have suggested that an ethical review board will have to screen cases proposed for fetal therapy and will have to set guidelines for human fetal research to minimize risks as much as possible.

Fetal therapy is a proven technique. In the next decade, the procedures will become far more sophisticated than they are today. The limits of this therapy can hardly be imagined. Even though there will be ethical questions concerning the procedures, the advances should reduce the probability of stillbirths and of mental and physical birth defects.

FOR DISCUSSION AND REVIEW

1. What are the major events in the development of a typical vertebrate embryo, from fertilization to morphogenesis? What is the significance of each?

2. What are the primary germ layers? When are they formed? What is their ultimate fate?

3. What are extraembryonic membranes? What are their functions? List the four extraembryonic membranes of a chick embryo. Give the location and function of each.

4. Describe the major events in the development of a human embryo, from fertilization to implantation. Give precise times and locations of these events, and explain their significance in relation to a typical menstrual cycle.

5. What is the signficance of the process of implantation?

6. List the extraembryonic membranes of a human embryo, giving the location and function of each.

7. What is the function of the placenta? The umbilical cord?

8. Describe the major events in human development during each of the three trimesters.

9. What is a fetus?

10. What is the indifferent sex stage? When does it occur? Describe the transformation of male and female genitals from this stage.

11. What are homologous structures? Give several examples.

12. What are the major phases of childbirth? Give the details, sequence of events, and average duration of each.

13. How do twins develop? Triplets? Quadruplets?

14. What is a teratogen? List several teratogens and their effects on fetal development. (Check magazines or newspapers for references to teratogens not mentioned in this book.)

15. What are human pseudohermaphrodites? What developmental events cause them to form?

SUGGESTED READINGS

BEHOLD MAN, by Lennart Nilsson. Little, Brown: 1978.
   A fascinating photographic essay by a famous photographer. The chapters detailing human development are particularly relevant to this chapter.
A CHILD IS BORN, by Lennart Nilsson. Delacorte: 1977.
   An unusual and fascinating account of human development, documented with color photographs of live human fetuses. The photography has earned Nilsson many awards and an international reputation.
UNDERSTANDING PREGNANCY AND CHILDBIRTH, by Sheldon H. Cherry. Bantam Books: 1973.
   One of many popular accounts (but perhaps the best) of what prospective parents should know about pregnancy and childbirth. Includes developmental events, a diet for mothers, options at the hospital, and many other important topics.
VERTEBRATE DEVELOPMENT, by Howard Manner. Kendall Hunt: 1975.
   A simple, clearly written, and detailed account of the development of selected vertebrates.

SECTION FIVE

GENETICS

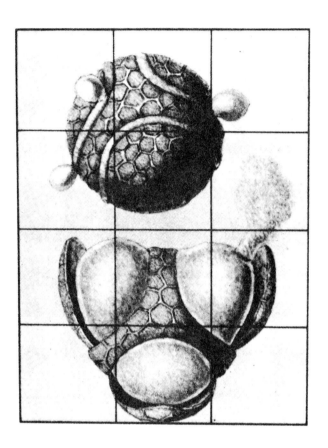

Franz Bauer, a superb botanical illustrator of the Eighteenth century, was unrivaled in rendering the minute detail of plant structures.

13

MENDEL AND THE PRINCIPLES OF GENETICS

PREVIEW QUESTIONS

After reading this chapter, you should be able to answer the following questions:

1. Who was Gregor Mendel, and why did he become famous?

2. What are Mendel's first and second laws?

3. Name several important discoveries about genetics that were made after Mendel's work became widely known.

During Charles Darwin's lengthy career in science in the 1800s, the accepted theory of inheritance was the theory of acquired characteristics. Made popular by the Chevalier de Lamarck, the theory stated that organisms transmit to their offspring the characteristics they have developed to cope with their environment. Darwin did not accept Lamarck's theory, even though he realized that his own theory of evolution required some explanation of how new variations in a species are passed on to their offspring.

Because no satisfactory theory existed, Darwin reintroduced the theory of pangenesis. According to this theory, cells of an organism produce *pangenes,* or *gemmules*—tiny hereditary particles—that are carried in the blood to the reproductive cells. Each gemmule recalls the part of the body from which it came and therefore can re-create that aspect of the organism in offspring.

Darwin's idea of pangenesis was not new. In fact, it was a sophisticated version of a view proposed by Aristotle around 300 B.C. Aristotle believed that semen and menstrual fluid were purified blood and that these pools of blood became mixed during intercourse. The blood carried recollections of the donors' bodies. When mixed, it formed the basic makeup of a new individual, one that blended the parents' characteristics.

This idea sounds wild given what we know today, but it was the only accepted point of view for two thousand years. Even today, vestiges of Aristotle's hypothesis persist in our language. We commonly speak of "blue bloods" (royalty) or of people with "bad blood." We refer to our relatives as "blood relatives," implying that our inheritance is in the blood.

Darwin was never really satisfied with pangenesis, because he saw a weakness in it. If new individuals arose from a blend of gemmules, then the adaptive characteristics of a species (which were the focus of Darwin's work) would become more and more diluted each generation. If new individuals had characteristics intermediate between their parents, how were highly adaptive traits maintained in populations? Gregor Mendel provided the answer.

GREGOR MENDEL

Gregor Mendel (Figure 13–1) was born in 1822 near Brünn, the modern city of Brno, Czechoslovakia. He was the only son of poor peasant farmers, so he was expected to work alongside his father as they tilled, cultivated, planted, and harvested their crops. He developed a profound interest in and fondness for plants.

Mendel was an excellent student. In the village school, he was frequently at the top of his class. His scholastic ability gave his mother hope that he might become something other than a farmer. His parents arranged, probably at great sacrifice, for him to attend a nearby college for two years. There he studied both philosophy and science. But what career could a promising scholar pursue? A religious post was almost the only choice at the time. Mendel entered the Augustinian monastery in Brünn, where he was ordained a priest in 1847.

As a monk, Mendel was appointed to be a teacher. He was stocky, healthy, and always cheerful, and he was well liked by his young pupils in the lower grades. He also was a substitute teacher in the local secondary school. When he attempted the examination that would license him as a permanent secondary teacher, however, he failed. This disappointment encouraged him to ask his superiors for permission to leave the monastery and travel to Vienna to attend the university.

Mendel studied botany and other sciences for two years at the University of Vienna. When he returned to Brünn, he failed the teacher certification examination two more times. For the next decade and a half, he remained only a substitute teacher. However, Mendel had more genius than his biographers generally credit him with. He designed a perfect experi-

Figure 13–1

Gregor Mendel, the "father of genetics." (Courtesy The American Museum of Natural History.)

ment, an atypical thing for any scientist of the time to do. Most proceeded by blind trial and error, so most discoveries were accidents.

In 1856, while teaching, Mendel began his investigations in plant breeding, a lengthy and intensive series of experiments that occupied the next eight years of his life. Mendel selected two varieties of garden peas for his experiments. One variety produced tall plants with yellow rounded seeds. The other produced dwarf plants with green wrinkled seeds. He planted the seeds in separate plots in the garden of the monastery. He cross-fertilized the plants himself, a technique he had learned in Vienna. The seeds from the **hybrids** (the offspring of two varieties), like their parents, were also planted separately.

It would have been difficult to keep a record of every plant throughout its life history, so Mendel kept records of only the possible combinations of specific characteristics. He used simple mathematics to express the results of various crosses among the pea plants and their offspring. He tallied each combination that he actually observed, and he expressed the relationships of various combinations as ratios. Using a statistical method before statistics had even been developed, Mendel raised and recorded the combination of seven physical factors in 12,980 pea plants.

Mendel belonged to a local scientific society. The members frequently met to talk about scientific matters and to read and discuss their experiments. Mendel presented the results of his experiments with his pea plants during two lectures to the society, which published his results in its 1866 recorded proceedings. It is ironic that Darwin never knew of Mendel's work. Mendel was an ardent fan of Darwin and read each of his books as soon as it became available.

Meanwhile, in Germany, an already famous biologist named Karl Nägeli was also breeding plants and thinking about heredity. Darwin's newly published *Origin of Species* stimulated Nägeli to begin a series of experiments on plant breeding using a plant known as hawkweed. Mendel wrote to Nägeli, asking for advice on his own experiments. Nägeli thought that independent recombination was impossible and that every strain, if bred long enough, would become true. Therefore, he encouraged Mendel to grow more peas. Mendel never did. He evidently thought that eight years and 12,980 pea plants were sufficient to support his hypothesis.

At one point, Nägeli asked Mendel for some of his pea seeds so that he could plant them in the gardens at Munich. Mendel promptly sent them, but they were never planted. In 1886, Nägeli published his work on plant breeding but never mentioned Mendel.

Mendel began to experiment on his own with hawkweed. He found it an impossible plant to work with. It was very small and difficult to cross-fertilize. It is now known that many of the seeds of the hawkweed are produced without fertilization, a fact that would have confused any experimental results.

In 1868, Mendel became abbot of his monastery and from that time on had little time for science. Also, his eyesight became progressively poorer, which would have made further experimentation difficult for him. He died in 1884, probably from smoking too much—he smoked twenty cigars a day for most of his life. He died a monk, not the "father of genetics," as he was later proclaimed. He had conducted one of the most important experiments in the history of science, but only a few members of a local scientific society knew anything about it.

MENDELIAN GENETICS

What had Mendel discovered in the monastery garden that would alter the course of science and affect future generations? What he discovered had little or no significance in 1866, when he published his results, and probably seems to have little significance to anyone today. But it is important to remember that no one before Mendel had bred organisms while keeping careful records of the expression of certain characteristics in each generation.

What Mendel discovered was the systematic inheritance of what he called factors. Inheritance, as he saw it in his garden peas, was not only systematic but predictable. The rules he established to allow prediction of factors in future generations are today referred to as Mendel's principles, or Mendel's laws, and they were the foundation of modern genetics. All animals and plants studied since Mendel have been found to operate by Mendel's laws. The genetic predictions made possible because of these laws are commonly called **Mendelian genetics.**

The Law of Segregation

No one knows for sure why Mendel picked pea plants, but pea plants were locally available to him. He obtained thirty-four varieties from local seedsmen. The varieties differed in specific ways. For example, some varieties were taller than others.

The ripe peas in the mature pods of certain varieties also differed. Some were smooth and round; others were wrinkled. Some mature pea seeds were green inside; others were yellow. Some peas had gray seed coats surrounding the pea seeds; others had white seed coats.

The pods on the mature plants of certain varieties differed too. Some mature plants had full pods, called inflated pods; others had constricted pods, in which the shape and placement of the pea seeds were visible from the outside. The color of the pods varied, some being yellow and others green. Finally, the flower arrangements on the plants differed. Some varieties of plants had axial flowers (flowers located anywhere along the length of the pea vine). Other varieties had only terminal flowers (flowers at the end of each stem, not along the full length of the stem).

The characteristics of the peas used in Mendel's time are not much different from the characteristics of many varieties of peas available today. If you are a garden enthusiast, you can select tall garden pea plants or dwarf ones, for example.

Mendel planted the seeds of the peas in a small garden (approximately 20 by 120 feet) next to the monastery at Brünn. The first ones he planted constituted the **parental generation (P₁)**. Plants of different varieties were cross-pollinated, or, more simply, crossed. Peas are normally self-fertilizing, so Mendel had to cross-fertilize the plants himself. He removed the stamens from one plant—for example, from the variety producing yellow seeds. He then pollinated the pistils with pollen collected from another plant—for example, the variety producing green seeds.

Mendel carefully collected the seeds of the parental plants and planted them in separate plots in his garden. When these plants grew, they constituted the **first filial generation (F₁)**, which Mendel called **hybrids.** To his surprise, some of the characteristics of the parent plants were not expressed in the hybrid plants. For example, even if one of the parent plants had yellow seeds and the other green seeds, the hybrid plants all had yellow seeds. The results must have puzzled Mendel at first, but he continued his work.

Mendel did not cross-fertilize the F₁ plants. Instead he let them self-fertilize. When the seeds from this generation were mature, he planted them again in separate plots. These **F₂** plants had some of the characteristics of both grandparents. Also, factors that had seemed lost in the F₁ generation were again expressed.

The results of these experiments clearly did not support the concept of pangenesis, or blending inheritance. They seemed instead to support an entirely different principle. The factors, as Mendel called them, did not become blended together but remained discrete.

There was only one way to interpret the results. Mendel proposed that a factor for a particular characteristic was inherited from each parent. He further proposed that one of the factors is always more **dominant** than the other and that offspring display only the more dominant characteristic. However, continued breeding would produce new generations that would express the less dominant factors, called **recessive** because neither parent contributed a dominant factor. This explanation is now called the **principle of dominance.** Mendel's garden peas demonstrated many dominant and recessive factors. (See Box 13–1.)

BOX 13–1

MENDEL'S PEAS AND GENETIC SYMBOLS

Mendel described many varieties of garden peas during the years that he conducted his experiments. Each variety had a different combination of seven dominant and recessive characteristics.

Today, we express the dominance and recessiveness of specific characteristics by symbols. The most popular symbols are upper- and lowercase letters. An uppercase letter (for example, A) is used for the dominant characteristic, and a lowercase letter (a) is used for the recessive characteristic. The letter chosen is not arbitrary. It is usually the first letter of the name for the recessive condition.

Following are the seven characteristics of pea plants used by Mendel and the proper letter designating each (Mendel determined which were dominant and which were recessive):

Tall (D) versus dwarf (d) plants
Round (W) versus wrinkled (w) seeds
Yellow (G) versus green (g) seeds
Green (Y) versus yellow (y) pods
Gray-brown (WH) versus white (wh) seed coats
(note the addition of an extra letter to avoid confusion with W and w)
Inflated (C) versus constricted (c) pods
Axial (T) versus terminal (t) flowers

Today, if an organism has two dominant factors, it is said to be **homozygous dominant.** If it has two recessive factors, it is said to be **homozygous recessive.** If it has one dominant factor and one recessive factor, it is said to be **heterozygous.**

One of Mendel's experiments will provide an example of dominance and recessiveness. Of the many crosses that Mendel made, one was between plants that produced only smooth, round peas (homozygous dominant) and plants that produced only wrinkled seeds (homozygous recessive). (See Figure 13–2, which shows the shapes of the peas in their pods.) Mendel used pollen from the flowers of plants that were known to produce round seeds to pollinate the flowers of plants that were known to produce wrinkled seeds. He also did the reverse and, of course, obtained the same results. The offspring from this

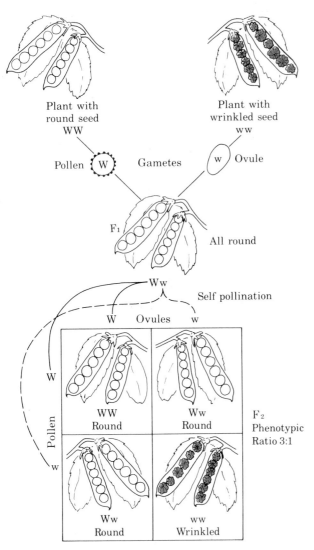

Plant with round seed WW — **Plant with wrinkled seed ww**

Pollen (W) Gametes (w) Ovule

F₁ All round

Ww Self pollination

W Ovules w

Pollen W

WW Round — Ww Round

w

Ww Round — ww Wrinkled

F₂ Phenotypic Ratio 3:1

Figure 13–2

An example of one of Mendel's many crosses with pea plants. In this instance (a monohybrid cross), only a single characteristic—the shape of the pea seeds—was considered. The seeds were either round (dominant) or wrinkled (recessive). The first generation (F₁) demonstrates Mendel's principle of dominance. The kinds of individuals in the second generation (F₂) demonstrate Mendel's principle of segregation.

cross, the F₁ generation, all inherited both a dominant and a recessive factor. These individuals were heterozygous. Thus Mendel's F₁ pea plants all produced round seeds because round is dominant over wrinkled.

Mendel conducted many experiments in which he followed the distribution of single characteristics.

When he planted the seeds from the F₁ plants, he let the plants pollinate themselves. Surprisingly enough the distribution of the characteristics in the second generation, the F₂, was consistently the same. Figure 13–2 shows the result. Three of the plants inherited one or two dominant factors. Only one inherited two recessive factors. The consistent ratio was 3:1. That is, of every four plants, three expressed the dominant characteristics and one the recessive.

The results Mendel obtained were exactly what he would have predicted on the basis of his principle of dominance. He didn't know about meiosis or gamete production. He simply examined his data and interpreted it the only way he could. His observations had been careful, and his interpretation of them led him to suggest an entirely new theory of inheritance. Mendel also deserves credit for his courage, because what he proposed was contrary to the thinking of all the scientists of his time. Could a monk in his garden change the viewpoint of the entire scientific community? Mendel did.

The occurrence of two factors, their separation in the pollen and ovules, and their transmission un-

---

CONCEPT SUMMARY
MENDEL'S CONCLUSIONS

---

1. The factors responsible for the characteristics studied were paired.
2. Prior to reproduction, the pollen and ovules had only one factor for each characteristic.
3. If one of the factors inherited was dominant and the other was recessive, the dominant trait would be expressed and the recessive trait would be suppressed.
4. If these individuals were allowed to self-pollinate, then in the next generation a fourth of the individuals would have two dominant factors, a fourth would have two recessive factors, and half would have one dominant factor and one recessive factor (see Figure 13–2). Thus three out of four would display the dominant trait, and only one (the one with two recessive factors) would display the recessive trait.
5. The factors were not blended but rather were transmitted as discrete, unchanged units. A recessive factor could be paired with a dominant and not be immediately expressed. It could, however, be expressed in the next generation, unchanged from the characteristic possessed by the grandparents.

---

changed to future generations is now referred to as Mendel's principle of segregation.

Mendel did not know anything about genes. However, the "factors" that he hypothesized are now called **genes.** Today, we know that genes are carried on **chromosomes.** Chromosomes in cells are paired, and the pairs are called homologous chromosomes. One member of each pair is inherited from the male parent, the other from the female parent. Genes on paired chromosomes are also paired. If one chromosome of a pair has a gene influencing eye color, the other also has a gene influencing eye color—at the exact same location. These paired genes are called **alleles** (meaning "parallels").

The appearance of an organism, regardless of its alleles, is its **phenotype** (*pheno* means "to appear"). The particular arrangement of alleles on homologous chromosomes constitutes the **genotype** of the organism. For example, the F_1 individuals in Figure 13–2 all produced round seeds, which is the phenotype. However, each of the round-seeded individuals carried an allele for wrinkled seeds, a recessive trait. Therefore, each pea plant carried both the dominant and recessive alleles. This description is of the plant's genotype, or its genetic composition.

Principle of Independent Assortment

Mendel's first experiments all considered just one set of factors (alleles) at a time, a condition referred to as **monohybrid crossing.** Mendel thought he understood the inheritance of the seven characteristics that he studied one at a time. But what might be predicted if he considered more than one set of factors at a time? Mendel had purebred lines of peas that produced yellow, round seeds, two dominant characteristics. Other pea plants produced only green, wrinkled seeds, two recessive characteristics. What would happen if plants of these two pure-breeding types were cross-pollinated? Would yellow, round and green, wrinkled always be linked together? Or would they separate, as Mendel hypothesized, and recombine to create new phenotypes?

Mendel crossed these two pure lines. Figure 13–3 shows the results. Today we call a genetic experiment considering two traits at once a **dihybrid cross.** As might be predicted, Mendel's F_1 individuals expressed only the dominant alleles and were thus all yellow, round. These plants self-fertilized to produce the F_2

individuals. Mendel found that a total of 315 F_2 plants were yellow, round. He also found that 32 F_2 plants were green, wrinkled.

As he probably hoped, Mendel also saw combinations of alleles that were not present in either the grandparents or the parents: 101 plants produced yellow, wrinkled peas, and 108 plants produced green, round peas. The numbers of these four phenotypes were as follows: 315 yellow, round plants, 101 yellow, wrinkled plants, 108 green, round plants, and 32 green, wrinkled plants. The ratio was $315:108:101:32$, or approximately $9:3:3:1$. (Ratios are always expressed by first arranging numbers in order from highest to lowest. Then the smallest number is divided into each of the larger numbers.)

Mendel concluded that the two factors were distributed independently of one another but that each set of factors was distributed as it had been in the monohybrid crosses. In other words, the ratio of yellow to green was approximately $3:1$, as was the ratio of round to wrinkled.

Today, we know that Mendel's principle of **independent assortment,** as illustrated in this experiment, is based on the random assortment of chromosomes during meiosis▫ and their distribution into gametes. Mendel was lucky enough to have selected inherited characteristics whose alleles were located on separate chromosomes. He undoubtedly would have become very confused if some of the characteristics he had selected were located on the same chromosomes (linked).

Figure 13–3 not only displays the results of Mendel's dihybrid cross but also demonstrates a common method of genetic computation. The checkerboard in the figure is called a **Punnett square** after R.C. Punnett, a colleague of William Bateson, who devised the method. Letters indicating the possible alleles in male gametes are arranged down the side of the checkerboard, and those possible in female gametes are arranged across the top. The number of squares in the checkerboard depends on the number of possible allelic combinations in gametes. The letters in each square express a predicted genotype of an individual in the next generation, a combination of the male gamete at the end of that row and the female gamete at the top of that column. There is no particular reason for it, but geneticists always show the capital letters of alleles before the lowercase letters. For example, the combination of G and W and g and w is always written GgWw.

Figure 13–3

An example of one of Mendel's more complicated crosses, which considered two characteristics at once (a dihybrid cross). In this instance, the shape of the pea seeds (round is dominant, wrinkled is recessive) and their color (yellow is dominant, green is recessive) are considered. The kinds of individuals of the second generation (F₂) demonstrate Mendel's principle of independent assortment. The shape and color of the pea seeds are inherited independently. The ratio of phenotypes in the F_2 generation was 315:108:101:32, or approximately 9:3:3:1.

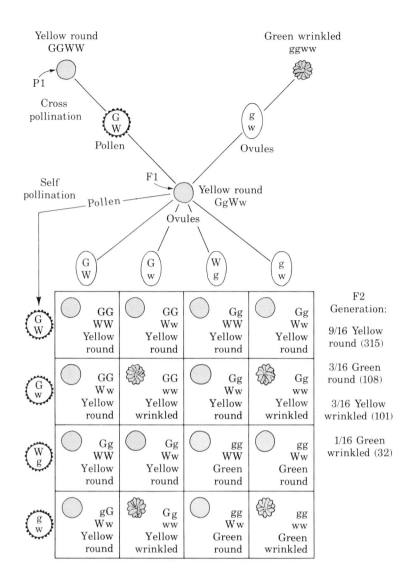

GENETICS AFTER MENDEL

In the second half of the nineteenth century, a Dutch botany professor named Hugo De Vries began some experiments in plant breeding. Like many scientists of the time, De Vries had been stimulated by the work of Darwin. He hypothesized that evolution could occur in jumps as well as by the accumulation of slight variations over time. He began to grow two strains of the evening primrose. After crossing and growing more than fifty thousand plants, De Vries believed that he had created eight new species. (In reality, they were only stable strains of the same species.) In 1900,

De Vries published a book that described his experiments and explained his theories of inheritance. De Vries had read Mendel's paper and recognized its importance, and his references to Mendel brought Mendel to the attention of many biologists.

In the same year, William Bateson rode a train to London on his way to address the Royal Horticultural Society. During the trip, he read Mendel's paper. Bateson, a cautious scientist and plant breeder, was so impressed by Mendel's work that he incorporated the major points of Mendel's ideas into his speech. Immediately afterward, Bateson had Mendel's original paper translated into English and published. Thus

Mendel became known to the entire scientific world. Bateson was the strongest supporter of the new Mendelian theories of heredity and was responsible for the birth of genetics as a field of study. He gave the field much of its terminology, including the words *genetics, homozygous, heterozygous,* and F_1 and F_2 *filial generations.*

Mendel's fame was established for all time by the work of Thomas Hunt (T. H.) Morgan, an American who began his work in genetics around 1908, after he had established a career as an embryologist. The work of De Vries was a prime topic of discussion among most scientists at the time, and the subject of mutations intrigued many brilliant minds.

Morgan believed that the investigation of mutations might be a new approach to the study of genetics, but he was skeptical of Mendel's claims, because there was no current evidence to support his theory of inheritance. Morgan was the first to make extensive use of the fruit fly *(Drosophila).* He bred this little fly for a year without any significant findings. Finally, a fly that had white, rather than red, eyes appeared in his cultures. Within another year, forty different kinds of flies with changed physical characteristics had been reared. Various crosses among these flies proved that characteristics were inherited in the fashion predicted by Mendel—now called Mendelian inheritance. This was modern-day support for Mendel's theories, although more evidence was to come.

Morgan not only substantiated Mendel's findings but also made several significant findings of his own on how organisms inherit characteristics. Morgan found that the various mutations of flies reared in his laboratory were associated with the four pairs of chromosomes possessed by *Drosophila.* His work conclusively proved the **chromosomal theory of genetics,** which states that chromosomes are the elements that transmit inherited characteristics. He also found that there were more mutations than chromosomes, thereby showing that chromosomes are the carriers of genes, which cause the expression of individual characteristics.

However, Morgan found that some characteristics did not segregate as Mendel predicted. These characteristics, expressions of genes on the same chromosome, are said to be **linked.** When one character is expressed, so is the other. Two of these linked characteristics one day appeared separately. Morgan proposed that this had happened because a chromosome had been broken and had rejoined with another chromosome. He thus identified the phenomenon of **crossing over.**

Morgan made one other important contribution. He was one of the first to encourage scientists to work together as teams on the same biological problem. Thus many other scientists became engaged in *Drosophila* genetics and together made other important contributions to the understanding of heredity.

In 1940, geneticists George W. Beadle and Edward Tatum were conducting genetic experiments at Stanford University. They became dissatisfied with *Drosophila* and switched to the common bread mold, *Neurospora crassa.* They grew thousands of tubes of the mold and exposed the cultures to X-rays in the hope of causing mutations. Some of the affected cultures failed to grow on a gelatin slab that had the basic nutrients for promoting growth of the mold. Beadle and Tatum found that these nongrowing strains needed vitamin B_6 to grow. The molds were unable to manufacture the vitamin from the basic nutrients in the gelatin. This discovery established the **one-gene, one-enzyme principle.** The X-rays had changed one gene, a gene that produced the enzyme that promoted the manufacture of vitamin B_6. Subsequently, genetics and cellular chemistry became one and the same study.

While Mendel was growing his peas in Germany, a physician named Griedrich Miescher found that the nuclei of many cells contain an unusual substance. He called it nuclein, and it later came to be called **nucleic acid.** The relationship of this substance to chromosomes and genes would become one of the most important discoveries in science. The relevance of nucleic acids was discovered slowly, because many technical advances had to come first. Among them were the development of sophisticated microscopes, techniques of staining and examining cells, and biochemical identification and separation of substances (chromosomes, genes, and so on).

Miescher's magic substance was found to be deoxyribonucleic acid, or **DNA**▫. Geneticists James Watson, Francis Crick, and Maurice Wilkins were awarded the Nobel Prize in 1962 for their work in describing the chemical nature and structure of DNA. Their discoveries and many discoveries made by scientists who followed them established the fact that the hereditary material DNA is found in the cells of living organisms and that the principles of genetics and evolution apply to all organisms.

SUMMARY

1. Gregor Mendel provided the first experimental evidence of the inheritance of physical characteristics. He conducted a nearly perfect genetic experiment using the common garden pea.

2. Although Mendel attempted to provide the scientific community with his data, he was not successful in his lifetime. He did not become known as the father of genetics until decades after his death.

3. Mendel established the basic principles of inheritance that became the foundation of modern genetics. He proposed the following:

— a. A pair of factors controls the expression of inheritable characteristics.

— b. Offspring inherit one member of a pair of factors from the male parent and the other from the female parent.

— c. Each factor is transmitted as a discrete unit. Factors do not blend or become diluted in future generations.

— d. One member of a pair of factors may dominate the other. If so, only the dominate factor will be expressed. However, the recessive member will remain a discrete unit to be expressed in future generations.

— e. Paired factors separate during the formation of pollen and ovules. Offspring inherit one member of each pair from each parent. The separation is known as the principle of segregation.

— f. Each pair of factors segregates independently of any pairs of other factors, a phenomenon known as the principle of independent assortment.

4. Mendel's experiments were publicized at the turn of the twentieth century by William Bateson, who also coined much of the current terminology in genetics.

5. Thomas Hunt Morgan provided modern evidence to support Mendel's original experiments. He was the first to use the fruit fly as a tool in genetic experimentation.

6. Morgan conclusively proved the chromosomal theory of genetics. Furthermore, he found that the inheritance of some traits could not be predicted by Mendel's laws because the traits were linked on the same chromosome.

7. George W. Beadle and Edward Tatum are credited with the one-gene, one-enzyme principle of inheritance.

8. James Watson, Francis Crick, and Maurice Wilkins established the chemical basis of heredity when they described the DNA molecule.

FUTURE FOOD

The work of Gregor Mendel caused a revolution not only in science but also in agriculture. Since the beginning of agriculture, cultivated food plants had been improved only by selection. That is, people saved seeds from plants that displayed desirable characteristics and planted them the next season. Most of these plants were adapted to very small geographic regions and could be grown successfully nowhere else.

Although there are more than 300,000 species of flowering plants, humans have selected fewer than 400 as food plants and have cultivated only 150. Today, a few major food plants support the world: wheat, corn, rice, potatoes, sugar cane, beets, beans (particularly soybeans), coconuts, and bananas.

Mendel's work demonstrated that plants could be hybridized—indeed, bred like animals—and that the resulting offspring would have characteristics of each parent. During Mendel's time, there were few attempts at plant hybridization. Even when hybridization was accomplished, the results were a mystery, because the principles of heredity were unknown.

When Mendel's work was rediscovered, it caused a wave of scientific research in plant hybridization, most of it conducted by U.S. land grant colleges. Considerable prog-

ress was made between 1910 and 1940. New plants, notably soybeans and sorghum, were also introduced into the United States from other countries. Because of hybridization, they became important crops. Many new varieties of corn and wheat appeared. They were disease-, frost-, and drought-resistant, and they produced more grain. The availability of commercial fertilizer doubled and even tripled yields in the Midwest.

As a result of this progress there was so much corn in storage just before World War II that it was cheaper to buy corn than to grow it. As a result, the bottom fell out of the market and people began to feed corn to their cattle.

The "green revolution" is an attempt to grow more productive and nutritious grains to help feed the growing populations of the world. One of the first new grains developed as a result of the green revolution was a hybrid wheat. This wheat is much shorter than conventional varieties. Therefore, more of the plant energy goes into making grain rather than straw. The shorter plants are also more resistant to the hail and winds that topple taller varieties.

Another miracle grain is wheat that grows in the tropics. Genetic engineering of this hybrid made it resistant to both drought and flood and independent of temperature.

Also, flowering is not timed by the length of the day, as it is in temperate regions.

The first rice was IR–8, developed in 1966. This grain is also independent of temperature and day length, so two or three crops can be grown each year, instead of only one. Yields of IR–8 are 300 percent greater than yields of conventional types of rice.

High-lysine corn is another new grain. Humans have been unable to live exclusively on corn because two essential amino acids, lysine and tryptophan, occur in it in only small quantities. A diet restricted to corn would eventually produce malnutrition, unless the amino acids were consumed in other protein sources. In 1964, the first high-lysine hybrid corn was produced. Its lysine content was 70 percent greater than in regular corn. Today, corn can be a good source of protein. High-lysine corn is now grown in many areas and is the mainstay of the diet of many Colombian Indians, who benefit from their improved nutrition.

The green revolution has had its failures too. But it must continue if the world is to be fed. Feeding the burgeoning population is the most serious problem that has ever faced humankind, and its most serious moments most likely will come in the next decade.

FOR DISCUSSION
AND REVIEW

1. Why did Mendel begin his experiments? What was his profession?

2. Why did Mendel's important contributions to science remain obscure for nearly thirty years?

3. Who rediscovered Mendel's valuable experiments? What role did this person play in the development of the field of genetics?

4. Who provided the modern information that proved that Mendel's original experiments were correct? What kind of organisms did he use? How has that organism proved useful in the study of genetics?

5. Explain the chromosomal theory of genetics.

6. What is the significance of the one-gene, one-enzyme principle of inheritance? Who provided the evidence for this principle? When?

7. Define the following terms: *dominant, recessive, phenotype, genotype, allele, principle of segregation, principle of independent assortment.*

SUGGESTED READINGS

GENETICS IS EASY, by P. Goldstein. Lantern: 1967.
An easy-to-read introduction to genetics.

A HISTORY OF GENETICS, by A.H. Sturtevant. Harper & Row: 1965 (hardcover).
One of the most complete histories; considerable additional detail on all aspects.

14

TYPES OF INHERITANCE

It was not long after the rediscovery of Mendel's principles of heredity that scientists began to find examples of inheritance that could not be explained by Mendel's laws. In short, heredity is much more complex than Mendel realized. This chapter illustrates patterns of inheritance that can be explained by Mendel's laws as well as a number of other types of inheritance for which new rules have been constructed. In all cases, inheritance is predictable once the mechanisms are completely understood.

MENDELIAN INHERITANCE

Mendelian inheritance describes by simple schemes of dominance and recessiveness how characteristics in animals and plants are inherited. Mendel's peas are the classic example.

Fruit Fly Inheritance

More than likely, laboratory work is part of your biology course. If so, you will probably have the opportunity to observe firsthand the inheritance of characteristics that are predictable by Mendel's laws. For this purpose the fruit fly, *Drosophila melanogaster,* is commonly used. The fruit fly has been an important tool in exploring how characteristics are inherited. There are hundreds of known mutants for the fruit fly. In fact more is known about the inheritance of this seemingly insignificant fly than about the inheritance of nearly any other organism.

If you do study fruit flies, it is likely that you will study the inheritance of some simple dominant and recessive traits. Figure 14–1 illustrates the inheritance of one physical characteristic of fruit flies—the

PREVIEW QUESTIONS

After reading this chapter, you should be able to answer the following questions:

1. When the results of a genetic cross are analyzed, the data are usually expressed as a phenotypic or a genotypic ratio. What is the difference between the two types of ratios?

2. What are linked alleles?

3. How is sex inherited?

4. Give an example of each of the following: sex-linked inheritance, sex-limited inheritance, and sex-influenced inheritance.

5. What are multiple alleles?

6. Name a human trait that is influenced by gene interaction. How is this trait inherited?

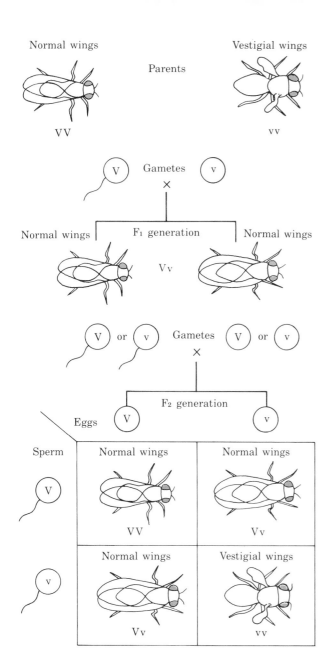

Figure 14-1

A monohybrid cross. When normal-winged flies (VV) are crossed with vestigial-winged flies (vv), all of the F_1 offspring have normal wings (Vv), but all carry the recessive allele (v) for vestigial wings. When the F_1 offspring are bred together, three-fourths of the F_2 generation have normal wings and one-fourth have vestigial wings, a phenotype ratio of 3:1. The genotype ratio is 1 VV to 2 Vv to 1 vv, or 1:2:1.

wings—in a **monohybrid cross.** The natural characteristics of the fruit fly are often referred to as **wild-type** characteristics, because they are the adaptive traits possessed by flies that are wild or in nature.

The mutant flies that are studied in the laboratory would probably never survive in nature, because there would be strong selection against them. Normal wings are a wild-type characteristic and are dominant over most other wing conditions. One common mutant almost completely lacks wings, a trait known as vestigial wings. A fly without wings would have little success in the wild, although it can be propagated in the safe, nonvariable environment of a culture bottle in the laboratory.

As Figure 14–1 shows, if you cross normal-winged flies with vestigial-winged flies, all of the flies of the **first generation** (F_1) will be normal-winged. Like Mendel's peas, all of the F_1 flies will be **heterozygous,** expressing the dominant trait for normal wings and carrying the recessive trait for vestigial wings. The lower portion of Figure 14–1 shows what happens when the F_1 heterozygous flies breed among themselves. Again, the resulting **second generation (F_2)** flies show a distribution of characteristics similar to the distribution Mendel found with his peas. There are three normal-winged flies for every vestigial-winged fly, a ratio of normal to vestigial of 3:1. This ratio is called the **phenotype ratio** because it is computed on the basis of how the flies look. However, as Figure 14–1 illustrates, some of the F_2 winged flies are carriers of the recessive trait for vestigial wings, just as all of the F_1 (heterozygous) individuals are.

The **Punnett square** method of predicting the gene combinations of the F_2 flies, illustrated in Figure 14–1, clearly indicates the carrier (heterozygous), noncarrier (homozygous), and vestigial (homozygous) flies. Among the F_2 flies, 25 percent have both alleles for the dominant condition of normal wings, 50 percent have one dominant and one recessive allele (and thus express the dominant normal wings but carry the vestigial ones), and 25 percent have two recessive alleles and vestigial wings. The ratio of the genetic composition of the F_2 flies is 1:2:1, a ratio known as the **genotype ratio.**

Fruit flies can also be used to demonstrate the predictable results of a **dihybrid cross** using simple Mendelian characteristics (dominant and recessive). (See Box 14–1.) The two characteristics used for the cross in Figure 14–2 are again normal wings and vestigial wings, but added to this are the traits red eyes

BOX 14–1

SOLVING GENETIC PROBLEMS

Solving many genetic problems is not only easy but fun. The first step is always to write down in symbols the information that is available. Say you are told that an individual is heterozygous for two physical characteristics. You could represent the genotype as AaBb or use any other symbols. Perhaps you know only that the individual is dominant for two traits. Then all you can do initially is to deduce the genotype, which you can represent as A_ B_. Whatever the case, first write down the genotype or phenotype as best you can, using appropriate symbols.

The second step in solving any genetic problem is to deduce which alleles will be carried by the gametes, the eggs and sperm cells that form during meiosis. What gametes will be produced by the genotype AaBb? If you have had algebra, you can compute the answer in seconds, because the gametes are all of the possible combinations of the two sets of letters:

AaBb = AB

AaBb = Ab

AaBb = aB

AaBb = ab

Algebraically:

$$A(B + b) = AB + Ab$$

$$a(B + b) = aB + ab$$

The first letter of the genotype is paired with each of the second letters. If the second letter of the genotype is different from the first, then it too is paired with each of the second set of letters.

The third step is to arrange the possible female gametes across the top of a Punnett square. Then arrange the possible male gametes along one side. A Punnett square does not always have eight or sixteen boxes. It can have any even number of boxes, depending on the number of possible gametes. For example, a male who has the genotype AABB will produce only AB gametes. Therefore, the Punnett square will have only one row.

The fourth step is to combine male and female gametes in the various boxes of the Punnett square. Each new combination is a possible genotype of the offspring.

Is there an equal chance of producing all of these offspring? No. The odds are computed in the fifth step. Count the number of phenotypes represented in the Punnett square and then the number of individuals who will have each phenotype. (Each box represents an individual.) Arrange all the numbers for each phenotype in a series—the largest number first, the next largest second, and so on. Then divide each of the larger numbers by the smallest number. The result is a ratio, the phenotype ratio (something to something to one) It states how probable each phenotype is. The phenotypes in Figures 13–3 and 14–2 maintain the typical ratio of $9:3:3:1$. One phenotype may be represented in nine of sixteen boxes, which means that there are nine chances out of sixteen (or a 56 percent chance) that a particular set of parents will have offspring of that particular phenotype. If only one box out of sixteen is of a particular phenotype, the chances that it will appear are one out of sixteen (6.5 percent).

The sixth and last step is to compute the genotype ratio. Find all of the different genotypes in the Punnett square. Then count how many cells have the same genotypes. Arrange and compute a ratio as you did for the phenotype ratio. In Figures 13–3 and 14–2, the ratio is $4:2:2:2:2:1:1:1:1$.

If you follow these rules, you can solve any genetic problem.

and black eyes. Wild fruit flies always have bright red eyes, a dominant characteristic. Many other recessive eye colors are possible, among them black or sepia.

Figure 14–2 shows that the **P₁ (parent)** flies are of two different phenotypes. The P₁ males have normal wings and red eyes, and the P₁ females have vestigial wings and black eyes. The flies are chosen this way so that the males will not be at a disadvantage when

courting females, because wing displays are important courtship gestures. The P₁ flies have been obtained from pure breeding lines in the laboratory. They have been bred again and again to make sure that the dominant males are homozygous and carry no recessive traits and that the homozygous recessive females carry no dominant alleles.

The F₁ flies of the dihybrid cross will all be hetero-

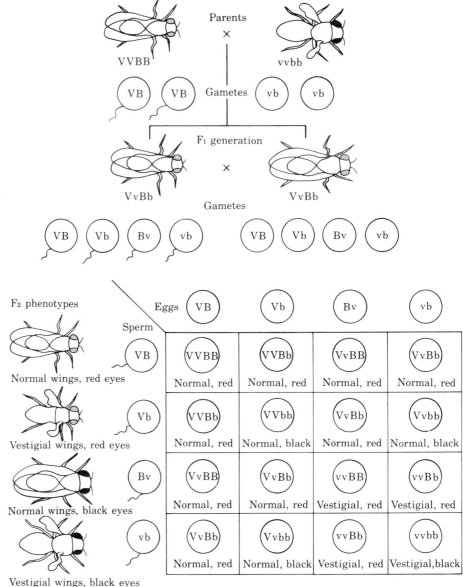

Figure 14–2

A dihybrid cross. When wild-type flies with normal wings and red eyes (VVBB) are bred with flies with vestigial wings and black eyes (vvbb), all of the F_1 generation have normal wings and red eyes (VvBb), but all carry the recessive alleles for vestigial wings (v) and black eyes (b). When the F_1 flies are bred together, nine out of sixteen of the F_2 offspring have normal wings and red eyes, three out of sixteen have vestigial wings and red eyes, three out of sixteen have normal wings and black eyes, and one out of sixteen has vestigial wings and black eyes, a phenotype ratio of 9:3:3:1.

zygous. All will have normal wings and red eyes, and all will carry the recessive alleles for vestigial wings and black eyes. What will happen if these flies are allowed to freely breed and create an F_2 generation? The phenotypes and genotypes of the F_2 generation can be predicted with a Punnett square, as illustrated in Figure 14–2. The Punnett square in this instance has sixteen boxes. Nine of the boxes predict wild-type flies with normal wings and red eyes. Three of the boxes predict flies with normal wings but black eyes. Three of the boxes predict vestigial-winged flies

with red eyes. Only one box predicts a fly with vestigial wings and black eyes. The phenotype ratio, in this instance 9:3:3:1, is more complicated than for a monohybrid cross. The genotype ratio is 4:2:2:2:2:1:1:1:1.

The F_2 flies in this example have an assortment of genetic characteristics that is different from those of either their parents or grandparents. This illustrates Mendel's principle of independent assortment.

What can be predicted if three characteristics are considered in a trihybrid cross? Obviously, the mathematics becomes complex. The Punnett square will have sixty-four cells. Because most species inherit thousands or tens of thousands of alleles, it is almost impossible to predict the entire range of phenotypes of an organism.

Human Inheritance

Many human characteristics are known to be inherited as Mendelian traits—simple conditions of dominance and recessiveness. For example, dark eyes are generally dominant to blue eyes, and dark hair is generally dominant to blond hair.

However, there are many intermediate phenotypes in humans, such as hazel eyes and red hair. Predictions based on pedigree analysis□ are not always accurate, especially when it is assumed that these traits (eye and hair color, for example) are simple dominant and recessive. Actually most human characteristics are controlled by many genes. It is known, for example, that red hair is controlled by alleles different from those of either dark or blond hair.

This does not mean that humans have only a few characteristics that are inherited as simple Mendelian dominants and recessives. The list for humans at present includes 583 defects inherited as simple dominants and 466 as recessives, plus another 635 suspected simple dominant and 481 suspected recessive traits—for a grand total of 2,165.

The inheritance of an earlobe characteristic is one example of simple inheritance in humans. Human earlobes are of two configurations (see Figure 14–3). Some earlobes hang in a pendulous fashion, a dominant trait. Others are attached to the superficial tissues of the side of the face and do not hang free, a recessive trait. Often, people with the latter trait look almost as though they lack earlobes. People with pendulous earlobes are said to have "free" earlobes (A). The others are said to have "attached" earlobes (a). Inheritance of this condition can be predicted using a Punnett square.

Suppose you have attached earlobes. Is it possible that both of your parents have free earlobes? Of course it is. Both of your parents may be carriers or heterozygous for this trait (Aa). Suppose that you and your sister and father have free earlobes but that your mother has attached earlobes. Can you determine your genotype? Your mother must be aa, so you and your sister are Aa. But your father may be either AA or Aa. What we have just done is known as a pedi-

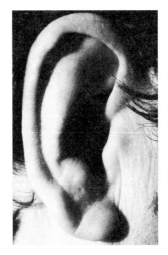

Figure 14–3

Free and attached earlobes of humans. Free earlobes (left) are dominant over attached earlobes (right), which are recessive. (Photos by Tom Ardelt.)

gree analysis, a form of phenotype and genotype detective work using Mendel's laws (see Box 14–2).

Have you ever tasted phenylthiocarbamide? This bitter substance, abbreviated PTC, cannot be tasted by everyone. Pedigree analysis has shown that the ability to taste it comes from a dominant allele inherited in simple Mendelian fashion. Nontasters are homozygous recessive. About 75 percent of the people in this country are tasters. Furthermore, this Mendelian characteristic is also shared by many of our primate cousins. The distribution of tasters and nontasters among chimpanzees is about the same as it is among humans. Nontasting chimpanzee parents produce only nontasting offspring. Other primates have also been checked. All marmosets seem to be tasters, and all spider monkeys seem to be nontasters.

The evolutionary significance of this characteristic is unclear at the present time. However, substances closely related to PTC are found in vegetables belonging to the cabbage family. These substances may be related to thyroid disease in humans. Perhaps at one time in our evolution, tasting was an adaptation that told us to avoid certain detrimental foods. Today, however, there is no correlation between the ability to taste PTC and the like or dislike for cabbage-related vegetables.

Some people have the **Rh factor,** a protein on the surface of the red blood cells. Such people are said to be Rh-positive, or Rh$^+$. Those who lack the antigen

BOX   14–2

PEDIGREES

If you have ever had a registered dog, you have probably seen its pedigree. The pedigree provided by the American Kennel Club shows the dog's parents, grandparents, aunts, uncles, cousins, and brothers and sisters. In some cases, great-grandparents or great-great-grandparents are known. The pedigree usually notes all the names and relationships of the dog's ancestors. Often, award winners and champions are also noted.

A genetic pedigree is similar to a dog pedigree. However, it traces the distribution of a genetic characteristic among ancestors. Such a pedigree can be of great value in genetic counseling. You can construct a pedigree for your own family. It is not only an educational experience but also fun.

A few rules of construction make pedigrees uniform, so they can be interpreted by anyone. The pedigree in Figure 14–9, showing Queen Victoria's family and the distribution of hemophilia among her descendants, can be used as an example. Each row on a pedigree chart represents a separate generation. The oldest known generation is labeled I, the next oldest II, the next III, and so on. The following symbols are used to denote the characteristics of each individual in the pedigree:

○   Female

□   Male

○—□   Spouses

● or ■   Positive genetic trait

⊙ or ⊡   Carrier of recessive trait

Identical female twins

Identical male twins

Fraternal twins (female/male)

Fraternal twins (female/female)

Fraternal twins (male/male)

If there are many descendants and the pedigree is very crowded, unafflicted siblings can be shown with a single symbol that has in it the number of individuals of that sex. For example, ③ indicates three females. A stillborn child is shown with a symbol smaller than that for a living child. The first child of a marriage appears on the extreme left and the other children to the right, in order of birth. Finally, a question mark indicating doubt about a designation, may appear at any point in a pedigree.

With this information, interpret the pedigree given here. Is the trait dominant or recessive? Can you identify the carriers?

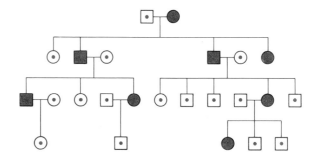

Properly constructed pedigrees can give considerable information about the inheritance of a trait. For example, a simple dominant trait is expressed in siblings of every generation, the incidence among males and females being about equal. A recessive trait, on the other hand, may be represented in siblings in every other or every third generation. A characteristic that is most frequently expressed by either males or females is probably a sex-linked characteristic. A pedigree showing a trait distributed only among males indicates Y-linked or a rare X-linked recessive characteristic. What can you conclude about the inheritance of the trait illustrated here?

are considered Rh-negative, or Rh⁻. Surveys have shown that about 85 percent of the U.S. population is Rh⁺ and 15 percent Rh⁻.

The inheritance of the Rh factor is somewhat complicated. However, from family pedigrees, it appears that the Rh factor is caused by a simple dominant al-

lele. The dominant and recessive alleles are usually designated as D and d, respectively. Rh⁺ persons can be either DD or Dd; Rh⁻ persons are dd. Recent research has shown that three closely linked gene loci are involved, each coding for a specific **antigen** (protein) associated with the Rh factor. The dominant allele D is the most common, and its coded polypeptide is considered the strongest antigen.

Most people's acquaintance with the Rh factor is from its association with complications during pregnancy. The most frequent question asked in an introductory biology laboratory by female students is whether they can find out if they are Rh^+ or Rh^-. The Rh factor problem that can arise during pregnancy is a problem only for an Rh^- woman who might carry an Rh^+ fetus.

Generally, the placenta□ is reasonably good at separating the blood of mother and fetus. But late in pregnancy and during delivery, the weight and activity of the fetus may cause some minor breaks in the capillary walls of the placenta. Some of the fetal red blood cells may enter the bloodstream of the mother. When the mother is Rh^- and the fetus is Rh^+, the Rh antigen on the fetal red blood cells triggers the production of antibodies by the mother. The effect is as though the mother had received a blood transfusion of Rh^+ blood. Because this occurs late in pregnancy, the mother has only a short time to produce antibodies. The few that are produced pose no threat to the child.

However, if an Rh^- woman's second pregnancy involves another Rh^+ fetus, the result can be serious. The antibodies produced by the mother are then sufficiently concentrated to cause an antigen-antibody reaction in the fetus. The affected fetuses are often born anemic, with jaundice, abnormal blood cells, and severe liver and brain damage. They may be stillborn or may live only a few weeks. This disorder is called erythroblastosis fetalis (see Figure 14–4).

Problems with the Rh factor during pregnancy occur only if an Rh^- woman marries an Rh^+ man. If the man is Dd, the chances of having an Rh^+ fetus are 50/50. If the man is DD, every child will be Rh^+. At one time, one of the most important questions dis-

Figure 14–4

Development of an Rh^+ fetus within an Rh^- mother. The placenta is not always a perfect barrier. If the blood of an Rh^+ fetus crosses into the mother's circulation, the mother produces antibodies against it. There is usually no problem for an Rh^+ fetus with the first pregnancy. However, during the second pregnancy, the mother produces many more antibodies, which can destroy the red blood cells of the Rh^+ fetus and cause its death. This condition, called erythroblastosis fetalis, can be prevented if a preparation called Rhogam is given to the mother after each pregnancy.

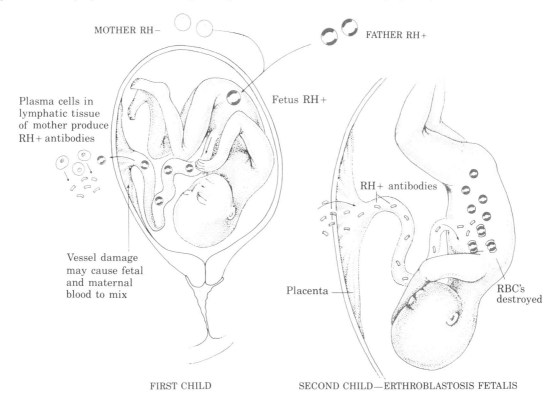

MOTHER RH− FATHER RH+

Plasma cells in lymphatic tissue of mother produce RH+ antibodies

Fetus RH+

Vessel damage may cause fetal and maternal blood to mix

RH+ antibodies

Placenta

RBC's destroyed

FIRST CHILD

SECOND CHILD—ERTHROBLASTOSIS FETALIS

cussed during a marriage proposal was "What is your Rh factor?" Rh⁻ women searched for Rh⁻ men to avoid any complications. However, since the early 1960s, the situation has changed. Rh⁻ women can now marry whomever they choose because of a substance called Rhogam.

After the first Rh^+ birth, the Rh^- mother is given an injection of Rhogam, which is a mixture of Rh antibodies. The antibodies react with any fetal Rh^+ red blood cells (containing antigens) that may have entered the woman's circulation. She therefore does not become sensitized and does not produce antibodies of her own. (It is important to stop the woman's own immune system from producing these antibodies.) Thereafter, the Rh^- woman is given Rhogam after every birth or abortion. (During an abortion, blood from the fetus may also seep into the maternal circulation.) Erythroblastosis fetalis is prevented by these measures, so Rh^- women need not worry about the Rh factor of a future husband. However, the treatment works only for women who have not yet been sensitized. Thus it is important to know a woman's chances of having an Rh^+ fetus.

Rhogam is produced by volunteer men who are Rh^-. Each donor is given the Rh antigen and allowed to produce antibodies. Later, a pint of blood is taken from the volunteer, and the antibodies are filtered from the serum. The final mixture is Rhogam.

Many simple dominant and recessive alleles cause malformations of the human body. Most of these involve the hands, feet, arms, and legs. One of the first known examples of a dominant human allele was the condition known as brachydactyly. The inheritance of the dominant allele produces hands with short, stubby fingers and fused terminal bones. Another dominant allele calls a premature halt to the growth of the long bones of the body, causing the condition known as achondroplasia (see Figure 14–5). Human dwarfs have short arms and legs but normal-sized torsos. They are four feet or less in height, with disproportionate bodies.

A recessive allele, when homozygous, affects the release of growth hormone from the pituitary gland, producing human midgets. Midgets are also four feet or less in height, but they are perfectly proportioned, because there simply has been no growth since early childhood. Human midgets may soon become nearly nonexistent. When the condition is detected early, the individual can be given injections of growth hormone and will then grow normally.

Figure 14–5

Human dwarfs have a condition known as achondroplasia, which is inherited as a simple dominant trait. The arms and legs are short, and often there are abnormalities of the joints. (Notice the woman's elbow and the man's fingers.) The head is often enlarged, and the forehead is high and flattened. The torso is of normal proportions. (Wide World Photos.)

MECHANICS OF NON-MENDELIAN INHERITANCE

Mendel was lucky to select the common variety of garden peas for his experiments. It was also a stroke of genius or luck that the characteristics he chose were not linked. (Alleles that are linked are carried on the same chromosome. Mendel picked characteristics whose alleles were on separate chromosomes.) However, many characteristics have more complex modes of inheritance, and Mendelian inheritance has proved inadequate to the task of explaining them.

Linked Alleles

Even in fruit flies, there are many examples of linked alleles—different traits on the same chromosome. For example, the alleles for normal and vestigial wings are on the same pair of chromosomes as are the alleles for gray (dominant) and black (recessive) body color. If we use Mendel's laws, we would predict that a dihybrid cross between VVBB and vvbb flies will produce only heterozygous F_1 flies, VvBb. (If you cannot make that prediction, see Figure 14–2, which explains the reason for this prediction.)

What can we predict for the F_2 generation? In order to simplify the prediction, let us assume that the F_1 flies are crossed with their vvbb parents. This will eliminate the need for a sixteen-box Punnett square. A four-box Punnett square is all that is needed. Going by Mendel's laws, we can predict that a cross between VvBb and vvbb will produce 25 percent VvBb, 25 percent Vvbb, 25 percent vvBb, and 25 percent vvbb. (Construct a Punnett square if you cannot deduce this result.) In fact, however, the actual result is 40 percent VvBb, 10 percent Vvbb, 10 percent vvBb, and 40 percent vvbb. How is this result possible? Is it predictable?

The second question is the easier to answer. Yes, the result is predictable and repeatable. Notice that most of the offspring have phenotypes just like those of the parents—normal wings and gray color or vestigial wings and black color. That is the first clue to an explanation. The alleles are located on the same chromosome (linked). Therefore, where one goes, so does the other. The chromosomes carrying the alleles VB and vb are independently assorted, but the alleles themselves are not. Thus the prediction from Mendel's law of independent assortment fails.

If the alleles are linked on the same chromosome, how do the genotypes Vvbb and vvBb and the corresponding phenotypes of normal wing–black body and vestigial wing–gray body come about? During meiosis□, homologous chromosomes come to lie side by side, a process called **synapsis.** During synapsis, sections of chromosomes from one homologue and the genes they contain may be exchanged for similar sections from the other homologue, an event called **crossing over□.** Alleles that are linked can become unlinked and be switched to separate chromosomes during crossing over. Alleles that are close to each other on the same chromosome ("closely linked") are less likely to become separated during crossing over

than are alleles separated by greater distances. If crossing over occurs among closely linked alleles, it is highly probable that some will remain linked and stay in the same arrangement as they were on the original chromosome.

How far apart are linked alleles when they cross over? Geneticists use the distribution of offspring to make that prediction. The results of the fruit fly example show that about 20 percent of the flies have genotypes that came about by crossing over (Vvbb and vvBb). If crossing over did not occur, we would expect all F_1 flies to have normal wings and gray bodies or vestigial wings and black bodies. The 20 percent figure is a repeatable result. Therefore, geneticists say that the V and B alleles are twenty map units apart, or twenty crossing-over units (originally called centimorgans after T.H. Morgan, who first deduced that crossing over could be a natural genetic phenomenon).

The results of genetic crosses involving linked alleles have allowed geneticists to map the approximate locations of genes on chromosomes. The position of a gene on a chromosome is called its **locus.** Gene maps are extremely complete for the chromosomes of *Drosophila.* For example, the X chromosome□ (one of the sex chromosomes) of *Drosophila* is approximately sixty-six map units long and contains approximately five hundred genes located a discrete number of map units apart, as determined by their frequency of crossing over.

Incomplete Dominance

Another type of inheritance that is not entirely predictable by Mendel's laws is **incomplete dominance.** In this case, the dominant allele is not completely dominant, so when an individual is heterozygous, the phenotype is often different from that of either the homozygous dominant or the homozygous recessive individuals.

Perhaps one of the best examples of incomplete dominance occurs among a special breed of cattle known as short-horns. These cattle can be white, red, or roan. (Roan coloration is essentially a blending of the colors white and red.) When red cattle (WW) are mated with white (ww), all of the offspring are heterozygous and roan color (Ww). At first inspection, this example might appear to contradict Mendel's laws and to favor the older concept of blending inheritance. But when heterozygotes are bred (Ww ×

Ww), the offspring produced are in the ratio of one red (WW) to two roan (Ww) to one white (ww). The appearance of red offspring and white offspring shows that the characteristics for coat color are discrete, as predicted by Mendelian inheritance. The appearance of roan offspring shows that the allele for red (W) is not completely dominant over the allele for white (w).

Pink carnations are another example of incomplete dominance. When red carnations (WW) are crossed with white carnations (ww), all of the F₁ flowers are pink (Ww).

X AND Y CHROMOSOMES

In special slide preparations, chromosomes can be viewed through a microscope. Human chromosome counts are usually made using white blood cells that have been cultured and treated in specific ways. Similar techniques are used for the cells of many other animal species. Once the cells under study are prepared, the chromosomes can be counted. Each chromosome in a cell has a **homologue,** an identical chromosome with complementary alleles. Homologous chromosomes are also physically identical. Therefore, chromosomes can be matched or paired by their size, shape, and physical configuration. The final profile of matched chromosomes is called a **karyotype** (*karyo* means "nucleus," so the karyotype indicates the type of chromosomes in the nucleus). Karyotypes have been determined for thousands of animal and plant species. The karyotype technique plays an important role in the clinical determination of various human genetic disorders.

Figure 14–6 illustrates a normal human karyotype. Humans have a total chromosome number of forty-six, or twenty-three pairs. When human karyotypes are made, twenty-two matching pairs, the **autosomes,** can be identified. However, in human males, the chromosomes of one pair are not physically identical. One chromosome is much larger than the other. This same pair of chromosomes can be identified in human females, but there they are perfect matches. Because of their relationship to sex, these two chromosomes are called the **sex chromosomes** (see Box 14–3). In the human male, the sex chromosomes include a large **X chromosome** and a much smaller **Y chromosome.** In the human female, they include two large X chromosomes and no Y chromosome.

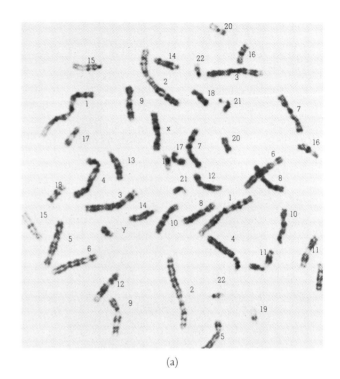

(a)

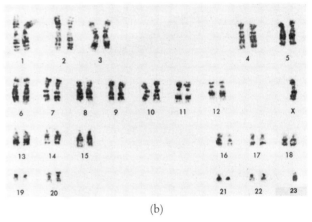

(b)

Figure 14–6

Human chromosome pairs. When white blood cells are treated in a special way, the chromosomes become apparent (a). Once photographed, chromosomes can be cut from the prints and matched up as homologous pairs. The result is a karyotype (b). Humans have twenty-three pairs of chromosomes (2*n* = 46). The twenty-three homologous pairs have been matched in this photograph. One member of each pair is maternal in origin, the other paternal. Notice the difference in the shape of the X and Y chromosomes. Is this example the karyotype of a male or a female? (Photo courtesy of Prenatal Diagnosis and Cytogenetic Laboratory, University of California, San Francisco.)

MORE GENETIC SYMBOLS

Some recessive conditions are known to deviate from normal. Therefore, they are abbreviated by the first letter of the abnormality. For example, vestigial wings in the fruit fly, a recessive and abnormal characteristic, is represented by a v. Normal wings, a dominant trait, is then represented by a V. A heterozygous fly is Vv. Similarly, albinism in humans is a recessive trait represented by a. The normal condition is represented by A.

It is not always clear whether some characteristics are truly normal. For example, is it normal for humans to have free earlobes or attached ones? We don't know. We do know, however, that the condition of attached earlobes is recessive. For that reason, attached earlobes are usually represented as a and free earlobes as A. What about the ability to taste phenylthiocarbamide? The significance of the adaptation for tasting or not tasting the substance is unclear. Here again, nontasting is the recessive trait and for that reason is usually represented as n (with tasting being shown as N).

When the inheritance of a particular trait is well understood, other symbols, such as superscripts, are often used. For example, the cause of sickle-cell anemia, a recessive trait, is perfectly understood.

Hemoglobin (Hb) affected by the sickle-cell trait is called hemoglobin S. Because the trait is known to involve a particular molecule, Hb^S is used to represent the recessive alleles and Hb^A to represent the normal, or dominant, alleles. However, S and s could be used to specify the same thing.

Superscripts become particularly useful when dealing with sex-linked inheritance, where the distribution of the allele and the sex of the offspring must be considered at the same time. Take, for example, human color-blindness. The genotype of a normal male can be expressed as $X^C Y$. The "Y" indicates the Y chromosome of a male. It has no gene for color vision. The genotype of a female carrier can be expressed as $X^C X^c$. The superscripts indicate that the alleles are carried on the X chromosome. The ratios of offspring from a mating of the two genotypes can be computed by the usual Punnett square:

	X^C	X^c
X^C	$X^C C^C$	$X^C X^c$
Y	$X^C Y$	$X^c Y$

The results: one normal female, one female carrier, one normal male, and one color-blind male. The superscripts have allowed computation of both sex and the distribution of alleles for color-blindness.

Sex Determination

In humans and most other mammals, sex determination involves the Y chromosome. Its presence means that the sex is male, and its absence means that the sex is female (see Figure 14–7). The Y chromosome is usually thought of as inert–that is, as having few or no genes. However, some researchers have suggested that an important gene (or genes) on the Y chromosome in mammals produces a chemical inducer early in embryonic development that stimulates a ridge of tissue to begin differentiating into testes. The Y chromosome may have no other function.

The sex of an offspring is determined by chance. There is a 50 percent chance that an egg will be fertilized by a sperm cell carrying an X chromosome and a 50 percent chance that it will be fertilized by a sperm cell carrying a Y chromosome. Therefore, the distribution of the sexes is expected to be in the ratio

of 1:1. If you are one of three boys in a family and your mother is pregnant again, the chances of her having another male child are 50/50. Her chances of having a female child are also 50/50. The previous birth of three male children will not influence the sex of a future child.

The U.S. population is 51.5 percent male and 48.5 percent female, a very slight deviation from the expected 50/50. Consider the families that have only two children. By chance, some of them will have two girls, some two boys, and some one girl and one boy. Is the ratio of males to females among these families also 1:1? The data clearly show that it is. Among two-children families, 23.6 percent have two girls, 26.5 percent have two boys, and 49.9 percent have one girl and one boy, the male/female ratio in two-children families is 51.5 percent to 48.5 percent, or 1.06:1.

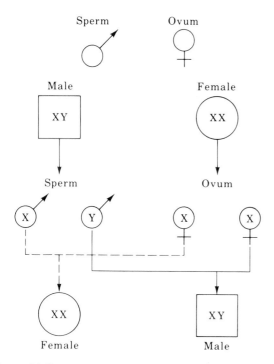

Figure 14–7

The determination of sex. Males of many species (including humans) have only one X chromosome and a much smaller Y chromosome, whereas females have two X chromosomes. The presence of the Y chromosome produces a male, and its absence produces a female.

Dosage Compensation

Chromosomes are in part composed of **genes,** which are coded messages for the manufacture of specific proteins. Because males have only one X chromosome and females have two, one might think that the total amount of coded information (DNA) will be different for males and females. It has been shown in the fruit fly, however, that the single X chromosome inherited by males creates twice the amount of protein as either one of the X chromosomes inherited by females. The male X chromosome works harder than its female counterpart, and its total output is equal to that of two female X chromosomes. This equalizing pace is known as dosage compensation.

Dosage compensation also occurs in humans, but the mode of compensation is the reverse of that for fruit flies. In human females, only one of the X chromosomes is active at any one time, and its dosage equals that of the single male X chromosome.

In 1949, Dr. Murray Barr was studying nerve cells

of cats. He discovered a clump of chromatin material in every nucleus of the female nerve cells. The clump was not associated with the usual, more centrally located, larger mass of chromatin. The small clump could not be found in cells taken from males. Because it occurred only in females, Barr called it sex chromatin. Since then, it has become more commonly known as the **Barr body,** after its discoverer (see Figure 14–8). We now know that the Barr body is the inactive X chromosome of females and that it is characteristic of all normal mammalian females. This is the basis of the "sex determination test" that is sometimes used at athletic events.

Dosage compensation in humans and all other mammals is interesting for yet another reason. The two female X chromosomes can be represented as X^1X^2. In one cell, X^1 may be active, and in another cell, X^2 may be active. X^1 and X^2 may also have dif-

Figure 14–8

The Barr body in the white blood cell of a human female (*arrow*). To prevent females from producing twice as much genetically coded protein as males, one of the two X chromosomes of females remains inactive. (It is called a Barr body after its discoverer.) As a result, the amount of protein made by the X chromosome of females is the same as that of males, a phenomenon called dosage compensation. (Courtesy Carolina Biological Supply Company.)

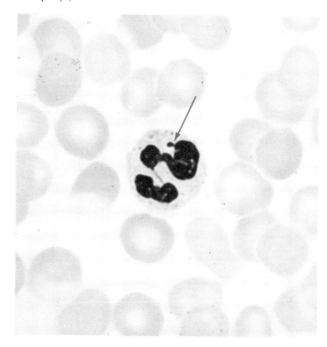

ferent alleles, so some cells may have a genetic message that is different from that of others, depending on which X chromosome is active. Some individuals have two major categories of cells, some coding for one set of genetic messages, others coding for other genetic messages. Such individuals are called **mosaics** (patchworks, much like quilts).

Female mosaics are often difficult to identify. However, among cats, a calico coat provides an excellent example. The X chromosomes of calicos can be either X^Y, which codes for a black coat, or X^y, which codes for a yellow coat. Females that are $X^Y X^y$ have some cells with an active X^Y and other cells with an active X^y and hence a coat that is partly black and partly yellow. A third allele located on an autosome (not on a sex chromosome, where X^Y and X^y are located) inhibits all color, and cells that contain it produce a white coat. When this third allele combines with X^Y and X^y, the result is the mixture of yellow, black, and white called calico.

There are human female mosaics too. A rare condition known as anhydrotic dysplasia reduces the number of sweat glands normally present in the skin. It is carried on the X chromosomes. Heterozygous women have some areas of the skin with the normal number of sweat glands and other areas with no sweat glands. They are true mosaics. In one **instance**, a pair of twin girls heterozygous for the condition had different patterns of skin without sweat glands. Apparently, the twins had a different distribution of cells with active X^1 and active X^2.

Human X chromosomes are also associated with protein factors involved in normal blood clotting. For simplicity's sake, the allele for this protein will be called BC, for blood clotting. A comparison of males who are $X^{BC}Y$ and females who are $X^{BC}X^{BC}$ shows that they produce the same amount of BC protein—further proof that females have only one active X chromosome. However, females who are $X^{BC}X^{bc}$ produce less of the BC protein, because only some of their cells have an active allele that codes for it.

SEX-LINKED INHERITANCE

Thomas Hunt Morgan was one of the first scientists to raise fruit flies and to identify natural mutants. Among the first mutants he found was a fruit fly with white eyes instead of red. When he crossed white-eyed flies with red-eyed flies, the results did not con-firm what he expected. Crosses of these traits produced females that were all red-eyed and males that were all white-eyed. Morgan reasoned that these results could be explained only if the allele for white eyes was associated with sex determination. That, in fact, turned out to be the case. The allele for white eyes (w), which is recessive to the allele for red eyes (W), is carried on the X chromosome. Because male flies have only one X chromosome, they should have white eyes (X^wY) more frequently than do females, who have two X chromosomes (X^WX^w). Any time a male fly inherits X^w, it has white eyes. However, a female fly may inherit X^w and still have red eyes. In general, then, **sex-linked traits** are more frequently expressed by males. Females are often carriers of the recessive trait and pass it to their sons.

Hemophilia

The disease known as hemophilia is perhaps one of the most famous examples of sex-linked inheritance in humans. **Hemophilia,** often called bleeder's disease, is caused by a recessive allele on the X chromosome. The dominant allele, X^H, codes for the normal antihemophilic globulin, an important protein factor that aids in the clotting of blood. The recessive allele, X^h, fails to code for this protein.

A hemophiliac must live a well-guarded life. A minor injury, such as a cut or bruise, could result in death through uncontrollable internal or external bleeding. A normal person's blood clots in less than ten minutes, but a hemophiliac's blood does not clot for thirty minutes to an hour. Therefore, some of the things we think of as common procedures, such as having a tooth pulled, are hazardous to hemophiliacs.

Hemophilia has often been called the "royal" disease, because it has occurred frequently among the royal families of Europe. Evidently, it all began with a naturally occurring mutation in Queen Victoria. Her children married into the royal courts of many countries. Two granddaughters, both carriers, married into the Russian and Spanish royal families, and both had hemophiliac sons. Figure 14–9, a pedigree of Queen Victoria's family, shows which females were carriers and which males were afflicted.

Perhaps one of the most famous hemophiliacs in the royal family was Czarevich Alexis, the son of the last czar of Russia, Nicholas II. The boy was coddled all his life and often suffered from severe bouts of bleeding. His parents, especially his mother, were willing to enlist the services of anyone who could help their son. Rasputin, a monk, claimed to have supernatural powers that could cure him. The boy did improve, but no one knows whether it was because of Rasputin's powers or because of the constant flock of physicians attending him. As a result of his ability to cure, Rasputin became a powerful figure in the royal court. His influence led to much unhappiness and helped contribute to political unrest. Eventually, the Russian monarchy fell. During the Russian revolution, in 1918, Alexis, his parents, and four sisters were executed.

King Edward VII was Queen Victoria's son. He did not inherit hemophilia. Thus the current royal family of England, which is descended from him will never have to fear inheriting hemophilia.

Female hemophiliacs are rare. They result from the marriage of a male hemophiliac and a female carrier. (Can you deduce why?) The odds against such a union are estimated to be about 1 in 50 million in random matings.

Today, hemophilia is not as dreaded as it once was. Those who suffer from it are now often given the missing proteins by transfusion from normal compatible blood donors. With adequate blood proteins from the transfusions, the blood of hemophiliacs clots as fast as it would if they did not have the disease. However, the cost of the transfusions is about $20,000 per year. And despite all therapeutic measures, if hemophiliac males by chance marry carrier women, they may propagate hemophiliac sons and daughters or carrier daughters.

Color-Blindness

Another sex-linked characteristic is color-blindness, the most common form of which is red-green. Those who are afflictd have difficulty distinguishing red

Figure 14–9
Queen Victoria and her descendants. This pedigree illustrates the inheritance of hemophilia in several royal families.

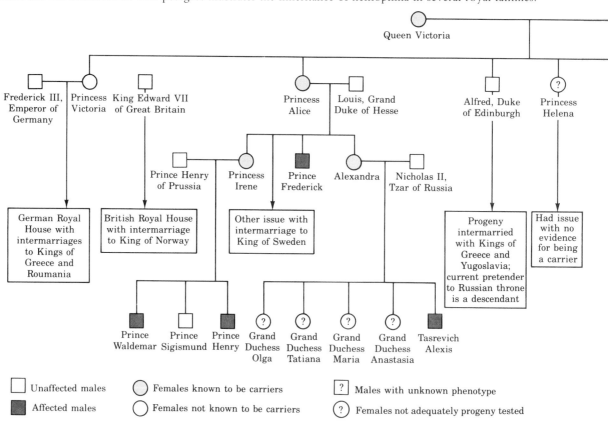

from green. They perceive green as red. There are three kinds of color receptors in the retina of the eye. All are **cone cells** that are sensitive to either red, green, or blue. People who are red-green color blind have defective green cones. As can be predicted, more men than women suffer from color-blindness (6 percent of U.S. men versus 0.5 percent of U.S. women). Women are often carriers. The inheritance of this form of color-blindness (see Figure 14–10) is identical to that for hemophilia.

There are two other forms of color-blindness, both much rarer. One is green-red color-blindness. People with this form have a difficult time distinguishing between green and red. They perceive red as green because the red cones are defective. Finally, a very rare form of color-blindness (occurring in fewer than 1 percent of men) results in insensitivity of all three of the cones. Individuals with this type of color-blindness see everything in shades of gray.

SEX-LIMITED INHERITANCE

Some characteristics are expressed only in males or only in females, a condition of **sex-limited inheritance.** Breast development and beard growth are obvious expressions of sex-limited inheritance. Their mechanisms are poorly understood, but they are indirect and involve hormones. Current research suggests that the specific characteristics of maleness and femaleness may not be associated with the sex chromosomes at all but may instead be the expression of autosomes. The sex chromosomes, however, undoubtedly influence the expression of the autosomes.

Research has demonstrated the existence of one particular allele that codes for hairy ears (see Figure 14–11). This allele appears to be part of the Y chromosome. Therefore, the trait is expressed only by men, usually after they reach thirty years of age. The outer rim of the external ear becomes covered with

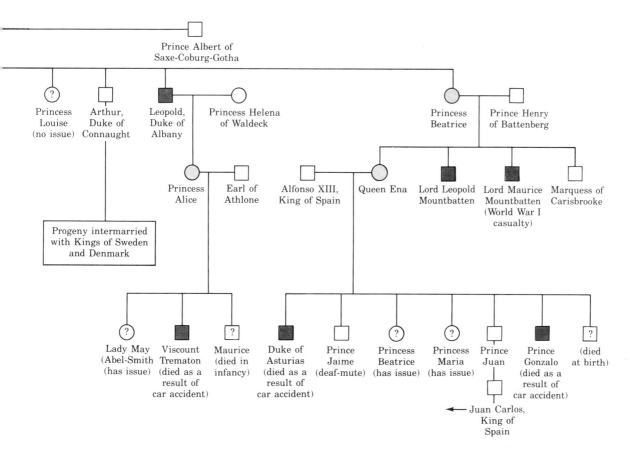

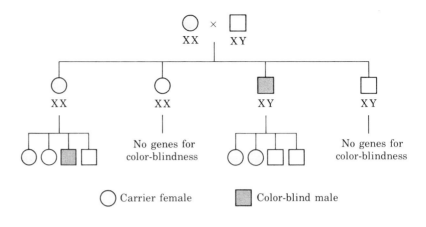

Figure 14–10

A pedigree for the inheritance of color-blindness. A carrier female marries a normal male. Half of their daughters are carriers, and half of their sons suffer color-blindness. The carrier daughter marries a normal male. Again, half of their daughters are carriers, and half of their sons suffer color-blindness. The carrier daughter's color-blind brother marries a normal female. Here, too, half of their daughters are carriers, and half of their sons suffer color-blindness. This inheritance is typical for sex-linked characteristics.

Figure 14–11

Hairy ears of some human males. The Y chromosome is thought to be generally inert, although some believe that it may code for inducer substances that cause differentiation of testes. It also carries at least one physical characteristic—hairy ears, as shown in these photographs. Because the trait is carried on the Y chromosome, it is expressed only in men. (Photographs by Mr. S. D. Sigamoni, Photography Department, Christian Medical College, Vellore; courtesy of the University of Chicago Press.)

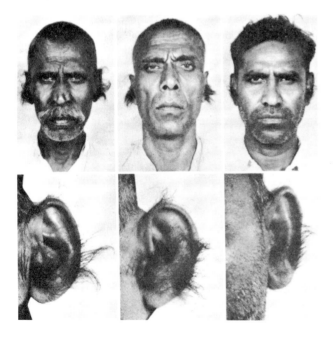

either a heavy or a sparse coat of long, stiff hair (not to be confused with the hair that often surrounds the opening of a man's ear canal). The hairy ears trait is rare in the West, but relatively common in several Asiatic countries, particularly India and Sri Lanka.

SEX-INFLUENCED INHERITANCE

Traits that are expressed differently in males and females are called **sex-influenced characteristics.** Even when both sexes have the same genotype, the characteristic is expressed differently.

For example, why are (for the most part) only men bald? Patterned baldness of men is quite common. It begins in the twenties or thirties, when the hair on the crown of the head thins, and proceeds to a loss and thinning of hair in every direction from the crown. The result may be a completely bald head, hair only on the sides of the head, or a bald crown. Middle-aged women may have generally thinning hair, but they are rarely bald, except as a side-effect of serious illness.

Baldness is inherited as a simple Mendelian trait. The allele B causes baldness. The allele b does not. However, the expression of B differs according to the sex of the inheritor. In men, B causes baldness. In women it does not. Women who are Bb never become bald. Women who are BB also never become bald, but they have very thin hair when middle-aged. On the other hand, men who are BB or Bb do become bald. Only bb men have normal hair. Figure 14–12 illustrates the inheritance of human baldness. It should help dispel the myth that a man who has a bald father will necessarily also be bald.

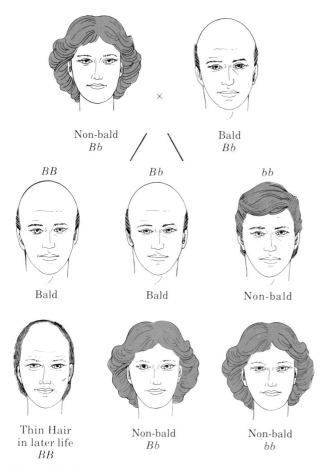

Figure 14–12

The inheritance of baldness, a sex-influenced characteristic. In men the allele B causes baldness. In women it does not. Men who are BB or Bb will go bald. Women who are Bb will never go bald. Even BB women will only have thin hair. If you are male and have a bald father, are you doomed to lose your hair?

Other characteristics seem to dominate one sex and not the other, but the details are unknown. For example, women's index fingers are about the same length as their ring fingers, whereas men's index fingers are shorter than their ring fingers. There seems to be no exception. Check your own fingers.

MULTIPLE-ALLELIC INHERITANCE

So far, most of the examples in this chapter have considered traits involving only two alleles at the same position on homologous chromosomes. The alleles

may be identical (homozygous) or different (heterozygous). Now, however, consider a diploid organism that has two alleles, A and B. A mutation changes B to C. If the resulting AC organism reproduces, the population will have three possible alleles—A, B, and C—although any one individual can possess only two of them. Inheritance controlled by more than two alleles is referred to as **multiple-allelic inheritance.** The inheritance of human blood types is an example.

Blood transfusions date back to the fifteenth century. From time to time, the transfusion of blood from other humans or animals became popular, but it was usually abandoned quickly. Generally, more patients were lost than saved, and most died of agonizing side-effects. Today, we know why transfusions were unsuccessful. Patients must receive blood that is chemically compatible with their own. Compatibility is based on blood type, a concept first suggested by Karl Landsteiner in 1900. He found that blood could be classified into several distinct groups. Today those groups are the blood types A, B, AB, and O.

The four blood types are related to the inheritance of specific **antigens**—usually large molecules of protein or polysaccharides. When an antigen enters the human bloodstream, the body's immune system produces **antibodies** to fight it. The antibodies are manufactured not only in response to invading antigens but also as a natural part of the body's defense mechanism. They are produced according to an inherited genetic code. The body also produces antigens according to a genetic code. The initial production of antibodies may not prevent measles or mumps, but from then on the antibodies are part of the immune system. If the antigen invades the body again, it will quickly be destroyed by the antibodies.

The typical antigen-antibody reaction causes a clumping of the antigens, thereby immobilizing them and preventing any toxic effects. That is exactly what happens when a human is given a blood transfusion with incompatible blood. The new red blood cells clump together and become immobilized. As a result, the patient goes into shock and often dies.

Two known antigens, A and B, are associated with red blood cells. These antigens are polysaccharides on the surface of the cells. Both are inherited. If you are blood type A, you have inherited the A antigen. Your body also produces antibodies against the B antigen, called anti-B (antibodies against B). If you are blood type B, you have inherited the B antigen, and your body has naturally (genetically) produced A antibodies (anti-A). If you are blood type AB, you have inherited both antigens. However, you cannot have any antibodies, because antibodies of either category would cause your red blood cells to clump together. Finally, if you are blood type O, you have inherited no antigens, but you produce both A and B antibodies (anti-A and anti-B) Table 14–1 helps illustrate the relationships among blood types, antigens, and antibodies.

What about blood transfusions? People with blood type O are called **universal donors,** because people of any other blood type can accept their blood. (Another type of blood antigen complicates the business of blood transfusion. A universal donor must not only be blood type O, but must also lack the Rh antigen.) Type O blood has no antigens, so A, B, and AB blood will not react against O blood cells. There is nothing for them to react with.

On the other hand, people with blood type AB are **universal recipients.** AB is a rare blood type, but individuals who have it can accept blood from any other person. They have no antibodies, so an antigen-antibody reaction is impossible.

The inheritance of blood type is by multiple alleles. The alleles A and B are both dominant. When both

Table 14–1

Major Human Blood Types, Antigens, and Antibodies

BLOOD TYPE	ANTIGENS ON RED BLOOD CELLS	ANTIBODIES IN SERUM
O	None	Both anti-A and anti-B
A	A	Anti-B
B	B	Anti-A
AB	Both A and B	None

are inherited, they are **codominant** (both expressed, as in blood type AB). The third allele, O, is recessive to both A and B. Therefore, if your phenotype is blood type A, your genotype can be either AA or AO. If your phenotype is O, your genotype can be only OO.

Blood can also be typed on the basis of other inherited antigens. For example, an antigen system known as the M and N system exists. M and N are considered weak antigens, and they elicit little, if any, antibody production. We seldom know whether we have M, N, or MN blood, because knowing is not important for blood transfusions.

POLYGENIC INHERITANCE

Many plant and animal characteristics are inherited not by single pairs of genes (alleles) but by several pairs of genes at different loci. The alleles may be located at separate sites on the same chromosome or even on separate chromosomes. Nevertheless, these multiple genes work together in the expression of a single characteristic. This type of inheritance is called **polygenic inheritance.** A few examples of characteristics controlled by many genes are human body weight, height, intelligence, skin color, and behavior patterns; plant height and fruit size; and milk production in cows and egg laying in chickens.

Any characteristic of a plant or animal species with continuous variation, where the trait is distributed in a typical bell curve, is likely to be controlled by polygenic inheritance. For example, if you plotted on a graph the height of the first thousand or so men that you found, the distribution of their heights would approximate a bell curve (see Figure 14–13). Most men

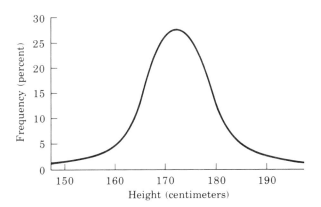

Figure 14–13
A bell curve that demonstrates continuous variation in height among human males. Characteristics that display continuous variation are often controlled by many genes. This kind of inheritance is called polygenic inheritance.

would be about 173 centimeters tall, fewer would be 170 and 175 centimeters, fewer yet would be 168 and 178 centimeters, and so on.

Skin Pigmentation

The color of human skin (pigmentation) is an example of continuous variation and a polygenic characteristic. At present there is no agreement on how many alleles at different loci are involved in pigmentation. Estimates range from two to six, but most authorities

believe that three or four adequately explain the continuous variation of human skin color.

Pigmentation is the product of special cells in the epidermis of the skin that manufacture a dark granular pigment called **melanin.** The amount of melanin produced by these cells is genetically coded. For example, assume that skin pigmentation is controlled by three sets of alleles at different loci, called B, C, and D. These alleles are codominant (all are dominant and expressed⊃). In an additive fashion, they increase the amount of melanin in the skin. The recessive counterparts of these alleles are b, c, and d. Recessiveness is also additive, but it does not code for any pigment. For example, a very dark black person may have the genotype BBCCDD, whereas a very fair white person may have the genotype bbccdd. The black person has six times more melanin than the white person. There are many possible intermediate genotypes.

How many genotypes are possible for three pairs of alleles at different loci? All possible arrangements of the three factors produce twenty-seven different genotypes (3 × 3 × 3 = 27) and seven phenotypes (you can compute this using a Punnett square). What if there were four genes involved? There would then be sixty-four genotypes and nine phenotypes. Either of these instances surely accounts for the many variations of skin color among humans.

Figure 14–14 illustrates the variation in human skin color that would occur if it were regulated by two sets of alleles instead of three. Five possible pheno-

Figure 14–14
The inheritance of human skin color. Most authorities accept the view that three or four pairs of alleles are involved in skin pigmentation. For simplicity, this illustration assumes only two pairs of alleles. The alleles are additive. The more dominant alleles there are, the more melanin is produced and the blacker the person is. If the dominant alleles were not additive, then BbCc and BBCC would be the same phenotype. But they are not. BBCC is twice as black as BbCc, because that person has twice as many dominant alleles.

PHENOTYPE:	Black	Dark	Intermediate	Light	White
GENOTYPE:	*BBCC*	*BBCc* *BbCC*	*BbCc* *BBcc* *bbCC*	*Bbcc* *bbCc*	*bbcc*

types would exist ranging from individuals with very dark skin to individuals with very light skin. Notice the variety of genotypes for each of the possible phenotypes.

GENE INTERACTIONS

Albinism

Some humans are albinos (meaning "white"). They lack any form of skin pigmentation, having extremely white skin and yellowish white hair and even lacking pigment in their eyes (see Figure 14–15). The lack of pigment allows the blood that bathes the capillary networks of the skin to show through, with the result that many parts of an albino's skin, including the eyes, may look pink. **Albinism** is not a fatal disorder. Albinos can live as long as anyone else. However, they often have eye trouble and frequently become blind. They sunburn easily and therefore must take precautions to remain out of the sun. They also have a high incidence of various forms of skin cancer because of their great sensitivity to ultraviolet radiation.

Albinos are unable to manufacture the pigment melanin because they lack an enzyme that converts the amino acid tyrosine to melanin. Albinism is inherited as a simple recessive trait (aa), but the amount of melanin actually produced in the skin is determined by another series of genes at different loci. Thus an individual who is BBCCDD for melanin could still be an albino if also aa, because no enzymes would be manufactured to promote melanin production. Most humans are AA and therefore manufacture the necessary enzymes for melanin production. Some are heterozygous (Aa) and carry the disorder.

The frequency of albinism varies. In Europe and the Soviet Union, about 1 out of every 100,000 is albino. In the United States, about 1 out of every 10,000 is albino. Worldwide, about 1 out of every 20,000 is albino. Albinism is most frequent among the Indians of the southwestern United States (specifically the Hopi, Jemes, and Zuni), where it affects about 1 out of every 200. It is estimated that 12 percent of the Hopis are carriers.

The albino Indians are admired by their tribes, and it is often considered good luck to have an albino around. Often, albino children are thought to be symbols of good deeds done in the past. Many women hope to have albino children. Although the number of albinos who marry is low, it is enough to keep the frequency of the recessive condition high in the population. The chances of intermarriage of carriers (Aa × Aa) is great. Carriers have a 25 percent

Figure 14–15

A sister and brother who are both albinos, demonstrating the inheritance of a recessive allele. Albinos lack an enzyme involved in the production of melanin, a normal skin pigment. (Courtesy A. M. Winchester, from *Heredity, Evolution, and Humankind*, West, 1976, p. 304.)

chance of conceiving albino children. Albinos who marry albinos have only albino children.

Pleiotropy

Many genes are known to have multiple, or pleiotropic, effects. Their primary effect can be thought of as a main effect, and other effects can be thought of as side-effects. For example, human albinos have no melanin pigment (main effect), but they also are nearly always nearsighted and have astigmatism (side-effects). The connection between skin pigmentation and eye structure is not known. It seems likely that the enzyme lacking in albinos plays some role in eye development.

One of the classic examples of the pleiotropic effects of genes occurs in *Drosophila*. White eyes are mutants of the normal red eyes, a trait that is sex-linked□. White-eyed males differ from red-eyed males not only in eye color but also in the shape of their reproductive organs.

The brevity of this discussion of **pleiotropy** is not meant to imply that pleiotropy is rare. In fact, it may be an extremely common phenomenon. Many of the human genetic disorders are pleiotropic, as Chapter 16 will show.

SUMMARY

1. Mendelian inheritance is inheritance of characteristics by simple dominant or recessive alleles. There are many examples of Mendelian inheritance among humans.

2. When genes occur on the same chromosome, they are said to be linked. The distance between linked genes can be experimentally determined.

3. Incomplete dominance is a type of inheritance in which neither allele in a heterozygous individual is dominant. The phenotype of the heterozygote is different from that of either of the homozygotes.

4. The sex of an organism is determined by the inheritance of sex chromosomes. In humans, XX codes for femaleness, and XY codes for maleness.

5. A human female (XX) has twice the number of X genes as a human male (XY), and the Y chromosomes contain little or no genetic information. To compensate for this imbalance, most females have one inactive X chromosome. Thus the number of active X genes is equal in males and females—a phenomenon called dosage compensation. The inactive X chromosome, which is often visible with a microscope, is called a Barr body.

6. Sex-linked traits are more frequently expressed by males. Females are frequently carriers who pass these traits to their sons. Hemophilia is an example.

7. Sex-limited characteristics are expressed only in males or in females, depending on the characteristic.

9. Multiple-allelic inheritance occurs when three or more alleles occupy the same position on homologous chromosomes. Blood group inheritance (A, B, AB, or O) is an example.

10. Many characteristics of organisms are determined by polygenic inheritance—that is, by several pairs of alleles. Skin color in humans is an example.

11. Some alleles affect or control the expression of other alleles, a phenomenon called gene interaction.

GENES AND CANCER

Is cancer inherited? Most forms are not. However, if someone in your family has had cancer, your chance of also having cancer is slightly higher than average. Therefore, there must be some relationship between cancer and heredity. (Your chance of having cancer also increases with age.)

There are a few kinds of inherited cancer. One rare form, which causes malignancies in the skin (xeroderma pigmentosum), is inherited as an autosomal recessive. Another type of cancer causes tumors in the retinas of eyes. It is inherited as a dominant trait. The malignant tumors of the retina begin during fetal development and generally are not detected until the child's vision becomes bad enough to require treatment—usually when the child is a year or two old. About 1 out of every 25,000 children inherits this form of cancer.

A commonly inherited nonmalignant tumor is the so-called polyp on the wall of the colon, which is inherited as a dominant trait. Various environmental conditions may cause polyps to become malignant.

There are other, extremely rare, forms of inherited cancer. Unfortunately, those who have inherited forms of cancer are more prone to develop other forms of cancer, such as leukemia and bone or brain cancer.

Still other forms of cancer are known to be caused by chromosomal anomalies. A chronic form of leukemia (myelogenous leukemia) is always associated with a deletion of part of chromosome 22 and the subsequent fusion of the remainder with chromosome 9. Another form of leukemia (nonlymphocytic leukemia) is always associated with the doubling of chromosome 1. Recent research has shown that mouse cells containing human chromosomes produce tumor-producing clones when exposed to a tumor-inducing virus, but only if the human chromosome 7 is present. Clones without chromosome 7 do not produce tumors, which suggests that certain virus-induced tumors in humans may depend on chromosome 7 for expression.

This evidence suggests a strong link between genes and cancer. The link is becoming more and more obvious. Cancer cells are normal cells that have been transformed. The transformed cells divide rapidly and lack the contact inhibition of normal cells. (Normal, nonmalignant cells do not divide when they come in contact with other cells.) The most popular and widely supported cancer theory suggests that all cancer cells come about by mutation of normal cells, an idea that proposes a direct relationship between cancer and DNA.

Many elements in the environment can cause mutations (changes in DNA). They are called mutagens. The list of known mutagens grows longer every day: air pollutants, herbicides, pesticides, chemical contaminates in food, cigarette smoke, various kinds of drugs, the charred proteins of a too-well-done steak or hamburger, radiation, and so on. Mutations may produce cell lines that are malignant or cancerous. In these cases, mutagens are called carcinogens. Many forms of cancer are probably environmentally produced, but induction is through the substance of inheritance, DNA.

Other evidence implicates genes in cancer. Three unusual genetic disorders—Fanconi's anemia, Bloom's syndrome, and Louis-Bar syndrome—are caused by inheritance of a simple Mendelian recessive allele. Those who have one of the diseases are homozygous recessive. They also have a higher than normal incidence of leukemia and other forms of cancer, and their chromosomes show a greater than average tendency to break. Heterozygotes carrying the recessive condition are more prone to malignancies than average but less so than those having the disorder.

Cancer is also associated with an individual susceptibility to the various forms of the disease. Many normal persons spontaneously produce cancer cells, but their immune systems quickly recognize these abnormal cells and destroy them. If the immune system becomes depressed for some reason, the cells may proliferate and eventually cause a malignancy. Such environmental factors as various drugs, physiological stress, and diet can depress the immune system. There is growing evidence that some forms of disease susceptibility are inherited.

Some viruses are known to induce cancers ranging from forms of breast cancer to papilloma (a wart or mole-like cancer that infects the skin or mucous membranes). Instead of killing cells, as many viruses do, these viruses transform cells, making them cancerous. The mechanisms are not clearly understood, but it seems that the invading viruses do not reproduce new viral particles or viral proteins. Instead, the invaded cell is transformed into a cell that lacks contact inhibition, and the resulting rapid division of the cell causes a malignancy. It is believed that the viral DNA becomes incorporated into the DNA of the host cell. The host cell then continues its own normal metabolism, but the viral DNA directs some protein synthesis. These viruses, called sym-

biotic viruses rather than virulent (causing the death of the host cell) viruses are implicated in many forms of cancer.

There is thought to be a strong link between environment-induced and virus-induced cancer. Both change or modify genetic messages. Current theory suggests yet another link. In some cases, symbiotic viral DNA becomes incorporated with host DNA but does not transform the host cell. The viral DNA may remain dormant for a long time, even years, until some environmental stimulus, such as a mutagenic or carcinogenic substance, induces its

activation. The result is cell transformation and eventual malignancy. The substances classified as mutagens or carcinogens may not cause direct transformation of cell lineages but may do just as much damage by stimulating symbiotic viruses.

We have made progress in determining the causes and treatment of cancer, but much remains to be learned. For instance, the connection between genes and cancer is still unclear. The next decade of research may make the relationship clearer. If so, new methods of treating cancer will result. Perhaps cures may even be found.

FOR DISCUSSION AND REVIEW

One of the best ways to familiarize yourself with the types of inheritance is to solve genetic problems. The following problems reflect the sequence of the text, so you may wish to reread sections of it as you attempt to solve the problems.

1. Normal-winged flies, carriers of the vestigial-wing trait, are crossed with homozygous recessive flies that have vestigial wings. How will the F_1 generation appear? What will be the phenotype ratio? The genotype ratio?

2. Flies homozygous for normal eyes are crossed with flies homozygous for black eyes. How will the F_1 generation appear? What are the phenotype and genotype ratios?

3. Flies heterozygous for normal wings (carriers of the vestigial trait) and homozygous for black eyes are crossed with flies heterozygous for both traits. What proportion of the F_1 flies will be homozygous recessive? Homozygous dominant? What proportion will have black eyes? Normal or red eyes?

4. You have attached earlobes, but both your parents have free earlobes. Is that possible? Explain.

5. You are a PTC taster, your brother is not, your father is, and your mother is not. What is your gen-

otype for this trait? (This is a pedigree analysis problem.)

6. You have free earlobes and are a nontaster. You marry someone who has free earlobes and is a taster. Will all of your children be tasters? Will they all have free earlobes? Can you expect to have a child with attached earlobes who is a taster? (Pedigree analysis again.)

7. Pink carnations are crossed with pink carnations. What color will the F_1 carnations be? What is the phenotype ratio? Explain.

8. Your parents have three boys and two girls. What are the chances that their next child will be a boy? A girl?

9. A man with hemophilia marries a woman who is normal (does not carry the defective gene). What are the chances that they will have an afflicted son? An afflicted daughter?

10. The female offspring in Question 9 will all be carriers. What implications does this have if they marry normal men?

11. A woman who suffers from red-green color-blindness marries a normal man. What are the chances that their daughters will be color-blind? Their sons?

12. A man has hairy ears. If he has daughters, they too will have hairy ears. Explain why this would be impossible.

13. A couple in their early twenties have the following alleles for baldness. The man is BB but shows no signs of balding, and the woman is bb. Will the man become bald? Will the woman have thin hair when she gets older? What are the chances that some of their sons will have normal hair and need never fear balding? Should their daughters be prepared to wear wigs? Explain.

14. In a recent paternity suit, a young woman (blood type A) had given birth to a child with blood type O. She proposed to the court that a young man with blood type AB was the child's father. Is that possible? What if the young man had blood type A? Type O? Type B?

15. A black man (BBCC) marries a woman with lighter skin (BbCc). The woman hopes that her sons will be as black as her handsome husband. Explain the possible skin colors that could result from this marriage. What is the possibility of a child as black as the man? What percentage of the children can be expected to be darker than the woman? Lighter?

SUGGESTED READINGS

GENETICS, 2nd ed., by Monroe W. Strickberger. Macmillan:1976.
Sophisticated and comprehensive coverage of genetics, intended for the more advanced student.

GENETICS: A SURVEY OF THE PRINCIPLES OF HEREDITY, by A. M. Winchester. Houghton Mifflin: 1977.
An easy to read and up-to-date discussion of the principles of genetics.

HUMAN GENETICS, 3rd ed., by A. M. Winchester. Merrill: 1979.
A very easy to read discussion of many aspects of genetics, including basic principles. All examples are related to humans.

PROBLEMS IN GENETICS WITH NOTES AND EXAMPLES, by D. Harrison. Addison-Wesley: 1970.
A good way to test your knowledge; some of the problems are difficult.

15

MOLECULAR GENETICS

It took more than a hundred years to fully explain the phenomena that Mendel discovered. Geneticists still study the expression of inheritable characteristics. However, modern genetics is based on study of the behavior of specific chemical molecules that are the material of heredity.

By the time Mendel's contributions were finally recognized, other scientists had discovered chromosomes and were studying their movement and distribution during mitosis▫ and meiosis▫. Of course, they soon found that the "factors" described by Mendel were located on the chromosomes. Chromosomes are now known to be composed of a substance known as **deoxyribonucleic acid** (**DNA**). The molecular arrangements of specific sections of DNA are the **genes.** They can be copied by an intermediating compound, a nucleic acid called **ribonucleic acid** (**RNA**). Ultimately, the RNA can be used to produce proteins. The whole process is summarized as

DNA→RNA→Protein
(gene)

Thus the instructions for the synthesis of proteins are coded in the DNA molecule.

DNA

You do not have to be a biochemist to understand the molecular structure of DNA. In fact, you do not need any appreciable knowledge of chemistry. So the following discussion need not be intimidating. If you read it carefully and patiently, you should be able to comprehend this complicated topic.

PREVIEW QUESTIONS

After reading this chapter, you should be able to answer the following questions:

1. What is DNA, and what are some of its unique properties?

2. What is the relationship between DNA and RNA?

3. How do cells synthesize proteins from amino acids?

4. What is a mutation?

5. What is genetic engineering, and how might it affect your future?

DNA is composed of a double string of many smaller units, known as nucleotides, just as proteins are composed of a series of amino acids and complex carbohydrates are composed of strings of simple sugars. Each nucleotide has three parts: a five-carbon sugar molecule, a phosphate group, and a nitrogenous (nitrogen-containing) base.

The sugar portion of the DNA molecule is deoxyribose sugar. It is called a pentose sugar because it has five (pent) carbon atoms. Figure 15–1 illustrates deoxyribose and ribose sugars, which resemble crowns or short pyramids and should be thought of as three-dimensional.

The deoxyribose sugar is attached to a phosphate group. Biochemists often number the carbon atoms in various compounds. Carbon atoms 3 and 5 of the deoxyribose sugar molecule are located on the same side. A phosphate group is attached to each of these carbons (see Figure 15–2). The phosphate attached to carbon 5 projects upward and chemically bonds with carbon 3 of the deoxyribose sugar in the nucleotide directly above. The phosphate attached to carbon 3 projects downward and chemically bonds with carbon 5 of the deoxyribose sugar of the nucleotide immediately below.

Carbon 1 of the deoxyribose sugar is attached to a **nitrogenous base.** In the case of DNA, there are only four possible nitrogenous bases (see Figure 15–3): adenine (A), guanine (G), thymine (T), and cytosine (C). **Adenine** and **guanine** are double-ringed structures called **purines. Thymine** and **cytosine** are single-ringed structures called **pyrimidines.** The combination of the sugar molecule, phosphate group, and nitrogenous base completes the basic structure of a **deoxyribonucleotide.**

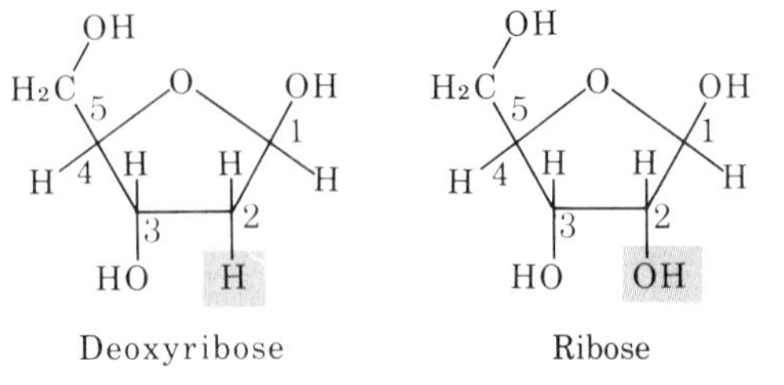

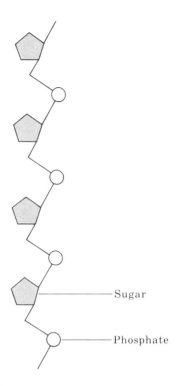

Figure 15–2
The attachment of phosphate to deoxyribose sugar molecules.

Figure 15–3
The pyrimidines, cytosine and thymine, and the purines, adenine and guanine. These compounds are the nitrogenous bases of DNA molecules.

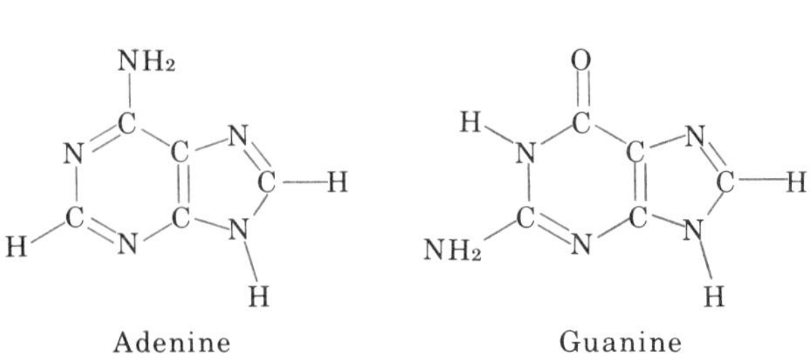

Cytosine

Thymine

Adenine

Guanine

Figure 15–1
The molecular structure of deoxyribose sugar and ribose sugar, both five-carbon sugars. They differ only in that ribose sugar possesses one more oxygen atom (shown as a shaded OH) than deoxyribose sugar has.

Deoxyribose

Ribose

Imagine that the DNA molecule resembles a staircase. Alternating sugar molecules (S) and phosphate groups (P) form the banisters of the staircase. The actual molecular arrangement of the sugar molecules and phosphate groups can be shown two-dimensionally:

```
    P
     \
      S
     /
    P
     \
      S
     /
    P
```

Figure 15–4 illustrates the same configuration but incorporates the actual molecular structure of both the sugar and the phosphate groups.

A nitrogenous base, either a purine or a pyrimidine, is attached to the inside of each sugar molecule. The nitrogenous bases can be visualized as the steps of the staircase. Their arrangement can be represented as

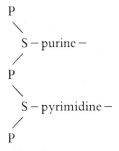

```
    P
     \
      S — purine —
     /
    P
     \
      S — pyrimidine —
     /
    P
```

To complete the DNA molecule, a complementary strand is added, because DNA is a double nucleotide chain. The other banister is also composed of alternating sugars and phosphate groups. Here too, either a purine or a pyrimidine projects from each sugar. The steps throughout the staircase are of the same

Figure 15–4
One strand of DNA, showing the relationships among deoxyribose sugar (S), phosphate (P), and nitrogenous bases (B).

The backbone

The base

Phosphate + sugar + base = nucleotide

There are four bases:

A = Adenine

C = Cytosine

T = Thymine

G = Guanine

S = Sugar

P = Phosphate

length (see Figure 15–5). A purine is always linked (by hydrogen bonds□) to a pyrimidine, because any other combination would produce steps of different width. Two pyrimidines would be much shorter than two purines. Actually, the molecular requirements are even more precise. *Adenine is always paired with thymine. Guanine is always paired with cytosine.* The structure of DNA is something like this:

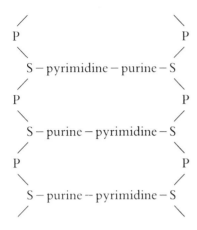

Figure 15–5
The ladderlike structure of DNA.

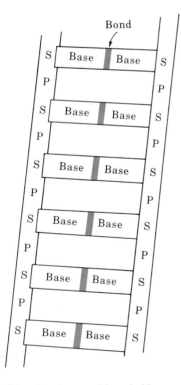

DNA molecule resembles a ladder

Substituting thymine or cytosine for the pyrimidines and adenine or guanine for the purines produces

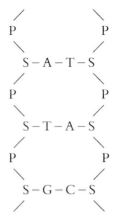

(Compare with Figure 15–6, which shows the actual molecular structure.) For the final configuration of DNA, the banisters are twisted so they cross each other, forming a double spiral. The DNA spiral is often called a **double helix,** or spiral staircase. Figure 15–7 shows the spiral structure of DNA and the location of the sugars, phosphate groups, and nitrogenous bases.

It should be obvious now that when one half of the DNA molecule is known, the other half can be predicted without error. Each half is the complement of the other. More precisely (and borrowing terminology from photography), one half represents a negative, the other a positive. For example, if the bases of one half of the molecule are

A-
T-
C-
G-
T-

then the bases of the other half will have to be

A-T
T-A
C-G
G-C
T-A

Francis Crick and James Watson predicted the detailed structure of DNA described here. Their prediction became known as the Watson-Crick model. In

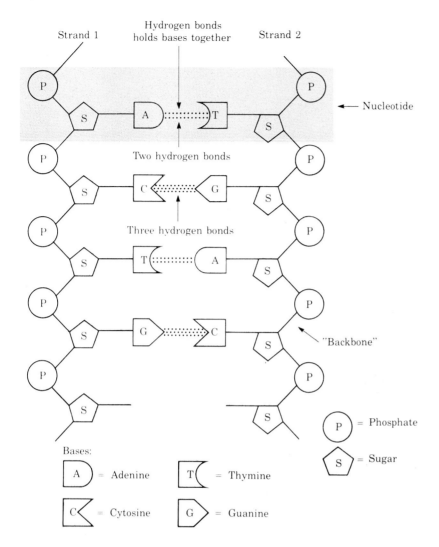

Strand 1 — Hydrogen bonds holds bases together — **Strand 2**

← Nucleotide

Two hydrogen bonds

Three hydrogen bonds

"Backbone"

(P) = Phosphate

(S) = Sugar

Bases:

A) = Adenine   T(= Thymine

C< = Cytosine   G> = Guanine

Figure 15–6
The complete DNA molecule, showing both
nucleotide strands. One strand is
complementary to the other.

1962, they shared the Nobel Prize with Maurice Wilkins for their discovery.

Replication

The Watson-Crick model predicted not only the structure of DNA but also that this remarkable molecule must be able to reproduce itself, a process referred to as **replication.**

Watson and Crick proposed that the parent DNA molecule essentially unzips during replication. This notion has since been confirmed in a variety of organisms. The hydrogen bonds between the nitrogenous bases are broken. As a result, the double helix begins to unzip and unwind. The unzipping creates two separate parent strands of DNA. Each becomes the mold, or **template,** for the production of a daughter strand.

The unzipping exposes the chemical bonds on the purines and pyrimidines that become templates. The nucleoplasm is a vast reservoir of free nucleotides from which each A on the parent strands is matched

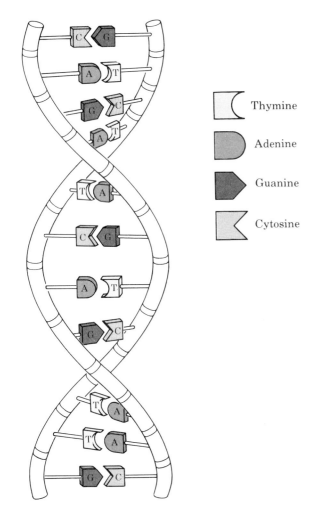

Figure 15-7
The DNA molecule twisted into a spiral called a double helix.

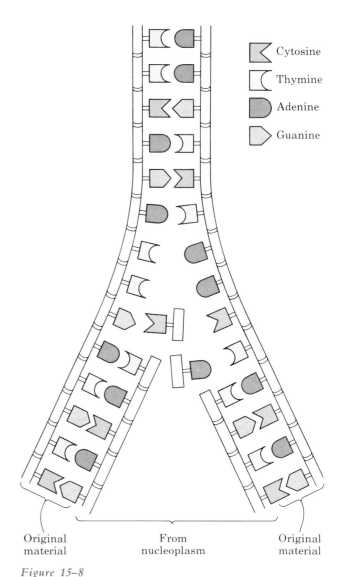

Original material | From nucleoplasm | Original material

Figure 15-8
Replication of a DNA molecule. The double helix uncoils and then unzips, separating base pairs. Complementary base pairs are attracted from the nucleoplasm and attach themselves to the exposed nucleotides. When replication is completed, each daughter strand is composed of one strand of the parent DNA and one new strand.

with a T nucleotide, each C is matched with a G nucleotide, and so on (see Figure 15–8). The enzyme DNA polymerase helps the nucleotides link up. The process of pairing bases during replication is called **complementary bonding.**

After each daughter strand bonds to a parent strand, the molecules twist again into a double helix. When the process is completed, two strands of DNA have been formed. Each is identical to the other and to the original parent molecule.

DNA replication occurs in cells in interphase, during the S (synthesis) period between the two growth periods, G_1 and G_2.

DNA and Chromosomes

When cells are not in the process of cell division the chromosomes are not tightly coiled and are too long and thin to be seen except with an electron microscope. In electron photomicrographs chromosome strands look like strings of beads. The bead-like structures, called nucleosomes, are not in contact with each other, although they are connected. Each chromosome strand is composed of a long continuous molecule of DNA which is associated with a variety of proteins, including histones. The DNA molecule extends through the nucleosomes and the strands connecting them. Each nucleosome is a concentration of histones with a length of DNA wound about it. A far grater number of DNA base pairs are present in nucleosomes than in the sections of DNA connecting them.

RNA

Another compound in living cells is ribonucleic acid (RNA). The structure of RNA is similar to that of DNA. RNA is composed of a string of **ribonucleotides,** each of which is composed of a five-carbon sugar molecule, a phosphate group, and a nitrogenous base.

The pentose sugar of each nucleotide in RNA is **ribose sugar** (see Fgure 15–1). Ribose sugar differs from deoxyribose sugar in having one additional atom of oxygen. The phosphate group of a ribonucleotide is exactly the same as that for a deoxyribonucleotide. And like DNA, one of four nitrogenous bases, two pyrimidines and two purines complete each nucleotide of RNA. The purines of RNA are adenine and guanine. The pyrimidines, however, are cytosine and **uracil** (U). Unlike DNA, RNA exists in living cells usually as a single strand rather than as a double strand.

There are three kinds of RNA. They differ primarily in their molecular size. The larger the molecule, the greater the number of ribonucleotides it includes. The different forms of RNA play unique and vital roles in protein synthesis.

The smallest RNA is called **transfer-RNA,** abbreviated **tRNA.** There is a specific class of tRNA for each of the twenty amino acids involved in protein synthesis. Each tRNA attaches to a specific amino acid.

Another RNA is **ribosomal-RNA,** abbreviated **rRNA.** The rRNA molecule is larger than tRNA and therefore contains many more ribonucleotides. It composes the **ribosomes** of the cytoplasm. Ribosomes are not pure rRNA. They are also composed of associated proteins.

The last type of RNA is the largest—**messenger-RNA,** abbreviated **mRNA.** As its name implies, mRNA carries the genetic message from the DNA in the chromosomes to the sites of protein synthesis, the ribosomes. The details of this process are discussed in the remainder of the chapter.

PROTEIN SYNTHESIS

A great variety of **proteins** exist in every organism. They are required for growth and replacement of cells; for building enzymes, hormones, antibodies, and other important elements; and for numerous other functions that maintain the organism in working order. All proteins are basically chains of **amino acids.** They figure prominently in any discussion of molecular genetics.

The Genetic Code

Francis Crick proposed that the sequence of nucleotides within the DNA molecule constituted what had previously been referred to as "genes" and that the sequence was in fact a code, not unlike Morse code. The code, according to Crick, dictated the sequence of amino acids in a specific protein.

DNA is composed of four types of nucleotides, which differ only in the types of nitrogenous bases they contain. The bases are adenine, thymine, cytosine, and guanine. Sequences of these nucleotides can code for hundreds of thousands of specific proteins. How is this coding accomplished?

First of all, there are approximately twenty known amino acids. The hundreds of thousands of proteins differ only in the number, kind, and sequence of these amino acids. Coding, then, is much simpler than it might first appear. However, if each nucleotide coded only one specific amino acid, then all proteins could contain only four kinds of amino acids—obviously not enough to account for the known variety of proteins.

What if a code word is just two nucleotides in the DNA molecule? This will code for only sixteen different amino acids (four nucleotides, two at a time, or $4^2 = 16$), which is still not enough. The next possibility is for the code to include three nucleotides ($4^3 = 64$). However, sixty-four possible code words is more than enough. Nevertheless, it was postulated early in the search for the genetic code that three nucleotides would code for a specific amino acid. Later research has verified this early supposition. The three nucleotides in a genetic code word (such as GAC and TGT) are referred to as a **codon.**

If there are sixty-four code words but only twenty amino acids, then there can be several different code words for the same amino acid. This has been shown to be true. These repeated codes, universal among living organisms, are called redundancies (unnecessary repetitions). For example, UUA, UUG, CUU, CUA, and CUG all code for the amino acid leucine. Some codons do not code for any amino acids. UAA, UAG, and UGG signal the end of a message ("stop"), the end of a protein, or the end of a gene.

A gene appears to be a series of nucleotides that code for a specific amino acid sequence. Some proteins are large and complex sequences of amino acids, and the gene for each of them is equally long and complex. Other proteins are simpler, shorter sequences of amino acids, and therefore their genes are shorter.

Transcription

DNA, with its genetic code, is located in the nucleus of cells within chromatin and chromosomes. The protein-synthesizing machinery of each cell is located in the cytoplasm. The information on DNA must therefore be transported to the cytoplasm. This occurs when the nucleotide sequence along the DNA molecule is copied into a strand of mRNA, which then communicates directly with the protein-synthesizing machinery of the cytoplasm. Each code word or codon is copied into mRNA by a process called transcription.

DNA is the template from which mRNA is manufactured (see Figure 15–9) with assistance from the enzyme RNA polymerase. Only one strand of the DNA molecule acts as a template. It is believed that one strand of the pair always serves this purpose. The opposite strand is complementary and therefore has a series of different bases. It also has a genetic message different from that of the other strand.

For transcription, the DNA molecule must be unzipped to expose the template strand. Complementary ribonucleotides, which abound in the nucleoplasm, are attached to the exposed bases of the DNA molecule. (RNA does not possess the base thymine but substitutes uracil for it.) If one nucleotide sequence on the DNA molecules is AAA, the complementary codon on a strand of mRNA will be UUU.

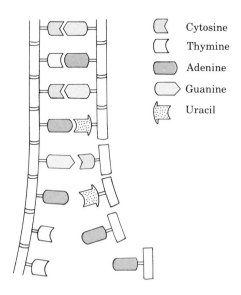

	Cytosine
	Thymine
	Adenine
	Guanine
	Uracil

Figure 15–9

Transcription of the genetic code from a DNA molecule to a strand of mRNA. mRNA does not contain the base thymine but substitutes uracil for it.

TAT will transcribe as AUA, GCG as CGC, and so on. Then the mRNA, with a faithful reverse copy of the genetic code, separates from the DNA template and passes through minute pores in the nuclear membrane and into the cytoplasm.

Translation

The process of joining amino acids in sequence to form a polypeptide (a small protein) from the information on an mRNA molecule is known as **translation.** First an amino acid must become attached to a specific tRNA. Then the various amino acids must be linked together into a chain. (Each of the amino acids was previously attached to a specific tRNA.) The second (and final) step occurs on ribosomes.

It is now known that tRNA is manufactured on the DNA template, much as mRNA is. tRNA, the smallest kind of RNA, has a complex molecular structure. It consists of several loops and an open end (the receptor arm) to which a specific amino acid can be attached (see Figure 15–10). The two-dimensional structure of tRNA looks like an aerial view of a cloverleaf on an interstate highway. For that reason, its molecular configuration is referred to as a cloverleaf.

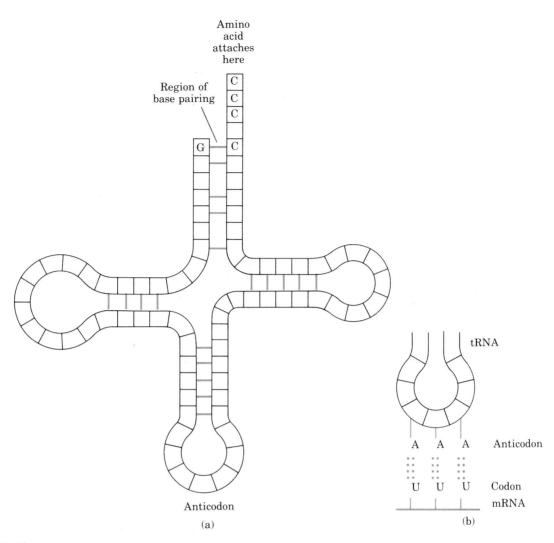

Amino
acid
attaches
here

Region of
base pairing

tRNA

A   A   A   Anticodon

U   U   U   Codon

mRNA

Anticodon

(a)

(b)

Figure 15–10

The cloverleaf molecular configuration of tRNA. (a) The open end is the end that attaches to free amino acids. (b) The anticodon loop attaches to a codon on mRNA.

There are as many different tRNAs as there are amino acids, because each tRNA is specific for a particular amino acid. One of the loops on a tRNA cloverleaf is known as the anticodon loop (see Figure 15–10). A nucleotide sequence on this loop (a triplet code) chemically binds to a specific codon on a molecule of mRNA. This is why there is a different tRNA for each amino acid: One end of the tRNA can read only one codon on mRNA, and that codon codes for a specific amino acid. In Figure 15–10, the anticodon loop of the tRNA is AAA, which will attach to mRNA each time the codon on that molecule is UUU. (The original nucleotide sequence on the

DNA molecule, then, was AAA. Do you see why it has to be so?)

The participants in protein synthesis are the mRNA, which contains codons for coding the amino acid sequence for a polypeptide or protein; the tRNA; and the ribosomes in the cytoplasm which are composed of rRNA.

When mRNA has copied the genetic information of a gene from the DNA molecule, it diffuses into the cytoplasm (see Figure 15–11). Once in the cytoplasm, the molecule of mRNA becomes chemically bonded to a ribosome. The site of bonding is exactly two codons wide because each ribosome is large enough to

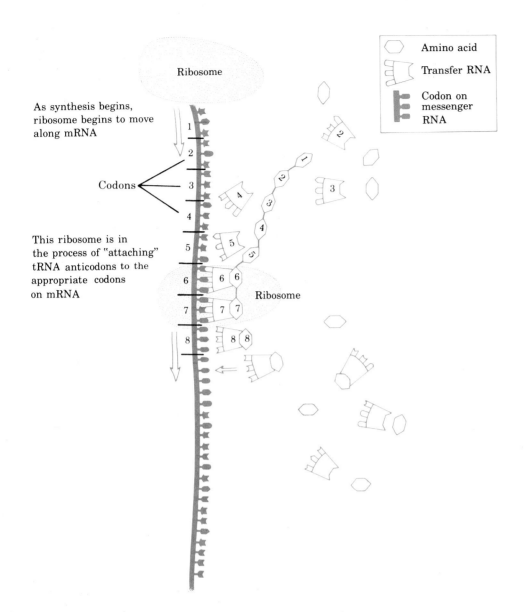

Figure 15–11
Translation of the genetic code into strings of amino acids and finally into polypeptides and proteins.

bond only that distance. In Figure 15–11, codons 6 and 7 are bonded to mRNA. While reading the information on the mRNA molecule, the ribosome moves from codon to codon. For example, in Figure 15–11, just a short time before, codons 5 and 6 were chemically bonded with mRNA. These codons coded for the tRNAs 5 and 6. The ribosome then moved and now is chemically bonded to codons 6 and 7, which code for tRNAs 6 and 7. The next move would include codons 7 and 8, and so on. The tRNA mole-

cules are attracted only to the codon being read, and they chemically bond to the codon that is complementary to the nucleotide sequence in the anticodon loop. (Figure 15–11 shows the tRNAs with the same number as the codons to avoid confusion.)

Each tRNA has attached to it a specific amino acid. As the tRNA arrives at the codon being read, it carries that amino acid with it. (Again, Figure 15–11 uses the same numbers for amino acids as it did for the tRNAs and codons.) In our example, as tRNA 5 moved away from codon 5, its amino acid was chemically bonded to the the existing chain 1–2–3–4–, making the chain 1–2–3–4–5. As tRNA 6 leaves codon 6, amino acid 6 will be added to the chain, making the chain 1–2–3–4–5–6. The process will continue until a polypeptide is formed. The synthesis stops when a stop codon is reached. The mRNA, ribosome, and polypeptide separate. The mRNA disintegrates or is read again by another ribosome. The ribosome may read other mRNA molecules. The polypeptide is freed into the cytoplasm of the cell.

An analogy should explain protein synthesis more clearly. Suppose we organize a human bucket brigade to act out the process of protein synthesis. We need twenty student volunteers, each one representing a different tRNA molecule. We also need twenty buckets, each one a slightly different color from any of the others. Each color represents a specific amino acid. Each student is given a bucket of a specific color. This illustrates the specificity of tRNA for a particular amino acid.

A long roll of paper is now unrolled in the hallway. The paper was previously marked off into sections one meter long. Each section is a make-believe codon. Each codon has the name of one of the twenty students written on it. This illustrates the codon-anticodon association of a codon for a specific tRNA. As the paper is unrolled, each codon displays a student's name. The student steps forward, stands on the codon, and places the bucket to the side of the codon. The student named on the next codon does likewise, and so on. In a while, we have a chain of buckets of different colors, which represents the polypeptide. We could make this analogy even more complete by having other students tie the handles of the buckets together with string, which would represent the chemical bonds in the polypeptide chain sequence.

This discussion describes a single ribosome reading a single molecule of mRNA and producing a single polypeptide. In some cells, however, the mRNA may be read by several ribosomes simultaneously, each translating at a different point along the mRNA. Such complexes, called *polysomes,* produce several identical polypeptides, one after the other.

Protein synthesis is complicated, requiring several enzyme-dependent systems. The description here is an oversimplification of the process. Because of its complexity, it may seem that protein synthesis would be slow. But that is not the case. Hemoglobin, the red blood pigment, is manufactured by cells that are precursors to red blood cells. Hemoglobin is a complex protein. It consists of four polypeptide chains, each having about 140 amino acids. Nevertheless, the cells that manufacture hemoglobin can make a molecule in about three minutes.

The speed of protein synthesis is less important than its accuracy. A mistake can be fatal to the organism. Therefore, the system must be nearly perfect. Perfection and accuracy are achieved by the nucleotide pairing between each tRNA (carrying a specific amino acid) and mRNA held chemically by the rRNA.

CONCEPT SUMMARY

EVENTS OF PROTEIN SYNTHESIS

1. Transcription. The nucleotide sequence of DNA is copied as a strand of mRNA.
2. mRNA leaves the nucleus and becomes chemically bonded to a ribosome.
3. Meanwhile, tRNA molecules are bonding with specific amino acids.
4. Translation. tRNA molecules with their attached amino acids align themselves two at a time along the mRNA by matching their anticodons with complementary mRNA nucleotide sequences. The amino acids are thereby arranged in their correct sequence.
5. As the amino acids are moved into the correct sequence, they are chemically bonded together.

The One-Gene, One-Enzyme Principle

A gene is an information-carrying sequence of codons on the DNA template. Genes first become transcribed by **mRNA** and then are translated to a polypeptide chain at the ribosomes in the cytoplasm. Research has shown that most proteins, including enzymes, are products of genes. Thus the structural proteins that maintain the integrity of cells, muscles, bones, and all other body components are the expression of genes.

During the past decade, researchers have proved the universality of the **one-gene, one-enzyme** (polypeptide) **principle.** The principle was originally proposed by George W. Beadle and Edward Tatum, who shared the Nobel Prize in 1958 for discovering this relationship.

Most molecular biologists now view the one-gene, one-enzyme principle in a broader sense. They speak of the one-gene, one-polypeptide principle, because all proteins, as well as enzymes, are polypeptides. Furthermore, the sequence of codons that specifies a particular polypeptide on the DNA molecule is now called a **cistron** rather than a gene.

Mutations

Even a minor change in the sequences of the nitrogenous bases in DNA will change the genetic message. Any alteration in the genetic message is called a **mutation.** Some changes are more serious than others. For example, an **enzyme**□ has an active site and inactive sites. If the change in the amino acid sequence involves only the inactive sites, it will probably have no significant effect on the organism. However, if it involves an active site, the enzyme may become only half as active or even totally incapable of any activity. Either situation could have far-reaching consequences for the organism.

Sometimes, as a result of a mutation, a whole segment of a chromosome, including many genes, becomes altered. For example, a segment may be lost during replication (called a deletion), or it may become repeated or duplicated. A portion of one chromosome may become transferred to another, a process known as a translocation. Sometimes, a chromosome segment is turned end over end, an error called an inversion. All these chromosome alterations change the genetic code and produce mutations.

The same phenomena that change the structure of chromosomes may occur within a gene on a chromosome. The changes are called gene mutations or point mutations. If a base pair is deleted or inserted, the entire gene may be misread. Changes in reading are referred to as frame-shift mutations. For example, suppose that a segment of DNA has the code CAT/ TAG/TAC. What will happen to the code if the A is deleted in the first codon? The deletion will cause a frame shift, and the whole gene will be misread, becoming CTT/AGT/AC__.

Substitution of a new base pair, which would cause a mistake in reading, is called a base-pair substitution. For example, if the sequence of bases being read from the DNA is ACC, then transcription will yield UGG. However, a base-pair substitution causing ACC to become ATC will yield a different codon, UAG, instead of UGG. UGG codes for the amino acid tryptophan, but UAG translates as a "stop" codon. Thus, no matter which codons follow, translation will stop

at this point, and only a short chain of amino acids will be produced. Geneticists have called this kind of base-pair substitution a nonsense mutation.

What if the substitution had occurred elsewhere? Consider the possibility of a base-pair substitution that changes the DNA from ACC to ACA. This will transcribe on mRNA as UGU, which translates as cysteine. Mutations causing amino acid substitutions are called missense mutations. Other base-pair substitutions may result in no change in the amino acid sequence. They are an example of the duplication (redundancy) of codons for the same amino acids.

How often do mutations occur? Natural, or spontaneous, mutations are reasonably rare. It is estimated that they occur at the rate of about once per person per generation. This means that you probably have at least one gene different from those you inherited from your parents.

What causes mutations? Spontaneous mutations are thought to be the result of simple chemical errors during DNA replication. Others may be induced by radiation of various types. Ultraviolet radiation in sunlight is one suspect.

There is growing concern that chemical agents in the environment may be increasing normal mutation rates. Many chemicals are suspect. They range from alcohol to caffeine, automobile exhaust to cigarette smoke, marijuana to nicotine, LSD to aspirin, and food preservatives and additives to air and water pollutants. Even the hamburger is suspected of harboring **mutagens** (mutation-inducing chemicals) and **carcinogens** (cancer-inducing chemicals).

Sickle-Cell Anemia

Many coding defects are inherited. One of the best-known examples of a human point mutation involves **hemoglobin,** which is a large protein and the red pigment carried by red blood cells. Normal hemoglobin, type A, is coded by the gene Hb^A. There are many kinds of abnormal hemoglobin. One of the best known is hemoglobin S, which is coded by the gene Hb^S. People who are Hb^{S}Hb^S have a disorder known as sickle-cell anemia. Whereas normal red blood cells are disk-shaped, the red blood cells of these individuals are curved into a sickle shape (see Figure 15–12). The sickle-shaped cells rupture frequently. The ruptures destroy the oxygen-carrying capacity of the cells. In addition, the sickle-shaped cells often become stuck sideways in small arterioles, blocking the usual

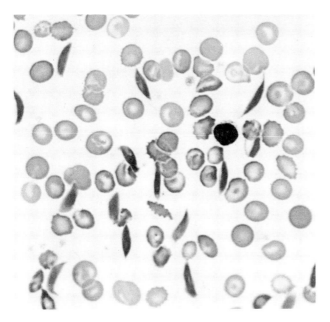

Figure 15–12
The blood cells of someone with sickle-cell anemia. Normal red blood cells are disk-shaped and contain normal hemoglobin, Hb^A. The red blood cells of people with sickle-cell anemia are sickle-shaped because they contain abnormal hemoglobin, Hb^S. Carriers of the sickle-cell trait can be detected because they have some sickled red blood cells. (Courtesy Carolina Biological Supply Company.)

flow of blood cells to tissues and causing irreversible damage because of oxygen depletion. Individuals who are Hb^{S}Hb^S usually die prematurely.

People who have the genotype Hb^{A}Gb^S are heterozygous for the sickle-cell trait. They have some normal red blood cells and some sickle-shaped cells. These individuals have a greater resistance to malaria than do people with normal cells. In tropical regions of the world, where malaria is prevalent, this genotype confers an advantage. The majority of heterozygotes are blacks whose ancestors lived in the tropics. These individuals suffer some muscle pain and shortness of breath, particularly at higher elevations, but otherwise may never know of their condition. Volunteer screening centers have been set up in many cities to detect sickle-cell carriers in the hope of reducing the frequency of Hb^{S}Hb^S offspring.

The difference between Hb^A and Hb^S is a single amino acid changed by a base-pair substitution. Hemoglobin consists of four polypeptide chains. Two of these chains are called alpha chains. The other two are

called beta chains. Alpha chains have 141 amino acids each. Beta chains have 146 amino acids each. One of the normal beta chains of Hb^A possesses the amino acid glutamic acid at position 6. Hb^S possesses the amino acid valine at position 6. The difference is the effect of a substitution in base pairs. How this seemingly minor change causes such dramatic expression in the phenotype is currently unknown. It is known, however, that the mutation from Hb^A to Hb^S is a recurrent one. That is, it has occurred in many individuals at many different times. Another type of sickle-cell anemia is described in Box 15–1.

Genetic Engineering

Since 1973, molecular biologists have been manipulating genes in test tubes. They are now able to create new breeds of DNA. These breeds are commonly referred to as **recombinant DNA,** which means that existing DNA is recombined in new or different ways. This area of study is one of the most exciting in molecular genetics today. Its application is called **genetic engineering.** Scientists are capable of adding genes to or deleting genes from different organisms. In fact, entirely new organisms may be engineered. Research in this area has stirred considerable controversy, and its opponents have made some frightening predictions such as the accidental creation of new dangerous disease organisms.

The techniques of recombining DNA are interesting and not all that difficult (see Figure 15–13). The organism most often used is the bacterium□ *Escherichia coli,* but other bacteria have been used as well. Bacteria have extra DNA in each cell, and it is not connected to the chromosome. The DNA occurs in rings called plasmids. With special laboratory techniques, the plasmids of bacterial cells are separated from the cells. Then the rings of DNA are cut open. Next the DNA sequence of a foreign gene is mixed with the opened plasmids. The donor DNA can be a real gene from another organism, or it can be artificially prepared with the proper nucleotides.

Plasmids that are opened will eventually heal closed again. Some will attach the foreign gene onto the already existing DNA. Many will heal without incorporating the foreign DNA. The plasmids that incorporate the foreign DNA are heavier because of it, and, again, they can be separated out by simple laboratory procedures. (Recombining DNA is also called

ANOTHER TYPE OF SICKLE-CELL ANEMIA?

Researchers have recently described a new form of sickle-cell anemia. Most sufferers have "sickling crises," a period when many of the red blood cells assume odd shapes and clog blood vessels. The resulting pain and tissue damage may become so intense that the sufferer dies prematurely. However, this reaction is not true of all who suffer from sickle-cell anemia. Researchers have identified individuals aged twenty-three and twenty-five who have sickle-cell anemia (Hb^SHb^S) but who have never had the usual symptoms of the disease.

The reason some victims of sickle-cell anemia have more severe reactions and symptoms than others is currently being studied. Researchers have shown that individuals with severe symptoms have deficient structural proteins in the membranes of their red blood cells. Sufferers with mild symptoms have fewer membrane anomalies.

Thus sickle-cell anemia may not be entirely the result of a single base-pair substitution in the DNA code for hemoglobin molecules. Genetic errors involving the proteins of red blood cell membranes may make sickle-cell anemia more severe.

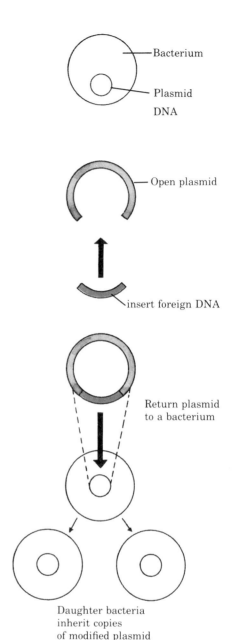

gene splicing. You can see why.) The cell walls of other bacteria are dissolved. When these bacteria are cultured with the plasmids that have the new genes, the bacteria eventually incorporate the plasmids. Bacteria reproduce rapidly, so it does not take long to create a whole new race of bacterial cells, called a clone. Each clone has a new DNA code—a gene.

What are the implications of recombinant DNA? Most are positive. For example, the gene that codes for insulin has been removed from a human pancreas cell and incorporated into a plasmid, and a whole new clone of bacteria is now rapidly and cheaply producing human insulin. Prior to recombinant DNA, all insulin was extracted from the pancreas glands of livestock when the animals were slaughtered. This source of insulin was expensive, and many people developed allergic reactions to it because of its impurities. Synthetic insulin, taken as tablets, helped ease this problem somewhat. Unfortunately, however, the insulin does not work for everyone.

Figure 15–13

Recombinant DNA. Bacterial plasmid DNA is removed from a bacterium and DNA from another source is incorporated into it forming recombinant DNA. The modified plasmid is next placed in another bacterium thus altering its genetic composition. The descendants of the modified bacterium will all contain the copies of the modified DNA.

Another compound being made by bacterial cells is the pituitary growth hormone. This hormone is important in the treatment of children who do not grow and who would become midgets without the hormone. Such children produce too little or none of the hormone themselves. Before recombinant DNA, the only growth hormone available was that extracted from the pituitary glands of human corpses.

The limits to recombinant DNA can hardly be imagined. There is a great deal of excitement about it.

Recombinant DNA also has economic ramifications. Companies have been established to clone bacteria to manufacture many products. Indeed, an entire new industry is beginning. Both scientists and financiers want to invest in this industry. Can people buy and sell bacteria? Can they copy a laboratory's clone? The U.S. Court of Patent Appeals recently ruled that new bacterial strains can be patented, which gives some protection to their creators.

Some aspects of recombinant DNA are less than positive. What if too many bacteria are given the gift of fixing nitrogen⁼? How are we going to limit their numbers? There is the possibility of too much of a good thing. Plants can actually be killed by too many nitrites.

Recombinant DNA could be very important in the study and prevention of disease. But what might happen if some of the engineered disease-causing organisms ever escaped? The implications are obvious. The probability of escape might be remote, but similar things have happened before. The last person on earth to die of smallpox got the disease from a virus that escaped from a laboratory.

Because of the escape problem, many people are opposed to recombinant DNA research. Actually, the problem is not so much the research as the organism used in it: *E.coli,* an extremely common bacterium that exists in the intestinal tracts of all humans and many other animals as well. If one of the recombinant strains ever did get away, it might run rampant and never be controlled. It could spread whatever new genetic messages scientists had given it, good or bad, around the globe. This is why the enthusiasm of some scientists for continued work with recombinant DNA has been curbed.

However, science has solved one of its own problems. If *E. coli* is so common and can exist nearly everywhere, why not create a bacterium that has a different life-style? This has now been done. A new,

weaker strain of *E.coli* that was made by adding and deleting various genes is now being used. This strain can be killed by ultraviolet light or detergent. Furthermore, it must have the pyrimidine thymine in its culture medium in order to live. Even if turned loose, it cannot survive outside the laboratory. Normal laboratory hygiene will also eliminate any strays. Outside the laboratory, ultraviolet light from the sun will destroy any survivors. The new strain has lessened some of the fears of recombinant DNA research. However, bacteria mutate frequently, and any natural mutation might allow the bacterium to again survive in the world outside the laboratory.

SUMMARY

1. Deoxyribonucleic acid (DNA) is composed of nucleotides, each containing a five-carbon sugar molecule, a phosphate group, and a nitrogenous base. The five-carbon sugar molecule of DNA is deoxyribose sugar, and the phosphate group is always the same. The nitrogenous base, however, is either a purine or a pyrimidine.

2. There are only two purines (double-ringed compounds): adenine (A) and guanine (G). The DNA pyrimidines (single-ringed compounds) are thymine (T) and cytosine (C).

3. Nucleotides pair to form a double-stranded molecule.

4. The DNA molecule is analogous to a staircase. The banister is an alternation of phosphates and deoxyribose sugars. The stairs are paired nitrogenous bases. In addition, the staircase is spiral (more specifically, it is a double helix).

5. The nitrogenous bases always pair a purine with a pyrimidine. The possible pairings are adenine-thymine, thymine-adenine, guanine-cytosine, and cytosine-guanine. If one base is known, the other can easily be derived.

6. The unique characteristic of DNA is its ability to reproduce itself, a process referred to as replication. During replication, the double helix unwinds, and the strands of the molecule unzip. When base pairs are separated, complementary nucleotides from the nucleoplasm pair with them, a process called complementary bonding.

7. Chromosomes, which appear as obvious rod-

shaped structures at the time of cell division, consist of DNA molecules and protein.

8. Ribonucleic acid (RNA) has a structure similar to that of DNA. It is composed of nucleotides formed from the five-carbon sugar ribose, a phosphate group, and a nitrogenous base.

9. The nitrogenous bases of RNA are the same as those for DNA, except that the pyrimidine uracil is substituted for thymine.

10. Unlike DNA, RNA is a single-stranded molecule.

11. There are three kinds of RNA: messenger-RNA (mRNA), ribosomal-RNA (rRNA), and transfer-RNA (tRNA).

12. The genetic code consists of sequences of nucleotides, called codons.

13. There are approximately twenty amino acids and sixty-four codons. Therefore, some codons are redundant. Each codon translates for a specific amino acid.

14. A gene is a series of nucleotides that codes for specific codons of the mRNA. Each codon specifies an amino acid. The amino acids eventually link to become a specific protein.

15. A cell must decode its genetic information. During transcription, the DNA nucleotide sequence is copied by a strand of mRNA, which then enters the cytoplasm.

16. mRNA containing the information for protein synthesis is produced by a process called translation. mRNA and tRNA become associated at the ribosome, tRNA shuttles amino acids to the mRNA template. Ultimately, a protein is produced.

17. A mutation is a change in a genetic message. There are several types of mutations: deletions, translocations, inversions, and point mutations, which include frame-shift mutations and base-pair substitutions.

18. The disorder known as sickle-cell anemia is caused by a point mutation.

19. Through genetic engineering, genes are transferred from one cell or organism to another. The cell or organism receiving the new gene may incorporate it into its own hereditary material and begin to produce proteins that it formerly was unable to produce.

THE NEXT DECADE

NEW APPLICATIONS OF GENETIC ENGINEERING

To many molecular biologists, the future of genetic engineering looks bright. Now that human insulin and growth hormone are being synthesized by bacteria, and interferon (a human antivirus protein) is being synthesized by yeast, numerous additional applications of gene splicing and gene transfer are being explored.

Throughout the world, genetic engineering labs are accomplishing amazing feats. In France, new, improved vaccines against rabies and other diseases are being developed. In Japan, efforts are being made to produce contraceptive vaccines against human sperm and ova. This could lead to an effective new approach to birth control. In Wales, cells are being induced to produce human blood proteins. Some day, a human blood substitute may be developed.

So far, microorganisms are being used to produce many products, but it may become common to transfer genes from one advanced complex organism to another. Recently, strains of giant mice have been produced by transferring rat genes that regulate growth hormone production into the fertilized eggs of mice. The mice that developed from these zygotes produced as much as eight hundred times the amount of growth hormone they would ordinarily produce, and they grew much larger than normal.

Experiments such as these could eventually pave the way for improving crops and livestock. Cows could give greater quantities of milk. Cattle could grow faster and larger, providing a greater supply of beef. New forms of disease-resistant crops

and animals could be produced. There is optimism that the genes in nitrogen-fixing bacteria that allow the bacteria to take nitrogen from the air and change it to a form that plants such as legumes can use as a natural fertilizer can be harnessed. Bacteria might then be engineered to produce inexpensive sources of nitrogen fertilizer. Perhaps nitrogen-fixing genes could even be incorporated into the cells of many important food plants, such as corn and other grains. If this could be accomplished, much less fertilizer would be needed to grow these crops.

The possibilities for genetic engineering applications seem endless. During the next decade, we may witness revolutions in medicine and agriculture because of them.

FOR DISCUSSION AND REVIEW

1. Compare and contrast the structures of DNA and RNA in regard to sugar molecules, phosphates, purines, pyrimidines, and molecular structure.

2. Draw a nucleotide of a DNA molecule, showing the spatial relationships among the phosphate, the deoxyribose sugar, and a purine or pyrimidine. How is the next nucleotide connected? What is the relationship of this nucleotide to the nucleotide of the complementary strand?

3. In DNA, adenine is always bonded to thymine, and cytosine is always bonded to guanine. Why?

4. Distinguish among replication, transcription, and translation.

5. Consider the codon ACT on the DNA molecule. What would be the complementary codon after replication? After transcription?

6. Consider the following base-pair sequence on a DNA molecule:

A-T
C-G
G-C
T-A
T-A
C-G

How many codons are involved? Rearrange the message as if (a) a deletion had occurred, (b) a translocation, (c) an inversion, (d) a point mutation, and (e) a frame-shift mutation.

7. It is thought that there are sixty-four codons, but their combined effect is to code for only twenty amino acids. Therefore, many codons code for the same amino acids. Why is this redundancy necessary?

8. What is the relationship between a chromosome and the DNA molecule?

9. List the various forms of RNA. How do they differ? What is the function of each?

10. Detail the process of protein synthesis, beginning with a strand of mRNA and ending with a string of amino acids. Make sure you include steps involving mRNA, rRNA, ribosomes, tRNA, and amino acids.

SUGGESTED READINGS

The molecular details of DNA are commonly found in any general biology, botany, zoology, or genetics textbook. The approach and amount of detail vary from source to source.

THE DOUBLE HELIX, by James P. Watson. Mentor: 1968.
A best seller written by one member of the team that deciphered the structure of DNA.

THE GENETIC CODE, by Isaac Asimov. Mentor: 1962.
An explanation of the world of DNA written for popular consumption.

IMPROVING ON NATURE, by Robert Cooke. Quadrangle New York Times Books: 1978.
Explores the brave new world of genetic engineering. Worth reading once the molecular detail presented in this chapter is understood.

PLAYING GOD: GENETIC ENGINEERING AND THE MANIPULATION OF LIFE, by June Goodfield. Harper & Row: 1979.
An account of scientists, society, DNA, and human ethics.

16

HUMAN GENETICS

The study of human inheritance seems to fascinate students in introductory biology courses. The reason for the interest is obvious. Information about human genetics has grown considerably in the past two decades. The focal point of the study of human genetics is similar to that of genetics in other organisms. The emphasis is not on what is normal but on the inherited characteristics that deviate from normal—**mutations**▫.

The list of human mutations is long. Most of them have a deleterious effect, causing physical and often mental crippling. Hemophilia and sickle-cell anemia▫ are two examples. Important sociological, psychological, and economic consequences are involved in all these disorders.

We hope to prevent or cure maladies like these. In order to do so, however, we must analyze the pattern of inheritance. This chapter is an introduction of some of the more common inherited disorders. (More information about them is available in any of the two dozen or more textbooks devoted entirely to human genetics.)

POINT MUTATIONS

Cystic Fibrosis

One of the most common human genetic disorders is cystic fibrosis. In the United States, about 5 of every 10,000 babies have the disease, which is caused by a simple recessive allele. The people who have it are designated cc, whereas normal people are either CC or Cc. The disease is common because many people are carriers. It is estimated that 1 white person out of

CAUSES OF SOME IMPORTANT HUMAN GENETIC DISORDERS

Point mutation — Cystic fibrosis, antitrypsin deficiency, muscular dystrophy, Tay-Sachs disease, phenylketonuria

Chromosomal anomaly

Autosomal trisomy — Down's syndrome, Edward's syndrome, Patau's syndrome

Sex chromosomal trisomy — Klinefelter's syndrome, trisomy-X, XYY males

Sex chromosomal monosomy — Turner's syndrome

every 25 is a carrier. However, the disease is rare among blacks and orientals.

Cystic fibrosis is a malfunction of certain pathways in protein metabolism. Many internal organs are affected, including the respiratory and digestive systems. Those with the disease have greater than normal amounts of mucus, particularly in the respiratory passages. They often have difficulty breathing and are constantly threatened by secondary complications and infections, such as pneumonia. Victims also may be malnourished because of the defects in the digestive system. Food is not completely digested, and most fats and proteins are not absorbed by the small intestine. The disease begins early in childhood and inevitably leads to an early death.

Treatment, which is effective to a degree, consists of special diets, constant supplies of antibiotics, and inhalants and decongestants for the respiratory system. Without treatment, death usually occurs before age five. With treatment, life can be prolonged for another twelve to sixteen years. Some women with the disease have lived long enough to reproduce. However, almost all the males who have it are sterile.

Prevention of cystic fibrosis can be accomplished only if carriers can be identified. Known carriers can then try not to marry other carriers. Biochemical tests that detect heterozygotes are available, but they are complex, expensive, and unreliable.

A disorder similar to and even more common than cystic fibrosis is antitrypsin deficiency. It affects about

HIGHLIGHT
GENES AND HEART DISEASE

Probably few subjects have had more written about them in the past decade than heart disease. Coronary heart disease is one of the major killers in the West.

There is no agreement at present on the cause of this disease. It may take many more years of research to determine the circumstances that produce diseased hearts. Diet, life-style, and stress have been implicated as potential causes. Today, millions of people are attempting to avoid heart disease by losing weight, jogging, and reducing the amount of saturated fats in their diets.

National awareness of the dietary connections to coronary heart disease have changed many of the habits of consumers and manufacturers. Foods that contain no fats or only polyunsaturated fats have become the mainstays of diets. There was a time when margarine did not exist and butter was on the table at every meal, when corn oil was unheard of and lard greased every biscuit, cake, and frying pan. Despite the current attention paid to diet, many people will still die of coronary heart disease. More and more research is indicating that various genetic errors may play an equal role with diet in this disease.

A first symptom of coronary problems is high blood pressure, or hypertension. Recently, a gene called HYP-1 (HYP for hypertension) was identified in rats. The gene directs the synthesis of a specific enzyme that is essential to the production of certain hormones in the adrenal glands. An abundance of these hormones is known to cause hypertension in humans. A salt-free diet, often prescribed for humans, did not curb hypertension in these rats.

Another factor that plays a role in coronary heart disease is fat deposits in arteries, called atherosclerosis. Recently, a protein (the expression of a gene) that promotes the cellular process leading to the formation of plaques in blood vessels was identified. The protein is a product of blood platelets. Other research has shown that at least some plaques are the result of cellular cloning. A single genetic error in a cell in the lining of a blood vessel produces a clone of cells with the same error. The result is a plaque blocking the blood vessel. Researchers claim that the accumulation of fats in these plaques is secondary and that it is not involved in the initiation of the plaques.

Until additional information is available, watching the diet may be one of the kindest things we can do for our bodies. Until we can watch our genes, that is.

6 of every 10,000 infants. About 1 out of 21 people are carriers. This disease is also inherited as a recessive trait. Its symptoms are similar to those of cystic fibrosis but less serious. Cirrhosis of the liver is common, as is emphysema. Victims may live reasonably normal lives and become middle-aged, and most reproduce.

Muscular Dystrophy

Another relatively common inherited disease is muscular dystrophy. One form of the disease strikes early in childhood, usually around the age of six. It is called Duchenne muscular dystrophy after its discoverer, Benjamin Duchenne, who described it in 1868. The first symptom is a lack of muscle coordination that may be seen in an inability to get up off the floor or in a sudden inability to negotiate stairs. The disease progresses rapidly. Within a few years, the muscles atrophy (shrink) to almost nothing. The child becomes skinny and is nearly incapable of any movement or body coordination (see Figure 16–1). Death usually occurs in the early teens.

Duchenne muscular dystrophy is inherited as a sex-linked recessive allele□. Therefore, the disease is much more common among boys than girls. Heterozygous females may show some minor symptoms of the disease. Boys who have the disease never live to reproduce, so the disorder is perpetuated by heterozygous females. The incidence of carriers is about 1 in every 20,000 females. However, 3 males in every 10,000 have the disease. How can so many boys have the disease when there are so few female carriers? It has been shown that about a third of all afflicted males have inherited the disease as a new, recurrent mutation from one of their parents.

The other major form of muscular dystrophy is adolescent muscular dystrophy. The symptoms of the disease do not develop until the individual is an adolescent. One type is inherited as a dominant allele, the other as a recessive. Both are rare in comparison to Duchenne muscular dystrophy.

Tay-Sachs Disease

The first symptoms of Tay-Sachs disease, which is inherited as a simple recessive trait, appear at about six months of age. As the disease progresses, there is an increased accumulation of fat in all body cells, but particularly in the cells of the spinal cord and brain. The affected infant begins to lose general body coordination. Seizures and blindness are common. Finally, a complete breakdown of motor and neural functions causes death before age five.

Tay-Sachs disease was first called familial infantile idiocy because of its effect on the brain. A British

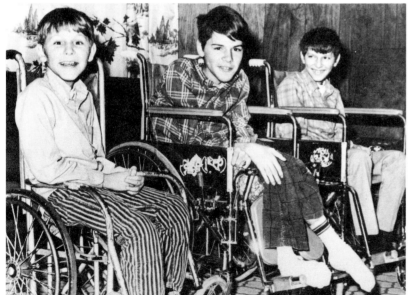

Figure 16–1
Muscular dystrophy, a sex-linked recessive trait, is more frequently expressed in boys than in girls. Pictured are three brothers who all suffer from the disease. They are ages 10, 16, and 13 (from left to right). The oldest (in the middle) shows more severe symptoms than do his brothers. Unfortunately, within a few years, the younger boys will be as seriously afflicted as he is. (Wide World Photos.)

ophthalmologist, Warren Tay, and an American neurologist, Bernard Sachs, both described the disease before the turn of the century. Thus it became known as Tay-Sachs disease. The disease is far more common in some genetic groups than in others. Among the Jews in Central Europe (called Ashkenazi), the incidence is approximately 1 in 2,500 to 6,000 births—a high rate. It is estimated that about 1 in every 40 Jews and 1 in every 400 non-Jews is a carrier.

Like so many other human genetic disorders, the disease is caused by the lack of an enzyme. In this case, the enzyme is hexosaminidase, which is necessary for normal fat metabolism. The enzyme can be detected in blood samples. Carriers (Tt) produce only half as much of the enzyme as noncarriers (TT). Therefore, they can be detected by relatively simple screening procedures. A screening program was set up in the Washington, D.C., area in 1971. About 8,000 carriers were detected. Both husbands and wives of several hundred couples were heterozygous. The risk that these couples will have a child with Tay-Sachs disease is high (1 out of 4, or 25 percent). In addition, 2 out of 3 of their normal children are likely to be carriers.

Amniocentesis can show if a fetus is homozygous recessive for the disease (tt) (see Box 16–1). If it is, the parents may decide on a therapeutic abortion. The disease is incurable, and infants who have it will be totally helpless and blind by the age of five. Even so, the Jewish community is not in agreement about therapeutic abortion for Tay-Sachs fetuses. Abortion is against its religious principles. There is also some debate about whether abortions will actually lower the frequency of the recessive allele. The homozygous recessive infants never reach reproductive age. They are born because their parents are carriers. Therapeutic abortion may only encourage parents who are carriers to try to have another child, and two-thirds of their normal children could be carriers. Increasing the number of carriers increases the incidence of children born with the disease.

Phenylketonuria

The chemical deficiency that causes phenylketonuria **(PKU)** is probably the best understood of all human genetic disorders. It is most common among whites with a Northern European ancestry and least common among blacks and orientals. PKU, which is in-

herited as a simple recessive trait (pp), affects about 1 of every 10,000 children born in the United States.

Phenylalanine is one of the essential amino acids. The body cannot manufacture it, so people must get it daily in certain foods, such as fish, cheese, and eggs. Individuals who have PKU lack the enzyme phenylalanine hydroxylase. This enzyme is necessary for the conversion of phenylalanine to tyrosine, an amino acid. Tyrosine is the precursor to the melanin pigment in skin and to the hormone thyroxin which is manufactured in the thyroid gland. It is also a necessary substance in the production of specific adrenal hormones.

An individual without phenylalanine hydroxylase accumulates an excess of phenylalanine, which becomes thirty times more abundant than other amino acids in the blood. The overabundance of phenylalanine and an associated chemical, phenylpyruvic acid, causes severe brain damage and retardation. Most individuals suffering from PKU have to be institutionalized. PKU victims often have fair skin and yellowish-red hair because of the lack of tyrosine.

PKU can be detected at birth. A simple test is required by law in most states. In the usual test, the Guthrie test, a drop of blood from the newborn is placed on a culture of mutant bacteria. Small amounts of phenylalanine stimulate the growth of these bacteria. The test is very sensitive, because the newborn PKU victim has only a minute elevation in the amount of phenylalanine. The disease can easily be detected when an infant is a few days old, because of these elevated phenylalanine levels. The earlier the treatment begins, the better the results.

Newborns with PKU are put on a diet that is extremely low in phenylalanine. The diet is expensive and difficult to prepare. However, it eliminates the buildup of phenylalanine and the dreaded symptoms of phenylketonuria. Once the brain is fully formed, at about six years of age, the child can begin a normal diet without fear of physical impairment or retardation. However, there is some evidence that the PKU diet should be continued for several more years. Many children who abandon the diet early develop serious behavioral disorders. Although treatment is effective, the intellect of PKU sufferers is average at best, and most PKU children have below-average IQs. Without treatment, however, they would be severely retarded.

Will detection and early treatment eradicate PKU? Obviously not. Those who survive PKU (children

BOX   16–1

AMNIOCENTESIS

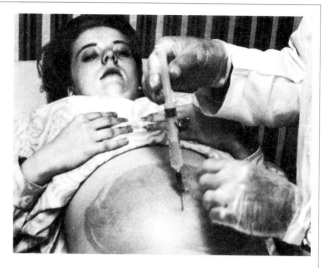

A relatively simple procedure during which a sample of amniotic fluid is taken from a developing fetus is known as **amniocentesis.** The procedure is conducted to determine whether the fetus suffers any genetic disorders or other serious detectable diseases.

Amniocentesis is usually performed on women who have been genetically counseled about the advisability of the procedure. Most of the women requesting it fear Down's syndrome. They are usually women over forty or women who already have a child with the syndrome. (The recurrence rate is about 15 percent.) The woman must be fifteen to seventeen weeks into the pregnancy. If the procedure is carried out any earlier, there will be insufficient amniotic fluid for a sample. If it is done much later, therapeutic abortion, if advised, will be less safe.

For amniocentesis, the abdomen of the patient is sterilized, and the position of the fetus and placenta is determined by an ultrasound scan. To avoid damage to the placenta or the fetus, the obstetrician chooses a course where the placenta is thin and the fetus is not in the way. A long hypodermic needle is inserted through the woman's abdominal wall into the uterus and finally into the amniotic cavity, and the sample of amniotic fluid is withdrawn. The procedure is essentially painless and is done on an outpatient basis.

The amniotic fluid contains cells that have been sloughed off from the fetal skin, respiratory system, and urinary tract. It may also contain various chemicals. The cells are cultured and later analyzed for karyotype (pairs of chromosomes) and many other biochemical properties. It takes about three weeks to complete the cell cultures and final analysis.

Currently, more than two dozen disorders can be detected by amniocentesis. For example, all of the point mutations and the chromosomal anomalies described in this chapter can be detected. Two fetal disorders whose genetic basis is unknown—spina bifida and anencephala—can be detected by leakage of specific fetal proteins. The occurrence of these disorders is as frequent as 1 in 500 births. Spina bifida, failure of the spinal cord to close, is nearly always fatal. Anencephala, a nearly total lack of the brain and part of the head, is always fatal.

The sex of the fetus can be determined by the absence or presence of a Barr body (the inactive X chromosome of females). Amniocentesis is rarely performed just to determine the sex of a fetus, however.

Amniocentesis may not reveal some defects, so fetal blood samples are also required in certain instances. The samples are obtained by fetoscopy, a procedure in which a small, rigid fiber optics scope with a needle is inserted into the uterus, and blood is withdrawn from the fetus. Fetal blood can be biochemically analyzed and used to predict many potential blood disorders. Among the disorders are abnormal or insufficient hemoglobin, blood clotting disorders, and blood platelet disorders. Fetal blood analysis can also predict Duchenne muscular dystrophy.

Amniocentesis is estimated to be about 99.4 percent accurate, but mistakes may be made. In one instance, a child with Down's syndrome was born after amniocentesis had shown a normal fetus. In another case, a metabolic disorder diagnosed by amniocentesis failed to develop. Furthermore, the sex of the fetus has been incorrectly identified from time to time.

Is amniocentesis safe? Generally yes, although there is about a 5 percent risk of a problem with the procedure. In about 2 to 5 percent of the cases, no fluid is obtained, and the procedure has to be repeated. In 5 to 10 percent of the cases, the cultured cells fail to grow, and again the procedure must be repeated. In about 50 percent of the cases, it is impossible to avoid the placenta, and the maternal/fetal barrier is broken, which produces bleeding. Finally, there is always the risk of technical and clerical error. However, all parents take risks. Of all newborns, 3 to 5 percent have some kind of genetic defect and another 1 to 2 percent are mentally retarded. Thus any risk resulting from amniocentesis must be weighed against the potential risk of not detecting fetal problems.

who are not treated die by age thirty) will be normal, healthy adults who can reproduce. All of their offspring will be carriers, so the incidence of PKU babies will ultimately increase. Adults who were born with PKU and were given a special diet until they were teenagers are known as phenocopies. They have a true genetic disorder, but treatment allows them to mimic normal PP or Pp individuals.

Female PKU phenocopies acquire another set of problems when they become pregnant. They still lack the enzyme for the metabolism of phenylalanine. This lack affects the developing fetus. Therefore, pregnant PKU women must once again go on a special diet to avoid the ingestion of too much phenylalanine. If this precaution is not taken, the child will be brain-damaged at birth. The child will most likely be Pp. (Simple chance dictates that the female phenocopy will probably not marry a male carrier.) If the mother does not take dietary precautions, however, the child will still be born with all the symptoms of PKU. In effect, the Pp child becomes pp—a reverse phenocopy.

CHROMOSOMAL ANOMALIES

Human genetic disorders are caused by more than point mutations. Entire chromosomes may be involved. During meiosis, gametes may acquire less or more than the normal haploid chromosome number. An unusual chromosome number (a chromosomal anomaly) is relatively rare, but it happens because of nondisjunction.

Nondisjunction occurs when one or more homologous chromosomes stay joined during the first meiotic division and move to the same cell during anaphase. It also occurs when sister chromatids fail to separate during anaphase of the second meiotic division. (See Chapter 5 for a review of the events of meiosis.) The results of nondisjunction at the first or second meiotic division are slightly different (see Figure 16–2). After nondisjunction, gametes either have an extra chromosome, because there are two copies of one chromosome, or lack one chromosome entirely.

What happens at fertilization? The gamete with two copies of the same chromosome acquires a homologue to this pair, so the zygote has three rather than two of these chromosomes. This is called a **trisomy**

(tri means "three"). The gamete missing a chromosome acquires a homologue to the missing chromosome, so the zygote has only one chromosome rather than a pair. This is called a **monosomy** (mono means "one"). The result is that the embryo will either be short an entire chromosome or have an extra one in every cell.

Monosomics are usually lethal. The monosomic fetus dies early in development and is aborted. Many human disorders are the result of trisomics, however. Most trisomics involve the sex chromosomes, but some involve the autosomes.

Autosomal Trisomics

Not all trisomics involve the sex chromosomes. In fact, one of the best-known involves one of the other twenty-two pairs of chromosomes, called **autosomes.**

Down's Syndrome

About 1 of every 650 infants is affected by Down's syndrome. The child with Down's syndrome has forty-seven chromosomes instead of forty-six because there are three copies of chromosome 21 (an autosome). Therefore, this syndrome is often called trisomy-21.

The disorder was first described by Langdon Down in 1866 and later was named in his honor. Down thought that the afflicted children were similar to mongoloids, and he concluded that the disease was a throwback to an ancient ancestor. The syndrome is still frequently called mongolism. The fact that Down's syndrome was caused by a trisomy was not known until 1959.

You may have seen children or adults with Down's syndrome (Figure 16–3). Afflicted individuals are always rather short and chunky. They also have poor muscle tone. As a result, their faces are rather flat, their bodies appear bulky, and they may walk peculiarly. Their hands are short, and their fingers are short and stubby. They are clumsy when using their hands. Frequently, they have what is called a simian crease, a single prominent crease across the palm. Normal palms have three or four major creases. One distinguishing trait of a child with Down's syndrome is an extra fold of skin in the corner of each eye, called the epicanthal fold, which gives the child an oriental appearance. The flat face and broad, flat nose add to the oriental likeness. The mouth cavity seldom

NORMAL MALE MEIOSIS
(formation of the sperm)

Primary spermatocyte

FIRST MEIOTIC DIVISION

SECOND MEIOTIC DIVISION

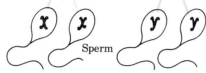

Sperm

Sperm		Ovum		Normal male		Sperm		Ovum		Normal female

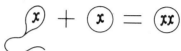

NONDISJUNCTION OF SEX CHROMOSOMES IN MALE MEIOSIS

Primary spermatocyte   Primary spermatocyte

NONDISJUNCTION AT FIRST MEIOTIC DIVISION   FIRST MEIOTIC DIVISION

SECOND MEIOTIC DIVISION   NONDISJUNCTION AT SECOND MEIOTIC DIVISION

Sperm

Sperm		Ovum		Male (Klinefelter's syndrome)		Sperm		Ovum		Male (XYY syndrome)

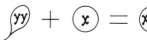

(a)   *(b)* (autosomes omitted)

Figure 16–2
Nondisjunction during the formation of sperm. The upper portion of this illustration shows the normal distribution of X and Y chromosomes during meiosis. The lower portion shows the distribution of X and Y chromosomes. (a) When there is nondisjunction during the first meiotic division, gametes are formed with both X and Y chromosomes and without sex chromosomes. (b) When there is nondisjunction during the second meiotic division, gametes are formed with either two X chromosomes or two Y chromosomes and without sex chromosomes. Nondisjunction may also occur during the formation of gametes in females. (After Ashley Montagu. 1968. "Crime and Chromosomes" in Psychology Today 2(5):p 47.)

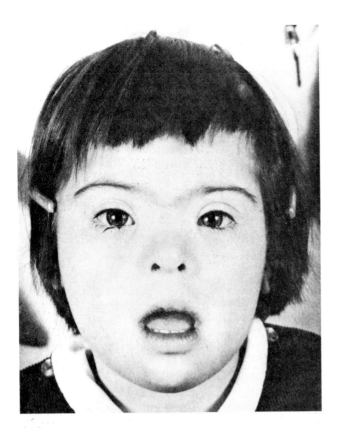

Figure 16–3

A 3½-year-old girl who suffers from Down's syndrome, or trisomy-21. The flat face, epicanthal folds of the eye, and thick furrowed lips are all characteristics of this chromosomal anomaly. (Courtesy of M. Bartolos; from M. Bartolos and T. A. Baramki, *Medical Cytogenetics*, Williams and Wilkins, 1967.)

grows to normal size, so the tongue generally protrudes from the mouth. The tongue is also oversized and rough in texture. Speech is slurred and often difficult to understand. The teeth are frequently abnormal and irregular in shape.

A child with Down's syndrome may also have at least one deformed ear, abundant neck skin, a gap between the first and second toes, and a smaller than average head. All Down's children are mentally retarded, most having an IQ of 25 to 50, which classifies them as severely retarded. (For a discussion of the inheritance of intelligence, see Box 16–2.) Pubic hair and body hair are sparse, and adult males lack beards. Females who live long enough can reproduce. Males produce motile sperm but in such small numbers that the males are always sterile.

In addition to their abnormal physical appearance,

BOX 16–2

THE INTELLIGENCE DEBATE

What is intelligence? Is it inherited? These questions were seriously debated just a few years ago. Not only do most people associate intelligence with the intelligence quotient (IQ), but to many people intelligence and IQ are the same thing. This may not necessarily be so, however.

IQ is measured by an examination. Early this century, a French schoolteacher named Alfred Binet developed an examination to determine the potential success of French schoolchildren. Based on the school curriculum, the exam became a good predictor of children's success in school. The standard Binet test for IQ has been modified many times, but it still uses the same principle.

Does the IQ test actually measure intelligence? No. It measures only achievement and a person's awareness of the values of Western society. The test is based primarily on verbal and mathematical skills and immediate recall. But intelligence includes many other things: specialized skills, experience, reasoning, and individual talents, for example. One test cannot measure all these things. In terms of natural selection, intelligence is best measured as all the skills and cunning that allow a person to survive. This concept is nearly impossible to measure in a complex technical society.

You have probably taken at least one IQ test. (A few years ago, nearly every schoolchild in the United States was required to take an IQ test.) The score is based on age. Students are evaluated on whether they perform above or below the level of their age group. For example, a ten-year-old who performs like any other ten-year-old has an IQ of 100. A ten-year-old who performs at the level of twelve-year-olds has an IQ of 120.   **(Continued, Facing Page)**

many children with Down's syndrome have heart deformities. They also have a tendency to contract infectious diseases, particularly those involving the lungs and respiratory system.

During the 1930s, the average life span of a child with Down's syndrome was nine years. Most died of respiratory diseases. The discovery of antibiotics curbed most of these fatal infections. Thus, by 1947 life expectancy was sixteen years; by 1963 it was

The intelligence debate began when extensive data about IQ scores began to appear. The main issue was that the average IQ of whites in the United States was 100, whereas the average IQ of blacks was 85. However, many blacks scored higher than whites. In fact, northern blacks scored somewhat higher than southern whites. The debate centered on the basis of the fifteen-point difference between the averages of blacks and whites. There were two points of view.

One group of authorities maintained that the difference was due to the depressed economic and educational environment of most blacks. This group said that blacks simply did not have the opportunities afforded most whites.

The other group of scholars interpreted the fifteen-point difference as a difference in innate, or genetically determined, intelligence. There is no doubt that various aspects of intelligence are inherited. From a large variety of facts, geneticists have decided that the heritability of intelligence is 80 percent. This means that 80 percent of the variation in basic intelligence is inherited and that the remaining 20 percent is determined by environment, say, by the kind of schools attended, the socioeconomic class, and so on. Scientists on this side of the debate contended that only 20 percent of the fifteen-point difference in IQ scores could be explained by the environment and that the remainder was a basic difference in innate intelligence. Various mathematical calculations made by these scholars suggested that if blacks and whites had exactly the same environment, the average IQ of blacks would be raised to only 88.

The debate was scholarly, based on observation. Unfortunately, many nonscientists seized onto the issue and misinterpreted the purpose of the debate. Some used the facts to conclude that the white race was superior to the black, an obviously unscientific use of the information.

Fortunately, the debate has had its good side as well.

A great deal of effort and money has been expended to eliminate discrimination and to afford all races equal opportunities. Affirmative action programs, for example, were designed to give the disadvantaged and minorities a better chance to succeed.

Is intelligence inherited? As stated earlier, about 80 percent of the variation in intelligence is inherited—as a polygenic trait. The genes involved do not give more or less intelligence, but instead have subtle effects on the development of brains and nervous systems.

The evidence for the heritability of intelligence is considerable and varied. For example, it is known that the children of a severely retarded parent can be normal. This suggests that the parent has a recessive defect that is not expressed in the children. On the other hand, the children of parents who are mildly retarded are frequently also mildly retarded. This also suggests a genetic basis. The information is similar to that obtained from a pedigree for any physical characteristic.

Studies of identical and fraternal twins have provided the best evidence for the heritability of intelligence. Identical twins have remarkably similar IQs. The concordance (agreement or relationship) is about 80 percent. This fact is the origin of the 80 percent heritability index for intelligence. Fraternal twins have less concordance; their IQs are no more alike than are the IQs of any other brothers and sisters—just what would be predicted for a physical characteristic.

One problem with these findings is that identical twins have an unusual psychological propensity to do like things and to be alike. A hypothesis was that twins could influence each other to the point that they developed the same IQs. Some identical twins who have been reared apart have been studied to test this possibility. These twins still have a concordance in intelligence between 60 and 80 percent, so it seems that their separate environments have little effect on their intelligence.

twenty-one years, and today it is about twenty-seven years. Some adults, however, reach the age of forty or fifty.

At one time is was common for parents of children with Down's syndrome to put them into mental institutions. However, these institutions are often overcrowded, and the children seldom receive the attention they need. The children can be trained, and they become obedient and loving people who respond positively to their entire environment. The awareness of Down's syndrome and its cause has removed much of the stigma that at one time surrounded these children. Today, they are often raised by their parents, although they remain retarded and childish all their lives. Training a Down's child is time-consuming, and the parents have to make great sacrifices if they raise the child at home. Many communities have associations of parents who have children with Down's syndrome, and they help each other and advise each other about proper care.

Trisomy-21 does not usually run in families. It is more frequent among women who are older when they get pregnant. It usually occurs because of nondisjunction during egg production. As mentioned earlier, the overall risk of having a Down's child is about 1 out of 650. It is only 1 out of 3,000 for women who are twenty years old, but it is 1 out of 800 for women thirty years old, 1 out of 500 for women thirty-five years old, and 1 out of 40 for women over forty-five. Recent clinical data have shown that in about 20 percent of the cases the extra chromosome was inherited from the father. Thus, although the risk is highest for older women, the cause can be attributed to either parent.

Down's syndrome is also caused by chromosome 21 becoming fused with chromosome 15 (a translocation). This form of Down's syndrome is less severe. It is inherited and does run in families. If the translocation occurs in the gametes of a parent, four kinds of offspring are equally probable: normal, Down's, a carrier who can produce Down's children, and an incompatible lethal combination that never develops.

Other Autosomal Trisomics

There are only two other trisomics of live infants involving autosomes. They are trisomy-18 (Edward's syndrome) and trisomy-13 (Patau's syndrome). In both syndromes, the newborns live only a few weeks, and many are stillborn. The infants suffer severe external and internal deformities and severe mental retardation.

Other trisomics involving autosomes have been identified. In all cases, the fetus dies early in development. Such abnormalities are referred to as lethal chromosomal anomalies.

Sex Chromosomal Trisomics

Many chromosomal anomalies are associated with chromosomes 23, the sex chromosome pair (X and Y□.)

Klinefelter's Syndrome

Individuals with Klinefelter's syndrome are XXY males rather than the normal XY males. This anomaly occurs in 1 of every 600 male births.

The symptoms of Klinefelter's syndrome are not detected until near the time of puberty. Klinefelter's males have sparse pubic hair; small scrotums; small, underdeveloped testes that never produce sperm; and normal-sized or smaller-than-average penises. The small testes cause a deficiency in the male hormone testosterone. As a result, the hips of Klinefelter's males are wider than those of normal males. The men show some obvious breast development, their body hair is sparse, their beards do not develop, and their voices are high-pitched. In addition, they are often much taller than average males, and their arms and legs are often long in proportion to their torsos. Many Klinefelter's males live normal lives. Some marry, but they are sterile. In a number of cases, there is some associated mental retardation. The average IQ of Klinefelter's males is about 90.

XXY males are the most common example of Klinefelter's syndrome, but some males are XXXY, XXYY, or even XXXYY. The greater the number of X chromosomes, the more severely retarded the individual.

Trisomy-X

As the name implies, trisomy-X produces females who have three X chromosomes and are thus XXX. Cases of females who are XXXX or even XXXXX have been reported, but the XXX condition is the most frequent. It occurs in 1 of every 500 female births.

Those affected by this syndrome were once called "super females," but the name caused some confusion. Females with trisomy-X have no "super" characteristics. Instead they have supernumerary X chromosomes. Their extra X chromosomes may come from their fathers, who produced abundant numbers of XX sperm through nondisjunction. Also a normal X-containing sperm cell may fertilize an XX egg produced by nondisjunction.

XXX females show no abnormalities. They appear as healthy, normal, and fertile, although some have a somewhat depressed mental ability. Most marry and have children. XXXX and XXXXX females are extremely retarded, show abnormal sexual development, and are sterile.

XYY Males

About 1 in every 500 to 700 male births is an XYY male rather than an XY. Most of these males appear to be normal, although at maturity most are tall (over six feet). Many have IQs of less than 100. XYY men are fertile.

Despite the apparent normality of XYY men, the condition has become quite famous. A study done in Scotland in 1965 demonstrated that many XYY males were in either mental or penal institutions because of their unusually aggressive personalities. The frequency of XYY men not in institutions was about 1 in 400, but surveys made in prisons showed frequencies as high as 1 in 25.

The Scottish study was followed by a great deal of publicity. Soon afterward, the XYY male condition became known as the "criminal syndrome." It was supposed that the extra Y chromosome gave these males an extra dose of maleness that was expressed as excessive and uncontrollable aggression. However, XYY males not in institutions are frequently of average height and are well-adjusted and mild-mannered, with no history of hostility. Obviously, then, having an extra Y chromosome does not make the possessor a criminal.

Many other studies were launched to determine the effects of the extra chromosome, and the debate continues today. Some authorities have suggested that the aggressiveness may be environmentally induced, because XYY children have a combination of tallness and slow mentality that may promote attacks by others. One study showed that XYY men were not so frequently found in institutions as was previously reported. However, other studies have shown that the number of XYY men in prisons in both the United States and Europe is ten times higher than in the general population and that most such men are confined to maximum security prisons.

Several criminal cases before various U.S. courts have involved XYY defendants who have asked that their chromosome condition be entered as evidence of why they committed crimes. Some have had their sentences reduced because of their XYY condition. Similar cases have occurred in other parts of the world.

Sex Chromosomal Monosomy

Because of nondisjunction, some individuals have a missing chromosome, a condition termed monosomic. As with trisomics, the affected individual (if born) often suffers a complex of physical and mental abnormalities.

Turner's Syndrome

Females with only one X chromosome instead of the usual two have Turner's syndrome. They are represented as XO instead of XX. This is the only known nonlethal monosomic syndrome among humans.

XO females (Figure 16–4) are rarely detected until the time of puberty. Usually, they are taken to a phy-

Figure 16–4

The woman in this photograph has Turner's syndrome. She is XO, which means that she has only one X chromosome rather than two. Birth control pills supply her with hormones that maintain normal breast and hip development. She is sterile. A common characteristic of afflicted females is that they are always quite short. This woman is standing next to a man of average height. Despite their symptoms, women with Turner's syndrome are often bright and well adjusted. (Courtesy of A. M. Winchester, from Heredity, Evolution, and Humankind, West, 1967, p. 165.

sician when they are teenagers because they do not start to menstruate. They never will. In addition, they lack pubic hair and breast development. Their breasts and nipples remain small. The nipples are spaced wide apart, causing what is known as a shield chest. XO women are usually very short. In most cases, the hairline on the back of the neck is quite low and may extend down to the shoulders. The skin of the neck is often webbed. If not webbed, it is extremely loose and can be pulled away from the underlying tissues as though it were rubber. The women are always sterile.

XO women suffer no mental deficiencies. Most have average intelligence, and many are very bright. They all suffer a peculiar defect in vision, however. They have faulty spatial perception. If they are shown a hexagon or an octagon and asked to draw what they have seen, the figure they produce seldom resembles what they were asked to draw.

Turner's syndrome occurs in about 1 of every 2,500 female births. However, the incidence is actually much higher. Approximately 20 percent of all spontaneous abortions are of XO fetuses. It is not known why so many XO fetuses are aborted. Those who have the condition can be treated if the condition is diagnosed soon enough. Hormone therapy, including estrogens, causes normal breast development and enlargement. Many women who receive therapy live full and meaningful lives.

SUMMARY

1. Cystic fibrosis is one of the most common human genetic disorders. It is caused by a simple recessive allele. Treatment is available, but it is only partly effective. The only prevention is identification of carriers.

2. Duchenne muscular dystrophy is a sex-linked recessive condition. Therefore, more males than females are afflicted. There is no treatment.

3. Tay-Sachs disease affects general body coordination. It is caused by the lack of a specific enzyme (which is coded by a simple dominant allele). There is no cure for this disorder.

4. Amniocentesis is a procedure for analyzing amniotic fluid withdrawn from around a developing fetus. Examination of cells in the fluid can determine whether the fetus is afflicted with many inherited disorders.

5. Phenylketonuria (PKU) is inherited as a simple recessive trait. Afflicted individuals lack the enzyme phenylalanine hydroxylase. Treatment is possible, but without it a child becomes severely mentally retarded.

6. Individuals afflicted with a genetic disorder that is curable may appear normal even though they still possess the disease-causing alleles. They are known as phenocopies.

7. Many human disorders are caused by chromosomal anomalies. The most common of them result when homologous chromosomes do not separate during meiosis, a situation called nondisjunction.

8. Down's syndrome is a common human disorder resulting from nondisjunction. The afflicted person inherits three of chromosome 21. Thus the disease is often called trisomy-21.

9. Nondisjunction may also create sex chromosome trisomics. For example, people with Klinefelter's syndrome are XXY males (possessing some female characteristics and sterile). "Super females" are commonly XXX and more rarely XXXX or XXXXX (normal and fertile, but with depressed mental abilities). XYY males are exceptionally tall and often have histories of aggression and criminality.

10. Nondisjunction may also create individuals who lack a chromosome (monosomics). For example, females with Turner's syndrome are XO. They have underdeveloped breasts and other physical deformities, and they are sterile.

THE NEXT DECADE

GENETIC COUNSELING

In the next decade, many couples will be advised to, or will elect to, visit a genetic counselor. Therefore, we need to know as much about this vital service as possible.

More and more medical complexes and urban areas are establishing genetic counseling centers. The counselors at these centers are often physicians with a specialty in a particular field of medicine. The most frequent specialty is pediatrics. Genetic counselors may also be doctors of philosophy, with advanced degrees in genetics, zoology, or some other branch of science. Some counselors even have Ph.D.s in one of the social sciences, such as psychology or sociology.

In any event, genetic counselors must have a complete knowledge of genetic principles and human genetics, and they must be compassionate. It is difficult to tell a couple that there is a 75 percent chance their next child will die of a genetic disorder within a year or two of birth or that their child may be more vegetable than human.

Many people seek the professional services of genetic counselors. About half need advice about a genetic defect that is caused by a single pair of alleles. Approximately a quarter are concerned about chromosomal abnormalities (the majority about Down's syndrome). The rest seek an understanding of genetic disorders caused by multiple genes. However, many people who need advice do not seek it for religious or moral reasons. Others do not know that it is available.

If you go to a genetic counselor, you will first receive a complete medical diagnosis of the problem or potential problem. You will most likely be given a complete physical. Samples of your blood will be analyzed for a large number of enzymes to help determine whether you have a genetic disorder or might be a carrier. Some of your white blood cells will be cultured. From them a karyotype will be constructed to determine your chromosome constitution.

In some cases, tests on a single individual may be sufficient for a diagnosis. More frequently, the same tests are done on spouses and on any children in the family. In some instances, close relatives may be asked to submit to the tests. If genetic defect in the fetus of a pregnant woman is suspected, amniocentesis may be included among the tests.

The genetic counselor may also prepare a family pedigree for the suspected disorder. Suppose that muscular dystrophy is feared. The genetic counselor will attempt to find out who in the family has had the disorder and will plot this information on a typical pedigree. Sometimes nongenetic diseases mimic certain genetic disorders and they may confuse a pedigree. For example, the family may recall that an uncle had muscular dystrophy when in fact he had some other disorder. Nevertheless, the pedigree can be extremely helpful to the genetic counselor. The medical tests conducted on the members of the family supply infor-

mation that will improve the accuracy of the pedigree.

When as much information as possible has been gathered, the genetic counselor can advise people on the probable risk of their future children having a particular genetic disorder. The risk may be nonexistent or 100 percent or somewhere in between. The genetic counselor will also advise several courses of action to reduce the risks. Only advice is given. The counselor never makes decisions for people.

Genetic counselors provide a necessary service, one that will improve and spread in the next decade. Nearly every family has members with some genetic defect, whether diabetes, colorblindness, hypertension, or muscular dystrophy. It is the goal of genetic counseling to reduce the number of humans with such defects.

There is a need for more genetic counselors and for more people to consult them as they would consult other physicians or dentists. It has been reported that 10 to 25 percent of the children admitted to pediatric hospitals have genetic disorders. Currently, less than 10 percent of the families of these patients seek or receive genetic counseling. We frequently hear that regular checkups prevent certain illnesses. In the next decade, this slogan may apply to genetic counseling.

FOR DISCUSSION
AND REVIEW

1. Describe the inheritance, symptoms, and treatment of cystic fibrosis.

2. The next time there is a telethon for muscular dystrophy, notice that nearly all of those afflicted are males. Why?

3. Children with Tay-Sachs disease die by the age of five. They therefore never reproduce. Why, then, is the disease so prevalent?

4. Describe in detail the metabolic interruption that occurs in the case of PKU. How is PKU treated?

5. What is a phenocopy?

6. What is the intelligence debate? Will it be resolved? What are the consequences of the debate?

7. What are the recognizable characteristics of a child with Down's syndrome? What causes Down's syndrome? Is there a cure? Explain.

8. Compare and contrast (causes, symptoms, and so on) Klinefelter's syndrome, trisomy-X, XYY males, and Turner's syndrome.

SUGGESTED READINGS

ELEMENTS OF HUMAN GENETICS, 2d ed., by L. L. Cavalli-Sforza. W. A. Benjamin: 1977.
Good coverage of most inherited human disorders and of other topics of biological interest.

THE GENETIC CONNECTION, by David Hendin and Joan Marks. Signet, New American Library: 1979.
A wealth of information and a reference all students should have. Covers everything from birth defects to genetic counseling.

HUMAN GENETICS, by Norman V. Rothwell. Prentice-Hall Biological Sciences Series:1977.
Excellent and detailed coverage for those who crave more information.

HUMAN GENETICS, by Sam Singer. W. H. Freeman: 1978.
Introduction to the principles of genetics, beautifully simple and detailed.

HUMAN HEREDITY AND BIRTH DEFECTS, by E. Peter Volpe. Pegasus: 1971.
Easy to read and comprehend and with complete coverage of birth defects.

KNOW YOUR GENES, by Aubrey Milunsky. Avon Books: 1977.
Written to be understood by everyone. Good coverage of many diverse topics.

SECTION SIX

EVOLUTION

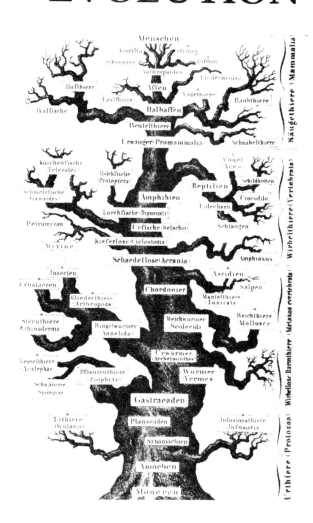

Ernst Haeckel, a Nineteenth Century German biologist and natural philosopher, drew this "evolutionary tree" that extends Darwin's theory, and suggests the descent of humans from single-cell organisms!

17

DARWIN AND THE EVIDENCES OF EVOLUTION

PREVIEW QUESTIONS

After reading this chapter, you should be able to answer the following questions:

1. Who were Charles Darwin and Alfred Wallace? Why are they famous?

2. What are the important points of the theory of the origin of species by natural selection?

3. What types of evidence support the concept of evolution?

Charles Darwin, the father of evolution, was born in England on February 12, 1809, the same day that Abraham Lincoln was born. Darwin was part of the English aristocracy. His father, Robert, was a successful country doctor, as was his grandfather, Erasmus. His mother, Susannah Wedgwood, was the daughter of the Wedgwoods who began the Wedgwood Pottery Company, makers of the famous china that bears their name. Charles Darwin (Figure 17–1) was the youngest of four children, three of whom were girls. His mother died when he was eight, and he was raised thereafter by his sisters. He was a favorite nephew of the Wedgwoods and spent considerable time in their home. Later he would marry one of their daughters, Emma.

Darwin had no money worries, so he spent all his leisure time in pursuit of skills that only the aristocracy could afford. He was a marksman and a horseman, and he enjoyed a hunt. He was also interested in flowers, rocks, birds, spiders, beetles, and other natural things, and he had extensive collections of some of them. Once, when he had collected a beetle in each hand, he spotted a third that he did not have in his collection. He popped one of the beetles into his mouth to free a hand to capture the third. The beetle in his mouth released a foul-tasting acid, and he was forced to spit it out. Thereafter Darwin's father paid an assistant to help Charles in his collections.

DARWIN'S DISCOVERY

It may be hard to believe, but Darwin, a genius in many ways, was a poor student. He was sent to a day

Figure 17–1
Charles Darwin as a young man. (Courtesy of the American Museum of Natural History.)

school that emphasized languages and the classics. Darwin loathed both but found a minor interest in mathematics.

It is doubtful that Charles Darwin would be eligible to enter a university today. Nevertheless, he began his college studies at the University of Edinburgh. He was expected to complete the requirements in medicine and to become a physician like his father and grandfather. However, he found early in his medical career that he could not tolerate the sight of blood. This was a disappointment to his father, who decided that if his son could not stomach a medical career, he should become a clergyman. At that time, it was common for sons of the aristocracy to become clergymen.

Darwin transferred to Cambridge University to begin his theology studies. He loved Cambridge. In his autobiography, he stated that his three years there were the happiest of his life. There is little wonder that Darwin liked Cambridge. It was the school for the "horsey set," and he liked nothing better than to ride and hunt. When he retired to his room, he would practice his marksmanship by pretending to shoot bits of flying paper instead of studying. A good share

of his time outside classes was spent with his friends. He joined the Glutton Club, whose members stayed up late at night playing games, drinking, and listening to music. He was very popular and a guest at the best affairs.

During Darwin's college years, he developed only one serious interest. He liked Professor John Henslow's botany class. Henslow was a botanist and a clergyman, a combination of professions that appealed to Darwin. Henslow interested Darwin not only in natural history but also in geology, and Darwin became a member of Henslow's Friday evening discussion group. When Darwin graduated, in 1831, he had the distinction among students of being a friend of his professor, a rare alliance at the time.

After his spring graduation, Darwin looked forward to taking his holy vows in the fall. He anticipated a summer vacation of traveling and hunting. However, Henslow had nominated Darwin to fill the post of naturalist aboard HMS *Beagle* during a two-year map-making excursion. Darwin had never dreamed of being a professional naturalist, but he was pleased with the prospect.

Darwin's father thought the job would be just another folly. After all, Charles did not make it as a medical student, and this was probably an excuse to get away from the Church. His father forbade him to accept the post. Reluctantly, Darwin wrote a letter turning down the offer.

Putting thoughts of being a naturalist behind him, Charles went to his uncle Wedgwood's home for the first day of partridge season. He told his uncle about the offer. His uncle thought it was a great opportunity and advised Charles to take advantage of it. (There is some belief that his uncle wanted to get Charles away from Emma, at least until he was ready to settle down.) Darwin and his uncle went to Charles's father and made their case, and Robert Darwin gave in. So Charles wrote a second letter, accepting the offer and expressing the hope that he was not too late to get the position.

The final hiring decision was left to the captain of the *Beagle*, Robert Fitz-Roy, who would have to share a cabin for two years with the man he chose. The captain liked Darwin and chose him. Fitz-Roy was impressed with Darwin's religious training. Even though the *Beagle* was on an official mapping mission, Fitz-Roy hoped to find evidence that supported the Book of Genesis. He wanted to prove the existence of Divine Creation and thought he and Darwin would find evidence of the Great Flood and the first appearance of all created things. Darwin had no reason to doubt the Bible, and he enthusiastically shared Fitz-Roy's objectives.

The Beagle

On December 27, 1831, the *Beagle* left the port of Plymouth. Darwin's greatest fear turned to reality when he became seasick. He was sick for several weeks, until he put his feet on ground for the first time in the Cape Verde Islands. (Some accounts of Darwin's voyage allege that he never did fully recover.)

Darwin's early interest in geology led him to take along a copy of Charles Lyell's *Principles of Geology*, which helped him understand the volcanic origins of the Cape Verde Islands. It also helped him realize what his adventure might mean. For the first time, he took a serious interest in something.

The voyage took longer than anticipated. Darwin did not return to England until nearly five years later. The *Beagle* had sailed most of the coasts of South America, New Zealand, and Australia, around the tip of Africa, and back home again in a complete circumnavigation of the world (see Figure 17–2).

Darwin learned much during his voyage. Seeing his first tropical jungle in Brazil, he was awed by its beauty and by the diversity of plants and animals. He spent considerable time watching the insects of the jungle. He observed wasps stinging prey and carrying them away as food for their larvae, and he watched a horde of army ants devastating everything in its path. He learned that life was a struggle, the task to prey or to be preyed upon. He collected many insects and noticed that many were camouflaged to avoid being eaten.

On his trip, Darwin also saw black slavery for the first time. His family had been among the first in England to protest slavery, but he had never actually observed it. Now Darwin and Fitz-Roy had their first disagreement. Fitz-Roy thought that slavery was part of the scheme of things because of the numerous references to slaves in the Bible. Darwin was furious. The argument might have damaged their relationship, but fortuitously Darwin remained ashore for

Figure 17–2
The route HMS *Beagle* followed around the world during her five-year voyage.

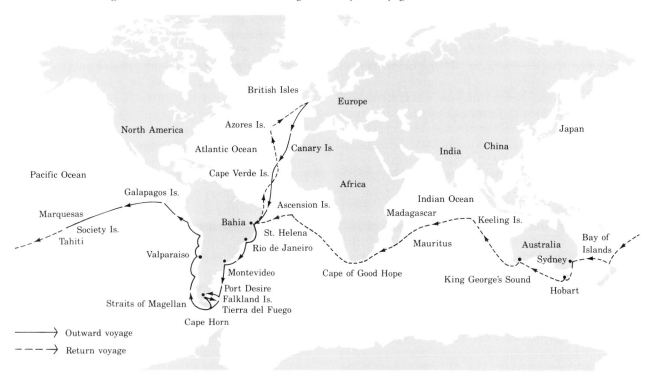

several months while Fitz-Roy charted the northern coast of Brazil.

In Argentina, Darwin unearthed a number of fossil remains of giant animals. He knew that there were living animals similar to these ancient beasts, but smaller. He began to wonder whether animals constantly changed—whether those that failed to adjust to the environment died. If he was right, then the creatures alive today, he reasoned, must be different from those God originally created. He began to doubt whether God could have created all the plants and animals he had seen.

Fitz-Roy made his own interpretation Darwin's giant fossils. He concluded that the beasts had been too large to board the Ark and had drowned in the Great Flood. Darwin had also found marine seashells among his fossils, and he thought that the land had once been sea bottom. Fitz-Roy quickly concluded that the sea had risen during the Great Flood to trap both seashells and giants. The frequent exchanges of ideas and contradictory theories between Darwin and

Fitz-Roy were instrumental in convincing Darwin of the truth of his ideas. Fitz-Roy was his sounding board.

Perhaps Darwin's greatest moment was the day the *Beagle* made port in the Galapagos Islands, September 15, 1835 (Figure 17–3). The average person would not be impressed with the Galapagos Islands. They are nearly barren volcanic land. But Darwin could not believe the immense variety of animals and plants that he saw and the unbelievable tameness of the animals. He found giant tortoises (*galapagos* is Spanish for "tortoise"), marine and land iguanas, and twenty-six species of land birds. Darwin quickly recognized that he was surrounded by animals that were similar to the specimens he had collected or observed earlier in South America.

Darwin also found out that the animals were different from island to island, even though the islands were only forty or fifty miles apart. For example, the tortoises of one island had different shell shapes from those of other islands. Darwin was particularly fasci-

Figure 17–3
A view across one of the Galapagos Islands. Volcanic islands such as the Galapagos are desolate and look nearly devoid of life. The view here is typical. (T. A. De Roy; Bruce Coleman, Inc.)

The little finches of the Galapagos today are known as Darwin's finches. In fact, the Galapagos species on which Darwin based much of his theory of natural selection are today reasonably undisturbed and in surroundings similar to those they had when Darwin first observed them. That is because some of the islands today are a wildlife sanctuary, visited by thousands of tourists each year (most of them biologists). The only permanent inhabitants are those working at the Darwin Foundation.

Figure 17–4
The woodpecker finch. There are fourteen species of finches called Darwin's finches on the Galapagos Islands. There are no woodpeckers on the Galapagos Islands, so this little finch occupies the niche commonly occupied by woodpeckers in other geographic areas. However, the woodpecker finch does not have the woodpecker's long tongue, which is used for probing deep into trees for grubs and insects. As a substitute, the woodpecker finch uses a cactus spine as a tool for probing. (Bruce Coleman, Inc.)

nated with the variety of finches. On one island, these birds had strong bills, for crushing seeds. On another island, their bills were more slender, for catching insects. On yet another island, the bills were adapted for feeding on fruits. One species of finch, the woodpecker finch, used a cactus spine as a tool to search for food in the holes of trees (see Figure 17–4).

What Darwin witnessed in the Galapagos was evidence of the formation of species. The islands were (and are) relatively new, being the result of volcanic upheavals. At first no life existed on them. Birds probably flew over them, and the seeds in their droppings began the vegetation. Some seeds may have floated in from the mainland. The tortoises could have come on tree rafts, or they may have been marine tortoises first. As each species arrived, it adapted to the food and to the habitat.

As the species adjusted, they underwent change. There were no woodpeckers on the Galapagos Islands. That is why one of the little finches was able to capitalize on the woodpecker's mode of existence, using cactus spines instead of a long pointed bill. The opportunities for the pioneer species were different from island to island, and so were their adaptations. Thus they diversified into new species. Darwin later wrote that the Galapagos Islands brought him nearer to "that great fact—the mystery of mysteries—the first appearance of new beings on this earth."

Once again Darwin presented his new ideas to Fitz-Roy. Fitz-Roy contended that it was Divine Wisdom that created a finch for every niche.

The *Beagle* arrived home on October 2, 1836. Darwin was not the same carefree youth who had left

five years before. While away, he had written that a man who wastes one hour will never discover the value of life. He could no longer tolerate idleness. He had become a disciplined scientist.

The next two years were among the most productive years of Darwin's life. He edited a five-volume work, *Zoology of the Voyage of the H.M.S. Beagle,* and published in 1839 the journal he had kept while aboard the *Beagle.* In that same year he married his cousin, Emma Wedgwood (who had had the best of everything, including piano lessons from Chopin). Darwin and his bride lived a few years in London. However, Darwin was becoming progressively more ill (later he would be nearly an invalid), and he hated the city. His father bought him a home in Down, sixteen miles from London, where he lived until his death, in 1882. Today the house is a museum, library, and mecca for biologists from around the world.

After the voyage and for the rest of his life, Darwin was very ill. He constantly complained of fever, dizziness, paralysis, and aches and pains. He never knew the true nature of his disease. Authorities today believe that he suffered from Chagas' disease, which is caused by a parasite spread by bloodsucking black bugs. Darwin had written about the irritating bites of these bugs when he was in South America.

Because of his illness, Darwin rarely left Down, even to go to nearby London. He never attended scientific meetings or gave lectures. However, he was constantly represented in the mainstream of science by his friends and colleagues—Thomas Huxley, Joseph Hooker, John Henslow, and Charles Lyell. They believed that Darwin's findings were enormously important.

Darwin never held a paying job. His services aboard the *Beagle* were free. He was never a professor, consultant, or profit-making author, but his contributions were considerable. During his life, he wrote nineteen books and published dozens of scientific papers on many subjects, including barnacles, earthworms, orchids, plant movements, vegetables, behavior, and humans. His most important contribution was *On the Origin of Species by Means of Natural Selection.* Later in his life, Darwin commented that he was surprised by his own achievements.

The Origin of Species

Darwin's theory of evolution was not clear in his mind while he was on the *Beagle.* He collected a lot of evidence, but it did not fit together then. He was haunted, however, by the conviction that life on earth was changing constantly through heredity and environment.

In 1838, Darwin read Thomas Malthus's *An Essay on the Principle of Population,* which proposed that more individuals were produced than could be supported by the environment. Darwin began to realize that he was closer to the truth about the origin of species than he thought.

Darwin could have had a difficult time accepting his own ideas. He had come from a religious family and had a degree in theology. Christians then believed that the earth and all life on it were created on October 23, 4004 BC, a Sunday, at 9:00 AM. Darwin now challenged that belief. He wrote that his faith in

religion eroded slowly, but when it did he felt no distress and never doubted that he was correct. It may have taken Darwin nearly twenty years to come to this conclusion (Figure 17–5).

In the summer of 1858 Darwin received a letter from Malaya from Alfred Russel Wallace, who sent with it an essay entitled *On the Tendency of Varieties to Depart Indefinitely from the Original Type.* The essay stated exactly what Darwin had been thinking for twenty years. In fact, Wallace had formulated the same theory as Darwin had about the creation of species. Darwin's friends Lyell and Hooker thought that he and Wallace should report their theory together. A joint paper was presented to the Linnaean Society the following month.

Darwin's *On the Origin of Species by Means of Natural Selection* was published the next year, in 1859. The first edition, totaling 1,250 copies, was sold out the first day. The book underwent seven editions, the last in 1871. It did not interest everyone, but it did create considerable interest among scientists, many of whom had been thinking along similar lines. The average Englishman did not comprehend the magnitude of Darwin's theory. Most of the popular attention Darwin received was from misinterpretations of his book. Many thought that he had implied that humans were descended from apes. This erroneous interpretation caught the interest of the press, which pub-

Figure 17–5
Charles Darwin as an old man. (Courtesy of the American Museum of Natural History.)

lished numerous cartoons mocking Darwin (see Figure 17–6).

The only organized objection to Darwin's theory came from the Church. The clergy debated the issue at the meetings of the British Association for the Advancement of Science held June 28, 1860, in Oxford. The Church was represented by Bishop Samuel Wilberforce, an eloquent speaker who had earned the title Soapy Sam. The Church intended to smash Darwin at the debate. Darwin was ill and did not attend. He was represented by his friends Professor Henslow, Thomas Huxley, and Joseph Hooker. The auditorium was packed with thrill seekers, those truly interested, and students who could hardly wait to see what would happen. Bishop Wilberforce eloquently put down Darwin and his ideas of evolution, calling the ideas a "casual theory."

Wilberforce turned to Huxley, also sitting at the speakers' table, and asked him whether it was his "grandfather or grandmother that had descended from an ape." Huxley, already perturbed, stood and

Figure 17–6
Cartoons of this sort were common in English magazines and newspapers when Charles Darwin first published his theory of evolution. This cartoon, from the magazine *Punch,* shows an ape wondering if he may also be a human and perhaps our brother. Darwin's critics did not fully understand his theory. (The Bettman Archive.)

said that he "would have rather descended from an ape than from a cultivated man who prostituted the gifts of culture and eloquence to the service of prejudice and falsehood." It was blasphemy to criticize the Church, and to insult the clergy was nearly treason. Huxley's retort caused a great stir in the audience. Some cheered, some snickered, and one woman fainted. Huxley's response ended the debate, and the audience filed out in confusion.

During the confusion, a man went to the podium holding a Bible. He declared that the only truth was in the book he held. Few heard him, and fewer recognized him. It was Captain Fitz-Roy. He realized that he had been instrumental in helping Darwin come to his conclusions about the creation of species, and that was the greatest sadness in his life. He had originally hoped that together they would prove the Bible, not disprove it.

The Scopes Trial

Charles Darwin did not live to see the end of the debate. Sixty-five years later, in Dayton, Tennessee, a high school teacher named John Scopes agreed to be arrested to challenge a law that prohibited the teaching of evolution. Scopes was defended by Clarence Darrow from Chicago, a famous defense attorney. The prosecution was represented by the even more famous William Jennings Bryant, who had run for president of the United States on three occasions. Scopes was found guilty and fined $100. Later, the case was dismissed by a higher court. However, it was well publicized and was popularly called the "monkey trial."

The monkey trial didn't change anything in Tennessee. In fact, the teaching of evolution was unlawful in that state until 1967. Tennessee was the last state to abolish such a law.

The Scopes trial came alive again in 1972. The U.S. premier of the film *The Darwin Adventure* was shown in the same courtroom in which the famous trial had occurred.★ Charles Darwin's grandson flew from England to speak after the film. After his speech, a group of religious fundamentalists rose and attempted to debate the issue of evolution all over again. What had

★*The Darwin Adventure* is an excellent biography of the life and times of Charles Darwin. It was filmed on location around the world. An equally interesting film is *Inherit the Wind,* a Hollywood dramatization of the Scopes trial.

Darwin discovered that rocked the world for more than a hundred years?

THEORIES OF EVOLUTION

Darwin's ideas were not entirely original. Others had given thought to evolution before him. Darwin's own grandfather, Erasmus Darwin, had theorized about evolution. Erasmus was an eccentric but good scientific thinker. He was an enormous man with a ravenous appetite. He was married twice and had eleven legitimate children and one illegitimate child. He belonged to a scientific group known as the Lunar Club (whose members were called "lunatics"), which investigated new scientific ideas. Evolution was one of them.

Erasmus never came to any conclusion about evolution, but he did put some of his ideas into a poem, "The Botanic Garden." In the poem he described how all living things came from a common ancestor, the "living filament." He published two other works on the same subject. Charles Darwin believed that his grandfather's ideas never influenced him.

Jean Baptiste Lamarck (1744–1829), a French botanist and zoologist, formalized the first theories of evolution. In 1809, he proposed "the theory of the inheritance of acquired characteristics." His theory is often called the theory of use and disuse. He contended that an animal's anatomy did not dictate its habits. Instead, the habits were dictated by the use of specific organs by the animal's ancestors. Lamarck saw evolution as the inheritance of acquired characteristics. For example, he perceived sightless moles as being the result of ancestors who had no use for eyes. He believed that swimming birds acquired web feet and wading birds acquired long legs because of their need. He thought that the sons of blacksmiths inherited the muscular arms of their fathers. His most famous example was the long necks of giraffes (see Figure 17–7). He believed that their long necks were acquired because each generation had stretched farther and farther to reach the leaves on trees. Lamarck went so far as to predict that if several children were deprived of one eye and then interbred, they would produce a one-eyed race. August Weismann, a German biologist, tested Lamarck's ideas much later. He cut the tails off mice each generation for twenty-two generations, but the offspring of every generation continued to be born with tails.

The principle of evolution that we know today is not difficult to comprehend. If Darwin and Wallace had not made the discovery, it would have been made by other scientists—but perhaps much later. *The Origin of the Species* stated the following provisions about **evolution:**

— Darwin had been impressed by Malthus's paper, which proposed that the numbers of animals, and humans, could increase geometrically (2–4–8–16 and so on) but that the resources necessary to support life could increase only arithmetically (2–3–4–5–6 and so on). In simple terms, more life is produced than the planet can support. Most organisms produce thousands, if not millions, of gametes. Not all of them fuse to form new individuals, but many more begin an existence than can be supported by the environment.

— From these facts, Darwin proposed that the organisms that survived were the result of **natural selection,** or survival of the fittest. He did not propose a physical struggle among organisms for an existence— a common misinterpretation of the principle of evolution. Instead he proposed that organisms would compete for space, food, shelter, water, and other resources. Only those organisms best adapted to the environment would survive. The remainder would perish.

— Darwin proposed that new variations of characteristics were created every generation but that only the individuals that successfully adapted would survive. Over a long period of time, the environment would change and these successfully adapting individuals would become a new species.

— Finally, the organisms that survived would reproduce and pass to their offspring the characteristics that allowed them to survive. Darwin referred to this process as heredity, but he did not know how organisms inherited characteristics. This was one of the weakest points of Darwin's theory, simply because the principles of genetics were unknown to him at the time.

A more modern statement defining the principle of evolution is not much different from what Darwin proposed. Today we say that new species develop from earlier forms by the transmission of slight variations in successive generations. The most technically precise definition of evolution is changes in allele frequency that occur in populations of organisms in time. More simply, evolution is the process by which

CONCEPT SUMMARY

MAJOR POINTS OF THE DARWIN AND WALLACE THEORY ON THE ORIGIN OF SPECIES BY NATURAL SELECTION

1. Organisms produce more offspring than the environment can support.
2. Offspring must therefore compete for limited environmental resources.
3. The members of a species vary considerably in their ability to compete for resources.
4. Those that are best adapted (most fit) to compete survive to reproduce and pass these successful traits to their offspring.
5. Offspring that inherit favorable traits are in turn better adapted to compete for resources and to survive.

the genetic makeup of species changes in concordance with a changing environment.

THE EVIDENCES OF EVOLUTION

What we know today about the origin of life and its evolution into current forms is the product of careful scientific investigation. Because many of the previous life forms no longer exist, however, scientists have had to rely on other evidence—most notably fossils and comparisons of living creatures.

Fossils

If you have ever been to a museum of natural history, you have undoubtedly passed display cases containing **fossils**—bonelike impressions or remains of organisms (see Figure 17–8). The number of fossils displayed was probably enough to overwhelm you, but it represents only a minute fraction of the life that has been present on the earth in past millennia. This fraction of fossils is referred to as the fossil record.

The process of fossilization is complicated and poorly understood. Fossil bones are not bones at all. During fossilization, the entire bone or just a portion may be replaced through mineralization, or mineral replacement. Scientists do know that few whole animals or plants ever become fossils. In fact, it is rare to find all of the remains of a single organism. The complete skeletal remains that are viewed in museums are often composites of many individuals of the same species. Many of the parts are made of plaster and are used to help paleontologists visualize and reconstruct the whole organism.

Many dead organisms that might become fossils are eaten by scavengers, who dismember them and scatter their bones. The bones are frequently crushed by larger animals and thus pass into oblivion as dust. Plants and animals that escape scavengers may be fossilized, but usually only in two specific situations. First, they may become buried in sediments. Sedimentation occurs most readily in water, so some of the best fossils are obtained from land masses that were once ocean bottoms. Second, organisms may become buried in volcanic dust, where they fossilize beautifully. There are other minor means of fossilization also. Organisms may become trapped in tree pitch, which later hardens as amber (see Figure 17–9). They may be trapped in molten tar or preserved by freezing in polar ice caps. Most organisms never become fossils, because they die where none of these situations exist.

Bones comprise much of the existing fossil evidence. Teeth, made of enamel, are even more resistant to environmental deterioration. The shape and ridges of teeth precisely identify the diet of an extinct species. (The soft parts of animals decompose and are recycled through the ecosystem.) Some animals left footprints or body impressions. The shapes and contours of hard egg shells in soft sediments have also

Figure 17–7
(On facing page.) A comparison of Charles Darwin's theory of evolution and Jean Baptiste Lamarck's idea of the inheritance of acquired characteristics. According to Lamarck, animals who acquired characteristics during their lifetime would pass them on to their offspring. His most famous example is illustrated by the giraffes on the left (a). Those with short necks had to stretch and even attempt to climb trees to reach food. Such exercises, according to Lamarck, made their necks longer. The trait was passed on to the next generation of giraffes, which likewise had to stretch their necks to make them longer. This continued until the long-necked giraffes of today evolved. Darwin's theory of evolution (b), on the contrary, would contend that giraffes with short necks would perish, because they could not reach enough leaves to sustain life. Only giraffes with necks long enough to reach leaves would survive—the survival of the fittest. Natural selection still monitors giraffes, eliminating all but those with the longest necks. As a consequence, most giraffes today have very long necks.

(a) Lamarck's explanation

(b) Darwin's explanation

Figure 17–8

A bed of fossil sea lilies, or crinoids. These creatures had hard armor plates that left a permanent impression in sediments. The study of fossils helps reconstruct past events in the history of the earth. The modern-day relatives of crinoids are the starfishes and sea urchins. (Courtesy Field Museum of Natural History, Chicago.)

been faithfully preserved. The shells of Foraminifera, crabs, insects, and the like and of turtles and mollusks, the scales of fish and reptiles, and the spines of dinosaurs are only a few kinds of animal armor that often become fossils.

There are more plant fossils than animal fossils. The reason is based on a fundamental characteristic of all plant cells—the possession of a **cell wall**. The nonliving wall of plant cells is hard and resists decomposition better than the softer living plant parts do. As a result, the shape of the living plant becomes preserved in stone as an impression. (For further information about fossils, see Box 17–1.)

Rock Strata

You may have seen the Grand Canyon or some of the areas of "table rocks" in the southwestern United States (Figure 17–10), where rock strata are tourist attractions. Even if not, you have undoubtedly driven along superhighways whose construction caused entire hills to be cut through, exposing the layers of different kinds of rock. Fossils in these layers progress usually from simpler forms of life to more complex forms as you move from deeper to shallower rock layers. The oldest rocks contain simple, mostly microscopic fossils. Younger rocks show fossils similar to organisms existing today.

Each rock layer represents a different time in the earth's history. Some periods may be as brief as a few years, others as long as tens or hundreds or thousands of years. In some cases, the veins of earth (called varves) may represent only seasons. For example, during a rainy season, great quantities of sediment may be deposited, whereas very little is left during

Figure 17–9
Millions of years ago, this insect became trapped in tree pitch. Later the pitch hardened. Now it is transparent amber, and the insect inside is almost perfectly preserved. (Courtesy Field Museum of Natural History, Chicago.)

Figure 17–10
As sediments are deposited, they often form distinct layers, called rock strata, that may differ in color and thickness. Each stratum represents a period of time in the history of the earth; the strata on top are newer than those below. When rivers cut through the strata, they create valleys of spectacular view, such as the Badlands of South Dakota (shown here) and the Grand Canyon. (Courtesy Field Museum of Natural History, Chicago.)

CHAPTER 17 *Darwin and the Evidence of Evolution*

BOX 17–1

DATING FOSSILS

Once a fossil is located, dusted, and brought back to the laboratory or museum, it must be identified. Paleontologists (fossil experts) specialize in specific animal or plant groups, such as extinct birds, reptiles, or mammals. It is often easy for paleontologists to quickly decide what type of animal or plant has been found. Their major question is always how long ago the creature lived.

Fossils can be dated in several ways. One way is called the carbon-14 method (commonly written ^{14}C). The stable form of carbon, and therefore the carbon that is most frequently cycled through food chains and webs, is carbon-12. However, part of the carbon that is cycled is carbon-14. ^{14}C, an unstable isotope, gives off beta particles, which make it radioactive. Because of its instability, ^{14}C is slowly converted—that is, it decays—to the next possible stable form, which is nitrogen-14.

The rate of change for ^{14}C is precisely known. It takes exactly 5,600 years for half of a specific amount of ^{14}C to be converted to ^{14}N. Thus the half-life for ^{14}C is 5,600 years. For example, if you had 2 micrograms of ^{14}C, in 5,600 years you would have only 1 microgram; in 11,200 years you would have only 0.5 microgram; and in 22,400 years, only 0.25 microgram.

The ratio of ^{12}C to ^{14}C in specific tissues, such as bone, is known for animals alive today. By biochemical assay, the ratio of ^{12}C to ^{14}C in a fossil is measured as the amount of beta particle emission. The result can be converted into years of age. However, when the specimen being dated is much over 50,000 years old,

the technique loses its accuracy, because little (or none) of the original ^{14}C is left. Thus this method is useful for dating only very recent fossils.

A similar method for dating fossils involves the conversion of potassium-40 (^{40}K) to argon. When an organism is trapped in volcanic ash, argon gas cannot escape. The decay rate can therefore be computed when the fossil is found. This method allows dating in millions of years, because the half-life of ^{40}K is 300 million years.

When fossils are still older, the conversion of uranium-238 to lead-206 can be measured. The half-life is 4.5 billion years. This method has been particularly useful in dating rocks and determining the age of the earth's strata.

Fission-track dating, a new method, is like uranium-238 dating in reverse. If uranium-238 is placed in an atomic reactor, the explosions that accompany its decay create a series of fine etchings in glass. These etchings can be observed with a microscope. The same process happens naturally in active volcanoes. A glass specimen taken from volcanic sediment shows fission tracks that can be counted. The specimen is placed in an atomic reactor so that the remaining ^{238}U will be used. During its decay, a new set of etchings is created. The total number of etch lines is proportional to the original amount of ^{238}U and, therefore, proportional to age.

These methods of fossil dating, singly or in any combination, allow paleontologists to determine the age of fossil remains and rocks with reasonable accuracy. Of course, the methods are not perfect, so some degree of error is expected. However, an error of a few thousand years, or even of tens of thousands of years, is negligible when one considers fossils that are several million years old.

---

the drier part of the year. The deposits made during wet versus dry seasons often have different colors and textures.

These strata allow paleontologists to estimate when the various layers were deposited. For example, in the Olduvai Gorge in Africa, where our human ancestors lived, three-hundred feet of stratified rock and ash are exposed. They represent about 2 million years of sedimentation.

If the studied strata of the earth's crust do not contain volcanic ash, it is nearly impossible to date them. The ratio of potassium-40 to argon is best preserved in volcanic ash. When one layer can be dated, then the age of fossils above and below can be estimated, which eliminates the need to date each fossil. For ex-

ample, one volcanic ash layer may be 1 million years old, and a lower layer may be 2 million years old. The fossils above the first layer are less than 1 million years old. The ones below the second layer are older than 2 million years. Those in between are between 1 and 2 million years old. The position of the fossils between the layers is also an indication of the age differences. Those positioned near the 1-million-year layer are younger than those farther down.

Continental Drift

There is a great amount of evidence that the evolution of life on earth is also demonstrated by the past movements of the continents. Even today, the hard crust of the earth is suspended on a softer layer that is in constant motion. The theory of **plate tectonics** explains that the continents and parts of the ocean floors are like plates slowly floating on a semiliquid gel.

At one time, about 230 million years ago, there was only a single, very large continent, Pangaea (see Figure 17–11). (Pan means "all"; geo means "earth.") The movements of the softer, slowly moving layers under Pangaea broke up this single continent and formed the continents we know today. Figure 17–11 shows how nicely the east coast of South America fits, like a piece of a jigsaw puzzle, with the west coast of Africa. Similar matches can be made between the other continents on the earth.

Continental drift has helped explain the distribution of living organisms around the world. For example, much of the wildlife of Australia is unique. Animals commonly found on other continents are absent in Australia. Marsupial species, mammals that carry their young in pouches, are abundant in Australia but sparse in the rest of the world. (The opossum is the only marsupial in North America.) The reason for the abundance of marsupials in Australia is that they evolved after the break-up of Pangaea, when Australia became an isolated island (see Figure 17–11).

Their ancestors inhabited other continents but became extinct because of competition, which was nearly nonexistent on the island of Australia.

Certain fish on the east coast of Africa have close relatives in India. How did these fish get to India? They are freshwater fish, so they could not have swum the oceans. Continental drift explains this dilemma. India was at one time part of Africa. Continental drift also explains why large flightless birds are found only in South America (Darwin's rhea), Africa (ostriches), and New Zealand (kiwi). Their common ancestor was from Gondwana (see Figure 17–11).

Continental drift also explains the geology of continents, such as the position of rock strata and mountain ranges. The continents still move at the slow but constant rate of a few millimeters per year. That movement is enough to cause observable results. Africa is moving northward and colliding with the Middle East, making the northern tier of the Middle East one of the most earthquake-prone areas on earth. A similar phenomenon is occurring along the California coast, near the San Andreas Fault. A serious earthquake there is inevitable. Like a conveyor belt, a slow ooze of inner earth emerges along midoceanic ridges and is pulled inward again along oceanic trenches or

Figure 17–11

Pangaea. (a) About 230 million years ago there was only one large continent on the earth, Pangaea. (b) About 180 million years ago, it began to break up into two large land masses, a northern mass called Laurasia and a southern one called Gondwana. As these land masses continued to break apart (about 65 million years ago), the continents as they are today were established.

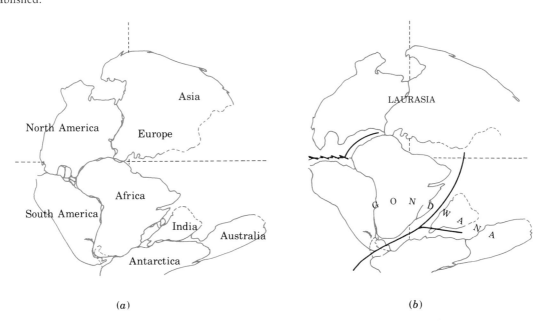

(a)

(b)

below mountain ranges. The continents ride on this movement.

The Geological Time Scale

By using various dating techniques, a history of the earth has been constructed. The chart of the earth's history is known as the geological time scale. It is represented in abbreviated form in Figure 17–12.

The history of the earth is divided into major eras, each with a descriptive name. The eras shown in Figure 17–12 are the Paleozoic (*paleo* meaning "old" and *zoic* meaning "animal"; hence "old animal"), the Mesozoic (*meso* meaning "middle"; hence "animal in the middle of the past"), and Cenozoic (*ceno* meaning "new"; hence "newer forms of animal"). Representatives of each era are shown in Figure 17–12.

Inspection of the figure reveals that the genus *Homo* (the human being) is rather new on earth. One of the great scientific mysteries is not how we evolved□ but how much longer we will last on earth as a viable genus. Figure 17–12 also corrects an erroneous assumption of filmmakers. Dinosaurs and humans did not roam the earth at the same time. They were separated by at least 70 to 75 million years.

In summary, fossils found in the crust of the earth offer the best evidence that the earth has been inhabited by many forms of animal and plant life, many of which are no longer living. The fossil record also provides irrefutable evidence that life was not created all at once. Living species have changed as a result of changing environments, just as Darwin proposed.

Comparative Studies

Evidence of evolution is not based solely on the existence of fossils in the crust of the earth. Comparison of **extant** (living) species can be just as important as studying **extinct** (past) species in understanding the process of evolution. The comparing and contrasting of extant species is known as the comparative approach. (This method is also used to interpret the behavior of organisms.) Similarities and differences among living organisms provide information about origins and degrees of relatedness.

Comparative Anatomy

There are more species of insects on the earth than of any other living group of animals. Despite their large variety, insects have similar anatomical features.

Whether they are dragonflies, honeybees, or beetles, they all have three body regions—head, thorax, and abdomen—and three pairs of legs on the thorax. No other animals have these characteristics. Thus it has been concluded that all insects are related and that they have evolved from a common ancestor.

Although frogs, snakes, birds, and dogs look different, they share characteristics that make them vertebrates. They all have gill slits (often only embryonic ones), a vertebral column, and a dorsal hollow nerve cord. Therefore, all vertebrates are related, and they inherited certain anatomical structures from a common ancestor.

Because the majority of vertebrate fossils are bones and teeth, the comparative anatomy of these structures is studied by paleontologists. The skeletal elements of all tetrapods (four-legged vertebrates) are remarkably similar. Tetrapods have two major skeletal parts: the **axial skeleton,** composed of the skull and the vertebrae, and the **appendicular skeleton,** which supports the appendages (legs and arms). Comparison of tetrapod skulls makes it obvious that, through evolution, the number of separate bony elements in the skulls has decreased. Many have fused. Thus the more recent tetrapods have fewer bony elements in their skulls.

The trunk contains the supporting skeleton, or vertebral column. The vertebrae of all tetrapods, from front to back or top to bottom, are classified as cervical, thoracic, lumbar, and caudal. The number of vertebrae in each region may vary from group to group, but each division is always represented. The vertebrae all have similar spines for the attachment of tendons and muscles. In addition, all tetrapods have two special cervical vertebrae, the first two below the skull. These vertebrae are called the atlas and the axis. The way they are put together allows the tetrapod to swing its head from side to side.

The appendicular skeleton consists of the bony elements that support the limbs. All tetrapods have a pectoral region (shoulder), from which the forelimbs extend, and a pelvic region (hip), from which the hindlimbs extend. The bony elements of the limbs of all tetrapods are remarkably similar. Figure 17–13 illustrates this similarity. The human forearm and hand have the same basic bony elements as do the wing of a bird, the forelimb of a lizard, and the forelimb of a frog. Some animals that have evolved in the tetrapod lineage lack hindlimbs, but they still have pelvic girdles and may also possess some of the bony elements

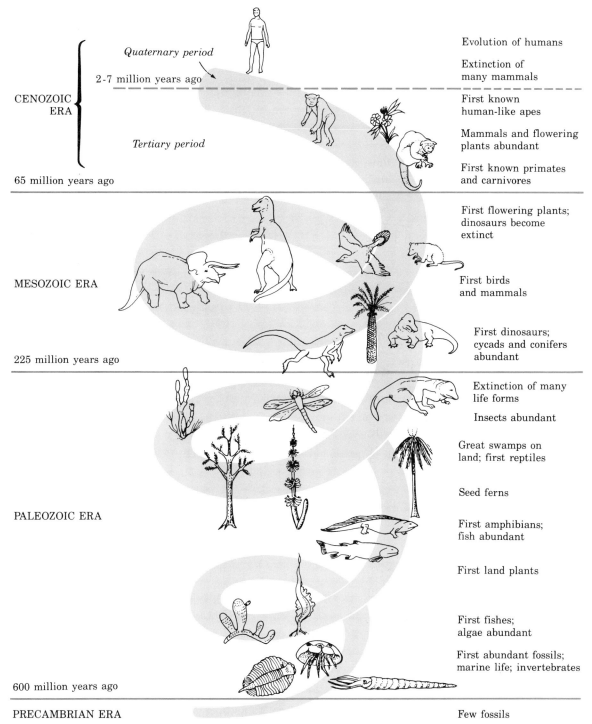

CENOZOIC ERA

Quaternary period

2-7 million years ago

Tertiary period

65 million years ago

Evolution of humans

Extinction of many mammals

First known human-like apes

Mammals and flowering plants abundant

First known primates and carnivores

MESOZOIC ERA

225 million years ago

First flowering plants; dinosaurs become extinct

First birds and mammals

First dinosaurs; cycads and conifers abundant

PALEOZOIC ERA

600 million years ago

Extinction of many life forms

Insects abundant

Great swamps on land; first reptiles

Seed ferns

First amphibians; fish abundant

First land plants

First fishes; algae abundant

First abundant fossils; marine life; invertebrates

PRECAMBRIAN ERA

Few fossils

Figure 17–12

A simplified geological time scale showing the major eras of the geological past and periods of the Cenozoic era. The spiral includes the forms of life prevalent at the time and new forms of life that evolved during the various eras. However, life did not evolve in a single linear pattern. Ancient groups gave rise to the separate lineages in existence today. Other groups, such as the dinosaurs, became extinct. A more realistic representation of the evolution of life would be a complex system of branching and rebranching lines, some leading from ancient ancestral forms to numerous groups of existing descendants and others ending in extinction some time in the past.

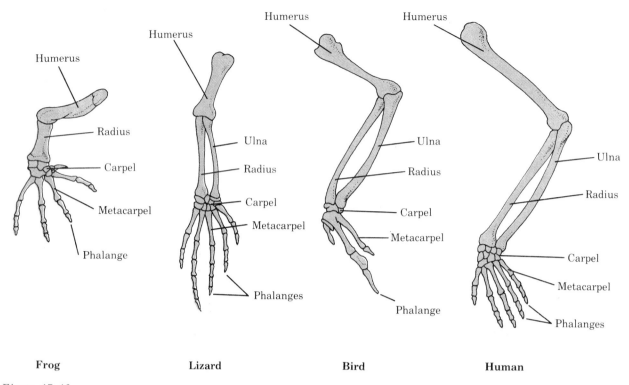

| Frog | Lizard | Bird | Human |

Figure 17–13

Homologous structures. Your own arm, the wing of a bird, the forelimb of a lizard, and that of a frog have the same bone elements and are therefore classified as homologous structures. Organisms that have such structures share a common ancestry.

of hindlimbs. These structures can be seen in the skeleton of some snakes and all whales.

Anatomically, comparative structures can be either homologous or analagous. Homologous structures are structures on different animals or plants that have a similar form and embryological development because they were inherited from a common ancestor. For example, the forearm and hand of a human, are homologous with the wing of a bird, and the forelimb of a lizard, or a frog. (see Figure 17–13). Analagous structures are structures that are different anatomically, because they did not evolve from a common ancestor, but that have similar functions. The wings of a bird are analagous to the wings of an insect. In both cases, they are used for flying. However, insect wings have no internal skeletal support, whereas bird wings do.

The comparative anatomist is equipped to explain the degrees of relationship among organisms, which give information about their evolution. (The scheme of classification in Appendix 2 is based on comparative anatomy.) If your hand were compared to similar structures in other tetrapods and rated by the number of similarities on a scale from 1 to 10 (least to most similarities), the following scores would emerge, compared to a bird wing, score 1; to a bat wing, 5; to a whale flipper, 6; to a monkey hand, 8; to a great ape (chimpanzee, gorilla, and orangutan) hand, 9. Therefore, humans and the great apes are more closely related than are humans and monkeys or humans and any of the other creatures. This example is an oversimplification, but it does illustrate anatomical similarities among closely related organisms.

Comparative Embryology

Shortly after Darwin published *On the Origin of Species,* one of his supporters, Ernst Haeckel, proposed the law of biogenesis. Haeckel and many biologists of his era became aware that the embryos of vertebrates were similar and followed similar developmental patterns. The similarities produced adults with similar anatomies.

Figure 17–14 illustrates the similarities among ver-

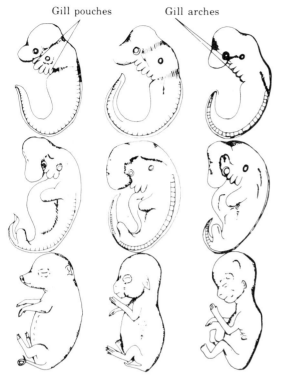

Hog   Rabbit   Human

Gill pouches   Gill arches

Figure 17–14

Comparative embryology. Another tool used in determining the relationships among organisms is the study of the similarities and differences in their embryonic development. In this case, the embryonic development of a human, a rabbit, and a hog are compared. The earliest stages look almost identical. Only minor differences are obvious during the middle stages. Definitive differences become more obvious during the later stages. Similarities in embryonic development is evidence that organisms are related.

tebrate embryos. Look closely at the mammals represented—hog, rabbit, and human. The early stages of development are nearly identical. The intermediate stages are quite similar. Only the final stages show the definitive characteristics of the species.

The comparison of embryos and their developmental processes is another means of measuring relationships. Relationships provide information about the past history or evolution of species. For example, vertebrate embryos that develop four-chambered hearts are more closely related to those that develop three-chambered hearts than to those that develop two-chambered hearts.

Vestigial Structures

Some structures in vertebrate embryos develop and then disappear. Other structures become part of the adult but apparently have no function. The latter are called **vestigial structures.** For example, one digit (actually the thumb) on each foot of cats and dogs never touches the ground because it is high on the foot, near the ankle. The structure has no obvious function, but it is a vestige of these animals' evolutionary history. Cats and dogs evolved from ancestors with five digits, all on the ground. The same is true for horses. They stand only on the tip of one toe, but vestigial digits can be found higher up the leg (see Figure 17–15).

Many vertebrates have a third eyelid, called the nictitating membrane. This lid can be seen in frogs, turtles, alligators, and many other aquatic animals. When they submerge themselves in water, the transparent membrane moves across the eye to protect the cornea while permitting sight. Vertebrates that do not have a nictitating membrane have a vestige of this structure, called the semilunar fold. Look into a mirror and examine the corner of each eye nearest the nose. You can see a small fold of skin, the semilunar fold, which is a vestige of a nictitating membrane.

We also have other supposed vestigial structures, such as tonsils and an appendix. The appendix is an extension of an upper region of the colon, called the caecum. In many vertebrates, the caecum is a large structure that houses a complex of microorganisms that digest cellulose. Thus it is a necessary digestive organ for herbivorous (plant-eating) animals. The human caecum, however, is quite small and no longer a chamber for cellulose digestion. The tiny appendix, is evidence that our ancestors may have been strictly herbivorous rather than omnivorous (plant- and flesh-eating). The appendix often becomes engorged with food and microorganisms. This engorgement may cause inflammation that requires the surgical removal of the appendix. However, neither the appendix nor the tonsils are entirely useless. Both are composed of lymphoid tissues, so they may contribute to the immune responses of the body.

Another example of a vestigial structure is associated with the ears. Many common mammals, including dogs and cats, are capable of moving their ears. Humans have vestigial muscles for ear movement. Most of us, however, never develop control of them. (Can you wiggle your ears?) Darwin described the ear

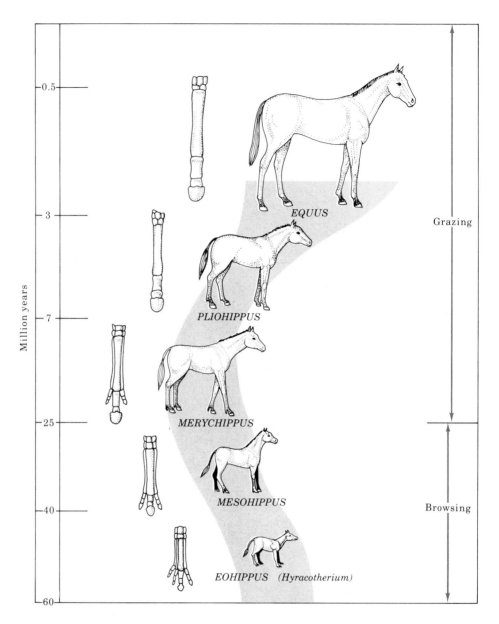

Figure 17–15
The horses of today have a single toe, on which they stand. They evolved from an ancestor that had five toes, which is typical of most mammals. As the middle toe became more and more important, the digits surrounding it became vestigial. Horses now have a few bony elements around their single toes that are vestiges of their past evolution.

Million years

-0.5

-3

-7

-25

-40

-60

EQUUS

PLIOHIPPUS

MERYCHIPPUS

MESOHIPPUS

EOHIPPUS (Hyracotherium)

Grazing

Browsing

point (now called Darwin's ear point), which is a point of cartilage in the rim of the outer ear of some humans (see Figure 17–16). He suggested that this lump of cartilage was a reminder of ancestors with pointed ears. Today we are left with only a vestige.

Comparative Protein Chemistry

Biochemists have developed a variety of sophisticated techniques that allow them to identify proteins, the amino acid sequences of proteins□, enzymes, and the actions of enzymes. The sequences of amino acids in enzymes and other proteins are direct translations of specific genes (alleles). Species sharing numerous biochemical traits are closely related. Therefore, the comparison of various proteins provides evidence for the degrees of genetic similarity among species.

Hemoglobin, the red blood pigment, has been studied extensively. So has the liquid portion of the blood, **plasma,** which is rich in various proteins collectively known as serum proteins. Many of these proteins are important in general homeostasis and immune reactions□. A recent survey of the plasma proteins of humans and chimpanzees shows amazing similarities. In fact, the investigator making the com-

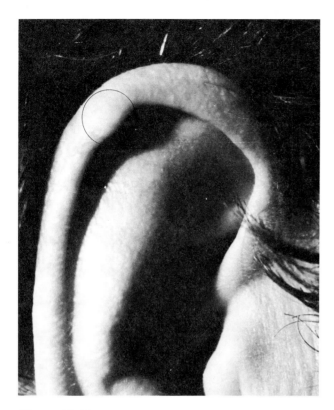

Figure 17–16

Darwin's earpoint. Many humans have a section of folded cartilage in the outer rim of the ear *(circled)*. Darwin suggested that this cartilage indicates our relatedness to animals that have pointed ears. Today we are left with only the vestige. (Photo by Tom Ardelt.)

parison concluded that, on the basis of blood proteins, humans *(Homo sapiens)* and chimpanzees *(Pan troglodytes)* should not be separate species but should constitute only subspecies.

The immune reaction is used to determine the relatedness of species and to provide evidence of evolution. Figure 17–17 illustrates the reaction of dog proteins with the antibodies from a wolf, a fox, and a cow. Such tests begin when some tissue of one animal, in this case the dog, is liquified and then injected into a rabbit. (A rabbit is used only because it is a convenient and cheap laboratory animal.) The muscle proteins of the dog act as antigens in the rabbit, and the rabbit manufactures antibodies for each of the antigens. Later, the serum of the rabbit is tested with other liquified muscle preparations. If a dog muscle protein preparation is mixed with the rabbit serum, a large number of specific antibodies for many

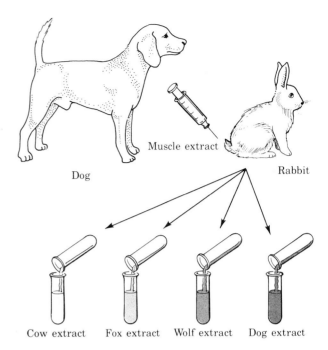

Cow extract   Fox extract   Wolf extract   Dog extract

Figure 17–17

Similarities in protein chemistry can be used to determine relatedness among organisms. In one type of test, dog muscle tissue is liquified and injected into a rabbit. The rabbit produces antibodies against the muscle proteins of the dog (antigens). Later, the antibodies in the rabbit's blood can be isolated. If the rabbit antibodies are then mixed with dog muscle proteins, many antigen-antibody complexes will form. If the rabbit antibodies are mixed with wolf muscle proteins, there will be fewer complexes formed. Even fewer will be formed with fox extract and fewer yet with cow extract. The test shows that the dog is more closely related to the wolf than it is to the fox or cow.

specific dog proteins will show up in the rabbit serum and the antibody-antigen complexes will precipitate (clot together). When the rabbit serum is added to wolf extract, the precipitate is less, because the wolf has proteins for which there are no antibodies in the rabbit serum. There are fewer antibodies for the fox extract and even fewer for the cow. The final comparison shows that the dog, wolf, and fox are more closely related to each other by protein chemistry (expression of genes) than they are to a cow. The dog, wolf, and fox are all **carnivores** (flesh eaters) and look alike, so we would expect some relatedness—which is verified by the results.

Biochemists have also devised tests that allow direct comparison of base pairs in DNA molecules from different species. Comparison of base pairs is in effect comparison of specific genes.

A great portion of our knowledge of human anatomy, physiology, and reproduction has been obtained by studying closely related animals, such as monkeys and apes, and many laboratory animals, such as mice, rats, rabbits, dogs, and cats. Almost all of the prescription and over-the-counter drugs that we consume were first administered to a variety of laboratory animals to see how they would react. Many of the hormones used to treat humans for various disorders are extracted from other animals—including cattle, pigs, and sheep—and they work perfectly well in humans. These facts became known because we are related through evolution to the animals used as our surrogates for research, and this is readily accepted by almost everyone. Yet, ironically, many people reject their relationship with other animals.

SUMMARY

1. Charles Darwin was born an aristocrat in England. He was to pursue a medical career but failed, so then he prepared for the clergy.

2. An unexpected opportunity arose. Darwin was offered the position of naturalist aboard HMS *Beagle,* a ship bound around the world on a mapping expedition. Captain Fitz-Roy selected Darwin, hoping he could help prove the accuracy of the Book of Genesis.

3. During the voyage, Darwin began to formulate his ideas about evolution. His visit to the Galapagos Islands played a major role in his understanding of the creation of species.

4. Darwin published his theory in 1859 in *On the Origin of Species by Means of Natural Selection,* twenty-four years after his voyage.

5. The major points in Darwin's theory are:
   a. Organisms produce more offspring than the environment can support.
   b. Offspring must therefore compete for limited environmental resources.
   c. The members of a species vary considerably in their ability to compete for resources.
   d. Those that are best adapted (most fit) to compete survive to reproduce and pass these successful traits to their offspring.
   e. Offspring that inherit favorable traits are in turn better adapted to compete for resources and to survive.

6. The theory of evolution was publicly debated in 1860 and continued to be a subject of controversy into the twentieth century.

7. Darwin's theory of evolution replaced Jean Baptiste Lamarck's ridiculed theory of the inheritance of acquired characteristics.

8. The strata of rock in the earth's crust contain fossils. Rocks and fossils can be dated and are thus evidence of the creation of species.

9. Other evidence of evolution includes similarities in the anatomy and embryological development of related species, vestigial structures, and the comparative protein chemistry of living species.

THE NEXT DECADE

SOCIAL DARWINISM

Charles Darwin established a theory that became not only the focus of biology but also a philosophy of life. That philosophy will be a legacy to future generations.

Unfortunately the main concepts of Darwin's principle of evolution were misunderstood and misused. Darwin's phrase *struggle for existence* was quickly interpreted to mean that humans are not created equal, that some are more fit than others. Much of human prejudice today has arisen from this prostitution of Darwin's ideas.

At the same time, some people interpreted Darwin's writings to imply that if a group struggled forcefully enough, it could outsurvive other human groups less willing to compete. Karl Marx, for example, proposed that Darwin's theory was the basis for the historic struggle between social classes. The almost surrealistic dream of Adolf Hitler for

the primacy of a "master race" began with a similar misconception.

The principle of evolution has often been used as the "natural" basis for imperialism and to comfort the conscience of warmongers. Darwin wrote that the best-adapted populations would increase, and this became the excuse for imperialism throughout the world. War, in that philosophy, is a convenient way to eliminate competitors, the unfit, and the unwanted. The advocates of war saw the substantiation of their ideas in animal aggression. They considered war the real manifestation of the struggle for existence. Middle Eastern countries have been engaged in a tense and often violent dispute for many years. They have many explanations for why hostility is necessary and inevitable, but the major reason for the crisis is that each country believes it is better fit than the others.

A common philosophy based on the idea of natural selection is that might makes right. This has been the banner of industry, business, and capitalism in general since the turn of the century. In fact, this philosophy is the basis of nearly all Western thinking. From kindergarten until the end of our adult lives, we are continually force-fed the notion that every endeavor in which we become engaged is just another struggle for existence. How many times have you heard: "It's a dog-eat-dog world out there." "If you are too weak to compete you won't make it." "Only the best will succeed."

Darwin did not intend to direct our destinies, yet he has. In fact, few people in history have affected our thinking, our philosophies, and our life-styles more than Darwin. Thus it seems reasonable to conclude that the philosophy of the next decade will not change appreciably.

FOR DISCUSSION AND REVIEW

1. Trace the highlights of Darwin's voyage aboard HMS *Beagle*. When did he begin to understand the evolution of species? What particular place that he visited supplied him with the most compelling evidence for his theory?

2. Why did Darwin wait so long before publishing his ideas? What contribution did Alfred Russel Wallace make to the theory of evolution?

3. Describe the Oxford debate. Who were the main players? Was the argument settled?

4. Compare and contrast Darwin's theory of evolution with Lamarck's theory of the inheritance of acquired characteristics.

5. Detail the provisions that were set forth in Darwin's *On the Origin of Species by Means of Natural Selection.*

6. What is the theory of plate tectonics? What does it propose?

7. Why do so few living organisms ever become fossils? How are fossils dated?

8. What is the evidence of evolution provided by rock strata? Comparative anatomy? Comparative embryology? Vestigial structures? Comparative protein chemistry?

9. Describe an antigen-antibody reaction.

SUGGESTED READINGS

DARWIN AND THE BEAGLE, by Alan Moorehead. Harper & Row: 1969.
A vivid biography of Charles Darwin, readable and beautifully illustrated.

THE EARTH, by Arthur Beiser and the Editors of Life. Life Nature Library, Time Incorporated: 1962.
An easily read and well-illustrated account of the earth from its beginnings. Includes a description of how fossils are formed.

EVER SINCE DARWIN, by Stephen Jay Gould. Norton: 1977.
A series of contemporary essays on Darwin, the theory of evolution, the origin of the earth and life, and science in a modern society.

THE ILLUSTRATED ORIGIN OF SPECIES, by Charles Darwin. Abridged and introduced by Richard E. Leakey. Hill and Wang: 1979.
The original text. A detailed account of the evidence compiled by Darwin for his theory of evolution. The text is the most important contribution to biological science. In this version Richard Leakey's comments are inserted in Darwin's text. Beautifully illustrated with photographs, maps, and diagrams.

18

THE MECHANISMS OF EVOLUTION

During the history of the earth, many species have evolved into new species. There is no doubt, then, that evolution has occurred since life originated. Is it still occurring? Obviously, it must be. However, the usual evolutionary process is extremely slow, so it is nearly impossible to see it happening in a single lifetime. In evolution, the crucial factor is how rapidly or slowly a species reproduces. A species of bacteria that can reproduce a new generation every hour will evolve much faster than humans do, because a human generation requires twenty years.

NATURAL SELECTION

Natural selection is the mechanism of **evolution.** It deals with genetic changes in populations, not in individuals. You as an individual will not change genetically and therefore will not evolve. Your genetic makeup was established at the time of your conception. Only part of your genotype will be passed to future generations. You will remain unchanged. You may never experience a mutation, but if you reproduce, you will affect the genetic makeup of future generations.

The production of gametes in your body and their subsequent union with gametes of the opposite sex will create new genetic combinations. The environment is the test of these combinations. Those that survive the environmental challenge may be carried into future generations. Those that fail to survive may never exist again.

PREVIEW QUESTIONS

After reading this chapter, you should be able to answer the following questions:

1. Explain natural selection. What are its different forms?

2. What is a species? What kinds of mechanisms keep species isolated and prevent closely related species from crossbreeding with each other?

3. By which means may new species develop?

4. What does the Hardy-Weinberg Law tell us about evolution?

There are three types of natural selection: directional, stabilizing, and disruptive. Figure 18–1 diagrams the differences among them. In addition, natural selection can be classified by the type of evolution that has occurred: divergent, parallel, or convergent.

Directional Selection

The most common form of natural selection is directional selection. Some phenotypes within a population are better adapted than others. Those that are best adapted are more likely to survive and to reproduce successfully than are other phenotypes. Reproduction increases the frequency of these phenotypes in the population. In other words, organisms with genetic combinations that are adaptable to some environmental change are favored. Those that are not adaptable are selected against and are lost from the population. Another way to describe directional selection is to say that it creates new species phyletically.

Directional selection operated in the past evolution of humans and continues to operate today. When early humans became hunters, certain phenotypes were more adapted than others for that kind of existence. Directional selection eventually developed a different type of phenotype than had been present before. Directional selection also operated during the evolution of the horse and of many other species.

The process of domestication is nothing more than directional selection, although in this case humans, not the environment, are now doing the selecting. Therefore, domestication is best referred to as artificial (directional) selection. Domestic animals have been selected for characteristics that make them valuable to human existence and economy.

Certain breeds of cattle, for example, were selected for milk production, others for meat production, and others for adaptability to hot, arid regions or cool, high elevations. Brahman cattle originated in Asia, where they were well adapted to hot humid weather. Various crosses of Brahman cattle with other breeds have produced hybrids that are heat resistant but retain the meat qualities of beef cattle. These animals survive well in weather typical of southern Florida, where other breeds of cattle often suffer from heat prostration. The famous Texas longhorns were selected for other characteristics. They produce good-quality meat without feedlot feeding, a trait not possessed by Hereford or Angus breeds.

Another form of directional selection in the process of domestication has been selection for specific behavior. The best strains of domestic animals are those that are the tamest and therefore the most easily maneuvered and worked. We have all heard stories about the aggressiveness of bulls, but at 4-H fairs many of the prize-winning bulls are as tame as household cats.

Stabilizing Selection

As indicated in Figure 18–1, during stabilizing selection the average phenotype is the best adapted, and selection favors the gene pool that exists. For example, nearly all house sparrows seem to be the same size. But in fact, house sparrows come in a range of sizes. During extreme conditions, such as severe winters, the larger and smaller sparrows do not survive. There is apparently some critical ratio between weight and wing length that is the most favorable phenotype.

The same phenomenon seems to exist in human populations. Statistical evidence suggests that human babies weighing between 3.2 and 3.6 kilograms at birth have a better chance of surviving and reaching sexual maturity than do babies weighing either under 3.2 kilograms or over 3.6 kilograms.

Figure 18–1

Types of natural selection. The best-adapted phenotypes in a population are always favored by natural selection. If they are the average and most frequent, the population changes little and is stabilized. If they are an extreme, the population changes in that direction. If there is more than one best-adapted phenotype, the original population is disrupted.

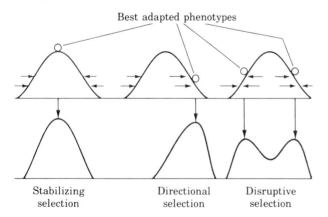

Best adapted phenotypes

Stabilizing selection   Directional selection   Disruptive selection

After stabilizing selection has been in force for a long time, there is essentially no change in the average phenotype. We often refer to plants or animals of this sort as living fossils or genetic relics. For example, *Latimera,* a coelacanth fish (Figure 18–2), was presumed to have become extinct 125 million years ago, although it was relatively common for about 2 million years before that time. In 1939, a living specimen was caught off the coast of eastern Africa. Since then, several others have been caught. The live coelacanths are essentially the same as the fossil remains.

The dawn redwood, *Metasequoia,* flourished 13 million years ago. Until recently, it was known only as a fossil, but live specimens have been found in China. A very primitive lizard-like reptile known as the tuatara *(Sphenodon)* that was common more than 72 million years ago still is living in New Zealand. Presumably, fossil specimens do not differ significantly from current ones.

Disruptive Selection

The exact opposite of stabilizing selection is disruptive selection. In this case, the extremes have more adaptable phenotypes than the averages do. As a result, the original population is disrupted into two or more separate groups that may later evolve into separate species (Figure 18–1). If disruptive selection results in the formation of many new species, it is called adaptive radiation.

Disruptive selection can be demonstrated in the laboratory. In a special apparatus, common fruit flies are tested to determine whether individual flies are positively or negatively phototactic (attracted to or

Figure 18–2

A living fossil. This fish, known as *Latimera,* was presumed to have become extinct about 125 million years ago. In 1939, a living specimen was caught off the eastern African coast. Since then, many others have been caught. Comparisons with fossil fish show that the fish has changed little.

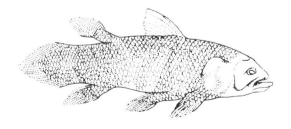

repelled by light). When a large number of flies are tested, most are found to be neutral to the light source, a few are positively phototactic, and a few are negatively phototactic. The individuals responding to light represent extreme phenotypes. The negatively phototactic flies are interbred, and the positively phototactic flies are interbred. New populations of flies are thus created. Each new population is again tested for response to light, and again only the extreme phenotypes are interbred. If this procedure is continued for a couple dozen generations, the final population of flies will nearly all be positively or negatively phototactic. This is a case of artificial disruptive selection.

Divergent Evolution

Disruptive selection results in divergent evolution. One of the most commonly cited examples of divergent evolution is the evolution of the Darwin's finches on the Galapagos Islands (see Figure 18–3). The Galapagos are volcanic islands that emerged from the sea about a million years ago. At first, they were lifeless, but eventually several species migrated to them. The ancestor of Darwin's finches is unknown, but it was undoubtedly a generalized (not specialized) finch. There were no other bird competitors on the Galapagos, so the original population of generalized finches became disrupted as extreme phenotypes and began to invade niches that had not previously been occupied. The early offshoots of the original population were disrupted again and again as adaptation continued (another case of adaptive radiation). This process went on until fourteen distinct species of finches evolved. Today they inhabit fifteen different islands. Some of the species are sympatric (found in the same area). Others are allopatric (found in different places). One of the most remarkable examples of the adaptations of the finches is the woodpecker finch. This finch, which lacks the long tongue of a typical woodpecker, exploits the niche usually inhabited by woodpeckers in other geographic areas by using a cactus spine to probe for insects.

Parallel Evolution

When a common ancestral stock has diverged (by disruptive selection) into two main lines, sometimes the two lines evolve similar structures and habits because they are exposed to similar environmental pressures.

(a)

(b)

Figure 18–3

Darwin's finches. The finches of the Galapagos Islands are an example of adaptive radiation. The first finches there were probably very generalized birds. Because they had no competitors (no other types of birds), these finches were able to radiate into a variety of niches. Today there are fourteen species of finches (thirteen are shown here), each occupying a separate niche. *(a)* These finches are all tree finches. Most are specialized for feeding on insects. The tree finch with the smallest beak has the habits of most warblers and is adapted to picking small insects from bushes. The largest tree finch, the one with the massive bill, is a vegetarian that eats fruits. *(b)* These finches are the so-called ground finches. Many have massive beaks that help them crush the seeds on which they feed. The species with longer and narrower beaks probe cactus flowers for nectar. The most specialized finch, the woodpecker finch, is shown holding a cactus spine, which it uses as a tool. (From *Biological Science: A Molecular Approach,* fourth edition, by the Biological Sciences Curriculum Study, D. C. Heath and Company, 1980.)

An example of such **parallel evolution** occurred among monkeys. New World monkeys evolved in Central and South America, whereas Old World monkeys evolved in Asia and Africa. The two are remarkably similar in appearance and habits. Nevertheless, they can be easily distinguished on the basis of certain specific characteristics. For example, while both New World and Old World monkeys have tails, the tails of New World monkeys are prehensile (capable of grasping), whereas those of Old World monkeys are not. Also, New World monkeys have three pairs of upper and lower premolar teeth, whereas Old World monkeys have only two pairs.

Convergent Evolution

When species have similar habitats and similar structures but are not closely related, convergent evolution is said to have occurred. The resulting similarities are a function of similar environmental pressures (see Figure 18–4). For example, wings have evolved independently several times—among insects, reptiles (all winged reptiles are now extinct), birds, and mammals (bats). The eyes of vertebrates and the eyes of cephalopods (octopuses and squids) are extremely similar in form and function, although these groups are not related by a common ancestor. The general body shape of sharks (fish), including fins, tails, and streamlined body shape, is very similar to the body shape of only distantly related aquatic organisms, such as porpoises (mammals) and extinct aquatic reptiles called ichthyosaurs.

SPECIATION

What is a **species?** In simple terms, it is a group of organisms that look similar and have similar anatomies and physiologies. A more complete definition is that a species has unique behavioral characteristics, that its members breed only with one another, and that they have a common ancestor. More precisely a species is made up of organisms that constitute a reproductively isolated aggregate. The important criterion for a species is that it must be reproductively isolated from other organisms.

Some populations of organisms are strikingly similar in form and structure but cannot interbreed. Therefore, they constitute separate species. For example, there is a very small population of Indian

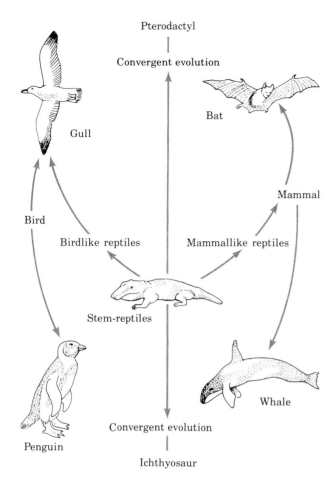

Figure 18–4

Convergent evolution. Convergent evolution is said to occur when animals that are not closely related become more and more alike during evolution. The animals evolve in similar ways because of similar environmental situations. Many millions of years ago, an ancestral reptile gave rise to two distinct lines in evolution—birdlike and mammal-like reptiles. Later, these ancestral stocks evolved into modern birds and mammals. The example at the top shows that the gull, the pterodactyl (prehistoric flying reptile), and the bat all evolved wings. The example below shows how representatives of these same groups became aquatic because of the similar environmental situations in which they independently evolved. (An ichthyosaur was a prehistoric aquatic reptile that had a large sail on its back.)

lions, which exist only in a wildlife sanctuary. These lions will never have the opportunity to breed with any other population of lions, because all other lions live in Africa.

Other organisms within a species have such ex-treme variations in appearance that the casual ob-server would place them in separate species. For ex-ample, various subspecies of gulls form a breeding ring (see Figure 18–5). Adjacent populations freely in-terbreed, but the two populations at the ends of the ring do not interbreed. By the definition just given, the end populations should constitute separate spe-cies—but they do not. Because the populations be-tween the extremes breed freely, genes flow between the two end populations through the chain.

The species concept is obviously complicated, and defining a species in specific terms is often difficult. There are always exceptions, and a certain amount of variation among organisms is expected. In practice, however, we have little trouble identifying most spe-cies. Nearly everyone recognizes that house mice, field mice, and kangaroo mice are separate species. Most of us can distinguish robins, cardinals, and crows. And obviously rabbits, squirrels, dogs, and cats are all separate species. A tribe of natives in New Guinea had 136 separate names for the birds in its jungles. When biologists studied the birds, they found 137 different species. The natives had confused only 2 species.

A species is an interbreeding population in which there is gene flow—the movement of genes from one part of a population to another. If the gene flow is interrupted, separate species may be formed. Species whose ranges overlap are called **sympatric species** (literally "sharing a homeland"). Species occupying different ranges are called **allopatric species** (literally "different homelands"). The lions of Africa and In-dia, therefore, are allopatric species. Many of Dar-win's finches live on the same islands in the Galapa-gos and are thus sympatric species.

Species-Isolating Mechanisms

Species are maintained through their isolation, which prevents them from breeding with other species.

Geographic Isolation

The integrity of species is often maintained simply because they are geographically separated from simi-lar organisms, so that there is no opportunity to in-terbreed. The Indian and African lions are an example of similar organisms separated by geographic bar-riers. The separation of the two populations of lions has been rather recent (since the agricultural revolu-tion, or in the past 10,000 years). Thus the lions of

Figure 18–5

A breeding ring of gulls. A, B, and C are separate subspecies. D is a separate species. Interbreeding of the gulls takes place in a circular pattern *(follow the arrows)*. C1 interbreeds with C2, C2 with C3, and so on. A1 interbreeds with A2, and so on again. However, at the ends of the chains, A2 does not interbreed with B3 and B4, even though they live side by side. Isolating mechanisms prevent their interbreeding. In that geographic area they are similar to separate species—*not subspecies*—because there is no gene flow.

one population look very much like those of the other population. These lions could reproduce if they were brought together, but this cannot occur naturally. They could interbreed only if humans transported members of one population to the other or mated them in a zoo. Therefore, these populations still belong to the same species. However, in time they may give rise to two separate species.

Another example of geographic isolation involves the squirrels that live on the north and south rims of the Grand Canyon. Millions of years ago, when the Grand Canyon was formed, a single population of squirrels became divided by this impassable barrier. The single population became two, and gene flow between them was halted. Today, the northern squirrels are called Kaibab squirrels and the southern are called Abert squirrels. The squirrels of the two populations do not look much alike (see Figure 18–6), and they are not known to interbreed. Some controversy exists over whether to consider the squirrels as distinct, separate species. However, under natural conditions they do not interbreed, so by definition they are classified as separate species.

Seasonal Isolation

Sometimes, organisms that appear similar are maintained as separate species because they breed during different times or seasons of the year. The isolating mechanism in this instance is time, and for that reason seasonal isolation is also called temporal isolation. For example, the toad *Bufo americanus* breeds early in May. A closely related species, *Bufo fowleri,* does not commence breeding until June, when *Bufo americanus* no longer is breeding. The two species may share the same pond, but the time of breeding prevents them from interbreeding. Biologists have been able to maintain both species of *Bufo* in the laboratory. By manipulating temperature and light conditions, the scientists can stimulate the two species to become sexually active at the same time. Under these conditions, the males of one species will clasp the females of the other species. The union of the two species can

Figure 18–6

Are these separate species? The Abert squirrel *(left)* lives on the southern rim of the Grand Canyon. The Kaibab squirrel *(right)*, commonly called the white-tailed squirrel, lives on the northern rim of the Grand Canyon. These two separate populations of squirrels are geographically isolated by the canyon and do not have the opportunity to interbreed. Therefore they constitute separate species. Some zoologists argue that they can interbreed if brought together. However, they never have the opportunity to interbreed in nature. (Abert squirrel, Al Lowry, Photo Researchers, Inc. White tailed squirrel © Sonja Bullaty/Photo Researchers, Inc.)

result in **hybrid** toads. The hybrids have a seasonal cycle different from either parent. They would therefore perish in nature unless they established a new breeding population—a new species.

Seasonal isolation also exists for plants. The common pine trees in California are two species, *Pinus radiata* and *Pinus muricata*. The two often stand side by side. However, they remain distinct species because the times of year when they shed pollen are different. *P. radiata* matures and broadcasts pollen in February. *P. muricata* does not release pollen until April. The pollen grains from one species are capable of fertilizing the ova of the other species, but this does not occur because of seasonal isolation.

Ecological Isolation

The species isolation known as ecological isolation is also referred to as **habitat isolation,** because similar species occupy different habitats and therefore have different ecological requirements. This form of species isolation is common among plants. For example,

there are two closely related species of the spiderwort *(Tradescantia canaliculata* and *T. subaspera).* One species thrives in open sun, the other in heavy shade. The different habitats cause one species to bloom earlier than the other, thereby isolating them. This too is seasonal isolation, but it is slightly more complex because the species have separate habitats. Common wild lilacs, *Ceanothus thyrsiflorus,* grow only on hillsides in good soil. A closely related species, *C. dentatus,* grows in dry, exposed, poor soil.

Some animal species are isolated ecologically. In Florida there are two closely related species of dragonflies, *Progomphus obscurus* and *P. alachuensis. P. obscurus* is found near lakes, where its eggs are deposited. *P. alachuensis* is found only near rivers and creeks.

Ethological Isolation

Sometimes, even though species breed at the same time and occupy similar habitats, their integrity is maintained because their behavior is different. The species are ethologically isolated. The behavior of the three-spined stickleback□ provides an example. There are several species of sticklebacks, and some of them are sympatric. Sympatric species do not interbreed because their courtship sequences are different.

Fireflies, common in the eastern United States, are obvious on hot, humid evenings. There are a number of species, and many are sympatric. The females attract males by signaling with a species-specific train of light flashes generated in the tip of their abdomens. (The next time you watch fireflies, see if you can detect differences in flashing rates. Each rate is characteristic of a different species.) Male fireflies are attracted to their own species-specific signal. One species of firefly gives two different signals. The first is species-specific and attracts a male mate. After mating, the female begins to emit a different rate of flashes. This rate is the species-specific signal of a sympatric species. The flashes attract males of the other species, which the female then devours.

Later in the evening, fireflies disappear and crickets begin their distinctive songs. Males sing songs that attract only females of their species, not females of other sympatric species. It was once thought that male crickets of one species sang several different songs. In fact, the songs were produced by males of different sympatric species that look identical.

Mechanical Isolation

Some separate species occupy the same habitats, breed at the same time, and even have similar behavior. These species are maintained because mechanically it is impossible for them to reproduce and hybridize.

For example, several species of guppy-like fish belonging to the family Poeciliidae inhabit the same areas of specific streams in Central America. The males almost continually pursue and court females, receptive or unreceptive, with great vigor. The males have an intromittent organ, the gonopodium. The males of closely related species have differently shaped gonopodia, so although males court females of a different species, they are unable to copulate with them. The gonopodia of males and the urogenital pores of females of the same species are like a lock and key—the gonopodium fits into only same-species females.

One obvious mechanical isolating factor among some species is size. Closely related species may be of quite different sizes, making mating between species nearly impossible. All domestic dogs belong to the same species, *Canis familiaris*. However, there are many separate breeds of the domestic dog, and the size discrepancies among some of the breeds creates a mechanical barrier to interbreeding. For example, it would be mechanically impossible for a female Chihuahua to mate with a male Saint Bernard—or, for that matter, vice versa. However, a Chihuahua could easily interbreed with a toy poodle, creating a mongrel, or a Saint Bernard could easily breed with a Great Pyrenees. Size here is not natural. It is a result of artificial selection by humans. Therefore all dogs are still considered to belong to the same species.

Mechanical isolating mechanisms are much more common among plants than among animals. The size and shape of many flowers become mechanical barriers to certain pollinators and encourage others. Various flowers are shaped for pollination by small bees, large bees, bumblebees, hawk moths, butterflies, long-tongued flies, beetles, or hummingbirds. Some very large flowers are pollinated only by bats. For example, two closely related species of sage require different pollinators. One species is pollinated by very small bees. The other is pollinated by much larger bees, because the flower will not open to admit the pollinator unless considerable weight is put on the lip of the flower. Small bees cannot enter, and their exclusion prevents interbreeding of the sage species.

Gametic Isolation

In some cases, the **gametes** (egg and sperm) are incompatible, so fertilization never results. The eggs of many species produce chemical substances that attract or direct the swimming movements of sperm to eggs. These fertilization substances are species-specific, and the sperm of one species are not attracted to the egg cells of other species.

Interspecific matings are known to occur among some fruit flies. When this happens, the reproductive tract of the female swells and kills the sperm by antigenic activity before the sperm reach the eggs. It is also relatively common for the pollen of one plant species to land on the pistil of another species. Often a pollen tube never germinates. But if a tube does form, it never reaches the ovary, thereby preventing union of egg and sperm. Among tobacco species, there are sixty-eight known combinations that are prevented by this mechanism.

Hybrid Inviability

If fertilization between two species occurs, the resulting hybrid individuals usually die before they complete development, keeping the species separate. In other words, the hybrids are inviable. Most often, the embryo undergoes only one or two cleavages and then dies. If development proceeds further, there are usually so many malformations that the organism dies before embryonic development is completed. For example, when certain species of goats are crossed with sheep, the hybrid embryos die in early development.

Hybrid Sterility

Sometimes a hybrid survives, but sterility maintains a barrier to gene flow between two closely related species. Perhaps the most well-known example is the mule, which is obtained by crossing a female horse with a male donkey. The mule is in many ways better than either the horse or the donkey. However, all mules are sterile. Therefore, the gene pools of the horse and donkey remain isolated.

An experiment started more than ten years ago has removed the barrier of sterility for one hybrid. American bison were crossed with beef cattle, and the resulting offspring were called beefaloes (see Figure 18–7). The hybrids have desirable characteristics. They do not have to be grain-fed (for good meat taste) prior to marketing, and there is less fat in the meat. They are also larger than cattle, providing more meat per animal, and they are ready for market sooner than cattle. The original beefaloes were sterile. However, this barrier was overcome by artificial insemination. Large amounts of semen from the hybrid were artificially introduced into female cattle. This backcross hybrid was fertile, and now beefaloes can propagate themselves.

The isolating mechanisms described in this section

Figure 18–7

A beefalo. Beefalos are a cross (hybrid) between American bison and beef cattle. The first hybrids were sterile, but artificial insemination created beefaloes capable of reproducing themselves. Beefaloes provide more meat because they are larger than beef cattle, and their meat has less fat. (Courtesy A. M. Winchester from *Heredity, Evolution and Humankind,* West, 1976, p281.)

are often categorized into premating and postmating mechanisms. Premating mechanisms include geographic, seasonal, ethological, and mechanical isolation. Postmating mechanisms include gametic isolation, hybrid inviability, and hybrid sterility.

The Origin of Species

The discussion of isolating mechanisms may suggest that the origination of a new species is nearly impossible. However, species evolve all the time, through several different mechanisms.

Hybrid Speciation

New species are formed, for example, when one or more of the isolating mechanisms fail to keep gene pools separate and the resulting hybrids are fertile. Among animals, hybrids are rarely fertile. When they are fertile, they seldom establish themselves as a successful species. There are exceptions, however. The beefalo is an example of a fertile hybrid (made fertile by artificial means) that has been successfully established. However, the breed is maintained by human supervision. It is not known whether it would survive in the wild.

Speciation by hybridization is more common among plants than animals. Interbreeding among plant species often causes dramatic changes in chromosome numbers. For example, the interbreeding of one species of a normal diploid chromosome number with another diploid species often produces a hybrid that has all the chromosomes of both parents and is thus tetraploid (*tetra* means "four").

Plants that have more than a diploid complement of chromosomes are known as **polyploids** (*poly* means "many"). Polyploid animals do not occur naturally because, for some unknown reason, the additional chromosomes are lethal. However, there are many common polyploid plants. Tobacco is a polyploid that evolved by the hybridization of two smaller species. Alfalfa is another example, a viable hybrid from two species of clover. Wheat, a crop important to the entire world, is also a polyploid species. The original wheat plant, very small and producing low yields of grain, had fourteen chromosomes. This species hybridized with goat grass, which also had fourteen chromosomes, to create a new species with twenty-eight chromosomes. Sometime later in the evolution of wheat, the tetraploid wheat species hybridized again with goat grass,

forming a hexaploid (forty-two chromosomes) that is now the common species of wheat that is grown the world over.

Many hybrid plants have more vigor than either of the parents, a condition referred to as **heterosis.** This is why hybrid vegetables are constantly created. The hybrid vegetables are most often heterozygous and express only dominant traits. All recessive traits are suppressed. As a result, the plants often grow taller and faster, produce more fruits, and so on. However, many hybrid vegetables are sterile. Those that are fertile are usually not propagated, because hybridization of the hybrids would allow expression of recessive traits, which are often undesirable. Therefore, new hybrid seed has to be obtained by continued crossing of the original parents. Hybrid corn seed is produced every summer in the Midwest by selective pollination.

A famous case of artificial plant hybridization was a cross made by the Russian biologist G. D. Karpechenko. He crossed cabbages and radishes, both of which have only eighteen chromosomes, hoping to create a plant with a cabbage top and a radish root—the best of both plants. Most of the initial offspring were sterile. However, some of the eggs and pollen produced a tetraploid (thirty-six chromosomes), a hybrid that was half radish and half cabbage. Unfortunately, the hybrid had a cabbage root and a radish top and was essentially worthless. However, it does propagate itself and is thus a new species.

Allopatric Speciation

New species arise most frequently when an original population becomes geographically isolated into two or more groups. Natural selection operates separately within each group. The resulting changes cause sufficient differences among the groups to establish them as separate species. Figure 18–8 illustrates a hypothetical case of allopatric speciation. The evolution of the Abert and Kaibab squirrels also fits the model of allopatric speciation.

The topography of the earth has been changing constantly since life was created. Mountain ranges have appeared and disappeared, rivers have cut enormous gorges and canyons, and volcanoes have changed the landscape. The continents were once all joined into a single land mass called Pangaea. The continents we know today were formed when Pangaea broke apart. Any of these events could provide sufficient geographical barriers for allopatric specia-

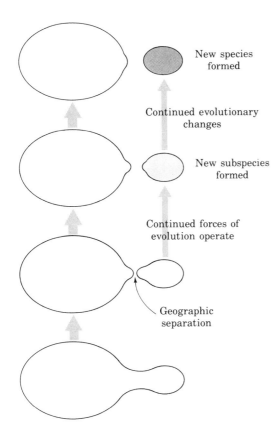

Figure 18–8

Allopatric speciation. The original population is shown at the bottom. A part of this population becomes geographically isolated from the main population, perhaps by the formation of a mountain ridge or a canyon. Eventually, the populations become entirely separated, and each constitutes a separate species.

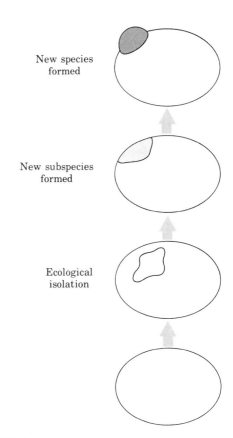

Figure 18–9

Sympatric speciation. A small segment of the original population (shown at the bottom) becomes isolated ecologically or ethologically. As the isolating mechanism comes into force, a new subspecies emerges. Given enough time, the subspecies will form a new species.

tion. Glaciation during the Pleistocene era in North America is one of the physical barriers that promoted some of the major differences between the flora and fauna of the western and eastern United States.

Sympatric Speciation

The reverse of allopatric speciation is sympatric speciation. Species form within an original population without physical isolation (see Figure 18–9). It is believed that the same isolating mechanisms that maintain these species after they form, such as ecological and ethological isolation, are the mechanisms that also create sympatric species. There is some debate among specialists whether species actually arise in this fashion. However, there is evidence that sympatric species do form.

Lake Malawi, one of the largest lakes in Africa, is known to harbor at least three hundred species of fish. Many of these fish belong to the family Cichlidae, called cichlid fish. Many hypotheses attempt to explain speciation among the cichlids, and most propose sympatric speciation. It is obvious that the cichlids in Lake Malawi today are ecologically (habitat) isolated, because they have evolved a great variety of feeding techniques. Among these techniques are scraping rocks, sifting sand, feeding on phytoplankton or zooplankton, and eating only fish scales, fish eyes, fish eggs, fish embryos, or insect larvae. There is also evidence of ethological isolation among many of the species. Most remarkable is the fact that speciation of these cichlids may have occurred during the past ten thousand years. Because of the apparent speed of creation of so many species, this is referred to as a case of **explosive evolution.**

Phyletic Speciation

New species are most commonly created phyletically. A single species undergoes continuous change through the forces of natural selection until it has changed so much that a new species is created and the original stock becomes extinct. This method of speciation played a major role in human evolution, which is described in more detail in Chapter 20.

THE HARDY-WEINBERG LAW

Evolution is most accurately defined as changes in gene frequency within a population. It is difficult to visualize gradual genetic changes within large populations. For the purpose of ordering and directing thinking about such elusive processes, scientists often use models—mental tools to facilitate analysis of biological problems. Models are always oversimplifications of real situations. However, if the model fits the situation, the scientist can explain the situation in simple terms. This often leads to a better understanding of natural processes. Models are used not only in biology but also in all applied and theoretical sciences.

The model that explains gene frequencies in a population is called the **Hardy-Weinberg law.** The concept of the model was first worked out in 1908 by Godfrey H. Hardy, an English mathematician, and Wilhelm Weinberg, a German physician. The model is credited to both men.

The Hardy-Weinberg law states that the gene and genotypic frequencies of a single trait in a large, random mating population—and in the absence of mutations (see Box 18–1), selection, and migration—will not change from generation to generation. In simpler terms, the model predicts that the **genotypes** of a single trait—homozygous dominant (SS), heterozygous (Ss), and homozygous recessive (ss)—have an equal ability to reproduce and will be represented in the same numbers generation after generation. In fact, what the model represents is a case where evolution does not occur, where there is no change in gene and genotypic frequencies. The state of no change becomes a base line from which actual change in populations can be measured and the rate of genetic change (evolution) can be mathematically computed.

CONCEPT SUMMARY

THE HARDY-WEINBERG LAW

The Hardy-Weinberg law states that gene and genotypic frequencies of a single trait will remain the same from generation to generation if the following conditions are met:

1. The population in question is large and isolated.
2. There is random mating within the population.
3. There is no selection for individual genotypes.
4. Mutations do not occur.
5. There is no migration of individuals into or out of the population.

The human trait known as sickle-cell anemia provides an example of how the Hardy-Weinberg model works. It is known that in some African populations the incidence of carriers (Ss) of sickle-cell anemia can often be 32 percent or more of the population. It is also known that 4 percent or more have the disease and are thus homozygous recessive (ss). These individuals usually die before reaching reproductive age. The frequency of any trait in a population is considered to be 100 percent. Therefore, the frequencies of the possible genotypes of the sickle-cell trait in a hypothetical African population can easily be computed:

$$SS = ?; \; Ss = 32\%; \; ss = 4\%\star$$
$$SS + Ss + ss = 100\%$$
$$\text{Therefore, } 100\% - 32\% - 4\% = 64\% \; SS$$

It is easier when making the next series of computations to use decimals instead of percentages to represent the genotypic frequencies:

$$SS = 0.64; \; Ss = 0.32; \; ss = 0.04$$
$$\text{Therefore, } 0.64 + 0.32 + 0.04 = 1.00$$

The Hardy-Weinberg model predicts that the genotypic frequencies given above will never change, that there will be no evolution. It also predicts that when the original population reproduces, a specific number of sperm cells will carry the S allele and a specific number will carry the s allele. Because the precise numbers of each are not known, they are symbolized algebraically by the letters p and q. The

---

$\star$The numbers have been approximated for convenience in mathematical computation.

MUTATIONS

What is the origin of the new phenotypes that become the objects of natural selection? These phenotypes come about by mutation (see Chapter 15 for a discussion of the mechanics of mutations). There are two types of mutation: micromutation and macromutation.

Micromutations can involve single genes or entire chromosomes. Any form of base-pair alteration within a gene produces a translation different from the original order. Micromutations may have only subtle effects on the phenotype, but their gradual accumulation may in time produce obviously new phenotypes. On rare occasions, the new phenotypes are more adaptive than the previously existing ones. However, most mutations are lethal.

Macromutations, which are also thought to be rare, involve major and dramatic phenotype changes within a few generations. By necessity, macromutations involve entire complexes of chromosomes. They may not occur at all among animals, but there are known instances among plants. For example, the tetraploid offspring of diploid parents create sufficient changes in the phenotype to establish reproductive isolation of the tetraploid in a single generation.

Mutations that occur naturally are referred to as spontaneous mutations. Among humans, the rate of spontaneous mutation is considered to be about two thousand mutations per million people. Mutations are induced by a number of natural agents, including radiation, ultraviolet light, and a number of chemicals. Agents causing mutations are called mutagens.

There is some concern that the by-products of modern technology may increase the natural incidence of mutation. For example, radiation exposure from X-rays, nuclear generating plants, and atomic bomb blasts are known to increase mutation rates. A number of common substances are also known mutagens. Among them are various dyes (red dye #2 was banned by the Food and Drug Administration in 1977 because of suspected mutagenic effects); saccharin (also banned by the FDA for the same reason); caffeine and theobromine, which are found in coffee, tea, and chocolate; and many other organic and inorganic chemicals.

A long list of chemicals added to food is also under suspicion of being mutagenic and carcinogenic (cancer-causing). There has been extensive research on the fluorocarbon gases used in aerosol cans to propel everything from whipped cream to drugs. The research has indicated that the escape of excessive fluorocarbon gas into the atmosphere may destroy the delicate ozone layer, permitting more ultraviolet radiation to penetrate our environment and rapidly increase the mutation rate among all organisms. It is estimated that all of these factors have increased the mutation rate among humans by about 125 mutations per million people, for a total of approximately 2,125 mutations per million.

---

frequency of sperm cells carrying the S allele is represented by p. The frequency of sperm cells carrying the s allele is represented by q. The total pool of sperm cells is expressed as p + q. The same symbols can be used for egg cells. Egg cells carrying the S allele are represented by p. Egg cells carrying the s allele are represented by q. Again, the total pool of egg cells is p + q.

The model assumes that fertilization will be random and that all eggs will be fertilized. The frequency of the alleles in the zygotes (the next generation) can be computed by calculating $(p + q) \times (p + q)$, which is the same as $(p + q)^2$. When solved, this is expressed as $p^2 + 2pq + q^2$. The frequency of zygotes that are homozygous dominant (SS) is p^2. The frequency of heterozygous (Ss) zygotes is 2pq. The frequency of homozygous recessive (ss) zygotes is q^2.

$$p^2 + 2pq + q^2 = 1.00$$
$$SS + Ss + ss = 1.00$$

In our example, 4 percent of the African population was afflicted with sickle-cell anemia. Even if no other information were available, the other two genotypic frequencies could be computed using the new formula:

$$q^2 = 0.04$$
$$q = \sqrt{0.04}$$
$$q = 0.2$$

The value of q is the gene frequency of s, and it represents the frequency of the recessive allele in a population—in this case, 20 percent. Also, p is a gene frequency, and it represents the frequency of the dominant allele (S) in the population. Therefore:

$$p + q = 1.0$$

If the frequency of q is 0.2:

$$p = 1.0 - 0.2$$
$$p = 0.8$$

Thus the frequency of S in the population is 80 percent. It is simple to compute the genotypic frequencies after p and q are known:

$$p^2 = 0.8 \times 0.8 = 0.64$$
$$2pq = 2 \times 0.8 \times 0.2 = 0.32$$
$$q^2 = 0.2 \times 0.2 = 0.04$$

In other terms:

$$0.64 + 0.32 + 0.04 = 1.0$$

These genotypic frequencies are exactly the same as those obtained by the original computation.

What does the Hardy-Weinberg law predict about the genotypic frequency of the next generation? From the first generation, 64 percent of the population (both males and females) contribute only gametes carrying S (because they are homozygous dominants). Half of the gametes produced by the heterozygotes (Ss), or 16 percent, carry only S; and half (also 16 percent) carry only s. The individuals contributing only gametes carrying ss, because they are homozygous recessive, are 4 percent. Therefore, from the entire potential pool of gametes, the frequency of S is 64 + 16, or 80 percent. The frequency of s is 16 + 4, or 20 percent (p = 0.8; q = 0.2). The genotypic frequency of the next generation is thus (repeating the computations):

$$0.64 + 0.32 + 0.04$$
$$\text{SS} \quad \text{Ss} \quad \text{ss}$$

The Hardy-Weinberg law predicts no change. What about the third and fourth generations? The same computations are made, and the genotypic frequency remains the same. There is no change and therefore no evolution.

Examination of the real situation shows that change does occur in the expression of the sickle-cell trait and that evolution is therefore occurring. One reason is that the ss individuals are afflicted with sickle-cell anemia, and many die before they reach reproductive age. Individuals who do not reproduce make no contribution of s alleles to the next generation. There is, then, selection against individuals who are ss. The genotypic frequency in the next generation is actually (mathematical computations omitted for simplicity):

$$0.71 + 0.29 + 0$$
$$\text{SS} \quad \text{Ss} \quad \text{ss}$$

Does this mean that the sickle-cell trait ends with this generation? No. The Ss individuals also carry the trait, and marriage among them produces new ss individuals. By selection, the frequency of p and q have obviously changed. They are now p = 0.84 and q = 0.16. Thus the frequency of ss is predicted to decrease in the next generation, from 4 percent to about 2½ percent (0.16 × 0.16 = 0.0256).

Will the sickle-cell trait eventually decline to zero? No again. There is a strong selection in favor of Ss individuals because they are more resistant to malaria. As long as the population is exposed to malaria, the frequency of Ss individuals is predicted to remain nearly the same or to increase. The positive selection for Ss keeps the frequency of s rather high, and ss individuals continue to be produced. Spontaneous mutation of S to s will also prevent the frequency of s from ever becoming zero.

Other factors affect the genotypic frequencies of future generations. What would happen if 10 percent of the Ss individuals emigrated from Africa to the United States? Their absence from the population in Africa would obviously change the frequencies of S and s and the genotypic frequency of the next generation in Africa. They would also influence future generations in the United States.

Nonrandom mating would increase the number of Ss and ss individuals in the United States, a situation that was predicted before blacks were voluntarily screened for carriers (Ss). Carriers are advised against marrying other carriers, because 25 percent of their children will be afflicted with sickle-cell anemia. This, then, is another kind of selection—selection against carriers.

It is unlikely that the carriers of the sickle-cell trait will ever become part of a very small isolated population. However, gene and genotypic frequencies can drastically change by chance in small populations. The effect of chance is called **genetic drift** or the founder effect. Often a small group of individuals form an inbreeding colony because of religious, philosophical, or other reasons. Inbreeding increases the gene frequencies of many characteristics, even those that might be deleterious. These groups often differ markedly in the frequency of certain characteristics compared to the original group they came from.

The religious sect known as Dunkers immigrated to Pennsylvania from Germany in 1719. The original group consisted of twenty-eight people. The Dunkers have strict rules about not marrying outside the sect. Thus the frequencies of blood types among Dunkers today is markedly different from those in the rest of the U.S. population and from the region in Germany where they originated. Nearly 60 percent of the Dunkers have blood type A, and only 3 percent have type B—as compared to 40 percent and 11 percent in the rest of the U.S. population. Obviously, many of the original twenty-eight immigrants had blood type A, and that fact established the high frequency observed today.

Another religious sect in Pennsylvania is the Amish. A specific kind of dwarfism occurs among members of this sect. The disorder causes short stature, extra digits on hands and feet, and other deformities. There are as few as a hundred cases of this disorder in the world today, and more than half of them are among the Amish. All of the Amish who suffer the disorder can be traced to a single pair of ancestors who came to Pennsylvania in 1744. It was simply chance that these two individuals married and became part of the small inbreeding Amish sect.

Another interesting example of the founder effect concerns night blindness, a disorder caused by a dominant allele. A butcher who lived in Montpellier, France, four hundred years ago, had night blindness. In 1907, it was found that 135 inhabitants of this village of several thousand suffered from night blindness, as compared to 1 in 300,000 in the rest of France. Most of the butcher's descendants remained in the village, and the frequency of the dominant allele increased by this chance effect.

Huntington's disease is a fatal disease caused by a rare dominant gene. The disease strikes its victims with essentially no warning and usually not until about age forty. However, some individuals have become afflicted when they were as young as fifteen or as old as seventy. The disease causes a slow degeneration of the nervous system. There is no cure for it. Afflicted individuals first lose their balance, weave when they walk, and have uncontrollable twitches in the arms, legs, and face. The symptoms worsen until death. The incidence of the disease is about 1 in 10,000 people. However, in the Australian state of Tasmania, the rate is 1 in 2,083 people. Nearly all of these cases trace back to a single Englishwoman who settled in Tasmania in 1848. Most of the cases of Huntington's disease in the United States can be linked to another Englishwoman, whose three sons (all afflicted) immigrated to this country.

In summary, the Hardy-Weinberg law provides a mathematical model from which evolution can be measured and predicted. The model allows measurement of the major influences affecting the genetic makeup of populations, including selection, migration, mating behavior, mutation, and isolation of a small population.

EVOLUTION IN ACTION

The various methods of speciation and the patterns of natural selection and evolution were formulated by biologists after intensive examination of the fossil record and other evidence that supports the concept of evolution. These processes are known to have occurred in the past, and they are assumed to be the active forces in evolution today. As mentioned earlier, evolution proceeds very slowly, so it is difficult to appreciate its progress during a lifetime. However, biologists have discovered situations in which evolution has occurred very rapidly, so we have been able to witness some evolution in action.

The peppered moth, *Biston betuloria,* is common throughout England. The moth naturally occurs in two forms (see Figure 18–10): the peppered form, which is mostly white with a peppering of dark spots, and the melanic form, which is entirely black (*melanic* referring to the pigment melanin, which causes the dark coloration of the moth). When a species occurs naturally in different colors, sizes, or forms, it is said to display polymorphism (*poly* means "many"; *morph* means "form"). The polymorphic nature of the peppered moth was first reported in 1848 in England. By 1895, in industrial areas of England, the melanic form far outnumbered the peppered form, accounting for nearly 98 percent of the population. The reverse was true in nonindustrial areas. These data were interpreted to mean that selection favored the melanic moth in industrial areas and the peppered moth in nonindustrial areas.

Many biologists were skeptical that selection could have operated so quickly. In the 1950s, H.B.D. Ket-

(a)   (b)

Figure 18–10

Peppered and melanic moths. *(a)* The melanic moth (entirely black) is extremely conspicuous as it rests on the white and black background of lichens on a tree trunk. It is equally obvious to bird predators. Because of excessive predation, it is rare where lichens are found on trees. The peppered moth is inconspicuous on the same background. (Can you find it?) Because of its protective coloration, it is most common where lichens exist. *(b)* Industry in England created air pollution, which killed tree lichens in industrial areas. Tree trunks devoid of lichens are dark in overall coloration. In this situation, the peppered moth is more conspicuous than the melanic moth. (Courtesy, Mrs. H. B. D. Kettlewell.)

tlewell of Oxford University conducted experiments to demonstrate the nature of selection among the two forms of the moth. In carefully controlled experiments, equal numbers of peppered and melanic moths were released into an area of English countryside where industry was absent. The trunks of the trees in this region are covered by lightly colored lichens. After release, observers with binoculars tried to determine the fate of as many moths as possible. Several species of birds began to prey on the moths when they rested on tree trunks. Of the 190 moths that were observed to be eaten by birds, 26 were peppered and 164 were melanic. Figure 18–10 gives an idea of how conspicuous the melanic moth was on a lightly colored background of lichens.

The same experiment was conducted in the midland of England, where extensive industrialization has caused most lichens to die and trees to become darkened by soot. The results of this experiment were the reverse of the previous one. The birds ate many more peppered moths than melanic moths. Again, Figure 18–10 gives an idea of how the moths appeared in industrial areas.

The moth example is often referred to as one of industrial melanism because the pollution of industry gave melanic moths a dark background that camouflaged them from predators. The explanation of the phenomenon is simple. The melanic moth was originally represented in the population as a rare phenotype (an extreme). Before industrialization, it was heavily preyed on by birds, and few of the moths survived to reproduce. But industrial air pollution and soot gave unexpected protection to melanic moths, allowing more to survive. In the industrial setting, the melanic phenotype was more adaptable than the peppered phenotype, so fewer and fewer peppered moths survived to reproduce.

During the past decade, England has become an example to the rest of the world by cleaning up its serious problems of air pollution. The air is becoming cleaner and lichens are returning to the trees. It was predicted that the new air standards of England would shift selection again in favor of the peppered moth, and it is happening.

THE HUMAN SPECIES

Every human on the earth today belongs to a single species, *Homo sapiens*. (Chapter 20 details human evolution and the time when there was more than one sympatric species of humanlike organisms.) However, large sectors of the population vary noticeably in skin color, body size, hair texture, and so on. This is not unusual. Many species are represented by populations that are genetically different and often partially isolated from other populations. Such groups are called **races.**

There are, of course, several races of *Homo sapiens*. The members of one race are more closely related genetically to each another than to members of other races, but they are members of a single species and therefore have the potential to interbreed. As far as anyone knows, interbreeding between any two human races will produce fertile offspring.

Nevertheless, there are barriers to interbreeding. Perhaps the purest races left on the earth are the Australian aborigines and the African Bushmen/Hottentots, because they have been isolated so long and because even today they are reluctant to breed with other races. Modern technology, however, has increased the mobility of our species, and today it is possible to travel to any part of the globe. This increased mobility has reduced one of the barriers—geographical isolation—that created human races in the first place. Thus interracial marriages are becoming more common. However, another barrier, a social one, often prevents mixed marriages. Some groups still consider interracial marriages socially unacceptable. Thus the integrity of most races on the earth today will last a long time.

Even though humans comprise a single species, there is considerable evidence for selection among specific populations. Sickle-cell anemia is an example. Carriers of the sickle-cell trait (heterozygous individuals) have an advantage in certain environments, because they have a higher resistance to malaria than is afforded noncarriers, who are homozygous dominants. (Individuals who are homozygous recessive for the sickle-cell trait usually die before sexual maturity.) In areas where the incidence of malaria is high, carriers have a better chance of survival than noncarriers, so selection favors the carriers.

However, the use of pesticides for killing mosquitoes has nearly eradicated malaria in some parts of the world. When the pressure of malaria is relieved, the carriers seem to be at more of a disadvantage than the noncarriers. In temperate regions, particularly at higher elevations, carriers often cannot get sufficient oxygen to the tissues and thus suffer from extreme muscle pain and shortness of breath. Most carriers are blacks who acquired the trait from ancestors. The ancestors benefited by being carriers, but the current disadvantages conferred on carriers and the general decline in malaria should reverse selection for this trait.

Many communities in the United States now have screening centers to detect carriers. They advise that carriers mating with each other have a high chance of bearing children with sickle-cell anemia. The screening programs should reduce the frequency of homozygous recessive and heterozygous offspring.

SUMMARY

1. Natural selection is the process of evolution wherein organisms that are best adapted to their environment and are most successful in reproducing pass their genotypes to succeeding generations. It can be one of three different types: directional, stabilizing, or disruptive.

2. When some phenotypes within a population are better adapted than others, they are more likely to survive. This constitutes directional selection.

3. When the average phenotype is the most adaptable and selection favors it, stabilizing selection is in force.

4. When extreme phenotypes within a population are more adaptable, the original population breaks into separate groups. This constitutes disruptive selection, which in turn results in divergent evolution. When many new species are formed, adaptive radiation has taken place.

5. Parallel evolution occurs when a common ancestral stock diverges into two main lines, each of which evolve similar structures and habits because they are exposed to similar environmental pressures. New World and Old World monkeys are examples.

6. Convergent evolution occurs when species have similar habitats and structures but have not evolved from a common ancestor. For example, wings have evolved independently several times among animals not closely related.

7. A species is a group of organisms that constitute a reproductively isolated population.

8. Species integrity is maintained by one or more of the following isolating mechanisms: geographic, seasonal, ecological, ethological, mechanical, gametic, hybrid inviability, and hybrid sterility.

9. Isolating mechanisms can be organized into premating and postmating types.

10. New species are formed in several ways when one or more of the isolating mechanisms fail to maintain the gene pool. Speciation can occur by hybridization, which is more common among plants than animals. Allopatric speciation occurs when an original population becomes geographically isolated into two or more groups. Sympatric speciation occurs within a population without physical isolation. The species become isolated instead by ecological and ethological mechanisms.

11. Phyletic speciation may be the most common method of creating new species. A single species undergoes continuous change until a new species is formed.

12. The Hardy-Weinberg law provides a mathematical model from which evolution can be measured and predicted.

13. The frequency of the melanic and peppered moths in England varies. The melanic form is more common in industrialized areas and the peppered form in rural areas. Pronounced changes in the frequency of occurrence of these moths have taken place during this century, an example of evolution in action.

14. Human beings comprise a single species but exhibit considerable variation, as do members of other species.

THE NEXT DECADE:

UNNATURAL SELECTION?

What you have just read in this chapter are the provisions of natural selection. In essence, natural selection is the normal biological process by which the genetic makeup of a population slowly changes as a result of environmental change. What, then, is unnatural selection? Just the reverse. The environment is controlled, but genetic changes are random, sporadic, and not linked to environmental change. Ultimately, unnatural selection lowers genetic fitness.

Human beings have been unnaturally selected since the discovery of fire, about 1½ million years ago. Fire changed our diets and our way of life. For the first time, we were able to control our environment.

Environmental threat today is less than it has ever been. We have shelter from environmental extremes. Too cold? Turn up the heat. Too hot? Turn on the air conditioning. Have an infection? Take antibiotics. A pain? Take pain killers. Want disease resistance? Use a vaccine. Despite our efforts, though, people are still killed by natural forces, such as lightning, floods, and earthquakes.

Myopia, or nearsightedness, provides a good example of unnatural selection. Myopia is an inherited disorder. How would our ancestors have coped with it 1 or 2 million years ago? Males would have been poor hunters and would have made fatal mistakes. Females would have had difficulty caring for offspring. More than likely, some environmental circumstance, such as preda-

tors, would have eliminated them and the children they hoped to protect. This is the process of natural selection. But today if you are myopic, you obtain prescription glasses that allow you to see as well as those who do not have the defect. What would happen to you if you had no glasses? You would be accident prone, and you would be unable to drive a car or do simple chores. You would be generally unfit. Miscalculations might result in a serious or fatal injury.

Another example of unnatural selection is diabetes. The genetic basis of childhood diabetes cannot be questioned. Diabetes is inherited as a polygenic trait. More than 4 million people in the United States suffer from it. They are maintained in the population because they inject themselves with insulin. Without these injections they would die. With the injections, most diabetics live to pass the trait to their offspring.

The human race is far from perfect. We harbor many genetic defects that are maintained in the population because of scientific and medical technology. Most of us have one or more genetic defects that will be passed on to our offspring in future decades. The harboring of genetic defects in human races is called genetic load, a term that describes the burden of defective mutations.

Could genetic load be reduced? Would it improve our lot? The answer is definitely yes to both questions. We have the ability to control our genetic fitness by applying ge-

netic knowledge to our own reproduction. Some people would be prohibited from reproducing because of their genetic load. Others with less of a load would be encouraged to reproduce. This theoretical ability to improve the human race is known as eugenics. The subject is not new. In fact, the earliest suggestion was made by Plato, who proposed that the better specimens of humans should be encouraged to mate and the worst should be limited in their mating.

Controlling human reproduction has been attempted many times in our history and has been endorsed several times in this century. For example, in the recent past, one of our states chose to sterilize ten thousand mentally retarded patients to prevent them from passing their defects to offspring. However, great concern was raised about the individual rights of these patients. Even if retarded, many of them could have had a more enriched and rewarding life by raising and loving a child. Many would have had normal children. Undoubtedly we will hear mention of eugenics in the next decade.

We will also hear serious discussion of euthenics which uses the tools of our highly technical environment to lessen the impact of our genetic load. Examples are drugs for controlling genetic diseases, such as insulin for diabetics, and glasses for those of us with poor vision.

What will be the consequences of

our genetic load in the next decade? Predictions have been made, but they forecast much farther into time than the next decade. For example, if a genetic disorder occurs once in every 15,000 births, it will take about 50 years for that rate to double. If the defect occurs once in 100,000 births, the rate will double in about 140 years. In general, it will take 200 to 300 years for our genetic load to double.

Medical advances in determining genetic defects in individuals and in fetuses may help reduce our genetic load. Genetic engineering, whereby new genes are inserted into cells, may become a reality and so reduce our genetic load in the next decade. However, even if the load is reduced, spontaneous mutations will still add to it. Most authorities believe that genetic load is no real threat to humankind. It may be of some concern, but it will not be our doom.

FOR DISCUSSION AND REVIEW

1. Draw diagrams that explain natural selection by (a) directional, (b) stabilizing, and (c) disruptive selection. Which is the most common?

2. Compare and contrast parallel and convergent evolution. Give examples of each.

3. Define *species*.

4. Give examples of each of the following isolating mechanisms: geographic, seasonal, ecological, ethological, mechanical, gametic, hybrid inviability, and hybrid sterility.

5. What is the difference between a premating and a postmating isolating mechanism? Arrange the list of isolating mechanisms into premating and postmating mechanisms.

6. Compare and contrast (with examples) allopatric, sympatric, and phyletic speciation.

7. What is a beefalo? How did it originate? Is it a new species? Will it survive?

8. Explain how hybrids can give rise to new species.

9. What are the stipulations of the Hardy-Weinberg law?

10. Suppose that 9 percent of the individuals in a population are homozygous recessive for a specific trait. What is the predicted frequency in the next generation of homozygous dominants? Heterozygotes?

11. The gene frequencies within a population are $q = 0.4$ and $p = 0.6$. What is the predicted frequency of homozygous dominants in the next generation? Homozygous recessive? Heterozygotes?

12. Explain genetic drift. Give examples of its occurrence.

13. Explain how the frequencies of the peppered and melanic moths changed as England became industrialized. Will the frequencies remain constant? Explain.

14. Give an example of selection occurring in human populations.

SUGGESTED READINGS

EVOLUTION OF REPRODUCTION, by G. R. Austin and R. V. Short. Cambridge University Press: 1976.
Examines the probable influences on the evolution of mammalian reproduction.

HEREDITY, EVOLUTION, AND HUMANKIND, by A.M. Winchester. West: 1976.
A good account of the concepts and mechanisms of evolution.

THE HOT-BLOODED DINOSAURS, by Adrian J. Desmond. Warner Books: 1977.
This book may not seem to pertain to this chapter, but the concepts of this chapter and Chapter 17 are used in its evolutionary detective story. If dinosaurs fascinate you, this is everything you ever wanted to know about them.

HUMAN VARIATION AND HUMAN MICROEVOLUTION, by Jane Underwood. Prentice-Hall: 1979.
Applies evolutionary concepts and mechanisms to the human species.

19

OUR PRIMATE HERITAGE

All primates, including humans, share certain characteristics, which are described in this chapter. In the classification scheme in Appendix 2, primates fall into the kingdom Animalia□, the phylum Chordata, and the subphylum Vertebrata.

The next level of grouping, the class Mammalia, includes the order **Primates,** which consists of two suborders. The suborders are **Prosimii** (*pro* means "before," and *simi* means "ape"; thus "primates before apes") and **Anthropoidea** (*anthro* means "man"; thus "appearing like man"). The prosimians include lemurs, lorises, and tarsiers. Representative prosimians often are in large zoos (see Figures 19–1 and 19–2). The anthropoids include monkeys, apes, and humans.

The suborder Anthropoidea is divided into three superfamilies: **Ceboidea,** or New World monkeys; **Cercopithecoidea,** or Old World monkeys; and **Hominoidea,** apes and humans. The evolution of Old World and New World monkeys is a case of parallel evolution□. Despite the similarity between them, the two groups of monkeys are different.

New World monkeys have prehensile tails (capable of grasping). They lack opposable thumbs on their hands, but their big toes are opposable. Their nostrils open to the side of the face because of their broad, flat nose (they do not have a protruding nose). Finally, they have three pairs of upper and lower premolar teeth.

Old World monkeys may have tails, but the tails are never prehensile. The monkeys have opposable thumbs on their hands and opposable big toes on their feet. Their nostrils open downward or forward, and they have a small, protruding nose. Finally, they have only two pairs of upper and lower premolar teeth.

PREVIEW QUESTIONS

After reading this chapter, you should be able to answer the following questions:

1. What are primates?

2. To which primate group do humans belong?

3. What anatomical features possessed by primates have contributed significantly to their success?

4. List some of the events that may have taken place during the evolution of primates.

Figure 19–1

The Asian tree shrew, a primitive prosimian whose ancestors radiated into a diverse group of advanced mammals—the primates—70 million years ago. The tree shrew of today is about the size of a rat and climbs trees much like a cat does. (It does not have fingers that can grasp the branches on which it travels.) The Asian tree shrew is similar in many ways to ground shrews, which are not primates but insectivores. Insectivores are the ancestors of primates. (Courtesy of the San Diego Zoo).

The Hominoidea are the apes—gibbons, siamangs, orangutans, chimpanzees, and gorillas—and humans. It is relatively common to refer to these creatures collectively as hominoids. Two of the important families of the Hominoidea are the Pongidae and Hominidae. (Gibbons belong to the family Hylobatidae.) The Pongidae are the orangutans, chimpanzees, and gorillas. The Hominidae include humans and their recent ancestors. The term **hominid** is often used to refer to humans and their ancestors.

EVOLUTION OF THE PRIMATES

The ancestors of the primates were mammals, members of the order **Insectivora**. Insectivores today are represented by shrews, moles, and hedgehogs. These mammals burrow in the ground and live on a diet almost exclusively of insects. The extinct insectivores

Figure 19–2

A tarsier. Tarsiers are more advanced than tree shrews but less advanced than monkeys and apes. They live in Indonesia and the Philippines. Notice the obvious primate characteristics: fingers and toes capable of grasping and large forward-placed eyes with binocular vision (but tarsiers are nocturnal and color-blind). Prosimians such as tarsiers were abundant for nearly 20 million years but declined in number when forced to compete with a new group of primates, the monkeys. Some authorities believe that ancient tarsiers may have been the direct prosimian ancestors of modern monkeys. (Courtesy of the San Diego Zoo.)

that gave rise to primates were much larger than present-day insectivores and apparently were ratlike creatures. They had been successful for nearly 100 million years, searching for insects on the floors of lush tropical forests. These early insectivores were good candidates for radiating into other niches because they were generalized rather than specialized. Above their habitat was a niche that was little exploited: the canopies of trees, inhabited only by birds and snakes.

The first insectivores climbed into trees 75 million years ago and began an **arboreal** (tree-living) existence. Moving among tree branches on clawed feet, with their noses pointed downward in search of insects, many undoubtedly had only a precarious hold in their new habitat. However, over millions of years, many adaptations were developed by the process of natural selection to fit them to their environment.

Early prosimians evolved from the insectivores. Prosimians are well adapted for life among the trees. Some of their most primitive living members are the Asian tree shrews (Figure 19–1), which are rats-sized and have large eyes and clawed feet for climbing. They are unable, however, to firmly grasp branches with their forelimbs or hindlimbs. All other prosimians show more advanced characteristics (Figure 19–2). Their eyes are more forward and their visual fields overlap, but they are color-blind. Their hands and feet are developed for grasping, and their thumbs and big toes are slightly opposable, although they have little ability to move individual digits. The prosimians are primarily olfactory animals, and their sense of smell is well developed. Most of them use urine to mark territory□ and to attract mates.

About 50 million years ago, there was an abrupt decline in the number of prosimians, presumably because of competition from newcomers. Monkeys, which had diverged earlier from the prosimian stock, were fierce competitors who were better adapted for arboreal existence. The prosimians possessed the rudiments of those adaptations, which were fully developed only among monkeys and apes—the Anthropoidea. These characteristics will be described in detail in this section of the chapter.

Opposable Thumbs

Moving from tree limb to tree limb became much easier when hands and feet were capable of grasping. Most primates have hands and feet that can grasp because their thumbs and big toes have rotated outward and around the hands or feet to a position where they face (oppose) the other digits. Hence the term opposable thumbs.

Human hands are excellent examples of this characteristic. (The big toes of humans are not opposable. They once may have been, but selection did not favor retention of this characteristic.) The thumb can touch the end of the other four fingers. When contact is made, a circle is formed by the two digits. This gives humans the ability to grasp objects.

The human grip is considered a precision grip. It is the most advanced grip for primates. A precision grip of lesser dexterity is characteristic of gorillas, chimpanzees, and many of the monkeys (see Figure 19–3). This grip allowed early hominids to develop and use tools. It allows humans to hold pencils to write or paintbrushes to paint. Chimpanzees can also hold paintbrushes, but their lack of a true precision grip prevents the control of the brush that is characteristic of human painters.

Locomotion

Primates do considerable vertical clinging to and leaping among branches of trees. This can easily be seen in zoos. One form of locomotion (moving from place to place) used by many primates is brachiation, hand-over-hand swinging from branch to branch (Figure 19–4). The anthropoids that make the most extensive use of brachiation are the gibbons and siamangs. In zoos they brachiate hour after hour and never seem to get tired. At other times, primates move by walking on all fours, a form of locomotion known as quadrupedalism (see Figure 19–5). This is how dogs and cats move, and it is the form of locomotion used by crawling children. Many anthropoids are capable of walking on the hindlegs, a style of locomotion known as bipedalism. All primates are capable of bipedalism to some extent, but only humans use it exclusively.

Clinging, leaping, and brachiation demand considerable strength and dexterity of the forelimbs, or arms. Primates are the only animals that can fully stretch the arms above the head—an adaptation for

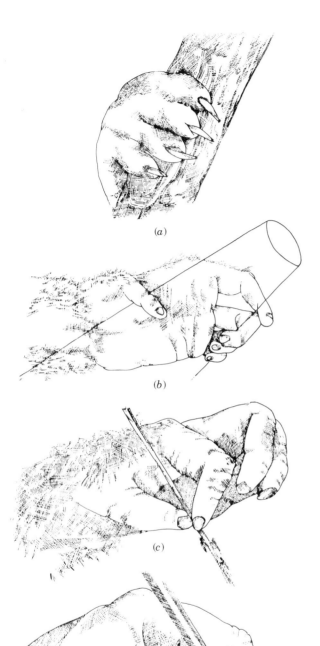

(a)

(b)

(c)

(d)

Figure 19–3

The hands of primates. Primates are characterized by digits on the hands and feet that are capable of grasping. (a) The primitive tree shrews could grasp little. They depended instead on their claws for leverage when climbing. (b) Most primates—for example, the orangutan—not only grip with their digits but also have opposable thumbs and toes, which improve grasping. (c) The chimpanzee has a more opposable thumb, which gives it an even more precise use of its hands. This chimpanzee is using a small twig as a tool, expertly probing for tasty termites with thumb and forefinger. (d) The human hand has a thumb that is completely opposable to each of the other digits, which gives humans a true precision grip. The precision grip allows the delicate and precise manipulation needed for drawing and writing.

Figure 19–4

Brachiation, a mode of locomotion common among many primates, involves hand-over-hand swinging. All apes are capable of brachiating. It requires grasping hands and feet and stereoscopic vision. Notice the opposable toes of this chimpanzee. (Courtesy of the San Diego Zoo.)

Figure 19–5
Quadrupedalism (walking on all fours) is a typical form of locomotion for primates. This gorilla demonstrates a specialized form of quadrupedalism—knuckle walking. Much of the gorilla's body weight is supported by its legs, but the anterior portions of its body are supported by the knuckles. Notice that the arms are longer than the legs. This kind of adaptation also allows many primates to walk bipedally, supporting all of their weight on the hind legs. The only exclusively bipedal primates are humans, although many other primates walk on their hind legs for short periods of time. (Russ Kinne/Photo Researches, Inc.)

hanging from trees. Furthermore, the primate forelimb can be rotated in a complete circle. To demonstrate this, hold your arm straight down along your side. Then raise it until it points directly in front. Next, move it upward until it is overhead. Continue the circle backward until the arm is in the original starting position. Other mammals are incapable of such complex movement. Another important primate characteristic can also be demonstrated. Hold your arm out in front of you, palm up. Then turn your hand 180° until the palm faces downward. Repeat. What you have demonstrated is the unique ability of primates to rotate the hand and lower arm. This is possible because of the way the bones are joined below the elbow.

The grip and rotating arms and hands give the average primate an amazing ability to move among trees. The precision and accuracy of these acrobatics is impressive. Primates rarely misjudge the distance and location of the next object to be grasped, even when they move at great speed. But the system is not perfect. There is no way to judge the safety of the next branch. Will it be strong enough to hold the weight of the primate? Is it rotten? There have been reports of primates falling to their death after grasping a rotten branch.

Perception

Vision□ is the dominant sensory mechanism of primates. They must be able to accurately compute the distance to the next branch while rapidly moving. **Stereoscopic vision** (depth perception or 3-D vision) gives them this ability. Some other mammals have stereoscopic vision, but it is never as well developed as it is in primates.

Primate eyes are large and positioned close together in the center of the face—rather than to the side, as is the case in many other mammals. The largeness of the eyes increases the visual field over that of other mammals and allows more light to enter each eye. The position of the eyes allows the visual field of each to overlap considerably. Look directly at some object. Cover your left eye with your hand. Turn your head to the right slowly while still looking at the object. You can turn your head about 45 degrees to the right and still see the object in the visual field of your right eye. The overlap of visual fields gives you nearly perfect distance perception.

Some of the optic fibers from the left and right halves of one eye pass to opposite sides of the brain. The fibers separate before entering the brain in a region known as the optic chiasma (see Figure 19–6). As you perceive an object, the brain computes distance by the slightly different information coming from the two eyes.

Anthropoids have another important visual characteristic—color vision. They eat insects, seeds, and even meat when they get the chance, but an important part of their diet is fruit. Color vision is an important adaptation for locating brightly colored fruits among the drab greens of the rainforest and jungle. Anthropoids also have a good sense of smell, although not as good as that of some prosimians.

The grasping hands of anthropoids were the tools for gathering colorful fruit. Forearm dexterity gave primates another important ability. They no longer had to run among branches with their noses down.

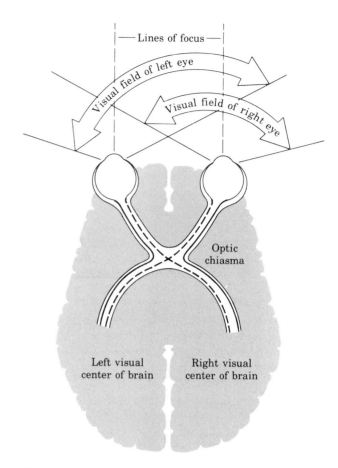

Figure 19–6

Stereoscopic vision. Three-dimensional vision is important to primates, because they must make frequent judgments about distance as they swing and leap among trees. This type of vision is possible only when some of the optic nerve fibers from each eye communicate with both the left and right cerebral hemispheres. The outer optic fibers from each eye go to the cerebral hemisphere on the same side. However, the inner optic fibers travel to the cerebral hemisphere of the opposite side. The separation of these fibers occurs in a region known as the optic chiasma.

They could grasp objects and bring them to the nose or mouth for inspection, smelling, and tasting. Furthermore, the fleshy pads on the tips of the fingers, backed by nails instead of claws (see Figure 19–7), became increasingly sensitive and permitted the detection of textures and shapes of objects. Primates are the only animals that spend considerable time picking up and inspecting objects. These manipulative skills and the ability to discriminate among objects are important adaptations.

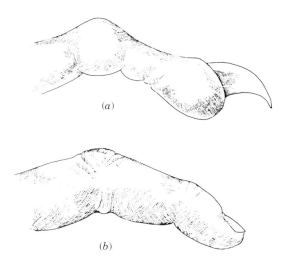

(a)

(b)

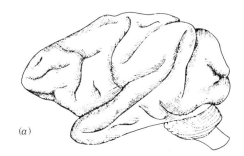

(a)

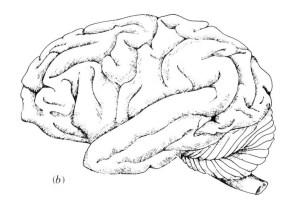

(b)

Figure 19–7

Claws and fingernails. (a) Animals that possess claws use them primarily for traction during locomotion. The fingertips are not particularly sensitive, because the claw is the probe that explores the environment. (b) The finger shown here is typical for most primates. The fingertip is backed by a fingernail, making the fingertip a sensitive pressure pad. This enables primates to determine the exact shape and texture of objects in their environment.

The Cerebrum and Behavior

Primates have enormous **cerebrums** in comparison to many other mammals (see Figure 19–8). The primate cerebrum also has a greater surface area because of a greater number of folds in its surface. The enlarged cerebrum was an important development for the complex coordination of the eye-limb movements of primates.

The large cerebrum is equally important in storing information—memory. It gives the primate a unique ability to recall previous experiences and to compare the outcomes of those experiences with some new situation. In humans, the same process is referred to as

Figure 19–8

The primate cerebrum, which is much larger than that of other mammals. As primates became more advanced, the size of the cerebrum increased as well. (a) The cerebrum of a monkey is considerably smaller than that of most apes, such as (b) the chimpanzee. (c) Humans have the largest cerebrums. The increased complexity and size of the cerebrum permits the storage of information or memory. Previous experiences can be compared with current experiences, an activity called thinking.

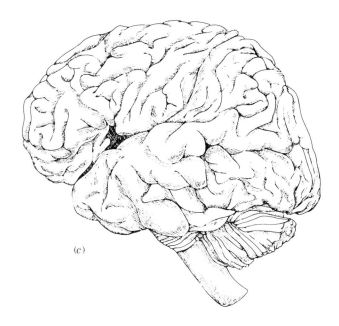

(c)

thinking. Memory, then, plays an important role in the life of primates.

The ability to learn changed the behavior of primates drastically. Their social behavior is among the most advanced in the animal kingdom. The majority of primates live in social groups (troops) that are maintained and protected by the most dominant individuals. The relationships and dependencies among individuals in primate troops are as complex as the relationships among humans. It is obvious that most primates recognize individuals and even remember them after a period of separation.

The offspring of primates depend more on their mothers than do the offspring of other mammals. A great amount of primate behavior is taught and learned during an early period of socialization. A majority of learned responses are acquired through imitation.

The increased infant dependency results in fewer births than for other mammals. Primates usually have single infants, rarely twins, and never litters. The births are spaced at least one year apart. In many species—for example, chimpanzees—they may be spaced several years apart. This allows females to gestate the fetus longer, reducing the risk of premature birth and other developmental hazards. Gestation periods for primates generally are longer than for most other mammals. (The elephant, however, holds the record for mammals, gestation is twenty-two months). Having single births drastically reduces the number of offspring that can be produced per unit of time. Primates have long life spans (chimpanzees may live forty to fifty years). The greater amount of time for reproduction compensates for the loss in reproductive capacity.

Increased infant dependency and living in groups promotes cooperation among members of primate groups. This is a form of advanced social behavior.

EVOLUTION OF THE HOMINOIDS

About 25 million years ago, the earth's climate changed. For millions of years, most land masses were covered with lush tropical rainforests. But then the earth began to cool, and the climate became more arid. As a result, the margins of the rainforests□ began to change. The cooler, more arid weather was too harsh for many of the delicate tropical plants exposed on the fringes of the forests.

In their place appeared new plants that had food reserves in underground storage organs and that could survive yearly periods of cooler weather by remaining dormant. These herbaceous perennial plants, which died out during cooler weather, formed an entirely different habitat, called grasslands□.

Grasslands also began to develop in the middle of forests. They continued to form and unite until, millions of years later, they were very large, as exemplified by the savanna of Africa today. The first grasslands had not been exploited by animals. The earliest invaders were probably insects, followed closely by rodents. A third group of pioneers—certain groups of primates—came down from the trees to explore the new habitat. The reason these primates left the forest is unknown. Perhaps they were forced to leave because of competition with other primates. Perhaps it was simply natural selection favoring expansion into less populated areas. These primates, the ancestors to all modern-day hominoids, began the lineage known as apes.

The fossil record shows that a number of primate species were common along the edges of the grasslands 12 to 25 million years ago. At one time, all the various primate fossils were classified as Proconsul, but now they are usually referred to as *Dryopithecus*. In what follows, we will use the old name for simplicity. That name was established when some of the first fossils of this primate stock were brought to London for identification early in this century. It seemed clear to the paleontologists examining the fossils that this hominoid resembled the modern-day chimpanzee. The London Zoo had a popular chimpanzee named Consul; it seemed only appropriate to name its ancestor Proconsul (*pro* means "before"; thus "before Consul").

Proconsul and its relatives were about 107 centimeters tall and weighed from 14 to 23 kilograms. They were much larger than any primate before them. (Asian fossils belonging to the same group are thought to have been as big as gorillas.) Proconsul primates could walk upright (bipedally). However, they probably went about on all fours more frequently and may even have walked on their knuckles, a habit of modern-day chimpanzees and gorillas. Fossil evidence suggests that they were also efficient brachiators.

The teeth of Proconsul clearly indicate that it was a

primate (hominoid), but not a close relative of humans (hominid). The shape of the jaw and the placement of the teeth are referred to as the dental arch. The dental arches of apes and Proconsul primates are elongate. Those of hominids are shaped more like a horseshoe, and the distance from front to back is approximately the same as the distance from side to side (see Figure 19–9).

Proconsul had long pointed canines, which suggests that it probably did not defend itself with stones, rocks, tools, and the like but rather with its long teeth, which it probably used in threat displays□. The grinding teeth, or molars, were also different from those of humans. Old World monkeys have molars with four grinding cusps. Hominids have molars with five grinding cusps. The characteristics of the teeth alone are sufficient to indicate that Proconsul was ancestral to hominoids, not to humans.

Most monkeys have little skin pigmentation under their dense coats of hair. Generally, the skin is glistening white. However, Proconsul is believed to have developed some pigmentation in the skin and probably considerable pigmentation on areas of the body that were not heavily covered by hair, such as the face, hands, and feet. The dark skin pigmentation (melanin) would have been a necessary adaptation. Proconsul primates most likely spent only the cooler

hours of the morning and evening in the grasslands and rested in the shade of the forest during the hotter times of day. They probably slept in trees at night. Even so, they were exposed to considerably more sunlight than were the primates living in the shade of the forest canopy.

Ultraviolet radiation stimulates the production of vitamin D in the skin. Certain amounts of vitamin D are necessary for survival. However, if too much ultraviolet radiation is absorbed by the skin, too much vitamin D is produced. Calcium and phosphorus levels of the blood are raised, kidney stones form, and the individual dies. (A lack of vitamin D, on the other hand, causes rickets and ultimately deformation of the limbs.)

Pigmentation in the skin prevents overexposure to ultraviolet radiation. To be successful, Proconsul had to have some pigmentation. (It has been estimated that in latitudes comparable to Denmark's, insufficient ultraviolet radiation reaches humans. Thus vitamin D must be included in the diet. Milk and other dairy products are good sources.

Proconsul continued an omnivorous diet (plant and animal matter), but with new foods incorporated into the daily menu. There were new and varied seeds in the grasslands. Proconsul troops probably also spent considerable amounts of time digging roots and tubers for food (a habit typical of modern baboons). Fruits were probably still obtained from the forests surrounding the grassland and from many of the new grassland plant species and bushes. Rodents were common, and it is reasonable to suspect that Proconsul occasionally chased and killed them and thus ate meat from time to time.

The fossil record suggests that Proconsul had few predators, a major reason for its success. The only serious predators of apes and humans are large cats, which did not appear on the scene until about 15 million years ago.

There were several varieties of ground or woodland hominoids 12 million years ago. The large number of fossils demonstrates their success. These primates have been lumped into three large taxonomic groups: *Dryopithecus (pithecus* means "ape"; also known as Proconsul*), Gigantopithecus,* and *Ramapithecus. Dryopithecus* is considered the ancestor to the Pongidae apes. *Ramapithecus* is the ancestor to humans (Hominidae—described in detail in Chapter 20). *Gigantopithecus,* a very large woodland ape from Asia, became extinct. It has no currently living relatives.

Figure 19–9

The dental arch. The fossil jaws of hominoids (apes) can easily be differentiated from those of hominids (humans) on the basis of the shape of the dental arch. (a) The dental arch of apes, including the modern gorilla, is long and almost rectangular. (b) The dental arch of all hominids, including contemporary humans, is shaped more like a horseshoe. The distance from front to back is nearly the same as the distance from side to side.

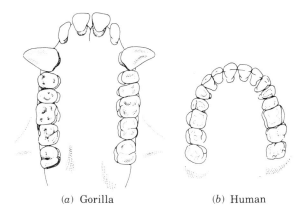

(a) Gorilla   (b) Human

The early divergence of ground apes into two distinct lines, the ancestors of apes and humans, is a significant occurrence. It is often erroneously assumed that humans evolved from apes and gorillas or that, at one time, humans looked like modern chimpanzees or gorillas. Both assumptions are wrong. First, modern apes have had a longer evolution than humans. They diverged from the woodland ape stock perhaps as early as 25 million years ago. The origin of hominids was not evident until 13 million years ago. Second, the long evolutionary history of the Pongidae has produced modern representatives that are well adapted to their habitats. They all dwell in forests, a habitat not occupied by early hominids. How could humans have possibly looked like chimpanzees and gorillas when they evolved at a different time and in a completely different habitat? What is true is that both modern apes and humans had a common ancestor about 25 million years ago. It resembled an ape more than a monkey but was distinctly different from modern apes.

SUMMARY

1. The ancestors of the primates were insectivores. Today they are represented by shrews, moles, and hedgehogs.

2. Ancestral insectivores gave rise to the prosimian group of primates about 75 million years ago. These primates exploited a new environment in the trees. They had forward-facing eyes and overlapping visual fields, but they were color-blind. Modern representatives are lemurs, tarsiers, and lorises.

3. About 50 million years ago, monkeys appeared. They were fierce competitors of the prosimians.

Monkeys, apes, and humans comprise the suborder Anthropoidea.

4. The anthropoids are exceptionally well adapted to their habitats. They have opposable thumbs that allow them to grasp objects, such as tree limbs and food. They are capable of complex locomotion, including brachiation, quadrupedalism, and bipedalism. Their arms are capable of rotating in a complete circle. Vision is their dominant sensory mechanism. They have both stereoscopic and color vision.

5. Primates have enlarged cerebrums, an important development for complex coordination and eye-limb movements.

6. Enlarged cerebrums also give primates the ability to store previous experience—memory.

7. Primate offspring have a greater dependency than do other mammal offspring. Complex behavior is taught and learned during an early period of socialization.

8. The hominoids evolved about 25 million years ago, when climatic change caused the formation of grasslands. The new, unexploited environment stimulated their evolution.

9. One of the earliest hominoids, Proconsul, was about 107 centimeters tall and weighed 14 to 23 kilograms. It walked upright. Its teeth and dental arches clearly indicate that it was a primate not closely related to humans.

10. There were several varieties of ground or woodland primates 12 million years ago: *Dryopithecus* (Proconsul), the ancestor to the apes (Pongidae); *Ramapithecus,* the ancestor to humans (Hominidae); and *Gigantopithecus,* a large ape that became extinct.

THE NEXT DECADE

PRIMATE PREDICTIONS

There may be less time than a decade to study our primate cousins in their natural habitats, because we have all but destroyed those habitats. In the next decade, we can look forward to hearing about more and more studies of primates in the wild. Many will be pioneer studies, and most will be our last chance.

All primates interest us. Those we are most interested in—in the hope that they may shed some light on our own beginnings—are the great apes. They are the orangutans, gorillas, and chimpanzees.

The longest and most detailed study of primates in nature is Jane Goodall's. She began studying chimpanzees in 1960 and continues to observe them today. More than two decades ago, as a very young woman, she entered the Gombe Stream Reserve on the shores of Lake Tanganyika in Africa. What was known about chimpanzees at that time had been deciphered from observations in zoos and laboratories. Most of the information was misleading.

Goodall's study had modest beginnings. She was alone in the jungle with a few camp helpers, who cooked and assisted her. Today, the area where she observed her first chimpanzee in the wild is a well-known research field station, where a dozen or more researchers work.

What Goodall discovered about chimpanzees in her first decade of study astonished the world. She found that chimpanzees have long-lasting social bonds, the strongest being between mothers and their offspring. Recognition of family members may last a lifetime. Each chimpanzee, adult or infant, is physically recognizable by any other. Goodall herself was able to recognize them all and gave them names.

Chimpanzees have almost a human body language. They stare, threaten by swinging their arms, hold hands, embrace, and even kiss. Their facial expressions are remarkably similar to ours. In addition, they make a number of discrete sounds—hoots, pants, and howls—that communicate their pleasures and displeasures.

Chimpanzees also make (as well as use) tools, a feat previously thought to be reserved for humans. They have been seen to pick twigs, remove all the leaves, and then stick their newly fashioned probes into termite nests. They transfer the termites from the probe to their mouths much as humans transfer rice grains on a chopstick.

Chimpanzees had been considered vegetarians. Goodall, however, watched them hunt small monkeys, kill their prey, and then share the bounty with other members of the troop.

To keep the chimpanzees nearby and to facilitate observation, Goodall gave them bananas. Bananas do not grow in the chimpanzee habitat, so their introduction was unnatural. True to what we see in zoos, though, the chimpanzees loved them. There was severe competition for the bananas, dominance became exaggerated, some chimpanzees became greedy, and fights frequently resulted. Goodall showed the world that chimpanzees in nature are more like humans than are any other animal.

Goodall's second decade with the chimpanzees (the 1970s) revealed unexpected behavior that tightened the link between human and ape. Chimpanzees move about in troops, each with its own home range. The borders are patrolled by dominant males, who fend off any intrusion by neighboring troops. On ten occasions, a lone male or a female with infant was violently attacked. Each attack was a cooperative assault by patrolling males from another troop who trespassed presumably to eliminate the local troop and assume control of its territory. What was of greatest surprise was that those attacked were beaten so badly that they were killed or died later of their wounds.

Still other violent behavior was observed. One female with two dependent offspring attacked another female, stole her infant, and killed it. The corpse was eaten by the attacking family. The same female was observed to repeat her violent act on two other occasions. War, gang war, murder, and cannibalism are behaviors rarely recorded for animal species other than humans.

Perhaps we have underestimated our primate heritage. Maybe an explanation for chimpanzee violence will be found in the next decade. Goodall is beginning her third decade of observing chimpanzees. What will she see? Will we find that chimpanzee society suffers the same ills as our own? Studies of primates in the next decade may give us not only valuable information about our past but also valuable insight into our future.

FOR DISCUSSION AND REVIEW

1. Distinguish among the Prosimii, Anthropoidea, Ceboidea, Cercopithecoidea, Hominoidea, Pongidae, and Hominidae. Give examples of each.

2. Trace the major evolutionary adaptations of the prosimians from their insectivore ancestors.

3. What characteristics made monkeys so successful?

4. What are the major events that led to the evolution of the hominoids? What did early hominoids look like? What were their general habits? Who was Proconsul?

5. What dental characteristics help differentiate other hominoids from hominids? What other characteristics help in this distinction?

6. Most monkeys have pigmented faces, hands, and feet but glistening white skin underneath the coat of hair. What is the adaptive significance of skin pigmentation?

SUGGESTED READINGS

THE EVOLUTION OF PRIMATE BEHAVIOR, by Alison Jolly. Macmillan: 1972.
A detailed treatment of comparisons between humans and other primates, with a focus on behavior.

MONKEYS AND APES, based on the television series Wild, Wild World of Animals. Time-Life Films: 1976.
A fascinating story of the evolution, behavior, and social systems of monkeys and apes. Excellent documentation and photographs.

THE PRIMATES, by Sarel Eimerl, Irven DeVore, and the Editors of Life. Time Incorporated: 1965.
The text and numerous illustrations explain the various groups of primates—their evolution and their relationships. Special attention is given to locomotion and behavior.

IN THE SHADOW OF MAN, by Jane van Lawick-Goodall. Houghton Mifflin: 1971.
The entire story of a young scientist's decade-long observation of chimpanzees in nature. Reads like a novel. Informative not only about chimpanzees but about primates in general.

20

HUMAN
EVOLUTION

Charles Darwin suggested more than a hundred years ago that Africa had to be the cradle of humankind because of the remarkable similarities between humans and the apes of Africa. Darwin seems to have been right. Most evidence collected to date supports the contention that human evolution began in Africa, although recent evidence indicates possible human origins in the Middle East and Asia. Many details about human history are still lacking (although there are no "missing links," as commonly supposed). Still, there is sufficient information to retrace human ancestry for the past 3 or 4 million years.

ARCHEOLOGICAL EVIDENCE

The first major discovery in the chronicle of human evolution was made in 1924. Miners working in an underground cavern in Taung (literally "place of the lion") in South Africa discovered a skull. The skull was nearly complete, which is a rare event in the process of fossilization. The skull and other artifacts in the cavern were sent to the famous archeologist Raymond Dart. After careful examination of the skull, Dart wrote an article, published in 1925 in the journal *Nature*, reporting that the skull had been a child's, that of a human ancestor. The skull was said to belong to a hominid called Taung child. Its teeth were those of a hominid and its face was flat, another hominid characteristic. The foramen magnum (an opening in the skull that permits the spinal cord to connect with the brain) was positioned not at the back of the skull, as might be the case in a prehistoric monkey, but at its base. This meant that the head was attached

PREVIEW QUESTIONS

After reading this chapter, you should be able to answer the following questions:

1. List some of the oldest hominid fossils collected so far.

2. What species of the genus *Homo* is believed to be the direct ancestor of *Homo sapiens?*

3. What are the evolutionary relationships of Neanderthals and Cro-Magnons to modern humans?

356

to the torso in such a way that the eyes pointed straight ahead. Taung child thus walked erect.

The discovery of Taung child was important. Like many other new scientific findings though, it went unnoticed—its importance unrealized—until many years later. However, Dart and a small group of archeologists knew that Taung child was important. They began digging in various parts of Africa, looking for more pieces to the puzzle of human evolution. They were rewarded. Within twenty years, a number of similar creatures were found at other sites in Africa, many accompanied by tools.

Other small prehumans and Taung child were classified as *Australopithecus africanus.* Larger, more robust creatures were classified as either *Australopithecus boisei* or *Australopithecus robustus,* depending on where the fossil remains were found (see Box 20–1). All these hominids were dated as living about 2 million years ago.

Olduvai Gorge

One of the most thoroughly worked sites in the search for human ancestry is an area in Tanzania (in Africa) known as Olduvai Gorge (Figure 20–1). The gorge is a gully about ninety meters deep and forty kilometers long, created by a seasonal river about 2 million years ago. Today its walls expose layer upon layer of rock and sand, which represent the history of this part of Africa for the past 2 million years. The

renowned archeologists Mary and Louis S. B. Leakey spent forty years in Olduvai Gorge looking for evidence of the human past. The Leakeys began their tedious work in the mid-1930s. Over their many years of labor, they found one of the most complete

Figure 20–1
A small shelf in Olduvai Gorge in Tanzania, Africa. The whole gorge is forty kilometers long and ninety meters deep. The rock strata along this portion of the gorge are obvious. The ninety meters of similar rock strata along the gorge contain the history of this region of Africa for over 2 million years. Paleontologists have been digging in Olduvai Gorge for four decades. The paleontologists shown here, like those before them, search for our origins. (Photo by Boyce A. Drummond III, Illinois State University.)

records of tool use by early hominids. Today their sons, Richard and Phillip, continue their work.

The tools of Olduvai Gorge fall into two distinct categories (see Figure 20–2). First are the choppers, which were extensively used by one type of hominid. They were made from round rocks about eight to ten centimeters in diameter. The final tool was fashioned by knocking off a chip or two from the stone, creating an exposed surface of rough and jagged edges. The hominids who created choppers are also believed to have used scrapers and hammerstones. These crude implements must have been satisfactory, because they were used for a million years.

The artifacts of the gorge also demonstrated the existence of a second culture, which coexisted with the first but was more advanced. The most important tool made by this type of hominid was the handaxe, a long instrument with a sharp edge. It looked something like a very large arrowhead. The user appar-

ently grasped the dull end and hacked away with the sharpened and pointed end. Other tools of this culture included scrapers, cleavers, chisels, choppers, and hammerstones.

The hominids using only choppers became extinct about a million years ago, but the fossil record shows that the hand-ax hominids coexisted with the first type for nearly half a million years. Olduvai artifacts have established that the two hominids lived side by side 1 to 1½ million years ago, although they maintained separate cultures.

Who were these hominids? More than twenty years of dedicated work by the Leakeys at Olduvai Gorge answered that question. In 1959, Mary Leakey found

Figure 20–2

The tools of our ancestors. *(a)* The chopper is a stone tool made from a small round rock eight to ten centimeters in diameter that had chips broken from it to create sharp and jagged edges. *Australopithecus africanus* made and used choppers. *(b)* A more advanced tool, the handaxe, is a longer stone tool with sharp edges. It resembles a large arrowhead. The user grasped the blunt end and used the sharper end as a multipurpose tool. *Homo habilis* ("handy man") made and used handaxes. These two hominids coexisted but had separate cultures. They used separate tools for nearly a million years, which indicates that they were satisfied with their own life-styles.

a skull fragment in the Gorge, the first evidence of who might have used the tools. More and more fragments were found. Nearly an entire skull was uncovered, but it was in many pieces. Its identity was not known until weeks later, after the fragments had been pieced together. The skull belonged to *Australopithecus boisei*. A few years later, an additional hominid was found and identified as *Homo habilis* ("handy man"). This was the advanced toolmaker. Later, fossils of the more primitive tool users, *A. africanus*, were also found in the Gorge.

What were the relationships among these hominids? It was assumed that the hominids probably interbred and that directional selection□ favored *H. habilis*. In other words, it was assumed that *Australopithecus* was slowly replaced by *H. habilis*. These were the human "roots."

Skull 1470

Hominid history was revised in 1977 because of new fossil evidence. The change may seem minor to some, but it places our *H. habilis* ancestors much earlier in history than previously thought and describes a more realistic relationship between *Australopithecus* and *Homo habilis*. The current interpretation of human history includes the same hominid ancestors, but it changes the time of their occurrence and their relationships.

In 1972, Richard Leakey discovered an important fossil skull along the eastern shore of Lake Turkana in northern Kenya. Like so many fossil skulls, this one was in many pieces. After weeks of patient reconstruction, the skull was identified as *Homo habilis*. But this skull was different. Its owner lived nearly 3 million years ago—1½ million years before the *H. habilis* whose skull had been found at Olduvai. This skull was therefore older than many of the existing australopithecine fossils. The skull was called skull 1470, which was the museum identification number assigned to it.

Other remains of 1470 indicated that this early version of *H. habilis* had the habit of always walking upright. The australopithecines had also walked erect, but their stride made them walk as we might walk when heading into a strong wind. (Contrary to popular notion, the australopithecines were not humpbacked, nor did they walk with a limp.) The ball on the joint of the upper leg of *H. habilis* was smaller than that of *Australopithecus,* and the bone attached to

it was longer and flatter. This minor change made *H. habilis* perfectly upright. Several fragments of *H. habilis* hands were also found near skull 1470. Reconstruction indicated that the hands were not much different from ours. They were quite dexterous and had a precision grip. All the evidence indicates that this hominid was quite small, about 120 to 140 centimeters (4 to 4½ feet) tall.

Since 1972, similar fossils have been found, some as old as 4 million years. The discovery of these fossils has led to new interpretations of the fossils previously found. It now seems that *Australopithecus* did not become *H. habilis*. The two were instead closely related contemporary cousins, the product of divergence from an earlier common ancestor.

THE EVOLUTION OF HOMINIDS

Ramapithecus, the earliest hominid to diverge from the hominoid line, appears to have been very successful around the margins of rainforests 9 to 12 million years ago. *Ramapithecus* differed from Proconsul (a hominoid) mainly because it possessed typical hominid teeth. Sometime between 4 and 9 million years ago, *Ramapithecus* diverged, creating the australopithecines and *Homo habilis,* first found in the fossil record about 4 million years ago. The reason for the divergence is unknown, but it could have been a climatic change that created new habitats to exploit.

What is more mysterious is the lack of fossils from the period between 4 and 9 million years ago. This period does not hide a "missing link." Either fossils have not been found yet or the climate did not favor fossilization during that time. But the time gap is of concern to specialists, because on one side of the gap is a single hominid—*Ramapithecus*—and on the other side are four closely related, coexisting hominids—"later" *Ramapithecus,* the two australopithecines, and *H. habilis.*

The later *Ramapithecus* was relatively small and upright and had a rather small cranial capacity. (The capacity of the skull, or cranium, is proportional to brain size. It is assumed that the larger the brain, the more advanced the hominid. The cranial capacity of the modern human averages about 1,400 cubic centimeters.) *Australopithecus boisei* was large, about 150 centimeters (5 feet) tall, and robust. It too was upright, and it has a cranial capacity of about 530 cubic centimeters. The smaller australopithecine, *A. africanus* (shown in Figure 20–3), was about 140 centimeters (4½ feet) tall, with a cranial capacity of 450 cubic centimeters. Its brain, however, was larger in proportion to body size than that of *A. boisei*. Finally, *H. habilis* was about 150 centimeters (5 feet) tall and more slender and upright than *A. boisei*. *H. habilis* had the largest cranial capacity, nearly 700 cubic centimeters.

About 2½ million years ago, these four hominids coexisted in the same community. It is assumed that they preferred a woodland habitat with some open grassland, like that of their ancestors. They had an advantage in direct access to water. Woodlands with open grasslands commonly border tropical lakes. Lake Turkana in Africa, where most of the fossil evidence now comes from, is a similar habitat.

Richard Leakey has offered the following explanation of the relationships among these hominids (see Box 20–2). The most common hominid was *A. boisei*. These creatures were large and had rather massive jaws, an adaptation for their feeding behavior. They probably ate mostly seeds and roots. Their jaws and broad molars acted as grinders for their hard, fibrous food. Males, females, and youngsters spent most of their time pulling roots with their hands, although they might have used sticks or rocks to dig with. They moved in troops, but each member was independent of the others. The protection of the troop was not the responsibility of specific individuals—each member looked out for itself—but there was safety in numbers. They climbed trees of the woodland at night to avoid predators while sleeping.

A. africanus shared the habitat. These hominids were smaller but more agile. Their troops favored more open grasslands but remained close enough to the woodland to have its protection. *A. africanus* troops are believed to have had more cohesiveness and to have been composed of more individuals than were *A. boisei* troops. They ate roots and seeds, but their diet also included berries, nuts, insects, and any small or injured animals that could be captured. In addition, they used scrapers to dig insects out of trees and decaying logs.

It is supposed that some of the larger members of each troop were always on the outlook for danger and thus were the watchdogs of the social unit. Youngsters were probably offered food by adults. Females were protected by males. The young were pro-

tected by both. They too took to the trees for sleeping. The behavior of *A. africanus* is thought to have been very similar to that of modern-day baboons.

Ramapithecus, although also present at the time, was relatively rare. The population is believed to have de-

Figure 20–3
Australopithecus africanus. (From Origins, by Richard E. Leakey and Roger Lewin. Dutton: 1977.)

THE REST OF THE STORY ABOUT THE EVOLUTION OF HOMINIDS

The scientists who find and study hominid fossils are called paleoanthropologists (*anthropology* is the study of humans, and *paleo* means "old"; hence scientists who study the humans of the past). Paleoanthropologists study, classify, and place hominid fossils in a meaningful sequence from past to present. They also hypothesize about the life-styles of these hominids. The term *story* in the title of this box is appropriate for the theories of paleoanthropologists because a story is a narrative that connects past events.

The narrative often gets confusing because each story has several versions. All the versions have the same plot. However, the hominids are often given different names, different relatives, different time periods, and even different life-styles. Another point of confusion is that the story supplied by any individual paleoanthropologist changes from time to time. Each time a new discovery is made, it provides more information about the story, so the story is modified (see "The Next Decade" at the end of this chapter for an example). Despite the confusion, however, the theories of paleoanthropologists have provided real excitement in science. The stories yet to come will be met with equal enthusiasm and excitement.

How does a paleoanthropologist hypothesize about the life-style and behavior of a fossil? Life-style is perhaps the easier to determine—through artifacts, the fossilized remains of animals and other articles found near the hominid fossil. Among these remains are animal bones, stones, and tools. Their arrangement gives an indication of life-style. One batch of artifacts may include rounded, natural stones. Another may include stones that obviously were beaten so they would have sharp edges (see Figure 20–2). This kind of evidence provides the information that one hominid used natural tools and therefore was advanced, while the other made tools and therefore was more advanced.

Artifacts tell much about life-styles, including the kinds of plants and animals eaten and the locations of campsites, fire pits, migratory trails, and territorial boundaries.

How can information about the behavior of a fossil hominid be acquired? Obviously, behavior cannot become a fossil, but artifacts often give a general indication of it. Archeologists can use artifacts to show probable behavior, such as tool making, meat eating, and hunting. Paleoanthropologists acquire certain behavioral information from cultural anthropologists. Cultural anthropologists are specialists who study current human cultures that exist in environments similar to those of the past. What interests them are the environmental adaptations of primitive cultures such as the aborigines of Australia, the Tasaday of the Philippines, the various peoples of the highlands of New Guinea, and a multitude of tribes in Africa and South America. The interactions of today's primitive people with their environment gives paleoanthropologists ideas of how fossil hominids may have dealt with their environment.

Similarly, primatologists provide information about the relationships between the structure and behavior of humans and other living primates. Others, such as pyschologists and linguists, provide information about the behavior and physiology of earlier hominids. This information may include the relationship of brain capacity to the development of languages.

The theories of paleoanthropologists are like all other theories. They are the best interpretation of the currently available facts that have general support. That is why the stories of different paleoanthropologists have the same plot.

The theory of hominid evolution presented in this chapter was proposed by Richard Leakey. It has been the topic of three popular books written by him and the featured story in many popular magazines. As you read his theory, recall the information from this essay. Leakey's ideas are based on the detailed interpretations of many scientists. However, as in the explanation of any scientific theory, his has the magic ingredient of speculation—the ingredient that brings out the taste of excitement.

clined because of competition with *A. africanus,* with whom they shared habits and some niches. This competition caused the extinction of *Ramapithecus* a few hundred thousand years later.

In this community, the troops of *H. habilis* were located nearer the lake. The members of *H. habilis* looked much like the australopithecines, particularly *A. boisei,* except that *H. habilis* lacked the massive jaws. *H. habilis* troops are believed to have lived in true campsites. The sites were temporary and perhaps used only a week or so. When food became scarce, the troops would move to another area, establish a campsite, and again search for food.

It is suggested that a primitive division of labor began in these early *H. habilis* camps. Although every camp member helped search for food, it is thought that the females spent most of their time collecting berries, nuts, tubers, and other plant materials. Two or three males at a time ventured away from the campsite to search for dead animal flesh, or carrion. The kills of predatory animals were often scavenged. Hunting for live animals was probably rare. Males were probably assisted frequently by females, and females would never pass up carrion during their collection trips. Although meat was eaten, and perhaps favored, plant materials were the major component of the diet.

H. habilis had an array of tools for digging plants, removing skins, cutting meat, and crushing animal bones so the marrow could be scooped out. The camp was the center of all activity—sharing, eating, sleeping. It is even believed that the members of the troops could communicate directly with one another through some form of primitive language. But the important characteristic of the *H. habilis* members is that they systematically collected, transported, and shared food. This is a major difference in behavior from that of the other hominids with whom they coexisted.

Ramapithecus became extinct early in the period of coexistence. *A. africanus* became extinct about a million years ago, probably because of competition with modern-day baboons, who now occupy the same niche. *A. boisei,* the larger individual, also became extinct, but somewhat later than *A. africanus. A. boisei* was extremely specialized. This probably contributed to its longevity. However, extinction may have come about because of overspecialization. *H. habilis* persisted, and it emerged about a million years ago as *Homo erectus.*

Homo erectus

As Figure 20–4 shows, *Homo erectus* ("erect man") did not look much different from modern humans. Its individual members were shorter than we are and had much larger molar teeth. The major differences are details of the face and head. *H. erectus* had a protruding jaw, large brow ridges, a flat face, a flattened

Figure 20–4

Homo erectus. (From Origins, by Richard E. Leakey and Roger Lewin. Dutton: 1977.)

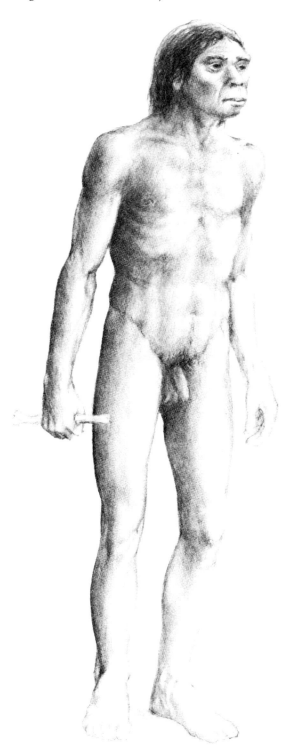

skull, and a domelike cranium to house a large brain. Cranial capacity was 775 to 1300 cubic centimeters.

It has been proposed that the most significant advancement of *H. erectus* was behavioral. *H. erectus* was a true hunting-gathering species, with a complete division of labor among adults. Males left campsites to hunt for meat instead of just hoping to find a dead carcass to scavenge from a predator. At first, the hunting bands probably chased some small animal until it was exhausted and then stoned or beat it to death. (The Hottentot Bushmen of Africa today will often follow their prey for days after it has been narcotized slightly by a poisoned spear. The animal finally drops of exhaustion miles from the original encounter and is killed by its hunters.)

Later in time, *H. erectus* members probably cooperated more when hunting—encircling the prey or using lures and traps to make their efforts more efficient. Food was transported back to the campsite. Females spent considerable time near the campsite searching for plant food, which still made up a large proportion of the daily diet. They also helped search for food, and some stayed in camp to care for youngsters.

There has been a great deal of speculation about *H. erectus*. How did its members behave? Were they the first "naked apes"? We will never know for sure.

Authorities believe that *H. erectus* individuals must have had considerably less body hair than did their immediate ancestors. Earlier hominids lived mostly under the woodland canopy, where they would have been well shaded from ultraviolet radiation. Their coats of hair afforded some warmth on cool tropical evenings, and the hair stood on end during the hot part of the day to promote cooling. They were exposed to direct sunlight for only brief periods of time.

Although *H. erectus* individuals existed in similar circumstances, they also hunted for long hours under the hot tropical sun of the open savannas. Specialists generally agree that it would have been a necessary adaptation for the number of hair follicles and the width of all hair to be reduced. Another important adaptation would have been an increase in the number of sweat glands all over the body. This would have made *H. habilis* less hairy than its ancestors, which earned this hominid the nickname "the naked ape." The adaptations would facilitate rapid cooling and prevent heat prostration when hunting. Today there are hunters on the savannas that have full coats of hair, such as lions and cheetahs. However, they

exert themselves for only a minute or two, because they overheat quickly. *H. erectus* is believed to have pursued prey for much longer periods of time.

The loss of hair created a problem. When the skin was covered with hair, it was lightly pigmented. Heavy pigmentation was typical only on faces, hands, and feet (as it is even now for nonhuman primates). It is assumed that when the hair was lost, melanin pigments accumlated in the skin as an adaption to prevent both excessive exposure of cells to ultraviolet radiation and excessive production of vitamin D (which causes kidney ailments). Because of the theoretical necessity of this adaptation, many specialists believe that *H. erectus* had black skin.

What kind of social behavior did *H. erectus* exhibit? The members of every troop were cooperative and sharing, and that was the key to their success. There is little doubt that individuals had separate identities and that various kinds of bonds existed, such as parental and pair bonds□.

Individual bonds may have been strengthened by the advent of several behavioral and physiological characteristics that humans possess today. First, the females of our species are the only primates that are sexually receptive at any time. All other female primates have estrus cycles□ and thus accept male mates only a few days each month. The continual receptivity of *H. erectus* females would have strengthened male-female bonds. Males were out hunting for days at a time, and their return was never predictable. When they did return, females were always receptive to sexual intercourse. The bonds created between males and females might have been the beginnings of stable pair bonds between specific males and females.

Second, the face-to-face posture of coitus is practiced by no primates other than humans. Perhaps this posture made sexual bonds stronger and more enduring. Sex became a social behavior and was thus used for more than just procreation.

Third, it is probable that females of a closely knit troop had synchronous menstrual cycles, as many women do now. Hence many females would conceive at the same time and many births would occur at the same time. This would facilitate cooperative care of the young by the mothers and would reduce infant mortality.

Fourth, like all primates, extinct or extant, there was a lengthy dependency of the young on their parents. Adult behavior and skills were acquired during this long period of socialization. Infant and juvenile dependency was probably longer for *H. erectus* than for its precursors, and this length favored higher survival rates.

Finally, the cooperative effort created some leisure time, which allowed more complex social behavior to evolve. Fire may have played a significant role in this development because it allowed social interaction to extend into the night. The use of fire evidently began late in Africa. It had a much earlier origin along the migratory routes of *H. erectus*.

CONCEPT SUMMARY

MAJOR POINTS OF RICHARD LEAKEY'S VIEWS OF HUMAN EVOLUTION

1. Several species of hominids coexisted in Africa 2½ million years ago.
2. All but *Homo habilis* became extinct. *H. habilis* persisted and evolved several important behavioral characteristics, including division of labor and systematic collection, transportation, and sharing of food.
3. *Homo erectus* evolved from *Homo habilis*.
4. *Homo sapiens* evolved from *Homo erectus*.

Migrations

The fossil record shows that the members of *H. erectus* were wanderers. They began to emigrate from Africa, first across the narrow land bridge to Asia and later northward into Europe. Why they began to migrate may never be known. What is known is that there were no sudden climatic changes in Africa that would have forced them to disperse into other regions. In fact, the climate in Europe was becoming much colder, which should have inhibited exploration.

H. erectus had the first tools to allow emigration, including cooperation, food sharing, and the ability to transport food. Carrying food led to carrying all other necessary commodities, including water and fire. Thus carrying is a unique hominid characteristic. *H. erectus* may have not carried fire from Africa, although fire was carried along the migratory route. A campsite of *H. erectus* unearthed near Nice, France, showed that fire had been used. The campsite is about a million years old. A significant aspect is that it had been rebuilt again and again by later arrivals.

The behavioral characteristics of *H. erectus* enabled

BOX 20–3

CAVE DWELLER CAPERS

Hundreds of accounts of human evolution appear in a variety of books. The details may differ, but the trends are essentially the same. In accounts of how early hominids behaved, however, there is considerable variation and perhaps even distortion.

One popular notion is that the australopithecines became extinct because they were the targets of hunting and belligerence on the part of the genus *Homo*. Some have even suggested that the early *Homo* used the leg bones of antelope and other animals as weapons, smashing the skulls of australopithecines and other members of genus *Homo* with the rounded ball joint. The fossil deposits (from supposed *Homo* campsites) that suggested this interpretation are now known to be from the dens of predatory cats, who left behind an assemblage of hominid skulls and other bones, the refuse of many satisfying meals.

It has also been suggested that Neanderthals became extinct because of murderous raids conducted by hordes of Cro-Magnons. Peking man (*Homo erectus*, found in the Orient) evidently opened the bases of the skulls of his fellows and removed their brains and ate them. This has been interpreted as evidence of the early origin of killing and cannibalizing other humans. A less violent interpretation is that this behavior may have been a ritual performed on the dead for some religious or superstitious purpose.

---

its members to explore any habitat. They were no longer tied to woodlands next to water. They could, and did, populate the Old World. Their larger cranium gave them the intelligence to handle unusual environmental problems. It also gave them an awareness of the world. Perhaps it was only human curiosity that drove them into the next valley and over the next mountain range (see Box 20–3).

It was surely inevitable that some populations of *Homo* would be left behind during major migrations. This was probably the fate of the Neanderthals.

Neanderthals

The first skull of a Neanderthal was found in 1856 in a cave near the German village of Neander by workers who were blasting rock. The significance of the find was not evident for decades. Even the scientific community was reluctant to accept the skull as being Neanderthal. It was first thought to be the skull of a Mongolian Cossack who had pursued Napoleon and his army only fifty years before. However, many other skulls and bones of Neanderthals were found over the next twenty years in other parts of Europe. Today there exists a fossil collection of over a hundred more or less complete individuals.

Neanderthal males were about 160 centimeters (5 feet, 4 inches) tall. Females were about 150 centimeters (4 feet 10 inches) tall. Neanderthals walked upright, as we do, and had low brows, protruding faces, receding jaws, and high-domed heads. If there was anything truly different about them, it was that they were much stockier than we are (see Figure 20–5). Their cranial capacity was 1,300 to 1,600 cubic centimeters. (Modern cranial capacity ranges from 1,200 to 2,000 cubic centimeters, with an average of 1,400. Thus many Neanderthals had larger brains than we do now.) Neanderthals existed half a million years ago but were most numerous from about 100,000 years ago. They became extinct 30,000 years ago.

Neanderthals were the legendary cave dwellers. However, they did not look like the unfortunate reconstructions that often were used to illustrate them. They have been portrayed as having heavy brow ridges and humped backs. This version of Neanderthals was reconstructed from the fossil remains of an old arthritic man. Figure 20–5 gives a more accurate illustration of their general appearance.

Neanderthals were adapted to a cold environment. They encountered the succession of glaciers that passed over most of the northern temperate regions of the world. In fact, at times they may have existed in an environment entirely of ice. They were not only skilled hunters but true predators, a specialization that did not occur among hominids before or after them. The Neanderthals did not kill animals just for the meat. They also fashioned the skin into clothing to protect themselves against the harsh climate. Natural caves became campsites that were illuminated and heated by fire.

Neanderthals had the leisure time and the basic intelligence to develop self-awareness. For the first time, hominids began to wonder who they were and how they fitted into the scheme of nature. It is thought that Neanderthals also had a primitive religion and may even have believed in some form of

Figure 20–5

Neanderthal man. (From Origins, by Richard E. Leakey and Roger Lewin. Dutton: 1977.)

afterlife. In Shanidar Cave in northern Iraq were found the remains of a Neanderthal man who apparently was buried in a ritual ceremony. The pattern of flowers around the corpse suggests that they were deliberately placed there. The fossil remains are only an orderly arrangement of pollen grains, but the pollen was sufficient to identify most of the species of flowers used in the burial ceremony. Many of the plants were herbs commonly used for medicinal purposes later in history. Neanderthals may therefore also have had some knowledge of disease and its treatment.

Neanderthals existed in other places besides northern Europe. Fossils have been found in France, Spain, the Middle East, and even China. They were probably reproductively isolated from the mainstream of hominid evolution, even though they are classified as *Homo sapiens* (along with Cro-Magnons, who were very similar to modern humans). Neanderthals are commonly placed in their own subspecies, *Homo sapiens neanderthalensis.*

Neanderthals must have been aware of Cro-Magnons, and some interbreeding may have occurred. However, Neanderthal genes were probably quickly absorbed by Cro-Magnons. Also, the extent of such interbreeding must have been limited because the two groups were ecologically separated. Neanderthals were dependent on the large herds of grazers found in northern regions to support their carnivorous existence. Cro-Magnons, on the other hand, were successful omnivores in more temperate regions. The fate of Neanderthals was probably the same wherever they existed. The time of their decline was near the beginning of a major period of glaciation, which is thought to have nearly exterminated their food—the large grazing herds in the north. Because of this and probably many other reasons, Neanderthals became extinct.

Cro-Magnons

The first fossil of a Cro-Magnon was found in southwest France near a large rock (Cro-Magnon) in 1868. Cro-Magnon emerged about 40,000 years ago. For an accurate description of what Cro-Magnons looked like, simply look at yourself in a mirror. We belong to the same species, and there are no major differences in appearance between us and Cro-Magnons.

Among the legacies left by Cro-Magnons are impressions of their artistic talents. The glaciation that ended the Neanderthals was not so severe for the Cro-Magnons because they populated more southerly areas. Nevertheless, it was cold enough for them to be forced into caves for protection. In these caves, they left paintings depicting their life and environment (see Figure 20–6). They most frequently painted animals, illustrating their partial dependence on the

Figure 20–6

Ice Age art. These cave paintings are from a cave inhabited by Cro-Magnon people about 25,000 years ago. The bison are now extinct, but the paintings provide a record of some of the large grazing herds that were present at the time. *(a)* One bison is a male. The position of the tail and the erect mane may indicate that it is displaying a threat to other bisons or to its human predators. *(b)* The other bison is a female lying down. A collection of similar paintings and other art forms of the Cro-Magnons has toured the museums of the world as "Ice Age Art." (American Museum of Natural History.)

(a)

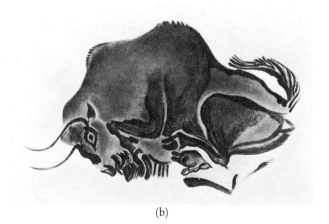

(b)

natural habits of game herds. More than a hundred cave paintings have been found in France and nearly half that many in Spain.

As the glaciers formed absorbed tremendous amounts of water, and the seas dropped as much as 122 meters. This created many new land bridges across vast stretches of what had previously been oceans. Human curiosity—or perhaps mutual competition—stimulated another major migration. This one established the major populations of the world. Human beings settled in many remote areas, each geographically isolated from the others. The isolation and thousands of additional years of continued evolution have created the obvious variations—the races—among human populations today.

Dispersion

Homo sapiens in Europe could not have dispersed and populated the globe by themselves. There is overwhelming evidence that the events discussed in this chapter did not simply happen once by chance. Instead, each stage in human evolution happened many times and in many locations. The fossil evidence suggests that the *H. erectus-sapiens* lineage existed in Europe, Asia, and Africa, as did the *Ramapithecus* lineage. The australopithecines, however, were unique to Africa. The transition from *H. erectus* to *H. sapiens* differed from region to region because of different environmental pressures, but the transition occurred at nearly the same time in three major centers.

The African population never left after the initial exodus of *H. erectus,* and no immigrants are known to have returned from Europe. This group of *H. sapiens* retained dark pigmentation as well as other anatomical adaptations and was the origin of the Negroid race.

The *H. erectus-sapiens* complex in Asia differentiated into the Mongoloid race. There is evidence that these people began to migrate northeast about 18,000 years ago. (They are called "early mongoloids," because they probably lacked the discrete mongoloid features of orientals today.) This assemblage is believed to have migrated into North America about 16,000 years ago via the Bering land bridge, which temporarily linked Asia with Alaska. The migrations occurred again and again—and migrations within the New World also occurred—until hundreds of geographically isolated Indian populations were established from Alaska to Chile. The last migration established the Eskimo and Aleut Indians. They have more mongoloid features than any other North American or South American Indians.

The European *H. erectus-sapiens* complex populated Eurasia and became the Caucasoid race. Much later, some migrated to the Americas.

What about the inhabitants of Australia? It is not known whether they descended from the Caucasians or the Mongoloids, although their features suggest Caucasian origins. The first arrivals in Australia, about 16,000 years ago, established what is today known as the Australian aborigine subgroup, perhaps one of the older isolated populations of *H. sapiens.* However, nearly all of the population of Australia today is made up of recent white settlers from Europe.

It should be obvious by now that our hominid ancestors were a restless breed. Migration of humans continues, and, considering the number of people now involved, its pace has accelerated. However, about 10,000 years ago, the need for extensive migration ended. Intelligence, cunning, and appreciation of nature led to the development of agriculture. Crops were nurtured, and the food supply became abundant. It was no longer necessary to be hunters and gatherers. Our herbivorous prey were herded rather than hunted. Plant materials were cultivated, not blindly sought. Our "roots" began with a woodland ape 12 million years ago. Today we represent the most intelligent form of life ever to inhabit the earth.

SUMMARY

1. The first major fossil discovery documenting human evolution was made in 1924. The fossil was called Taung child, and it walked upright and lived about 2 million years ago.

2. The artifacts and fossils of Olduvai Gorge in Africa revealed that about 1 to 1½ million years ago, two hominids lived side by side but maintained separate cultures. One was *Australopithecus africanus,* the other *Australopithecus boisei.* A third hominid, *Homo habilis,* an advanced toolmaker, was also discovered. It was assumed that the australopithecines were slowly replaced by the genus *Homo.*

3. In 1972, Richard Leakey discovered another important hominid fossil along the shores of Lake Turkana. Skull 1470 belonged to *H. habilis* but lived 3

million years ago. Other members of *H. habilis* lived 4 million years ago. This proved that the australopithecines did not become *Homo* but that they were closely related cousins, the product of divergence from an earlier ancestor.

4. *Ramapithecus* existed from about 9 to 12 million years ago.

5. "Later" *Ramapithecus, A. africanus, A. boisei,* and *H. habilis* were contemporaries 2½ million years ago. They could coexist because each had different adaptations and filled a different niche.

6. *Ramapithecus* was small, walked upright, and had a small cranial capacity. *A. boisei* was about 150 centimeters (5 feet) tall and robust, with a cranial capacity of 530 cubic centimeters. *A. africanus* was smaller, about 140 centimeters (4½ feet) tall, and had a larger brain in proportion to body size (cranial capacity was 450 cubic centimeters). *H. habilis* was about 150 centimeters tall, slender, and possessing the largest cranial capacity—700 cubic centimeters.

7. *Ramapithecus* became extinct first. *A. africanus* became extinct about a million years ago, presumably because of competition with modern baboons.

8. *H. habilis* persisted and evolved several important characteristics, including division of labor and the systematic collection, transportation, and sharing of food.

9. *H. erectus,* the descendant of *H. habilis,* looked much as we do, and had a cranial capacity of 775 to 1,300 cubic centimeters. They were true hunter-gatherers, with a division of labor. Males hunted, and females gathered food and cared for offspring.

10. *H. erectus,* the first "naked ape," had darkly pigmented skin. Bonds among individuals—males, females, and youngsters—were important to their survival.

11. *H. erectus* migrated from Africa into Asia and Europe. A portion of the population in Europe became the Neanderthals. Males were about 160 centimeters (5 feet, 4 inches) tall, females about 150 centimeters (4 feet, 10 inches) tall. Both were stocky and had a cranial capacity of 1,300 to 1,600 cubic centimeters. They were most numerous about 100,000 years ago and became extinct about 30,000 years ago.

12. *H. sapiens,* or Cro-Magnons, emerged about 40,000 years ago. They looked exactly like today's humans.

13. The *H. erectus-sapiens* complex existed in three centers of the world. One never left Africa, and it became the origin of the Negroid race. One populated Asia and gave rise to the Mongoloid race, some of which migrated into North America about 18,000 years ago. The final complex populated Eurasia and became the Caucasoid race.

LUCY'S LEGACY SHAKES THE HUMAN FAMILY TREE

It is a sure bet that in the next decade we will read and hear a great deal about human evolution. The facts in this chapter may not change appreciably, but interpretations of existing fossils and new fossil finds will make headlines. Undoubtedly, the human family tree will be modified again and again.

The family tree of any species, called a **phylogeny,** shows the evolutionary ascent of various members of the group in question. Phylogenies are commonly constructed for a large number of species. For example, an illustrated phylogeny for the horse appears in Chapter 17. Similar phylogenies are constructed by paleoanthropologists for humans. A recent physical anthropology text indicates that nine different phylogenies for humans were proposed in the last decade.

The story of human evolution given in this chapter would look like Illustration A if it were charted as a phylogeny. This is the phylogeny that most paleoanthropologists accept. It was proposed by Louis Leakey and is staunchly supported today by his widow, Mary, and their two sons, Richard and Phillip.

H. sapiens

Homo erectus   A. robustus

        A. africanus   A. boisei

Early Homo

Australopithecus africanus?

Illustration A

The phylogeny of Illustration A proposes that there are two major branches to the human family tree. The common ancestor for both branches is *Australopithecus africanus*. One branch includes other australopithecines, which are now extinct. (*A. robustus* was not described in this chapter. It was a hominid that existed in South Africa, 2,000 miles south of the site presented here.) The other branch includes various members of *Homo* that led to today's humans.

In 1978, paleontologist Donald Johanson, curator of the Cleveland Museum of Natural History, found a new player in the drama of our past. He named his discovery Lucy, and she became a headline in many publications. One article was entitled "Lucy: A 3.5 Million-Year-Old Woman Shakes Man's Family Tree." Johanson wrote a best-selling book, *Lucy: The Beginnings of Humankind*. The book was not only widely read but won Johanson the Pulitzer Prize for literature in 1981. Lucy's species, according to Johanson, is *Australopithecus afarensis*. (She was discovered in the Afar region of Ethiopia.)

Why was Lucy such a spectacular discovery? First, she and her kind existed in eastern Africa 2.9 to 3.8 million years ago. (Lucy is just part of a female skeleton, but other members of her family have also been found.) This makes them the oldest hominids ever discovered. What makes Lucy and her family hominids? Their locomotion. They walked upright, or were bipedal.

Many limb bones have been recovered, and they all suggest an upright walking hominid. Furthermore, actual fossilized footprints of these hominids were found by Mary Leakey and her party. (Andrew Hill actually found the footprints while he and other scientists were playing elephant-dung soccer after a hard day at the "dig".)

Lucy was only 107 to 122 centimeters (3½ to 4 feet) tall, and she was indeed female. The fossils of nearby males suggest the males were at least 152 centimeters (5 feet) tall. What is astounding about Lucy and her family is that from the neck up they resemble apes (hominoids), but from the neck down they are our ancient ancestors. They had very small brains compared to the brains of chimpanzees today. Their heads were ape-like, chinless, and with large, forward-thrusting jaws. Their faces were flat and chimpanzee-like.

What really astonished paleoanthropologists is that upright walking had always been considered an advanced trait—until the discovery of Lucy. It had been associated with hunting and toolmaking, both of which required a sizable brain. But Lucy and her family walked upright, had tiny brains, and were not toolmakers. Thus Lucy demonstrated that upright walking was a primitive hominid characteristic that preceded large brains and toolmaking. The combination of characteristics possessed by Lucy and her family must have been good adaptations, because the species lived unchanged for 1 million years.

Johanson determined that the discovery of Lucy changed our phylogeny. He proposed the phylogeny shown in Illustration B.

The phylogeny of Illustration B places Lucy and her kind (*A. afarensis*) as the common ancestor to the two major branches of the human family tree. One might think that that placement would not cause any disagreement among specialists. Wrong. It has sparked a lively debate among paleoanthropologists. For example, Mary Leakey takes the stand that Lucy is not unique, that she is simply an early *Homo* and is already accounted for in Phylogeny A. The debate is not just words. It also involves a great deal of work. Right now, paleoanthropologists are

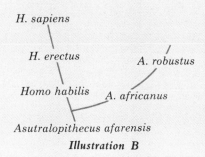

H. sapiens

H. erectus

A. robustus

Homo habilis   *A. africanus*

Asutralopithecus afarensis

Illustration B

reexamining many hominid fossils and artifacts. The fossils of Lucy and her family will be painstakingly scrutinized.

One thing is certain. The debate over Lucy will continue into the next decade. If Mary Leakey is right, then the human phylogeny will remain much as it appears in Phylogeny A. If Johanson is right, then Phylogeny B will be accepted.

Anyone who keeps up with current affairs in the next decade will hear more about Lucy. She could become as well known as the Lucy in the "Peanuts" cartoon strip.

FOR DISCUSSION AND REVIEW

1. Describe the significance of the discovery of Taung child and of skull 1470.

2. What hominids are known to have coexisted in Olduvai Gorge? What were their relationships? What kind of tools did they use?

3. What was the relationship of *Ramapithecus* to later hominids such as *Australopithecus* and *Homo?*

4. Describe the probable life-style along a lakeshore for *Ramapithecus*, *A. africanus*, *A. boisei*, and *H. habilis*.

5. What was distinctively different about *H. habilis* in comparison with the other hominids?

6. *H. erectus* is often referred to as the first "naked ape." Explain why this is probable. Was *H. erectus* black? Explain.

7. The bonds between individual *H. erectus* members are considered an important advance. Describe these bonds and their significance.

8. What is the evidence that *H. erectus* was a wanderer? Where did they go? How did they get there? Why did they wander?

9. Describe the proposed general life-style of the Neanderthals. Why did they become extinct?

10. Describe the probable life-style of the Cro-Magnons. Why were they so successful when the Neanderthals became extinct?

11. Describe the origins of races.

SUGGESTED READINGS

THE HUMAN SPECIES, by Richard M. Tullar. McGraw-Hill: 1977.
An excellent account of the human species, including its nature, evolution, and ecology. The book details the human primate ancestry, evolution, and questionable future.

INTRODUCTION TO PHYSICAL ANTHROPOLOGY, by Harry Nelson and Robert Jurmain. West: 1979.
For those who want additional detail on many topics, ranging from principles of evolution to human evolution.

LUCY: THE BEGINNINGS OF HUMANKIND, by Donald Johanson and Maitland Edey. Simon and Schuster: 1981.
Another view of the evolution of the human species.

THE NAKED APE, by Desmond Morris. Dell: 1967.
A provocative account of the origin and behavior of the naked apes—human beings.

ORIGINS, by Richard E. Leakey and Roger Lewin. Dutton: 1977.
Details an exciting theory of human origins. Well illustrated and readable. One of the finest and most interesting books on human evolution.

PEOPLE OF THE LAKE, by Richard E. Leakey and Roger Lewin. Anchor Press/Doubleday: 1978.
The theme is the same as in Origins—the documentation of human origins. However, this book is intended for a wider audience. It reads like a novel, and you will feel that you are experiencing the excitement of each discovery.

SECTION SEVEN

ECOLOGY AND BEHAVIOR

This drawing of a flea appeared in Robert Hooke's *Micrographia* (1665). This is one of the sixty plates in the book that were rendered with the aid of one of the earliest microscopes.

21

THE BIOSPHERE

PREVIEW QUESTIONS

After reading this chapter, you should be able to answer the following questions:

1. What is an ecosystem?
2. How extensive is the biosphere?
3. What are the names of the terrestrial biomes? List their characteristics.
4. Into what regions or zones do ecologists subdivide the oceans?
5. What kinds of freshwater environments exist?

Ecology (from *oikos,* meaning "house") is the study of the interrelationships of living organisms and their total environment. This environment includes the **abiotic environment**—physical factors such as temperature, rainfall, sunlight, and chemicals in the soil or water. It also includes the **biotic** or **biological environment**—the interrelationships of all the organisms in a given area or habitat.

The abiotic components affect the distribution of living organisms, which in turn can modify the physical environment. For example, bacteria decompose dead organic matter to produce inorganic substances necessary for plant growth, such as nitrates and phosphates. Furthermore, one species or group of species may provide the habitat for others. Trees provide the habitat for many species of birds and mammals. Coral reefs provide the habitat for fish and other marine animals.

A species' **niche** consists of the factors of the abiotic environment that affect the species and all interactions of that species with its biotic environment. For reasons explained elsewhere, no two species have the same niche in the same area. However, the niches of different species may be intimately interrelated. In a sense, every species on earth influences every other species in some way.

The members of a species in a given area are referred to collectively as a **population.** The area may be small—for example, a pond. It may be as large as the entire world; the human population, for example, is at home nearly everywhere on earth (see Chapter 24). All the populations of different species in a given area comprise a **community** (see Chapter 23). Some populations in a community are **autotrophic** (self-feeding) producers. Others are **herbivores** (plant eaters). Yet others are **carnivores** (meat eaters). And others are **decomposers** (microorganisms that reduce the complex chemical structures of dead organisms to simple compounds).

The community and all aspects of the physical environment associated with it comprise an **ecosystem.** Ecosystems are basic ecological units that are more-or-less self-sustaining (see Chapter 22). A small lake, an isolated few acres of forest, thousands of acres of grassland or desert, and a coral reef are examples of ecosystems. The abiotic environment of an ecosystem determines what kinds of organisms populate it, and these populations differ from those found in neighboring ecosystems.

Although the boundaries of an ecosystem may be clear-cut, they may also be vague. A lake has a fairly definite shoreline, but a desert may not be clearly set off from surrounding ecosystems. No matter how clearly defined it is, however, no ecosystem is truly independent and isolated from others. Many organisms pass from one ecosystem to another during migrations, and matter is carried from one ecosystem to another by wind and water. Because nutrients and energy often flow from one ecosystem to another, changes in one ecosystem may directly or indirectly affect another.

The **biosphere** is composed of all the ecosystems on earth. It therefore includes all living organisms.

AN OVERVIEW

Earth is often called the water planet because nearly three-quarters of its surface is covered by oceans. The oceans have an average depth of 4,000 meters, but the depth of ocean trenches may be greater than 9,000 meters. Life exists everywhere in the oceans, even in the near-freezing blackness of the greatest depths. The oceans and bodies of fresh water—ponds, lakes, streams, and rivers—make up the aquatic environment.

The remainder of the earth's surface is dry land. Most of the land masses (the terrestrial environment) occur in the Northern Hemisphere, as do most of the great bodies of fresh water. The heights of the land masses vary from sea level to nearly 9 kilometers, the height of Mt. Everest. Terrestrial organisms exist from sea level to elevations of approximately 6,700 meters, although spores of bacteria and molds have been collected high in the atmosphere. Therefore, the biosphere extends from the depths of the oceans to elevations greater than the highest mountains (see Figure 21–1). This distance is not as great as it

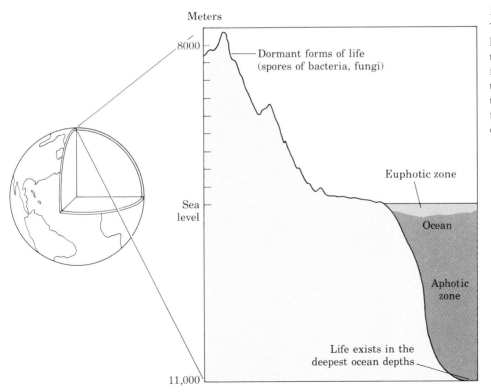

Figure 21–1

The extent of the biosphere. Living organisms are found throughout the biosphere—from the deepest ocean trenches to near the tops of the highest mountains. Spores from molds and bacteria exist even high in the atmosphere.

BOX 21–1

CLIMATE

The climate of a region is defined by its average weather conditions over a long period of time. Many factors—among them altitude, nearness to large bodies of water, and location on the continental land mass—may influence weather and climate. Ultimately, however, the major factors are the winds and the uneven distribution of sunlight reaching the earth's surface.

The sun radiates light energy evenly in all directions. Thus it would seem logical that, as the earth revolved around the sun, it would be uniformly heated by the sun's radiant energy. However, the earth rotates on an axis that is tilted at a 23.5 degree angle. This causes the Northern Hemisphere to be tilted toward the sun during the summer and causes the Southern Hemisphere to be tilted toward the sun during the winter. Summers are warmer than winters in the United States because when the North Pole is tilted toward the sun, the sun's rays strike the Northern Hemisphere more directly. The days are also longer, which provides more time for heating the earth's crust. During the winter, when the North Pole is tilted away from the sun, the sun's rays strike the Northern Hemisphere less directly (at an oblique angle). The days grow shorter, and, because there is less time for heating, the days are also cooler.

Near the equator, the earth receives solar radiation directly all year long, and the length of each day remains the same throughout the year. Thus the region around the equator is the warmest. Average temperatures decrease near the poles, which receive the least amount of solar energy. This differential in heating creates the winds that blow over the surface of the earth and in the upper atmosphere. The reason is simple. Warm air is less dense than cold air and therefore lighter. The wind patterns of the earth would be simple if the earth did not rotate.

Consider first the wind patterns that would exist if the earth did not rotate. The warm, light air in the vicinity of the equator would rise into the upper atmosphere and spread to the North and South Poles. As these winds blew toward the poles, the air would cool and become denser. Eventually, it would sink to near the earth's surface. At the same time, air already near the earth's surface would be moving from the North and South Poles and replacing the air that rises at the equator. These masses of air would then be

replaced by the air settling to the surface near the poles. Thus two gigantic gyres (circular currents) of circulating air would be produced, one in the Northern Hemisphere and one in the Southern Hemisphere. These air masses would be in constant motion. In the Northern Hemisphere surface winds would be blowing constantly out of the North, and in the Southern Hemisphere surface winds would be blowing constantly out of the South.

The earth does rotate, though, and the actual wind patterns around it are much more complicated than those just described. How are these patterns modified by the rotation of the earth?

Any point on the equator is at the earth's greatest diameter, and here the rotation of the earth reaches speeds of slightly more than 1,000 miles per hour. Since the earth decreases in diameter north and south of the equator, points at each succeeding latitude north and south have less distance to travel per rotation than at the next lower latitude. This situation produces a phenomenon, called the Coriolis effect, that causes water and air currents moving north from the equator to be deflected toward the east and those moving south from the equator to be deflected toward the west. (The Coriolis effect also directs currents of water in the same way. The next time you drain the sink or tub, notice whether the currents swirl to the right (east) or left (west). What would you predict?

The Coriolis effect prevents the warm air that rises above the equator from getting all the way to the North and South Poles. To see what actually happens, we will follow the events of the Northern Hemisphere only. (What happens in the Southern Hemisphere is a mirror image of what happens in the Northern Hemisphere.)

In the Northern Hemisphere, the warm equatorial air that rises and begins the northward movement is deflected to the east, forming a gyre. The surface winds in the gyre move in a southwesterly direction because of the deflection of the south-moving currents to the left. This gyre extends from the equator to the Tropic of Cancer, which is 23 degrees, 27 minutes from the equator. This gyre in turn induces a second gyre to the north, with the winds moving in the opposite direction. The surface winds move out of the west toward the northeast instead of the southwest, and winds in the upper part of the gyre move to the southwest instead of the northeast. A third gyre, the polar gyre, has surface winds once again blowing toward the southeast. (Diagram A depicts the general directions of the major surface winds in the Northern Hemisphere.) The United States and much of Canada

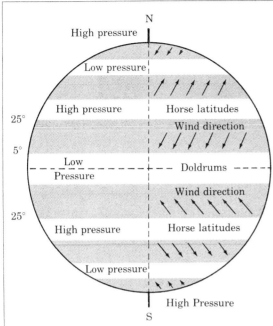

High pressure

Low pressure

High pressure — Horse latitudes

Wind direction

Low Pressure ---- Doldrums ----

Wind direction

High pressure — Horse latitudes

Low pressure

High Pressure

25°

5°

25°

N

S

Diagram A

fall within the gyre that has winds generally from the west.

Associated with the gyres are regions of high and low barometric pressure and regions of high and low precipitation. Generally, regions of rising air have low barometric pressure and high precipitation. The combination occurs because as the air rises and cools, water vapor condenses and falls back to earth as precipitation. This is true of the region along the equator called the doldrums. The high temperatures and high rainfall in this area provide the necessary general climate for tropical rainforests. The high pressure associated with the horse latitudes, shown on Diagram A, is the result of the sinking of cool air of two adjacent gyres. The air here has little moisture content. Most of the world's great deserts are found at this latitude, along the Tropic of Cancer and the Tropic of Capricorn (23 minutes, 27 seconds south of the equator), where there is little possibility of rainfall.

What about the climate of the United States? The northern part of the country is cooler than the southern part because reduced amounts of solar radiation are correlated with higher latitudes. The reasons for marked seasons were explained earlier. The rainfall patterns of the United States are associated with the general westerly flow of the wind across the country. Diagram B shows a section of the United States at 39 degrees north latitude. The West Coast of the country generally has high levels of precipitation because the moisture-laden wind blowing off the Pacific Ocean encounters a series of mountain ranges. The air is deflected upward and cooled, producing abundant rainfall. The land to the east of these mountain ranges is desert because the air now has little moisture left. As the wind descends from the Rocky Mountains, it again accumulates moisture from the earth's surface and from plants. This accumulation provides sufficient rain to support the extensive grassland biome of the central United States. Finally, nearing the East Coast, the air contains sufficient moisture to support deciduous forests (if other conditions are suitable).

Diagram B

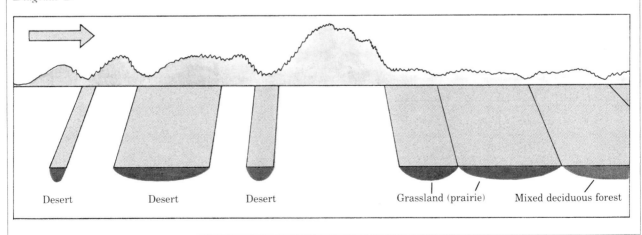

Desert   Desert   Desert   Grassland (prairie)   Mixed deciduous forest

seems—only 20 kilometers. Life actually exists only in a narrow zone near the earth's surface.

Living organisms are not uniformly distributed throughout the biosphere. Their distribution is limited by various factors in the biotic and abiotic environments. In terrestrial environments, the two most important factors are temperature and the availability of water. Both are important parameters of climate. (See Box 21–1.) On the basis of these two factors and other aspects of climate, terrestrial environments are classified into a number of biomes. A **biome** is a series of similar stable communities that extend over large geographical areas. Each is characterized by the presence of certain types of vegetation.

TERRESTRIAL ENVIRONMENTS

The following six biomes cover most of the land surface of the earth: tropical rainforests, grasslands, deserts, temperate deciduous forests, coniferous forests (taiga), and tundra (see Figure 21–2).

Tropical Rainforests

At low elevations near the equator, where there are high temperatures, high levels of rainfall, and intense light throughout the year, tropical rainforests are

Figure 21–2
The distribution of the original native vegetation of the world.

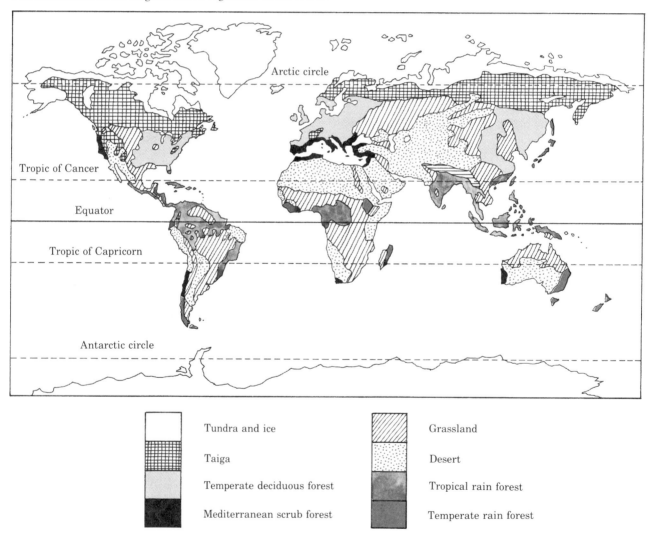

Tundra and ice		Grassland	
Taiga		Desert	
Temperate deciduous forest		Tropical rain forest	
Mediterranean scrub forest		Temperate rain forest	

found. The annual rainfall must be at least 200 centimeters (80 inches) for a rainforest to develop.

The characteristic vegetation of rainforests is broad-leafed evergreen trees, some 35 meters in height. These trees have shallow root systems, broad buttresslike bases, and branches only near their tops. The branches are often large, and they spread over a considerable area. They also bear numerous **epiphytes**—plants that cling to other plants for support. Many epiphytes are not rooted in soils. Instead they have aerial roots that absorb moisture. Orchids are an example.

The dense tree branches and epiphytes form canopies that effectively screen out much of the available light. Therefore, the plants that grow beneath the tall trees are shade-loving and do not survive in direct sunlight. Of various heights, these plants create different levels of canopies, each level screening out still more of the light (see Figure 21–3). So little light reaches ground level that few plants grow there, and the spaces between the bases of the trees are often bare. Most of the animals in rainforests live in the canopies and seldom, if ever, come to the ground. In South America, for example, the animals that inhabit the canopies include birds, monkeys, snakes, amphibians, tree sloths, insect- and fruit-eating bats, and, of

Figure 21–3
Profile of a tropical rainforest, showing the upper and lower canopies of vegetation that effectively reduce the amount of light reaching the forest floor.

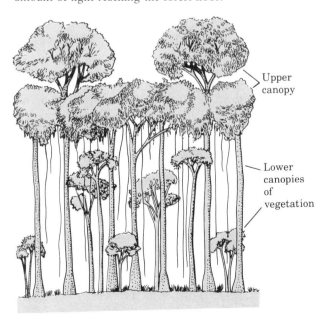

Upper canopy

Lower canopies of vegetation

course, countless species of insects. On the forest floor, several species of large rodents and wild pigs are common.

A greater variety of plant and animal species lives in rainforests than in most other habitats on earth. They make rainforests the most complex biome. Although hundreds of species of plants live in rainforests (compared to only a few dozen in other types of forests), population levels are low. A given area may contain only one or two of each species.

Surprisingly, the soil in rainforests is poor, because little or no **humus** (dead plant material) accumulates. After a plant or animal dies, it quickly decomposes. The nutrients released by decomposition are rapidly used by living plants or are leached from the soil by the heavy rains. The soil in tropical rainforests is poorly suited for agriculture when the forests are removed.

Grasslands

Extending from the tropics into the temperate regions are grasslands, including savannas. Rainfall varies from 25 to 125 centimeters (10 to 50 inches) per year. It is not distributed uniformly throughout the year, however, so grasslands have discrete dry and wet seasons.

Trees and shrubs grow in grasslands, but not in dense clusters, except along streams or in low, wet areas. Grasses abound, as do numerous species of nonwoody herbaceous plants. Grassland soil is especially rich because of the accumulation of plant debris that decomposes and provides humus. The humus comes not only from the leaves and stems of dead plants but also from the extensive root systems.

Grasslands have been converted into the richest farmland in the world. In the United States, most of the great prairies have been destroyed. Their land is now used for growing domesticated grasses such as corn and wheat.

Undisturbed grasslands support a variety of animals, including deer, antelope, and other herbivores. In the United States, large herds of bison roamed the Great Plains until their near extinction through hunting and destruction of their natural habitat.

Deserts

Where the annual rainfall is below 25 centimeters (10 inches) per year, deserts occur. Some deserts are in areas where precipitation is greater but unevenly distributed, the temperature is high, and long periods of drought occur.

Desert vegetation is sparse. Plants usually are spaced far apart, with bare ground between them. During rainy periods, a great many small plants suddenly grow, bloom, go to seed, and die back within a few weeks. Cactus and other plants generally associated with deserts grow most abundantly in warmer deserts, such as those in the southwestern United States. Sagebrush is the dominant plant species in the colder deserts, those in the northern states immediately east and west of the Rockies.

Animals that inhabit deserts, for example, kangaroo rats, are adapted to survive with little water. Some obtain water only from the plants they eat. Most are nocturnal and avoid the dehydrating effects of the hot sun.

Deserts can be converted into productive agricultural land with irrigation. However, the evaporation of irrigation water may eventually result in the accumulation of salt and other minerals in the soil, which limits its agricultural use.

Temperate Deciduous Forests

Where precipitation is abundant (75 to 150 centimeters, or 30 to 60 inches, per year) and evenly distributed throughout the year, deciduous forests occur. Moderate temperatures prevail, and there are distinct seasons—summer and winter. Deciduous trees lose their leaves at the end of the growing season and go dormant during the winter. The leaves that accumulate on the forest floor gradually decompose into humus, which provides nutrients for the wildflowers that bloom in abundance in spring, before the trees are fully leafed. Trees common in deciduous forests of the United States include maple, oak, beech, and hickory. Much of the deciduous forest that once covered most of the eastern United States has been destroyed by urbanization or agriculture. Chestnuts, once abundant, are nearly extinct because of chestnut blight fungus, brought to the United States from the Orient in 1904. Elms have been largely eliminated by Dutch elm disease. Animals common in deciduous forests include deer, raccoons, foxes, and numerous species of birds and insects.

Coniferous Forests

North of the deciduous forests, grasslands, and deserts, where precipitation is abundant (37 to 100 centimeters, or 15 to 40 inches, per year) but where winter temperatures are low, coniferous forests, called **taiga,** occur. These forests also extend southward at higher altitudes along such mountain ranges as the Appalachians in the eastern United States and the Rockies in the West. The dominant tree species are fir, spruce, and pine. These trees (conifers) retain their leaves (needles) throughout the year. Their dense foliage always shades the forest floor, so other plant life is sparse. Many of the animals inhabiting the taiga are migratory and spend winters further south. Others hibernate. Examples of animals associated with the taiga are moose, bears, and wolves. This biome is extensive only in the Northern Hemisphere, because there is little land mass in the Southern Hemisphere at latitudes suitable for its evolution.

Soils in the taiga contain little humus, because the rate of decomposition is slow and the summers are short. Therefore, the land has limited agricultural use. The taiga is, however, exploited for lumber.

Tundra

In the Northern Hemisphere between the taiga and polar icecap lies the tundra. It is the least complex of the terrestrial biomes, supporting only a few species of plants. Among them are mosses, lichens, and other low-growing plants, including certain species of grasses and sedges. Rainfall averages only 25 centimeters (10 inches) per year, and the growing season is little more than a month long. Only the upper 10 to 20 centimeters of the soil thaws. The soil beneath this layer, the permafrost, remains permanently frozen.

Few animals other than insects remain in the tundra during the long, dark winters, and even the insects go dormant then. The animals that inhabit the tundra during the short growing season include caribou, arctic hare, arctic fox, and numerous species of birds that migrate south during the winter, when temperatures get as low as −57 degrees Celsius.

The pattern of biomes is generally correlated with latitude. However, a similar pattern can be found as one moves from low altitudes to higher altitudes up a mountain (see Figure 21–4). A mountain climber near the equator might expect to find tropical rainforests at the base and deciduous forests at higher elevations, followed by taiga and then alpine tundra near the peak.

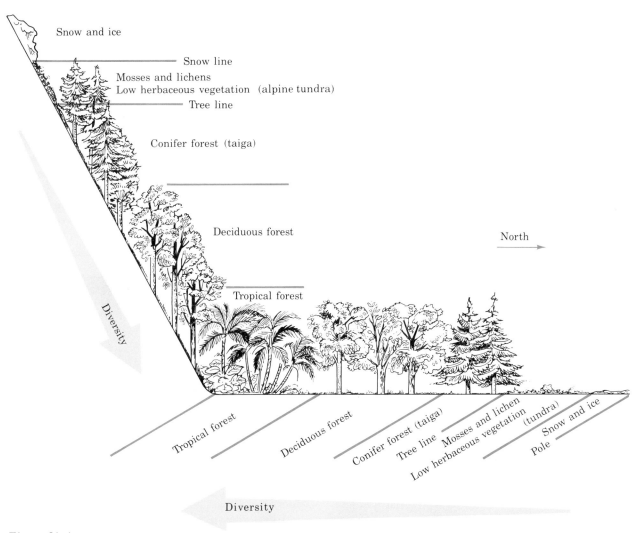

Figure 21–4

Distribution of terrestrial biomes correlated with latitude and altitude. A person traveling from the equator northward to the polar icecap would encounter the terrestrial biomes in the order indicated on the horizontal axis. (Grasslands occur from the lower to middle latitudes.) A person climbing a mountain near the equator would encounter similar biomes in the same order, but they would be distributed up the slope. The diversity of living organisms increases from low levels in colder regions to maximum levels in tropical regions.

AQUATIC ENVIRONMENTS

Aquatic environments include the oceans and the freshwater habitats of lakes, ponds, streams, and rivers. Freshwater and ocean environments are not always clearly separable. Where rivers flow into the ocean, fresh water and ocean water meet, and estuaries□ occur. This mixture, called brackish water, provides environmental conditions for complex communities.

Four major physical factors affect the distribution of living organisms in water: temperature, amount of light, salinity, and levels of dissolved nutrients. Salinity, the amount of dissolved salts in water, acts as a barrier separating freshwater and marine (ocean) organisms. Few species can move from one habitat to

CHARACTERISTICS OF TERRESTRIAL BIOMES

BIOME	TEMPERATURE RANGE	RAINFALL	DOMINANT VEGETATION
Tropical rainforests	Warm throughout year	High much of the year; in excess of 200 centimeters per year	Broad-leafed evergreen trees and epiphytes
Grasslands	Warm in tropics; warm to cold in higher latitudes	Low to moderate; 25 to 125 centimeters, not uniformly distributed throughout year	Grasses and other nonwoody herbaceous plants
Deserts	Warm in tropics; warm to cold in higher latitudes	Low; less than 25 centimeters per year; occuring during short time periods	Cactus and related plants in warm deserts; sagebrush in colder deserts
Temperate deciduous forests	Warm in summer; cold in winter	Moderate; 75 to 150 centimeters per year	Deciduous trees
Coniferous forests (taiga)	Cool in summer; cold in winter	Low to moderate; 37 to 100 centimeters per year	Coniferous evergreen trees
Tundra	Cool during brief summer; extremely cold much of the year	Low; averaging only 25 centimeters per year	Low-growing plants such as mosses, lichens, and grasses

the other. Salmon and Atlantic eels are two that can (see Box 21–2).

Oceans

The oceans of the world are interconnected bodies of water. However, life in them is not evenly distributed. A greater number of organisms live in the sunlit surface waters than in deep water, where great pressures occur, no sunlight penetrates, and temperatures may hover near freezing.

Oceans are dynamic. Great masses of water with slight variations in temperature, salinity, and density contain different communities of living organisms. These masses of water flow slowly and eventually mix, carrying with them pollutants such as DDT, which now can be found in all the oceans.

Light levels sufficient to maintain photosynthesis penetrate only the upper 100 to 200 meters of the oceans. This illuminated zone is called the **euphotic zone.** The water below is called the **aphotic zone.** Heterotrophs in the sea depend directly or indirectly

on the photosynthetic organisms in the euphotic zone.

Floating in the sunlit surface waters of the ocean are countless microscopic unicellular and multicellular organisms called **plankton. Phytoplankton,** microscopic photosynthetic protists, are transported by currents. Some "swim" feebly. Animal plankton, called **zooplankton,** occur at all depths but are most abundant in the euphotic zone, where many of them feed on phytoplankton or each other.

As Figure 21–5 indicates, ocean waters can be divided into several zones. The waters that cover the sloping and shallow continental shelves compose the **neritic zone,** which is divided into the intertidal and subtidal zones. The intertidal (or littoral) zone is alternately covered and uncovered by tidal water. Consequently, organisms living there are alternately hidden and exposed. The patterns and heights of tides—which are caused by gravitational attractions of the earth, sun, and moon—differ from one geographic region to the next. Intertidal shoreline communities are determined to a large extent by the size of the

BOX 21–2

OCEAN WATER AND FRESH WATER

The water that makes up the oceans if often called saltwater because of the high concentration of salt (sodium chloride) dissolved in it. Ocean water also contains a great variety of other dissolved organic and inorganic substances. Some substances are present in low concentrations, others in high concentrations. Some of those in low concentrations are essential for certain forms of sea life. Ocean water contains an average of 3.5 percent dissolved salts of various kinds. Fresh water also contains a wide variety of dissolved substances, including salt. It is a much less complex solution, though, and the concentration of dissolved substances is much lower than in ocean water. Fresh water usually contains only about 0.1 percent salt. The differences in the amount and kind of dissolved substances in fresh and ocean waters give them quite different properties as far as living organisms are concerned.

Because of the higher levels of dissolved substances in ocean water, it has a higher density than fresh water. Living organisms are thus more buoyant in ocean water than in fresh water. Differences in the concentrations of dissolved substances in ocean water and fresh water also create differences in osmotic properties (refer to Chapter 4), which are correlated with differences in the physiological characteristics of organisms living in the two environments. For example, freshwater fish do not drink water, but water enters their bodies by osmosis; as a consequence, these fish produce large quantities of urine to eliminate the water. At the same time, many important dissolved substances in the body fluids of the fish must be retained. Ocean fish, other than sharks and their relations, lose body water to the ocean by osmosis and therefore need to replace it. These fish drink sea water and excrete small amounts of urine. By drinking sea water, they also consume large amounts of salt. Ocean fish have special salt-secreting glands in their gills that rid them of the excess salt. Vertebrates that

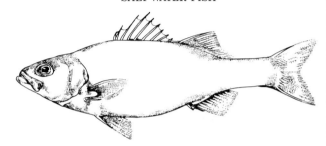

FRESH-WATER FISH

1. Drinks no water   2. Water enters body by osmosis   3. Produces large amounts of dilute urine

SALT-WATER FISH

1. Drinks salt water   2. Loses tissue water by osmosis   3. Gills excrete salt   4. Produces little urine

drink ocean water, such as sea gulls, have salt glands in their nostrils to eliminate excess salt.

Humans and most other terrestrial organisms need constant sources of water to replace water lost by the drying effects of the terrestrial environment and water lost in urine. They are physiologically equipped to drink only fresh water. Three-fourths of the earth's surface is covered by water, and earth is called the water planet, but fresh water is a limited resource and a precious commodity. Indiscriminate pollution and destruction of the sources of fresh water are foolhardy actions for creatures who can't live without it.

tides and the type of substrate (ground surface) in a particular coastal region. For example, rocky and sandy substrates exist on exposed, wave-swept coasts, and mud and silt are found in protected areas.

Benthic (bottom-dwelling) organisms on rocky shores either are attached (**sessile**) or cling tenaciously to the rocks in order not to be swept away by waves. The majority of benthic organisms (except for rooted vegetation and its epiphytes, found in sandy and muddy areas) burrow beneath the surface and are protected from waves during high tide and from drying during low tide (see Figure 21–6). The sub-

Figure 21–5

Classification of marine environments.

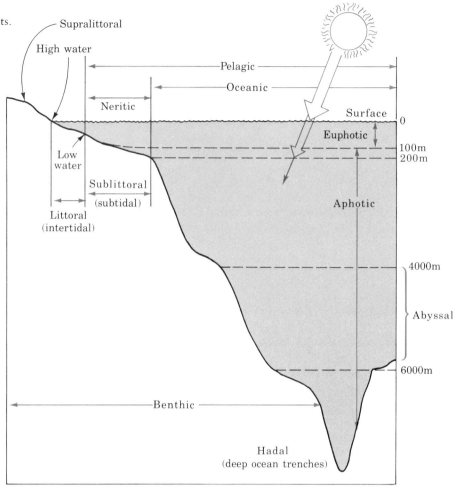

Figure 21–6

A mud flat at low tide. Although much of the mud flat appears barren, countless burrowing animals live beneath its surface, protected from dessication. (Jack Demit, © 1979, Photo Researchers, Inc.)

tidal (or sublittoral) zone is just below the tidal zone. Subtidal organisms are not affected by tides, but they also form communities related to the type of substrate. Plankton float and nekton swim freely in the neritic waters (see Figure 21–7). The **nekton** are strong swimmers and include most ocean-going fish.

The neritic water over the continental shelf and the **oceanic waters** that fill the deep ocean basins together form the **pelagic zone.** The oceanic waters are also subdivided into a number of zones based on depth. The most extensive of these zones, at depths between 4,000 and 6,000 meters, is the abyss. The bottom of the abyss is predominantly a flat plain covered with fine sediments and oozes, the skeletal remains of living organisms that sink to the bottom and accumulate over thousands of years. Organisms commonly found in the abyss include numerous species of burrowing invertebrates, a number of arthropods and sea cucumbers (leathery grublike relatives of starfish) that creep about on the surface of the substrate, and small, bizarre-looking fishes. The abyss is the largest single ecosystem on earth. Temperatures in it range from 4 degrees Celsius down to near 0 degrees Celsius. The only zone deeper than the abyss is the hadal region, which is limited to deep ocean trenches.

The numbers of marine organisms decrease with depth. However, even in the euphotic zone, these organisms vary in number and distribution. The crystal-clear tropical surface waters contain fewer organisms than do the more murky waters of temperate regions. In fact, it is the absence or presence of life that to a large extent gives these water masses their different appearances.

Why should the warm, sunlit tropical waters contain less life than the cooler waters to the north and south? Air temperatures in the tropics are high throughout the year. Surface water is likewise warm and thus lighter than the cooler, deeper waters. The surface layer floats on the cooler water beneath it, and little mixing occurs between the two (see Figure 21–8). Nutrient levels are low in the surface layer, because the phytoplankton rapidly recycle available nutrients. This layer also loses nutrients when plankton and organic material sink into the deeper zone.

Actually, the surface waters are separated from the deeper waters and kept from mixing with them by a thin layer of water, the **thermocline.** With depth, this layer decreases rapidly in temperature and increases rapidly in density (see Box 21–3). Surface waters in temperate regions support greater quantities of

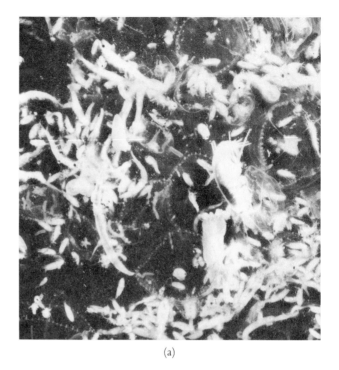

(a)

(b)

Figure 21–7

(a) Marine plankton consist largely of unicellular photosynthetic organisms and small animals that are able to swim strongly enough to maintain their vertical range in the water column. They are carried along in ocean currents. (R. Mariscal; Bruce Coleman, Inc.) *(b) Unlike plankton, nekton are strong swimmers, and many species migrate over long distances. (Courtesy John G. Shedd Aquarium.)*

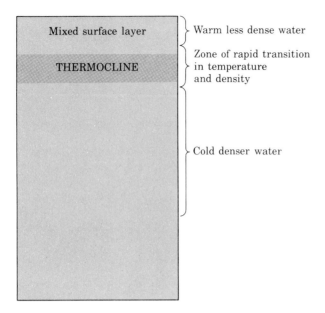

Figure 21–8
Stratification of a column of water in a tropical ocean. The upper layer of warm, less dense water is separated from the mass of deeper, colder, denser water by a thermocline. The two masses do not mix.

life because nutrients are renewed annually. During the summer, a thermocline exists. Phytoplankton growth during the summer uses most of the nutrients. But in the fall, as the surface waters cool, the thermocline is lost, and surface and deep waters begin to mix. The mixing that continues into the spring returns lost nutrients to the surface. During the spring, the thermocline reforms, and the cycle begins again. A sudden increase in the numbers of organisms occurs in the spring at about the time the thermocline is forming, and the increase begins the depletion of nutrients.

A phenomenon called upwelling brings cool, nutrient-rich waters to the surface in localized areas of both tropical and temperate regions. Upwelling can occur, for example, when offshore winds blow surface waters away from the shore, allowing the cool, deep, nutrient-rich water to rise to replace them (see Figure 21–9). Areas of upwelling are highly productive.

Estuaries

Areas of the ocean that are partly enclosed by land and that receive sizable amounts of fresh water from streams and rivers are called estuaries. Many estuaries

are shallow and often nearly empty at low tide. The substrate of estuaries consists mainly of fine sediments, such as sand, mud, and silt. In shallow estuaries, sufficient light reaches the bottom to support rooted vegetation, which is also abundant in shallow regions of deep estuaries. Estuaries are seldom depleted of nutrients for plant growth, because they receive renewed nutrient supplies from both fresh water and ocean tidal flows. Thus they are among the most productive environments on earth, and a large number of marine fish use them as nurseries.

Many estuaries are heavily polluted, because many of the major cities of the world—such as New York, San Francisco, and Seattle—are located on estuaries that serve as major seaports. Too often estuaries are used as dumping grounds for sewage and other wastes.

Freshwater Habitats

Freshwater environments range from large lakes to small, temporary ponds and from large rivers to tiny

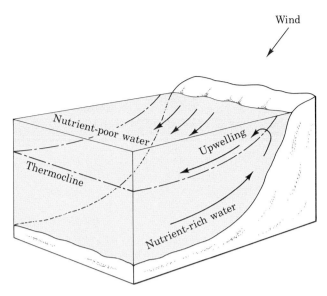

Wind

Nutrient-poor water

Upwelling

Thermocline

Nutrient-rich water

Figure 21–9
Upwelling brings cool, nutrient-rich water to the surface, where it mixes with or replaces nutrient-poor surface water. Areas of upwelling support highly productive communities of organisms. In this case, offshore winds move nutrient-poor surface water away from the coast. It is replaced by cooler water rising along the shore.

streams. Except for the large lakes, freshwater environments do not provide living organisms with the same degree of stability that oceans provide. (See Box 21–3.) They are subjected to greater temperature variations, they are more easily affected by pollution and flooding, and they may be reduced by evaporation during dry periods because they are smaller bodies of water. Variations in the kinds and quantities of substances dissolved in the water reflect the type of land mass surrounding it.

Freshwater habitats also differ from one another in levels of organic matter and oxygen. Lakes, ponds, and rivers that support a rich flora and fauna (plant and animal life) are **eutrophic.** Those that do not are **oligotrophic. Eutrophication** is the natural process that increases the levels of organic substances in fresh water and the capacity of the environment to support the increases. When humans add nutrients to rivers and lakes in the form of sewage, cultural eutrophication occurs. But excessive pollution ultimately leads to the destruction of the aquatic ecosystem□.

Freshwater habitats are subdivided into rivers and streams, which contain running water, and ponds and lakes, which contain still water. Lakes and large ponds may also have slow-moving currents, as do the

oceans. Rivers, streams, lakes, and ponds contain nekton, plankton, and benthic organisms, as do the oceans.

Lakes may be divided into three zones. The first, the **littoral zone,** is the shallow zone along the shore. It contains rooted vegetation. In many ponds and lakes, the vegetation within the littoral zone is arranged in three zones (see Figure 21–10). The zone closest to shore contains plants that extend above the surface of the water, the emergent vegetation. Cattails are an example. The next zone contains floating vegetation, including plants that have roots embedded in the substrate, such as water lilies. Finally, farthest from shore is the zone of submerged vegetation, plants that do not reach the surface. Living there are a variety of algae and rooted vegetation, such as *Elodea,* a common plant sold in pet stores for aquariums. The littoral zone also contains a rich variety of animal life; aquatic insects, snails, and worms are common, as are crayfish, clams, and fish.

The surface waters away from shore, where no rooted vegetation grows, constitute the second zone of lakes, the **limnetic zone.** This zone is penetrated by light and contains abundant phytoplankton and zooplankton during certain seasons. Fish are also present.

In deep lakes, the waters beneath the limnetic zone comprise the **profundal zone,** which receives little or no light and therefore supports no plant life. However, microorganisms such as bacteria and fungi live there, as do worms and the larvae of certain insects. In some lakes, low concentrations of oxygen may occur in this zone. The amount of oxygen influences the type of organisms that live there. Shallow lakes and ponds may lack a profundal zone.

In temperate regions of the world, the surface waters of large lakes become warm during the summer and form a less dense layer that floats on a deeper, colder, denser layer. These layers cool in the fall, and mixing occurs, as it does in oceans. Lakes and ponds in cold regions may freeze over in winter. When the ice is snow-covered, little light reaches the plants below. Photosynthesis stops, and oxygen levels drop as the surviving plants and animals use the oxygen for respiration. No oxygen can diffuse from the surface because of the ice. If the levels are severely depleted, fish and other active animals die in large numbers.

Communities of organisms in flowing water are quite different from those in standing water. In streams and small rivers, deeper pools of slow-mov-

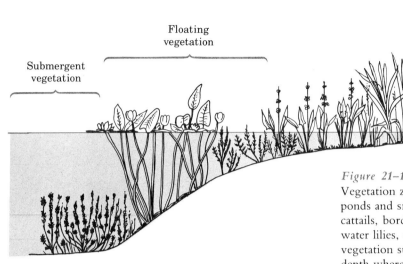

Submergent vegetation

Floating vegetation

Emergent vegetation

Figure 21–10

Vegetation zones commonly found in the littoral zone of ponds and small lakes. Emergent vegetation, such as cattails, borders the shoreline. Floating vegetation, such as water lilies, occurs in deeper water. Submergent vegetation such as fanwort or hair grass extends to the depth where light is no longer intense enough for photosynthesis.

ing water may occur between regions of more rapidly moving water called riffles or rapids. In riffles or rapids, water is aerated. Thus oxygen levels usually remain high.

To a large extent, the rate of flow of water determines the type of organisms that frequent specific areas of a stream. Rooted vegetation and burrowing organisms exist in quiet pools where sediment accumulates. Where water flows more rapidly, sediment usually does not accumulate, and benthic organisms must live beneath rocks or in crevices. Many stream dwellers possess specialized structures that allow them to hold fast to the substrate or to vegetation and thus avoid being swept downstream. Light reaches the bottom in sufficient quantities to support benthic plants (mostly algae) unless the water is very deep or carries a heavy load of silt.

SUMMARY

1. Populations of living organisms form communities that, in relationships with abiotic environments, constitute more or less self-sustaining ecosystems of living organisms interacting with their total environment.

2. Each species in a community has its own unique niche, or way of life, that cannot be occupied by any other species simultaneously.

3. All the ecosystems on earth comprise the biosphere, which contains all living organisms.

4. Terrestrial environments are represented as six major biomes, each characterized by specific temperature ranges, amount and distribution of annual rainfall, and vegetation. The major biomes are tropical rainforests, grasslands, deserts, temperate deciduous forests, coniferous forests (taiga), and tundra.

5. Aquatic environments are separated into ocean environments and freshwater environments. Ocean water is a complex solution of organic and inorganic substances, the most abundant of which is salt. Fresh water contains fewer dissolved substances and in general is a much less complex solution.

6. All aquatic life is subjected to less environmental fluctuation than is terrestrial life. However, freshwater habitats show greater variation in environmental factors than do ocean habitats.

7. Both freshwater and ocean habitats contain plankton, nekton, and benthic organisms.

8. Estuaries are areas of the ocean that are partly enclosed by land and that are diluted by freshwater runoff from adjacent land masses. They are among the most productive environments on earth.

9. Photosynthetic organisms in ocean and freshwater environments are limited to the sunlit surface waters. Waters below the depth penetrated by light contain no plant life. The heterotrophic organisms living there depend on organic matter produced in the sunlit euphotic (ocean) or limnetic (freshwater) zones.

FURTHER BIOME DESTRUCTION

The destruction of extensive tracts of natural biomes by humans is not new. It took less than two centuries for European and other immigrants to North America to convert the great prairie grasslands to monocultures, which caused the extinction or near extinction of dozens of animal and plant species. The countless flocks of passenger pigeons are gone, and the great herds of bison have been reduced to a handful of small, carefully cropped herds.

Now another biome is under attack—the great tropical rainforests that extend along the equator. According to the Food and Agriculture Organization of the United Nations (FAO), approximately 40 percent of the world's tropical forests have been destroyed in the last 150 years. Currently, they are being destroyed at an ever-increasing rate. An area of forest equivalent in size to the state of Massachusetts is being destroyed each month, and at this rate little rainforest will be left by the end of the century. Why is this diverse and productive biome being destroyed at such a fantastic rate, and what are the consequences of its destruction?

The why is answered by population pressures. As much as 90 percent of the world's population growth during the next twenty years will occur in tropical countries. According to the World Population Data Sheet for 1983, the population of Africa will increase 886 million by 2020. In Asia the population will likely increase by 1661 millon during the same period. In Latin America there will be an increase of 411 million. Where will the resources come

from to clothe and feed these millions? The people of the tropics are turning to the tropical rain forests for economic gain and agricultural land. Many experts believe it is inevitable that the rainforests will be destroyed. But the destruction of the tropical forest biome will produce little gain and may create serious problems, not only for the tropics but for the entire world.

It is generally believed that when tropical rainforests are cleared, the land rapidly loses its worth for agricultural purposes because it is rapidly converted into soil of a bricklike consistency, called laterite. Although this is true in many areas, there are a variety of other soil types in the tropic forests. They too, however, are marginal for agriculture. For example, in the Amazon basin, where only 4 percent of the soil has laterite characteristics, 87 percent of the soil has a pH of 5.3 or lower and contains toxic levels of aluminum. Much of the soil is deficient in trace elements, and 23 percent of it is waterlogged during at least part of the year.

Other sizable areas suffer moisture deficiencies during dry seasons. Even so, with careful management, some of the soil might be made productive. However, this would require wide use of fertilizers and other chemicals, a costly effort. Still other problems also must be faced. One of them is erosion. Of the world's supply of fresh water, 20 percent passes through the Amazon basin, and many deforested areas rapidly lose to runoff whatever soil exists. Also, in many areas where

converted land has been coached into agricultural use, plant pests are abundant and yields are often lower than desired. However, if plant strains better suited to tropical environments and with increased pest resistance could be developed, yields might be increased. There seem to be many ifs connected with tropical agriculture in converted forest areas.

Still another concern about the wholesale destruction of rainforests is the possibility of climate change, maybe for the entire world but certainly for the local areas of destruction. One potential cause of such change is the greenhouse effect. Certain chemicals in the atmosphere, such as CO_2 keep the sun's energy from being radiated back into space. It has been speculated that as the amount of CO_2 increases in the atmosphere, from the burning of fossil fuels, the average temperature of the earth will increase. The consequences of an increase as small as a few degrees could be disastrous. For example, the Great Plains of the United States and Canada, where most of the world's grains are grown, might become too hot for production to continue. Lands further north are less suitable for agriculture because of poor soil conditions and shorter growing seasons. A warming trend would encourage the polar ice caps to melt, and the sea level would rise. A rising of several inches would flood most of the world's major cities and much of the earth's rich, agriculturally important land. Scientists are now arguing whether this can or will happen.

How does climate change relate to

tropical rainforests? These forests act as a trap for CO_2. Plants continuously convert the CO_2 to the organic matter that is part of their structure. If tropical forests are destroyed and burned, great amounts of CO_2 used by plants will not be consumed. In addition, the destruction of the trees by burning will release more carbon dioxide into the atmosphere. Experts do not know if the destruction of tropical forests will increase CO_2 levels in the atmosphere enough to bring about the greenhouse effect, but certainly that increase coupled with the increase from burning fossil fuels adds to the danger.

Even if the world climate remains unaffected by rainforest destruction, local climates in areas where the forests are destroyed are more certain to change. But it is uncertain what the changes will be. Some predict an increase in temperature as well as a loss of water from the soil. Others predict either an increased or a decrease in precipitation. Mean temperature changes of one or two degrees and rainfall changes of as much as 10 percent have also been suggested. Perhaps in the next decade we will know with greater certainty what the trends will be.

Yet another interesting consequence of the destruction of tropical forests was discovered in Africa. The mosquito species of natural tropical rainforests do not carry human malaria. The species that do carry it prefer sunlit pools in which to deposit their eggs. They avoid the shady forests. When rainforests are destroyed, the newly created habitats contain numerous places for water to accumulate. These places become the preferred breeding grounds for the malaria-carrying mosquitoes. In several areas in Africa where forests have been cleared, malaria is on the increase.

One of the greatest concerns about the destruction of the world's tropical forests is that the habitats of millions of species of plants and animals are also destroyed, which will bring about the extinction of these species. Specialists estimate that half of the species of plants, animals, and other organisms live in the tropics. Rainforests are known to house great numbers of varieties of these species, many of which have not yet even been seen by biologists. These species represent an incalculable wealth of genetic material representing great biological variation. We do not know what new important crop species might be developed from yet unknown tropical plants, what drugs might be derived from unknown plants and animals, what numbers of species of great biological importance are being lost. And of course there still remains the ethical question: What right have we to bring about the extinction of hundreds of species of organisms? During the next decade, these issues will continue to be debated. Maybe, if we are lucky, the remaining tropical forests will come to be viewed as natural areas worth preserving. Maybe this complex biome with great biological potential for humankind will be saved. We can only hope so.

FOR DISCUSSION AND REVIEW

1. What constitutes an ecosystem?

2. Each species has its own niche. What is a niche?

3. What is the extent of the biosphere on the planet earth?

4. Name and characterize each of the major terrestrial biomes. Discuss human influences on each.

5. What is the abyss, and why are its environmental variations limited?

6. Why do freshwater habitats exhibit more environmental fluctuations than do ocean habitats?

7. Why are estuaries so productive?

8. Why are the surface waters in tropical regions generally less productive than the surface waters in temperate regions?

SUGGESTED READings

BIOLOGY OF MARINE LIFE 2d ed., by James L. Sumich. William C. Brown: 1980.
   An excellent introduction to marine ecology.

THE BIOSPHERE, by D. Flanagen. W. H. Freeman: 1970.
   A collection of articles on ecological processes in the biosphere.

TERRESTRIAL ENVIRONMENTS, by J. L. Cloudsley-Thompson. Croom Helm: 1975.
   An excellent detailed review of terrestrial biomes, including a brief chapter on freshwater habitats.

22

ECOSYSTEMS AND THE PHYSICAL ENVIRONMENT

An **ecosystem** must possess certain abiotic factors—energy, nutrients, water, oxygen, proper climate, and correct substrates and media—if it is to be sustained. Because an ecosystem is more-or-less self-sustaining, many of its chemical components must be recycled, often by living organisms. For example, if an essential mineral is in short supply in an aquatic ecosystem, **phytoplankton** productivity will level off or even decrease. As a consequence, **zooplankton** populations dependent on the phytoplankton for food will diminish. After death, decomposition of these organisms by bacteria and other microorganisms will renew the supply of essential minerals in the water, and phytoplankton growth and reproduction may resume for a while. Thus the abiotic and biotic environments are inseparable.

Variations in the physical and chemical aspects of the environment are common and normal. For example, temperature varies through predictable daily and annual cycles. The organisms within the ecosystem must be able to tolerate these variations. However, every plant and animal species has a limited range of tolerance. If the range is exceeded through gradual environmental changes, populations may become extinct, they may slowly acclimate to the changes, or they may migrate to places where conditions are more suitable. Specific environmental factors that limit the distribution of a species with restrictive tolerances are called **limiting factors.**

PREVIEW QUESTIONS

After reading this chapter, you should be able to answer the following questions:

1. How does soil type affect the distribution of plants?

2. What is primary productivity?

3. What is a food pyramid? Name the relationships that exist among the different trophic levels in food pyramids.

4. What is a biogeochemical cycle? Outline the important events of three such cycles.

WEATHER, CLIMATES, AND MICROCLIMATES

The distribution of terrestrial biomes is correlated with differences in precipitation, temperature, and so on. A climate is defined in terms of long-range averages for rainfall, humidity, temperature, amount of sunlight, and the like. It is not defined by the extremes in weather conditions that may exist on certain days. However, such extremes over short periods can have a major effect on organisms. For example, during an unusually warm spring, fruit trees may bloom earlier than normal. Heavy frosts later in the spring, however, may destroy the blossoms and the reproductive potential of the trees for that year.

Organisms are not uniformly distributed throughout ecosystems. Many species occupy habitats with specific microclimates. For example, certain insects are found only near the bases of grass plants, where there is sufficient shade, lower temperatures, and possibly higher humidity than higher up on the leaves of the grass plant, where different species may live. A single ecosystem, therefore, provides organisms with numerous microclimates, and species of the community have evolved to occupy them. Larger organisms are less dependent on microclimates but still occupy specific habitats within the ecosystem.

SUBSTRATES AND MEDIA

Organisms are surrounded by a medium—air or water. Even slight variations in the medium can affect the distribution of the organisms. For example, tiny organisms live in the spaces between sand grains on sandy beaches. They are surrounded by water with specific chemical and physical characteristics. Different species live at various depths in the sand. The depth for each species depends in part on the oxygen content of the water trapped between the sand grains. Another example is reef-building corals, which can survive only in ocean water above 18 degrees Celsius.

Many organisms depend on specific types of substrates, or ground surfaces. The soils required by plants differ in their ability to retain water and in their chemical composition. Organisms adapted to life on hard, rocky surfaces washed by the tides cannot survive on sandy beaches, because they cannot cling to loose sand. Even organisms adapted for burrowing in sand and mud substrates are restricted to specific areas

(see Figure 22–1). The amount of water retained between particles of sand or mud determines whether permanent burrows can be maintained. If the substrate retains too much water, the burrow walls collapse. The sizes of particles and the amount of water between them also determine the degree of difficulty that burrowing organisms encounter in digging.

In terrestrial environments, soil is the substrate that affects the distribution of plants and animals. The size of its particles is of primary importance. Sand particles measure 0.05 to 2 millimeters in diameter. Silt particles are from 0.002 to 0.05 millimeter in diameter. Clay particles have diameters less than 0.002 millimeter.

Most soils are composed of mixtures of sand, clay, and silt. Sandy soil does not retain moisture effectively and dries rapidly. Soils containing large amounts of clay retain large amounts of water and drain poorly, leaving little space for air among the soil particles. Few plants live in clay soils, because their roots take oxygen from the soil. Most plants grow best on **loam** soils, a mixture of nearly equal quantities of sand, clay, and silt.

Figure 22–1
Burrowing marine organisms construct tubes and burrows of several types. The distribution of burrowing species is determined to a large extent by the type of substrate and its water content. Many animals can dig and construct burrows only in substrates composed of particles in certain size ranges.

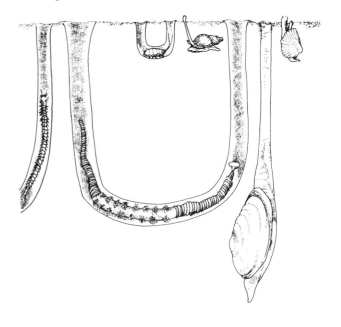

The nature of soils changes with depth. A typical vertical profile through a column of soil (see Figure 22–2) shows an upper layer, the topsoil, below which is the subsoil. Subsoil gradually merges with a layer of coarser parent material from which the finer soil particles are derived. Topsoil is of great importance to plants, because it contains the largest amount of **humus,** or organic matter. This humus, which is derived largely from the plants themselves, is decomposed by soil bacteria and other microorganisms to provide plant nutrients. Topsoil contains a complex community of soil organisms, including bacteria, fungi, algae, and other microorganisms and a variety of larger organisms, such as nematode worms, adult and larval insects, and earthworms.

Earthworms play an important role in soil ecology. They are substrate feeders. As they burrow through the soil, they ingest soil particles and organic matter. Much of the organic matter nourishes the worms. The undigested material, including soil particles, is brought to the surface and deposited as fecal castings. Earthworm activity mixes and redistributes the upper layers of soil and assists in distributing humus and other organic matter. Charles Darwin was one of the first scientists to study this behavior of earthworms and to appreciate its importance. He wrote an entire book about earthworms and their activities.

In addition to being food for soil organisms and a source of inorganic nutrients, humus is also important in water retention. It is spongy and soaks up water. Sandy soils retain more water if they are mixed with it. Humus also loosens clay soils, making them more suitable for plant growth. Many gardeners maintain compost piles, in which they place grass clippings, dead leaves, and other vegetable matter. This matter is left to decompose. Then it is used in the garden to enrich the soil. Gardeners who compost materials are manufacturing their own humus.

The chemical characteristics of soil also determine the distribution of plants. Plants require sizable quantities of nitrogen, phosphorus, potassium, sulfur, magnesium, and calcium for normal growth and reproduction. They also require small amounts of many other elements, such as iron and manganese, in order to survive. The latter elements are called **trace elements** or micronutrients. The absence of a critical trace element limits the distribution of some plants because it constitutes a limiting factor in the physical environment for the growth of plants.

ENERGY AND ECOSYSTEMS

Life on earth is possible because of sunlight. Not only does the sun provide heat energy to maintain the temperature of the earth, but it also provides light energy for **photosynthesis .** The organic substances produced by photosynthesis fuel the ecosystem. One of the most important biological principles concerns energy conversions through ecosystems.

Producers, Consumers, and Decomposers

The organisms within an ecosystem can be classified as producers, consumers, or decomposers. Producers (or autotrophic organisms) synthesize new organic

Figure 22–2
A vertical profile through a column of soil. The type and depth of the topsoil is an important factor in the distribution of plant species.

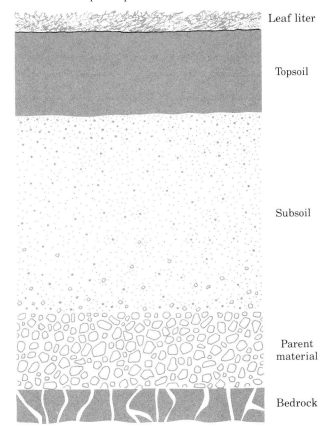

Leaf liter

Topsoil

Subsoil

Parent material

Bedrock

matter from inorganic nutrients. Photosynthetic organisms are producers that, in the presence of light and chlorophyll, convert carbon dioxide and water into simple carbohydrates. Carbohydrates can be converted into by these organisms other biologically important substances such as proteins and lipids. The conversions utilize mineral nutrients, such as nitrates and phosphates, from the soil.

In terrestrial environments, the most important producers are the rooted seed plants□. In aquatic environments, the algae are the most important (although rooted seed plants may appear in shallow waters). Microscopic phytoplankton living in the euphotic zone of the oceans□ are among the most important photosynthetic organisms on earth. One group of these microscopic organisms are the unicellular algae called diatoms which occur in both fresh and ocean water. They are often called "grasses of the oceans." (See Box 22–1.)

The producers' rate of production of new organic matter is called **primary productivity.** It is commonly expressed as the weight (usually in grams) of new plant material produced per unit area (usually in square or cubic meters) per unit of time (preferably a day or a year).

Methods of measuring primary productivity are not perfect. If different methods are used for the same habitat, different results are often obtained. In some instances, the differences are slight. However, general trends in productivity levels have been established. Over a period of a year, terrestrial productivity is greater in wet, tropical areas, such as rainforests, than it is in cooler or drier areas.

The productivity of a tropical rainforest is about 2,000 grams per square meter per year. That of a temperate grassland is only 500 grams per square meter per year. And that of a temperate deciduous forest is about 1,300 grams per square meter per year. Other estimates (in grams per square meter per year) are taiga, 800; tundra, 140; desert, less than 70; estuaries, 2,000; neritic phytoplankton, 350; and oceanic phytoplankton, 125. Agricultural land (including pasture and rangelands) is estimated to have a productivity of 650 grams per square meter per year. Coral reefs are believed to be among the most productive marine environments, comparing favorably with estuaries.

Consumers are organisms that feed on plants or on each other to satisfy their nutritional needs. They are heterotrophic organisms incapable of synthesizing organic compounds from inorganic precursors.

Organisms are arranged in different **trophic** (feeding) **levels.** Producers comprise the first level. Herbivores—animals that feed on plant material—constitute the second. Carnivores—animals that feed on the animals in the second trophic level—make up the third level. Additional trophic levels are possible, as illustrated in Figure 22–3.

After an organism dies, bacteria and fungi digest its body. Thus bacteria and fungi are known as decomposers. Organic and inorganic compounds from dead organisms become nutrients that primary producers incorporate into new living organic matter. Decomposers in any ecosystem play an important role in recycling nutrients and converting energy.

Food Pyramids

The trophic relationships among producers and consumers can be diagramed as a food pyramid. At the top of each pyramid is a carnivore that has no predators. Humans hold this terminal position in many pyramids.

Food pyramids can be analyzed in terms of differences in **biomass** (weight or volume of living organisms), energy content of different trophic levels, and numbers of organisms in each level. There is a marked decrease of all three at higher trophic levels. Figure 22–4 illustrates these relationships. When plants are eaten by herbivores, only about 10 percent of the energy stored in chemical bonds of organic molecules is transferred. About 90 percent of the energy is lost as indigestible plant material in animal fecal matter and in the form of heat. Similarly, 90 percent of the energy held in the tissues of animals in the second trophic level is lost when the animals are consumed by animals in the third trophic level, and so on. Actually, the amount of energy lost between trophic levels varies in different food pyramids. It depends on the nature of the food consumed.

Because energy is lost to each higher trophic level, the amount of living biomass supported in each level is less than that in the previous level. This may not seem obvious at first, because herbivores are usually much larger than the plants they feed on and predators are commonly larger than their prey. However, there are fewer organisms at each higher trophic level. A short food pyramid common in oceans illus-

BOX 22–1

THE LIFE OF A DIATOM

Unicellular or colonial photosynthetic organisms called diatoms are common in the oceans and in freshwater habitats. Most are planktonic, but benthic (bottom-dwelling) diatoms exist in shallow water. All are important primary producers of new organic matter.

When conditions are favorable, the rate at which diatoms can reproduce is remarkable. One diatom may reproduce up to twice a day by simple cell division (see Chapter 5). Over half a million diatoms may exist in a liter of ocean water.

The living diatom is enclosed in a glassy cell wall composed mostly of silicon. The wall is in two parts, resembling a pillbox bottom and lid. The lid is called the epitheca, the bottom the hypotheca. When the cell divides, one of the daughter cells inherits the epitheca, the other daughter the hypotheca. Each daughter produces a new hypotheca. The hypotheca of the mother becomes the epitheca of the daughter. From generation to generation, the offspring become smaller and smaller.

Obviously, there must be an end to the decreases in size. When a certain minimum size is reached, cells shed both halves of their cell walls and enter a sexual reproductive cycle that results in a new generation of larger cells. The larger cells secrete new cell walls and begin to divide again.

Diatoms must have a source of silicon to maintain their reproductive cycles, but silicon is not particularly abundant in ocean water. When sources become depleted, diatom reproduction is curtailed. Silicon thus is a limiting factor. The cell walls of many species of diatoms are elaborately covered with ridges, tubercles, and spines constructed of silicon. They are used by biologists to identify different species.

When diatoms die or are eaten, the protoplasm of the cell decomposes or is digested, but the cell walls do not

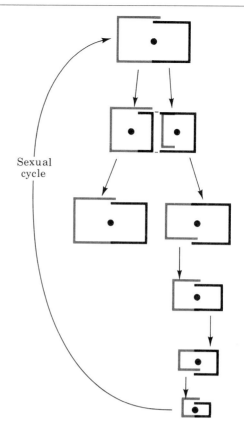

Sexual cycle

decompose. They become concentrated in animal feces and sink to the bottom of the ocean. Over long periods of time, the accumulated diatoms create diatomaceous ooze. This material is used commercially in the manufacture of polishes, detergents, and paint removers.

Diatoms are so small that only marine organisms with special feeding structures that strain plankton can feed on them. In oceans, arthropods known as copepods are important harvesters of diatoms. Copepods in turn are a major source of food for many fish, such as herring.

---

trates this principle. Phytoplankton (primary producers) are eaten by copepods (herbivores), which are eaten by herring (carnivores), which in turn, are eaten by larger fish (cod) or mammals (toothed whales). A single fin whale may eat as many at 5,000 herring per day. A herring may consume 7,000 copepods a day, and a copepod may filter as many as 130,000 phytoplankton cells per day. Simple multiplication shows

that 4.55 trillion unicellular phytoplanktonic organisms are needed each day to support a single fin whale.

Much of the phytoplankton eaten by a copepod is indigestible. Therefore, energy is lost. Fish cannot digest the shell of the copepod, so the chemical energy contained in it is lost to the herring. The total volume or weight of the whale is less than that of the small

Figure 22–3
Feeding levels in a food pyramid.

Hawks

Foxes

Chipmunks   Birds

Salamanders   Shrews

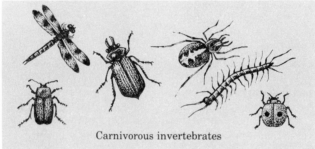

Carnivorous invertebrates

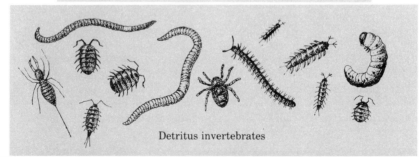

Detritus invertebrates

Leaf litter (detritus)

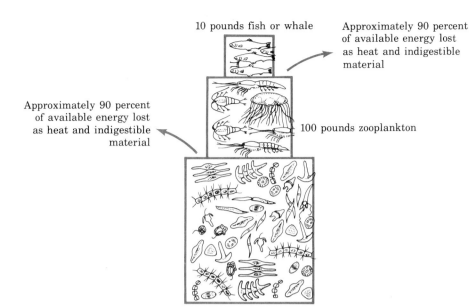

10 pounds fish or whale

Approximately 90 percent of available energy lost as heat and indigestible material

100 pounds zooplankton

Approximately 90 percent of available energy lost as heat and indigestible material

1000 pounds phytoplankton

Figure 22–4

Energy and biomass relationships among three trophic levels of a food pyramid. Although the exact percentage of energy lost between trophic levels varies with the species involved, as much as 90 percent may be lost. Consequently, a smaller biomass of living organisms is supported at each subsequent trophic level.

fish it consumed to reach its size. The total amount of chemical energy contained in its tissues may be 90 percent less than the chemical energy contained in the tissues of the herring.

The relationships in this food pyramid are clarified by the following formula: Approximately 1,000 kilograms of phytoplankton are necessary to produce 100 kilograms of copepods; approximately 100 kilograms of copepods produce only 10 kilograms of herring; and 10 kilograms of herring produce only 1 kilogram of a larger fish or mammal.

We eat both herring and larger carnivorous fish. Therefore, considering these data, it should be apparent that a diet of herring will support ten times as many people as will a diet of the larger fish that eat the herring. We can realize an even greater conservation of nutritional energy if zooplankton are harvested and used for food. The Soviet Union and other countries are now harvesting krill, tiny shrimplike arthropods, from which a substance called "ocean paste" is manufactured. Ocean paste is eaten by humans and fed to domestic animals. It is not very palatable, however.

Can you now see why most of the world's population lives on grains such as rice and wheat? Greater quantities of grains and other plants can be produced to feed the world's population because plants are at the bottom of the food pyramid.

Few predators feed exclusively on a single species

of prey. Most herbivores feed on a variety of plants. A fox feeds with equal enthusiasm on mice (in the second trophic level) and on birds (in the third trophic level). Many animals feed on different foods in different seasons or during different phases of their life cycle. Therefore, simple, direct food chains are uncommon. Many animals, including humans, are **omnivores.** They feed on a variety of plant and animal material. A more realistic picture of energy flow through an ecosystem is illustrated by interconnecting food pyramids, called **food webs** (see Figure 22–5).

Organisms that use a variety of food sources in a food web have a greater chance of avoiding starvation (and surviving) than do organisms that use a single food source. Single food sources may become extinct. Populations of a single-food species in an ecosystem may be subject to sizable fluctuations, due to environmental fluctuations such as short-range climatic changes. Organisms preying exclusively on this population would also suffer if the numbers of food organisms were low. That is why more complex food webs provide greater community stability within an ecosystem.

BIOGEOCHEMICAL CYCLES

Inorganic chemicals that are necessary for life, such as oxygen and carbon, are recycled within ecosystems.

TROPHIC RELATIONSHIPS AMONG LIVING ORGANISMS

Primary producers	Convert inorganic substances into organic substances by the process of photosynthesis; constitute the first trophic level of a food pyramid
	Examples: plants, unicellular and multicellular algae
Herbivores	Feed on primary producers; comprise the second trophic level of a food pyramid
	Examples: cattle, sheep, deer, copepods, and certain fish
Carnivores	Feed on herbivores or other animals; found in any trophic level above the second
	Examples: lions, wolves, and killer whales
Omnivores	Feed on both animal and plant foods; found in any trophic level above the second
	Examples: humans, bears, and raccoons
Decomposers	Utilize dead plant and animal matter for nutrients and energy; convert organic matter back to inorganic matter, which may be recycled by primary producers
	Examples: bacteria and fungi

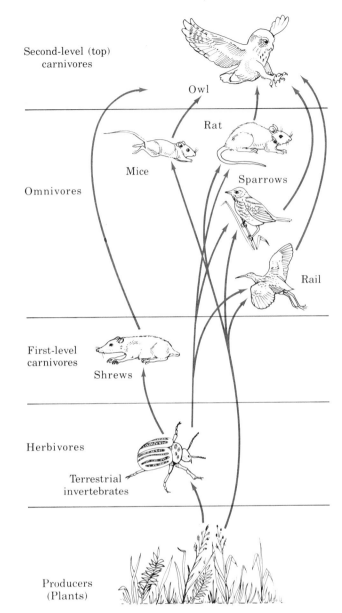

Figure 22–5
A simplified terrestrial food web. Most organisms feed on a variety of other organisms, often at different trophic levels. This creates a food web rather than a food pyramid. Complex food webs contribute stability to living communities and are a more realistic expression of the trophic relationships among living organisms.

Because these cycles often involve both biological and geological processes, they are called **biogeochemical cycles.** They include the water, carbon, oxygen, nitrogen, and phosphorus cycles.

Water Cycle

Living organisms are composed of 60 to 90 percent water, which is constantly cycled through their bodies. Terrestrial animals lose water through evaporation and excretion. They replace it with water contained in what they eat and drink. Plants absorb water from the soil but lose it through transpiration. Aquatic organisms obtain their water directly from the medium in which they live.

Oceans are the great reservoir of water for life. As Figure 22–6 shows, the heat of the sun causes water to evaporate. The water vapor that is produced rises into the atmosphere and forms clouds. When the clouds cool sufficiently, the water condenses and falls to earth as rain, sleet, hail, or snow. Some water is

(a)

(b)

(a) Tropical rain forests occur at low elevations near the equator where there are high temperatures and high levels of rainfall throughout the year. **(b)** A three-toed sloth is one of the animal species found in South America rain forests.

(a)

(a) Grassland biomes such as this African savanna and the prairies of North America extend from the tropics into temperate regions, but all have discrete wet and dry seasons. (b) The great cats such as these lions, are top predators in the food webs of African savannas. Their prey, grazing animals such as antelope, are abundant in undisturbed areas. (c) Scavangers, vultures and hyennas, finish consuming the carcass of an animal killed by lions.

(b)

(c)

(a) A sidewinder rattle snake is a native species in the deserts of the southwestern United States and northern Mexico. (b) Cacti are common plants in warm deserts (photo by Roger C. Anderson, Illinois State University). (c) Gila monsters are reptiles also found in the deserts of southwestern United States (photo by Richard S. Funk). (d) Deserts occur where the annual rainfall is below 25 centimeters per year. Desert plants are commonly spaced far apart with bare ground between them (photo by authors).

(a)

(b)

(a) Deer frequent the margins of deciduous forests (photo by Joseph E. Armstrong, Illinois State University). **(b)** This birdfoot violet is common in spring at the edge of deciduous forests in specific geographic areas (photo by Richard S. Funk). **(c)** A deciduous forest in autumn. Deciduous forests occur in temperate regions with distinct seasons and where there is abundant precipitation throughout the year. The autumn displays of color are known to everyone.

(c)

(a) Extensive coniferous forests (taiga) occur where precipitation is abundant but where winter temperatures are low. (b) Spruce grouse are common birds found throughout the year in undisturbed North American taiga.

(a) Certain species of caribou migrate northward into the tundra during the brief summer to feed and later return to more southern winter ranges. (b) The tundra in autumn. In the northern hemisphere, the tundra lies between the taiga to the south and the polar ice cap to the north. During the brief summer growing season, the low-growing tundra plant life blooms and reproduces. During this period only the upper 10 to 20 centimeters of the soil thaws; the soil below remains frozen.

(a) Zones of living organisms in the intertidal region of a rocky coast. Although the organisms are covered by sea water during high tide and exposed to the drying effects of air during low tide, life may be abundant here. Zones of different kinds of animals and algae develop in the intertidal region because these species have the ability to withstand exposure and because of predation during low tide. **(b)** Living coral. Coral reefs are among the most highly productive communities in tropical seas. **(c)** the oceans are dynamic and house complex communities of living organisms at all depths.

(a)

(b)

(c)

(a) A pond with emergent and floating vegetation along its margins. Submerged vegetation is below the surface in deeper water. (b) An unspoiled mountain stream (photo by authors). (c) Severe pollution of an aquatic environment. Humans too frequently have created environmental disasters that destroy natural communities.

(a)

(b)

(c)

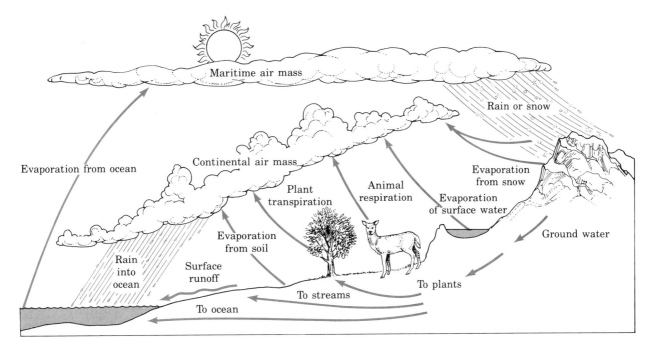

Figure 22–6
The water cycle.

absorbed by the soil. Most of it, however, flows into bodies of fresh water through drainage systems and eventually returns to the oceans. Plants absorb some water from the soil, but some also seeps deep into the earth's crust and is returned to the surface through springs and wells. The energy necessary to maintain the water cycle comes from the sun.

Carbon Cycle

Chains of carbon atoms make up most of the organic molecules that compose living organisms: proteins, carbohydrates, lipids, and nucleic acids□. The source of carbon for the synthesis of these molecules is carbon dioxide (CO_2) from the atmosphere, ocean water, and fresh water. Carbon moves through rapid short-term cycles (food pyramids and respiration) and through slower geological cycles, as Figure 22–7 shows.

During photosynthesis□, plants synthesize simple carbohydrates from water and carbon dioxide. These carbohydrates may be converted to other forms of organic molecules by plant cells. Animals feed on plants and on other animals to obtain the organic nutrients needed for their growth and reproduction.

Carbon dioxide is constantly recycled in ecosys-

tems by the respiratory processes of living organisms□. Plant and animal respiration degrades such compounds as simple sugars into water and carbon dioxide. This carbon dioxide is returned to the atmosphere and to water. When plants and animals die, decomposers use the dead bodies for their own nutritional needs. They too respire and release carbon dioxide. Decomposers play a major role in the carbon cycle.

Enormous amounts of carbon are also trapped for long periods of time in such natural materials as coal, petroleum, and carbonaceous rocks. Coal and petroleum represent the accumulated compressed and altered bodies of dead plants and animals that did not decompose. Carbonaceous rock, limestone, and chalk are often compressed skeletal remains. The chalk rocks of the white cliffs of Dover were formed from the accumulation of shells of the unicellular marine organisms called foraminifera (shelled relatives of the common amoeba).

Many other limestone and chalk deposits represent extinct coral reefs. During the life of a coral reef, millions of organisms secrete calcium carbonate shells and skeletons that remain after the death of the animals and plants that produced them. Carbonaceous rock and other deposits may eventually weather and

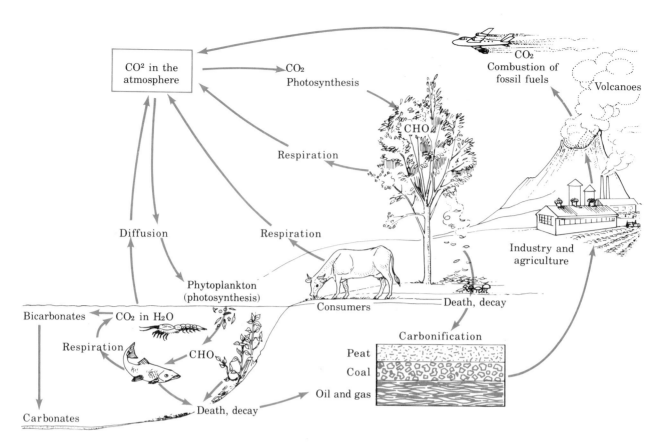

Figure 22–7
The carbon cycle.

erode by slow natural processes. Thus carbon dioxide is returned to ecosystems.

When fossil fuels (coal and petroleum) are burned, carbon dioxide is released back into the atmosphere. Humans have caused measurable increases in atmospheric carbon dioxide levels since the Industrial Revolution. These increases may have ecological consequences□.

Oxygen Cycle

Most living organisms require oxygen for cellular (aerobic) respiration□. Its source is molecular oxygen (O_2) from air or water. During cellular respiration, oxygen combines with hydrogen to produce water. The primary means of renewing and maintaining oxygen levels is photosynthesis□. During photosynthesis, plants free oxygen from water molecules. Some of this oxygen is used by the plant cells themselves, but much of it is released into the atmosphere and the water.

Oxygen plays yet another role. Ozone (O_3), another molecular form of oxygen, forms a screen in the upper atmosphere that absorbs large amounts of the ultraviolet radiation produced by the sun. If it were not for this protective shield, life would be restricted to areas protected from destructive radiation. In recent years, it has been discovered that fluorocarbons, the propellant gas in many aerosol spray cans and the coolant in refrigerators and air conditioners, react with ozone. They may reduce the critical ozone layer.

Nitrogen Cycle

Animals and plants require nitrogen for the synthesis of amino acids, proteins, nucleic acids, and nucleoproteins. Animals obtain nitrogen in the amino acids from the plants they eat or from the tissues of animals they prey on. Plants obtain nitrogen from the soil or, if they are aquatic, from the surrounding water.

The most abundant form of nitrogen is the gas N_2,

which comprises 78 percent of the atmosphere and is abundant in ocean water and fresh water. Few living organisms can use atmospheric nitrogen but certain species of microorganisms fix, or convert, N_2 to a form that can be used by plants and animals. Ammonia (NH_3), ammonium ions (NH_4^+), and nitrates (compounds that have the nitrate group NO_3^-) are the forms of nitrogen used by plants. The nitrogen cycle (see Figure 22–8), which produces these alternate forms, is composed of three interlinked processes: nitrogen fixing, nitrification, and denitrifica-

tion. Each of these processes is carried out by different groups of microorganisms.

Decomposing microorganisms enzymatically attack dead organic matter, breaking proteins into amino acids and further degrading amino acids into free ammonia (NH_3), which plants can use. However, although some plants can use ammonia, most require nitrates (NO_3^-). The conversion of ammonia to nitrates, called nitrification, requires two steps. The first is accomplished by the nitrite bacteria in soils. *Nitrosomonas, Nitrosocystis,* and *Nitrosospira* are the ge-

Figure 22–8
The nitrogen cycle.

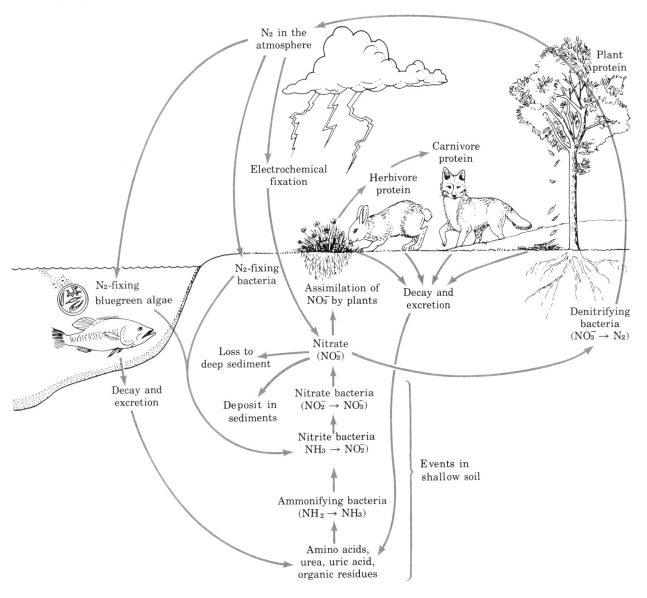

nus names of three of the important groups of nitrite bacteria. These bacteria convert ammonia into nitrite (NO_2^-). A second group of bacteria, which includes the genus *Nitrobacter,* is called the nitrate bacteria. These bacteria convert the nitrite into nitrate and release it into the soil, where it is absorbed by plants.

Nitrogen fixing is the process of converting free nitrogen (N_2) to ammonia. Bacteria belonging to the genera *Azotobacter* and *Clostridium* are among the few groups of living organisms that can fix nitrogen. Of special interest are the symbiotic□ nitrogen-fixing bacteria that live in the root nodules of members of the bean, pea, and alfalfa family of seed plants (see Figure 22–9). These bacteria belong to genus *Rhizobium.* About 90 percent of the nitrogen fixed by them is released into the plant tissues as amino acids. Some of the fixed nitrogen is also released from root nodules into the soil, enriching it for other plants. Farmers often rotate crops of alfalfa, beans, and peas with crops of plants that are incapable of fixing nitrogen. This practice conserves nitrates in the soil. Farmers also make extensive use of commercial fertilizers that contain large amounts of nitrates and of other chemicals, such as phosphates, needed for good plant growth. Free nitrogen is also fixed in the atmosphere by lightning, but the amount is only 10 percent of that fixed by organisms.

Usable nitrogen is lost from the soil through the activities of denitrifying bacteria. They convert it to gaseous nitrogen (N_2), which organisms other than the nitrogen-fixing bacteria cannot use. This process is called **denitrification.**

The quantities of nitrogen processed through the nitrogen cycle are impressive. As much as fifty to one hundred kilograms of nitrogen may be fixed by the bacteria in the root nodules of an acre of alfalfa in a single growing season. The nitrogen-fixing bacteria free in the soil may fix as much as twelve kilograms per acre per season.

Figure 22–9
Root nodules of a yellow wax bean, a leguminous plant. (Runkl/Schoenberger; Grant Heilman Photography.)

Phosphorus Cycle

Another element required by plants, animals, and microorganisms is phosphorus. It is an essential element in many biologically important chemicals, such as adenosine triphosphate (ATP)—the energy molecule—and DNA. The store of inorganic phosphorus is in the phosphorus-containing rocks in the earth's crust. As Figure 22–10 shows, weathering and erosion cause these rocks to slowly dissolve. Phosphorus, often in the form of phosphates (PO_4^{-3}), is carried away in fresh water to the sea. Animals obtain phosphorus from their drinking water and from plant and animal tissues. Plants absorb it from the soil and water in which they live. The concentration of phosphorus compounds in soil and water is usually much lower than is the concentration of nitrogen compounds. Therefore, plants deplete phosphorus supplies more rapidly than nitrogen supplies, and phosphorus commonly becomes a limiting factor for plant growth.

Great stores of phosphorus occur in some marine deposits. Enterprising industrialists hope some day to mine this phosphorus for use in commercial fertilizers. However, fertilizers rich in phosphorus have greatly increased the phosphorus levels in lakes, rivers, and streams. The result has been increased **eutrophication,** which depletes oxygen and thereby stifles aquatic life.

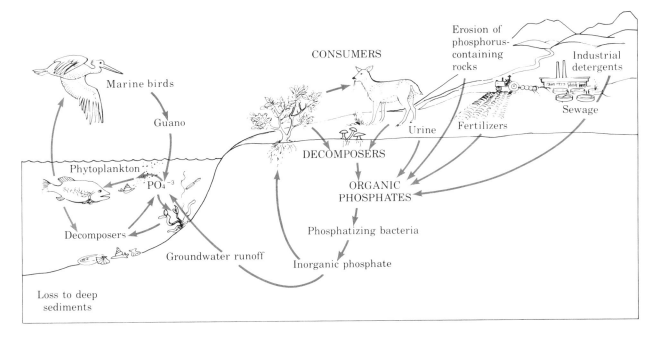

Figure 22–10
The phosphorus cycle.

SUMMARY

1. Any physical or chemical factor in an ecosystem that limits the distribution or reproductive abilities of a population is called a limiting factor.

2. Climate is defined in terms of long-range averages in rainfall, temperature, humidity, and so on. Organisms living in specific climates must be able to tolerate normal variations in these components.

3. Microclimates to which certain species are adapted may exist within a general habitat.

4. Variations in the physical or chemical components of the medium (air or water) in which a species lives affect its distribution.

5. Organisms with specific substrate requirements cannot exist where the essential characteristics are lacking.

6. Energy flow through an ecosystem follows predictable sequences. Light energy trapped by primary producers is converted to chemical energy in the producers' tissues. Primary producers represent the first trophic level of food pyramid.

7. Herbivores are the second trophic level. They feed on the organisms in the first level but realize only a small percentage of the available chemical energy in their food. As much as 90 percent may be lost as metabolic heat or in the form of indigestible material.

8. Carnivores, the third trophic level, are equally inefficient in energy transfer. Consequently, each trophic level contains a considerably smaller biomass than the preceding one.

9. Food webs are a more realistic representation of trophic relationships within a community than are simple food pyramids. Complex food webs contribute to community stability.

10. Inorganic substances essential for life are recycled through ecosystems. These cycles, called biogeochemical cycles, include the water, carbon, oxygen, nitrogen, and phosphorus cycles.

THE NEXT DECADE

OVERLOADED CYCLES

Humans have greatly altered the environment in countless ways. Many of these alterations have occurred over the last century. Some have gone unnoticed by the public until fairly recently. One of the unnoticed changes has been the subtle increase in the levels of many chemicals in the environment.

During the next decade, many of these compounds will reach critical levels, creating major problems for humans around the world. Even now, increases in the environmental levels of carbon dioxide (CO_2) and nitrogen compounds are causing scientists serious concern. They worry about potential dangers resulting from the imbalance of the carbon and nitrogen biogeochemical cycles.

Since the beginning of the Industrial Revolution, the amount of CO_2 in the earth's atmosphere has been on the increase. In 1860, the atmosphere contained 283 parts per million of CO_2. By 1960, it contained 330 ppm; 1980 levels were 340 ppm. These increases have resulted from the burning of fossil fuels, which has also increased dramatically during this period.

The concern about increased levels of CO_2 is related to what is called the greenhouse effect. The earth and its atmosphere are something like a greenhouse. Light energy from the sun penetrates the atmosphere and warms the earth, which in turn radiates longer wavelengths of heat energy back into space. CO_2 acts something like the glass in a greenhouse in that it lets in light but inhibits the radiation of heat into space. As the levels of CO_2 increase, more heat is retained in the atmosphere. This is the greenhouse effect.

During the last hundred years, the average temperature of the earth has increased by one degree. A 0.2 degree Celsius increase has occurred since 1960. A one-degree temperature increase may not seem like much, but there are signs that it may be enough to begin a serious chain of events. For example, as the earth's temperature increases, there is danger that the polar ice caps will melt and raise the sea level. If this were to happen, much of the lowlands of the world would be flooded, including many major cities. It is estimated that as much as 25 percent of Louisiana and Florida would be inundated, as well as 10 percent of New Jersey.

Is there any evidence that this melting has already started? Maybe. Scientists at Columbia University's Lamont-Doherty Geological Observatory have reported that data gathered by satellites indicate that the area of the Antarctic ice pack decreased by approximately 35 percent (2.5 million square kilometers) between 1973 and 1980. Other scientists have reported that the mean sea level is rising at the rate of approximately three millimeters per year. Many believe that these changes must be linked to the greenhouse effect.

What is even more disturbing is that the rate at which CO_2 accumulates in the atmosphere may be increasing because of two additional sources of CO_2. First, the great tropical rainforests of the world are being destroyed at an alarming rate (see Chapter 21's "Next Decade"). Plants use CO_2 partly to synthesize their own bodies. The destruction of the tropical rainforests reduces the amount of CO_2 taken up by plants. The plants themselves are also burned to create farmland, which produces additional CO_2 in the atmosphere. Second, if the earth's average temperature increases, so will the temperature of the oceans. The oceans hold an enormous amount of CO_2 in solution. The amount of gas that water can hold decreases as the temperature of the water increases. (The next time you heat a pan of water on a stove, watch the bubbles of gas form and leave the water.) The oceans have been a great stabilizer for the atmospheric CO_2. The amount of CO_2 in the atmosphere is currently in balance with the amount in the oceans. If the oceans become warmer and begin to give up CO_2, however, the earth may be in serious trouble, because the additional CO_2 in the atmosphere will increase the greenhouse effect.

There is no doubt that the levels of CO_2 in the atmosphere, the average world temperatures, and the changes in sea level will be closely monitored during the next decade. Newspapers and magazines will keep everyone informed of the changes taking place.

Another serious problem for the environment is the increased use of fertilizers for agricultural productivity. During World War I, the Germans developed a process to synthesize nitrates from methane gas and steam. This process became the basis for making artificial fertilizers. The U.S. Department of Agriculture reports that in 1978 alone U.S. farmers used over 45 million tons of fertilizers containing high levels of nitrogen and phosphorus compounds. Recent estimates suggest that 9 million metric tons of fixed nitrogen are being added to the ecosystems of the world each year. This amount is about 10 percent of the amount of nitrogen naturally fixed each year. Much of it is leached from the soil, ending up in rivers, streams, lakes, and groundwater.

Elevated nitrogen levels in water cause two serious problems. The nitrogen in water acts as a fertilizer, enhancing the growth of aquatic plants and algae. Many bodies of water become choked with such plant life. The rapid growth is called a bloom. Often, blooms can cause the environmental disasters such as the widespread death of fish and other organisms. The "death" of Lake Erie is an example. However, in fairness to farmers, it must be emphasized that not all the nitrogen in waterways comes from fertilizers. Another source is raw and processed sewage.

For humans, a potentially more serious problem than plant blooms is a medical one correlated with high levels of nitrites in drinking water. The consumption of too many nitrites can cause a condition called methemoglobinemia. Infants and children are especially susceptible. A person with this condition transports less oxygen than normal to the tissues because the hemoglobin has less affinity for oxygen. Methemoglobinemia is usually found in the children of migrant farm workers. However, the potential exists for it to become much more common, especially in rural areas, where well water may be contaminated by high levels of nitrogen compounds.

The problems associated with the nitrogen cycle should be solved more easily than those of the carbon cycle. Simply reduce the amounts of fertilizer used for agriculture. However, pressures on the agricultural industry around the world to produce more food for more people each year will ensure the continued use of high levels of fertilizers. Still, during the next decade, it should be possible to establish a better balance between human needs and the nitrogen cycle.

FOR DISCUSSION AND REVIEW

1. What is a limiting factor? Give several examples of limiting factors, and explain how they affect populations of living organisms.

2. Distinguish between climates and microclimates.

3. Discuss several examples of specific substrate requirements of living organisms.

4. Describe the process of energy flow through an ecosystem. What is the ultimate source of energy for all ecosystems?

5. Describe several food pyramids.

6. What is a food web? Why does the concept of a food web make more sense ecologically than does the concept of a simple food pyramid?

7. Outline the major events in each of the following biogeochemical cycles: water, carbon, oxygen, nitrogen, and phosphorus.

8. What influences, if any, have human populations had on each of the biogeochemical cycles?

SUGGESTED READINGS

CLIMATES OF HUNGER, by Reid A. Bryson and Thomas J. Murray. University of Wisconsin Press: 1977.
How a changing climate has affected humans, their environment, and their diet.
ECOLOGY, by Taylor R. Alexander and George S. Fichter. Western: 1973.
Part of the Golden Guide Series. A well-illustrated thumbnail overview of ecological principles for the general reader.
ECOLOGY, by Eugene P. Odum. Holt, Rinehart and Winston: 1975.
An excellent short introduction to the field of ecology, which Odum calls the link between the natural and social sciences.

23

COMMUNITIES AND THE BIOTIC ENVIRONMENT

PREVIEW QUESTIONS

After reading this chapter, you should be able to answer the following questions:

1. What is a community?

2. Explain the principle of competitive exclusion.

3. Describe the kinds of interactions that exist between predators and their prey.

4. What is symbiosis? In what forms does it exist? Give an example of each form.

5. Which kinds of communities are most stable? Least stable?

6. Define *ecological succession*.

A **community** is composed of all the different species of organisms living in a specific habitat. The organisms interact with members of both their own species and other species. Sometimes the relationships are obvious. (A deer in a community feeds on certain plant species, and a fox in the same community preys on mice and small birds.) In other cases, the relationships are subtle, and thorough studies are necessary to discern them. For example, for over a decade, Roger Paine of the University of Washington has been studying a community of intertidal organisms that live on the rocky coast of Mukkaw Bay, Washington. Many of the plants and animals that comprise this community—algae, barnacles, and mussels, for example—are sessile organisms. That is, they are permanently attached to rock surfaces. Other organisms—starfish and limpets (snails with conical shells), for example—creep slowly about the rock. They have special adaptations that allow them to cling tightly to it.

Paine has demonstrated several interesting relationships among these species (see Figure 23–1). The starfish feed on the mussels and limpets. The limpets feed on the algae. The barnacles and mussels filter plankton from the sea during high tide. When a starfish removes a mussel from a rock and eats it, space is made available for other organisms, such as algae. By eating limpets, starfish contribute to the survival of algae. That survival indirectly benefits the starfish, because the larvae of the mussels prefer to settle on the surface of algae—where they undergo metamorphosis—and later move onto rock surfaces. In short,

Space colonized by
Endocladia muricata

Mussel larvae
settle on *Endocladia*

Spores and sporelings
of *Endocladia* are
eaten by limpets

Mussels on and
around *Endocladia*
are eaten by *Pisaster*

Figure 23–1

An intertidal, rocky coastal community typical of the Pacific Northwest includes mussels, starfish, limpets, and algae. The ecological relationships among these organisms have been well documented by Roger Paine and others.

the starfish contribute to the survival of their preferred food species, the mussel.

If a species is removed from a community, unexpected and far-reaching effects often result. For example, when starfish were removed from selected areas of the Mukkaw Bay community, the mussel population expanded about two-thirds of a meter a year downward into the lower intertidal region, crowding out many other species. Starfish contribute to the stability of the entire community.

Humans have frequently altered natural communities by exterminating certain species or introducing new ones. Often this practice has adversely affected humans, directly or indirectly.

COMPETITION AND COEXISTENCE

Populations within a community compete for finite resources such as space and food. Two species cannot continue to exist if they compete for exactly the same resources in the same way. Another way of express-

ing this principle, which ecologists call the principle of **competitive exclusion,** is to say that two species cannot occupy the same niche. A **niche** is an organism's complete way of life, the sum of its existence. A complete description of an organism's niche would include physiological tolerances, food habits, behavior, relationships with other species, and structural adaptations.

The principle of competitive exclusion has been demonstrated in the laboratory. Two very similar species have been brought together and forced to compete for exactly the same resource. In a classic experiment, C. F. Gause, an ecologist, cultured two species of protozoans, *Paramecium caudatum* and *Paramecium aurelia,* separately and together. When they were cultured separately under the same set of controlled environmental conditions and fed with the same species of bacterium, both species flourished. However, when they were cultured together under the same environmental conditions and forced to compete for the same food, *Paramecium aurelia* survived and *Paramecium caudatum* did not. (See Figure 23–2.)

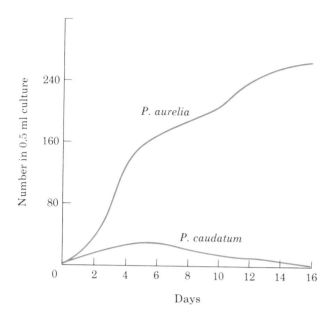

Figure 23–2

Population densities of two species of *Paramecium* through time when grown together. Notice that due to competition for the same food source the *P. caudatum* population died out while the population of *P. aurelia* increased.

If two species attempt to occupy the same niche in nature, one of two outcomes will be likely. First, one species will be unable to compete and will become extinct. However, one species may be able to survive under certain conditions and the other under other conditions. This has been demonstrated in a series of laboratory experiments. When two similar species of grain beetles, *Tribolium confusum* and *Tribolium castaneum,* are raised together and forced to compete for the same food (flour), one species dies out—as is predictable from the principle of competitive exclusion. However, the species that survives can be either one, depending on environmental conditions. *Tribolium confusum* survives when the two species are kept at 24 degrees Celsius and 30 percent humidity. *Tribolium castaneum* survives if conditions are warmer and more humid (34 degrees Celsius and 70 percent humidity). In nature, these species have different niches.

Second, if one or both of two species competing for the same niche evolve and adopt slightly different ways of exploiting common resources, both may survive. For example, two closely related species of cormorants, *Phalacrocorax carbo* and *Phalacrocorax aristo-*

telis, coexist on the same coastlines and seem to exploit the same niche. Cormorants feed on fish and other forms of marine life. However, examination of the stomach contents of these birds has revealed that they ingest different food organisms. Most of the stomach contents of *Phalacrocorax aristotelis* consisted of shrimp and flat fish that live on the ocean bottom. Most of the stomach contents of *Phalacrocorax carbo* consisted of fish species that swim above the bottom.

In a similar case, three species of insect-eating warblers forage for insects in the same spruce trees in Maine. Although they seem to use the same food source, careful study has revealed that they are not in competition. Each species feeds in a different area of the trees (see Figure 23–3). Blackburnian warblers feed among the outer portions of the upper branches. Black-throated green warblers feed among the middle outer branches. Yellow-rumped warblers feed among the lower branches.

Competition is an ecological phenomenon that can create a divergence of species through natural selection□.

TROPHIC RELATIONSHIPS

Energy flows through ecosystems via food pyramids and food webs. Some species in a community provide food for other species. Herbivores feed on plants and become food for carnivores and parasites. After death, all organisms furnish nutrients for countless microorganisms. Trophic (feeding) relationships among species affect the community structure.

Prey-Predator Interactions

A predator is an animal that attacks, kills, and feeds on another animal, its prey. Predators are usually, but not always, larger than their prey. Lions and other predatory cats can successfully kill antelope that are nearly the same size and weight as themselves. Often, small predators hunt in groups and cooperate in killing larger prey animals. For example, wolves hunt in packs and kill prey as large as moose.

The relationships of predator and prey species are in a delicate balance that ensures the survival of both species. If a predator caused the extinction of its prey population, it too would become extinct. Predators do influence prey populations by keeping them below the carrying capacity of the ecosystem.

Blackburnian warbler

Figure 23–3

Three species of warblers as illustrating the principle of competitive exclusion. The warblers do not occupy the same niche, although they eat similar foods and feed in the same trees, because they utilize different regions of the tree. The Blackburnian warbler feeds among the upper tree branches. The black-throated green warbler feeds among the middle outer branches. The yellow-rumped warbler feeds among the lowest branches.

Black-throated green warbler

Prey and predator populations usually display cyclic fluctuations (see Figure 23–4). As the number of prey increase, so do the number of predators. More predators mean increased predation and a consequent decrease in the prey population. A lack of prey decreases the predator population, and the cycle begins again. What keeps predators from consuming the entire prey population? A predator usually has more than one prey species, so pressures on a single prey species are lessened. Prey species also produce many more offspring than predators do, which contributes to their ability to maintain their population.

Predators have evolved efficient mechanisms to capture and kill prey. The prey have responded by evolving mechanisms to escape or hide from their predators. Studies of predators have confirmed that most of the prey that are killed are old or young or ill. Healthy adults most easily escape their predators. Predation thus benefits the prey species by regulating its population both qualitatively and quantitatively.

Herbivores and Plant Populations

Many herbivores are browsers, consuming part of a plant but leaving enough so that the plant can continue to grow and can replace the lost parts. Other herbivores are plant predators, consuming entire plants, seeds, or young seedlings. Most authorities believe that browsers seldom consume more than 10 percent of the total net plant production. The effects of a single browsing species, then, are not great. But the same plant species may be used by a number of different browsing species. In this case, the pressure on plant species may be considerable.

The herbivores that consume seeds or seedlings appear to limit populations of the food species to a much greater extent. In desert shrub plant communities, kangaroo rats consume almost 90 percent of the available seeds, whereas browsing predators eat only

Myrtle warbler

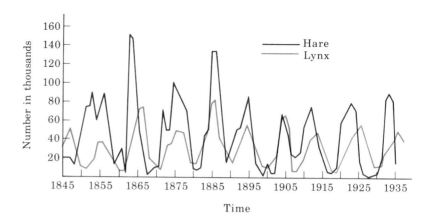

Figure 23-4

Population fluctuations of the snowshoe hare and the Canadian lynx from 1845 to 1935. Populations of the varied hare peaked approximately every ten years. Paralleling the prey (hare) population cycles were the population cycles of the predatory lynx. Increases in prey populations were followed by increases in predator populations. (Redrawn from "Fluctuations in the numbers of the varying hare [*Lepus americanus*]," by D.A. MacLulich. University of Toronto Studies, Biology Series, no. 43, 1937. Reprinted, 1974.)

7 percent of the cactus vegetation and 3 percent of the other vegetation in the community.

Many plants have evolved adaptations, such as spines and thorns, that discourage browsing by herbivores. Other plants have evolved chemical protection. Their tissues contain bad-tasting or bad-smelling chemicals that reduce their palatability or digestibility. Some plants are even toxic to many animals. For example, African elephants are serious plant predators, eagerly consuming plant foliage with thorns an inch long. The only tree or shrub that does not invite browsing is the elephant bane tree, which is ignored by all but the most naive elephants. It tastes bad and is therefore protected from potential predators.

Symbiosis

Referring to special relationships between two species, **symbiosis** means "living together" (*sym* meaning "together," *bios* meaning "life"). The relationships may be between plants, between animals, or between plants and animals. Three broad categories of symbiosis are recognized: parasitism, mutualism, and commensalism.

Parasitism

One type of symbiotic relationship involves a **host** and a **parasite.** The parasite lives with its host, derives nutrients from it, and thereby harms it. The degree of harm may be minor or so severe that the host dies. **Parasitism** is widespread and common. Few, if any, species are without parasites.

Parasitic relationships evolved gradually. The best-adapted parasites are those that harm their hosts the least. After all, when a host dies, so do the parasites it supports. Parasites are generally very specialized organisms that have evolved mechanisms to overcome the defense mechanisms of the host species and special structures to attach to or penetrate the host. Often, they have complex life cycles that ensure infection of other hosts.

External parasites (ectoparasites) cling to the outside of the host or are associated with body cavities such as the mouth, gill chambers, and cloaca. Fleas and ticks are examples of these parasites. Internal parasites (endoparasites) reside deep within the body of the host. Examples are tapeworms, which inhabit intestines; blood parasites, such as blood flukes; and the protozoan that causes malaria. Many internal parasites spend certain periods of their life cycles in different host species. The host species in which the parasite reproduces sexually is called the primary host. Other hosts involved in the life cycle are called secondary, or intermediate, hosts.

The life histories of two important parasites of humans will serve to illustrate some of the general characteristics of parasites. The two are the blood fluke (*Schistosoma mansoni*) and the liver fluke (*Opisthorchis sinensis*). The blood fluke is a common human parasite in parts of Asia and Africa. Since the completion of the Aswan High Dam in Egypt, built to retain irrigation water to increase agricultural productivity, the number of cases of schistosomiasis in Egypt has doubled. The irrigation ditches have provided additional habitats for the intermediate host of the blood fluke, a certain species of snail. Almost 80 percent of Egypt's agricultural workers are infected with blood flukes. Blood fluke infections are difficult to cure and are very serious. The worms and their eggs are carried throughout the body and cause considerable tissue damage to various organs including the brain and liver.

The adult male and female blood fluke worms live coupled together in pairs in the circulating blood of the human host (see Figure 23–5). The female releases eggs, each containing a worm embryo, into the small veins in the wall of the human's large intestine. The muscular contractions of the intestine move the eggs through the lining of the intestine and into its contents. The eggs pass from the body in the feces.

If the embryonated eggs find their way into a body of water, they hatch, and a free-swimming larva, called a miracidium, is released from each egg. The miracidium lives only a short time unless it locates and penetrates the correct intermediate host, a snail. In the snail, it changes into a larval form that reproduces asexually. Finally larvae called cercaria leave the body of the snail and swim freely in the water. When a cercarium contacts human skin, it creates a small opening in the skin and enters the bloodstream, where it develops into a sexually mature adult. The life cycle of *Schistosoma* is outlined in Figure 23–6. The free-swimming larvae do not feed, but in their parasitic stages, the blood flukes absorb nutrients from human blood and from the body fluids of snails.

Many internal parasites including *Schistosoma* have both asexual and sexual reproductive phases. These dual processes ensure the production of large numbers of offspring to complete the life cycle. Probably, few of the offspring survive, because the chances are slim that the host will be present in the right place at the right time. Often, parasitic life cycles follow the food pyramids of the host species. For example, the parasite may use the prey of a predator as an intermediate host. In some cases, the eggs or cysts that are the infectious stages are deposited on plants eaten by herbivores.

The life cycle of the liver fluke (*Opisthorchis*) illustrates the use of a food organism as an intermediate host. Adult liver flukes live in the upper bile ducts of humans and some other mammals that eat fish. The worms are small (about a centimeter long) and flat. Several thousand may be present in a single human. Heavy infections may block and damage bile and liver ducts and also damage liver tissue. The eggs pass down the bile ducts and enter the intestine, from which they are passed in feces. As is the case with the blood fluke, aquatic snails become infected with asex-

Figure 23–5
Schistosoma males and females circulate in pairs in the blood of their human hosts. The wider body of the male is folded to form a groove. The female is carried in the groove. (Macmillan Science Co., Inc., Chicago, IL 60620.)

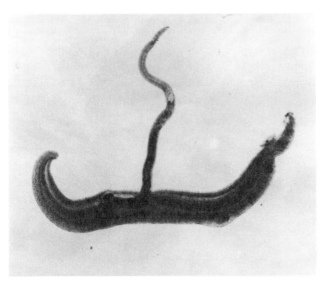

Figure 23–6
Life cycle of the human blood fluke, *Schistosoma mansoni.*

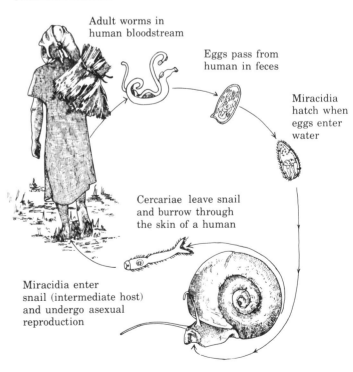

Adult worms in human bloodstream

Eggs pass from human in feces

Miracidia hatch when eggs enter water

Cercariae leave snail and burrow through the skin of a human

Miracidia enter snail (intermediate host) and undergo asexual reproduction

CHAPTER 23 *Communities and the Biotic Environment*

ually reproducing larvae. Eventually, cercaria are released from the snails. However, in this case, the cercaria bore into the muscles of certain fish instead of into humans. When humans eat raw fish—as is common in the Orient, where the liver fluke is abundant—the young flukes emerge from the cysts in the fish muscle and make their way into the bile ducts, where they mature.

Both *Schistosoma* and *Opisthorchis* have larval stages that require aquatic snails as intermediate hosts. Proper management of sewage and human waste keeps water supplies from being contaminated with these parasites in their infective stages. The use of untreated human wastes as fertilizer in the rice paddies of some countries facilitates the spread of liver flukes.

Cholera and typhoid fever, caused by bacteria, are two serious diseases transmitted in polluted water. On a world-wide scale, more human illness is associated with polluted water than with any other single environmental factor.

Other internal parasites have much less complicated life cycles. The common roundworm parasite *Enterobius* is an example. *Enterobius* species are found worldwide. The North American species is probably the most common worm parasite of school children. In certain U.S. school districts, as many as 50 percent of the children are infected. Male and female adult worms live in the human large intestine, where they mate. After mating, the females emerge through the anus at night and deposit eggs in the perianal region. They then reenter the large intestine, leaving the eggs to contaminate the bed clothing and hands of the individual. Humans are infected and reinfected by swallowing the eggs, which hatch in the small intestine. Juvenile worms migrate along the intestinal tract until they reach the large intestine, where they mature and complete the cycle.

Mutualism

A symbiotic relationship that benefits both species is termed **mutualism.** In fact, mutualism usually benefits both species to the extent that one cannot exist without the other. Mutualistic relationships may provide essential nutrients or protection from predators and parasites, or they may fulfill some other necessary function.

One complex mutualistic relationship has evolved between the thorny acacia trees in Central America and Africa and certain species of ants (see Figure 23–7). The ants hollow out and occupy the thorns on the acacia. Some acacias produce food for the ants in the form of plant juice at the base of the leaves or on nodules on the tips of the leaves. In exchange for room and board, the ants attack and kill other small insects that feed on acacias. By biting the larger insects and other herbivores, the ants discourage browsing. Acacia

Figure 23–7

A thorn of an acacia tree and one of the tree's symbiotic ants exemplify a mutualistic relationship. Both the ants and the tree benefit from the relationship. (Robert Mitchell/Animals Animals.)

shoots raised without their ant associates show markedly reduced growth and survival rates.

Another mutualistic relationship involves lichens, the small, often greenish growths on rocks and tree trunks. Although a lichen appears to be a single plant, it is actually a complex formed by alga and fungus species. The photosynthetic alga supplies nutrients for the heterotrophic fungus. The fungus secretes chemicals that dissolve the nutrients needed by the alga from the substrate on which the lichen grows. It also provides protection from dessication.

Coral reefs would not exist if it were not for the mutualistic relationship between the coral polyps and unicellular photosynthetic algae called zooxanthellae, which live in the cells of corals. Corals are colonial animals. Thousands of polyps contribute to the secretion of the skeleton that houses the members of each colony. Corals are also carnivores, voraciously feeding on planktonic animals small enough to be captured by their stinging tentacles. However, the clear tropical waters that surround coral reefs do not contain enough plankton to maintain the countless coral polyps and other filter-feeding animals on each reef.

The zooxanthellae carry on photosynthesis and produce organic nutrients that are released into the coral tissue to supplement the diet of the corals. They also create a chemical balance that facilitates the rapid secretion of coral skeleton by the coral polyps. Reef-building corals exist only in shallow water, where sufficient light is present to support photosynthesis of the zooxanthellae. The zooxanthellae use carbon dioxide and nitrogenous wastes produced by the corals. Recent studies have shown that the organic nutrients released by zooxanthellae are translocated from regions in the coral colonies where the algae are abundant to newly formed parts of the colony, where they are rarer.

Commensalism

When one species benefits and the other is neither benefited nor harmed, **commensalism** exists. Although many commensal relationships involve sharing food with another organism (the word *commensal* means "sharing the same table"), other types of relationships are equally common. The commensals may obtain shelter, protection, or transportation by associating with another organism. For example, epiphytic plants such as orchids are commensals that grow on the branches of trees. The orchids receive the benefit of being supported and held above ground level, where the light intensity is greater. They do no harm to the trees and derive no nutrients from the trees.

Occasionally, several commensals may associate with an organism at the same time. The innkeeper worm, *Urechis caupo,* is a large cigar-shaped worm that burrows in the muddy substrate of bays and estuaries. It pumps water through its burrow, collecting food particles and organisms in a mucous net that it secretes. When the net is filled with food, it begins to eat, but it rejects food particles that are above a certain size. The rejected particles are eaten by the commensal guests, which live in the burrow with the innkeeper worm. These guests include segmented scale worms, crabs, and certain small fish.

CONCEPT SUMMARY
FORMS OF SYMBIOSIS

Parsitism	A parasitic species lives with and benefits from its host, deriving nutrients from the host and causing it harm
Mutualism	Both symbiotic species are benefited to the extent that neither can live separately from the other
Commensalism	One of the symbiotic species, receives benefit, but the other is neither benefited nor harmed; commensals usually can exist in the absence of the other symbiotic species

COMMUNITY STABILITY

Communities are dynamic. The organisms that compose them grow, reproduce, die, and are replaced by the next generation. The population sizes fluctuate within the limits set by the complex trophic relationships and by the carrying capacity of the ecosystem. Organisms migrate among communities, sometimes randomly and sometimes in annual patterns. Yet communities, if undisturbed by humans or natural catastrophes, exhibit a kind of overall stability when their populations achieve a balance in sharing and using the resources of the ecosystem.

Ecologists agree that the most stable communities are the most complex ones. A finer degree of balance is achieved by their member populations, and a greater degree of efficiency in energy use is possible. Complex stable communities contain numerous species, and few dramatic changes occur in populations. Communities in this stable state are called **climax communities.** Ecological balance in them is maintained by interactions among the producers, consumers, and decomposers. Simpler communities tend to be less stable, because there are fewer relationships possible among the species to serve as checks and balances.

An extreme in community simplicity has been achieved by modern agricultural practices. In the United States, hundreds of thousands of acres of native prairies—grassland climax communities—have been converted to populations of only wheat, soybeans, or corn. The practice of growing a single crop species over a large area is called monoculture (see Figure 23–8). Such communities must be carefully managed and are expensive to maintain, because large amounts of fertilizers and pesticides are needed. Herbicides are used to eliminate competing plant species, and pesticides are used to eliminate plant parasites and

insects. Often, the pesticides eliminate both the pest and its natural predators. Even with good management, monocultures are in a precarious state. An outbreak of a disease organism or a pest that is resistant to pesticides and is without natural controls could sweep through the monoculture uninhibited. The failure of the potato crop in Ireland in the 1840s because of potato blight fungus and the spread of corn blight in the United States in 1970 are examples of such catastrophes.

Many natural communities have been drastically altered by human activities. Humans have introduced new species into stable community systems, causing instability. For example, humans introduced rabbits into Australia, where they had no natural predators. The subsequent population explosion of rabbits caused severe damage to the natural vegetation and affected populations of native herbivores. The edible African fish *Saratherodon* (Tilapia) *mossambicus* has been introduced into nearly thirty countries in Asia and South America as a source of protein. The behavior of this species when breeding disrupts the natural

Figure 23–8

A corn field, showing monoculture. (Courtesy Funk Seeds International.)

environment to such an extent that many populations of native fish and other organisms are destroyed. A large number of these populations are important and preferred foods in their countries. Species introduced into North America that have disrupted natural communities include English sparrows, starlings, and gypsy moths (described in Box 23–1). Natural calamities—fires, floods, droughts, and the like—also disrupt communities.

In time, disrupted communities usually return to the natural climax condition. (The rebirth of a community is called **ecological succession.**) For example, if a small farm in southern New England is abandoned and left undisturbed, the land will return to the natural climax community structure for that region—a beech-maple forest (see Figure 23–9). First, the fields will become populated with annual plants, such as ragweed. Gradually, these plants will be replaced by populations of perennial plants, which survive harsh winters. Woody shrubs will appear next. They will gradually be replaced by tree and shrub species that comprise the beech-maple climax forest.

During succession, a number of general changes occur, regardless of the species involved. The total **biomass** (living organic substance) and nonliving organic matter gradually increase. They reach a maximum when the climax community is established. The numbers of species also increase. At first, the increase is rapid, but later its rate slows. As the numbers of species increase, so does the complexity of food webs. Succession is a process of change that continues until the climax community is established and community stabilization is maximized. The species inhabiting the disturbed area also change. New species replace earlier ones.

Figure 23–9

Ecological succession. An abandoned plowed field first is colonized by annual plants, such as ragweed. They are gradually replaced by perennial plants, such as grasses. As the soil is modified by the plants, the field becomes more suitable for other species. Shrubs replace the grasses. Then they are replaced by beech and maple saplings, which mature to create a beech-maple climax forest.

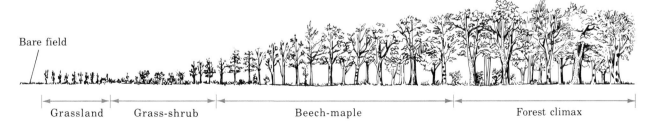

Bare field

| Grassland | Grass-shrub | Beech-maple | Forest climax |

What causes these changes? The activities of the organisms themselves alter the biotic and physical environment, making it more suitable for the next species to follow in sequence. For example, as the first annual plants die, their bodies contribute organic matter to the soil. The accumulation of organic matter changes the water retention capability of the soil and releases nutrients that alter the chemical nature of the soil. These changes favor populations of perennial plants, which become more successful than the annuals. The climax community species achieve a balance between the physical and biotic environments that eliminates further changes.

What has just been described is actually secondary ecological succession—a return to the original climax state after a disruption has occurred in an already existing climax community. Another type of succession, primary succession, is the establishment of new life in an area where life did not previously exist and the subsequent development of a climax community. For example, newly exposed bare rocky surfaces are soon colonized with lichens, which begin to erode the rock chemically. The eroded rock particles and wind-blown sediments fill in cracks and crevices in the rocks and become soil in which grasses and other small plants can take root. As these plants produce humus and as additional soil accumulates, further successional stages follow. The last stage is a climax community.

SUMMARY

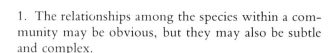

1. The relationships among the species within a community may be obvious, but they may also be subtle and complex.

2. Two species cannot occupy the same niche in a community. This is called the principle of competitive exclusion.

3. Competition for similar ecological resources in a community can lead to divergence of species through natural selection.

4. Prey-predator relationships benefit both species and result in cyclic fluctuations of both the prey and predator populations.

5. Herbivores are considered browsers if they do not kill the plants on which they feed. They are considered predators if they destroy the entire plant, seed, or seedling.

6. Three categories of symbiosis exist: parasitism, mutualism, and commensalism.

7. The most stable communities are climax communities. They are complex and contain numerous species whose interactions result in balanced populations and the highest efficiency of energy use.

8. If climax communities are disrupted, they return in time to their normal mature state through the process of secondary ecological succession.

9. Primary ecological succession begins with the colonization of a new area devoid of life and ends with the establishment of a climax community.

INTEGRATED PEST MANAGEMENT

During the next decade, a new kind of war will be waged. Called integrated pest management (IPM), it will combine, expand, and build on techniques used in the past.

What does *pest* mean in this context? The numbers of organisms in an undisturbed community are maintained at appropriate levels by a series of natural checks and balances. However, humans have upset many of these checks and balances, causing the extinction of some species and the expansion of others. Expanded populations that begin to cause human inconvenience or harm are considered pests. They attack crops and domesticated animals, destroy stored foods, harbor and transmit disease, and often disrupt and unbalance ecosystems in other ways.

Generally, three methods are used to control pests: mechanical or cultural, chemical pesticide, and biological. Mechanical or cultural control includes a variety of techniques. They range from collecting insect crop pests by hand to rotating crops. The intent is to disrupt the life cycles of certain pests. For example, some insects that feed on corn lay their eggs in the soil of the cornfields during the summer. The eggs overwinter, and the larvae hatch the following spring. If the farmer has planted soybeans instead of corn, the corn pest population will be eradicated because it will have no corn plants to eat.

Chemical pesticides have been broadcast into the biosphere since World War II. Now people are realizing that their widespread use has as many harmful effects as good ones. For example, chemical pesticides kill desirable species as well as pests. They have to be applied repeatedly, and many pests become immune to their effects. They may become concentrated in food, poisoning it. Finally, they are costly to produce, and their manufacture often increases pollution levels.

The final method, biological pest control, has had an increasing success rate. It uses natural enemies and diseases of pests to limit their population size, thereby reducing the need for pesticides. Biological control systems have many advantages. If they are well developed, they do not pollute the environment or endanger humans and nonpest species. They do not disrupt natural ecosystems. Furthermore, they usually provide permanent control of pest populations, because pests rarely develop resistance to natural controls.

Biological control is not a new idea. It has been practiced for centuries in limited ways. For example, the ancient Chinese used predatory ants to control insect pests in citrus trees. As detailed knowledge of the biology and life cycles of insects and other pests was gained through basic research, scientists applied this information in the development of control systems. An estimated three hundred species of insect pests have been brought under some degree of control by biological control systems.

Currently, lacewings and lady beetles (lady bugs)—predatory insects that feed on a variety of other insects—are raised commercially to rid crops of insect infestations. Lacewings have been especially effective in controlling bollworm populations in cotton fields. Garden centers also sell insect pathogens (disease-producing organisms). The most common of them are spores and toxins of the bacterium *Bacillus thuringiensis,* which is sprayed on vegetables and flowers to infect and kill pests. The caterpillars of at least twenty species of butterflies and moths are susceptible to the bacterium.

Scientists working in the field of biological control have found ingenious new ways of dealing with pests without harming other species. On several occasions, they have used the pest's own sex drive to eliminate the species. Screwworms (fly maggots that burrow into the skin of cattle and other browsing animals) became a major problem in the Southwest. Biologists solved the problem by raising millions of male screwworm flies and then irradiating them, making them sterile. When the flies were released into the environment, the sterile males mated with normal, fertile females, which then laid unfertilized eggs that could not develop. The screwworm population was significantly reduced in a single generation.

The females of many insects produce sexual attractants called pheromones. Even in very small amounts, the pheromones attract males from miles around. Either females themselves or the pheromones can be used to bait traps. Then the thousands of males lured to the traps by the scent can be destroyed.

During the next decade, new avenues of biological control will be explored. New resistant plant strains will be developed. Lethal genes may be bred into pest populations. Perhaps harmless species and pests with similar niche requirements will be placed in competition. By slightly modifying some environmental factors, giving an edge to the harmless species, we may be able to eliminate certain pest species. In any case, applied biological pest control will continue to come from knowledge gained from basic research. In order to develop effective control systems, scientists need to know as much as possible about the biology of pests and their natural enemies. If you hear of a scientist spending years studying what may seem to be an obscure insect or parasite, remember that the research may be paving the way for yet another success in biological control.

Now specialists are working toward making integrated pest management the main thrust in pest control. This system approaches pest control from an ecological point of view. It uses all the knowledge that can be gained about the biology of the pest, the crop, and the host, as well as about the abiotic environmental factors. In other words, integrated pest managers treat the pest and its host as an ecosystem. The aim is not to eradicate the pest but simply to keep its numbers low enough to eliminate the problem and avoid economic loss. Each situation is carefully monitored. If the pest's population size increases to the danger level, cultural and biological control systems are put into use.

Only if these systems fail are expensive and potentially dangerous pesticides used—and then only in the lowest amounts necessary to do the job. Successful IPM programs have demonstrated that as much as 50 to 75 percent less pesticide than usual is needed. (The agricultural companies producing chemicals are less than enthusiastic about IPM, of course.)

In addition to savings on pesticides, in several cases (such as cotton crop management in Texas) significant savings have been realized on reduced fertilizer use and a reduced need for irrigation water. Presumably, without the stress of supporting insect pest populations as well as their own bodies, the plants need fewer nutrients and less water.

Integrated pest management utilizes all aspects of the biology of both the pest and the crop. For example, the management of cotton pests became easier when strains of early-maturing cotton were introduced. These cotton plants mature before the insect pests do. By the time the insects are ready to feed on the plants, their food supply is gone.

Pest management systems are expensive at first because of the number of specialists who need to study the situation and the amount of monitoring that has to be undertaken. Ultimately, however, they can result in significant savings and increased productivity. IPM will become more and more common in the next decade, because it is the most economical and effective way to control pests while at the same time reducing the pollution and other side-effects of earlier pest control measures.

FOR DISCUSSION AND REVIEW

1. Select a community and describe the relationships that exist among the common species in it.

2. If two closely related species compete for the same ecological resources in a community, what will be the likely consequences?

3. In what ways do predators benefit their prey species?

4. Describe and explain the cyclic fluctuations in the sizes of prey and predator populations.

5. Which has the greater impact on populations of plant species—browsers or plant predators? Why?

6. Differentiate among parasitism, mutualism, and commensalism. Give an example of each.

7. What are the general characteristics of a stable climax community?

8. What kinds of natural and unnatural events can seriously disrupt a climax community? List several climax communities, and discuss the normal course of ecological succession following the disruption of each.

SUGGESTED READINGS

BIOLOGICAL CONTROL BY NATURAL ENEMIES, by Paul Debach. Cambridge Univeristy Press: 1974.
A discussion of alternatives to the use of pesticides.

COMMUNITIES AND ECOSYSTEMS, by Robert H. Whittaker. Macmillan: 1970.
Insights into how ecologists mathematically analyze ecological phenomena.

SYMBIOSIS, by Thomas C. Cheng. A Biological Sciences Curriculum Study Book. Pegasus: 1970.
An excellent and interesting introduction to the intimate relationships of species in close coexistence.

24

POPULATIONS AND HUMAN ECOLOGY

A **population** is any group of individuals of the same species in a given area or region at a specific time. We can express the size of a population by counting the number of individuals—for example, the number of humans in a certain city or county, the number of maple trees in an acre of deciduous forest, or the number of mice in ten square meters of grassland. Some ecologists express population size in terms of **biomass** (the weight or volume of organisms in a given area). The advantage of doing this is that population size can then be easily related to the flow of energy or material through an ecosystem.

Population sizes fluctuate. The causes of these fluctuations and their effects on other populations living in the same environment are one part of ecology. The only time a population is stable is when the number of deaths in the population equals the number of births and the number of individuals leaving the area equals the number of new arrivals.

Such stability is rare in nature. Usually, populations experience frequent increases and decreases. Sometimes, these changes are dramatic. Often, they are cyclic. For example, the ecologist D.A. Mac-Lulich has shown that populations of the snowshoe hare in Canada increase dramatically about every ten years. The increases are offset by intervening decreases. Changes in the populations of the Canadian lynx, a predator, closely follow the population changes of the hare.

Many studies have demonstrated predictable changes in populations under controlled laboratory

PREVIEW QUESTIONS

After reading this chapter, you should be able to answer the following questions:

1. What happens when a population exceeds the carrying capacity of its environment?

2. How do the factors that affect population size apply to human populations?

3. What is demographic transition?

4. List several major ecological problems facing human populations today. How can each of them be solved in the future?

5. Which forms of pollution pose the most serious threats for humans today?

conditions. These studies have used populations of microorganisms such as protozoans, unicellular algae, and bacteria, which are easily cultured under uniform conditions of light, temperature, nutrient supply, and so on.

First, there is a gradual increase in the number of individuals as the pioneers in the culture reproduce. During this period, few, if any, individuals die. While the amount of reproducing individuals is increasing and the death (mortality) rate is low, the size of the population increases rapidly in a short period of time. However, the rate of increase slows and the size of the population levels off as the reproduction rate decreases and, the death rate increases. Eventually the reproduction and death rates balance each other. Populations under these controlled conditions reach a certain stable level, called the asymptotic limit. This type of population growth pattern is called a sigmoidal or S-shaped, curve (see Figure 24–1).

Natural environmental conditions are seldom as stable as those in the artificial environment of a laboratory culture. Populations are not isolated from one another. They interact. No perfect asymptotic limit is ever realized. Rather, the natural population varies around a limit set by the natural carrying capacity of the environment. The **carrying capacity** is the limit to which a population can grow and be sustained over an extended period. If the population exceeds the carrying capacity of its environment, it decreases rapidly, because the resources needed to sustain the increased numbers—nutrients, energy, and so on—become exhausted. For example, when nutrient levels in the surface waters of temperate lakes and oceans are renewed in the spring, population explosions of phytoplankton, called blooms, occur. However, these large populations soon decrease because of the depletion of nutrients in the water and the browsing by herbivores.

Some other factors that control populations are predation, disease, and parasitism□. The environmental factors that limit population size act as **environmental resistance** to population growth. **Biotic potential** is the reproductive potential of a species. It allows the population to increase at its maximum rate under optimal conditions. Environmental resistance is balanced by biotic potential.

Self-regulating factors also exist within populations. They are related to **population density.** Every species has an optimal density in a given habitat. If the population density is very small, members are less likely than normal to find each other to reproduce. Thus, the population decreases. On the other hand, if the population density is too high, disease and parasitism become negative factors. Overcrowding may also adversely affect the ability to reproduce. As the number of individuals increases, so do behavioral and social interactions—particularly aggression□. Aggression often interferes with egg laying, care of the young, nest building, courtship, and copulation. This phenomenon has been observed in laboratory studies of many animals, including flour beetles (insects) and rats (mammals).

Perhaps the most famous of these studies was conducted by the ecologist John B. Calhoun during the 1960s. He placed several pairs of Norway rats in a quarter-acre enclosure and provided them with enough food and water to support a population of more than 5,000 individuals. Environmental factors were maintained to enhance population growth. As expected, the population increased rapidly. When it reached approximately 150, pregnant females frequently miscarried. If they delivered, they often failed

Figure 24–1

A sigmoidal population growth curve. A small population in an environment conducive to growth and reproduction will increase dramatically. However, it will level off as the percentage of older, postreproductive individuals in the population increases and as the population size reaches the maximum that the environment can support over an extended period—the environmental carrying capacity.

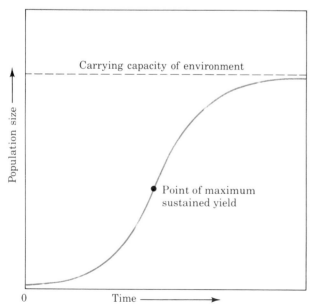

to produce milk for the young. Nest building and other normal activities were disrupted. Males attempted to copulate with animals of both sexes. Presumably because of these behavioral aberrations, the population did not increase further even though food and other resources were plentiful.

Another factor that influences changes in population size is the age structure of the population (see Figure 24–2). A rapidly growing population contains a large percentage of individuals of reproductive age. A population containing mainly individuals past reproductive age or individuals not yet old enough to reproduce will not expand rapidly. A population composed of many old individuals may actually decrease in size, because the mortality rate among them is greater than the birth rate. Populations tend to be most stable when there are nearly equal numbers of prereproductive individuals, reproductive individuals, and postreproductive individuals.

HUMAN POPULATIONS

Few environmental topics have received as much attention over the past few decades as (1) the rapid rate of human population growth, (2) what this rapid increase means in terms of food supplies to support the

Figure 24–2

Three forms of population age pyramids. *(a)* A broad-based, triangular pyramid represents a population containing large numbers of young people. It is a rapidly expanding population. The age structure of the Mexican population in 1976 approximated this form. *(b)* This form of population age pyramid represents a stable population with nearly equal numbers of prereproductive, reproductive, and postreproductive individuals. The population of Switzerland in 1947 approximated it. *(c)* The population age pyramid of a population decreasing in size is characterized by a narrow base, because there are fewer prereproductive individuals than there are individuals in the other two age categories. The population of US Blacks in 1980 had approximately this age structure. (Redrawn from a. John S. Nagel 1978. Mexico's Population Policy Turnaround. Population Bulletin 33(5):7. b. Arthur Haupt and Thomas T. Kane, Population Handbook, Washington, D.C.: Population Reference Bureau, 1978. c. John Reid, Black America in the 1980s. 1982. Population Bulletin 37(4):5.)

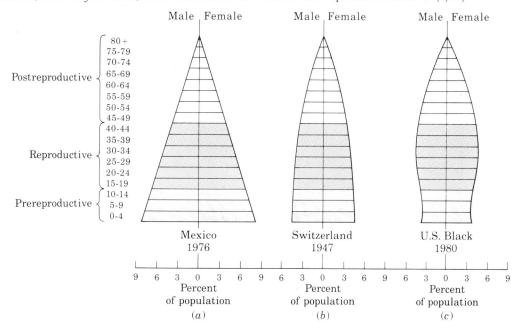

population, (3) what methods should be used to slow population growth, (4) the extent to which natural ecosystems are being altered to provide food for human consumption, and (5) the effect of crowding on the general quality of life. It would be foolish to assume that human populations can exceed the carrying capacity of the earth. Humans are living organisms. They are subject to the same environmental limitations as any other species.

Figure 24–3 depicts human population changes over the past 500,000 years. No one knows how many humans existed during most of this period. The best estimates of various authorities, however, seem to be that, prior to 12,000 to 30,000 years ago, the human population probably did not exceed 3 million. High mortality rates and inefficient methods of food procurement limited the population. As humans developed means to increase the carrying capacity of the environment, though, their population increased.

These cultural changes may have allowed the human population to increase to 5 million by about 30,000 years ago. By about 10,000 years ago, new agricultural practices had been developed. They al-

lowed the formation of more stable communities and still more efficient food production. Food supplies could also be stored. As a result of these activities, the number of humans increased to as many as 250 million by about AD 1. Specialists estimate that it took approximately another 1,600 years for this population to double.

Figure 24–3 depicts this population growth as a slow, steady increase. However, as mentioned earlier, natural populations tend to fluctuate around the limit set by the carrying capacity of the environment. Undoubtedly, such fluctuations existed in human populations. Disease, drought, and famine caused marked decreases in the population. Periods of good weather and low levels of disease promoted population growth. However, death rates remained high, and the overall population increased slowly. By modern standards, it might almost seem to have been stable.

The human population doubled during the period between 1 and 1650, again between 1650 and 1850, again between 1850 and 1930, and again between 1930 and 1975. It was estimated to be 4,585 million in 1982. The time necessary for the human population

Figure 24–3

Human population changes over the past 500,000 years. The human population increased very slowly until modern times. (Redrawn from Population Bulletin, February 1962, Population Reference Bureau.)

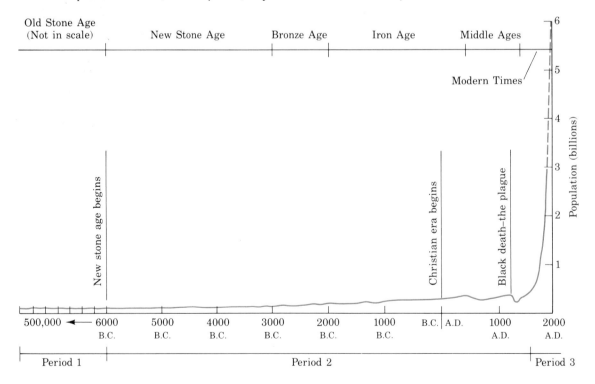

to double has decreased markedly. If current trends continue, it may reach a staggering 6.2 billion by the year 2000 and double that amount by 2020.

The rapid increase in the human population during the last 300 years is correlated with the increased efficiency of agricultural practices, the agricultural development of previously unexploited land areas, and the use of a greater variety of food resources—all of which have increased the carrying capacity of the environment. Since 1850 and the Industrial Revolution, even greater efficiency has been achieved, and modern systems of transportation, preservation, and distribution of food have come about. Of course, all these changes have required increased use of energy resources. In addition, during the past century, great advances have been made in the fields of health and medicine. These advances have increased life expectancy and reduced death rates.

Human life spans in industrialized countries, and to some extent in many underdeveloped countries, have increased. The most influential factor in the population boom, however, has been the decrease in death rates, particularly among infants and children. Birth rates have increased, and more individuals are living long enough to reproduce. Many of the diseases responsible for death among the prereproductive and reproductive age groups have been controlled.

For example, malaria, a disease caused by any of four species of protozoan parasites that infect human blood cells, has been fairly well controlled in the industrialized nations and in some of the underdeveloped countries as well. Until recently, one of the most dramatic reductions in malaria occurred in Sri Lanka (Ceylon). Prior to 1946, malaria was the chief cause of death there. After 1946, DDT was used extensively to eradicate the mosquito populations that transmit the disease. As a result, by 1970 the death rate dropped by nearly two thirds (from twenty-two deaths per thousand in 1946 to eight per thousand by 1970). Unfortunately, mosquito control has not been effectively continued and malaria is again becoming common in Sri Lanka and other tropical third world countries. Also, in many areas, mosquitoes are becoming resistant to DDT and other insecticides. Recently the World Health Organization estimated that 214 million people have malaria. They estimate that between 1972 and 1980 malaria cases increased by 4.7 million exclusive of Africa where reliable statistics are difficult to obtain.

The rates at which human populations increase are not uniform: 1983 estimates indicate that human populations in Africa, Latin America, and certain areas of Asia are among the most rapidly growing. Populations in Europe, North America, Russia, and Japan are growing much less rapidly. (See Table 24–1.)

As countries become industrialized, both birth rates and death rates decline. This phenomenon is called the demographic transition. (**Demography** is the study of human populations.) One of the most dramatic examples of demographic transition occurred in Japan, where the birth rate dropped a remarkable 46 percent between 1948 and 1958.

HUMAN ECOLOGY

We humans are a highly successful species. We occupy most of the world, and our populations continue to increase dramatically. Our ability to learn, share, and cooperate has allowed us to develop culture and technology and to exploit ecosystems as no other species has ever done. We have eliminated potential predators, controlled much disease, and diverted the energy flow of ecosystems into producing food species.

However, we may finally have exceeded the carrying capacity of the earth. We have exploited and brutalized natural ecosystems. We have polluted the

Table 24–1
Population Growth Rates and Doubling Times

REGION	ANNUAL INCREASE (Percent)	POPULATION DOUBLING TIME (Years)
Africa	3.0	23
Southwest Asia	2.6	26
Middle South Asia	2.3	30
Southeast Asia	2.1	33
East Asia	1.4	48
USSR	0.8	83
North America (United States and Canada)	0.7	95
Latin America	2.3	30
Europe	0.4	199

SOURCE: Data from the 1983 World Population Data Sheet of the Population Reference Bureau.

earth, water, and air with human and technological wastes. Our numbers have grown so large that food is no longer available for all of us. We are also rapidly using up the world supplies of nonrenewable resources, such as fossil fuels.

Is the human species doomed? The problems we face are complex. They will not be solved simply by applying science and technology to them. However, given favorable economic and sociological conditions, careful utilization of scientific knowledge can solve many of the problems. We should, for example, be able to limit population growth, reduce pollution, and develop alternate energy sources. Our success, however, will depend on our understanding the nature of the problems and on our willingness to cooperate with each other in resolving them. By conserving the remaining natural resources and by reducing pollution, we can gain time to develop new technologies that will bring us into greater ecological harmony.

Industrialized countries, such as the United States, will need to be world leaders in such efforts. These countries, which hold only a fraction of the world's population, use most of the food, energy, and other natural resources.

Food Supply and World Nutrition

Approximately half of the people in the world are hungry. Between 10 and 20 million die each year of starvation and malnutrition or of complications resulting from lack of food. These numbers account for a sixth to a third of all deaths each year.

Most of us have neither experienced real hunger nor seen a starving human. A starving person loses more than a third of total body weight before death because the body uses its own tissues as an energy source. Normal health requires adequate amounts of carbohydrates, fats, proteins, minerals, vitamins, and water. Carbohydrates (sugars and starch) are used in cellular respiration□ to provide energy for normal body functions. Excesses are converted to fat, which also can be used to produce energy and which acts as an insulator to retain body heat. Proteins□ are the most important structural molecules of the body. They also function as enzymes to catalyze the body's biochemical reactions. Vitamins□ play many roles. One of the most important is their relationship with protein molecules in activating enzymes. Minerals also are important. For example, calcium is needed for

forming bones and teeth, and iodine for producing hormones in the thyroid gland. Deficiencies in any of these dietary components can result in malnutrition. (See Box 24–1.) Deficiencies of calcium in a child's diet will produce rickets. Iodine deficiencies in both adults and children cause the formation of goiters.

In recent decades, food production and world population growth have barely kept pace. The widespread use of pesticides and fertilizers following World War II increased food production during the 1950s. But increased birth rates during the late 1950s and early 1960s erased any gain. The green revolution□ of the late 1960s helped increase food yields to some degree. In 1965, the United Nations Food and Agriculture Organization estimated that the world's food supply was sufficient to furnish each person with 2,420 kilocalories a day if the food could be distributed uniformly. Even with uniform distribution, however, there would be no food left over, and protein would still be deficient in most diets.

Demographers expect the world population to continue to grow. It may reach more than 9 billion by the year 2020. Most of the world's fertile agricultural lands are already under cultivation. Many people erroneously believe that deserts, forests, and other marginal agricultural lands can be brought under cultivation. Serious problems preclude this possibility, however. They include lack of water, high cost, high energy demands, and low soil fertility.

Irrigation

Most arid lands have no immediate source of water that can be used for irrigation. Where irrigation has been used, various problems have developed. For example, the evaporation rate from open irrigation canals and from soil is tremendous. Up to 99 percent of the water may be lost. Evaporation also concentrates dissolved salts in the soil and ultimately makes the land unusable for agriculture. This problem can be alleviated to some degree by installing underground drainage systems to drain away water and excess salts. However, in many areas, such systems cannot be built because of the high cost of materials and the energy needed for installation.

Pilot projects for other new types of irrigation systems that limit water loss have been successful, but they too are expensive. Crops can be enclosed in large greenhouses, which prevent excessive loss of water by evaporation. Humidity is 100 percent. This system, which requires less water than do open irriga-

UNICEF photo by FAO

People who live where food is abundant may still suffer from malnutrition by failing to eat a properly balanced diet. When the diet is grossly inadequate, the combined deficiencies of vitamins, minerals, energy food, and proteins result in serious malnutrition and eventual starvation.

Two conditions, marasmus and kwashiorkor, may occur during prolonged malnutrition. Marasmus develops when a diet does not supply enough kilocalories to sustain normal body functions. Adults and children suffering from marasmus first use body fat and then body proteins as energy sources. As a result, they become wasted and have low body temperatures and low disease resistance. Protein deficiency is an important aspect of marasmus even if the diet contains the minimum daily requirement of protein, because both dietary protein and body protein are used for energy.

A child may have a diet containing sufficient kilocalories from carbohydrates (for example, rice) and fats and still suffer a serious form of malnutrition. This form of malnutrition, known as kwashiorkor, develops in children whose diets are deficient in protein. A healthy adult needs at least a gram of protein per kilogram of body weight per day. Children and pregnant women need more.

In the Shanaian language, *kwashiorkor* means "the evil spirit that infects the first child when a second is born." When the first child is weaned after the second is born, its primary supply of protein (the mother's milk) is lost. If the new diet is too low in protein to meet the needs of the growing young body, kwashiorkor develops.

Kwashiorkor is common in Africa, Asia, South America, and Central America. The condition usually begins to affect children at about two years of age. They stop growing, their bellies become swollen, their skin ulcerates, and their hair color fades. Disease resistance is lowered, and many children develop dysentery (severe diarrhea). The dysentery causes further loss of amino acids, vitamins, and other nutrients, thereby worsening the condition. Often, children with kwashiorkor die if they become infected with such simple childhood diseases as measles. Children suffering from marasmus or kwashiorkor during their first year of life, a critical period for brain growth, suffer mental retardation if they survive, because they develop fewer brain cells than children with normal diets.

Millions of people in the world suffer from or die of malnutrition. At the same time, hundreds of thousands of people in industrialized countries suffer from and die of diseases associated with obesity.

tion systems, has proved successful on a limited trial basis. One such project on the Arabian Peninsula produced approximately a million pounds of vegetables in one year. However, the construction and maintenance costs of the facility, as well as the cost of fertilizer (which is needed in large quantities), make these crops extremely expensive.

Even when irrigation is possible, disruption of natural ecosystems can have many unexpected results. For example, in Egypt, the Aswan High Dam was constructed on the Nile River to provide hydroelectric power and water for irrigation. The dam did help reclaim arid lands, increase production of cotton, grain, and other crops, and produce tremendous electrical power. However, numerous unexpected developments have had far-reaching ramifications. The river water entering Lake Nasser above the dam loses velocity, and the heavy load of silt carried by the river settles to the bottom of the lake. In the past, during times of flooding, silt was spread as a natural fertilizer. Now large amounts of expensive fertilizer must be used along the river system.

Also, the river below the dam has a higher velocity. As a result, erosion has occurred along the river banks and in the river delta. Furthermore, the river water below the dam contains much less organic matter than before. As a consequence, approximately 18,000 fewer tons of sardines (a good protein source) are now caught each year in the Mediterranean. The lost nutrients once fed this resource.

The lack of annual floods in the lower Nile has promoted the accumulation of salts in soils along the river, decreasing their productivity. (The floods rinsed excess salts from the soil.) The number of humans infected with the blood parasite *Schistosoma* has also increased rapidly, because the irrigation system created new habitats for snails, an intermediate host. It may be years before the Egyptians know whether the construction of the dam has brought more benefit than harm.

Some irrigation projects encourage the establishment of desalinization plants along the ocean coasts of arid lands. Desalted ocean water can then be used for irrigation. Presently, few countries can afford such plants, and some environmentalists worry about the disposal of salt. However, if the system becomes feasible, the amount of land available for agricultural purposes will be increased by more than 5 billion acres. Unfortunately, extensive irrigation systems are unlikely unless the world's energy crisis is resolved.

What about converting more forest to farmland? The possibility of significant farmland increases by this means seems less likely than increases by converting arid lands. The taiga is too far north to serve as potential farmland; its growing season is too short. Most of the more temperate hardwood forests have already been destroyed. Those that remain are on mountainsides or on land that would be difficult to till.

The extensive tropical rainforests of South America and Africa are ecosystems with high rates of primary productivity. However, their soils are low in nutrients. All dead organic material is rapidly recycled through the ecosystem or leached from the soil. Application of large amounts of fertilizers is not always a solution to the problem, because many of the tropical soils rapidly change following the removal of natural vegetation.

In a process called laterization, the soils become compact and hard, forming a substance, laterite, that is firm enough to be used as building material. Tropical soils form laterite in only a few years when natural vegetation is removed. The temples and other structures in Angkor Wat, Cambodia, are constructed of laterite and have withstood erosion and natural destruction for centuries (see Figure 24–4).

Soil Erosion and Conservation

Nearly 3 million tons of the world's soil are eroded each year. The soil is carried away by the wind or washed into streams and rivers. The loss is due largely to poor agricultural practices. Even in the richest agricultural regions of the United States, the loss reaches nine to twelve tons per acre per year. In developing countries, where the land is more intensively used, erosion rates are twice as high. Even where soil conservation practices are well understood, as in the United States, they are often ignored. How-

Figure 24–4
A temple of Angkor Wat, built centuries ago from laterite. Soils in disturbed tropical rainforests may be converted to laterite and are unsuited for agriculture. (M. P. Kahl; Bruce Coleman, Inc.)

ever, there is a movement toward increased use of conservation practices, at least in North America. For example, more land is being plowed in the spring rather than the fall, and new types of plowing are being tried.

Green Revolution

World food supplies have been maintained not so much by increasing the amount of available land for agriculture as by increasing the efficiency of farming techniques (at great energy expense) and producing new plant hybrids that have higher yields and higher nutrient value. These projects have met with varying degrees of success.

In the mid-1960s, U.S. scientists developed a number of varieties of wheat and rice that produced higher yields and higher nutrient levels than most varieties of native grains. This green revolution was thought to be the solution to the world's food shortages. When raised under favorable conditions, these grains have been impressive. In a two-year period (1967–1968), the wheat harvest in Pakistan was increased by nearly 60 percent. Similar increases of wheat in India and rice in the Philipines followed.

However, the new varieties require large amounts of fertilizer and often are not resistant to plant diseases. Therefore, expensive pesticides must be used. Most farmers in underdeveloped countries cannot afford to raise these miracle grains. In addition, many native people find the texture and taste of the new grains unappealing. In some markets, these grains are cheaper than native grains because many people will not eat them. Also, in some localities, the large amounts of pesticides used in cultivating the high-yield rice have become concentrated in the food pyramid. Thus, fish, crabs, and other aquatic animals that were used as protein sources are now poisonous.

Aquaculture

Perhaps exploitation of marine resources can help solve the world's food and protein shortages. However, the populations of several important food species of aquatic animals have already been exploited to such a degree that fishing for them is no longer profitable. International cooperation in regulating fishing has been unsuccessful, and populations of herring, tuna, salmon, and other fish are being depleted rapidly. Several whale species have been hunted to near-extinction. Populations of the great blue whale are so low that the species will probably become extinct.

HIGHLIGHT

AMERICANS, MEAT, AND THE WORLD GRAIN SUPPLY

The average American consumes 110 kilograms (240 pounds) of meat each year, and much of it is beef. It takes a large amount of grain to produce high-quality beef—or any other meat for that matter.

During the first months after nursing is over, most male calves browse on the range but receive no grain. During the next year, they also browse, but now they receive some grain as well, so they will get accustomed to eating it before they are sent to feedlots. When the young steers are nearly two years old and weigh about 700 pounds, they are sold to feedlot operators, who will keep them for as long as nine months while fattening them for the market. The fattening process makes the meat well marbled, a characteristic prized by most Americans. In the feedlot period, each steer will consume approximately 2,500 pounds of grain and more than 300 pounds of soybeans, for a weight gain of about 2 pounds per day. Therefore, 16 pounds of grain and beans are needed to produce a single pound of beef. By regulating the amount of grain fed to cattle and other animals, U.S. agribusiness can effectively control the world grain supply and stabilize grain prices. If cattle were not fed grain, more grain would be on the world market, prices would probably be lower, and a greater supply of grain would be available for human consumption. But U.S. beef would not be as well marbled.

Although the oceans cover nearly three-quarters of the earth's surface, much of the open sea is not very productive. The most productive areas are coastal shelves. However, estuaries and coastal areas are being filled to create land for industrial sites. They are also being rapidly polluted by urbanization. Estuaries serve as nurseries for many food fish, and their destruction will contribute to further decreases in food supplies.

Another hope for increasing protein production in ocean and freshwater environments is aquaculture. Aquaculture of such shellfish as mussels and oysters (Figure 24–5) and such fish as catfish and trout has been successful. However, most species of marine organisms have specialized food requirements that often change with each developmental stage of the organ-

Figure 24–5
Oyster aquaculture in Oregon. Large numbers of oysters are grown attached to strings suspended from rafts. (Photo by authors.)

ism. This makes it impractical to cultivate the species in enclosed areas. It is ironic that underdeveloped countries, which suffer protein shortages, often export many of their fishery products to industrial countries.

It may be possible to exploit populations of fish and other aquatic organisms that have not been used in the past. However, marketing new food species is often difficult because of the unwillingness of people to use food with new textures and tastes. This may be difficult to understand at first, but it becomes easier if we think about including rat, dog, or cat meat in our own diets.

Nutrition

The deficient diets of many people in the world have been improved by nutritionists' combining of food products to produce more balanced diets. Soybean products are rich in protein and nutritious in general, but they have not been popular. However in parts of Asia, soy protein has been added to soft drinks, which are widely accepted. Essential amino acids can be synthesized by microorganisms such as yeast and added to flour and other products, improving their food value. Undoubtedly, modern technology will provide new solutions to specific problems of food production and nutrition. However, modern technology requires a large amount of energy, an expensive and increasingly rare commodity. Meanwhile, the world's population continues to increase.

Energy and Nonrenewable Resources

Environmental commodities that can be replaced are known as **renewable resources.** An example is food. More can be grown after a supply has been eaten.

However, some important resources—**nonrenewable resources**—occur in finite quantities; they are not replaceable. When the world's supplies of these commodities are gone, there will be no more.

Among the important nonrenewable resources are various metals, minerals, and fossil fuels—coal, natural gas, and petroleum. In recent years, public attention has focused on the rapidly diminishing supplies of fossil fuels, because we use these substances extensively to supply our energy needs. As fossil fuels become more and more scarce, they may end up being too expensive to use. This could happen before we have developed alternate sources of energy. If it does, the consequences will be incredible. Food production, processing, and distribution will be curtailed. People may have to live without electricity, heat, and gasoline. Exhaustion of any of the nonrenewable resources will have an enormous impact. This section of the chapter concentrates on fossil fuels and possible alternative energy sources.

Fossil Fuels

How long will the world's supplies of fossil fuels last? Predictions are difficult and estimates differ because many factors affect the rate at which fuels are used. As supplies dwindle and prices increase, the rate of use may slow significantly. Increased prices may stimulate further explorations, and new sources of fossil fuels may be found. (Major oil reserves were discovered recently in Mexico.) However, at current use rates, oil reserves may be used up in about a hundred years. Coal reserves may be depleted in three hundred to four hundred years.

The timing will depend greatly on the attitude of Americans, who represent about 5 percent of the world's population but who consume the greatest amount of energy in the world. The average American consumes 250,000 kilocalories of energy per day. That amount is double the amount used by individuals in other modern industrial nations. It is 12.5 times the amount used by people in advanced agricultural nations. It is 63 times the amount used by individuals living at primitive survival levels. If Americans were to develop strong energy conservation programs, considerable time would be bought for developing other energy sources. Time is important for developing the technology to utilize shale oil reserves and to economically and safely convert coal to forms that cause less pollution when burned.

The extensive oil shales and tar sands in Colorado,

Wyoming, Utah, and Alberta, Canada, could add another 25 percent to the world's petroleum supply. However, it is currently more economical to use other sources of oil. New technology may reduce the costs of oil production from shale, but environmentalists are justifiably concerned about the impact on the environment of mining these reserves. Thousands of acres of wild and beautiful areas will be disrupted, and rivers and streams that are now clear and unpolluted may be filled with sediments and pollutants.

Similar concerns are frequently expressed about coal mining in the United States. The least expensive way to mine coal is strip mining, a process that removes surface layers of topsoil, subsoil, and rock, exposing the coal beneath (see Figure 24–6). The mining is done in long strips that follow the coal deposit. When the coal has been removed from a strip, the surface covering the next strip is used to fill in the first one.

The unreclaimed remains of strip mines are an ecological abomination. Ugly piles of rock and unfertile earth remain to erode. Little, if any, vegetation colonizes this desolate terrain. Runoff from it may contain sulfur compounds that pollute streams and destroy natural ecosystems in the surrounding area. More than a million acres of land have already been strip mined. Much of this disturbed land has not been restored to its natural state. Approximately another 40 million acres of land could profitably be strip mined, but some of it is agriculturally important.

Land that has been strip mined can be reclaimed. The process is expensive, however, and takes many years for full restoration. Experimentation in using human waste to improve soil quality in strip-mined areas is underway.

Contour mining, a way of strip mining shallow coal deposits on hill and mountain slopes, is another common method of mining. It too defaces the countryside and creates many environmental problems.

Coal production in the United States will continue to increase as supplies of other fossil fuels continue to decrease. Coal is used to produce energy. It can also be converted to liquid fuel (oil) and methane gas. These conversions add to the cost of the resulting fuel, but oil and gas (especially gas) are more convenient as energy sources. Also, gas produces much less pollution than other fossil fuels do. Many of the coal deposits in the United States are of high-sulfur coal. When burned, this coal releases large amounts of sulfur compounds into the air. These compounds are

Figure 24–6
A strip coal-mining operation in Colorado. (EPA-Documerica, Bill Gilette photo; courtesy U.S. Environmental Protection Agency.)

among the most important air pollutants, causing serious health and environmental problems.

It makes sense economically and environmentally to develop alternative and cleaner sources of energy as rapidly as possible. Unfortunately, more effort seems to be expended in exploring for fossil fuels than exploring for alternative energy sources. Some possible alternatives to the great dependence on fossil fuels are nuclear power, solar energy, hydroelectric power, wind power, geothermal energy, and energy from garbage.

Nuclear Power

The great hope of the 1970s, nuclear power by nuclear fission, may be a fading dream. Once it was considered a clean, cheap source of power that would provide 21 percent of the world's energy by the year 2000. But only about a quarter of the 1,800 nuclear power plants projected to be operational by the end of the century have actually been built.

Some of the leading energy experts are saying that nuclear energy is dead. There are several reasons for

their viewpoint. One is the soaring costs of nuclear power plants and the lack of agreement on how to store the dangerous waste they produce. Another is the question of how long the necessary reserves of uranium (the fuel for nuclear power plants) will last. Yet another is increased public opposition to nuclear power because of 1979 accident at the Three Mile Island power plant. Also, power companies seem to have overestimated current needs for electrical power, which is the only form of energy generated by nuclear power plants. In the United States, many orders for nuclear power plants have been canceled, and almost no new orders have been placed. Nuclear power is under similar pressure in all the major industrial countries of the world except France and the USSR. It looks like nuclear power is far from being either cheap or clean.

A different source of nuclear power, nuclear fusion, may be available in the future. However, serious economic problems will probably delay its development and may even stop it. Nuclear fusion involves the fusion of the nuclei of two atoms instead of the splitting of atoms that occurs in nuclear fission. The result is the release of enormous quantities of energy but much less radioactivity than in nuclear fission. Scientists have not yet solved the problems associated with controlling nuclear fusion reactions. Some predict, however, that the problems will be solved by the end of the century if the necessary research is supported.

Solar Energy

Solar heating can be active or passive. Both are effective in saving the energy of fossil fuels, but active solar heating is more expensive and complicated. It uses solar panels to concentrate light energy and convert it to heat. The heat is then transferred to water or air to be circulated through a building by pumps or fans. The equipment is expensive, and there are maintenance costs.

Passive solar heating uses solar heat directly. It is cheap, simple, and maintenance-free. It depends on the design and position of the building and on the natural flow of energy. A building designed to maximize the use of passive solar heating should have two to three times as much insulation as other buildings and fewer windows, except on the south side. All windows should have double or triple panes, and those on the south side should have special blinds for shade. Shading is essential when the sun is high during the summer. However, the south windows should be exposed to the sun during the winter, when the sun is lower. The light energy entering the windows helps heat the inside of the building. In some cases, the east, west, and north exposures of the building are buried in soil, which increases the building's insulation and decreases its heat loss. It is predicted that by 1985 at least 1 million buildings in the United States will have passive solar heating, which will continue to increase in popularity.

Yet another use of sunlight is for heating water. Simple water heaters can be placed on the roofs of buildings. Such heaters are required by law in northern Australia. They are also common in Israel and Japan. If more efficient means of concentrating the sun's energy become available, solar energy will undoubtedly become one of the most important alternative sources of energy in the future.

Hydroelectric Power

Only about 4 percent of U.S. energy needs are met by hydroelectric power plants. It is doubtful that much more electrical power will be generated in this manner in the United States, because the country lacks appropriate sites for more dams and hydroelectric plants. The impact of a large dam on the ecosystems of any locality is great. It must be carefully studied before any new undertakings are considered. The geology of the area must also be carefully evaluated to be sure that the dam will be safe.

Hopes for the use of tidal energy from the oceans are limited because of the high cost of building and maintaining the facilities and the lack of suitable sites. Also, the impact of such facilities on bays and estuaries must be considered, since these marine communities are highly productive.

Wind Power

Windmills have been used for centuries as a source of power to grind grain and pump water. Unfortunately, the amount of wind and its duration varies widely from place to place and from season to season. Therefore, windmills probably are better suited for some areas than for others. In any case, wind power systems usually need conventional backup electrical systems or expensive storage systems. Nonetheless, experts feel that if the best wind sites around the world were utilized, many times the current amount of electricity could be generated.

Geothermal Energy

In some places, the heat deep in the earth's crust turns water to steam. The heated water comes to the earth's surface as hot springs and geysers. Communities in California, Iceland, New Zealand, and Italy have taken advantage of this energy source. Where available, it can be used to heat buildings and produce inexpensive electrical power. Unfortunately, few regions on earth are endowed with the type of geological formations that furnish geothermal energy.

Energy from Garbage

If fossil fuels become scarce, why not burn wastepaper, sewage, manure, cornstalks, and other combustible garbage? Americans produce more garbage per person than the people of any other country in the world. Half of this refuse is paper. The remainder is plastic, glass, metal, potato peelings, bones, and so on.

The problem with burning waste is the expense of gathering it and sorting the combustible from the noncombustible. Some material, such as plastic, produces large amounts of noxious air pollutants when burned. If everyone were willing to sort the garbage, separating paper from other solid wastes, communities could burn the combustible wastes and realize a savings in fossil fuels. In France, an electrical company has been burning waste for fifty years. However, it seems unlikely that waste will be used extensively to produce electrical energy in this country until the public makes a greater commitment to conservation than it has in the past.

Conservation

The energy problems facing the world are serious and may quickly become critical. Industrial nations have not responded to the energy shortage as quickly and efficiently as they should. (It seems that the only thing that has stimulated energy conservation so far has been the increased cost of fuel.) The facilities for using alternative energy sources to replace or supplement fossil fuels are still undeveloped. More time is needed for research, development, and installation of such facilities. In the meantime, nations and individuals, especially the people of the United States, must do everything possible to conserve energy in order to stretch the supplies of fossil fuels.

Each individual or family should attempt to establish and implement an energy savings program. So should each community and nation. Much can be done without dramatically altering one's life-style. Simply reducing excessive use of energy (turning off unneeded lights, for example) creates a significant savings. Other items that should be included in an energy conservation plan include increasing energy efficiency (for example, driving smaller, more energy-efficient cars). If everyone contributes in some way, the energy savings in the United States alone could make a great difference to the energy problems of the entire world.

Pollution

Contamination that leads to instability and harm to the biotic community of any part of an ecosystem is termed **pollution**. Pollution adversely affects humans as well as other species. The use of more and more fertilizer and pesticides to increase food production leads to water pollution and the concentration of poisons in the food pyramid. Human wastes accumulate in large amounts and become a source of pollution, as do industrial and agricultural wastes and the wastes

from burning fossil fuels. Sometimes these pollutants affect only local areas, but many are distributed worldwide.

Of course, every species leaves wastes, including its own dead bodies, in the environment. When the ecosystem is in balance, these wastes are recycled and the integrity of the system remains intact. Humans, however, have greatly altered natural ecosystems. They have produced such an overwhelming amount and variety of by-products that the natural ecosystem may no longer be able to accommodate the load.

Air Pollution

The major sources of air pollution are industrial wastes and exhausts from automobile and other internal combustion engines that burn fossil fuels. Therefore, air pollution is greatest near large urban industrial centers (see Figure 24–7). However, air pollutants are distributed worldwide in the major air currents above the earth. Probably no place in the atmosphere lacks at least some level of contamination.

Pure air (if there is such a thing) contains the following gases (from most to least abundant): nitrogen (N_2), oxygen (O_2), inert gases (mostly argon, Ar), carbon dioxide (CO_2), methane (CH_4), and hydrogen (H_2). In terms of relative quantities, nitrogen represents 78 percent of the gases in the air and oxygen 21 percent. The remaining gases account for only 1 percent. Air also contains water vapor, which varies widely in concentration. In addition to these components, pure air contains small amounts of a few other gases, such as carbon monoxide (CO), nitrogen oxide (NO), and ozone (O_3) produced by natural phenomena, such as lightning and solar radiation. They occur in small concentrations in the lower atmosphere and are recycled.

Pollution has significantly increased the levels of oxides of nitrogen, carbon dioxide, carbon monoxide, ozone, and particulate matter, such as soot. This is the air we breathe. Also of great importance has been the atmospheric accumulation of sulfur dioxide (SO_2) and sulfur trioxide (SO_3). These compounds are hazardous to health and create other serious environmental problems as well. (See Box 24–2.) Various organic compounds, such as hydrocarbons, are also present in the air we breathe.

Each air pollutant has its own effect on the tissues and cells of plants and air-breathing animals. The pollutants also interact to produce more complex and less understood problems. When polluted air is in-

Figure 24–7
Industrial air pollution in Pittsburgh, Pennsylvania, around 1960. (Grant Heilman Photography.)

haled, the larger particles of soot become entangled and trapped in mucus secreted by cells lining the upper respiratory passages. Laden with particulate matter, this mucus is driven upward along the walls of the air passages by the beating of microscopic cilia borne on the epithelial cells that line the respiratory passages. (Heavy smokers may not have many of these cilia left.) Eventually, the mucus is coughed up and swallowed or spit out. Finer particulate matter is carried into the air sacs of the lungs and engulfed by special cells lining the walls of these chambers. Air pollutants adhering to these particles are likewise taken up by these cells.

Some of the polluting gases, such as sulfur dioxide and sulfur trioxide, dissolve in water and enter tissues, causing burning and irritation. Others, such as carbon monoxide and nitrogen oxides, dissolve in tissue water and diffuse into the bloodsteam. In the blood, they combine with the respiratory pigment, hemoglobin, reducing the ability of the blood to carry oxygen. Some air pollutants combine to form compounds that may be carcinogenic (cancer inducing). Breathing heavily polluted air contributes significantly to such respiratory problems as bronchitis and emphysema, and aggravates asthmatic conditions.

Air pollution increases in industrial areas and cities when there are thermal inversions. In these inversions, a warm air mass moves over colder air, inhibiting the usual upward flow of air currents. (See Figure 24–8.) Inversions trap air masses and allow the

BOX 24–2

ACID RAIN

Two of the most objectionable components of air pollution are sulfur dioxide (from factories, metal smelters, and coal-burning power plants) and nitrogen dioxide (from automobile exhaust, power plants, and anything else that burns fossil fuels). Sulfur dioxide can be transformed into a strong acid, sulfuric acid (H_2SO_4), within several days after its release into the atmosphere. Nitrogen dioxide can be converted to nitric acid (HNO_3). Both compounds may be carried long distances by the winds before falling back to earth as acid rain or snow.

Of the many areas of the world seriously affected by acid precipitation, two are Sweden, and the eastern United States and Canada. Both regions are downwind from heavily industrialized areas.

The effects of exposure to acid rain over several years can be serious. Acid rain changes the pH of small lakes and disrupts their ecosystems. Entire populations of fish species that are sensitive to pH changes (trout, for example) have been destroyed, as have other species of aquatic life. This destruction greatly reduces the diversity of species in lake communities. The lowered pH of waterways also causes harmless chemicals in lake bottom deposits to be converted into toxic chemicals. Such conversion places further stress on these communities. For example, certain mercury compounds are converted to the highly toxic methyl mercury.

More than 300 lakes in New York State's Adirondack Mountains alone have been affected by acid rain and are in various stages of destruction. Canadian specialists predict that as many as 48,000 Ontario lakes will be dead within eighteen to twenty years if acid rain is not stopped.

Plants also are harmed by acid rain. Trees and food crops are stunted and damaged. Whole forests in Sweden have been stunted.

If the United States builds its proposed 350 coal-burning power plants, scheduled to be completed by 1995, there could be a 10 to 15 percent increase in atmospheric acid. The United States exports more acid rain to Canada than the other way around, so the Canadian government is particularly concerned about the problem. The Canadian and U.S. governments started negotiations about the problem in 1980.

In the same year, the Environmental Protection Agency began to require all new coal-burning power plants in the United States to remove as much as 75 percent of the sulfur dioxide from their smoke emissions with special devices called scrubbers. Existing power plants, however, do not have to comply with the order. Each of the older plants emits almost seven times more sulfur dioxide than a new plant does.

Acid rain is a serious problem. It will have to be dealt with at a time when we are turning back to coal to help us through the energy crisis.

usual air pollutants to concentrate dramatically. If the condition persists for several days, pollution may build to a level that is a serious health hazard. (As a result of a 1952 air inversion in London, 4,000 more people than usual died.) Most seriously affected are older people and people with histories of heart or respiratory problems.

Certain air pollutants are associated with specific industries. For example, hydrogen fluoride (HF) is associated with industries that produce aluminum, and arsenic compounds are given off by smelters. When fluorine and arsenic compounds settle on vegetation, herbivores may be poisoned. Fluorine compounds cause cattle to lose weight. They also cause the formation of abnormal bones and teeth.

Each year, millons of dollars of damage to vegetation is caused by air pollution. Where sulfur dioxide is abundant, it can cause the total destruction of plants. Many air pollutants cause discoloration and curling of leaves and petals. Because lichens will not grow on trees in areas that have air pollution, they serve as good indicators of local air pollution. Where there are lichens, the air is clean.

In time, air pollution may cause a change in the earth's temperature. The average temperature of the earth has remained fairly constant for thousands of years because there is a balance between the amount of solar energy the earth receives and the amount of energy (infrared radiation) it reflects and radiates into space. If the earth and its atmosphere were to absorb more energy, the earth would heat up. If the earth were to reflect or radiate more energy, it would cool.

The amount and variety of gases in the earth's atmosphere affect this balance. For example, the gases in the atmosphere that absorb infrared radiation are water vapor, ozone, and carbon dioxide. Since 1850, the level of carbon dioxide in the atmosphere has increased from approximately 280 parts per million to

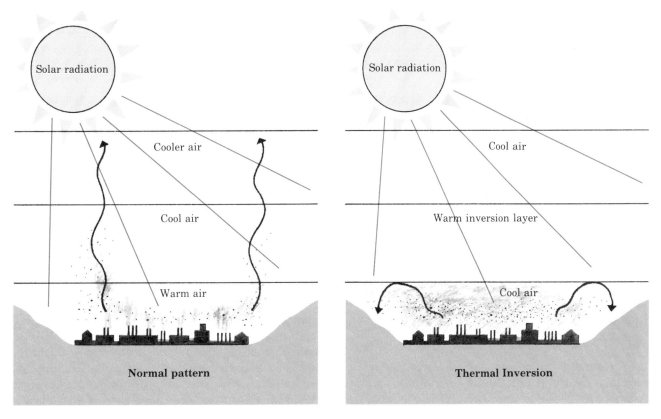

Figure 24–8
Thermal inversion. When a warm layer of air forms above a cool layer of surface air, the surface can not rise as it does under normal conditions, and pollutants become concentrated often forming health hazards.

340, an increase of over 12 percent. About a quarter of this increase has occurred in the past decade. If the present rate of increase continues, the amount of carbon dioxide in the atmosphere will double by the year 2020. As carbon dioxide accumulates in the atmosphere, it absorbs more and more of the reflected infrared radiation. This could cause an increase in temperature referred to as the **greenhouse effect.** An increase of one or two degrees may not cause any noticeable harm to ecosystems. However, any further increase could have serious consequences by changing the world's climate□. Melting polar icecaps and glaciers could cause sea levels to rise, flooding most of the major population centers and fertile lands.

Burning fossil fuels has been considered the primary cause of increased atmospheric carbon dioxide. However, it is now thought that only about half of the carbon dioxide results from such burning. The remainder results from forest destruction. Forests represent an enormous store of carbon-containing compounds. They contain 90 percent of all the carbon held in vegetation. When forests are destroyed, much of the carbon contained in plant tissue is released into the atmosphere as carbon dioxide by the respiratory activities of decomposing microorganisms□.

Increased particulate matter in the atmosphere may reverse the warming effect caused by carbon dioxide. All kinds of particulate matter are being thrown into the air—from steel mills, jet aircraft, automobile exhausts, volcanic activity, and plowed soils. Particulates produce a layer of dust that shades the earth from direct sunlight and reflects solar energy back into space. This may cause the earth's crust to cool. Much of the particulate matter is associated with what we call smog. In large urban areas, smog may decrease the amount of sunlight reaching the ground by 30 percent and may reduce ultraviolet radiation by 90 percent.

Two other serious air pollutants are fluorocarbons from aerosol cans and nitrogen oxides from automobiles and power plants. These substances may decrease the ozone layer in the stratosphere, which shields the earth's surface from much of the ultraviolet radiation of the sun. Any increase in ultraviolet radiation would have a wide range of effects on living organisms. For example, it would increase the incidence of skin cancer and of mutations□ and would adversely affect plants.

The extent to which air pollution may affect humans and natural ecosystems is just beginning to be appreciated. It would be sensible for us to act quickly to reduce air pollution and to seek clean energy sources that will not contribute to the problem.

Water Pollution

All naturally occurring water—oceans, lakes, streams, rivers, ponds, springs—contains varying amounts of dissolved substances. There is no such thing as pure water in nature. Water pollution is caused by adding substances to natural water that make the water less suitable for human use and that disrupt aquatic ecosystems. The most important sources of water pollution are sewage and industrial wastes.

When human populations were small and there was little, if any, industrialization, organic waste and sewage could be dumped into rivers or other bodies of water. There it would eventually recycle, as do the wastes from all other organisms. The organic matter in sewage decomposes and eventually becomes nutrients that plants use. However, the addition of human wastes to water introduces into the ecosystem a large variety of microorganisms, many of which cause human diseases such as cholera, typhoid fever, and parasitic infections. Consequently, water-borne diseases have plagued humans throughout history.

As human populations grew, the amount of sewage became too great to be processed by aquatic ecosystems. Not only did water become unfit to drink, it became foul and bad smelling. The reasons are not complex. As the amount of organic matter increases in water, the microorganisms that use the nutrients in it rapidly increase in number. The inorganic nutrients released by decomposition serve as nutrients for algae. They too reproduce rapidly and use more and more of the available oxygen. Eventually, there is not enough oxygen to support the aerobic (oxygen-using) organisms in the ecosystem, and they die. Their bodies contribute to the organic mass and further complicate the situation. Eventually, only anaerobic microorganisms survive. Many produce methane gas and hydrogen sulfide, a gas with an unpleasant odor that is associated with polluted water.

Following the Industrial Revolution and the rapid growth of urban centers, water pollution became so great that sewage treatment plants were developed. Sewage was first passed through a series of screens that strained out large objects and then passed into large settling tanks, where particulates settled out. Straining and settling constitute primary sewage treatment.

Secondary sewage treatment, which is common today, aerates the material from settling tanks, keeping the oxygen level high while aerobic bacteria and other microorganisms consume the organic matter. Finally, the fluid is treated to kill any bacteria. Chlorine has commonly been used for this purpose. However, it has been shown that chlorine combines with chemicals in the fluid to produce substances believed to cause cancer. Therefore, chlorine is being phased out, and other chemicals, such as ozone (O_3), or ultrasonic energy treatments are used instead.

The treated liquid is released into streams and other bodies of water. The solid fraction that has settled out is burned or buried. Primary and secondary sewage treatment remove up to 90 percent of the objectionable organic matter in human sewage. However, the fluid that leaves the sewage plant is rich in nutrients, nitrates, and phosphates. These often cause increased growth of algae and other undesirable organisms in bodies of water.

The most modern sewage treatment plants (around Lake Tahoe) include a third process, tertiary sewage treatment), before sewage is released into natural waters (see Figure 24–9). This treatment removes much of the dissolved nutrients by ion-exchange systems and reverse osmosis. Tertiary treatment is particularly important for industrial sewage, which contains a

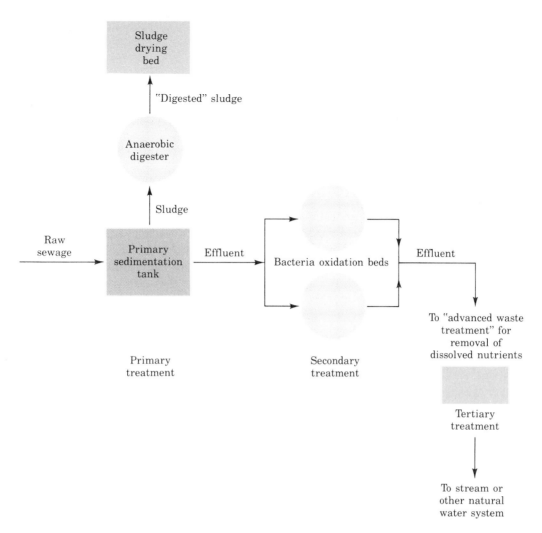

Figure 24–9

A flowchart diagraming the relationships among primary, secondary, and tertiary sewage treatment.

wide variety of dangerous chemicals. However, it is expensive, and few communities can afford it.

Although the majority of American communities use both primary and secondary treatment (20 percent of U.S. communities have only primary treatment), many systems are inadequate, because storm sewers are often connected to sanitary sewers. During heavy rains, sewage plants cannot accommodate the influx. Retaining gates are opened, and raw sewage pours into natural waters. The problem is most serious when many communities use the same river for sewage disposal. Less than a decade ago, this overuse "killed" many rivers, including the Illinois and the Mississippi. The situation has improved somewhat today because of new sewage plants, but the problem is still serious.

One of the most publicized ecological disasters was the near destruction of Lake Erie during much of this century. Approximately 11 million people lived around Lake Erie. Some 1.5 million gallons of sewage from communities and 10 million gallons of wastes from industry, flowed into the lake each day. Much of this sewage was untreated or had secondary treatment at best. At one time, some of the rivers flowing into Lake Erie contained so much organic matter and gases that they were fire hazards.

Fortunately, thanks to the concerted efforts of communities and cities along the lake to improve waste disposal methods during the last fifteen years, Lake Erie now has improved water quality. Many of the beaches that were closed for years because of water pollution have been reopened. Fish populations have increased to the point that sport fishing is again a common recreational activity even near large cities.

Another problem with sewage disposal is the use of detergents. Soap is biodegradable (bacteria can digest it), but it leaves residues in water that contain minerals. The residues do not dissolve. They cause "ring around the tub" and "tattletale gray" laundry. Enterprising chemists solved this problem in the early 1960s by developing synthetic detergents. The detergents do not leave residues, but they are nonbiodegradable. Although consumers liked them, rivers and streams became foamy because bacteria could not degrade the compounds. Today, biodegradable detergents are available. However, they contain phosphates, which enrich natural waters and disrupt ecosystems.

The quantity and variety of waste products entering the ecosystem are incredible. Radioactive wastes, heavy metals (such as mercury), pesticides, and polychlorinated biphenyls (PCBs, used in the plastics industry) are only a few of the toxic materials that contaminate water supplies.

Sewage ultimately reaches the oceans, where marine organisms are also drastically affected. For example, sea urchin populations in polluted areas along rocky shores have increased significantly, because added nutrients have increased kelp growth. Sea urchins feed voraciously on kelp, and their large populations destroyed so much kelp that populations of numerous other organisms that use kelp beds as a habitat have been affected.

Over a million tons of oil are spilled into the ocean each year, and this amount will probably increase as more offshore oil wells are constructed. Oil spills affect organisms along shorelines, and floating oil slicks harm plankton in the open sea. Some methods of cleaning up oil spills cause some fractions of crude oil to sink to the bottom and smother benthic organisms.

Polychlorinated biphenyls, pesticides (such as DDT), and radioactive wastes are widespread in the ocean and have become concentrated in food pyramids. It would be foolish to assume that the oceans are immune to environmental pollutants. They are threatened as severely as small streams are.

Pesticide and PCB Pollution

Since World War II, millions of tons of pesticides have been broadcast into the environment. A large number of insect pests are killed, but microorganisms cannot decompose many of the pesticides. As a consequence, they have accumulated in ecosystems.

The benefits of pesticides are obvious. Epidemics of diseases such as typhus (transmitted by lice) and malaria (transmitted by mosquitoes) have been controlled. Farming practices have been improved, and yields have been increased. Unfortunately, pesticides cause as many serious environmental problems as they solve. They kill not only pests but also many beneficial species. Some beneficial species are natural predators of the exterminated species.

Two major groups of insecticides have been used extensively: chlorinated hydrocarbons and organophosphates. DDT, dieldrin, endrin, aldrin, and chlordane are chlorinated hydrocarbons. Parathion, malathion, and diazinon are organophosphates. Chlorinated hydrocarbons are stable compounds, but some, such as DDT, may be altered biologically into other toxic compounds that persist in the environment for years. They kill a wide variety of organisms, including vertebrates. Organophosphates are biodegradable and do not persist in the environment, but they too are toxic, even to humans.

A great variety of herbicides are used to kill plant pests such as weeds. Two widely used herbicides are 2,4-D and 2,4,5-T, both of which are biodegradable and do not persist in the environment. However, they are toxic to humans and other animals.

The following example illustrates one of the unexpected consequences of using pesticides. DDT and other chlorinated hydrocarbons were used to control pests in a valley in Peru where cotton was a major crop. During the first four years of their use, cotton production rose from 440 to 650 pounds per acre, but it dropped to 350 pounds per acre in the fifth year and remained low thereafter. The pesticides had killed a wide range of insects, including not only the pests but their natural predators as well. After a number of generations, the pest insects developed a resistance to the pesticide. Their predators did not. In the absence of natural predators, populations of the resistant insect pests increased and caused greater destruction to cotton crops than before the use of pesticides. As many species of insects become resistant to the commonly used pesticides, we respond by developing new and more potent pesticides, and the war goes on.

Because chlorinated hydrocarbons such as DDT persist in the environment for years, they are transported in water and air currents to all parts of the world. Now they are concentrated in every food pyramid in every ecosystem on earth. Even animals inhabiting Antarctica contain sizable quantities of DDT and other pesticides in their tissues. This problem should be of great concern to humans, because we, like other predators, are at the top of the food pyramid. Organisms at the higher trophic levels concentrate the largest amounts of pesticides in their tissues.

Chlorinated hydrocarbons are not very soluble in water, which usually contains only 1 to 2 parts of pesticide per billion parts of water. However, these compounds are very soluble in fats. Aquatic microorganisms absorb them in fat and oils, where they accumulate to concentrations many times greater than in water. Zooplankton that feed on countless contaminated phytoplankton cells concentrate the pesticide still further in their tissues. Their levels may be as high as 1 part pesticide per million parts of fat, 1,000 times more concentrated than in water. Fish feeding on zooplankton further concentrate the pesticide to levels as high as 3 to 8 parts per million. Birds feeding on fish concentrate the compounds further—up to 3,000 parts per million. The increased accumulation of toxic substances in the food pyramid is referred to as **biological magnification** (see Figure 24–10).

Many species of predatory birds—eagles, hawks, pelicans, and the like—have shown serious adverse effects from this accumulation. It frequently interferes with eggshell production in many birds. The shells are thin and are easily broken by the bird's weight during incubation. If the shells aren't broken, the accumulated pesticides often adversely affect the developing embryos. As a result, the populations of many birds have decreased. For example, some pelican populations in the Gulf of Mexico nearly became extinct. Pesticides on vegetation can also be concentrated by herbivores such as cattle. By eating meat and drinking milk, humans concentrate pesticides in their tissues.

Human tissues may contain levels of DDT as high as 5 to 27 parts per million. Human milk may contain levels higher than 0.117 part per million, exceeding by 70 percent the tolerance levels for milk set by the United Nations. Therefore, infants begin to accumulate pesticides in their tissues with their first suckle. DDT may induce liver cancer and upset reproductive

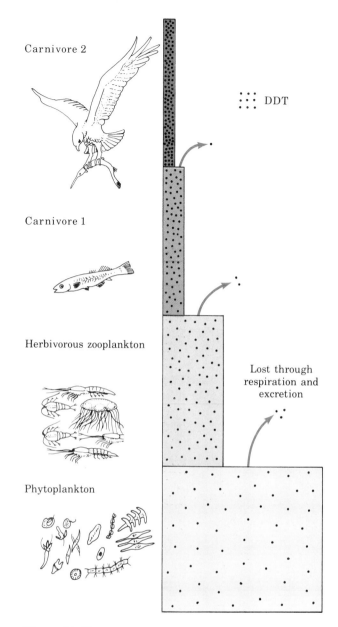

Figure 24–10
Biological magnification of the pesticide DDT in the tissues of organisms in an aquatic food pyramid.

processes in mice. How will it affect humans? There is insufficient evidence at present, but there is little doubt that the effects will be undesirable.

DDT and other pesticides also have an adverse effect on soil bacteria that are important in the nitrogen cycle. They may eventually decrease soil fertility and lower productivity.

If DDT and similar pesticides are so dangerous, why hasn't something been done about them? In some cases, something has been done. For example, in 1970, the Environmental Defense Fund (a group of scientists and lawyers supported by funds from private citizens and organized to bring lawsuits on environmental issues) brought a suit that resulted in a ban on the use of DDT in the United States. The ban went into effect in 1972. Since then, there have been noticeable improvements. For instance, there has been a gradual increase in the reproductive success of many species of birds affected by toxic DDT. Unfortunately, the United States still manufactures DDT and exports it for use in other countries.

PCBs are not pesticides, although they are chemically similar to DDT. Until the mid-1970s, PCBs were used extensively in industry to soften plastics and other polymers, and as coolants in large electrical transformers, and as additives to pesticides, adhesives, and duplicating paper. PCBs become concentrated in food pyramids and may have the same effects as pesticides on the health and well-being of organisms. They have become so concentrated in Lake Michigan salmon that the salmon should carry the label "If eaten, may endanger human health."

Other Infamous Toxins

Countless toxic pollutants are released into the environment, some accidentally and some intentionally. Three that have received considerable publicity are polybrominated biphenyls, agent orange, and dioxin.

Polybrominated biphenyls (PBBs) are fire retardants that were accidentally mixed with cattle and chicken feed in Michigan in 1980. Thousands of cattle and chickens had to be destroyed after their tissues were found to contain the chemical. Humans also became contaminated from drinking milk and eating beef and chicken before the mixup was discovered. PBBs have become widely distributed in Michigan and have been found in the waters of Lake Michigan.

Agent orange, a defoliant herbicide, is actually a mixture of two herbicides—2,4-D, which may or may not be harmful to humans, and 2,4,5,-T, which has been shown to cause birth defects in animals. Agent orange was widely used during the Vietnam War, between 1962 and 1970, to defoliate the jungle so the enemy could be found. After 1970, the Department of Agriculture outlawed its use in public areas such as parks, recreation spots, irrigation ditches, and lakes. It was still widely used, however,

in forest management and rice culture and to kill brush on grazing lands. In 1979, it was banned. By 1980, over 1,200 Vietnam veterans exposed to agent orange filed claims with the Veterans Administration, claiming that their exposure resulted in such conditions as liver, muscle, and skin disorders, testicular cancer, loss of sex drive and emotional disorders.

Dioxin, an unavoidable contaminant in 2,4,5-T and one of the most deadly toxins known, may be one of the factors that causes the ill effects of agent orange. Dow Chemical Company and other manufacturers of 2,4,5-T claim that dioxin is introduced into the environment through all forms of combustion, not just from 2,4,5-T. However, the Environmental Protection Agency disputes this claim.

Radiation

Every living organism is exposed to naturally occurring radiation from radioactive substances in the earth, from cosmic rays, and from other background radiation. Since World War II, the production and use of radioisotopes, X-rays, and other forms of radiation has greatly increased. As nuclear power plants become more common, environmental pollution by radioactive substances may become even greater due to improper waste storage, nuclear plant accidents, and possibly even sabotage.

Thermal Pollution

Water is used to cool heat-producing systems such as energy plants. This practice raises the temperature of natural waters when the heated water is released into them. Aquatic organisms require more oxygen in warm water than in cold water, but warm water holds less oxygen. Therefore, the organisms in unnaturally warmed waters are subjected to stress and often die.

Noise Pollution

Long-term exposure to loud noise can result in hypertension and hearing loss. Even guinea pigs suffer serious hearing loss after exposure to loud rock music.

Food Pollution

Most commercially processed foods contain food additives, substances added to prevent spoilage and staleness. Food colorings are added to enhance the appearance of food. Large amounts of sugar are also added to many foods, because people like sweet food.

Another cause of concern is the hormones and other compounds fed to cattle and other domesticated animals to induce rapid growth and, supposedly, improve the quality of meat. How will the concentrations of these compounds affect us?

The Future of Homo sapiens

Will humans, through polluting the world, cause their own extinction? Many predict it, but no one really knows. We do know that humans have caused major disruptions to ecosystems, and we do know the dangers of the continuing population spiral. If we have not exceeded the carrying capacity of our environment, we are dangerously close. What can we do to improve our chances for survival? It is no longer sufficient to assume that scientists and politicians can and will do something about it. It is up to each of us. With our understanding of biology, we can begin active programs for protecting ourselves and our environment. We can become concerned and informed citizens. We can make lists of what we can do. If we try, we will come up with very long lists.

SUMMARY

1. The sizes of natural populations of organisms fluctuate continuously and often in cycles, as is the case in prey-predator relationships. However, the size of a population cannot long exceed the carrying capacity of the habitat.

2. The carrying capacity of a habitat is the limit to which a population can grow and be sustained over an extended period.

3. Although the human population has increased dramatically during the past three hundred years and is still increasing rapidly, it may have exceeded the earth's carrying capacity for the species.

4. The impact of the human population on the biosphere has been great. While striving to obtain more food and to improve the quality of life, humans have drastically altered major ecosystems, polluted the environment, and used up most of the nonrenewable resources.

5. During recent decades, increased food production and human population growth have barely kept pace. Between 10 and 20 million people die each year of malnutrition and starvation.

6. Most of the world's fertile agricultural lands are presently under cultivation. The conversion of arid lands to agricultural use by irrigation presents serious problems. The conversion of rainforests to agricultural land is ongoing but presents numerous problems, including poor soil fertility and erosion.

7. In order to increase food supplies from the ocean, alternate food organisms will have to be used. Many populations of marine species currently used for food have been reduced by overfishing. Aquaculture has had few successes in increasing food production, but research is continuing.

8. Supplies of the world's nonrenewable resources, such as natural gas and petroleum, are seriously short. Humans will have to seek alternate energy sources, such as nuclear power, solar energy, and wind power.

9. Humans have extensively polluted the biosphere with air pollutants from automobile and industrial emissions and with water pollutants such as human sewage and industrial wastes. This heavy pollution has long-term negative effects on human health, agricultural productivity, and possibly even the world's climate. The quality of life is decreased because of pollution.

10. The widespread use of pesticides has deleterious effects on all living organisms, because many of the nonbiodegradable pesticides and the residues of biodegradable pesticides may be concentrated in food pyramids, reaching toxic levels in the tissues of top predators, such as humans.

THE NEXT DECADE

POPULATION OF THE UNITED STATES

According to the 1983 World Population Data Sheet of the Population Reference Bureau, the population of the United States was 234.2 million at midyear. Other 1983 demographic data for the United States follow:

— Birth rate: 16 per 1,000 individuals
— Death rate: 9 per 1,000 individuals
— Population growth rate: 0.7 percent annually
— Number of years to double the population: 95
— Population projection for the year 2000: 268 million
— Infant (under one year old) mortality rate: 11.4 deaths per 1,000 live births
— Population under 15 years: 23 percent
— Population over 65 years: 11 percent
— Life expectancy at birth: 74 years
— Urban population: 74 percent

How does the United States compare with the rest of the world? Czechoslovakia, Portugal, Australia, and New Zealand have the same birth rates as the United States. Most European nations, Japan, Canada, and Cuba have slightly lower birth rates. Most of the remaining countries have birth rates considerably higher than that of the United States. Lower death rates than that for the United States occur in 57 countries. Many of these countries have rapidly expanding populations

composed of high percentages of young people. Few countries have life expectancies equal to or greater than that of the United States. The people of Hong Kong, Iceland, and Japan have the highest life expectancies in the world: 76 years. In general, the countries of Africa have the lowest life expectancies. Afghanistan, Chad, and Ethiopia show the very lowest: 40 years.

The general fertility rate for a population is computed by dividing the total number of births in a year by the total number of women of reproductive age. In the United States, the fertility rate declined from 1910 until the mid-1930s. Following the Great Depression, it began to increase, except for a brief drop during World War II. After the war, the fertility rate rapidly increased, peaking in 1957. This peak is called the postwar baby boom. After 1961, the fertility rate gradually dropped, reaching an all-time low of just under 1.8 in 1976. Since then, it again increased, to 1.9 in 1982, and then decreased to 1.8 in 1983. This rate is still below the 2.1 births per woman needed to replace the population (called zero population growth). Children of the baby boom years reached reproductive age in the late 1970s. So far, the fertility rate of this generation has continued to remain low.

How will the population size of the United States change in the future? No one can say for sure, because economic and social changes influence fertility rates. The 1982

World Population Data Sheet predicts that the U.S. population will reach 268 million by the year 2000, and will double by the year 2078 assuming that the population growth rate continues. But will it?

Five leading demographers predict that U.S. women of childbearing age will reach a record 55 million by the mid-1980s and that annual births will reach nearly 4 million before the end of the decade and then decline. Death rates are expected to continue to decline. Predictions are that the life expectancy for men will be 71.9 years by 1990 and 72.9 years in the year 2000. The life expectancy for women will be 80 years in 1990 and 81.1 years in 2000.

A factor that may have a significant impact on the U.S. population is the predicted increase in legal and illegal immigration. It is difficult, for two reasons, to determine exactly how many immigrants are in the United States. First, although records are kept of how many legal immigrants enter the country each year, no one keeps score of how many people emigrate (leave to settle elsewhere). Second, the number of illegal immigrants can only be estimated. The best guess of the number currently living in the United States is between 3 and 6 million. This estimate is based on the fact that the 1980 census showed 5.5 million more people living in the United States than were predicted.

During the 1970s, an average of 420,000 legal immigrants arrived in the United States each year. In 1980, the number jumped to over 800,000 because of the influx of large numbers of Haitians, Cubans, and Southeast Asians. In 1981, approximately 700,000 legal immigrants were admitted. What percentage of the total U.S. population growth do these figures represent? During the 1970s, legal immigration alone accounted for more than 20 percent of the population growth. In 1980, it jumped to 33 percent. In 1981, it dropped slightly, to 31 percent. If the demographers are correct, legal and illegal immigration is going to significantly affect the U.S. population during the next decade.

FOR DISCUSSION AND REVIEW

1. Define the term *population,* and discuss how changes in ecological factors affect population size.

2. What happens to populations that exceed the carrying capacity of their habit?

3. Outline the changes in human population size over the past 500,000 years, and discuss the probable reasons for the changes.

4. List several ways by which the world's food supply might be increased. Discuss the feasibility of each plan.

5. As the world runs out of fossil fuels, alternate energy sources will need to be developed if society as we know it is to continue. List several possible alternate energy sources and the advantages and disadvantages of each.

6. Calculate your energy use for a day. For example, how many lights do you use and for how long? How much heat or air conditioning? How much hot water? How much gasoline? Don't forget the energy used by others for your comfort—public transportation, lights and heat in the classroom, and so on. How could you reduce your energy use by 10 percent? By 25 percent? By half?

7. What harmful effects do air and water pollution have on ecosystems?

8. What can be done to reduce pollution?

9. What is biological magnification of pesticides in food pyramids?

SUGGESTED READINGS

THE COUSTEAU ALMANAC, by Jacques-Yves Cousteau and the staff of the Cousteau Society. Doubleday: 1981.
A fascinating encyclopedia of environmental facts and issues.

ENVIRONMENT AND MAN, by Richard H. Wagner. Norton: 1978.
How humans affect their environment.

THE POPULATION BOMB, by Paul Ehrlich. Ballantine Books: 1978.
The consequences of unlimited populations growth.

POPULATION HANDBOOK, by Arthur Haupt and Thomas T. Kane. Population Reference Bureau: 1978.
An excellent introduction to demography and population dynamics.

SCIENCE, 9 May 1975.
An entire issue devoted to thought-provoking articles about the world's food problems.

THE SEA AGAINST HUNGER, by C. P. Idyll. Crowell: 1978.
An updated discussion of the contribution of the oceans to the world's food supplies.

SILENT SPRING, by Rachel Carson. Fawcett Books: 1962.
The classic that brought about widespread recognition of the harm of indiscriminate use of pesticides.

25

THE BIOLOGY OF BEHAVIOR

PREVIEW QUESTIONS

After reading this chapter, you should be able to answer the following questions:

1. How does an ethologist analyze the behavior of an animal?

2. What are the various types of releasing stimuli? Give an example of each.

3. List some of the ways that your Umwelt differs from that of an animal such as a dog or cat?

4. What are territories? What functions do they serve? How are they maintained?

The special branch of biology that deals with observations and interpretations of animal interaction—**ethology**—is a relatively new area of scientific investigation. The word itself is derived from the Greek *ethos,* meaning "habit" or "character." Ethology is the study of habit or character, or the biology of behavior. It began as a discipline in the mid-1930s, but it did not become common until after World War II.

Ethologists study all sorts of animal life, among them microscopic animals, invertebrates, fish, amphibians, reptiles, birds, and mammals (including humans). They interpret behavior from an evolutionary viewpoint, asking questions about its adaptive significance and about its evolutionary relationships among closely related species. This kind of questioning, sometimes called the comparative approach, often brings out detailed comparisons of the behaviors of closely related species.

MODAL ACTION PATTERNS

Ethologists record units of behavior. The units have been given a variety of names, but most commonly they are called fixed action patterns (FAPs). Today, they are better known as **modal action patterns (MAPs).** Many specialists objected to the term *fixed* because it implied that behavior was inflexible. *Modal* implies a range of variation that can be seen for all behavior patterns.

Some familiar MAPs are illustrated in Figure 25–1, which shows the tails of dogs and wolves. Each position is a MAP. Tail position is a useful barometer of emotion for many animals. Examine Figure 25–1. You should be able to assign an emotion to each position of the tail. Figure 25–2 illustrates the facial

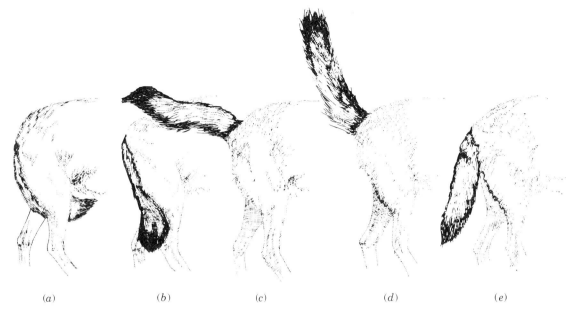

The tails of dogs and wolves, barometers of emotion for: (a) submission, (b) subordination, (c) threat, (d) dominance, and (e) neutrality. Tail position is one means of animal communication. Animals also show changing facial expressions that are linked to tail postures.

expressions of dogs and wolves. Dogs, wolves, coyotes, foxes, and hyenas belong to a single order of mammals, the **carnivores,** which have many MAPs in common.

As ethologists observe MAPs, they must be able to describe them in enough detail that other observers will be able to identify them when they are performed again. For example, the grooming behavior of a dog may consist of many MAPs. Among them are hind leg scratch to flank, hind leg scratch to ear, tongue lick to forepaws, tongue lick to genitals, tooth bite to hair, and forepaw wipe.

MAPs are often classified into functional categories, such as those used for grooming, aggression, sex, and parental care. However, it is important in classifying MAPs not to assign the supposed function prematurely. The terms used here to label grooming postures of the dog are all descriptive. None suggests function. Each posture, however, may have some maintenance function, such as grooming the fur or relieving an itch.

It is also important that the descriptive terms used for MAPs not be anthropomorphic (*anthro* means "man," and *morphos* means "form"; literally "in the

Figure 25–2
Facial expressions of dogs and wolves are: (a) neutral; (b) threatening or attacking—erect ears, staring eyes, display of teeth, and often an accompaniment of low growling; (c) conflict between attacking and running away—lowered position of ears signaling fear but display of teeth showing tendency to attack; (d) appeasement, the opposite of threat. (Try matching the tails of Figure 25–1 with these facial expressions.)

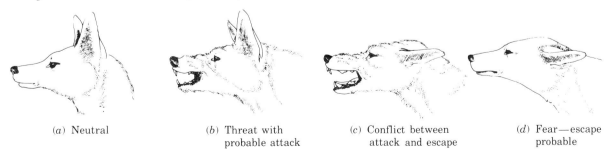

(a) Neutral   (b) Threat with probable attack   (c) Conflict between attack and escape   (d) Fear—escape probable

form of man"). Humans too often interpret the behavior of other organisms in the context of their own experiences and emotions. Common anthropomorphisms used in reference to other animals are love, jealousy, selfishness, anger, and loneliness. Animals have emotions, but there is little evidence that their emotions are like ours.

When MAPs for a given species have been observed, described, and named, they constitute an **ethogram**—a catalog of MAPs. Each ethogram is species specific. That is, it is unique for a given species. The ethograms for the social behavior of a dog and a cat would be quite different. However, the ethograms for a coyote and a wolf would be similar because of their close relationship.

A sequence of MAPs is comparable to the sequence of words in a sentence. Like a sentence, it conveys a message. If a few words in a sentence are out of sequence, the message is confused. If the words are completely jumbled, the message is lost. MAPs, however, seldom occur in rigid sequences, such as 1-2-3-4-5. If they did, the animal would have no flexibility in communication. Instead, MAP sequences vary in the same way that sentence structure varies. Although sentences can be constructed in a variety of ways and still convey the same basic message, there are rules governing sentence structure. One of the challenges for ethologists is to determine the rules of MAP sequencing.

Often, sequences of MAPs are performed as a series of repeated phrases, such as 1-2-1-2-1-2-1-2-3-1-2-3-2-3-2-3-2-3-4-4-3-4-3-4-3-4-5-3-4-5-3-4-5-5-5-5-5, which may start over again. The repetition not only gets the message across but also encourages the receiver to respond. It may also stimulate a specific motivation, such as readiness to mate.

In their natural habitats, animals cannot afford to behave conspicuously all the time, because they would attract predators. Thus many MAPs are performed only during certain times of the day or year. For example, some of the most conspicuous MAPs are those involved in mate selection and courtship, but many animals mate only once a year.

RELEASING STIMULI

Animals respond to other members of their species by performing specific MAPs or sequences of MAPs. Because MAPs (as well as colors, shapes, and smells)

often release or stimulate a performance by a receiver, they are called **releasers.**

The famous ethologist Konrad Lorenz proposed that organisms respond to releasers because their nervous systems possess an **innate releasing mechanism (IRM).** He believed that each species has programmed, genetically encoded mechanisms in its nervous system that respond only to specific environmental stimuli. These stimuli release adaptive behavior. In essence, Lorenz viewed the IRM as a lock and key. The right key (the right releaser) can turn a specific lock (the IRM) to unlock a specific behavior pattern. Despite the apparent abstractness of Lorenz's model, it does have structural and functional counterparts in the nervous system.

Releasers affecting IRMs are generally thought of as visual releasers if the organisms visually perceive them. Many visual releasers have become extremely exaggerated during evolution. A series of MAPs may now be performed more slowly, certain actions may be temporarily frozen, and the anatomical areas involved may acquire bright coloration or other special structures to make the releaser even more obvious. Obvious releasers are called **displays.** The peacock that spreads his beautiful tail feathers to attract potential mates provides a dramatic example of a display. Similar displays are common among other birds and among fish, reptiles, and mammals.

Visual displays are the best-known characteristics for most species of animals. Releasers may involve senses other than vision, however. They may, for example, also be sounds or chemical scents called **pheromones**.

Visual Releasers

Visual communication can best be illustrated with the classical example of the three-spined stikleback fish. Male and female sticklebacks are identical, except during the breeding season. In the spring, when breeding begins, the male acquires a "nuptial coloration." His neck and belly become bright red. The female's belly becomes swollen and distended with maturing eggs.

As males acquire their red bellies, they also begin delineating a small territory. There are always more male sticklebacks than available territories. This situation causes keen competition and frequent bouts of aggression among the males. Once a territory is acquired, the male begins constructing a nest. (See Figure 25–3.) He carries bits and pieces of aquatic plants

Figure 25-3

The courtship sequence of the three-spined stickleback. The male stickleback digs a depression in the substratum, which becomes the base of his nest. After nest construction, the male engages in certain MAPs. (a) He defends his territory and courts females. His red belly advertises his courtship desire and becomes a sexual releaser. (b) When a female approaches, the male performs his ritualized zigzag courtship dance. The female appeases the male by standing head up. The courtship sequence eventually positions the female in the nest of the male. (c) The male pushes at the posterior of the female, which causes her to release eggs. The male then chases the female from the nest and enters it himself to fertilize the newly laid eggs. After repeating the courtship sequence with several different females, the male begins his parental tasks of caring for the eggs. (d) Within the nest the male "fans" the eggs, increasing the oxygen supply for the developing embryos and dispersing their waste products.

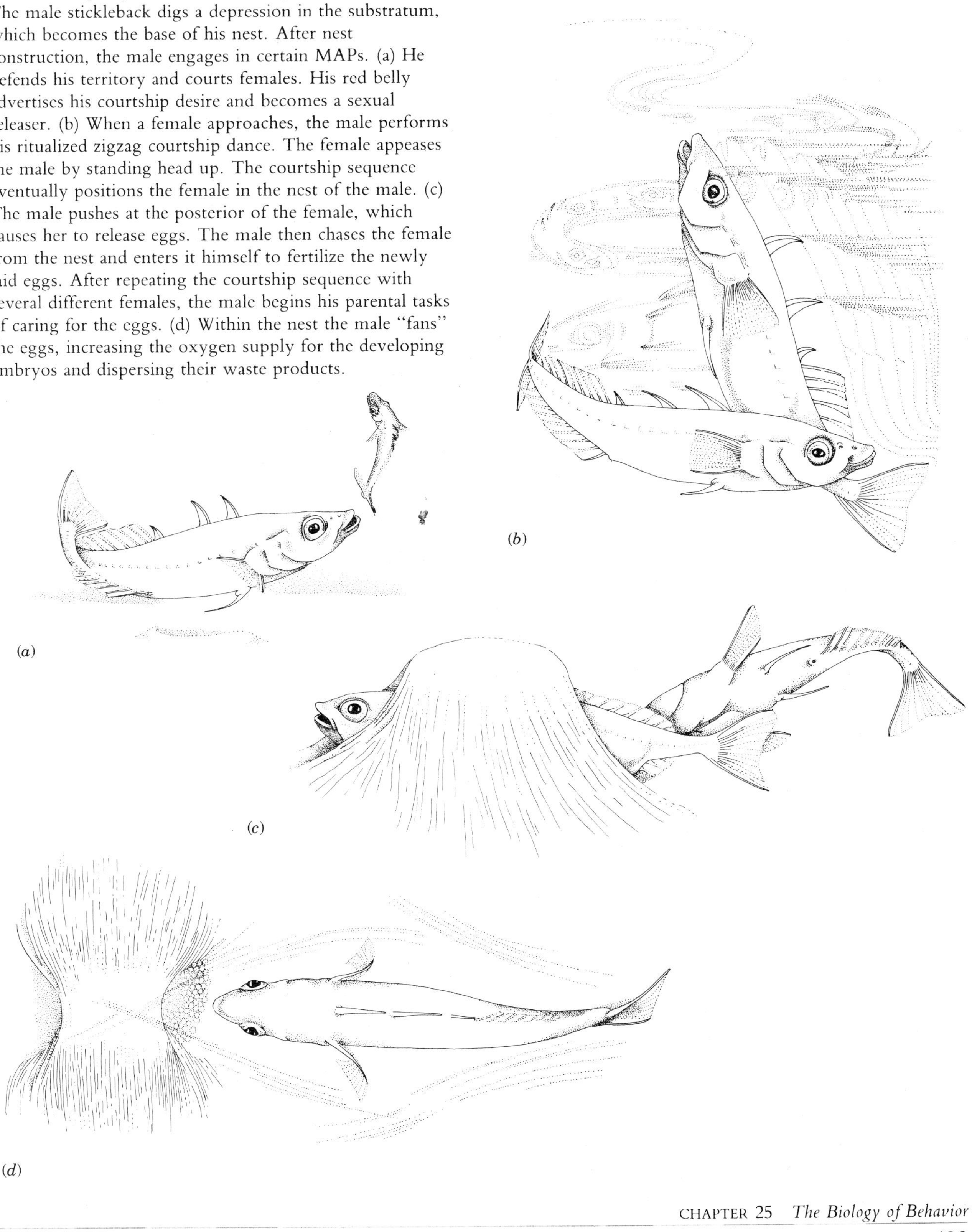

(a)

(b)

(c)

(d)

to the center of the nest, where he has dug a depression. After each deposit of plant material, the male glides over the nest, releasing a glue-like substance produced by his kidneys. That substance holds the vegetation together. When the glued plant material approximates a miniature haystack, the male forces himself through the accumulation, creating a tunnel with an entrance and an exit.

The male is constantly challenged by other males for possession of his territory. An intruding male is quickly recognized by his red belly. Threat, counterthreat, and tests of strength quickly determine the victor and owner of the territory. Previous possession ranks high in the final decision.

The male's activities motivate females to approach. Rather than attacking an approaching female, the male courts her. (Some of the MAPs involved in the courtship are shown in Figure 25–3.) The first response of the male is generally the performance of a zig-zag courtship dance (a display). The male approaches the female and retreats, approaches and retreats, creating a zig-zag motion. The female responds to these overtures by specific MAPs that are appeasing to the male, such as standing head up and remaining motionless. Failure of the female to submit to the male may cause the male to attack her.

The male leads the female to his nest. When she is positioned inside, he prods her posterior region with his snout. The prod is a specific stimulus that causes the female to spawn. She releases several hundred eggs. After spawning, the male becomes aggressive toward the female, chasing her away from the nest. As she retreats, the male quickly assumes her position within the nest and releases sperm (milt) over the eggs, fertilizing them.

This sequence of behavior is never rigid. Some MAPs occur more often than others, and many are repeated again and again. The female often interrupts the sequence at some point (even in the nest) and flees from the male. Most often, the male pursues the female and begins the sequence again by performing the zig-zag dance.

After the reproductive sequence is completed, the male may parol his territory, attack an approaching male, or court another female and repeat the entire sequence. Males continue to be sexually active for several days and may court several females. When the nest is filled with eggs, the male stops courting. His behavior changes abruptly, becoming parental. The male swims into the nest and, with fanlike movements of his fins, aerates the eggs until they hatch. His care of the developing embryos is never interrupted by renewed courtship with another female. Eggs hatch into free-swimming larvae that stay with the male and are protected by him. (It is a biological mystery that the eggs all hatch at once, beause they may have been laid and fertilized as much as a week apart.)

Why does a male attack an approaching male but court an approaching female? Obviously, the presence or lack of the red belly must be an important releaser. Models have been used to test this hypothesis (see Figure 25–4). Males in sexual readiness were presented with various models. One of the models was a life-sized stickleback as it might look outside the breeding season. Others were of fish that lacked specific details but had red bellies. The red-bellied models were attacked much more frequently than was the lifelike mimic without a red belly. Red, then, is a releaser that provokes attack. Other characteristics of the male seemingly go unnoticed.

Can the eliciting of behavior be so simple? It would be rare in the natural habitat for a male stickleback to see the color red anywhere but on another male, so in this case simplicity is efficiency. However, in this experiment, the severity of threat and consequent attack were a function of the position of the model. A model in a head-down posture provoked a more severe attack than did one with its head horizontal. Therefore, when one releaser was superimposed on another, it created a more intense response.

A similar series of experiments was carried out with models that simulated females. The lifelike mimic was used again as a control. Other models were a variety of fishlike shapes without detail but with swollen lower surfaces, suggesting the swollen abdomens of females (see Figure 25–4). The models with the swollen profiles elicited more courtship behavior (zig-zagging) from the males than the lifelike mimic did. This demonstrated that the swollen abdomen releases male courtship behavior.

Auditory Releasers

Sounds can also be behavioral releasers. You are probably aware that it is the female mosquito that bites. The beating wings of the female mosquito pursuing your blood create a distinct sound. This sound is not detectable in detail by human ears. To male mosquitoes, however, it is a sexual attractant that fa-

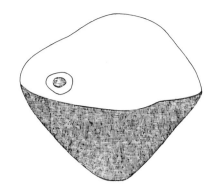

MALE MODELS

FEMALE MODEL

Figure 25-4
Stickleback models.

HIGHLIGHT

HUMAN VISUAL RELEASERS

Do members of the human species respond to one another by visual releasers? The answer is yes.

One of the first things a baby does socially is smile. (a MAP). The first time a baby smiles, mom, dad, brother, or sister smiles back. Each smile commands another, until a rigid stimulus-response chain develops. For the rest of life, each person will smile more frequently at a smiling face than at any other facial expression.

The dozen or so facial expressions studied by Charles Darwin are all examples of visual releasers and MAPs. The smile is a universal human gesture, signaling appeasement and friendliness. The grimace, a sign of anger or threat, is easily understood by all races. Children born blind have the same set of facial expressions as sighted children, although they have never seen a face.

The natural releasers that identify masculinity in most societies are tallness, broadness, wide shoulders, narrow hips, and facial hair. Femininity is identified by shortness, broad hips, breasts, and rounded faces.

cilitates mating. (Sounds of similar pitch also attract male mosquitoes, as is illustrated in Figure 25–5).

The songs of birds, perhaps some of the most obvious animal sounds in the environment, are extremely complex. The function of individual notes or trills is generally unknown. However, experiments have shown that the songs are extremely important in establishing territory and in advertising for a mate. Songs of birds of the same species that have neigh-boring territories may be slightly different and may function as the "signatures" of individual birds.

Birds of certain geographical regions sing a "dialect" of their species-specific sound that differs from the dialect of birds of another region. Therefore, in many instances, bird songs function as an isolating mechanism□. In general, birds instinctively recognize their own species' songs even before they are old enough to sing themselves. Many birds sing, however, only if they listened to other birds singing when they were nestlings.

Some of the most eerie and fascinating sounds of the animal kingdom are made by whales. The humpback whale and other species have been studied in their natural breeding grounds. Sound recordings made there reveal that the whales sing extremely complex songs that may last as long as thirty minutes—only to be repeated again and again, sometimes for days on end. (An album, "The Songs of the Humpback Whale," has been produced by the American Museum of Natural History.) Whales make low-frequency sounds that are audible to human ears. Wa-

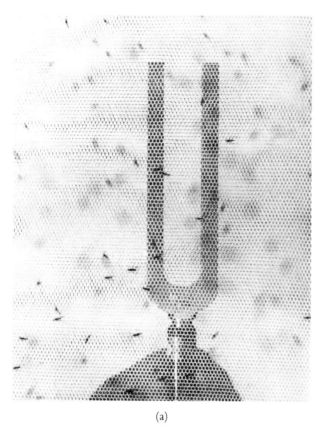 

(a)   (b)

Figure 25–5

Attracting mosquitoes. (a) A few resting mosquitoes ignore the prominent silhouette of a tuning fork. (b) When the tuning fork is activated, dozens of males converge on it. Male mosquitoes are attracted to females by the sounds of their wing beats. The tuning fork produces a similar frequency, and males eagerly approach. (Photos courtesy of Edwin R. Willis, Illinois State University.)

ter is a particularly good transmitter of sound. Whales may therefore hear other whales singing from distances of five hundred miles or more.

Pheromones

Chemical or pheromone, communication was first demonstrated in insects when the mating behavior of male cockroaches was shown to be released by a special chemical substance produced by the receptive female. Males that sense the presence of this sexual pheromone through their antennae begin searching frantically for the source—the female. They are attracted by the pheromone, not by the visual image of the female. Although their searching activities appear unordered, they seek increasing concentrations of the pheromone. As they get closer to the female, the concentration of the substance becomes greater. The tar-

get of the male cockroach is maximum concentration—the source of the substance.

The female silkworm moth *(Bombyx)* releases a sexual attractant called bombykol. Tests conducted in Tokyo demonstrated that a single female could attract males from a distance of 3 to 5 kilometers. The males have enormous antennae that selectively sieve the air to detect the sex pheromone. A similar pheromone has been demonstrated for the gypsy moth.□

Hundreds of pheromones in insect communication have been identified. They are classified according to function—sex pheromones, alarm pheromones, and trail pheromones, for example.

Pheromones also occur among vertebrates. In fact, probably all animal species produce chemical releasers. Unlike insects, vertebrates have not evolved special receptor organs to detect only pheromones. Instead, vertebrates sense pheromones through their

461

HIGHLIGHT

HUMAN PHEROMONES

The search for human pheromones is recent. The pheromones have not yet been identified, but there is considerable evidence that they are produced. These substances affect human behavior just as they affect the behavior of other animal species.

A number of human body structures seem to have no function. For example, axillary hair (hair in the armpits) and pubic hair seem to have no use whatsoever. They have been considered vestiges of ancestral fur coats. Close examination, however, reveals that these hairy areas (including scalps) contain specialized sweat glands called sebaceous glands. The secretions from these glands have more proteins and oils than do the watery fluids that pass from other sweat glands. They are even more abundant and more oily in males than in females. The secretions cling to hair, creating scent traps. The secretions of each hair patch have a different smell. It is also significant that the axillary, pubic, and head hair patches of humans—upright, naked apes—are located where scent can be dispersed by currents of air. If humans walk or run without clothes, eddies of air rush between the legs, under the arms, and across the head, increasing the dispersal of the scents.

It has been shown experimentally that women are particularly sensitive to the smell of musk, a common pheromone of many vertebrates that is often used as a territorial marker. Sensitivity to musk is strongest in sexually mature women at the time of ovulation. This suggests that, in the human past, males produced a sex pheromone that stimulated receptive females. The recent trend toward musk aftershave lotions thus echoes an ancient practice. For the past three thousand years, humans have anointed their bodies with musk substances, often extracted from deer or boars. (But beware the man wearing musk aftershave lotion, because he will probably attract more flies than females. Flies respond to musk because it leads them to a variety of mammals on which they feed and lay eggs.) Prepubescent females are not sensitive to musk. They prefer sweet, fruity smells—smells not preferred by sexually mature women.

Women who live together develop synchronous menstrual cycles. The original data, collected on college campuses, showed that the menstrual cycles of young women living closely together in dormitories and sororities changed from an average of eight days apart to only five days apart. A 'synchrony pheromone'' is thought to be secreted with underarm perspiration. Some women seem to be able to dominate or drive other women's menstrual cycles. A dominant woman's underarm perspiration was painted at random on the upper lips of women volunteers (half were painted with alcohol, half with alcohol and underarm secretion). Within four months, the beginning days of the menstrual cycles of those who received the secretion were reduced from 9.3 to only 3.4 days apart. The pheromone may be related to female sex hormones or to their metabolic by-products. Discrimination tests among men and women seem to suggest that women are sensitive to the compounds and men are not. When the evidence of synchronous menstrual cycles was related to freshman biology students, many of them doubted it. Three different groups of female students therefore decided on their own to keep personal records of when they menstruated. Each group lived together in a dormitory complex. The three groups all reported that within three months of cohabitation they were menstruating within a few days of one another.

Other observations suggest that women who date regularly or who wear or work with musk products have shorter menstrual cycles than other women. This suggests that male pheromones may have some influence on women.

---

nasal (olfactory) membranes. That is, they smell the pheromones. Olfactory membranes allow the detection of many chemical substances□, including species-specific pheromones. For this reason, pheromone communication among vertebrates is often called olfactory communication. Some fish, however, are known to taste pheromones as well.

Most mammals recognize potential mates in part by sex pheromones. In many cases, the female signals her receptivity by vaginal secretions that act as pheromones. **Copulins** is the name usually given to these secretions. It is common to observe male monkeys and apes in a zoo testing the receptivity of females first by visually inspecting the genitals and then by tasting and smelling the vaginal secretions.

The same substances are produced by the vaginal membranes of dogs, cats, and many other mammals. The pheromone produced by female dogs seems particularly potent. Females in **estrus** (heat) attract large numbers of males from considerable distances. If an absorbent tissue is pressed to the vulva of an estrous female and then thrown to males, the tissue commands all of the males' attention. The female that produced the pheromone is completely ignored if she no longer possesses the smell of the pheromone.

Many pets mark familiar objects with pheromones. Male gerbils have midventral (midabdominal) glands that produce a pheromone used to mark territory and familiar objects within the territory. The more dominant the male, the larger the gland and the more pheromone produced.

CONCEPT SUMMARY:
TYPES OF RELEASING STIMULI

TYPE OF STIMULUS	EXAMPLES
Visual releasers	Body form and pigmentation in fish; peacock tail displays; human male and female secondary sexual characteristics
Auditory releasers	Bird songs; frog calls; sounds produced by insects
Chemical releasers (pheromones)	Insect sex pheromones; fish alarm pheromones; copulins of mammals

---

THE UMWELT

People often assume that every animal species perceives the world in exactly the same way. This is not true. Each species has its own perceptual world, and it is often quite different from that of other species. A particular species' perception of the world is referred to as its **Umwelt** (a German word meaning "environment").

An example of contrasting Umwelts will illustrate this concept. The Umwelts of humans and dogs are quite different. For example, humans hear sounds in the range of 20 to 20,000 cycles per second. (Few people hear the entire range, and the range narrows with age.) Dogs perceive sounds well above 20,000 cycles and therefore hear things that humans cannot hear. A human who blows a special dog whistle hears nothing. But a dog hears the sound and responds by running toward it. The olfactory sense of humans is reasonably good, but that of dogs is much better. Humans are visually oriented; dogs are olfactorily oriented. Are there other differences in Umwelts? Dogs are color-blind, so their visual Umwelt is only black and white; humans perceive color. Humans have better three-dimensional sight than dogs. The Umwelts of dogs and humans therefore are much different. That is expected, of course, because each species has a different evolutionary history and is adapted for a different existence.

Stimulus Filtering

Within the Umwelt of an organism are far too many stimuli for the organism to respond to. Each organism therefore is capable of responding to only the stimuli that it perceives, a concept known as **stimulus filtering.**

As you read this passage, your nervous system is processing considerable information. You are comprehending the order of words in each sentence and then synthesizing the ideas of each paragraph. In terms of stimulus filtering, you are extremely busy. If you are taking notes, you are working up to full capacity. If you are absorbed in the ideas on this page, you are ignoring unconsciously, or by stimulus filtering, many other kinds of environmental information. If you stop reading for a few moments, you may suddenly realize that you are too cold or too hot or that you hear the refrigerator motor or voices in the hallway. All of this information was filtered out while you were reading.

Motivation

Another factor that may determine an animal's readiness to behave is motivation—the fluctuation in the physiology of an organism that influences its behavior. Psychologists use the term *drive*.

The nest-building and reproductive behavior of the three-spined stickleback fish occurs only during the breeding season. It is said that these fish are not "motivated" at other times. Almost all research has shown that motivation is related to the kinds and amounts of hormones circulating in the blood. Male sticklebacks in nature show sexual activity only when their bodies are producing certain hormones.

The production of hormones is often stimulated by environmental events, such as day length. As days get longer in the spring, the hypothalamus (part of the brain, connected to the pituitary gland) is stimulated to manufacture precursors to many hormones. Male sticklebacks display aggressive, nest-building, and courtship behavior at any time of the year if they are given injections of the male sex hormone testosterone.

If the gonads (testes and ovaries) of animals are removed, most of the animals lose their motivation for reproductive activity and related behavior, such as aggression and territoriality. Castration is often performed to tame such animals as horses, bulls, boars, and even cats. Replacement of the proper hormones often restores the missing behavioral patterns. For example, if a male chicken (cock) is castrated, it will no longer crow. If it is castrated at a young age, it will not grow the large comb typical of males. If the castrated cock (called a capon) is given injections of the male hormone testosterone, it will crow and grow a comb.

These examples illustrate that hormones affect behavior. Behavior can also affect hormone production. Courtship of ring doves involves a bowing-cooing display by males. As the female becomes motivated, she starts building a nest and eventually she lays eggs. The female does not have to participate in the courtship, however, to build a nest and lay eggs. If she sees a courting male through a glass window, her hormone production is equally stimulated.

Nature versus Nurture

The terms *instinct, innate, programmed,* and *preprogrammed* are frequently used in describing behavior. Such terms refer to the idea that much behavior is an expression of the genetic composition of a species. Other terms often used in describing behavior include *learned, acquired,* and *nurtured.* These terms imply that behavior is acquired through experience in the environment. This dichotomy about behavior has created a debate called the nature-nurture argument. *Nature* implies that behavior is natural to the species. *Nurture* implies that it is guided by experience and learning.

For scientific purposes, the argument is settled. Exclusive reliance on either point of view is incorrect. The truth is a compromise. Some of the behaviors of a species are learned. Others are expressions of genes or instinct. An ethologist has convincingly shown that the nut-burying behavior of red squirrels does not have to be learned. The first time a squirrel has the opportunity to bury a nut, it does so perfectly with a complicated sequence of MAPs□. However, red squirrels have to learn the behavior of opening nuts. Each experience doing so makes them more proficient.

As a general rule, learning probably plays a greater role in the behavior of birds and mammals than of lower vertebrates, such as fish. Obviously, learning plays an important role in human behavior, but it would be naive to assume that humans have no instincts.

Imprinting

Perhaps the best definition of **imprinting** is that of a behavior acquired or learned at some specific time. In short, imprinting is a phenomenon that incorporates both nature and nurture. It involves sensitive, or critical, periods during development or during an individual's lifetime. These periods are the times when a particular experience will encourage the development of a specific behavior. If a period is missed, the animal will not develop the specific behavior associated with it even if the experience is repeated at a later time.

Konrad Lorenz formalized the ideas about imprinting when he discovered that geese, shortly after hatching, would follow a human as readily as they would follow the mother goose. The more the young geese followed the human, the stronger their attachment to humans. In fact, many of them ignored other geese entirely and even sought humans as sexual companions when they were mature. Lorenz referred to this phenomenon as sexual imprinting.

Many pets become sexually imprinted on humans. Canaries that are "hand" raised may show the normal courtship and copulation behaviors when adult but may repond sexually only to human hands. Most dogs are socially imprinted on humans and often attempt to copulate with them by mounting their legs.

Research with dogs has shown that the critical age for dog socialization is about six to eight weeks. If you want a lovable companion and friend, your dog should be in the company of humans during this critical period. The critical period for cats is ten to twelve weeks. For chickens it is one day after hatching.

TERRITORY

The acquisition and defense of space by animals is known as **territoriality.** Most territories are fixed. That is, they are in the same general location and stay about the same size (see Figure 25–6). However, to some degree, they constantly fluctuate in size. Territories may be occupied by only one individual, by two, or perhaps by many, as is the case with harems, herds, prides, and troops. A pride of lions may consist of a dozen or more members that occupy the same territory. Robins, red-winged blackbirds, squirrels, and many other species maintain fixed territories.

The territories of some animals are moving, or floating (see Figure 25—7). For example, troops of baboons defend their sleeping trees and the areas in which they roam for food. As they move from sleeping sites to feeding ranges, their territorial requirements move with them, but their movements throughout the day cause their requirements for space to change. When they approach a waterhole to drink, all territoriality vanishes. They tolerate individuals from other troops only at waterholes.

Finally, all organisms, whether occupants of a territory or not, have an invisible form of individual territoriality called individual distance. Each individual requires and defends a critical ring of space around itself. The amount of individual space varies, depending on whether the animal is dominant or subordinate, male or female, mature or immature, and so on. It is often the violation of individual distance that causes one cow in a herd to charge and butt another cow or one sea lion in a harem to bark violently at another sea lion.

The sizes and functions of territories vary greatly. For many animals, the territory may be quite large, including shelter, a food source, room for a mate to be courted, and room for care of offspring. Among these animals are many fish, lizards, songbirds, and small mammals (such as raccoons, rabbits, and squirrels), as well as lions, wolves, and baboons. The territory for other animals may be smaller, acting only as a nest site. Food and other necessities are obtained elsewhere. This type of life-style is typical of many species of colonial nesting birds, such as gulls, terns, gannets, and flamingos.

There is thus no universal way of using a territory. The defended space is used differently by different species. Sex and reproduction are more frequently involved with territoriality than is any other kind of behavior.

Territorial Marking

If territories are to be defended, they must have rec-

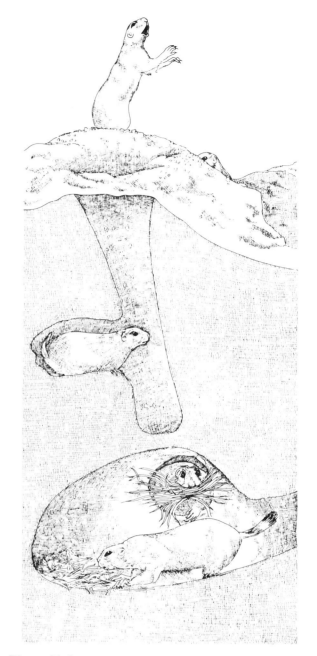

Figure 25–6
A fixed territory. The fixed territories of prairie dogs are a common sight in the plains of the United States. They often border one another in large clusters called prairie dog towns. The boundaries of each territory are rigid throughout the year, although the inhabitants may change from time to time. Typically, there are one male, several females, and several young pups. Each territory has a pronounced mound near the center. It is the entrance to the underground burrow system.

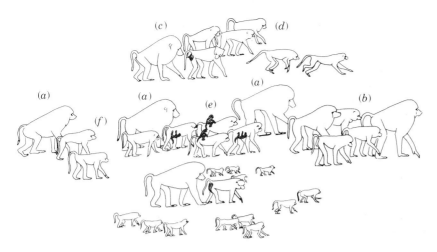

Figure 25–7

Floating territories. It is common for many animals that form troops or herds to have territories that move with the social group. Here a troop of baboons moves as a cohesive social unit, protecting its territorial needs wherever it goes. However, all territoriality disappears when the baboons visit waterholes, because this is a resource used by many animal species at any given time. The positioning of the members of this troop of baboons is typical. (a) Dominant males protect the core of the troop or bring up the rear. (b) Younger and less dominant males lead. (c) The most dominant males consort with estrous females. (d) Juveniles play around the periphery. (e) Females with young are better protected than (f) females without young.

ognizable boundaries. All species mark the perimeters of their territories.

Territorial boundaries can be visually marked by trees, meadows, rocks, streams, rivers, and outcroppings. Most animals mark them by scent with special **pheromones**, which are usually secretions from specialized glands. For example, hamsters rub their flank glands on the walls of their burrows, nests, and other surroundings in their territories. The midventral glands of gerbils, the chin glands of rabbits, the perorbital glands of deer, and the cheek glands of elephants are used in a similar fashion. Many mammals have glands that produce musk, a distinctive odor in small-mammal houses of zoos.

Functions of Territories

There have been many hypotheses to explain the adaptive functions of territoriality. Among some of the more popular are the following. Territories (1) provide for an unequal distribution of resources (more for the dominant, less for the subordinate); (2) reduce the time necessary for aggression; (3) regulate population size; (4) encourage efficiency in exploiting the environment, because they are familiar areas; (5) deter predators (again, because of familiarity); (6) ensure a food supply (where applicable); (7) reduce physical contact among species members, reducing the spread of diseases and some parasites; and (8) advertise reproductive capability and stimulate it in others. There is considerable evidence to support each of these contentions.

One of the more obvious adaptive functions of a territory is selection of the "best of fit." Although the term is impossible to define, it suggests that individuals holding territories are better, in a biological sense, than those not holding territories. Perhaps the more fit individuals have larger territories; perhaps they are bigger or more colorful, make louder noises, make more noise, look fiercer, perform sequences of MAPs more skillfully because they have more experience, and so on. In the final analysis, the factors that make up fitness are inherited. Therefore, the most fit pass their biological advantages on to their heirs.

HUMAN TERRITORIALITY

Many parallels exist between the territoriality displayed by other animal species and the possession and protection of property by humans. But do humans have a "territorial imperative," as suggested by ethologists Konrad Lorenz, Desmond Morris, and Robert Ardrey? A definitive answer is difficult, but observation of human behavior supports this hypothesis.

During human evolution, territoriality had an obvious advantage. If humans had not evolved this instinct, it is doubtful that they would have survived. Early humans moved about in small bands or groups, frequently setting up temporary camps or territories. These areas became home bases for hunting sojourns extending into nearby ranges. Campsites were well-protected areas that contained food stores, shelter of some sort, and often water. Fossil evidence suggests that lakeside or riverside campsites were preferred. The residing tribe defended the area of the campsite, its resources, and its occupants. The infants and women at the sites were secure. There was obviously competition with other human tribes for these campsites, and they were defended against human and other animal intrusion.

What about humans in modern societies? Are we territorial? Do we defend and protect our homes, food stores, family members, and personal belongings? Certainly we do. But is our defensive behavior learned from our culture, or is it a biological characteristic of being human? Research comparing modern and primitive cultures indicates that all humans, regardless of culture, display the same protective behavior—which is called territoriality. Obviously, then, it must be at least partly biological.

Like animals of all other species, humans also have individual space, territory that goes with them everywhere. Several levels of individual space exist. First is intimate space—space reserved near our bodies for intimate relations with other humans, frequently involving physical contact. Intimate space is usually reserved for parents, siblings, and mates. Others who attempt to intrude are discouraged, often physically. Second is casual space—space extending around us about an arm's length. This is the ring of privacy we reserve for ourselves while talking to friends and acquaintances of our own peer groups. Last is social space—a larger space around us for interacting with someone not in our social class, perhaps the biology professor or the minister or counselor. This distance is usually six feet or more from our bodies.

You can demonstrate and observe human territorial behavior with one or more of the following exercises:

1. The next time you go to the library, enter the stacks and see how close you can get to someone. Select a complete stranger. If you get much closer than the average casual distance accepted by most students, the person on whom you have intruded will probably move away from you or even leave the stacks because of your presence.

2. The next time you have coffee with a friend, sit directly across from the person. While talking, carefully but deliberately move objects (ashtray, cup, spoons, napkins, and so on) away from you and closer and closer to your friend (this works best if you slide the objects inch by inch across the table). Usually, the intrusion will elicit predictable territorial defense behavior. Your friend may become flustered or angered and may talk louder than before, possibly unaware of the cause of this change in behavior.

3. Go to a restaurant or cafeteria. Select a table at or near the center of the room. Sit there as long as you can, doing whatever you like. How do you feel? Uncomfortable? Probably. Get up and take a seat along the wall. Position your chair so you face the wall. How do you feel now? More uncomfortable? Probably. If others are sitting in similar positions, notice how often they turn their heads to see what is going on behind them. Finally, turn your chair so that your back is to the wall and you are facing most of the room. How do you feel now? Comfortable? Probably. What you have just demonstrated is human preference, in any strange room, for a seat along the wall, the back against the wall.

4. Go to a waiting room early in the morning before anyone else arrives. (Try the waiting room of the dean's office or your department's main office.) Take a corner seat. (You probably would have anyway.) Where does the next arrival sit? Probably not as far away from you as possible but at least half as far as possible. How about the third arrival? The third person will probably attempt to sit about halfway between you and the second arrival. Get up and move closer to either of the others. What happens? The closer person may find an excuse to get up and move, creating the same space barrier that originally existed between you. However, as more and more people come in, spatial relationships are no longer maintained.

AGGRESSIVE BEHAVIOR

Aggression is most frequent when territories are being established, often at the beginning of the breeding season. The first male robins and red-winged blackbirds returning to compete for territories each year are particularly aggressive.

Threat

The prelude to aggression is **threat,** which is performed in a similar fashion by all animals, from fish to humans. During threat displays, animals make themselves look as large and conspicuous as possible.

Threat is well demonstrated by dogs and wolves (see Figure 25–8). They raise their bodies upward by stretching their legs, which gives them a fraction of an inch to several inches in increased height. Simultaneously, they hold their tails straight out or up, which makes them look larger. Their hair becomes erect (a response called **piloerection**), but not all over the body. Typically, the hair that becomes erect is that along the shoulders, on the back of the neck, and along the full length of the spine to the tip of the tail. This produces an illusion of largeness. The head is held up, ears erect, mouth open, and canine teeth exposed. Low growling may be heard.

An element of threat common to all animals is that the threatener stares at its opponent. Four kinds of behavior may be displayed after threat: confrontation, appeasement, diversion, and displacement.

Confrontation

Probably the rarest outcome of mutual threat is **confrontation**—physical combat between the animals, often erroneously called a fight. Confrontation is often limited to the times of territorial establishment and mating (which are frequently concurrent).

When confrontation does occur, it is usually a test of strength, with little or no physical damage to the participants. It is uncommon for animals in nature to hurt or kill one another. It is true that animals may receive minor wounds and even bleed during a confrontation, but the wounds are rarely life-threatening.

Dogfights, for example, are tests of strength. The dogs commonly align laterally, head to tail, showing threat and occasionally smelling the genital region of the opponent. During confrontation, there is a great deal of growling and display of teeth. Confrontation

Figure 25–8

The posture of a dominant wolf or dog as it threatens a subordinate. The dominant's ears and tail are upright. The animal stares. Its mouth is open, its teeth are exposed, and it emits a low growl. All are common elements of threat. The threatening individual makes itself as large and conspicuous as possible, a common indication of threat throughout the animal kingdom.

among dogs, wolves, and all other carnivores is an attempt by one animal to throw the other to its side on the ground. The stronger animal, using its paws, often jumps on top of the weaker one, forcing it to the ground. Grasping the nape of the neck in the jaws is another act of dominance. It is common for a dominant wolf to gently grasp the nape of the neck of a subordinate wolf as a gesture of dominance.

It is generally agreed that only humans use weapons in confrontations. The assumption that other animal species possess and use weapons is an anthropomorphism. For example, the enormous antlers carried by many species of deer may appear to be menacing weapons. But they are used in ritualized confrontations in which the deer lower their heads and lock antlers to give them leverage to push. If the antlers are freed, the deer repeat their charge, locking antlers and pushing again. (Some deer become so entangled antler to antler that they are unable to free themselves and die of starvation.) Deer could attack the unprotected flanks of opponents and split them open with their antlers, but this behavior is rarely observed. Horns and antlers act as signals of rank, with the older, more dominant males having the largest ones.

They are often also signals of gender, since the females of most species are hornless.

Appeasement

More frequent than confrontation is **appeasement,** the antithesis (reverse) of threat. Appeasement postures signal "I do not threaten." When they are the result of mutual threat, they add the message "I give up." The postures are attempts by animals to make themselves look as small and inconspicuous as possible so as to turn off threat.

In the appeasement posture of a dog or wolf (see Figure 25–9) all elements of threat are reversed. The animal slouches to the ground, legs flexed. Often, the chest and belly are pressed to the ground. The tail is tucked between the legs and held against the lower abdomen. The head is dropped, the ears are lowered, and the hair becomes sleek. Appeasement postures are universal from fish to humans, although their forms differ from species to species.

Jane Goodall's study of chimpanzees in Africa has shown that chimpanzees often extend the open palm as a gesture of appeasement while lowering the body. The palm may be touched or grasped by a more dominant individual. In fact, chimpanzees perform a hand-shaking ritual as an appeasement gesture. They bow and kiss as appeasement gestures and when greeting one another. They also smile in appeasement.

Diversion

Appeasement behavior turns off threat behavior and the hostility of the dominant animal (albeit slowly).

Subordinate animals not only turn off threat but often turn on behavior that is incompatible with aggression. That behavior confuses the purpose of the threat and creates a **diversion** that stimulates the dominant animals to perform some other behavior.

It is common for subordinates to perform juvenile behavior in an attempt to stimulate parental behavior from the dominant animal. A dog lies down in front of a more dominant dog, rolls over, and exposes its belly. This position is the same as the position that young puppies display when inviting grooming from their mothers.

Many animals feed their young by putting food morsels into their mouths. Songbirds, for example, feed insects to their hungry young. Parental carnivores eat enormous amounts of food and then regurgitate it later for their offspring. Young carnivores beg for food from a parent by pushing their muzzles against the edge of the parent's mouth. It is common for adult carnivores to perform juvenile food-begging displays as a diversion. Adult robins squat before other adults, flap their wings, and gape in the ritual of juvenile food begging. Likewise, adult carnivores may lower their bodies and push their muzzles against the mouth of the dominant as a gesture of diversion.

Displacement

Mutually threatening animals near confrontation may be caught between two conflicting drives—to attack and to flee. It is impossible to do both. If the drive to perform one act, say to attack, is essentially equal to the drive to perform the other, then neither behavior is displayed, because they inhibit each other. How-

Figure 25–9
Appeasement. A wolf or dog displaying appeasement lies before a threatening and more dominant wolf or dog. In appeasement, the reverse of threat, the animal makes itself look as small and inconspicuous as possible. It not only appeases by the inconspicuousness of its body but also performs a ritualized food-begging gesture toward the muzzle of the dominant as a form of diversion to thwart any potential attack.

ever, the neural conflict releases the inhibition on other, less urgent behavior, which is then performed.

The behavior displayed during such conflict situations, called **displacement** behavior, comes in a variety of forms. Chickens may suddenly give up their defense and start pecking at the ground as if eating. Stickleback fish may retract their spines and begin fanning as if incubating eggs. Primates may interrupt a possible confrontation to pick at their skin with their fingers or teeth. Displacement behaviors are thus normal behaviors that seem out of place during aggression.

CONCEPT SUMMARY:
RESPONSES TO THREAT BEHAVIOR

TYPE OF RESPONSE	COMMENTS
Confrontation	Combat between two animals—a test of strength, usually with little or no physical damage; rarest outcome of mutual threat
Appeasement	Animal makes itself look as small and inconspicuous as possible, signaling "I do not threaten" or "I give up"; much more frequent than confrontation
Diversion	Subordinate animal performs a behavior incompatible with aggression, confusing original purpose of threat
Displacement	During mutual threat, the drives to attack and to flee are both inhibited; neural conflict releases inhibition of another behavior, which is performed

---

DOMINANCE HIERARCHIES

The elements of aggression—territoriality, threat, confrontation, appeasement, diversion, and displacement—are obviously complex. However, their interaction creates stable communities of organisms. The stable group is usually referred to as a **dominance hierarchy.**

Dominance hierarchies were originally called **peck orders,** because the initial observations were made

with chickens. Chickens peck one another as their main form of confrontation. When chickens are grouped together, a dominance hierarchy emerges after an interval of time (see Figure 25–10). The most dominant individual (called alpha after the first letter of the Greek alphabet) pecks but is not pecked. The most submissive, or least dominant, individual (called omega after the last letter of the Greek alphabet) is pecked by all but pecks none. Specific characteristics of individuals are associated with their dominance position. For example, if the comb and wattle of a dom-

Figure 25–10

Peck orders. Dominance hierarchies, originally called peck orders, are common among most animals. In the hierarchy of this small flock of chickens, the most dominant chicken, the alpha, is at the top. Next in order are two equally matched beta individuals. Lower in the hierarchy is gamma. The most subordinate is omega. Alpha does most of the pecking. Omega does none but receives the majority of pecks from those above it in the hierarchy. What would happen in this particular hierarchy if beta 1 became dominant over beta 2?

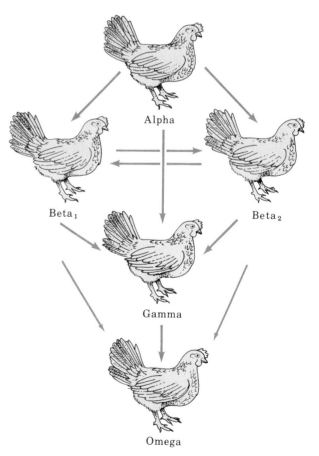

Alpha

Beta₁   Beta₂

Gamma

Omega

inant hen are removed, she loses her position in the peck order. Once the order is established, the frequency of pecking decreases, more food is consumed, the chickens gain weight, and the hens lay more eggs.

Primate dominance hierarchies have been the most extensively studied hierarchies. In general, there is always a dominant male, the control male, who dictates the movements of the primate troop. Whenever he moves, every member of the troop follows. The dominant individual may be assisted in his role by one or more high-ranking males. Together they form a high-ranking group. Younger males are always subordinate to the higher-ranking males but often have more sexual freedom, because they are not considered sexual rivals.

Female primates often have their own dominance hierarchy within the troop, but all are subordinate to the most dominant males. Females in estrus are more dominant than females not in estrus. Those with offspring are more dominant than those without them (see Figure 25–7).

Youngsters also have a dominance hierarchy. The most dominant are the offspring of the most dominant females. They acquire their status because they are assisted by their dominant mothers any time they scream or whimper during encounters with other youngsters. Parental care lasts for several years for most primates. It is likely that individuals learn their rank within the troop during this time.

SEXUAL BEHAVIOR

Various aspects of sexual behavior have been described in other chapters. The examples in those chapters illustrate the adaptations that have come about to ensure the fertilization of gametes. Here the focus is on the behavior that brings the sexes together prior to fertilization. The behavior of sexual reproduction can be divided into four phases: pair formation, courtship, precopulatory behavior, and copulation.

Pair Formation

The first behavior displayed by organisms during or near the time of reproduction is **pair formation.** During pair formation, two important pieces of information are exchanged: gender and sexual receptiv-

ity. Sexual releasers relay this information. For example, crickets chirp, birds sing, mosquitoes vibrate their wings, sticklebacks display their red and distended bellies, red-winged blackbirds display their red wing patches, dogs secrete vaginal pheromones, and peacocks display their enormous train of feathers as sexual releasers.

Pair formation usually occurs quickly if the male and female look different, because gender is quickly determined. For example, only male sticklebacks have red bellies, only peacocks (not peahens) have beautiful trailing feathers, and only male deer have antlers. It is difficult for humans (but probably not for the species) to distinguish males and females of some animals, because they look alike. Examples are doves, lizards, rabbits, and gulls. Sexual releasers for these animals include a variety of behavioral displays, sounds, and pheromones.

At the end of pair formation, a bond exists between a female and a male. A mate has been chosen.

Courtship

Once a pair of organisms has determined sexual readiness, courtship establishes behavioral synchrony (a meshing of behavior) between a male and a female in preparation for mating. Generally speaking, vertebrate males are larger and more aggressive than females. They also maintain and protect a territory. Although the male usually works at attracting a potential female mate, the female must approach a larger and more aggressive animal. Thus the initial phases of courtship involve considerable displays of dominance and threat by the male. They are reciprocated by appeasement and diversionary gestures of the female. The exchange of threat and appeasement between male and female is slowly replaced by submission and appeasement on the part of both.

The initial state of threat and aggression is overcome by diversionary behavior, which often involves parental and juvenile responses. In the courtship of domestic chickens, the cock pecks at the ground for food. The hen approaches and pecks at the same spot. This is exactly the relationship between a hen and her chicks, as she shows her young where and at what to peck. The female robin not only crouches and begs for food from a prospective mate but also makes sounds identical to those of a young robin begging for food. The male then feeds the female.

Precopulatory Behavior

Following courtship, a brief bout of precopulatory behavior ensures physiological synchrony of a mated pair in preparation for fertilization of gametes. Sexual releasers, which play a crucial role during pair formation and courtship, play a minor role now.

Precopulatory behavior involves close physical contact of the mated pair and mutual sexual stimulation. For example, a male dog rapidly licks the vulva of an estrous female. The licking and tasting of the vaginal discharge is sexually stimulating to the male, and his actions sexually stimulate the female.

As part of precopulatory behavior, it is common for females to position themselves in a way that will facilitate copulation. Female mammals of many species (dogs, cats, horses, and monkeys, for example) firmly plant their legs, arch their back, and deflect their tail from their vulva to accommodate intromission and the weight of the male during copulation. This posture is called **lordosis.**

Copulatory Behavior

For most animals the act of **copulation** is brief, being intended only to fertilize gametes. Both male and female are behaviorally indifferent to each other. If you have had hamsters or gerbils as pets, you have probably witnessed how quickly these creatures copulate. Copulation among primates may involve only a half a dozen or so pelvic thrusts from the male.

SUMMARY

1. Ethology, the biology of behavior, is a relatively new branch of scientific investigation.

2. The units of behavior are called modal action patterns (MAPs). Sequences of MAPs are comparable to the words of a sentence conveying a discrete message.

3. A complete catalog of MAPs for any animal, referred to as an ethogram, is species-specific.

4. Combinations and sequences of MAPs stimulate, or release, the performance of a behavior by a receiver. These MAPs are called releasers.

5. In many instances, releasers have become extremely exaggerated during evolution and now are performed more slowly or are temporarily frozen. Such releasers are called displays.

6. Organisms respond to many types of releasers. Releasers may be visual displays, sounds, or chemical substances called pheromones.

7. The sequence of behavior involved in the courtship of the three-spined stickleback fish is primarily a sequence of visual releasers.

8. Bird songs are obvious examples of auditory releasers. Perhaps the most unusual sounds made by any animal are the eerie sounds produced by whales.

9. Pheromones are discrete substances left in the environment as chemical messages. They were first discovered among insects (sex pheromones) but are now known to be equally common among vertebrates. In fact, all species probably produce one or more of them. Pheromones are classified according to function, such as sexual, alarm, or trail.

10. The perception of the world by each different species is referred to as its Umwelt. The world perceived by humans is quite different from the world perceived by dogs.

11. There are too many stimuli within each species' Umwelt for an organism to respond to all of them. The organism therefore responds to what is appropriate at any one time. The ability to respond to selected stimuli is called stimulus filtering.

12. An animal's readiness to behave is determined by its motivation—from the kinds and amounts of hormones circulating in the blood.

13. The nature-nurture argument was about whether all behavior is genetically programmed (nature) or learned by experience (nurture). The argument is generally considered settled now. Behavior is usually a combination of both factors.

14. Imprinting is an animal's learning of a specific behavior response at a genetically determined time. The exact time is known as the critical period.

15. The defense of space by aggression is known as territoriality.

16. Territories are marked by visual landmarks and by pheromones.

17. Threat is the prelude to animal aggression. A threatening animal makes itself look large and conspicuous.

18. Confrontation—actual physical combat between animals—may occur after threat.

19. Appeasement behavior occurs more frequently

than confrontation behavior. Appeasement signals are the opposite of threat signals.

20. Diversionary behavior is also displayed to avoid threat. It occurs when an animal turns on a behavior that is incompatible with threat.

21. Animal aggression and territoriality establish socially stable communities of animals. The communities are known as dominance hierarchies or peck orders.

22. Sexual behavior is divided into four phases. (a) Pair formation displays sexual releasers that help identify sex and sexual receptivity. (b) Courtship involves behavioral synchrony between a male and female in preparation for mating. (c) Precopulatory behavior ensures physiological synchrony of a pair prior to mating. (d) The act of copulation is the last phase.

Displays are everywhere common among animals as a means of communication. **(a)** The gorilla is hunched, arms folded, mouth open, grunting as it stares at other gorillas. This behavior is a mild form of threat—this individual is displaying his dominance to other gorillas in the troop. Many animals display with color and are often said to be in display dress. **(b)** The peacock spreads its beautiful tail feathers as it struts and stomps its feet. This peacock is performing a sexual display to attract pea hens who watch his display from the cover of the jungle. If a pea hen is attracted, she will approach and after a brief courtship the peacock will mount the female and inseminate her (photo by authors). **(c)** The display of the male frigate bird is similar. Males have very large vocal sacs that are bright red. Males inflate these vocal sacs to make them more conspicuous, and as they are slowly deflated they make distinct sounds—both the sexual display of the vocal sac and the sounds attract females for mating (photo by Boyce A. Drummond III, Illinois State University).

(c)

(a)

(b)

Aggression among animals is best defined as defensive behavior among individuals of the same species. It begins with a threat. Threat is a common display, and it is performed similarly among all animals; the animal makes itself look as large and conspicuous as possible. Birds may elevate feathers and or raise their wings as the owl **(a)** is doing. Many animals threaten with facial displays as is illustrated by the bearded lizard **(b).** The lizard threatens by opening its mouth and lowering a spiny fold of skin under its chin to create a "beard" that makes it look larger and fiercer. The lion-tailed macaque **(c)** threatens by opening its mouth wide to display its sharp teeth and by shrieking.

Following threat, confrontation may occur. The impala **(d)** when confronting one another lower their heads, lock horns, and push in a test of strength; the contest is a ritual and physical injury is rare. It is a test of genetic fitness.

(a)

(b)

(c)

(d)

Agression as a test of strength may also be seen among fish **(a).** Another response to threat may be appeasement which communicates, "I do not threaten you". Appeasement is the reverse of threat—the performer makes itself look small and inconspicuous. **(b)** Juvenile gulls have dark plumage in comparison to adults. They must frequently display appeasement to adults; this juvenile signals appeasement by lowering the body. Many animals prostrate the body, lower the head, cover the eyes and hunch downward as signals of appeasement, actions illustrated by an Allen's monkey **(c).**

(d) There are many other types of social behavior that are often confused with aggression. The bobcat chasing a mouse is not aggression but rather the behavior between predator and prey. The bobcat obtains its food in this manner; his behavior does not necessarily improve his genetic fitness among bobcats.

(d)

(a)

Courtship behavior among animals has many common displays found in other social contexts, such as aggression. It is common for potential mates to perform displays of dominance and appeasement. **(a)** The terns are involved in courtship, and one bird offers the other a gift of fish, an act of appeasement. **(b)** An albatross pair during courtship shows both dominance and appeasement. One of the birds struts with head held high, while the other has lowered the body in act of appeasement (courtesy of Boyce A. Drummond III, Illinois State University). **(c)** Many animals court from a ceremonial spot called a lek. A male grouse struts in sexual display on a lek to which females are attracted for mating. **(d)** Some animals form harems when reproducing. Elephant nose seals during the breeding season exist in harems composed of one male, the harem master, and several females. This male is trumpeting his dominance over the females in the harem and advertising his dominance to any other bull who may attempt to take his place. Bulls frequently engage in exhaustive tests of strength to determine which will be the harem master.

(b)

(c)

(d)

THE NEXT DECADE

SOCIOBIOLOGY

In the next decade, we will hear more and more about human behavior because of a new scientific discipline—sociobiology. The new discipline has combined the concepts and theories of ecology, ethology, and evolution. It uses their framework to explain all behavior biologically. It also strives to explain all human behavior as a result of complex genetic evolution over several million years.

Many sociobiologists consider their work the completion of the Darwinian revolution. Darwin's theory of evolution (see Chapter 17) proposed that the organisms that survive and reproduce are those that are better adapted. Sociobiologists believe that the tactics (behavior) for survival are passed to offspring in genes. They view all social interaction—all behavior, including conformity, spite, altruism (Good Samaritanism), and aggression—as evolving by natural selection and therefore as being coded in the genes.

Sociobiology was formally launched in 1975, when E. O. Wilson of Harvard University published his lengthy book *Sociobiology: The New Synthesis*. (Wilson won the Pulitzer Prize in 1979 for his popular account of sociobiology, titled *On Human Nature*.) Since then, many other books have been published on the topic, some intended for professional consumption, some written for the layperson.

At the core of sociobiological thinking is the "selfish gene" concept—the concept that all genes attempt to propagate themselves at any cost. All life may in fact exist

solely to serve DNA and the genes. All life forms are like huge spaceships carrying a payload of millions of genes that direct every motion of their carriers. For some genes, the trip will never be over, because the goal of genes is perpetual life. They guide their organisms for the ultimate satisfaction of being transported into a new spaceship and continuing their journey. They can be transported into newer and newer carriers only if each carrier reproduces. If the carrier fails to reproduce, the genetic payload is lost. Selfish genes try to make each new carrier better than the previous model so that it will have a better chance of competing with other carriers of its species, giving it an edge on survival and reproduction—and, for the genes, a new carrier.

Does this explanation sound like science fiction? It is an extreme interpretation, certainly, but it summarizes one aspect of sociobiology. At least it shows how important genes are to all life processes, including behavior.

Are all genes selfish? What about those of heroes who save others from drowning or burning? We often do things for others that do not seem selfish. This form of behavior, known as altruism, is displayed by humans and many other animal species. Sociobiologists interpret it as selfish in the following way. If you save someone from drowning and enough other people do likewise, then it is highly probable that someone will save you if you fall into a lake. Therefore, it is your selfish

genes that call for you to save a drowning person. In simpler terms, if you scratch my back, some day I will scratch yours. When the good deed is returned in kind, it is known as reciprocal altruism.

There are many examples of altruism in the animal kingdom, and they can all be interpreted as gene selfishness. It is common for many birds to give alarm calls to flock members when a predator (or other danger) threatens. Giving the alarm is altruistic, but is it selfish? Yes. When the alarmer gives its call, the entire flock freezes and all noise ceases. If the flock were not alarmed, the noise of foraging and chirping would help the predator locate it. Calling the alarm is a selfish act that saves the caller from predation. But why not fly away and not bother to call? The sight of a single bird in flight would draw the predator to the bird.

What if you save your own child instead of saving a stranger? Is that act selfish? Yes. Selfish genes are always on the alert for such situations. Saving or helping a relative is called kin selection. There are copies of your genes in your children and in other close relatives. Therefore, helping them survive and reproduce perpetuates many of your own genes. You can transmit your genes directly by reproducing or indirectly by helping a close relative. If you choose not to marry or to have children, your selfish genes insist that you help a close relative who has a better chance of reproducing. Your act could be as simple as helping to send your sister's children through school. Your act helps propagate many of your genes.

Sociobiology proposes that all human social behavior is selfish. It is in the interest of selfishness even for a mother to love a child. With such a view, it is not surprising that sociobiology has as many critics as it does supporters. Most of the critics believe that sociobiology is a form of genetic capitalism, because it defends the current structure of societies as natural and inevitable. (Most of the critics are Marxist. The philosophy of Marxism takes issue with selfishness.) One of the constant battles of civilization has been to convince people to control or modify their behavior for the benefit of the group. If we become convinced that our "bad" behavior is natural and inevitable, will we be less likely to want to correct it? This may be a question for the next decade. Many scholars believe that much of human behavior is learned and passed on by culture, not by genes. Therefore, they believe that sociobiology cannot predict human destiny.

The debate over sociobiology will continue into the next decade. It is an intellectual battle and will have little effect on personal lives. However, the word *sociobiology* will appear often—on television and radio and in newspapers and magazines. You or your children may take a class in sociobiology, and there may someday be university departments of sociobiology. One sociobiologist predicts that anthropology, economics, law, political science, psychology, and psychiatry will someday all be branches of a larger discipline—sociobiology.

FOR DISCUSSION AND REVIEW

1. What is a modal action pattern? Observe any common animal (your dog, birds in the yard, animals in the classroom or zoo) and attempt to describe several MAPs. Describe a specific sequence. What message was conveyed?

2. Compare and contrast a MAP and a display. A display and a releaser.

3. Observe some common animals interacting. Identify the visual releasers. The auditory releasers. Do the animals show any behavior that might be directed by pheromones? Explain.

4. Describe the essential features of your Umwelt. If you can locate the information, compare and contrast your Umwelt with that of another species.

5. Explain the following statement: The better the student, the better the ability to filter stimuli.

6. How do hormones affect behavior?

7. Explain the nature-nurture argument. How has it been settled?

8. Define *imprinting*. Why are household pets usually imprinted on their human companions?

9. Again, observe some common animal. Does it have a territory? What are its territorial boundaries? How is the territory marked? How is the animal benefited by having a territory? What MAPs does the animal use for threat behavior? For appeasement behavior? For diversionary behavior?

10. Describe threat behavior. Give several examples of animal threat displays (other than those described in this chapter). Do they all conform to your definition of threat? Explain.

11. Describe appeasement behavior. Give several examples of it, using animals not described in this chapter.

12. What are diversionary and displacement behaviors? Why are they important alternatives to confrontation?

13. What are the four phases of sexual behavior? Describe the functions of each. Which phase would you predict to be the longest? The shortest?

14. Select an animal species and describe the behavior of the male and female through the four sexual phases.

15. Explain how the principles of ethology apply to human beings.

SUGGESTED READINGS

ANIMAL BEHAVIOR: ITS DEVELOPMENT, ECOLOGY, AND EVOLUTION, 2d ed., by Robert A. Wallace. Goodyear: 1979.
A leading textbook on animal behavior for beginners.

Examines behavior from developmental and adaptive perspectives.

THE EXPRESSION OF THE EMOTIONS IN MAN AND ANIMALS, by Charles Darwin. University of Chicago Press: 1965.

A reprint of the original 1872 version. It is both interesting and fun to read sample sections and to see the illustrations. However, it is difficult reading because it is written in archaic English.

AN INTRODUCTION TO ANIMAL BEHAVIOR, 3d ed., by Aubrey Manning. Addison-Wesley: 1979.

An excellent introduction to the principles of ethology, with additional chapters on motivation and learning.

KING SOLOMON'S RING, by Konrad Lorenz. Crowell: 1952.

A delightful introduction to animal behavior written by the father of ethology, a Nobel laureate. Whimsical and entertaining. Includes many topics, from buying to pitying animals.

THE OXFORD COMPANION TO ANIMAL BEHAVIOR, edited by David McFarland. Oxford University Press: 1982.

This interesting volume is arranged like an encyclopedia. You can look up nearly any topic about animal behavior. For example, you could find additional information about any topic in this chapter.

APPENDIX 1

THE METRIC SYSTEM

Units of length

Kilometer (km)	= 1,000 meters
Meter (m)	= 100 centimeters
Meter (m)	= 1,000 millimeters
Centimeter (cm)	= 0.01 meter
Millimeter (ml)	= 0.001 meter
Micrometer★ (μm)	= 0.000001 meter
Nanometer★★ (nm)	= 0.000000001 meter
Angstrom (A)	= 0.0000000001 meter

★A micrometer is often called a micron
★★A nanometer is often called a millimicron

Conversions between English and Metric Systems

1 km = 0.62 mile	1 mile = 1.609 km
1 m = 1.09 yards	1 yard = 0.94 m
1 m = 3.28 feet	1 foot = 0.305 m
1 cm = 0.394 inches	1 inch = 2.54 cm

Units of mass (weight)

Metric ton (t)	= 1,000 kilograms
Kilogram (kg)	= 1,000 grams
Gram (g)	= 1,000 milligrams
Milligram (mg)	= 0.001 gram
Microgram (μg)	= 0.000001 gram

Conversions between English and Metric Systems

1 t	= 0.9842 ton	1 ton	= 1,0160 t
1 kg	= 2.205 pounds	1 pound	= 0.4536 kg
1 g	= 0.0353 ounce	1 ounce	= 28.35 g

Units of volume (for liquids)

Kiloliter (kl)	=	1,000 liters
Liter (l)	=	1,000 milliliters
Milliliter (ml)	=	0.001 liter
Microliter (μl)	=	0.000001 liter

Conversions between English and Metric Systems

1 kl	= 264.17 gallons	1 gallon	= 0.0038 kl
1 l	= 1.06 quarts	1 quart	= 0.95 l
1 l	= 2.18 pints	1 pint	= 0.47 l
1 ml	= 0.034 fluid ounces	1 fluid ounce	= 30.0 ml

Conversion of temperature between Celsius and Fahrenheit

$$°C = 9/5 \ °F + 32$$

THE MAJOR PHYLA OF LIVING ORGANISMS

This is a selected list of only the most important phyla. Students who wish to review what a phylum is and how organisms are classified should consult Chapter 1 of this text. Those interested in seeing a complete list of the phyla of the living world can consult *Five Kingdoms: An Illustrated Guide to the Phyla of Life on Earth,* by Lynn Margulis and Karlene V. Schwartz (W.H. Freeman, 1982).

KINGDOM MONERA

The cells of organisms belonging to the kingdom Monera lack mitochondria, plastids, and nuclear membranes. Therefore, they contain no well-defined nuclei and are called prokaryotes (pro means "before"; *karyo* means "nucleus"). If flagella are present, they do not contain the complex fibrillar structures found in the flagella of other organisms.

Phylum Schizophyta

Bacteria (Schizophyta)□ are a diverse group of organisms. Their classification is based on cell structure, physiological characteristics, composition of DNA, and ways of staining when exposed to certain dyes. The classification of bacteria changes as new information about them is acquired.

Phylum (Division*) Cyanophyta

Some species of blue-green algae (Cyanophyta) are unicellular. Others are filaments formed of chains of cells. These algae contain chlorophyll and other red and blue pigments, which give them their characteristic blue-green color. They carry on photosynthesis. They survive in environments where few other organisms can live. For example, some of these algae live in hot springs where water temperatures reach 70 degrees Celsius. Some fix nitrogen. Blue-green algae may grow rapidly and produce blooms in water polluted with phosphates.

KINGDOM PROTISTA

The organisms belonging to the kingdom Protista include unicellular, colonial, and multicellular forms. The cells are eukaryotic, containing well-defined nuclei enclosed by a nuclear membrane and mitochondria and sometimes plastids. However, the cells of the multicellular varieties do not form the tissues typical of true plants and animals.

Some members of this kingdom are photosynthetic□, but many are heterotrophs. The heterotrophic protists are commonly called protozoans. The photosynthetic or autotrophic ones are commonly called algae.

At one time, protozoans and algae were regarded as separate phyla, each containing several classes of organisms. Now it is common to consider these classes as phyla. Furthermore, organisms commonly called algae have been placed in three different king-

---

*Botanists refer to the phyla of algae, fungi, and plants as divisions.

doms: the blue-green algae in Monera and the remainder in Protista and Plantae. Botanists commonly consider Euglenophyta, Chrysophyta, and Pyrrophyta as algae and the remainder as protozoans.

Phylum (Division) Euglenophyta

Unicellular photosynthetic organisms that swim through water with movements of long, whiplike flagella□ are **euglenoids.** They have no cell walls and are able to change shape. Two varieties of chlorophyll (a and b) are found in their chloroplasts, but some euglenoids are able to survive as heterotrophic organisms when their chlorophyll is lost under experimental or natural conditions. Exposure to ultraviolet radiation, for example, may cause such a loss.

Phylum (Division) Chrysophyta

The golden algae and diatoms are in the phylum Chrysophyta. Their golden color comes from a yellow-to-brown pigment that disguises the green chlorophyll. The organisms also possess two kinds of chlorophyll (a and c). Most species in this phylum are diatoms□.

Phylum (Division) Pyrrophyta

Most of the Pyrrophyta are called dinoflagellates. Like the diatoms, they are important primary producers in the oceans. They are unicellular and possess two flagella oriented at right angles to each other. The beating of these flagella causes the cells to spiral through water. Certain dinoflagellates are symbiotic organisms□ living in the tissues of reef-building coral. Many dinoflagellates have a cellulose cell wall. Others are naked. Many are bioluminescent (producing light). When these species are abundant in ocean surface waters, the sea glows.

Certain species of dinoflagellates produce deadly toxins. When these species are abundant, they discolor the seawater, creating what is called a red tide. Toxins produced during these outbreaks poison and kill numerous species of marine animals. Clams and mussels that feed on these dinoflagellates become toxic and unsuitable for human consumption.

Phylum Zoomastigina

Members of the phylum Zoomastigina are unicellular heterotrophs. Some are colonial. Although most are free-living, some are parasitic. African sleeping sickness in humans and a related disease in cattle and game animals are caused by members of the genus *Trypanosoma.* Some species in this phylum are symbiotic and live in the digestive tracts of termites, cattle, and other animals, where they contribute to their host's digestion of food.

The members of the phylum possess one or more flagella□ with which they swim. Species living in fresh water have contractile vacuoles□ to expel water that moves into the cytoplasm by osmosis□.

Some members appear to be identical to certain members of the phylum Euglenophyta except that they lack chlorophyll. The resemblance suggests that one of the phyla evolved from the other or that they share a common ancestry.

Phylum Sarcodina

The phylum Sarcodina includes a number of important groups of organisms, three of which are the foraminifera, the radiolaria, and the amoebae. The fossilized calcareous skeletons of foraminifera formed the White Cliffs of Dover and other chalk deposits. The glasslike skeletons of radiolaria also form important marine deposits. The amoebae are naked, and some are parasitic. Certain species of amoebae cause dysentery in humans.

All members of Sarcodina move or gather food by the extension of cytoplasmic appendages called pseudopodia (false feet). The extension of a pseudopod is brought about by a change in the viscosity of the cytoplasm in the region of formation. After the pseudopod has been extended, the viscosity of the remaining cytoplasm changes, and this too flows forward. Eventually, the entire organism has advanced. Food organisms are surrounded by pseudopodia and enclosed in fluid-filled bubbles in the cytoplasm, called food vacuoles. Here is where the food is digested.

Phylum Sporozoa

All members of the phylum are parasitic and without means of locomotion—except for the male reproductive stages, which bear flagella and can swim. The life cycles of most Sporozoas are complex. Spores are transmitted from host to host. Often, they are enclosed in coats that prevent them from drying out during transfer. Sometimes, they are transmitted from host to host in a second host, called a vector.

For example, sporozoans that cause malaria in humans and other animals are transmitted from host to host by mosquitoes that inject the spores when they bite. The sporozoans absorb nutrients from the body fluids of their hosts. Because they are not under osmotic stress, they lack contractile vacuoles□.

Phylum Ciliophora

The members of the phylum Ciliophora all possess hairlike outgrowths of their cell surfaces, called cilia□. Movement of the cilia causes movement of the organism and assists in food gathering. Cilia create water currents that sweep food particles into the cytostomes (cell mouths) of these creatures, where they are taken into food vacuoles for digestion. Most ciliates exist as single individuals, but some species form colonies.

The ciliates are the most complex members of the kingdom Protista and show the greatest degree of cellular specialization. They possess more than one nucleus. Usually, they have one large macronucleus and one or more micronuclei. Ciliates reproduce by fission, as do most of the Protista, but many also reproduce sexually by conjugation. Most ciliates are freeliving, but some are parasitic.

KINGDOM FUNGI

Although many biologists classify the fungi as members of the kingdom Protista, R. H. Whittaker, a noted ecologist, has elevated them to separate kingdom status. Many biologists are adopting this system of classification.

Most fungi exist as long filaments called hyphae. The walls of the hyphae are composed partly of chitin, a strong, stiff substance. The cytoplasm within a hypha is continuous and multinucleate. That is, the cytoplasm is not subdivided into specific cells, each containing a nucleus. Sometimes the hyphae are partially subdivided by cross-walls. Hyphae often form masses called mycelia.

Fungi lack plastids and chlorophyll. They absorb nutrients through their hyphal walls. Often, they secrete digestive enzymes into their environment to digest complex organic matter into molecules small enough to be absorbed. Some fungi—for example, the mushrooms produce complex reproductive structures.

Phylum (Division) Phycomycetes

The members of the phylum Phycomycetes produce reproductive spores in special sacs called sporangia. The hyphae lack cross-walls. The phylum includes the water molds and a wide variety of terrestrial molds. Some water molds cause infections on fish and are a problem in fish hatcheries. A species that infects potato plants caused the Irish potato famine of 1845 and 1846. The common bread mold produces black sporangia filled with spores.

Phylum (Division) Ascomycetes

The members of the phylum Ascomycetes produce two kinds of spores, one asexually and the other (an ascospore) sexually. Many species exist only as hyphae and mycelia (for example, the powdery mildews); others give rise to sizable fleshy reproductive bodies (for example, the edible morels and truffles.) One important group are the unicellular yeasts, which are essential in baking and the production of alcohol. Some are also used in the production of certain vitamins.

Ascomycetes caused the chestnut blight that eliminated most of the North American chestnut trees and the Dutch elm disease that destroyed most North American elm trees. The genus *Penicillium,* an important member of the phylum, is the source of the antibiotic penicillin. Certain species of *Penicillium* are used in the manufacture of blue, Roquefort, and Camembert cheeses.

Phylum (Division) Basidiomycetes

The phylum Basidiomycetes includes the rusts, smuts, mushrooms, puffballs and shelf fungi. The life cycles of these organisms are often complex. The rusts and smuts are important plant pathogens. Some of them infect wheat and other grains. Many of the fleshy fungi, such as mushrooms and puffballs, are edible, but some are toxic. Wild mushrooms should not be eaten unless the edible species are known—and even then mistakes can be made. Only a few genera of mushrooms are grown commercially for human consumption. The fleshy part of the basidiomycete is the reproductive part, which produces spores. The remainder of the organism, the mycelium, is beneath the surface of the substrate and survives much longer than the transient reproductive part.

Phylum (Division) Myxomycophyta

The slime molds comprise the phylum Myxomycophyta.

KINGDOM PLANTAE

Most plants are multicellular organisms. Most are also capable of photosynthesis, because they contain chlorophyll in plastids within their cells. Plant cells possess well-developed cellulose cell walls and contain large vacuoles in their cytoplasm. Most organisms in the kingdom Plantae possess well-defined tissues, the higher ones having special vascular tissues. Plants generally are fixed, nonmotile organisms.

The Rhodophyta, Phaeophyta, and Chlorophyta are usually called algae. Although they are structurally much simpler than the other plant phyla, they are included in the plant kingdom because many of them are multicellular and some show tissue development to a limited degree. Also, they are believed to be more closely related to the higher plants than to the algae classified in the kingdoms Monera and Protista.

Phylum (Division) Rhodophyta

The red algae—phylum Rhodophyta—include both unicellular and multicellular forms. Nearly all are marine organisms, and the multicellular forms are commonly called seaweeds. Although members of this phylum contain chlorophyll, most do not appear green because red pigments disguise the chlorophyll. These algae can grow deeper in the oceans than most other plants. In the deep waters, the red pigments facilitate the absorption of light energy and transfer it to the chlorophyll. Certain red algae are edible. Some are commercially important as sources of agar, a gelatinous substance used to solidify the media used to grow microorganisms such as bacteria.

Phylum (Division) Phaeophyta

The phylum Phaeophyta contains the brown algae. A brownish pigment disguises the chlorophyll contained in the cells of these conspicuous plants. The members of the phylum, primarily marine, are known as seaweeds and kelp. Some kelp species may reach a length of up to thirty meters and are among the most rapidly growing plants. Each kelp plant has a holdfast that attaches the plant to the sea bottom, a long stemlike stipe, and leaflike fronds. Many of these large, heavy plants also possess gas-filled floats that allow them to float upward in the water. Kelps are harvested and used as food, as a source of iodine, and as fertilizer.

Phylum (Division) Chlorophyta

The phylum Chlorophyta are the green algae. Although species of green algae live in ocean water and fresh water and on moist soil, most are freshwater forms. The phylum includes unicellular, colonial, and multicellular forms. Most biologists believe that green algae were ancestors to the higher plants and that multicellular forms evolved from colonies of unicellular forms as the individuals became more and more interdependent.

Phylum (Division) Bryophyta

The mosses and liverworts are both in the phylum Bryophyta. They are regarded as primitive land plants from which the higher plants may have evolved. None of the plants in this phylum have vascular tissue. All are limited to damp areas, where sufficient water is present to allow the flagellated sperm cells to swim to the female reproductive structures that house the eggs. Liverworts are flat, low, leaflike plants that encrust the substrate on which they grow. Mosses are erect little plants that grow up from horizontal filamentous growths called protonema.

Phylum (Division) Tracheophyta

The plants in the phylum Tracheophyta are vascular plants. They possess true roots, stems, and leaves. Their cells are specialized as vascular tissue, which transports water and nutrients throughout the plant. This phylum includes the largest and tallest plants on earth. Some of the plants in the phylum are the club mosses, horsetails, and ferns, none of which produce seeds, and the well-known seed-producing plants— the gymnosperms and the angiosperms. The gymnosperms include the cycads, ginkgoes, and conifers (pines and spruces, for example).

The angiosperms—the flowering plants—represent the largest single plant group. They are subdivided into the Dicotyledoneae and the Monocotyledoneae. The monocots include the grasses, lilies, and orchids. The flowers of these plants usually have three petals and other floral structures, and the veins (vascular tissue) in the leaves are usually parallel. The dicots include the remainder of the flowering plants, most trees and shrubs, and many smaller herbaceous plants. The floral parts of these plants usually occur in fours or fives, not threes, and the veins in the leaves are netlike or arranged in some other nonparallel pattern. Most edible plants are angiosperms.

KINGDOM ANIMALIA

The animals are nonphotosynthetic (heterotrophic) multicellular organisms that possess a variety of tissues□ and undergo a characteristic pattern of embryonic development□. Animal cells lack the cell walls and large vacuoles that are typical of plant cells. Most animals have the power to move and therefore have evolved muscle and nervous systems. Animal movement is associated primarily with food procurement but serves many other functions as well. The animal kingdom contains approximately thirty phyla, but only the nine largest and most important are included in this survey.

Phylum Porifera

The animals in the phylum Porifera are the sponges. They are aquatic (mostly marine) sessile animals with bodies formed of loosely organized cells. The cells are not arranged into the tissues characteristic of other animal phyla. Sponges are often brightly colored. Their bodies are supported by internal skeletons. In some, the skeletons are composed of small calcareous spicules of various shapes. In others, the spicules are siliceous and glass-like. In addition to spicules, many sponges possess fibrous skeletons. These skeletons are composed of networks of tough fibers of a protein called spongin, which is chemically similar to the collagen fibers found in the connective tissues of other animals. Sponges that are fished commercially have extensive spongin skeletons. After being caught, these sponges die and decompose, leaving behind the fibrous skeleton, which is washed, dried, and sold. However, most sponges sold today are artificial plastic varieties.

Although sponges have no digestive system, they filter tiny food particles from the water in which they live. The body of a sponge contains one or more large openings called oscula (singular: osculum) and numerous microscopic pores called ostia. Water passes in through the ostia. It then goes through a series of canals, pores, and chambers. It leaves the body through the oscula. Some of the canals and chambers are lined with cells called choanocytes, each of which bears a single flagellum. The beating flagella maintain the flow of water. Also, the choanocytes engulf food particles that are in the water flowing over them. These food particles are digested in tiny food vacuoles.

Sponges reproduce both sexually and asexually. Only the flagellated larvae are motile. They are able to swim to new locations before settling to the substrate and transforming themselves into new adult sponges.

Phylum Cnidaria

Also called Coelenterata, the phylum Cnidaria contains corals, sea anemones, jellyfish, and the like. Most of these relatively simple animals live in the ocean, although a few, such as hydra, live in fresh water. Some species—the reef-building corals, for example—form complex colonies. Cnidarians have a simple digestive cavity enclosed by a body wall. The wall is composed of two cell layers separated by a layer of jellylike material called mesoglea. The digestive cavity has a single opening to the outside—the mouth—which is commonly surrounded by one or more rows of tentacles. The tentacles bear cells containing stinging capsules called nematocysts. When a food organism contacts one of the tentacles, the nematocysts discharge tiny threads, some of which inject deadly toxins into the prey. The others stick to and entangle the prey.

Phylum Platyhelminthes

The free-living and parasitic worms called flatworms (which are named after their body shape) make up the phylum Platyhelminthes. Most free-living flatworms, such as planaria, are aquatic. The flukes and tapeworms are important parasites□ of humans and other animals. Free-living flatworms and flukes possess simple digestive tracts. Tapeworms, which lack a digestive system, live in the digestive tracts of other animals and absorb nutrients digested by their hosts into their tissues. The long ribbonlike bodies of tapeworms are subdivided into sections called proglottids. Each section contains a complete set of male and female reproductive systems. Tapeworms attach themselves to the lining of the digestive tract of the host by a structure called a scolex, which often bears hooks.

Phylum Nematoda

The nematodes or roundworms, occur almost everywhere—in water, in soil, and as parasites in the bodies of both plants and animals. They are among the most abundant animals on earth, but because of their small size, they usually go unnoticed. Many parasitic nematodes infect humans. Among them are hookworms, filarial worms, ascarid worms, and pinworms. Ascarid worms are found in many animals, including dogs and cats. The animals are often already infected when they are born and thus need to be wormed. Nematodes are elongated worms that have a mouth at one end and an anus (posterior opening of the digestive tract) at or near the opposite end. The digestive cavity or tube is separated from the body wall by a fluid-filled cavity. The body plan of these worms is regarded as a tube-within-a-tube type.

Phylum Mollusca

The mollusks include such diverse creatures as snails, slugs, clams, oysters, squid, octopuses, and chitons. Most mollusks live in the oceans, but many live in fresh water or on land. Although they show great variation in form, all mollusks have the same basic body plan. That plan has been modified during the evolution of the different groups. The ancestral mollusk was probably snail-like, with a well-developed head that bore the mouth and sense organs, a broad muscular foot on which to creep about, a conical-shaped visceral mass containing the internal organs, and a conical shell covering the visceral mass. At the posterior end of the animal, beneath the shell, a fold of tissue, called the mantle, created a cavity that contained a pair of gills. Water circulated through the cavity.

Modern mollusks display various modifications of this plan. For example, clams have evolved large bivalve shells that enclose the entire animal. The broad muscular foot has been adapted for digging, and the head is greatly reduced. The gills in most clams are greatly enlarged and are used not only to obtain oxygen from the water but also to strain food particles from the circulated water. Squids and octopuses are streamlined for rapid locomotion and predation. Consequently, during evolution, their shells were re-

duced or lost. Their foot became modified into tentacles to capture food. Their head became fused with the foot region and now bears well-developed sense organs. Their eyes are nearly as complex as human eyes.

Phylum Annelida

The segmented worms, or annelids, include the earthworms, leeches, and a wide variety of marine worms called polychaetes. The latter possess numerous bristlelike structures, called chaetae, on certain regions of their bodies. The head is well-developed in many polychaetes, but it is small and rather inconspicuous in earthworms and leeches. Behind the head, the body is divided into segments by ringlike constrictions called annuli. The digestive tracts and major blood vessels pass through each of the segments, extending from the head region to the posterior region of the body. Each segment, however, possesses its own identical set of excretory tubules, certain nerves and muscles, and other structures. Many annelids continue to grow throughout life by adding new segments at the posterior end of the body just in front of the terminal section.

Phylum Arthropoda

Most of the living species of animals are in the phylum Arthropoda. The largest of the animal phyla, it includes the insects, the crustaceans, (crabs, lobsters, shrimp, and copepods), the arachnids (spiders and scorpions), and the millipedes and centipedes. Arthropod specialists are currently debating whether to divide this single large phylum into three separate phyla.

The body of an arthropod has a rigid external covering called an exoskeleton. The presence of this protective covering affects all aspects of arthropod life. For example, in order to grow, an arthropod must periodically shed its old exoskeleton and produce a larger one. This process is called molting. Prior to molting, the epidermis of the arthropod secretes a new soft exoskeleton beneath the old one. A split or tear occurs in the old one, and the animal works its way out through the opening. While the new exoskeleton is still soft and flexible, the arthropod expands it by taking in large amounts of air (if it is terrestrial) or water (if it is aquatic). After expansion, the new exoskeleton hardens. The air or water is eliminated as tissues grow and fill in the additional space. The continuous exoskeleton must be modified to form joints that allow such appendages as legs and mouth parts to flex and bend.

The largest group of arthropods is the insects. They include flies, beetles, butterflies, ants, wasps, bees, moths, termites, cockroaches, crickets, grasshoppers, fleas, lice, and others. The body of any insect is composed of three regions: (1) the head, which bears the simple and compound eyes, the mouth parts, and the antennae; (2) the thorax, which bears the three pairs of legs that characterize this group (insects belong to the class Hexapoda), and the wings, if any are present; (3) the abdomen, which contains the anus and the reproductive openings. Like annelids, arthropods are segmented. Each body region is composed of a specific number of segments. For example, the thorax is composed of three segments, each bearing one of the three pairs of legs.

Phylum Echinodermata

The echinoderms are exclusively marine animals. Sea urchins, starfish, brittle stars, and sea cucumbers are examples. The echinoderms have pentamerous (five-part) radial symmetry. That is, many of their body parts occur in fives and are arranged around a central body region. For example, a starfish has five rays, or arms, that extend from a region of the body called the central disk. Each arm contains a pair of digestive glands that are attached to the stomach in the central disk and a pair of gonads that open through five pores on the top of the central disk. Although the adult echinoderms have radial symmetry, their larvae are bilateral symmetry and must reorganize their body parts before becoming adults.

Echinodermata literally means "spiny skin." Many (but not all) echinoderms possess a well-developed skeleton with numerous spines that extend outward to give the animal an obvious or subtle spiny appearance. The spines are covered by the epidermis, or skin, and therefore are really internal structures.

The unique feature of echinoderms is their water vascular system, a system of fluid-filled canals that follow the radial pattern of the echinoderm. In the starfish, radial canals of the water vascular system extending along each of the five rays give rise to rows of numerous structures called tube feet, which lie in a groove along the lower side of each ray. Each tube foot bears a tiny sucker that allows it to attach itself to hard objects, and each can act independently of the others. Tube feet can be extended by internal fluid pressure, can attach to the substrate, and then can contract, pulling the animal along. Therefore, tube feet are the means by which many echinoderms move. They are also used in gathering food.

Phylum Chordata

The phylum that humans belong to is Chordata. The most important subgroup within the phylum is Vertebrata—animals that have a backbone or vertebral column. The major groups of vertebrate animals are the cartilagenous fish (such as skates, rays, and sharks), which possess skeletons made of cartilage rather than bone; the bony fishes, with bony skeletons; the amphibians (frogs, toads, salamanders, and the like); the reptiles (turtles, snakes, lizards, and the extinct dinosaurs); the birds; and the mammals, which all have hair and nurse their young.

These forms and the other minor groups of chordates, the sea squirts and cephalochordates, have the following three characteristics in common during at least one stage in their development: (1) pharyngeal gill slits, (2) a notochord, and (3) a dorsal hollow nerve cord.

The gill slits are well developed in the minor groups of chordates and in fishes. They are also present in larval amphibians, such as tadpoles. However, they are reduced, modified, or lost in most other groups of chordates. Humans have gill slits during a short period early in their embryonic life.

Except in the case of cephalochordates, lamprey eels, and hagfish, the notochord is also an embryonic or larval structure. It is a rod of special tissue that forms beneath and parallel to the nerve cord and vertebral column. Presumably, it acts to support the developing animals, but it disappears late in development in humans and most other vertebrates.

The dorsal hollow nerve cord is enclosed within the vertebral column and the skull of vertebrates. The brain and spinal cord represent the central nervous system in these animals. They contain fluid-filled cavities and therefore are regarded as hollow.

The chordates are among the largest and most successful animals ever to have lived. They are found almost everywhere on earth, from the deepest ocean depths to the highest mountains. They occur in deserts and near the poles of the earth. Winged insects, birds (including the reptiles from which they are descended), and flying mammals (bats) are the only animals that have evolved the power of flight. Humans and the other great apes are the most intelligent animals living.

GLOSSARY

The glossary includes neither the units of measure found in Appendix 1 nor the taxonomic groups or terms used exclusively in Appendix 2. Only boldface terms from the text are included.

Abiotic environment (Greek: *a*, "not"; *bio*, "life"). The physical factors of an environment, such as temperature, amount of rainfall or sunlight, and chemicals in the soil.

Acetylcholine. A neurotransmitter substance secreted at the ends of many neurons.

Acid (Latin: *acidus*, "sour"). A substance that releases hydrogen ions in water.

Actin (Greek: *aktis*, "a ray"). One of the two types of protein filaments found in muscle.

Active immunity (Latin: *immun*, "free"). The ability to resist infection that is gained by the buildup of antibodies caused by previous exposure to antigens. Also called acquired immunity.

Active transport. Transfer of a substance through a cell membrane against a concentration gradient, a process that requires the expenditure of energy by the cell.

Adaptation (Latin: *adaptare*, "to fit"). The accumulation, through the course of evolution, of characteristics that make organisms better suited to their environments.

Adaptive radiation. The evolution from a common ancestral group of several distinctly different groups with different characteristics.

Adenosine diphosphate (ADP). A substance formed when ATP (adenosine triphosphate) is used by a cell to supply energy. ATP → ADP + phosphate + energy.

Adenosine monophosphate (AMP). A substance formed when ADP (adenosine diphosphate) is used by a cell to supply energy. ADP → AMP + phosphate + energy. This reaction occurs only if ATP (adenosine triphosphate) is in short supply.

Adenosine triphosphate (ATP). An organic compound—composed of adenine, ribose, and three phosphate groups—that is a primary source of usable energy in cells. ATP is formed from ADP (adenosine diphosphate) and phosphate during cellular respiration.

Adipose tissue (Latin: *adip*, "fat"). A form of connective tissue composed of fat cells.

ADP. *See* Adenosine diphosphate (ADP).

Aerobic respiration (Greek: *aeros*, "air"; *bios*, "life"). Respiratory processes requiring oxygen.

Aggression. An attack by one member of a species on another member of the same species; may occur in connection with behaviors such as maintaining a territory or defending young.

Alkaline (Arabic: *alkali*, "soda ash"). Referring to substances that release hydroxyl (OH^-) ions in water; basic—the opposite of acidic.

Allantois (Greek: *allas*, "sausage shaped"). One of the extraembryonic membranes of birds, reptiles, and mammals. A pouch extending from the posterior region of the digestive tract that stores metabolic waste; forms part of the placenta in mammals.

Allele (Greek: *allelon*, "of one another"). An alternative form of a gene located at the same locus (position) on a chromosome.

Allopatric species (Greek: *allo*, "other"; *patria*, "fatherland"). Populations of two or more closely related species that occupy different (although often adjacent) geographic regions and do not interbreed, thereby maintaining their reproductive isolation.

All-or-none principle. The property of muscle cells or nerve cells to respond either maximally or not at all.

Alternation of generations. The inclusion of alternting haploid gamete-producing generations (gametophytes) and diploid spore-producing generations (sporophytes) in the life histories of plants. Haploid spores produced by meiosis germinate, producing the gametophyte stage. The union of haploid gametes gives rise to diploid sporophytes.

Alveolus (Latin: diminutive of *alveus*, "hollow"). The saclike endings of the air passages in the lungs of humans and other mammals.

Amino acids. The building blocks of proteins. Compounds that contain an amino group ($-NH_2$) and a carboxyl group ($-COOH$).

Amniocentesis. Removal of a small amount of amniotic fluid from the amnionic sac that encloses a fetus during development.

Amnion (Greek: *amnos*, "lamb"). One of the extraembryonic membranes of birds, reptiles, and mammals; forms a fluid-filled sac in which the developing embryo floats.

Amplexus (Latin: *amplex*, "to embrace"). A copulatory embrace performed by amphibians that ensures that the female's eggs will be fertilized by the male as they are laid.

Amylase (Latin: *amylum*, "starch"). An enzyme that catalyzes the digestion (hydrolysis) of starch.

Anaerobic respiration (*an*, "without"; Greek: *aeros*, "air"; and *bios*, 'life"). A respiratory process of cells and organisms, such as certain bacteria, that can continue in the absence of oxygen.

Analogous structures (Greek: *analogos*, "proportionate"). Structures that may have similar functions and appearances but that differ in how they develop embryologically. Example: a bird's wing and an insect's wing.

Anaphase (Greek: *ana*, "up"; *phasis*, "phase"). The stage of mitosis when the chromatids of each chromosome separate. During anaphase of the first division of meiosis, pairs of homologous chromosomes separate; during anaphase of the second division of meiosis, the chromatids of each chromosome separate as they do during mitosis.

Androgens (Greek: *andros*, "man"; *gennan*, "to produce"). Male sex hormones.

Animalia (Latin: *animali*, "animal"). The kingdom that includes all heterotrophic multicellular organisms. These organisms have well-defined tissues (for example, muscle tissue, nervous tissue).

Annual plant (Latin: *annus*, "a year"). A plant that completes its entire life cycle in a single growing season and then dies. Seeds produced by the plant produce the next generation during the following growing season.

Antagonistic muscles. Muscle pairs that work in opposition to one another. Example: the biceps muscle that flexes the human arm and the triceps muscle that extends it.

Antennae (Latin: *antenna*, "a sailyard"). Sensory structures that project from the heads of many arthropods. Singular is antenna.

Anterior (Latin: *ante*, "before, in front of"). The front end of an organism.

Antheridium (Greek: *anthus*, "flower"). The structure that produces male gametes in lower plants such as mosses. Plural is antheridia.

Anthropoidea (Greek: *anthropos*, "man"; *oeides*, "like"). The suborder of order Primates that includes monkeys, apes and humans.

Anti- (Greek: *anti*, "against").

Antibody. A protein produced by the body in response to foreign substances such as bacteria, which act as antigens. Antigens stimulate the production of antibodies, which react with them and destroy or neutralize them.

Anticodon. The set of three nucleotides on transfer RNA molecules that combines with a complementary set of three other nucleotides, the codon, on messenger RNA molecules during the synthesis of proteins.

Antidiuretic hormone (Greek: *diouretikos*, "promoting urine"). The hormone, produced in the hypothalamus of the brain and released by the pituitary gland, that regulates water reabsorption by nephrons in the kidney.

Antigen (Greek: *gennan*, "to produce"). A foreign substance, usually a protein or polysaccharide, that stimulates the production of antibodies.

Anus. The posterior opening of the digestive tract.

Aorta (Greek: *aorte*, "aorta"). The largest artery in the human body, which conducts blood from the left ventricle.

Aphotic zone (*a*, "without"; Greek: *photos*, "light"). The mass of ocean water below the depth to which light penetrates. Because there is no light, plant life cannot grow in this zone.

Apical meristem. The growing tip of a plant's root or stem, composed of meristematic tissue.

Appeasement. The antithesis of threat. During appeasement behavior, an animal attempts to make itself as small and inconspicuous as possible.

Appendicular skeleton (Latin: *appendic*, "hang to"). The part of the skeleton that includes the bones found in the arms, legs, hands, feet, wrists, ankles, shoulders, and pelvis.

Appendix (Latin: *append*, "an appendage"). A small, blind, tubelike sack found at the junction of the large and small intestines.

Applied research. Research that has a direct and obvious application to human welfare.

Arboreal (Latin: *arbor*, "a tree"). Referring to animals adapted for living in trees.

Arch- (Greek: *archos*, "first or earliest").

Archegonium (Greek: *archegeonos*, "the first of a race"). The structure that produces female gametes in lower plants such as mosses.

Archenteron (Greek: *enteron*, "intestine"). The initial embryonic undifferentiated digestive cavity produced during gastrulation and lined with endoderm.

Arteriole (Greek: *arteri*, "an artery"; Latin: *iol*, "little"). An artery that is less than 0.3 mm in diameter.

Artery. A blood vessel that conducts blood away from the heart to the tissues. The walls of arteries are thick and elastic.

Asexual reproduction. Any reproductive process that does not involve the union of gametes. Examples: binary fission, budding.

Asters (Greek: *aster*, "star"). The radiating fibrils that appear at the ends of the spindle during mitosis.

Atheriosclerosis. The buildup of fatty masses in the walls of arteries. A form of arteriosclerosis, hardening of the arteries.

Atom (Greek: *atomos*, "indivisible"). The smallest unit of an element. If an atom is subdivided, the characteristics of the element are lost.

ATP. *See* Adenosine triphosphate (ATP).

Atrioventricular node. A group of cells responsible for stimulating the ventricles to contract at the same time.

Atrium (Greek: *atrion*, "hall"). A chamber of the heart that receives blood from a vein and passes it next into a ventricle.

Auditory canal (Latin: *audi*, "hear"). The outer ear duct through which sound from the external environment is transmitted to the eardrum.

Auto- (Greek: *autos*, "self or same").

Autonomic nervous system. The portion of a vertebrate nervous system that is not under voluntary control. Autonomic nerves innervate structures such as the heart, intestine, and kidneys.

Autosome (Greek: *soma*, "body"). Any chromosome other than a sex chromosome.

Autotroph or autotrophic (Greek: *autos*, "self"; *trophos*, "feeder"). An organism that is able to synthesize organic compounds from inorganic substances. Example: photosynthetic plants.

Auxins (Greek: *auxein*, "to increase"). Growth-regulating plant hormones.

Axial skeleton (Latin: *axi*, "axis"). The part of the skeleton that includes the bones found in the skull, around the larynx, in the inner ear, in the vertebral column, in the sternum, and in the ribs.

Axon (Greek: *axon*, "axle"). The part of a neuron that conducts a nerve impulse away from the nerve cell body.

Bacillus (Latin: *bacillum*, "a little stick"). A rod-shaped bacterium.

Bark. The plant tissues exterior to the bark cambium in the stems of woody plants.

Barr body. The inactive X chromosome in mammalian females.

Basal body. A cell organelle located at the bases of cilia and flagella that has a structure identical to that of a centriole.

Base. A substance that releases hydroxyl ions (CH^-) in water. Basic is the opposite of acidic.

Basic research. Research performed for its own sake, without regard to any obvious direct application to human welfare. Nonetheless, basic research has produced knowledge that has greatly benefited humans.

B-cell. The type of lymphocyte that produces antibodies for the immune system.

Benthic or benthos (Greek: *benthos*, "bottom of the sea"). Referring to organisms living on or burrowing into the bottom of lakes and oceans.

Bicuspid valve (Latin: *bi*, "two"; *cuspi*, "point"). The two-leaved valve located between the left ventricle and atrium of the heart.

Bile. The secretion of the liver that is stored in the gallbladder and released to emulsify lipids in the small intestines of vertebrates.

Bio- (Greek: *bios*, "life").

Biogeochemical cycle (Greek: *geo*, "earth"). The flowing and cycling of chemicals through the biosphere through and between living organisms and their nonliving environment. Examples: the water cycle, the nitrogen cycle.

Biological clock. The rhythmic cycle of activities performed by an organism that is often linked with natural-cycle phenomena such as day and night and the tides. Biological clocks are innate and may continue in the absence of apparent environmental cues.

Biological control. The use of natural predators or disease organisms to control a pest species.

Biological magnification. The concentration of pesticides in the tissues of organisms in a food chain or web. The concentration of pesticides increases at each higher trophic level.

Biomass. The total weight of all the organisms in a particular area or habitat.

Biome. The worldwide community complexes that develop in regions characterized by specific temperature ranges and amounts and distribution of precipitation. Examples: tropical rainforests, grasslands, coniferous forests.

Biosphere. The region of the earth's surface, including the water and atmosphere, that supports and is occupied by living organisms.

Biotic or biological environment (Greek: *bio*, "life"). The interrelationships of all organisms in a given area or habitat.

Biotic potential (Greek: *bio*, "life"; Latin: *poten*, "powerful"). The capability of a species to increase its population size.

Bladder. An elastic sac that serves as a reservoir for fluids such as urine.

Blastocoel (Greek: *blastos*, "germ"; *koilos*, "hollow"). The central fluid-filled cavity of a blastula.

Blastocyst (Greek: *kystis*, "bladder"). The modified blastula of mammals, which consists of two cell masses—an inner mass that becomes the embryo and a peripheral mass that becomes part of the placenta.

Blastopore (Greek: *poro*, "opening"). The opening of the archenteron of a gastrula to the outside.

Blastula. A spherical ball of cells containing a fluid-filled cavity produced by cleavage of a fertilized egg, the zygote.

Blood cell. A blood corpuscle that can be either an erythrocyte (red blood cell) or leukocyte (white blood cell).

Bone. A type of connective tissue consisting of bone cells (osteocytes) embedded in a matrix of material containing mainly calcium and phosphorous compounds.

Bone marrow. Spongy, highly vascularized tissue, found in the cavities of bones, in which red and white blood cells originate.

Bowman's capsule. The cup-shaped end of a nephron, which encloses the glomerulus.

Brain. The often extremely complex enlarged anterior portion of the nervous system of bilaterally symmetrical animals. In humans it is the center not only for body coordination but for logical thought processes and stimulus perception.

Brain stem. The portion of the brain that includes both the hindbrain and the midbrain and that controls and coordinates many unconscious body functions, such as digestion, respiration, excretion, and circulation.

Bronchiolus (Greek: *bronch*, "windpipe"; Latin: *iol*, "little"). A small branch of a bronchus.

Bronchus (Greek: *bronchos*, "windpipe"). A branch of the trachea that conducts air to a lung. In each lung, the bronchus branches into numerous bronchioles. Plural is bronchi.

Budding. The form of asexual reproduction during which a small part of a parent individual develops into a new individual with the same genetic composition as the parent.

Bursa (Latin: *burs*, "purse"). A fluid-filled saclike structure found in many synovial joints, such as the knee.

C3 plants. Plants whose first molecule produced by the dark reaction is phosphoglyceraldehyde (PGAL). This molecule contains three carbon atoms.

C4 plants. Plants whose first molecule produced by the dark reaction contains four carbon atoms (for example, malate).

Cambium (Latin: *cambiare*, "to exchange"). Plant meristematic cells that divide and produce cells that differentiate into new plant tissues, such as xylem and phloem.

Capillary (Latin: *capillaries*, "hairlike"). A thin-walled blood vessel in tissues; the site of the exchange of substances between the blood and the tissue fluid. Capillaries connect arteries with veins.

Carbohydrate. Organic compounds, such as sugar, starch, and cellulose, in which there are twice as many hydrogen atoms as oxygen atoms.

Carbon cycle. The biogeochemical cycle for the reutilization of carbon atoms that includes the biological processes of photosynthesis and respiration.

Carcinogen. A chemical substance or any environmental factor that causes or induces cancer.

Cardiac muscle (Greek: *cardi*, "heart"). The type of muscle tissue found in the heart.

Carnivore (Latin: *carno*, flesh"; *vorare*, "to devour"). An animal that feeds largely or exclusively on the flesh of other animals.

Carotenoid pigment. Yellow and orange plant pigments that absorb light energy and transfer it to chlorophyll molecules. Thus, they are accessory pigments for photosynthesis.

Carpel (Greek: *carpus*, "a fruit"). Collectively, the ovary, style, and stigma of an angiosperm flower. A flower may have one or more carpels, which may occur singly or be fused together.

Carrying capacity. The environmental factors that limit the population size of a species over a long period of time.

Cartilage (Latin: *carilago*, "cartilage"). A type of resilient connective tissue that, along with bone, forms the vertebrate skeleton.

Catalyst (Greek: *katalysis*, "dissolution"). A substance, such as an enzyme, that regulates the rate of a chemical reaction but that is not used up by the reaction.

Cell. The basic unit of life, which consists of a nucleus surrounded by cytoplasm that is enclosed by a semipermeable cell membrane. Plant cells possess a nonliving cell wall exterior to the cell membrane.

Cell cycle. The period from the beginning of cell division until the beginning of the division of a daughter cell. The cell cycle consists of the mitosis phase, the G1 phase, the S phase (when DNA replicates), and the G2 phase.

Cell division. A process by which one cell gives rise to two cells. *See also* Mitosis.

Cell membrane. The semipermeable membrane that encloses the cytoplasm of a cell.

Cell plate. A structure that forms on the equatorial plate of a dividing plant cell, where new cell membranes and the middle lamella will form.

Cell sap. The complex solution that fills the large water vacuoles in plant cells.

Cell theory. The theory that all organisms are composed of one or more cells and cell products and that all cells originate from preexisting cells.

Cell wall. A nonliving, fairly rigid cell layer external to the cell membrane of plant cells, composed primarily of cellulose. The cells of fungi and bacteria also possess cell walls.

Cellular differentiation. The transformation of nonspecialized cells into specialized cells that occurs during embryological development and growth in animals and plants.

Cellular respirtaion (Latin: *respirare*, "to breathe"). The process by which cells use oxygen and produce water and carbon dioxide while converting the energy of food molecules into high-energy chemical bonds of compounds such as ATP. Some cells are able to respire without the presence of oxygen (*see* Anaerobic respiration).

Cellulose. A complex carbohydrate that is the main constituent of the walls of plant cells.

Central nervous system. The portion of the nervous system represented by the brain and spinal cord. It coordinates all sensory information.

Centriole (Latin: *centrum*, "center"). A cellular organelle in animal cells, usually located near the nucleus on the outside of the nuclear membrane. The centriole gives rise to the spindle during animal cell division.

Centromere (Greek: *kentron*, "center"; *meros*, "part"). The region of a chromosome to which spindle fibers attach and that holds the two chromatids together.

Cerebellum (Latin: *diminutive of cerebrum*, "brain"). The portion of the vertebrate brain involved with the coordination of body movements and equilibrium.

Cerebral cortex. The outer layer of the cerebrum in the brains of mammals; in it are found the centers for voluntary movements and conscious sensations.

Cerebrum (Latin: *cerebrum*, "brain"). The largest portion of the human brain and a major portion of all vertebrate brains, consisting of two cerebral hemispheres joined together.

Cervix (Latin: *cervix*, "the neck"). The juncture of the uterus and the vagina in humans.

Chemical bond. The electrical attraction between two atoms of a molecule that holds the atoms together.

Chloro- (Greek: *chlōros*, "green").

Chlorophyll. The green pigment found in plant and algal cells that is necessary for photosynthesis. Different types of chlorophyll (such as chlorphyll a, b, and c) are found in various combinations in different types of photosynthetic organisms.

Chloroplast. The membranous organelle of plant cells that contains chlorophyll and in which photosynthesis occurs.

Cholesterol (Greek: *chole*, "bile"; *stereos*, "solid"). A crystalline compound ($C_{27}H_{44}O$) that is found in gallstones, nerve tissue, blood, and bile.

Chorion (Greek: *chorion*, "a skin or membrane"). The outermost extraembryonic membrane of birds, reptiles, and mammals. In mammals, it forms part of the placenta.

Chorionic gonadotropin. The hormone produced by the placenta that maintains pregnancy and that is excreted in the urine and forms the basis of pregnancy tests.

Choroid layer (Greek: *choroid*, "like a membrane"). The vascular pigmented layer of the eye found inside the sclera.

Chrom- (Greek: *chrōma*, "color").

Chromatid. One of the two strands of a chromosome that are joined by a centromere but that separate during anaphase of mitosis.

Chromatin. The nucleoprotein of chromosomes that stains darkly with certain stains.

Chromatin threads. The uncoiled, thin strands of chromosomes found in the nucleus during interphase.

Chromoplast. A pigment-containing organelle (plastid) of plant cells.

Chromosomal anomalies. Abnormalities of chromosomes, including variations in their number, form, and size, that result in genetic defects and changes of cells and organisms possessing them.

Chromosomal theory of genetics. The theory that chromosomes are the elements that transmit inherited characteristics from one generation to the next.

Chromosomes (Greek: *soma*, "body"). The threads or rods of chromatin in the nucleus of cells that bear the genes in a linear order.

Chyme (Greek: *chy*, "juice"). The semifluid contents of the stomach and intestine.

Cilium (Latin: *cilium*, "eyelash"). Tiny hairlike extensions of the surfaces of some cells. Each encloses a ring of nine microtubules surrounding two centrally located microtubules. Plural is cilia.

Circadian (Latin: *circa*, "about"; *dies*, "a day"). Referring to biological rhythms and behaviors that follow approximately 24-hour cycles.

Cisterna (Greek: *cist*, "box"). A fluid-filled cavity found between parallel membranes of a golgi complex.

Cistron. The sequence of nucleotide bases in DNA that determines the order of amino acids in a peptide chain.

Class. A taxonomic subdivision of a phylum, composed of one or more related orders.

Cleavage. The cell divisions of a fertilized egg (a zygote) leading to the formation of a blastula.

Climax community. A stable community structure established through ecological succession.

Clitoris (Greek: *kleitoris*, "a small hill"). The female homolog of a penis.

Clone (Greek: *klon*, "twig"). A line of cells or a population of individuals descended asexually from an original cell or parent individual. All have the same genetic composition.

Coccus (Greek: *coccus*, "a berry"). A bacterium with a spherical shape.

Cochlea (Greek: *kochlias*, "snail"). The portion of the inner ear that contains the sensory receptors for hearing.

Cochlear duct (Latin: *cochl*, "snail shell, spiral"). The membranous sensory structure that contains the auditory receptors.

Codominant (Latin: *co*, "together"). The condition of two different dominant alleles composing the genotype and both being expressed in the phenotype.

Codon. A series of three nucleotides on a molecule of messenger RNA that specifies a single amino acid.

Coelom (Greek: *koilia*, "cavity"). The body cavity of higher animals, which may contain the internal organs and which is completely lined with mesoderm.

Coenzyme. A substance that is required along with an enzyme to catalyze a biochemical reaction. Some coenzymes are vitamins, such as riboflavin and thiamin.

Collagen (Greek: *kolla*, "glue"). A protein material found in connective tissue that when hydrolyzed, is converted to a gelatin.

Collecting tubule. The duct that drains kidney filtrate from the distal convoluted tubule to the pelvis of the kidney.

Collenchyma tissue. The plant tissue composed of elongate cells with thick secondary cell walls that strengthen growing plant tissues.

Colloid. A substance whose particles (groups of molecules), when dispersed in a solvent, remain uniformly suspended but that not form a true solution.

Colon. The large intestine of vertebrates.

Colony (Latin: *colonus*, "farmer"). A group of organisms, often physically connected, that may exhibit some degree of specialization for the division of labor.

Columnar epithelium (Greek: *column*, "pillar"; *epi*, "upon"). A type of epithelial tissue in which the cells are taller than they are wide.

Commensalism (Latin: *cum*, "together"; *mensa*, "table"). The form of symbiosis in which one member is benefited but the other is neither harmed or benefited.

Community. The populations of interacting species in a given large or small area or habitat.

Competitive exclusion. The concept that two species cannot occupy the same niche.

Complete flower. A flower that possesses sepals, petals, pistils, and stamens.

Cones. The cone-shaped photosensory cells in the retina of the eye, which are sensitive to bright light and which are also responsible for color vision.

Confrontation. The rarest outcome of mutual threat, a test of strength between two combatants during which little or no physical damage is done.

Conjugation (Latin: *conjugatio*, "a blending"). The form of sexual reproduction during which two unicellular organisms exchange genetic material.

Connective tissue. The type of animal tissue in which relatively few living cells are suspended in a large amount of nonliving intercellular material, the matrix. Examples: fibrous connective tissue, cartilage, bone, blood.

Contractile vacuole. An organelle that regulates the water content of unicellular organisms such as protozoa, which live in hypotonic media.

Control. A standard against which observations and results from experimental work can be checked to determine their validity.

Controlled experiment. An experiment in which a set of results or observations are compared between a control and an experimental group.

Convergent evolution (Latin: *cum*, "together"; *vergere*, "to incline"). The evolution of structures with similar form and function in two or more distinctly different groups of organisms, often occupying similar habitats.

Copulation (Latin: *copul*, "bond or link"). The sexual union of a male and female.

Copulins. Sexual pheromones.

Cork. A portion of the bark of woody plants, a secondary tissue composed of nonliving cells with cell walls infiltrated with a waxy material.

Cornea. The front transparent covering of the eye.

Coronary artery. An artery carrying oxygenated blood to heart muscle.

Coronary circulation. The system of blood vessels that supply and drain the heart tissue itself.

Corpus luteum (Latin: *corpus*, "body"; *luteus*, "yellow"). A mass of yellow endocrine cells that develop, following ovulation, at the site of the ruptured graffian follicle in a human ovary. The cells produce estrogen and progesterone.

Cortex (Latin: *cort*, "bark, shell"). The widest layer of the root, found just inside of the epidermis and extending to the endodermis.

Cotyledon (Greek: *kotyledon*, "shaped like a cup"). An embryonic leaf of a seed plant.

Covalent bond. A chemical bond formed between two atoms that share electrons to complete their outer orbitals.

Cranial nerves (Greek: *crani*, "skull"). The portion of the peripheral nervous system that consists of pairs of nerves that carry stimuli directly to and from the brain of vertebrate animals.

Cristae (Latin: *crista*, "a crest"). Folds or projections of the inner membrane of a mitochondrion.

Crossing over. The exchange of corresponding sections of chromatids between homologous chromosomes that occurs during synapsis in the first meiotic division.

Cubodial epithelium (Greek: *cubo*, "cube"; *epi*, "upon"). A type of epithelial tissue in which the cells are approximately as tall as they are wide.

Cyte, cyto- (Greek: *"vessel or container"*). Current use: pertaining to cell.

Cytochrome (Greek: *chroma*, "color"). An iron-containing protein that transfers electrons in redox reactions in the electron transport system, a part of aerobic cellular respiration.

Cytokinesis (Greek: *kinesis*, "motion"). The division of the cytoplasm of a cell during telophase of mitosis or meiosis.

Cytology. The study of the form and function of cells.

Cytoplasm. The living matter of a cell external to the nucleus.

Dark reaction. One of two periods in photosynthesis; chemical energy derived by the light reaction is used to synthesize new organic molecules from carbon dioxide and hydrogen ions.

Daughter cells. The two cells that result from the process of cell division.

Daughter chromosomes. Chromatids after they have separated during anaphase of mitosis or anaphase II of meiosis.

Deciduous (Latin: *decidere*, "to fall off"). Referring to plants that shed their leaves during certain seasons.

Decomposers. Organisms, such as bacteria, that break down dead organic matter into inorganic molecules that can serve as plant nutrients. Thus, they play a critical role in many biogeochemical cycles.

Dehydration synthesis. A chemical reaction that bonds together two organic molecules, such as amino acids, and during which a molecule of water is formed.

Demography (Greek: *dem*, "people"; *graphi*, "write"). Study of human populations.

Dendrites (Greek: *dendron*, "tree"). The processes of a neuron that conduct a nerve impulse toward the nerve cell body.

Deoxyribonucleic acid (DNA). The genetic material of the nuclei of cells, having a molecular form of a double helix composed of two spiralling chains of sugar molecules (deoxyribose) alternating with phosphate groups. The two chains are linked together by purine and pyrimidine base pairs.

Deoxyribose. The five-carbon sugar in DNA.

Dermis (Greek: *derma*, "skin"). The connective tissue layer beneath the epidermis in the skin of vertebrates.

Diaphragm. The muscle used in breathing that separates the chest cavities from the abdominal cavity.

Diastole. The relaxation phase following ventricular contraction, during which the ventricles fill with blood.

Diastolic pressure. The lowest pressure in the arteries, occurring when the ventricles are relaxed.

Differentiation. Development of an unspecialized cell into a more mature and specialized form.

Diffusion (Latin: *diffundere*, "to pour out"). The movement of substances from regions of high concentration to regions of lower concentration caused by the random movement of molecules in solution or suspension.

Digestion. The conversion of complex food molecules to simpler molecules by enzymatic action, a process often called hydrolysis.

Dihybrid cross. (*Di*, "two"; Latin: *hybrida*, "a mongrel"). The mating of two individuals during which the distribution of two sets of genetic characters are analyzed.

Dioecious (Greek: *di*, "two"; *oikos*, "house"). Referring to animals with separate male and female individuals and plants that have male and female flowers on separate individuals.

Diploid or diploid number (Greek: *diploos*, "twofold"). Twice the number of chromosomes found in gametes. A chromosomal composition of a pair of each of the different chromosomes typical of the species.

Directional selection. Selection in a changing environment for individuals possessing phenotypes that are best adapted to the environmental changes.

Disaccharide. A sugar composed of two monosaccharides. For example, the disaccharide maltose can be digested to produce two molecules of glucose.

Disassociate. To split the ionic bonds holding atoms together when a molecule is placed in solution.

Displacement. Behavior displayed by an animal when the drives to attack and to flee are essentially equal and inhibit each other. The behavior that results from the inhibition of two opposing drives.

Display. A visual releasing mechanism. Example: the display of a peacock's tail that serves to release the sexual responses of peahens.

Disruptive selection. Selection for phenotypes that are less common than the average phenotype.

Distal convoluted tubule (Latin: *dist*, "distant"). The duct that connects Henle's loop to the collecting tubule in nephron.

Divergent evolution. The development of at least two different phenotypes as the result of disruptive selection.

Diversion behavior. The responses to threat that divert the behavior of the threatening individual. For example, a threatened individual may begin to perform juvenile behaviors, such as begging for food.

Dizygotic twins. Fraternal twins that develop from two eggs ovulated near or at the same time. Each embryo is implanted separately in the uterine wall and possesses its own extraembryonic membranes.

DNA. *See* Deoxyribonucleic acid (DNA).

Dominance hierarchy. Referring to animals in the same group or family that have established a series of dominance relationships among themselves, ranging from the least dominant individual to the most dominant; Peck order.

Dominant. The phenotypic expression of an allele over all other allelic forms at the same locus.

Dorsal (Latin: *dorsal*, "back"). The back of an individual; the opposite of ventral.

Double fertilization. A characteristic of angiosperms; a sperm fuses with an egg, forming a diploid zygote, while another sperm fuses with two polar nuclei, producing a triploid endosperm nucleus.

Double helix. Descriptive of the form of DNA molecules. *See* also Deoxyribonucleic acid DNA.

Duodenum. The first portion of a vertebrate small intestine, where much of the digestion of food takes place.

Dynamic steady state (Greek: *dyn*, "power, energy"). The concept that the internal systems of an organism are kept in a constant state of change within certain set limits through time.

Ear. The organ of equilibrium and hearing in vertebrate animals.

Eardrum. The thin, flexible sheet of connective tissue that separates the outer and middle ear.

Eco- (Greek: *oikos*, "house or home").

Ecological succession. The gradual changes in the structure of a community that ultimately lead to the establishment of a stable climax community.

Ecology (Greek: *logos*, "discourse"). The study of the interactions between organisms and their abiotic environment.

Escosystem. The abiotic environment and all the organisms composing communities that interact with it and one another.

Ectoderm (Greek: *ektos*, "outer or outside"; *derma*, "skin"). One of the three embryological germ layers; ectoderm gives rise to the skin and its derivatives and to the nervous system.

Ejaculation (Latin: *ejacul*, "throw out"). The expulsion of seminal fluid through the urethra.

Ejaculatory duct. The duct connecting the seminal vesicle to the urethra in the mammalian male reproductive system.

Electrocardiogram (Greek: *electr*, "electricity"; *card*, "heart"). A recording of the electrical events occurring in the heart.

Electron. A negatively charged subatomic particle that orbits the atomic nucleus of an atom.

Electron transport system. A series of electron donors and acceptors that are a part of the aerobic cellular respiratory process. During successive series of redox reactions, energy is released; it then becomes incorporated into ATP molecules. Other electron transport systems occur in photosynthesis.

Element. A pure substance composed of a single kind of atom. The approximately 100 known elements comprise all matter in the universe.

Embryo (Greek: *en*, "in"; *bryein*, "to swell"). The early developmental stage of an organism contained within a seed, egg, or reproductive system of its mother.

Embryology (Greek: *en*, "in"; *bryein*, "to swell"; *logos*, "discourse"). The study of the development of an organism's fertilized eggs or other cells that give rise to new individuals.

Embryo sac. The female gametophyte of an angiosperm.

Endergonic reaction. A chemical reaction that requires an external supply of energy for its continuation.

Endo- (Greek: *endon*, "within").

Endocardium (Greek: *end*, "inner"; *card*, "heart"). The layer of epithelial cells lining the inside surface of the heart chambers.

Endocrine glands (Greek: *krinein*, "to separate"). Glands and cells that secrete products called hormones directly into the bloodstream. Hormones carried in the blood reach specific target sites (tissues or organs) that respond physiologically to the hormonal stimulus.

Endoderm (Greek: *derma*, "skin"). One of the three embryological germ layers of animals, formed during gastrulation and giving rise to the lining of most of the digestive system and its outgrowths, such as the liver and the pancreas.

Endodermis. The single layer of cells that separates the cortex from the vascular tissues in young plant roots.

Endometrium. The lining of the uterus in mammals, which thickens and then is sloughed off during menstruation.

Endoplasmic reticulum (Latin: *reticulum*, "network"). An elaborate system of membranes in the cytoplasm of cells that forms complex systems of channels and compartments. Parts of the endoplasmic reticulum bear ribosomes.

Endosperm (Greek: *sperma*, "seed"). The triploid (3n) tissue in the seeds of angiosperms that nourishes the developing embryo.

Energy. The ability to do work.

Energy of activation. The energy input necessary to start a chemical reaction.

Environmental resistance. Factors in the environment of a species that act to limit population size.

Enzyme (Greek: *en*, "in"; *zyme*, "leaven"). A protein molecule that acts as a catalyst for a chemical reaction in a cell or organ of an organism.

Epi- (Greek: *epi*, "upon or over").

Epicardium (Greek: *epi*, "upon"; *card*, "heart"). The outermost layer of the wall of the heart.

Epicotyl (Greek: *epi*, "upon"; *cotyl*, "cavity"). The part of a plant embryo above the cotyledons that gives rise to the stems and leaves.

Epidermis (Greek: *derma*, "skin"). The outermost layer of cells of a plant or animal.

Epididymus (Greek: *didymos*, "testis"). A complex of coiled tubules, adjacent to a testis, in which sperm are stored.

Epiglottis (Greek: *epi*, "upon"; *glott*, "tongue"). The structure near the base of the tongue that covers the opening of the larynx during swallowing.

Epiphyte (Greek: *phytum*, "a plant"). A plant that lives on (is mechanically supported by) another plant but does not derive nourishment from it. Example: many orchids and vines.

Epithelial tissue (Greek: *epi*, "on"; *thele*, "nipple"). A tissue layer that covers surfaces and lines cavities of organs and ducts. An epithelium always has one free surface.

Equational division *See* Meiosis II, or equational division.

Equatorial plane or plate. The position of alignment of chromosomes during metaphase of mitosis.

Erythrocytes (Greek: *erythros*, "red"; *kytos*, "vessel"). A hemoglobin containing red blood cells.

Esophagus (Greek: *esophag*, "esophagus"). The tube that connects the pharynx to the stomach.

Essential amino acids. The amino acids that the body

cannot synthesize and that must therefore be included in the diet for good health.

Estrogen. The female sex hormone that promotes the development of secondary sex characteristics and stimulates the growth of the uterine wall during the menstrual cycle. It is produced by ovarian follicles.

Estrus (Greek: *oistros*, "driven mad, in a frenzy"). A period of sexual receptivity and intense sexual drive in many female mammals.

Ethogram. A complete description or catalog of the modal action patterns of an animal or its behavioral repertoire.

Ethology (Greek: *ethos*, "habit"; *logos*, "discourse"). The study of habit; more conventionally defined as the biology of behavior.

Eu- (Greek: *eu*, "true").

Eukaryotic cells (Greek: *karyon*, "kernel"). Cells with nuclei enclosed by a membrane or envelope and possessing cell organelles such as golgi bodies and mitochondria.

Euphotic zone (Greek: *phos*, "light"). The thin upper layer of ocean water that contains sufficient light for photosynthesis.

Endosperm (Greek: *sperma*, "seed"). The triploid (3n) tissue in the seeds of angiosperms that nourishes the developing embryo.

Energy. The ability to do work.

Energy of activation. The energy input necessary to start a chemical reaction.

Environmental resistance. Factors in the environment of a species that act to limit population size.

Enzyme (Greek: *en*, "in"; *zyme*, "leaven"). A protein molecule that acts as a catalyst for a chemical reaction in a cell or organ of an organism.

Epi- (Greek: *epi*, "upon or over").

Epicardium (Greek: *epi*, "upon"; *card*, "heart"). The outermost layer of the wall of the heart.

Epicotyl (Greek: *epi*, "upon"; *cotyl*, "cavity"). The part of a plant embryo above the cotyledons that gives rise to the stems and leaves.

Epidermis (Greek: *derma*, "skin"). The outermost layer of cells of a plant or animal.

Epididymus (Greek: *didymos*, "testis"). A complex of coiled tubules, adjacent to a testis, in which sperm are stored.

Epiglottis (Greek: *epi*, "upon"; *glott*, "tongue"). The structure near the base of the tongue that covers the opening of the larynx during swallowing.

Epiphyte (Greek: *phytum*, "a plant"). A plant that lives on (is mechanically supported by) another plant but does not derive nourishment from it. Example: many orchids and vines.

Epithelial tissue (Greek: *epi*, "on"; *thele*, "nipple"). A tissue layer that covers surfaces and lines cavities of organs and ducts. An epithelium always has one free surface.

Equational division *See* Meiosis II, or equational division.

Equatorial plane or plate. The position of alignment of chromosomes during metaphase of mitosis.

Erythrocytes (Greek: *erythros*, "red"; *kytos*, "vessel"). A hemoglobin containing red blood cells.

Esophagus (Greek: *esophag*, "esophagus"). The tube that connects the pharynx to the stomach.

Essential amino acids. The amino acids that the body

cannot synthesize and that must therefore be included in the diet for good health.

Estrogen. The female sex hormone that promotes the development of secondary sex characteristics and stimulates the growth of the uterine wall during the menstrual cycle. It is produced by ovarian follicles.

Estrus (Greek: *oistros*, "driven mad, in a frenzy"). A period of sexual receptivity and intense sexual drive in many female mammals.

Ethogram. A complete description or catalog of the modal action patterns of an animal or its behavioral repertoire.

Ethology (Greek: *ethos*, "habit"; *logos*, "discourse"). The study of habit; more conventionally defined as the biology of behavior.

Eu- (Greek: *eu*, "true").

Eukaryotic cells (Greek: *karyon*, "kernel"). Cells with nuclei enclosed by a membrane or envelope and possessing cell organelles such as golgi bodies and mitochondria.

Euphotic zone (Greek: *phos*, "light"). The thin upper layer of ocean water that contains sufficient light for photosynthesis.

Eustachian tube (*Eustachio*: "an Italian anatomist"). A tubular passageway connecting the middle ear cavity to the pharynx and permitting air pressure on both sides of the tympanic membrane (eardrum) to equalize.

Eutrophic (Greek: *trophos*, "feeder"). A body of water with a rich nutrient supply that supports numerous living organisms.

Eutrophication (Greek: *eu*, "well"; *trophos*, "feeder"). A natural process whereby a body of water gains increasing levels of organic material through time.

Evolution (Latin: *e*, "out"; *volvere*, "to roll"). The continuous genetic change in organisms that results from their adaptation by natural selection to ever-changing environments.

Ex- (Latin: *ex-*, "out").

Excitatory neuron (Latin: *excit*, "arouse"). A neuron that, when stimulated, induces an impulse to start in another neuron, such as a motor neuron.

Excretion (Latin: *cernere*, "to separate or sift"). The elimination of metabolic waste by organisms.

Exergonic reaction. A chemical reaction releasing energy.

Experimental group. The set of individuals compared to the controls to determine the effects of a treatment.

Explosive evolution. The development of many species from one or a few ancestral lines in a short period of time.

Extant. Presently living.

External respiration. The process of drawing air into the lungs, where oxygen can diffuse into the blood and carbon dioxide can diffuse out of the blood.

Extinct. No longer living.

Extraembryonic membranes. The chorion, amnion, allantois, and yolk sac of birds, reptiles, and mammals. Membranes that enclose, protect, or support, both mechanically and physiologically, a developing embryo within an egg or in the uterus of the mother.

Eye. An organ of vision or sight.

F_1 **(first filial) generation.** The offspring of a cross of two parent organisms.

F₂ (second filial) generation. The offspring resulting from a cross between two members of an F₁ generation.

Facilitated diffusion. The process by which molecules unable to pass through a cell membrane by themselves are carried across a cell membrane by a carrier molecule from a region of high concentration to a region of low concentration.

Fallopian tube. A mammalian oviduct—the duct through which the ovum passes from the ovary to the uterus.

Family. A group of closely related genera.

Fat. A substance present in some tissues that is composed of carbon, hydrogen, and oxygen and is not soluble in water.

Fatty acid. A weak organic acid and a major component of simple lipids of plants and animals.

Fermentation (Latin: *fermentum,* "leaven"). A form of anaerobic respiration that usually produces alcohol or lactic acid.

Fertilization (Latin: *fertilis,* "to bear or produce"). The fusion of a sperm cell with an egg, or ovum, to produce a zygote, which develops into a new individual.

Fetus (Latin: *fetus,* "fruitful or pregnant"). An unborn offspring following the major events of embryological development. In humans, the unborn child is called a fetus after approximately three months of development.

Fibrin (Latin: *fibr,* "fiber"). A fibrous protein that helps form blood clots.

Fibrous connective tissue. Connective tissue containing networks of protein fibers and used to bind cells together.

Filament. The structure that supports the anther.

Flagellum (Latin: *flagellum,* "whip"). A long, hairlike process that extends from a cell and is often used in locomotion or feeding. Although longer and fewer in number than cilia, both types of cell processes have the same internal arrangement of microtubules. Plural is flagella.

Flower. The reproductive structure of angiosperms, which may contain the reproductive structures of one or both sexes.

Fluid mosaic model. The presently accepted model of membrane structure in which protein molecules are embedded in a lipid bilayer.

Follicle stimulating hormone (FSH). The hormone produced by the pituitary gland that stimulates the development of ovarian (Graffian) follicles.

Food chain, food pyramid, food web. A series of organisms at different trophic levels representing producers, herbivores, and carnivores, through which energy and materials pass in a community or ecosystem. The members of each trophic level above the producers feed on the trophic level below them.

Forebrain. One of the three major parts of the brain. Its greatest part consists of the cerebrum.

Fossil (Latin: *fossilis,* "to dig up"). Any recognizable remains of an organism preserved in the earth's crust. It may be represented as a footprint, skeletal part, or even feces.

Frond (Latin: *frond,* "leaf"). A fern leaf.

Fruit (Latin: *fructus,* "fruit"). A mature, ripened ovary or group of ovaries of an angiosperm, containing a seed or seeds.

FSH. *See* Follicle stimulating hormone (FSH).

G₁ phase. The phase of the cell cycle immediately following mitosis, during which considerable cell growth and synthesis of new cell organelles occur.

G₂ phase. The phase of the cell cycle that immediately follows the S phase and precedes mitosis, during which the synthesis of proteins and other substances (but not DNA) occurs in preparation for cell division.

Gallbladder. The saclike structure that stores and concentrates bile.

Gamete (Greek: *gametē,* "wife"). A haploid sex cell, commonly called an egg or sperm, that unites with another such cell to produce a diploid zygote.

Gametophyte or gametophyte generation. The haploid gamete–producing generation in the life cycle of plants.

Ganglion (Greek: *gangli,* "knot"). A mass of nerve cell bodies that produces an enlargement, or swelling, on an otherwise smooth nerve of relatively constant diameter.

Gastrula (Greek: *gaster,* "stomach"). The stage of early embryological development in animals following the blastula. During the formation of a gastrula, an inner layer of cells, the endoderm, forms; it initiates the formation of the digestive system.

Gastrulation. The process of forming a gastrula from a blastula, often by the invagination of the wall of the blastula.

Gene (Greek: *gennan,* "to produce"). The biological unit of heredity; a self-replicating sequence of nucleotides in the DNA of a chromosome that controls the development of a specific trait of an organism.

Gene frequency. The mathematical expression for the frequency of a particular allele in a population.

Gene pool. The total of all the alleles of all the genes that occur in all individuals in a population.

General senses. The sensations of touch, pain, heat, cold, pressure, and body position.

Genetic code. The system of three base-pair sequences that occurs in DNA molecules and that determines the order of amino acids in protein molecules synthesized by cells.

Genetic drift. Random changes in the gene pool of small populations caused by change rather than by natural selection.

Genetics. The study of heredity.

Genotype. The total genetic composition or hereditary assortment of all the genes of an organism. *See also* Phenotype.

Genotypic ratio. The ratio of the different genotypes resulting from a cross of two parents.

Genus. A group of closely related species. Plural is genera.

Gill. The respiratory structure of an aquatic animal through which the exchange of oxygen and carbon dioxide occurs; commonly a thin-walled area of the body wall with an increased surface area to facilitate gas exchange.

Gland. An organ or cell that secretes one or more substances to the outside of itself.

Glomerular filtrate. The fluid portion of the blood that passes from the glomeruli of the kidneys into the nephrons.

Glomerulus (Latin: glomus, "ball"). The knot of capillaries enclosed by each Bowman's capsule in a vertebrate kidney. Plural is glomeruli.

Glucose ($C_6H_{12}O_6$). A six-carbon sugar commonly found in living organisms. Glucose is metabolized as a source of energy by cellular respiratory processes.

Glycerol. A three-carbon compound that combines with fatty acids to produce simple lipids.

Glycolysis (Greek: *glykys,* "sweet"; *lysis,* "solution"). The conversion of a molecule of glucose into two molecules of pyruvic acid during cellular respiration. This conversion results in the net gain of two APT molecules.

Golgi complex. The cytoplasmic organelle found in eukaryotic cells that plays a role in the synthesis of complex substances and in cellular secretion of a variety of these substances.

Gonad (Greek: *gone*, "seed"). Ovary or testis, a gamete-producing organ.

Gonorrhea. A contagious venereal disease caused by a bacterium.

Granum. The portion of a chloroplast that houses chlorophyll molecules and is the site of the light reaction of photosynthesis. Plural is grana.

Greenhouse effect. The accumulation of carbon dioxide in the atmosphere that may result in the gradual increase in the temperature of the atmosphere and the earth's surface.

Guard cells. One of the two cells that open and close a stoma in a leaf's surface and thereby regulate the flow of gases and water loss from a leaf.

Gustation (Latin: *gust*, "taste"). The sensation of taste.

Habitat (Latin: *habere, habitus*, "to hold"). The area in an environment lived in or occupied by a species of an organism.

Habitat isolation. A mechanism of speciation wherein one species gives rise to other species as a result of adaptations to separate, nonidentical habitats.

Haploid (Greek: *haploos*, "single"). Referring to a cell that possesses only one member of each pair of homologous chromosomes. Gamates and the cells of the gametophyte generation of plants are haploid.

Haploid number. The number of chromosomes found in gametes and plant cells of the gametophyte generation; equal to one-half the diploid number.

Hardy-Weinberg law. The mathematical expressions for the frequencies of a pair of alleles in a population.

Hemo-, Hemato- (Greek: *haima*, "blood").

Hemoglobin. The iron-containing pigment in red blood cells that combines with and facilitates the transport of oxygen.

Hemophilia. "Bleeder's disease," a hereditary disease. The blood of people with hemophilia does not clot property, and many hemophiliacs die of blood loss.

Henle's loop. The duct that connects the proximal and distal convoluted tubules of kidney nephrons.

Hepatic portal circulation (Greek: *hepa*, "liver"). A system of blood vessels that carry blood from the digestive system to the liver.

Herbaceous (Latin: *herba*, "grass"). Any nonwoody plant.

Herbivore (Latin: *herba*, "grass"; *vorare*, "to devour"). An animal or other organism that eats mostly plants or other photosynthetic organisms.

Hermaphrodite. Any organism that is both male and female or that possesses both male and female sexual organs.

Hetero- (Greek: *heteros*, "other or different").

Heterochromatin (Greek: *hetero*, "different, other"; *chroma*, "color"). Densely staining areas within the nucleus, composed of DNA and protein.

Heterosis. Hybrid vigor. When members of two unrelated strains are mated, their offspring often are more vigorous and better adapted than their parents.

Heterotrophic or heterotroph (Greek: *trophos*, "feeder"). An organism that must feed on other organisms or organic matter produced by other organisms because it cannot synthesize its own organic compounds from inorganic substances, as autotrophic organisms can.

Heterozygous (Greek: *zygos*, "yolk"). Referring to an organism or cell that possesses two different alleles for a trait at the same loci, or positions, on a pair of homologous chromosomes.

Hexose (Greek: *hexa*, "six"). Any six-carbon sugar.

Hindbrain. One of the three major parts of the brain. It contains three major parts: the cerebellum, medulla oblongata, and pons.

Homeo-, homo-, homol- (Greek: *homos*, "same or similar").

Homeostasis (Greek: *homeo*, "like, unchanging"; *stasis*, "standing"). The maintenance of a stable environment inside an organism that differs from and is independent of normal changes in the environment in which the organism lives.

Hominid (Latin: *homini*, "man"). Modern-day humans as well as fossil humanlike relatives. This group does not include modern apes.

Hominoidea. The super-family to which humans and the great apes belong.

Homologous chromosomes. The pair of similar chromosomes in diploid organisms that synapse during the first meiotic cell division; chromosomes that bear alleles or genes at corresponding locations, or loci.

Homologous structures. Structures of animals that develop in a similar fashion and therefore are similar. Example: the arm of a human and the wing of a chicken.

Homologues. Homologous chromosomes.

Homozygous (Greek: *zygos*, "yolk"). Referring to an organism or cell that possesses the same alleles for a trait at the same loci, or positions, on a pair of homologous chromosomes.

Homozygous dominant. The case of an organism possessing two dominant alleles for the same trait.

Homozygous recessive. The case of an organism possessing two recessive alleles for the same trait.

Hormone (Greek: *hormaein*, "to excite or set in motion"). The secretion of an endocrine gland or cell that is carried in the blood and other body fluids to a different part of the body, where it stimulates or affects other cells or an organ.

Host. An organism on or in which a parasite lives.

Humus (Latin: *hum*, "ground"). Dead plant material on or mixed in the soil.

Hybrid. The offspring of two parents with different genotypes or the offspring of parents of different species.

Hydrogen bond. A weak chemical bond formed between a hydrogen atom and a more negatively charged portion of another molecule. Example: the hydrogen bonds that connect the hydrogen atoms of one water molecule with the more negatively charged areas occupied by the oxygen atoms of two other water molecules.

Hydrolysis (Greek: *hydra*, "water"; *lysis*, "loosening"). The enzymatic splitting of one molecule into two through the addition of water. The hydrogen ion attaches to one of the two molecules produced and the hydroxyl ion (OH^-) to the other.

Hymen (Greek: *hymen*, "membrane"). The membrane that partially occludes the vaginal opening.

Hypertonic (Greek: *hyper*, "above or over"; *tonos*, "tension or tone"). Having more solute molecules and fewer solvent molecules (water) than does a hypotonic solution separated from it by a semipermeable membrane. In this situation water molecules move from the hypotonic solution through the semipermeable membrane into the hypertonic solution because of osmotic pressure.

Hypo- (Greek: *hypo*, "under or less than").

Hypocotyl (Greek: *hypo*, "beneath"; *cotyl*, "cavity"). The part of a plant embryo found below the cotyledons.

Hypothalamus (Greek: *thalamos*, "inner chamber or room"). The floor of the brain beneath the cerebral hemispheres, which contains centers of the autonomic nervous system that regulate the body's water balance, temperature, sleep, and appetite. The hypothalamus also produces substances that stimulate and inhibit the pituitary gland.

Hypothesis (Greek: *thesis*, "arranging or setting down"). An assumption or proposal based on accumulated evidence that can be further tested and thereby supported or rejected by experimentation or by gathering further evidence or information. An aspect or step in the scientific method.

Hypotonic. Having fewer solute molecules and more solvent molecules than another solution with which it is compared. *See also* Hypertonic.

Ileum (Latin: *ile*, "intestine"). The third part of the small intestine, which joins the jejunum to the large intestine.

Immunity. The ability of a living organism to resist and overcome infection.

Imperfect flowers. Flowers that possess either stamens or pistils but not both.

Implantation (Latin: *in*, "into"; *plantare*, "to set"). The attachment of a developing embryo to the endometrium of the uterus in mammels.

Imprinting. A brief period of rapid learning that occurs within critical periods, usually following hatching in birds and birth in mammals. A genetically determined time to learn a specific response. Example: the following response of ducks and geese first demonstrated by Konrad Lorenz.

Incomplete dominance (codominance). A blending in the expression of traits determined by two alleles. Example: a cross between red carnations and white carnations produces pink carnations.

Incomplete flower. A flower that lacks at least one of the following: sepals, petals, pistils, or stamens.

Incus (Latin: *incu*, "an anvil"). The middle bone of the three bones found in the ear.

Independent assortment. *See* Mendel's second law.

Inferior vena cava (Latin: *infer*, "low"; *ven*, "vein"; *cav*, "hollow"). The major vein that drains blood from the posterior part of the body back to the heart.

Inhibitory neuron. A neuron that, when stimulated, reduces the chance of an impulse occurring in a neuron with which it synapses.

Innate releasing mechanism (IRM). A genetically coded, programmed system of neurons that respond to only specific environmental stimuli.

Inner ear. A composite of interconnected fluid-filled structures that contain detectors for sound and balance.

Instinct. An innate behavioral pattern not based on previous experiences. Examples: the sucking or nursing response of humans, human facial expressions.

Insulin (Latin: *insul*, "island"). A hormone that is secreted by the islets of Langerhans in the pancreas and that controls the amount of dissolved sugar in the blood.

Integumentary system. The skin and its derivative parts.

Internal receptors. Sensory cells that provide information about pressure, touch, temperature, position, movement, and pain.

Internal respiration. The utilization of food to produce energy within cells.

Interneuron (Latin: *inter*, "between"). A nerve cell that synapses with two other nerve cells, commonly between a sensory neuron and a motor neuron.

Interphase. The period of a cell cycle between mitotic or meiotic cell divisions. The period of the cell cycle during which DNA is replicated. Interphase is commonly divided into the G_1, S, and G_2 phases.

Intrauterine devices (IUDs) (Latin: *intra*, "within"). Plastic coils or loops inserted into the uterus to serve as a form of birth control.

Invagination (Latin: *in*, "in"; *vagina*, "sheath"). The inward folding of a layer of tissue or cells. During gastrulation, one side of a blastula invaginates to form a two-layered embryo, the gastrula.

Ion (Greek: *ion*, "going"). An electrically charged atom or group of atoms produced by lowing or gaining electrons.

Ionic bond. A chemical bond formed by the electrical attraction between a positively charged ion and a negatively charged ion.

Iris (Greek: *iri*, "iris of the eye"). The pigmented tissue in the eye that surrounds and controls the size of the pupil.

Iso- (Greek: *isos*, "equal").

Isogamy (Greek: *gamos*, "marriage"). Sexual reproduction associated with the union of two gametes alike in size and form.

Isolating mechanisms. Mechanisms that prevent closely related species in the same area from interbreeding. These may be morphological, behavioral, or physiological in nature.

Isotonic. Referring to a solution having the same concentration of solute and solvent molecules as a solution being compared with it. *See also* Hypertonic; Hypotonic.

Isotope (Greek: *topos*, "place"). Any of a number of possible forms of a chemical element. All forms have the same number of protons and electrons but differ in the number of neutrons.

Joint. A connection between adjacent bones.

Jejunum. The middle portion of the small intestine, which connects the duodenum to the ileum.

Karyotype (Greek: *karyon*, "nut or nucleus"; *typos*, "type"). A description or illustration of the number, size, and shape of the chromosomes in the cells of an organism.

Keratin (Greek: *keratos*, "horn"). A water-resistent, horny protein found in the epidermis, nails, hooves, horns, claws, feathers, and hair of vertebrates.

Kidney. An excretory organ of vertebrates that also functions in maintaining the body's water and solute balance.

Kingdom. A major taxonomic unit of the living world, which includes a number of related phyla that share a few basic characteristics. Examples: kingdom Animalia, kingdom Plantae.

Krebs cycle. A series of biochemical reactions in aerobic cellular respiratory, which follows glycolysis.

Kwashiorkor. A severe protein deficiency of humans caused by a lack of dietary protein.

Labia majora (Latin: *labi*, "lip"; *major*, "larger"). Folds of fatty tissue that enclose the vulva area of female mammals.

Labia minora (Latin: *labi*, "lip"; *minor*, "smaller"). Folds of tissue lying within the folds of the labia majora.

Lamella (Latin: diminutive of *lamina*, "a plate or leaf"). A layer or thin sheet or plate. Plural is lamellae.

Large intestine. The final portion of the intestine, which includes the colon and the rectum.

Larva (Latin: *larva*, "ghost"). An immature stage in the life cycle of an animal that bears little or no resemblance to the adult form. Example: a frog's tadpole. Plural is larvae.

Larynx (Greek: *laryn*, "gullet"). The cartilaginous structure at the upper end of the trachea that contains the vocal cords; the voice box.

Law. A scientific statement or conclusion that is invariably true under precisely stated conditions. Example: Mendel's first and second laws of genetics.

Lens. The transparent structure within the eye that focuses light rays onto the retina.

Leucocytes (Greek: *leukos*, "white"; *kytos*, "vessel or cell"). White blood cells that help defend the body from infection.

Leucoplast (Greek: *leukos*, "white"; *plastes*, "molder"). A colorless organelle used for starch storage in plant cells.

LH. *See* Luteinizing hormone (LH).

Ligament. A white band of tough, fibrous connective tissue that connects one bone to another.

Light reaction. The first stage of photosynthesis—a series of biochemical reactions that result in the incorporation of light energy into chemical energy in the chemical bonds of ATP.

Limiting factor. Any environmental factor that prevents an organism from living in a certain place or that may set a limit on its population size where it does exist.

Limnetic zone (Greek: *limm*, "lake"). Lake surface water that is located away from the shore. No rooted vegetation occurs in the limnetic zone.

Linkage. The tendency for two or more genes located on the same chromosome to be inherited together.

Lipase (Greek: *lipos*, "fat"). An enzyme that digests (hydrolyzes) simple fats into molecules of fatty acids and glycerol.

Lipid (Greek: *lipos*, "fat"). Any of the fatty organic compounds such as fats, waxes, and oils.

Littoral zone (Latin: *littor*, "the seashore"). The area of a lake along the shore that contains rooted vegetation.

Liver. A larger glandular organ that produces bile and performs many other functions.

Locus (Latin: *locus*, "place"). The position of a gene on a chromosome. Plural is loci.

Lordosis (Greek: *lord*, "bent backward"). A position assumed by some animals that facilitates copulation.

Lung. One of a pair of organs used in external respiration.

Luteinizing hormone (LH). A hormone secreted by the pituitary gland that helps stimulate ovulation and the formation of the corpus luteum.

Lymph (Latin: *lympha*, "lymph"). The fluid portion of blood that filters through the walls of capillaries and collects in lymph channels for return to the circulatory system. It contains a variety of leucocytes.

Lymph node (Latin: *lymph*, "water"; *nod*, "swelling"). The structure interposed in a lymph vessel that functions to filter lymph and produce lymphocytes, a type of white blood cell.

Lymph vessel (Latin: *lymph*, "water"). The passage that carries lymph and is part of the lymphatic system.

Lymphatic system. The system of channels and lymph nodes through which lymph passes on its return to the circulatory system.

Lysosome. A cytoplasmic organelle in the form of a membrane-enclosed vesicle containing a variety of hydrolytic enzymes.

Malaria (Italian: *malari*, "bad air"). A disease caused by several species of protozoans that infect red blood cells.

Malleus (Latin: *malle*, "hammer"). The first bone of the three bones found in the ear.

Marine (Latin: *mare*, "the sea"). Referring to organisms living in the oceans.

Matter. Any substance that occupies space and has mass.

Medulla oblongata. The most posterior portion of the brain, which connects with the spinal cord. The medulla controls autonomic responses such as respiration and circulation.

Megaspore (Greek: *megas*, "large"; *spora*, "a seed"). The haploid spore that produces the female gametophyte in plant life cycles.

Meiosis (Greek: *meiosis*, "to make smaller"). The two successive nuclear divisions that occur during the formation of gametes and spores and that reduce the number of chromosomes from the diploid number to the haploid number. Each cell resulting from meiotic divisions receives one number of each pair of homologous chromosomes.

Meiosis I, or reductional division. The first cell division of meiosis, during which homologous chromosomes separate and produce two haploid cells.

Meiosis II, or equational division. The second cell division of meiosis, during which the two chromatids of each chromosome separate to form daughter chromosomes.

Melanin (Greek: *melas*, "black"). A dark pigment that occurs in special pigment cells, called melanocytes, in the skin of many animals.

Melanocyte (Greek: *melan*, "black"). A pigmented cell containing melanin. Melanocytes are found in the bottom layer of the skin.

Mendelian genetics. The series of predictions resulting from the laws of principles discovered by Gregor Mendel that govern the inheritance of some traits.

Mendel's first law. The factors that determine hereditary traits are inherited as units, some being dominant and some recessive; although they may occur as pairs in organisms, the two can segregate in the next generation. The law of segregation.

Mendel's second law. The inheritance of one pair of factors occurs independently of the simultaneous inheritance of other factors or traits. Each pair of factors "assorts independently" of other factors. The law of independent assortment.

Menopause (Greek: *men*, "month"; *pausis*, "cessation"). The time during middle age of human females when the recurring menstrual cycle stops.

Menstruation (Latin: *menstrualis*, "monthly"). The cyclic discharge of blood and disintegrated uterine lining through the vagina of human females and of certain other primates.

Meristematic tissue. Generalized, undifferentiated plant tissues that give rise to new cells. Example: apical meristem and the cambium.

Mesoderm (Greek: *mesos*, "middle"; *derma*, "skin"). The primary germ layer of an embryo that develops between the ectoderm and the endoderm. Mesoderm forms the muscles, the connective tissues, the circulatory system, and most of the reproductive and excretory system.

Messenger RNA (mRNA). The nucleic acid that transcribes the genetic code of genes. mRNA moves from the nucleus, where it is formed, to the ribosomes, where it serves as a template for ordering the amino acid sequence during protein synthesis.

Metabolism (Greek: *metaballein*, "to change or alter"). The total of all the physical and chemical processes of a cell or organism that maintain its living state.

Metamorphosis (Greek: *meta*, "after or over"; *morphosis*, "form"). The often abrupt transition from one stage in the life cycle of an animal into another. Example: the change from a tadpole to a frog.

Metaphase (Latin: *meta*, "middle"; *phasis*, "form"). The stage of mitosis and meiosis when the chromosomes are aligned on the metaphase, or equatorial plate.

Micro- (Greek: *mikros*, "small").

Microbody. A cell organelle; a small membrane-bound, fluid-filled vesicle containing a mixture of enzymes usually used to break down amino acids and lipids.

Microfilament. Fibrous structure located within cells that may support structures and that may be involved in cellular movements and contractions.

Microspore (Greek: *spora*, "seed"). The spore that becomes the male gametophyte, the pollen grain, in the life cycle of plants.

Microtrabecular system. A complex system of fine filaments and fibrils that extend throughout the cytoplasm of cells.

Microtubule. A cytoplasmic organelle in the form of a hollow cylinder formed by thirteen filaments arranged in a circle; found in most eukaryotic cells.

Midbrain. One of the three major parts of the brain, functioning to transmit information from the forebrain to the hindbrain.

Middle ear. The air-filled cavity containing the three bones of the ear.

Middle lamella. The layer between the cell walls of adjacent plant cells that is rich in pectin and glues the cells together.

Mimicry (Greek: *mimos*, "to imitate"). The physical copying by one organism of another organism or nonliving thing that serves as an adaptation for survival. Example: some insects look like leaves or twigs.

Mineral. A dietary inorganic substance needed for good health. Examples: calcium, iron.

Mitochondrion (Greek: *mitos*, "thread"; *chondrion*, "granule"). The cytoplasmic organelle that contains most of the enzymes involved in cellular respiration. The site of ATP formation during respiration. Plural is mitochondria.

Mitosis (Greek: *mitos*, "thread"; *osis*, "condition or state"). The nuclear division associated with the division of somatic cells, during which the chromosomes divide to provide each daughter cell with a complete set of chromosomes—the same complement as that of the mother cell.

Modal action pattern (MAP). The recognizable and repeatable behavioral event or action of an animal that can be assigned a function.

Molecule (Latin: *molecula*, "a little mass"). The smallest chemical unit of a substance composed of two or more atoms.

Monera. The kingdom that includes bacteria and blue-green algae. The cells of members of this kingdom lack nuclei and most membranous cytoplasmic structures.

Mono- (Greek: *monos*, "single").

Monoecious (Greek: *oikos*, "house"). Having both sexes in the same individual; in the case of animals, being hermaphroditic; in plants, having both stamens and pistils on the same individual but in separate flowers.

Monohybrid cross. A mating between two individuals during which an analysis of the distribution of a single genetic trait is made.

Monosaccharide (Greek: *sakcharon*, "sugar"). A simple sugar such as glucose, a six-carbon sugar, or ribose, a five-carbon sugar.

Monosomy (Greek: soma, "a body"). The condition in which one chromosome is missing from a normal diploid complement.

Monozygotic twins (Greek: *zygotes*, "yolked together"). Identical twins—genetically identical because they both came from a single fertilized ovum after it divided into two cells, or two cell masses, each of which developed into one of the twins.

Mons veneris (Greek: *mon*, "single"; Latin: *vener*, "pertaining to venus"). An elevated pad of tissue above external female genitalia; in adult women, covered with pubic hair.

Morphogenesis (Greek: *morphe*, "form"; *gennan*, "to produce"). The development of a particular body part or organ in regard to size, form, and other detailed structural features.

Morula (Latin: *morula*, "a little mulberry"). A solid ball of cells produced by the cleavage of a zygote; precedes the blastula in human development.

Mosaic. An individual who possesses patches of tissue that differ genetically.

Motivational state (Latin: *mot*, "move"). A mode of existence in which certain types of behaviors predominate, depending on the kinds and amounts of hormones in the body.

Motor neuron. A nerve cell that transmits nerve impulses from the central nervous system to a voluntary (skeletal) muscle.

Motor unit. A motor neuron and the muscle fibers it stimulates.

M phase. The part of the cell cycle during which mitosis and cytokinesis take place.

Multicellular organism. Any organism whose body is composed of more than one cell.

Multiple-allelic inheritance. Traits regulated by more than two alleles. Example: blood type.

Muscle fiber. A multinucleated cylindrical elongated muscle cell.

Muscle tissue (Latin: *muscul*, "muscle"). A type of tissue that causes movement by being able to contract.

Mutagen. A substance or physical factor that causes changes in the structure of DNA—mutations.

Mutation (Latin: *mutare*, "to change"). A change in the form of a chromosome or the chemical structure of a gene that is inheritable.

Mutualism. A symbiotic association of two organisms in which both are benefited and each is unable to survive without its partner.

Myelin sheath (Greek: *myel*, "spinal cord"). A complex layer of lipids laid down by Schwann cells that surrounds neuron axons and protects and insulates them from other neurons.

Myo- (Greek: *mys*, "muscle").

Myofibril. A contractile element in a muscle fiber composed of overlapping myofilaments of actin and myosin. *See also* Sliding filament model.

Myosin. Protein filaments in muscle fibers that overlap with the thinner actin filaments. *See also* Sliding filament model.

NAD. *See* Nicotinaminde adenine dinucleotide (NAD).

Natural selection. The process of evolution, where organisms that are best adapted to their environments and are the most successful in reproducing pass their genotypes to succeeding generations.

Nekton (Greek: *nektos*, "swimming"). The aquatic organisms, such as most fish, that are strong, effective, and capable swimmers able to move against water currents and tides.

Nephron (Greek: *nephros*, "kidney"). The function and structural unit of the kidneys of vertebrate animals.

Neritic zone. The seawater that covers the continental shelves.

Nerve. A smooth bundle of nerve cells, or nerve cell processes, and associated connective tissue that extend from the central nervous system to various parts of the body.

Nerve impulse. A temporary wave of depolarization of the electrical charges, associated with the nerve cell membrane, sweeps quickly along a neuron.

Nerve tissue (Latin: *nerv*, "nerve"). A type of tissue that is able to carry and transmit nerve impulses.

Neura-, neuro- (Greek: *neuron*, "nerve").

Neural tube. An embryonic tubular invagination along the dorsal surface of a vertebrate embryo that closes over and forms the brain and the spinal cord.

Neuron (Greek: *neur*, "nerve"). A functional nerve cell.

Neurotransmitter chemical or substance. A chemical secreted by the axonal ending of a neuron that diffuses across the synapse to another neuron; in sufficient quantities, it may induce the postsynaptic neuron to transmit a nerve impulse.

Neurulation. The embryological formation of the neural tube and central nervous system following gastrulation of an embryo of a vertebrate animal.

Neutrons Along with protons, subatomic particles that comprise the atomic nucleus of atoms; unlike protons, neutrons are uncharged particles.

Neutrophil. A type of leukocyte that stains readily with neutral dyes.

Niche. The total way of life of a species; the unique way in which a species uses the resources of its environment. No two species can occupy the same niche at the same time for an extended period.

Nicotinaminde adenine dinucleotide (NAD). A coenzyme in biological redox reactions that plays an important role in respiration and photosynthesis.

Nitrogen cycle. The biogeochemical circulation and reutilization of nitrogen atoms, involving plants, animals, and important groups of microorganisms that play key roles in nitrogen fixation and nitrification.

Nitrogen fixation. The biological processing of atmospheric nitrogen into nitrogen compounds usable by plants. Some nitrogen-fixing microorganisms live symbiotically in plant roots.

Nitrogenous base. One of the purines or pyrimidines that form a part of DNA and RNA molecules.

Nondisjunction. The failure of homologous chromosomes to separate during meiosis; this causes half of the gametes to possess two, instead of one, member of a homologous pair; the other half of the gametes possess none of the chromosomes of a particular homologous pair.

Nonrenewable resources. Commodities in the environment that are not replaceable once used, such as fossil fuels.

Nucle- (Latin: *nucleus*, "kernel").

Nuclear membrane. The double membrane that encloses the nucleus in cells of eukaryotes.

Nucleic acid. Acidic organic compounds that contain the genetic code—DNA or RNA.

Nucleoid. The region of the cytoplasm of a bacterial cell that contains the nucleic acid molecule.

Nucleolus. A spherical body composed of RNA, DNA, and protein in the nucleus of cells. Believed to be the site of manufacture of ribosomal RNA (rRNA).

Nucleoplasm. The protoplasm that is contained in a cell's nucleus.

Nucleotide. A section of a DNA or RNA molecule composed of a phosphate group, a pentose sugar, and a pyrimidine or purine.

Nucleus. A cellular organelle that contains the genetic material of the cell.

Oceanic water (Greek: *ocean*, "the ocean"). The water that fills the deep ocean basins.

Oligotrophic (Greek: *olig*, "few"; *trophos*, "feeder"). A body of water with a poor nutrient supply and therefore low in nutrients necessary for productivity.

Omnivore (Latin; *omni*, "all"; *vorare*, "to devour"). An animal that eats both plants and animals.

One-gene, one-enzyme principle. The concept that each gene can give rise to only one polypeptide chain.

Ontogeny (Greek: *on*, "existing"; *gennan*, "to produce or originate"). The embryological, larval, and juvenile development of an individual organism from zygote to adult.

Oo- (Greek: *oion*, "egg").

Oocyte (Greek: *kytos*, "vessel"). A cell that undergoes meiosis and gives rise to an egg, or ovum.

Oogenesis (Greek: *gennan*, "to produce"). The origin and differentiation of an ovum, or egg, by the process of meiosis.

Oogonium (Greek: *gone*, "generation"). The cell in an ovary that differentiates into an oocyte.

Ootid. A haploid cell produced by meiosis that will differentiate into an ovum.

Optic nerve (Greek: *opt*, "eye"). A nerve that transports information from the sensory cells in the eye to the brain.

Orbit (Latin: *orbi*, "circle"). The socketlike cavity of the skull that contains an eye.

Order. A taxonomic group consisting of a number of closely related families of organisms.

Organ *(Greek: organon, "a tool").* A body structure or part with defined functions that is composed of several tissues. Examples: heart, stomach, kidney.

Organelle. A recognizable and structurally distinct protoplasmic structure in cells. Examples: mitochondria, Golgi bodies, chloroplasts.

Organ system. A group of organs anatomically ordered to complete a major body function. Examples: the digestive system, the excretory system.

Organic compounds or substances. Chemicals containing carbon; also substances produced by living organisms.

Organism. A living thing or individual—a microbe, fungus, animal, or plant.

Organogenesis (Greek; *gennan*, "to produce"). The embryological development of organs and organ systems.

Osmosis (Greek: *osmosis*, "impulsion"). The movement of water or some other solvent through a membrane separating two solutions. The solvent moves from the solution lower in solute content into the solution with higher solute contact. Although the solvent can pass through the membrane, the solute cannot. Thus, the membrane is semipermeable.

Osmotic pressure. The internal force exerted by cellular contents that influences the flow of water through a cell membrane.

Outer ear. The part of the ear that consists of the pinna and the auditory canal.

Ov-, ovi- (Latin: *ovum*, "egg").

Oval window. The membrane-covered opening of the inner ear that is in contact with the stapes.

Ovary (Latin: *ovaria*, "ovary"). The female gonad that produces eggs, or ova.

Oviduct. A duct through which ova, or eggs, pass to a uterus or to the exterior of an animal.

Oviparous. Referring to animals that lay eggs, such as birds and reptiles.

Ovulation. The release of an ovum, or egg cell, from an ovary.

Ovule. A seed plant's egg cell.

Ovum. The female gamete or sex cell. Plural is ova.

Oxidation. The loss of electrons by an atom or molecule during a redox chemical reaction.

Oxygen debt. The depletion of oxygen from a tissue as the result of exercise.

Pair formation. The establishment of a permanent or temporary relationship between a male and female in anticipation of mating.

Palisade layer. A layer of columnar chlorophyll-containing cells beneath the upper epidermis of a leaf.

Pancreas (Greek: *pan*, "all"; *kreas*, "flesh"). A complex gland that secretes a mixture of digestive enzymes into the duodenum and that contains endocrine cells that produce insulin.

Parallel evolution. The development of similar structures and habits caused by similar environmental pressures in two or more different lineages of organisms.

Parasite (Greek: *parasitos*, "one who eats at another person's table"). An organism that lives on or in the body of another organism, the host, from which it derives nutrients and to which it often causes some degree of harm.

Parasitism. A form of symbiosis in which one member is benefited at the expense of the other member.

Parasympathetic nervous system. The division of the autonomic nervous system of vertebrates that is, in a sense, antagonistic to the other division, the sympathetic nervous system. Generally, the parasympathetic system slows body functions and the sympathetic system speeds them up, as during an emergency situation.

Parenchyma tissue (Greek: *en*, "in"; *chein*, "to pour"). A generalized plant tissue composed of thin-walled cells with large vacuoles. The cells may be photosynthetic and may store materials and furnish the plant with support.

Parental generation (P_1). The two parental organisms that are crossed to produce offspring.

Parthenogenesis (Greek: *parthenos*, "virgin"; *genesis*, "production"). The development of an unfertilized egg, or ovum, into an adult.

Passive immunity (Latin: *immun*, "free"). The ability to resist infection because of the transference of antibodies from one organism to another.

Peck order. *See* Dominance hierarchy.

Pelagic zone (Greek: *pelag*, "the sea"). The water that lies over the continental shelves plus the water that fills the ocean basins.

Penis. The male copulatory, or sex, organ of mammals and some birds.

Pentose. A five-carbon sugar.

Pepsin (Greek: *peps*, "digest"). An enzyme that breaks down complex proteins to form shorter, simpler proteins.

Peptide bond. A covalent chemical bond that unites amino acids in the primary structure of a polypeptide or protein.

Perennial plant (Latin: *per*, "through"; *annus*, "a year"). A plant that lives for more than one year and blooms during each blooming season.

Perfect flower. A flower that contains both male parts (stamens) and female parts (pistils).

Pericardial cavity (Greek: *peri*, "around"; *card*, "heart"; Latin: *cav*, "hollow"). The cavity that is surrounded by the pericardium and in which the heart is located.

Pericycle (Greek: *peri*, "around"). A thin layer of parenchyma cells in plant stems immediately outside the primary phloem.

Peripheral nervous system (Greek: *peripher*, "outer surface"). A system of cranial nerves and spinal nerves that transmit information to and from the brain.

Peristalsis (Greek: *peri*, "around"; *stalis*, "contraction"). The rhythmic contractions of the smooth muscle of the intestinal tract that move food through the system.

Petiole. The stalk portion of a leaf that attaches it to a stem.

Phagocytosis (Greek: *phagein*, "to eat"; *kytos*, "hollow vessel"). The process by which a cell engulfs, or eats, particulate matter. Example: white blood cells phagocytize bacteria.

Pharynx (Greek: *pharyn*, "throat"). The part of the digestive tract that connects the mouth cavity to the esophagus.

Phenotype (Greek: *phainein*, "to show"; *typos*, "type"). The visible, or outward, expression of the genetic composition of an individual.

Phenotypic ratio. The ratio of the different phenotypes resulting from a cross of two parents.

Pheromone (Greek: *phorein*, "to carry"). A substance secreted by an organism into its environment that induces a behavioral, physiological, or developmental response by other members of the same species.

Phloem (Greek: *phloos*, "bark"). A vascular tissue in higher plants that is composed of sieve tubes and companion cells. Phloem conducts substances formed in the leaves to other parts of the plant.

Phloem fibers (Greek: *phloe*, "bark of a tree"). Conductive tissue cells that transport food material from the leaves to the stems and roots.

Phosphorus cycle. The biogeochemical cycle that cycles the element phosphorus throughout the biosphere.

Photolysis (Greek: *photo*, "light"; *lys*, "loose"). The dissociation of water or other compounds caused by the action of light.

Photophosphorylation (Greek: *photos*, "light"). The production of ATP (adenosine triphosphate) that occurs in the light reaction of photosynthesis.

Photoreceptors. Light-sensitive cells (rods and cones) within the retina.

Photosynthesis. (Greek: *photos*, "light"; *syn*, "together"; *tithenai*, "to place"). The physiological process in plants and many protists and monerans that converts light energy to chemical energy and at the same time produces new organic substances from inorganic precursors such as water and carbon dioxide.

Photosystem. A series of redox reactions that comprise a part of the light reaction of photosynthesis.

pH scale. A scale of 1 to 14 used to indicate the degree of acidity or alkalinity of a solution: 1 through 6 is acid, 7 is neutral, and above 9 is alkaline, or basic.

Phylogeny (Greek: *phylon*, "tribe or race"; *genesis*, "generation"). The evolutionary history of a group of organisms and the evolutionary relationships of that group to other related groups of organisms.

Phylum (Greek: *phylon*, "tribe or race"). A group of closely related classes of organisms that comprise a unique major, though not necessarily large, taxonomic subdivision of living organisms. Plural is phyla.

Phytoplankton (Greek: phyt, "a plant"; *planktos*, "wandering"). The minute autotrophic organisms that float freely in both fresh water and ocean water.

Piloerection (Latin: *pil*, "hair"; erect, "upright"). The erection of hairs.

Pinna (Latin: *pinn*, "wing"). The external part of the ear that can be seen.

Pinocytosis (Greek: *pinein*, "to drink"; *kytos*, "a hollow vessel"). The uptake or ingestion of droplets of fluid by a cell.

Pistil. The seed-producing structure of a flower, commonly consisting of a stigma, style, and ovary.

Pistillate. Referring to a flower that contains one or more pistils but is without stamens.

Pituitary gland (Latin: *pituitarius*, "producing phlegm"). The small endocrine gland attached to the hypothalamus at the base of the brain that secretes a large number of hormones, including those that regulate growth, sexual development, and sexual reproduction.

Placenta (Latin: *placenta*, "a flat plate or cake"). The structure that attaches a fetus to the uterine lining of the mother and is composed partly of fetal tissues and partly of maternal tissues. Gas, nutrients, and excretory wastes are exchanged between the fetal and maternal blood through the placenta.

Plankton (Greek: *planktos*, "wandering"). Mostly minute heterotrophic and autotrophic organisms that float freely in both fresh water and ocean water.

Plantae (Latin: *planta*, "plant"). The kingdom containing all plants.

Plaque. A material made of a combination of lipids and calcium that can accumulate in the walls of arteries, causing the disease called atherosclerosis.

Plasma. The fluid portion of blood, which is a clear, colorless, complex solution of many substances in water.

Plastid. A cytoplasmic organelle of plant cells that usually contains pigments. Example: chloroplasts.

Platelet (Greek: *plate*, "a flat surface"). A small cell fragment in the blood of mammals that functions in blood clotting.

Plate tectonics. The movement of sections of the earth's crust.

Pleiotropy (Greek: *pleion*, "more"; *tropos*, "turn"). The involvement of a single gene in the expression of several different traits.

Pleural cavity (Greek: *pleur*, "rib"; Latin: *cav*, "hollow"). The cavity in mammals in which a lung is located.

Polar body. A minute cell formed during oogenesis. During oogenesis, only one functional ovum is produced with most of the cytoplasm of the original oocyte. During each meiotic division, one large cell, the future ovum, and one small cell, the polar body, are formed.

Pollen or pollen grain (Latin: *pollen*, "fine flour or dust"). The male gametophyte generation of seed plants.

Pollination. The transfer of pollen from the anthers, where it is formed, to the stigma of the same or other flowers of the same species.

Pollution (Latin: *pollut*, "defiled"). Contamination in an ecosystem that leads to instability and harm to the biotic community.

Poly- (Greek: *polys*, "many").

Polygenic inheritance. The inheritance of traits that are determined by the interaction of several different alleles, such as skin color, height, and weight.

Polypeptide. A molecule formed by a few to many amino acids joined by peptide bonds.

Polyploidy (Greek: *ploos*, "fold"; *odes*, "resembling or like"). The occurrence of more than two complete sets of chromosomes in a cell.

Polysaccharide. A carbohydrate molecule composed of several simple sugars (monosaccharides) joined together. Examples: starch and cellulose.

Pons (Latin: *pons*, "bridge"). The lower portion of the hindbrain, which connects the cerebellum with the cerebrum.

Population. All of the members of a species living in a specific habitat or geographical region.

Population density. The number of individuals per area.

Posterior. The rear, or tail end, of an animal; the opposite of anterior.

Primary oocyte. A cell that undergoes meiosis to eventually give rise to an egg, or ovum. *See also* Oocyte.

Primary productivity. The rate of synthesis of new organic matter by autotrophic organisms inhabiting a specified area.

Primary spermatocyte. A cell that undergoes meiosis to produce sperm cells. *See also* spermatocyte.

Primates. The order of class Mammalia that contains prosimians, monkeys, apes, and humans.

Primitive (Latin: *primus*, "first"). An early nonspecialized stage in the evolution of a group of organisms; contrasts with advanced or specialized.

Primordial soup (Latin: *prim*, "first"). The original mixture of organic and inorganic matter from which life arose.

Principle of dominance. The concept of one allele being expressed in the phenotype to the exclusion of all other alleles for the same trait.

Principle of summation. The addition of excitatory and inhibitory stimuli that affect the initiation of an impulse in a motor neuron.

Pro- (Greek; *pro-*, "before, in front of, forward").

Profundal zone (Latin: *profund*, "deep"). The lake water under the limnetic zone that receives no light and thus supports no plant life.

Progesterone (Latin: *gestus*, "to bear or carry"). A hormone that is produced by the corpus luteum after ovulation and that helps regulate the menstrual cycle. Progesterone is also produced by the placenta and is necessary to maintain pregnancy.

Prokaryotic cells (Greek: *karyon*, "nut or nucleus"). Any cell of a member of the kingdom Monera; prokaryotic cells lack membrane-enclosed nuclei and other membranous cytoplasmic organelles, such as plastids and Golgi bodies.

Prophase (Greek: *phasis*, "an appearance"). The first stage, or phase, of mitosis or meiosis, during which the nuclear membrane disappears and the chromatin coils and condenses to become visible chromosomes.

Proplastid. The cytoplasmic organelle of a plant cell from which the various types of plastids, such as chloroplasts, are thought to differentiate.

Prosimians (Latin: *simia*, "an ape"). Lemurs, tarsiers, tree shrews, and other lower primates. The more primitive of the two primate suborders.

Prostaglandins. A number of lipid compounds that stimulate smooth muscle contraction and inhibit lipolysis, platelet aggregation, and gastric secretion.

Prostate gland (Greek: *prostates*, "one who stands before"). The gland that encloses the base of the urethra in male mammals and that secretes some of the seminal fluid.

Protein (Greek: *proteios*, "primary"). One of the major chemical constituents of living matter; macromolecules formed from complex, convoluted chains of amino acids bonded to one another by peptide bonds.

Prothallus (Greek: *thallus*, "a young shoot or twig"). The gametophyte stage in the life cycle of a fern.

Protista (Greek: *protist*, "very first"). The kingdom containing algae (except blue-green algae), fungi, slime molds, and protozoa; individuals have a true nucleus and chromosomes.

Protonema (Greek: *nema*, "a thread"). The early gametophyte generation of a moss plant; a thin filament of cells that germinates from a spore and gives rise to the leafy moss gametophyte.

Protons (Greek: *protos*, "first"). Positively charges subatomic particles that, along with neutrons, comprise the bulk of the atomic nucleus.

Protoplasm (Greek: *plasma*, "something that is molded"). Living cellular substance or material.

Proximal convoluted tubule (Latin: *proxim*, "nearest"). The duct that connects Bowman's capsule to the Henle's loop in the kidney.

Pseudohermaphrodite (Greek: *pseudes*, "false"; *hermaphroditos*, "a person who is both male and female"). A human or other mammal who is genetically a male or a female but who has external genital organs similar to those of the opposite sex.

Pseudostratified epithelium (Greek: *pseudo*, "false"; Latin: *strat*, "layer"; Greek: *epi*, "upon"). A type of epithelial tissue in which the cells lying next to the free surface appear to be in several layers; however, they are only one cell-layer thick.

Pulmonary circulation. The circulation of blood from the right ventricle through the pulmonary arteries to the lungs, where it is oxygenated, and then back to the left atrium via the pulmonary veins.

Pulse (Latin: *puls*, "beat"). The surge of blood sent into the circulatory system from the heart.

Punnett square. A checkerboard diagram used to determine all the possible combinations of alleles that may be produced by fertilization during the analysis of the inheritance of one or more pairs of alleles.

Pupil (Latin: *pupill*, "pupil of the eye"). The central opening of the eye, through which light enters to strike the retina.

Pure strain. A group of interbreeding organisms homozygous for certain traits and therefore always producing offspring with the same genotypes and phenotypes as the parents for those traits.

Purine. Organic molecules, called nitrogenous bases, that are essential components of nucleic acids and ATP. In DNA molecules, purines pair with pyrimidines. Examples: adenine, guanine.

Pyrimidine. Organic molecules, called nitrogenous bases, that are essential components of nucleic acids. In DNA molecules, pyrimidines pair with purines. Examples: cytosine, thymine, uracil.

Radial symmetry. The arrangement of the major parts of an organism around a central axis so that any plane perpendicular to and passing through the axis will divide the organism into two similar halves.

Radicle (Latin: *radic*, "root"). The part of the plant embryo that forms the root system of the plant.

Random assortment. The separation of the two chromosomes comprising homologous pairs and their random movement to opposite poles of a cell during the first meiotic cell division; the random movements of one pair in no way affect the separation of any other pair of homologous chromosomes.

Recessive allele (Latin: *recedere*, "to recede"). An allele whose phenotypic expression occurs only in the homozygous state; when coupled with a dominant form of the allele, only the dominant phenotype is expressed.

Rectum (Latin: *rect*, "straight"). The final portion of the large intestine of mammals, which terminates at the anus.

Redox reaction. An oxidation-reduction during which one atom or molecule loses one or more electrons (is oxidized) while another atom or molecule gains one or more electrons (is reduced).

Reduction. The addition of one or more electrons to an atom or molecule.

Reductional division. *See* Meiosis I, or reductional division.

Reflex (Latin: *reflexus*, "bend back"). An involuntary functional unit of the nervous system that involves a sensory neuron, an interneuron, and a motor neuron. Example: the knee-jerk reflex.

Releaser. A morphological structure or behavioral signal that results in a response or reaction by another member of the same animal species.

Releasing factors. Chemical substances produced in the hypothalamus of the brain that either stimulate or inhibit the activity of the pituitary.

Releasing mechanism. A hypothetical part of the brain that is selectively stimulated by a specific releaser. By analogy, a lock and key; the key is the releaser, and the lock is the releasing mechanism.

Renal (Latin: *renalis*, "the kidney"). Pertaining to a kidney.

Renal artery (Latin: *ren*, "kidney"; Greek: *arteri*, "artery"). The vessel that supplies blood directly from the aorta to a kidney.

Renal vein *(Latin: ren*, "kidney"; *ven*, "vein"). The vessel that carries blood from the kidney to the inferior vena cava.

Renewable resources. Commodities in the environment that can be replaced once used, such as food.

Respiration (Latin: *respirare*, "to breathe"). The process by which cells convert the chemical energy of nutrients such as glucose and other organic compounds into a usable form for cell functions, such as the high energy bonds of ATP. The process requires oxygen and produces carbon dioxide and water. *Respiration* is also defined as the uptake of oxygen and the elimination of carbon dioxide by an organism.

Retina (Latin: *rete*, "net"). The inner, light-sensitive layer of the eye that contains the rods and cones.

Rh factor. An antigen found in the blood of about 85 percent of the white population that was originally discovered in the blood of rhesus monkeys.

Rhizoid (Greek: *rhiza*, "root"). Rootlike structures of nonvascular plants.

Rhizome (Greek: *rhizoma*, "a mass of roots"). Modified underground plant stems often used for storage. Example: a potato.

Ribonucleic acid (RNA). A nucleic acid containing the pentose sugar ribose, the purines adenine and guanine, and the pyrimidines cytosine and uracil. RNA plays a key role in protein synthesis and is the genetic material of many viruses.

Ribose. The pentose sugar found in ribonucleic acid.

Ribosomal RNA (rRNA). The type of RNA found in ribosomes, the sites of protein synthesis.

Ribosomes. The sites of protein synthesis; small cytoplasmic granules composed of RNA and protein that occur freely in the cytoplasm or are attached to the membranes of the endoplasmic reticulum.

RNA. *See* Ribonucleic acid (RNA).

Rods. Light-sensitive cells found in the retina of the eye. Rods are especially sensitive to dim light and are responsible for black and white vision but not color vision.

Root. The lower portion of a vascular plant that anchors the plant in the soil and serves as the organ for the uptake of water and inorganic nutrients.

Rough ER. A type of endoplasmic reticulum that has ribosomes associated with it.

Round window. A membrane-covered opening near the base of the cochlea.

Saliva (Latin: *saliva*, "spittle"). The secretions of the salivary glands of the mouth that moisten and lubricate food and that contain an enzyme that begins the digestion of starch.

Salivary gland (Latin: *saliv*, "spittle"). A gland that secretes saliva.

Saturated fat. Fat containing fatty acids in which every carbon atom is bonded to four other atoms (no double bonds exist). Saturated fats solidify at room temperature and are easily converted to cholesterol, a compound implicated in human heart and artery diseases.

Schwann cell. A cell that produces part of the mylin sheath that may surround an axon.

Scientific method. The procedure used by scientists while conducting scientific investigations. It involves preparing and testing hypotheses—tentative explanations of the relationships among various phenomena. If a hypothesis cannot be supported through objective observations and experimentation, it is rejected, and an alternative hypothesis is formulated and tested.

Scientific name. A designation given to every species,

consisting of two Latin words. The first word is the genus. The second is a subdivision of the genus. The genus is always capitalized; the subdivision name is not; both are italicized.

Scion. The plant to which a stem of another plant is grafted.

Sclera (Greek: *scler*, "hard"). A tough, fibrous layer of connective tissue that surrounds the eye and is continuous with the cornea; the white of the eye.

Sclerenchyma tissue (Greek: *sclero*, "hard"; *enchyma*, "an infusion"). A plant tissue, composed of cells with thick secondary cell walls, that serves to support and protect the plant.

Secondary oocyte. The cell that results from the reductional division (meiosis I) of a primary oocyte.

Secondary sex characteristics. Characteristics that develop as an organism becomes sexually mature and that serve to distinguish between the sexes but that are not involved directly in reproduction. Examples: the mane of a male lion, the comb of a rooster.

Secondary spermatocyte. A cell that results from the reductional division (meiosis I) of a primary spermatocyte.

Secretion (Latin: *secretio*, "to secrete"). A substance, manufactured by cells of a gland, that is released from cells through cell membranes.

Seed. A mature ovule of a seed plant, which contains an embryo of the sporophyte generation, a seed coat, and often endosperm.

Segregation. The basis of Mendel's first law of genetics: during meiosis, the separation of alleles and chromosomes into different haploid gametes.

Semen (Latin: *semen*, "seed"). The sperm-carrying viscous fluid produced by the male reproductive system and deposited in the reproductive tract of the female during copulation.

Semicircular canal (Latin: *semi*, "half"; *circum*, "around"; *canal*, "duct"). One of three ducts within the internal ear that houses receptors associated with balance.

Semilunar valves (Latin: *semi*, "half"; *lun*, "moon"; *valv*, "folding door"). The valves found between the left ventricle and the aorta and between the right ventricle and the pulmonary artery.

Seminal vesicle. A gland associated with each of the male reproductive ducts that secrete a portion of the semen.

Seminiferous tubules (Latin: *semen*, "seed"; *ferre*, "to bear"). The coiled ducts in the testes where sperm are produced.

Semipermeable membrane. A membrane, such as a cell membrane, through which certain substances can pass but others cannot.

Sensory neuron (Latin: *sensi*, "feeling"). A nerve cell that leads from a receptor toward the central nervous system and transmits nerve impulses toward it.

Sepal. One of the four main parts of a flower. Sepals are the leaflike, usually green structures that enclose the other flower parts in a flower bud.

Sessile (Latin: *sessil*, "sedentary"). Referring to an organism, usually an animal or protist, that is attached to its substrate and is unable to move from one location to another.

Sex chromosomes. The X and Y chromosomes, which determine the sex of an individual. In mammals, males possess an X and a Y chromosome in each diploid cell; females possess two X chromosomes.

Sex-influenced inheritance. The inheritance of a genotype that is expressed differently in males than in females, although the inherited alleles are the same. Example: human baldness.

Sex-limited inheritance. The inheritance of genotypes for characters that are expressed only in one sex. Examples: breast development in females, beard growth in males.

Sex-linked inheritance. The inheritance of alleles on the sex chromosomes. Example: alleles for color blindness.

Sex-linked traits. Characteristics determined by alleles located on the sex chromosomes.

Sexual reproduction. The form of reproduction that involves the formation of haploid gametes and the subsequent union of a male gamete with a female gamete, producing a diploid zygote.

Simple epithelium (Greek: *epi*, "upon"). A type of epithelium tissue in which the cells lying next to the free surface are only one cell-layer thick.

Sinoatrial node. A specialized mass of cardiac muscle tissue that stimulates the heart to beat; the pacemaker of the heart.

Skeletal muscle (Greek: *skelet*, "skeleton"; Latin: *muscul*, "muscle"). The type of voluntary striated muscle that attaches to bones and is responsible for causing movement of the skeleton. Organisms have conscious control over this tissue. *See also* Striated muscle.

Skeleton (Greek: *skelet*, "skeleton"). A framework of bones and cartilage used to support and protect the body.

Sliding-filament model. The explanation of how striated muscle cells contract because of the relative movements of actin and myosin filaments in sarcomeres.

Small intestine. The region of the digestive tract where most digestion and absorption take place.

Smooth ER. A type of endoplasmic reticulum that does not have ribosomes associated with it.

Smooth muscle (Latin: *muscul*, "muscle"). The type of slow-contracting involuntary muscle tissue that is found in internal organs of the body. Organisms have no conscious control over this tissue.

Sodium pump. The active transport system, associated with the cell membrane of a neuron, that maintains the unequal distribution of potassium and sodium ions inside and outside the cells, thereby electrically polarizing the membrane.

Solute (Latin: *solvere*, "to dissolve"). The substance that dissolves in a solvent to form a true solution.

Solution. The uniform, permanent distribution of a solute (usually a solid or gas) in a solvent (usually a liquid).

Solvent (Latin: *solvere*, "to dissolve"). A fluid capable of dissolving solutes to form a true solution.

Somatic cell (Greek: *soma*, "the body"). Any cell of a multicellular plant or animal other than the gametes. Somatic cells are usually diploid.

Somatic nervous system. The portion of the nervous system that mediates conscious voluntary action (in contrast to the autonomic nervous system, which regulates unconscious involuntary actions, such as digestion and heartbeat).

Sorus (Greek: *sorus*, "a heap"). A cluster of sporangia on a fern's leaf. Plural is sori.

Special senses. The senses of taste, smell, vision, and hearing.

Speciation (Latin: *species*, "sort or kind"). The formation of new species through the process of evolution.

Species (Latin: *species*, "sort or kind"). A taxonomic subdivision of a genus, containing populations of similar organisms that interbreed but that do not usually breed with members of other subdivisions of the genus under natural conditions.

Sperm- (Greek: *sperma*, "seed").

Spermatid. A haploid cell, produced by meiosis, that differentiates into a spermatozoon or sperm cell.

Spermatocyte. A cell that undergoes meiosis and produces spermatids.

Spermatogenesis (Greek: *genesis*, "origin"). The process by which sperm are formed from spermatogonia.

Spermatogonium. A cell that becomes a spermatocyte, which in turn produces spermatids. Plural is spermatogonia.

Spermatophore (Greek: *phorein*, "to bear"). A structure that is produced by a male reproductive system and that serves as a packet, or container, for sperm.

Spermatozoan. A male gamete that is a haploid cell. Plural is spermatozoa.

Sperm cell. A spermatid that has completely matured to form a haploid male gamete. *See also* Spermatozoan.

Spermicides. Substances that kill or inactivate sperm. Some are used in birth control.

S phase. The period of the cell cycle during which DNA is replicated.

Spinal cord. The nerve tissue of the central nervous system that is enclosed within the vertebral column of vertebrates.

Spinal nerves. Any of the paired nerves that connect to the spinal cord.

Spindle. A structure, composed of fine filaments, that forms during prophase of mitosis and meiosis and within which the chromosomes are aligned during metaphase.

Spirillum. (Latin: *spira*, "a coil or spiral"). A bacterium with a coiled or spiral form. Plural is spirila.

Spongy layer. A layer of loosely organized cells below the palisade layer in the leaf of a plant. Spaces between the cells allow room for air circulation.

Spontaneous generation. The concept, finally disproved by Pasteur, that living organisms arise directly from nonliving matter. For example, it was once believed that flies were formed from decaying flesh and that mice were formed from old clothes.

Spor- (Greek: *spore*, "seed").

Sporangium. A structure in plants and many microorganisms that produces spores. Plural is sporangia.

Spore mother cell. A diploid cell in plants that undergoes meiosis to produce haploid spores.

Sporophyte generation, or sporophyte. The diploid spore-producing generation of plants that have alternation of generations.

Squamous epithelium (Latin: *squam*, "scale"; Greek: *epi*, "upon"). A type of epithelial tissue in which the cells lying next to the free surface are flat or scale-like.

Stabilizing selection. Selection for the most common phenotype.

Stamen (Latin: *stamen*, "anything standing upright, a thread"). The pollen-producing male organ of a flower.

Staminate. Referring to a flower that contains male organs (stamens) but not female organs (pistils).

Stapes (Latin: *stupe*, "stirrup"). The bone in the middle ear that abuts the oval window.

Starch. A carbohydrate with large molecules that is composed of a chain of many glucose molecules bonded together.

Stem. The usually upright portion of a plant that is usually above ground, that usually bears leaves, and, in the case of flowering plants, bears the flowers. Often stems can be horizontal, and some may occur underground (for example, rhizomes).

Stereoscopic vision. The type of vision that allows an animal to see objects in three dimensions.

Stigma (Greek: *stigma*, "a spot"). The part of the female structure of a flower on which pollen must be deposited during pollination.

Stimulus filtering. The process of tuning out unwanted or irrelevant signals. Example: a daydreamer may not hear the professor lecturing.

Stock. The plant to which a scion is grafted.

Stolon (Latin: *stoloni*, "a twig or shoot"). A horizontal plant stem such as the runner of a strawberry plant.

Stoma (Greek: *stoma*, "mouth"). The openings on the lower surface of a leaf through which gases pass. Each stoma is bordered by two guard cells, which act to open and close it. Plural is stomata.

Stomach (Greek: *stomach*, "stomach"). A muscular, saclike organ between the esophagus and small intestine in which food is stored and mixed and in which protein digestion begins.

Stratified epithelium (Latin: *strat*, "layer"; Greek: *epi*, "upon"). A type of epithelial tissue in which there exists more than a single layer of cells.

Striated muscle (Latin: *stria*, "streaked"; *muscul*, "muscle"). The type of muscle that attaches to bones and is responsible for causing movement of the skeleton. Organisms have conscious control over this tissue. In longitudinal section, the tissue appears striated. *See also* Skeletal muscle.

Stroma (Greek: *stroma*, "anything that is spread out, such as a blanket"). The substance inside a chloroplast in which the lamellae and the grana are suspended.

Style (Greek: *stylus*, "a pillar or column"). The stalk portion of a pistil that arises from the ovary and bears the stigma on its distal end.

Suspension. A temporary mixture of solid particles or liquid droplets in a gas or liquid. The particles are not in solution and will settle out in time. Examples: dust particles in air; oil droplets in water.

Sym, Syn (Greek: *syn*, "together").

Symbiosis (Greek: *Symbio*, "living together"). The close association between organisms belonging to different species. Symbiosis includes parasitism, in which one organism is benefited and the other harmed; mutualism, in which both of the partners are benefited; and commensalism, in which one member is benefited but the other is neither benefited nor harmed.

Sympathetic nervous system. The division of the autonomic nervous system that generally speeds up many body functions in response to emergency situations. The sympathetic system is in a sense antagonistic to the parasympathetic system.

**Sympatric species (Greek: *sym*, "with or together"; *pa*-

tria, "fatherland or habitat"). The species of living organisms that have overlapping ranges or habitats.

Synapse (Greek: *synapsis*, "conjunction or union"). The tiny gap between the axon of one neuron and the dendrite or cell body of another neuron.

Synapsis (Greek: *synapsis*, "conjunction or union"). The pairing of homologous chromosomes during prophase of the first meiotic division.

Syphilis. A human venereal disease caused by a bacterial infection transmitted by sexual contact.

Systemic circulation. The circulation of blood throughout the body—exclusive of pulmonary circulation, which refers to the circulation of the blood from the heart to the lungs, where it is oxygenated, and back again to the heart.

Systole (Greek: *systol*, "contraction"). The contraction phase of the heart muscles.

Systolic pressure (Greek: *systol*, "contraction"). The highest pressure obtained in the arteries, occurring when the ventricles contract.

Taiga. A coniferous forest where precipitation averages 37 to 100 centimeters per year and winter temperatures are low.

Target cell. A specific cell type on which a hormone has an effect.

Taxonomist (Greek: *taxis*, "to put in order or arrange"; *nomos*, "law"). A biologist whose specialization is the naming and classifying of organisms.

T-cell. A type of lymphocyte that functions in the immune reaction.

Telophase (Greek: *telo*, "an end"; *phasis*, "phase or form"). The final stage of mitosis and meiosis, during which the nuclei of the two future daughter cells are reorganized and cytokinesis occurs.

Template. A mold or pattern. For example, the genetic code in DNA molecules serves as the pattern for the synthesis of more DNA and RNA.

Tendon (Greek: *tebib*, "tendon"). An inelastic cord of fibrous connective tissue that connects muscle to either bone or cartilage.

Territoriality. The acquisition and defense of a space by an animal.

Territory (Latin: *terra*, "the earth"). An area occupied by one or more animals that is aggressively defended by them against intruders of the same species. Examples: nesting and food gathering territories of birds.

Testis, or testicle (Latin: *testis*, "a testicle"). A male gonad that produces sperm and that may also produce sexual hormones. Plural is testes.

Testosterone. The male sex hormone that is produced by the testes of higher vertebrates and that induces the development of male sex characteristics.

Tetrad (Greek: *tetra*, "four"). The configuration formed by a pair of homologous chromosomes during synapsis in the first meiotic division; in plants, the four spores formed from a spore mother cell during meiosis.

Theory (Greek: *theorein*, "to look at"). A generalization established by supporting the validity of a hypothesis through experimentation and observations.

Thermocline (Greek: *therm*, "heat"; *clin*, "slope"). A thin layer of water that separates a warm surface layer of water from a colder water mass beneath it.

Threat. The prelude to aggression, with animals looking as fierce and large as possible.

Tissue (Latin: *texere*, "to weave"). A group of cells with similar structure and function. Examples: muscle and epithelium, parenchyma and collenchyma.

Trace element. An element needed by plants and microorganisms in minute amounts for normal growth and development. Also called micronutrients.

Trachea (Greek: *trachelos*, "the throat"). The windpipe, or air tube, that conducts air from the pharynx of a mammal to the bronchii, which lead to the lungs. Also the respiratory ducts of insects and other arthropods. Plural is tracheae.

Tracheid (Greek: *tracheia*, "rough"). A long, tapering cell in the xylem of vascular plants that functions in conducting substances through the plants and that gives support to the plant.

Transcription. The process by which the genetic code of DNA is transferred by enzymatic action to messenger RNA. The code, or base sequence, in the RNA is complementary to that of the DNA that serves as the template.

Transfer RNA (tRNA). Small molecules of RNA that are specific for and attach to individual amino acid molecules prior to their assembly into proteins on the ribosome-mRNA complex.

Translation. The establishment of the correct amino acid sequence during protein synthesis according to the template of base sequences in mRNA.

Translocation. (1) The transportation of the photosynthetic products through the phloem in vascular plants. (2) The transfer of a piece of a chromosome to another chromosome.

Transpiration (Latin: *spirare*, "to breathe"). The loss of water vapor through the stomata of the leaves of plants.

Tricuspid value (Latin: *tri*, "three"; *cuspi*, "point"; *valv*, "folding door"). The valve separating the right atrium and right ventricle.

Trisomy (Latin: *tri*, "three"; Greek: *some*, "a body"). The abnormal occurrence of three homologous chromosomes, instead of two, of a particular type in a diploid cell.

Trophoblast (Greek: *trophos*, "feeder"; *blastus*, "a bud"). The portion of the blastocyst (blastula) stage in the early development of a human or other mammal that gives rise to the chorion and later becomes a part of the placenta.

Tuber (Latin: *tuber*, "a knot or knob"). A modified underground plant stem. Example: a potato.

Turgor (Latin: *turgere*, "to swell"). The rigid nature of a plant cell caused by the uptake of fluids. A loss of turgor results in wilting.

Umbilical cord (Latin: *umbilicus*, "navel"). The stalk, or cord, containing blood vessels that connects a mammalian embryo or fetus to the placenta.

Umwelt. The world as perceived and experienced by an animal living in it. The Umwelt of every species is different. Example: humans see color, dogs do not.

Unicellular organism (Latin: *uni*, "one"). A microorganism whose entire body is composed of only one cell.

Unit membrane. The common three-layered structure of all cell membranes, both external and internal, seen in electron photomicrographs.

Universal donor. An individual whose blood is type O.

Type O blood can be received by individuals having any type of blood.

Universal recipient. An individual with blood type AB, who can receive any type of blood.

Unsaturated fat. Any fat composed of molecules containing fatty acids with one or more double bonds. *See also* Saturated fat.

Urea (Greek: *ouron*, "urine"). The nitrogenous waste product of mammals, produced by the liver and other cells and excreted by the kidneys in the urine.

Ureter (Greek: *ourein*, "to urinate"). The duct, or tube, that carries the urine from a kidney to the bladder in humans and other mammals.

Urethra. The duct, or tube, that carries the urine from the bladder to the outside of the human body; also found in other mammals.

Urine. The fluid produced by the kidneys, containing nitrogenous waste products.

Uterus (Latin: *uterus*, "the womb"). The expandable portion of a female reproductive tract in which eggs are stored or in which a fetus develops.

Vacuole (Latin: *vacuus*, "empty"; *ole*, "ending"). A membrane-enclosed, fluid-filled space in a cell.

Vagina (Latin: *vagina*, "a sheath"). The terminal portion of the female reproductive duct. The male's penis is inserted into the vagina during copulation.

Vascular bundle. A group of supporting and vascular tissues in a plant. Example: the veins in a leaf.

Vascular tissue. Xylem and phloem in plants; tissue composed of tubes and vessels through which fluids pass.

Vas deferens (Latin: *vas*, "duct"; *deferens*, "carrying away"). The duct in the male mammalian reproductive system that conducts sperm from the epididymis to the urethra. Plural is vasa deferentia.

Vasectomy. A means of male contraception in which the vasa deferentia are cut and tied off, preventing the passage of sperm to the urethra.

Vegetative reproduction. A form of asexual reproduction in plants. Examples: the propagation of strawberries by runners, the propagation of potatoes from sections of tubers containing eyes.

Vein (Latin: *vena*, "a vein"). (1) A blood vessel that conducts blood toward the heart from the tissues. (2) A vascular bundle in the leaf of a plant.

Vena cava (Latin: *vena*, "vein"; *cava*, "a hollow or cave"). A large vein that opens into the right atrium of the heart and conducts blood to it.

Venous blood (Latin: *ven*, "vein"). Blood carried by the veins to the heart.

Ventral (Latin: *venter*, "belly"). The lower surface or undersurface of an animal; the opposite of dorsal.

Ventricle (Latin: *ventriculus*, "a small stomach"). A chamber of the heart that has a thick, muscular wall and that receives blood from an atrium or auricle and pumps blood from the heart. Also, a cavity in the brain or other organ.

Venule (Latin: *ven*, "vein"; *ule*, "small"). A small vein.

Vesicle (Latin: *vesicula*, "a little bladder"). A small vacuole in a cell.

Vestibule (Latin: *vestibul*, "a porch"). A cavity of the inner ear that houses sensory receptors associated with balance.

Vestigial structures (Latin: *vestigium*, "a trace of sign, such as a footprint"). An organ or other structure in an animal that is useless and underdeveloped and that is homologous with an organ or structure that was functional in an evolutionary ancestor.

Villus (Latin: *villus*, "shaggy hair"). One of the small, fingerlike projections in the lining of the small intestine of a human or other vertebrate into which digested food is absorbed.

Virus (Latin: *virus*, "a poison or stench"). A noncellular particle so small that it can be seen only with an electron microscope. A virus consists of a core of nucleic acid and a protein coat; it is parasitic and reproduces only while in other living cells.

Vitamin (Latin: *vita* "life"). An organic substance needed in a very small quantity for normal growth and development of an organism. Organisms are unable to synthesize most vitamins, which must be supplied in the organism's nutrients.

Viviparous (Latin: *vivus*, "alive"; *parere*, "to bring forth"). Giving birth to living young, as opposed to laying eggs. Most mammals are viviparous.

Vulva (Latin: *vulv*, "a covering"). The external portions of external genitalia of female mammals.

Wild type. The normal or most common phenotype of individuals comprising a species. The expected appearance of a species in nature or the wild.

X chromosome. One of the two types of sex chromosomes.

Xylem (Greek: *xylon*, "wood"). The vascular tissue in plants that conducts water and nutrients from the roots upward through the stems to the leaves and other structures.

Y chromosome. One of the two types of sex chromosomes.

Yolk. The stored nutrients in an egg, or ovum, that nourish the developing embryo.

Yolk sac. The extraembryonic membrane in many vertebrates that encloses the yolk and makes it available to the developing embryo. In humans and other mammals whose eggs may contain little yolk, the yolk sac serves as the initial site for blood cell formation.

Zooplankton (Greek: *zoo*, "an animal"; *planktos*, "wandering"). The mostly minute heterotrophic organisms that float freely in both fresh water and ocean water.

Zygote (Greek: *zygotos*, "yoked together"). A fertilized egg, or ovum; a diploid cell formed by the fusion of two haploid gametes.

INDEX

Single page references indicate where topics are mentioned; page ranges indicate expanded treatment of topics; boldface pages indicate accompanying illustration.

Abert squirrel 376–377
Abiotic environment 420
Abortion 240–241, 272
Abyss 431
Absorption, food 122
Abstinence 244
Acacia tree *458*
Acetylcholine 164
Achondroplasia 298
Acids *20*
   gamma aminobutyric (GABA) 164
   glutamic 164
   uric 155
Acquired characteristics 280
Acrosome 88, 254, 259
Actin 104
Active transport *73*
Adenine 316
Adenosine triphosphate, ATP 58
ADH, antidiuretic hormone 152
Adipose tissue 96
Adolescent muscular dystrophy 335
ADP 58
Adrenals 169
Adrenogenital syndrome 275
Afterbirth 269
African Bushmen/Hottentots 387
African elephants 456
Age structure *467*
Agent orange 486
Aggression *503–505*
Aging 173–174
AIDS (Acquired Immune Deficiency Syndrome) 246
Air pollution *478–482*

Albinism *310*
Alcohol consumption 274
Algae 440
All-or-none principle 105
Allantois
   human 261–262
   terrestrial vertebrate 258
Alleles 285
   linked 299
   multiple 307–308
Allergies 209
Allopatric speciation 380–*381*
Alternation of generations 87, *205–216*
Altruism 509
Alveolus (alveoli) *125*–126, 128
Amber *359*
Amino acids 22–*24*, 120, 321
Amish 385
Ammonia 155, 447
Amniocentesis 273, *377*
Amnion
   chick 257
   human 260, 261, *262*
Amnionic cavity 260
Amniotic fluid
   chick 257
   human 261
Amoeba *38*
AMP 58
Amphibians *220*
Amplexus 203
Amylase 189
   salivary 113–*114*
Anaerobic respiration 67–68
Analogous vs. homologous 364
Anaphase 80
Anatomy 362
Androgens 266
Anencephala 337
Anhydrotic dysplasia 303

Angiosperms 209–214
Animal behavior *490–511*
Animal pole 255
Animalia 10
Animals
   cold-blooded 148
   laboratory 368
   warm-blooded 148
Annelida 119, 154, 159, 521
Annual plants 183
Annual rings 183–*184*
Ants *458*
Antagonistic muscles 107–*108*
Anterior pituitary 170
Anther 210, 212
Antheridia 206, 208
Anthropoidea 391
Antibodies 140–142, *308*
Anticodon *323*
Antidiuretic hormone 152
Antigens 140–142, 307–*308*
Antihistamine 209
Antitrypsin deficiency 334
Anus 117
Anvil *167*
Aorta 134
Apes 391–402
Aphotic zone 432
Apical area 177
Appeasement *504*
Appendix 116, 365
Apple 201
Applied research 5
Aquaculture 473–*474*
Aquatic environments 421, *427–434*
Aqueous humor 166
Archegonia 206, 208, *217*
Archenteron 255
Arteries 130
Arterioles 130

Arthritis *102–103*
Arthropoda 119, 159, 171, 521
Artificial insemination *249*
Artificial organs
  heart 136
  kidney *151*
Ascomycetes 517
Asexual reproduction 88, *196–201*
  economic importance 199, 201
  limitations 199
Aster *79–80*
Astigmatism 166
Aswan High Dam 456, 471–472
Asymptomatic limit *466*
Atmosphere
  early earth 27–30
Atoms 14
Atomic nucleus 14–*15*
Atomic number 14
Atomic weight 14
ATP, adenosine triphosphate 57, *58*, 63–67
Atrioventricular (AV) node 133
Atrium (atria) 131, *132*
Auditory canal 167
Auditory releasers
  animal *494–496*
Australian aborigines 387, 414
Australopithecus afarensis 416
Australopithecus africanus 404, *407*, 416
Australopithecus boisei 404, 406, 407
Australopithecus robustus 404
Autonomic nervous system 158
Autosomes 300
Autotroph(ic) 56, 206, 420, 439
Auxin 191
Axis 213
Axon 161

Bacillus thuringiensis 463
Bacillus, bacteria 50–*51*
Bacteria 50–*51*
Balance 167
Baldness 306–*307*
Bark 178–180
Bartholin's glands 231
Barr body *302*
Basal body 44
Basal body temperature method 242
Base-pair substitution 326
Bases *20*
Basic research 5
Basidiomycetes 517
Bateson, William 286
B-cells 140–142
Beadle, George W. 287
Beagle, HMS 349–353
Bean plant *176*, 214–*215*
Bees
  drones 205

queen 205
reproduction 205
workers 205
Beef *473*
Beefalo *379*
Behavior *490–510*
  human, social *398*, *410*–411
  learned 499
Benthic organisms *429–430*
Bering land bridge 414
Best of fit 501
Betta splendens 203
Bicuspid valve 131–132
Big Bang theory 27
Bile 115
Billings method *242–244*
Binary fission 196–*197*
Biodegradable 484
Biogenesis, law of, 253, 364
Biogeochemical cycles *443–449*
Biological control, pests 463–464
Biological feedback 234
Biological magnification *485*
Biomass 440, 461, 465
Biome 424
  destruction *435–436*
Biosphere *420–436*
Biotic environment 420
Biotic potential *466*
Bipedalism 393
Birds 204
  songs 495
Birth control *238–244*
  chemical methods 241
  mechanical methods 238–*239*
  methods, list
  natural methods *241–244*
  surgical methods 240–241
Birth defects 273
Birthing stool 269
Bladder 150
Blastocoel 255, 259
Blastocyst 259
Blastodisc 257
Blastopore 255
Blastula 255
Blood 135, 138, 174
  cells 138–*139*
  clotting 138
  glucose level 119
  pressure 134
  transfusions 307–308
  type (human) 307–*308*
Bombykol 496
Bond, chemical 15–18
Bone 99–101
  development *100*–101
  marrow 100
Bottom-dwelling organisms 429
Bowman's capsule 150

Brachiation 393–*394*
Brachydactyly 298
Brain 155
  chemistry *164*
  stem 156
Breathing 126, 128
Breeding ring 375–*376*
Broadcasters 202
Bronchi 126
Bronchioli *125*–126
Browsers 455
Bryophyta 518
Bulbourethral glands 225
Budding *197–198*
Buds 210, 212
Bufo americanus 376
Bufo fowleri 376
Buffowing organisms *438*
Bursae 101
Bursitis 102

Cactus 426
Caecum 365
Calyptra 206
Cambium 177, 183
  cork 178
Canadian lynx 465
Cancer *90–91*, *312–313*
  lung 129
Capillaries 130
Capsule 206
Carbohydrates *24–26*, 117, 119–120
Carbon *15*, 21
Carbon-14 method *360*
Carbonaceous rock 445
Carbon cycle 445–*446*
Carbon dioxide 71, 445, 450, 478, 480
Carbon monoxide 478
Carboxyl group *24*
Carcinogens 312, 326
Carcinoma 90
Cardiac cycle 133
Cardiac muscle 103, *104*, 131
Carnations 300
Carnivore(s) 420, 491
Carotenoid pigment 59
Carrying capacity 454, 466
Cartilage 95, 100–*101*
Catalyst 21–*22*
Caucasoid race 414
Ceboidea 391
Celery 181
Cell(s) 5
  animal *38–46*
  cancer 46, 54
  companion *181*
  cycle *77–83*
  culture 54
  death 88–89
  division *77–82*

fusion 75
HeLa 54
megaspore mother 210–*211*
membrane *40*
organelles 50
permeability 68
plant *81–82*
plate *81*
pollen mother 210–*211*
sap 50
structure *36–51*
stone 178
study of 37–38
target 169
transformed 312
wall *46–48, 70, 81*
Cell division
bacterial (prokaryote) *51*
eukaryote 77–82
Cell theory 36
Cellular
differentiation 36, *88, 257*
organelles *38–46*
respiration *57, 63–68*
Cellulose 24, 46
Central nervous system *155–158*
Centriole 44, 79–80
Centromere(s) *79–80*
Cercaria 457, 458
Cercopithecoidea 391
Cerebellum 156
Cerebral cortex *157*
Cerebrum 155, *157*
and behavior *397–398*
Cervical caps 238–239
vertebrae 362
Cervix *232*
CG, chorionic gonadotropin 236
Chemical bonds *15–18*
covalent *17*
hydrogen *17–19*
ionic *16*
peptide *23–25*
Chemical reaction(s) *18–21*
endergonic 18, *21*
exergonic 18, *21*
redox 18
Chemoreceptors 128
Chick, development of 257–*258*
Chicken peck order 505
Chicken treading 204
Childbirth *266–269*
methods of *269*
questions about *272–273*
Chimpanzee, behavior of *401*, 504
Chlorinated hydrocarbons 484
Chlorophyll 48, 59
Chloroplast(s) *48–49*, 177, 183
Chlorophyta 518
Cholera 458

Cholesterol 26–27, 121–122
Choppers *405*
Chordata 521
Chorion
chick 257
human 261, *262*
Chorionic gonadotropin (CG) 236, 261
Choroid 166
Chromatid 79–*80*
Chromatin 44, 78
Chromoplast 48–49
Chromosomes 77, 285, 315, 320
daughter 80
homologous *300*
karyotype *300*
numbers 82
sex *300*
X 300
Y 300
Chromosomal anomalies *338–344*
Chrysophyta 516
Chyme 114
Cichlid fish 381
Cigarette smoking 129, 274
Cilia 43, *44*
Ciliophora 517
Circulation *130–135*
Circulatory systems *130–140*
open-closed 137–138
Circumcision 223
Cisternae 42
Cistron 326
Class, taxonomy 7
Classification 7–8
human *393*
Claws *397*
Clay soils 438
Cleaning wrasses 205
Cleavage
human *259*
vertebrate *255*
Climate *422–423*, 438
Climate changes 435–436, 450
Climateric 228
Clinical insemination 249–250
Clitellum 204
Clitoris 229
Cloaca 203
Clones 220
Cnidaria 118–119, 159, 519
Coacervates 28
Cocaine 160
Coal reserves 475
Coccus, bacteria 50–*51*
Cochlea 167–*168*
Cochlear duct 167
Cockroach, mating behavior 496
Codon 322
Coelacanth *373*
Coenzymes 21–*22*

Coexistence 453–*454*
Coitus interruptus 243
Collagen 95
Collecting tubule 151
Collenchyma *178*
Colloid 14
Colon 116–117
Color, spectrum *59*
Color-blindness *304–306*
Color vision 396
Commensalism *459*
Commensals 459
Communities 420, *452–462*
climax 460
stability 459–462
Comparative
approach 362–368
anatomy 362, 364
embryology 364–*365*
protein chemistry 366–*367*
Companion cells 181
Competition 453–*454*
Competitive exclusion, principle of 453
Complementary bonding 320
Composites 209
Compost 439
Conception 249–251
Condoms 238, 244
Confrontation *503*
Cones, of eye 166, 305
Cones, gymnosperm
ovulate 214
staminate 214
Conifers 214, 426
Conjoined twins 271
Conjugation 216
Connective tissue 95–*96*
fibrous 95–*96*
Constipation 117
Consumers 440
Conservatory 472, 478
Continental drift *360–362*
Contour mining 475
Contraception 238–244
Contractile vacuole *71*, 153
Controlled experiment 4
Controls 4
Convergent evolution 374
Copepods 441
Copulation 203, 507
Copulatory behavior 507
Copulins *229, 497–498*
Corals 438, 440, 459
Cork *178–180*
Cormorants 454
Corn, hybrid 380
high-lysine 289
Cornea, of eye 166
Coronary
arteries 134–135, *136*

bypass *136*
  circulation 134
Corpus luteum 236
Cortex 150
  plant 184
Cotyledons 213
Courtship 506
  animal *494–495*
Cowper's glands 225, 227
Cranial capacity 406
  nerves 158
Crick, Francis 287
Crickets 378
Criminal syndrome 343
Cristae, of mitochondria *43*
Critical period 499
Cro-magnon man *413–414*
Cross, common genetic *292, 294*
Crossing over, of genes 83, *85,* 202, 287,
  299
Cryptorchism 264
Cultural anthropologists 408
Curettage 241
Cuticle *182*
Cutin 177
Cyanophyta 515
Cystic fibrosis 333–335
Cytochrome 64
Cytokinesis *81*
Cytokinin 191
Cytology 37
Cytoplasm 40
Cytosine 316

Dark reaction *59–63*
Darwin, Charles 280, *348–357*
Darwin, Erasmus 355
Darwinism, social 369
Darwin's finches *352,* 373, *374*
Daughter
  cells 81
  chromosomes 80
Dawn redwood 373
DDT 484–486
Death, cell 88–89
Deciduous trees 183, 426
Decomposers 420, 440
Defecation 117
Dehydration synthesis 23
Deletion, chromosomal 326
Delivery *269*
Demographic transition 469
Demography 469
Dendrite 161
Denitrification 448
Denitrifying bacteria 448
Dental arch 399
Deoxyribonucleic acid (DNA) 77, 216,
  287, 312, *315–321*
Deoxyribonucleotide 316

Deoxyribose, structure of *316*
Depolarization, nerve 162
Depressants 160
Dermis *97*
Desalinization plants 472
Deserts 425–426, 440
Detergents 484
Development
  human *258–267*
  terrestrial vertebrate *256–258*
Developmental anomalies *272–274*
Diabetes 120, 389
Diaphragm *126*
  birth control
Diarrhea 117
Diastole 134
Diatomaceous ooze 441
Diatoms 440
  life cycle *441*
Dicot vs. monocot *213*
Dicotyledons 213
Diet 117
Differentiation, cellular *257*
Diffusion 69
  facilitated 72–73
Digestion *122*
  human 114
  plant 189
Digestive enzymes *114*
Digestive system *112–124*
Dihybrid cross 285, 292
Dinoflagellates 516
Dinosaur 362
Dioecious 209
Dioxin 486
Diploid 82, 201
Dissacharides 25–26
Dispersion, human 414
Displacement behavior 504–505
Display 492
Distal convoluted tubule 151
Divergent evolution 373
Diversionary behavior 504, 506
Division of labor 409
Dizygotic twins 270
DNA, deoxyribonucleic acid 77, 216,
  287, 312, *315–321*
  polymerase 320
  recombinant *327–329*
Dog, domestic 378
Domestication 372
Dominance
  codominance 308
  incomplete 299–300
  principle of 283
Dominance hierarchies *505–506*
Dominant 283
Dopamine 164
Dosage compensation *302*–303
Double fertilization *211,* 212

Double helix *318*
Doubling time 488
Down's syndrome 337, 338, *340–342*
Dragonflies 377
Drive 499
Drones 205
Drosophila 287–295
Drug, action 160
Dryopithecus 398–399
Duchenne muscular dystrophy *335*
Dunkers 385
Duodenum 114
Dutch elm disease 426
Dwarfism *298*
Dynamic steady state 147

Ear, human *167–168*
Ear point *365–367*
Eardrum 167
Earlobes 295
Ears, hairy *306*
Earthworms 126, 204, 439
Ecdysone 171
Echinodermata 521
Ecological isolation, species 377
Ecological succession *461*
  primary 462
  secondry 462
Ecology
  definition 436
  human *469–487*
Ecosystems 421, 437
Ectoderm *256,* 259
Ectoparasites 456
Ectotherms 148
Edward's syndrome 342
Egg(s) 202–205
Ejaculation 226
Ejaculatory ducts 225
Electrocardiogram (ECG) *133*
Electron transport system 64, 66–67
Electromagnetic spectrum 58–*59*
Electrons 14–*15*
Electron acceptor 60
Elements 14
  trace 439
Elodea 433
Embryos(s) 88, 206
  chick *258*
  human *259*
  plant 206, 208, 212
  vertebrate *253–278*
Embryo sac 210, 212
Embryology 253
Embryonic
  development *253–278*
  disc 260, 270
  leaves *213*
Embryophyta 206
Emphysema 129

Emulsification 115
Endergonic reaction 18, *21*
Endocardium 131
Endocrine system *168–171*
   glands *168–169*
Endoderm *256*, 259
Endodermis 184
Endometrium 234, 261
Endoparasites 456
Endoplasmic reticulum *39, 40–42*
Endosperm 210, 212, 214
Endotherms 148
Energy *57, 474–478*
   geothermal 477
   garbage 477
   hydroelectric 477
   nuclear 476
   solar 477
      active 477
      passive 477
   wind 477
Energy and ecosystems 439–443
Energy and its forms 57
Energy of activation *21*
Enterobius 458
Environments
   chemical 32–33
   fresh water 432–434
   marine, classification *428–432*
   terrestrial 424–426
Environmental resistance 466
Enzyme(s) 21–*22*, 23, 326
Epicanthal fold 338, *340*
Epicardium 131
Epicotyl 213
Epidermis *96–98*
   plant 177, 184
Epididymis 224, 226
Epigenesis 255
Epiglottis 125
Epiphytes 425
Epitheca 441
Epithelium 94–*95*
   ciliated 95, 129
   columnar *95*
   cuboidal *95*
   pseudostratified 95–*95*
   simple 94-*95*
   squamous *95*
   stratified 94–*95*
Equatorial plane 80
Equational division 86
Eras 362, *363*
Erectile tissue *224*
Erection 224
Erythroblastosis fetalis *297*
Erythrocytes 138
Escherichia coli 327
Esophagus 113
Estrogen 234, 241, 266

Estrus 231, 498
Estuaries 432, 440
Ethogram 492
Ethological isolation, species 377
Ethology *490–511*
Eugenics 389
Eukaryotic cells 50, 216
Euglenophyta 516
Euphotic zone 432
Eustachian tubes 167
Euthenics 389
Eutrophication 433, 448
   cultural 433
Evergreens 214, *215*, 424–425, 426
Evolution 7, *348–417*
   convergent 374–*375*
   divergent 373
   evidences *356–368, 403–406*
   explosive 382
   hominoid *398–400*
   human *403–417*
   parallel 373–374
   theories of 355–*356*
Excretion 153
   tubular 152
Excretory system 149
Exergonic reaction 18, *21*
Exocrine glands 168
Exoskeletons 108
Experiment, controlled 3
Experimental group 3
External fertilization *202–204*
Extracellular digestion 119
Extraembryonic membranes
   human 261–262
   terrestrial vertebrate 257
Eye *165–166*
   aging 173
   anterior chamber 166
   posterior chamber 166
   vitreous chamber 166

F_1, first filial generation 283, 292
F_2, second generation 292
FAD 66, *67*
Fallopian tubes 232
Family, taxonomy 7
Farsightedness 166
Fats 120–122
   saturated 26
   unsaturated 26
Fatty acids *26–27*, 121
   essential 121
Feces 117
Female, human reproduction *228–238*
Female pseudohermaphrodites 275
Fermentation 67–68
Fern, life cycle 207–*208*, 216
Fertility rate 488
Fertilization 201–205

double 210, 212
   cross 204
   external 202–203
   human 258–*259*
   internal *203*–204
   membrane 255
   plant 210
   vertebrate *254*–255
Fertilizer 448, 460
Fetal alcohol syndrome 274
Fetal therapy *277–278*
Fetoscopy 331
Fetus 263
Fiber(s) 178
   phloem 181
Fibrin 138
Filaments
   muscle 104–*106*
   stamen 210
Filter feeders 118, 452
Filtration 152
Fimbriae 235
Finches, Darwin's *373–374*
Fingernails 397
Fireflies 378
First trimester, human 262–*263*
Fishes 458
   freshwater 429
   ocean 429
Fission *197*, 216
   nuclear 32, 476
   track-dating *360*
Fitz-Roy, Robert 349
Fixed action patterns (FAPs) 490
Flagella 43
Flatworms 153, 159
Florigen 191
Flower(s) 209, 187
   complete 210
   imperfect 209
   incomplete 210
   perfect 209
   pistillate 209
   staminate 209
Flowering plant, life cycle 210–*211*
Fluid mosaic model 40–*41*
Fluorocarbons 383, 446, 482
Flukes, blood 456
   liver 456–*458*
Flies, fruit 373, 378
Follicles *282*, 234
Follicle stimulating hormone, FSH 227
Food, future 470
Food chains 6, 443
Food pollution 486
Food pyramid 440–443, 457
Food supply *470–474*
Food vacuole 72
Food web 443–*444*
Foramen magnum 403

Forebrain 155
Foreskin 223
Forests
  coniferous 426
  temperate deciduous 426, 440
Fossil fuels 446, 450, 475–476
Fossilization 356
Fossils 356, *359*
  dating *360*
Founder effect 384, 385
Freshwater habitats *432–435*
von Frisch, Karl 499
Frogs 203
Fronds 207
Fructose *25*
Fruit 187, 212
FSH, follicle stimulating hormone 227, 234–235
Fungi 8, 517
  chestnut blight 426

G_1 cell phase 78
G_2 cell phase 78
Galapagos Islands *351–352*
Galaxy 27
Gallbladder 115
Gallstones 116
Gamete(s) 82, 201
Gametic isolation, species 378
Gametophyte 206, 208, *217*
  generation 87, 205, 206, 209, 212, 216, *217*
Ganglion 163
Garbage, energy from 477
Gas, matter 13
Gastric juice 114
Gastroesophageal sphincter 113
Gastrula 255
Gastrulation 255–256, 259–*260*
Generative nucleus 210
Gene(s) 285, 315
  and cancer *312–313*
  flow 375
  frequency 382
  interactions *310*–311
  splicing 328
Genetic code 321–*322*
Genetic counseling *345*
Genetic disorders, human *334*
Genetic drift 384
Genetic engineering *327–329*
  applications *331*
Genetic load 389
Genetic problems 293
Genetic symbols 283, *301*
Genetic variability 201
Genetics
  chromosomal theory of 287
Genital folds 264
Genital tubercle 264

Genitalia, development 264, *265*
Genotype 199, 285
Genotypic ratio 292–*293*
  frequencies 382
Genus (genera), taxonomy 7
Geographic isolation, species *375–377*
Geological time scale 362–*363*
Geothermal energy 477
Germ layers
  primary *256*
German measles 272
Germination, of seed 214–*215*
Gestation 398
Gibberellins 191
Gigantopithecus 399
Gill slits 128, 362
Gills 128
Glans penis 223–224
Glaucoma 166
Glomerular filtrate 152
Glomerulus 151
Glucose 64–*65*, 119
  phosphate 189
Glycerol *26*
Glycogen 120
Glycolysis 64–*65*, 67
Golgi complex *39, 42*
Gondowana *361*
Gonopodium 204, 378
Gonorrhea 244–*245*
Goodall, Jane *401*
Gout 103
Grafting *201*
Grain beetles 454
Grana, of chloroplasts *49*
Granulocytes 139
Grasslands 425, 440
Grazers 455, 456
Greenhouse effect *435, 450, 481*
Green revolution 289, 473
Grip, precision 393
Grooming behavior 491
Growth, tree 183, *184*
Guanine 316
Guard cells 177, 183, 186, *188*
Guayule 193
Guppies 204, 378
Gustation 164
Guthrie test 336
Gymnosperms, life cycle 209, *214–217*
Gypsy moth *461*

Habitats
  aquatic *427–434*
  isolation, species 377
  terrestrial *424–427*
Hadal zone 431
Hair(s)
  human 173
  plant *182*

Hallucinogenic drugs 161
Hammerstones 405
Handaxe *405*
Haploid 82, 201
Hardy-Weinberg Law *382*–385
Hashish 179
Haversian canals *99*–100
Haversian system *99*–100
Hay fever *209*
Hearing, human 167–168
  aging 173
Heart 130
  aging 174
  disease 136, *334*
  sounds 132–133
Hectocotylus 204
HeLa cells 54
Helix 318, *319*
Hemoglobin 130–134, 138, 326
Hemophilia 303–*304*
Hemp 179
Henle, loop of 151
Henslow, John 349
Hepatic portal circulation 135
Herbicides *484–486*
Herbivore(s) 420, *455–456*
Heredity, see genetics and/or inheritance
Hermaphroditism 204
Hernia 224
Herpes 246
Heterochromatin 46, 78
Heterotrophs 56, 440
Heterosis 380
Heterozygous 283, 292
Hexosaminidase 336
Hindbrain 156
Histamine 209
Histones 321
Homeostasis 7, 147
Hominidae 392
Hominoidea 391
Homo erectus 409–411
Homo habilis 406, 408
Homo sapiens 387
  future of 487
Homologous vs. analogous *364*
Homologue 300
Homozygous 283
Honeybee 205
Hormonal anomalies 275
Hormonal control
  arthropod *171*
  flowering 191
  human growth 329
  plant growth *191*
  human reproduction 227–228, *234–237*
  sexuality 227–*228, 498–499*
Hormone(s) *168–171*
  activity 169–171
  list of human *170*

juvenile in arthropods 171
list of plant 191
Horse *366*
Host
    intermediate 456
    parasitic 456
    primary 456
Human, see the specific topic
Humus 425, 439
Humpback whale 495
Hunting-gathering habit 410
Huntington's disease 385
Huxley, Thomas 354
Hybrid 283
    inviability 378–379
    speciation 380–*381*
    sterility 379
Hybrid DNA 327–329
Hybridization 380
Hydra 197–198
Hydrochloric acid 114
Hydroelectric power 477
Hydrogen
    atom 15
    atmospheric 27–28, 478
    bonds 17–*19*
    fluoride 480
    pH 20
Hydrolysis 24–*25*, 113
Hymen 230–*231*
Hypertension 134
Hypertonic *70*
Hypocotyl 213
Hypothalamus 156, 168, 227–*228*, 234,
    266
Hypotheca 441
Hypothesis 3–5
Hypotonic *70*
Hysterotomy 241

ICSH, interstitial cell stimulating
    hormone 228
Ileum 114
Immune reaction *367*
    system 140–142
Immunity 142
    active/acquired 142
    nonspecific 140
    passive 140
Implantation, of human embryo *260–261*
Imprinting 499
Incomplete dominance 299–300
Incus 167
Independent assortment, principle of 285
Indifferent sex stage, human 264
Individual distance
    animal 500
    human 502
Inducers *257*
Industrial melanism *386*

Inferior vena cava *135*
Infertility 249–251
Inguinal canal 224
Inheritance of acquired characteristics,
    theory of *355, 357*
Inheritance, human *295–298*
    non-Mendelian 298–300
Innate releasing mechanism (IRM) 492
Inner cell mass 259
Inner ear 167–*168*
Innkeeper worm 459
Insect(s) 154, 362
Insectivora 392
Instinct 499
Insulin 168, 169, 328
Integumentary system 96–*98*
Intelligence 174, 266, *340–341*
Intercourse, sexual 258
Intermediate host, parasites 456
Internal fertilization 203
Interphase 78
Interstitial cell stimulating hormone,
    ICSH 228
Interstitial cells, testes 228
Intertidal 428
    organisms 452–*453*
Intestine
    small 114–*115, 117*
    large 116–117
Intracellular digestion *72,* 118–119, 189
Intrauterine devices, IUDs 239
Intromittant organ 204
Inversion 326
Involution 255
Ions 16
IQ *340–341*
Iris 166
Irradiation 274
Irrigation 470–472
Irritability 7
Islets of Langerhans 168
Isotonic 70
Isotopes 14–15
IUDs, intrauterine devices 239

Jejunum 114
Joints *101–103*
    cartilaginous 101
    disorders *102–103*
    fibrous 101
    synovial 101
Jojoba 193

Kaibab squirrel 376–*377*
Kangaroo rats 455
Karyotype *300*
Keratin 97
Kidney 149, 174
    artificial *151*
    transplants 151

Kingdoms *8*–10
Klinefelter's syndrome 342
Knuckle walking *395*
Krebs cycle *64–67*
Krill 443
Kwashiorkor 120, *471*

Labia majora *228*
Labia minora *228*
Labor *269*
Lacewings 463
Lacteal *117*
Lactic acid 68
Lady beetles 463
Lake Erie 483
Lake Malawi 381
Lake Turkana 406
Lamarck, Jean Baptiste 355
Lamaze method 269
Lamella or chloroplast *49*
Lamella, middle 48
Large intestine 116–117
Larva 89, 171
Larynx 125
Laterite 435, *472*
Laterization 472
Latimera 373
Laurasia *361*
Laws of genetics 282–284
Leaf veins 182
Leakey, Louis and Mary 404
Leakey, Richard 406
Leaves 177, *182,* 187
Learning 499
Leboyer method 269
Lens 166
Lenticels 177
Leukemia 90
Leucoplast 49
Leukocytes 138–*139*
LH, leutinizing hormone 228, 234–237
Lichens 459
Life
    characteristics 5–7
    search for 11
Life cycle(s), see specific organisms
Ligaments 101
Lignin 180
Light 57, 59
Light reaction *59–61*
Lightning 28
Lilacs, wild 377
Limiting factors 437, 441, 448
Limnetic zone 433
Limpets 452
Linkage, of genes 283
Lions, Indian 375
Lipid(s) *26–27*
Lipoproteins 121
Liquid, matter 13

Littoral zone 428, 433–*434*
Liver 115–116
Liver fluke, life cycle *457–458*
Loam 438
Lock-and-key theory 492
Locomotion *393–396*
Locus 299
Loop of Henle *150*, 151
Lordosis 507
Lorenz, Konrad 492
Lubrication 233
Lucy *416–417*
Lungs 126, 153
Luteinizing hormone, LH 228, 234
Lymph 139–140
   nodes 139–140
   vessels 139
Lymphatic system *138–140*
Lymphocytes *140–142*
Lymphoma 90
Lysosomes *39*, 42

M cell phase *78–82*
Macromutations 383
Macronucleus 216, *218*
Malaria 469
Male, human reproduction *222–228*
Male, pseudohermaphrodites 275
Malleus 167
Malnutrition 471
Malpighian tubules 154
Maltose *25*
Map units 299
Maramus *471*
Marijuana 179
Master gland, pituitary 168, 170, *228, 234*
Mating behavior, cockroach 496
Matter 13
Mechanical isolation, species 378
Medulla 150
   oblongata 126, 156
Megagametophyte 210, 212
Megaspore 209, 210, 212, *217*
Megaspore mother cell 210, 212, *217*
Meiosis *82–87*, 201, 205
   I *82–85*
   II *85–87*
Melanin 97–98, 309
Melanocytes 97–98
Membrane(s)
   cell *39*–40
   fluid mosaic model 40, *41*
   mucous 115–116
   nuclear *39*, 40
   transport 72, *73*
Memory *397*
Mendel, Gregory *281–285*
Mendelian genetics 282
Mendelian inheritance *291–295*

Mendel's Laws 282
Menopause 237
Menstrual cycle *234–238*, 497
Menstrual fluid 237
Menstrual phase 237
Meristematic tissue 177, 184
Mesoderm *256*
Metabolism 5–6
Metamorphosis 89
Metanephridia 154
Metaphase 80
Methane 475
Methemoglobinemia 451
Microbodies 42
Microclimates 438
Microfilaments 43
Micromutations 383
Micronucleus 216, *218*
Micropyle 210, 212
Microscope(s) *37–38*
   electron *37–38*
Microsporangia 212
Microspore 209, 210, 212, *217*
Microtrabecular system 43
Microtubules *43–44*
Microvilli 116–*117*
Midbrain 156
Middle ear 167
Middle lamella *48*, 81
Migrations, human 411
Miller, Stanley 29
Minerals 122, *124*
Miracidium 457
Miscarriage 272
Mitochondria *39, 43*
Mitosis *77–83*
Mixed nerves 158
Modal action patterns (MAPs) *490–495*
Models 382
Molecular genetics *315–332*
Molecule(s) 16
Mollusca 520
Monecious 208, 209
Monera 8, 515
Mongoloid race 414
Monkey trial 354
Monkeys
   New World 374, 391
   Old World 374, 391
Monocot vs. dicot *213*
Monocotyledons 213
Monoculture *460*
Monohybrid *283*
   crossing 285, *292*
Monosaccharides *24–26*
Monosomy 338, *339*
   sex chromosomal *343–344*
Monozygotic twins *270*
Mons veneris 228
Morgan, Thomas Hunt (T. H.) 287

Morphogenesis 256, 263
Mortality rate 466, 488
Morula 259
Mosaics *303*
Mosquito *494–496*
Moss, life cycle 206–*207*, 216
Motivation 498–499
   states 169
Motor unit 104
Mouth, human 113
Mouthbrooding fish 203
Movement *106–108*
Mucus method *242–244*
Mud flat *430*
Mule 379
Multiple allelic inheritance *307–308*
Muscle(s)
   action *104–108*
   antagonistic *107*
   cardiac 103–*104*
   contraction 104–*106*
   tissue 103–104
   skeletal 103–*105*
   smooth 103–*104*
   striated 104
   system *103–108*
Muscular dystrophy *335*
Musk *497*
Mussels 452
Mutagens 326, 383
Mutations 312, *326, 383*
   chromosomal 326
   missense 326
   nosense 326
   point *333–338*
Mutualism *458–459*
Myelin sheath 161
Myofibrils 104
Myofilaments 43
Myopia 389
Myosin 104

NAD, nicotinamide adenine dinucleotide 64–65
NADP, nicotinamide adenine dinucleotide phosphate 60
Narcotics 160, 272
Nasal cavities *125*
Natural childbirth 269
Natural selection 199, *355–356, 371–374*
   theory on origin of species *356*
Nature vs. nuture 499
Neanderthal man *411–413*
Nearsightedness 166
Negative feedback 234
Negroid race 414
Nekton 431
Nematoda 119, 520
Nephron *150*
Neritic zone 428

Nerve(s) 163
  cord *159*
  cranial 158
  impulse 162, 163
  mixed 158
  optic *165*
  spinal *158*
  tissue 96
  tracts 156
Nervous sytem 155–168, 266
Neural tube 256
Neuron(s) *161–163*
  excitatory 163
  inhibitory 163
  interneurons 161
  motor 104, *161*
  sensory *161*
Neurotransmitter *164*
Neurulation *256, 261*
Neutrons 14–*15*
Neutrophils 138, *139*
Newts 203
Niche 420, 453
Nictitating membrane 365
Night blindness 385
Nitrate 447, 451
  bacteria 448
Nitric acid 480
Nitrification 447
Nitrite bacteria 447
Nitrogen 446
Nitrogen cycle *446–447*
Nitrogen fixing bacteria 448
Nitrogen oxides 478, 480, 482
Nitrogenous bases *316*
Nobel Prize winners 3
Node(s), lymph 139, *140*
Noise pollution 486
Nondisjunction *338*
Nonrenewable resources *474-478*
Nonvertebrates 171
Norepinephrine 164, 169
Norway rats 466
Nostrils
Nuclear
  fission 476
  power 476
Nuclear membrane *39*, 40, 44, *45*
Nucleic acids 287
Nucleoid 50
Nucleolus 46
Nucleoplasm, 40, 44
Nucleosome 321
Nucleotides 316, 321
Nucleus *39, 44–46*
  endosperm 210, 212
  generative 210, 212
  macro- 216
  micro- 216
  polar 210, 212

sperm 210, 212
  tube 210, 212
Nutrition *117–124*, 474
  World *470–474*

Oat seedling *191*
Oceans *428–432*
Octupus 204
Oil
  reserves 475
  shales 475
  spills 484
Olduvai Gorge *404*
Olfaction 165
Omnivore 443
One gene-one enzyme theory 287, 325
Oocyte 87
Oogenesis 82
Ootid 87
Opisthorchis sinensis 456–458
Opposable thumbs *393–394*
Optic chiasma *396*
Orbit 165
Orbitals *15*
Orchids 459
Order, taxonomy 7
Organic compounds *21–27*
Organ(s)
  plant 182–186
  sensory 155
Organ systems
  animal 94–174
  plants 186–191
Organisms
  multicellular 5
  unicellular 5
  reproduction 196–*197*
Organogenesis 256, 263
Organophosphates 484
Orgasm 227, 233
Origin of life 27–30
Origin of species 353–354, *380–382*
Osmoregulation 153
Osmosis *69–70*
Osmotic pressure 70
Osteoarthritis 103
Osteoblast 100
Osteocyte 100
Outer ear *167*
Oval window 167
Ovary
  flower 210
  human 232
Oviducts 232
Oviparity 257
Ovists 255
Ovulate cone 214, *217*
Ovulation 235
Ovule 210, 212, 214, *217*
Ovum 87, 233, 254, 259

Oxidation-reduction reaction 18, 19
Oxygen 71
Oxygen cycle 446
Oxygen debt 68
Ozone (shield) 383, 446, 478

Pacemaker 133
Palasade layer 183
Palolo worms 202
P_1, parental generation 283
Pair formation 506
Paleoanthropologists 408
Pancreas 114, 119, 168
Pangea *361*
Pangenes 280
Pangenesis 280
Pap smear 232
Parallel evolution 273, 274
Paramecium 197, 216, 453–*454*
Parasites 456
  external 456
  internal 456
Parasitism *456–458*
Parasympathetic nervous system 158
Parathyroid glands *169,* 170
Parenchyma 177, 189
Parthenogenesis 205
Particulates 482
Pasteurization 30
Patau's syndrome 342
PBB's (polybrominated biphenyls) 486
PCB's 486
Peas 281–285
Peck orders *505*
Pectin 48
Pectoral region 362
Pedigree analysis 295, *296, 304*
Pelagic zone 431
Pelvic region 362
  inflammatory disease 245
Penis *223-224*
Pentose, structure of *316*
Peppered moth *385-387*
Pepsin 114
Pepsinogen 114
Peptide(s) 164
  bonds *23-25*
Perception *396*
Perennial plants 206
Pericardial cavity 131
Pericycle 186
Peripheral nervous system 155, *158*
Peristalsis 113
Permafrost 426
Permanent tissues 177
Pest management, integrated 463–464
Pesticide(s) 463, *484–486*
  DDT *484-486*
  pollution *484-486*
Petals 210, *212*

Petiole 182, 187
PGA, phosphoglyceric acid 61–62
PGAL, Phosphoglyceraldehyde 62–63
pH *20*
Phaeophyta 518
Phagocytosis *72*
Pharynx, human 125
Phenocopies 338
Phenotype 285
Phenotypic ratio *292–295*
Phenylalanine 336
Phenylketonuria, PKU 336, 338
Phenylpyruvic acid 336
Phenylthiocarbamide, PTC 295
Pheromones 203, 463, *496–498*
    human 229, 231, 497
Phloem *181*, 190
    fibers 181
Phosphate 316, *317*
Phosphoglyceraldehyde, PGAL 62–63
Phosphoglyceric acid, PGA 61–62
Phosphorus 448
    cycle 448–*449*
Photolysis 59–60
Photophosphorylation *60–61*
Photoreceptors 166
Photosynthesis *56–63*, 68, *186*, 188
Photosystems P-680 and P-700 60–61
Phototropism 191
Phycomycetes 517
Phyletic speciation 382
Phylogeny *416*
Phylum 7
Phytoplankton 428, 437, 466
Pills, birth control 241
Piloerection 503
Pinna 167
Pinocytosis *72*
Pinus radiata 377
Pinus muricata 377
Pistil 210–*212*
Pistillate 209
Pituitary 168, 227, 234
    growth hormone 329
Placebos 241
Placenta 236, 257, 261–*262*
Plankton 428, *431*
Planaria *198–199*
Plant(s)
    angiosperms, life cycle *209–214*
    annual 183
    biennial 183
    blooms 451
    C3, C4 *61–63*
    deciduous 183, 426
    digestion 189
    dioecious 209
    gymnosperms, life cycle *214–217*
    evergreens 424, 425, 426
    fern, life cycle 207–*208*

hormones 190–*191*
hydridization 289
monoecious 209
moss, life cycle 206–*207*
perennial 183, 206
predators 455
reproduction
    asexual *198–201*
    sexual *205–216*
saps 190
structure 175–186
systems 186–190
tissues *175–181*
Plantae 8, 518
Plaques 334
Plasma 95, 135, 138
    proteins 138
Plasmids 327
Plastids 48–*49*
Plate tectonics 361
Platelets 138
Platyhelminthes 119, 153, 519
Pleiotropy 311
Pleural cavity 126
Polar body 87
Polar nuclei 210
Polarized molecules *18*
Pollen (grain) 209, 210, 212, 214, *217*
    sac 212
    tube 212, 214, *217*, 378
Pollen mother cell 210, 212
Pollination 210
Pollution *478–487*
    air *478–482*
    food *485–486*
    noise 486
    pesticides *484–486*
    thermal 486
    water 482–484
Polychlorinated biphenyls 484
Polygenic inheritance *308–309*
Polymorphism 385
Polypeptides 23
Polyploid 380
Polysaccharides 25–26
Polysomes 325
Pongidae 392
Pons 156
Population(s) 420, *465–489*
    age pyramids *467*
    density 466
    growth *469*, 488
    human 467–469
    United States 488–489
Porifera 118, 519
Postmating isolation mechanisms 380
Potassium-40 360
Potato 198, *200*
Prairies 425
Precipitation 423

Precopulatory behavior 507
Predation, role of in population control
    454, 455, *456*
Prehensile tails 374, 391
Pregnancy 262-266
Premating isolating mechanisms 380
Premenstrual tension 237
Prepuce 223
Preputial glands 223
Prey 454
Prey-predators 454–*456*
Primary cell wall 46
Primary germ layers *256*
Primates *391–417*
    predictions 401
Primordial soup 27–28
Procaryotic cells 50
Proconsul 398–399
Producers 439–440
Productivity, primary 440
Profundal zone 433
Progesterone 236
Progestins 241
Prokaryotic cells 50, 216
Proliferation phase *234–236*
Promisians 391, *392*, 393
Prophase *79*
Proplastid 48
Prosimii 391, 393
Prostaglandins 171
Prostate gland 225, 226
Protein(s) *22–24*, 120
    complete 120
    incomplete 120
    synthesis *321–325*
Protenoid 29
Prothallus 208
Protista 8, *9*, 515
Protonema 207
Protonephridia *153–154*
Protons 14–*15*
Protoplasm 40
Proximal convoluted tubules 151
Pseudohermaphrodites 275
Puberty 234
Pulmonary arteries 130, *131*, 134
Pulmonary circulation 130, 134
Pulmonary veins 130, *131*, 134
Punnett square 285–*286*, 292
Pupil, of eye 166
Purines 316
Pyloric sphincter 114
Pyramid, food *440–444*
Pyrimidines 316
Pyrrophyta 516
Pyruvic acid 64–*65*

Quadrupedalism 393, *395*
Queen Victoria 303, *304*

Races, of modern man 387, 414
Radiation 486
Radicle 214
Radioactive element 14, 15
Rainforests, tropical *435*
Ramapithecus 399, 406–407
Random assortment 85–*86, 202*
Reactions, see chemical reactions
Reabsorption 152
Recapitulation, principle of 253
Receptor(s)
    external *163–168*
    internal 155
    sites on nerves 162
    stretch 163
Recessive 283
Rectum 117
Red blood cells 138
Redox reactions 18–19
Reduction division 82
Redundancy of DNA codes 322
Reflex
    knee-jerk 163
Regeneration 88, 198
Releasers *492–498*
    auditory *494–496*
    pheromone *496–498*
    visual *492–495*
Releasing factors 168
Renal
    artery 150
    corpuscle 151
    pelvis 150
    vein 150
Resources
    nonrenewable *474–477*
    renewable 474
Replication of DNA 319–*320*
Reproduction 7, *196–221*
    bacterial 51, 216
    human *222–238*
    viral 52
Reproductive structures 175
Reproductive isolating mechanisms
    375–380
Reptiles 148
Respiration
    aerobic *63–68*
    anaerobic 63, 67
    cellular (internal) 123, *124*
    external *125*, 128, 129
    plant *186, 188*
Respiratory systems 123, *125–130*
Retina 166
Rheumatoid arthritis 103
Rh factor *295–298*
    incompatibility *272*
Rhinitis 209
Rhizobium 448
Rhizoids 208

Rhodophyta 518
Rhogam 298
Rhythm method *242–243*
Ribonucleic acid (RNA) 315, 321
    messenger, mRNA 321
    ribosomal, rRNA 321
    transfer, tRNA 321
Ribonucleotides 321
Ribose, structure *316*
Ribosomes *41*, 321
Ribulose diphosphate 61, *62*
RNA (ribonucleic acid) 315, 321
RNA polymerase 322
Rods 166
Rock strata *358–360*
Root(s) 184–*186, 187*
    cap 184–*185*
    hairs 184–*185*
    nodules *448*
    tap 186
    tips 184–*185*
Rough ER *41*
Round window 167
Roundworms (nematodes) 119, 520
Rubella 272

S cell phase 78
Sagebrush 426
Salinity 427
Saliva 113–*114*
Salivary glands 113
Salt glands 429
Sarcodina 516
Sarcolemma 104
Sarcoma 90
Sarcomeres 104
Sarcoplasm 104
Saturated fats 26–27
Savannas 425
Scales, cone 214
Schistosoma mansoni 456–457
Schistosomiasis 456–457
Schizophyta 515
Schwann cells 161
Scientific method 3–5
Scientific name 8
Scion of graft *201*
Sclera 165
Sclerenchyma 177–*178*
Scopes trial 354
Scrapers *405*
Screwworms 463
Scrotum *223,* 224
Scurvy 117, 123
Sea anemone *220*
Sea urchins 484
Seasonal isolation, species 376–377
Sebaceous glands 497
Second generation (F_2) 292
Second trimester, human *263–265*

Secondary cell wall *236–237*
Secretory phase 236–237
Sediments 356, 358
Seed 187, 209, 212, 213
    coat 212, 214
Seed plant, life cycle *208–216*
Segregation, law of 282
Selection 199, 355–*356, 371–374*
    adaptive *373–374*
    artificial 378
    directional 372
    disruptive 373
    divergent 373
    gametic 378
    kin 509
    stabilizing *372–373*
    unnatural 389–390
Selfish gene concept 509–510
Semen 227
Semicircular canals 167
Semilunar fold 365
Semilunar valves 132
Seminal pool 233
Seminal vesicles 225, 226
Seminiferous tubules 224–*226*
Semipermeable membrane *69,* 70
Senses, human *163–168*
    general 164
    special *164–168*
Sensitive periods 499
Sepals *210,* 212
Sequence of development 262
Serosa 115
Serotonin 164
Serum proteins 366
Sewage *482–484*
    industrial 482
Sewage treatment
    plants 482
    primary *482–483*
    secondary 482–*483*
    tertiary *482–483*
Sex
    aging 174
    chromosomes *300–303, 339*, 342–344
    determination 301–*302*
    reversal 205
Sex-influenced inheritance *306–307*
Sex-limited inheritance *305–307*
Sex-linked inheritance *303–307*
Sexual arousal 229–230
Sexual behavior 506–507
Sexual imprinting 499
Sexual releasers 506
Sexual reproduction
    animal *201–205*
    plant *205–216*
    single-celled organisms 216, *218*
Sexually-transmitted diseases *244–247*
Shanidar Cave 413

Siamese fighting fish 203
Siamese twins *271*
Sickle-cell anemia 301, *326–328*, *382–385*, 387
Sieve
  elements *181*
  plates *181*
  tubes *181*
Sigmoidal (s) curve *466*
Silicon 441
Simian crease 338
Sinoatrial (SA) node 133
Skeletal muscle 103–*105*
Skeletal system 97–99
Skeleton
  appendicular *98*–99, 362
  axial *98*–99, 362
  hydrostatic *107–108*
Skin
  aging 173
  pigmentation *309–311*, 399
Skull 1470 406
Sliding filament theory 105–*106*
Smegma 223, 229
Smell 165
Smoking 129, 274
Smooth ER 41
Snail 457
Snowshoe hare 465
Sociobiology 509–510
Sodium 16
Sodium/potassium pump 73, 162
Soil 425, 438
  erosion 472
Solar energy 477
Solid, matter 13
Solute 14
Solution 13
Solvent 14
Somatic system 158
Sori 207
Space
  casual 502
  individual 502
  intimate 502
  social 502
Sparrows, house 372
Speciation 374–*376*, *380–382*
Species 7, *374*
  allopatric 375
  human 387
  isolating mechanisms *375–380*
  sympatric 375, 378
Sperm
  banks 249
  cells 87, 254
  fetilization by 258–*259*
  nuclei 210
  numbers 202
  production 225

Spermatids 86, 88, 225
Spermatocytes
  primary 82, 225
  secodary 82, 225
Spermatogenesis 82, *84*, 225–226
Spermatogonia 82
Spermatophore 204
Spermatozoa 88
Spermicides 241
Spermists 255
Spiderwort 377
Spina bifida 337
Spinal cord 156, *157*
Spinal fluid 158
Spinal nerves *158*
Spindle 79–80
Spirillum, bacteria 50–*51*
Sponges 118
Spongy layer *182*–183
Spontaneous generation 30
Sporangium 207
Spore mother cell 206
Spores 206, 207
Sporophyte generation 87, 205, 207, 209, 212, 214, 216, *217*
Sporozoa 515
Staminate cones 214
Stamens *210*
Stapes 167
Starch, phosphorylase 189
Starfish 452
Stems *187* 183–184, 187
  herbaceous 183
  woody *183–184*
Stereoscopic vision 166
Sterilization *240*
Stickleback, behavior 378, *492–494*
Stigma, of flower 210, 212
Stillbirths 272
Stimulants 160
Stimulus filtering 498
Stimuli, releasing *492–498*
Stirrup *167*
Stock of graft *201*
Stolon 199
Stomach 113
Stomata 177, 183, 186, *188*
Stone cell 178
Strawberries 199, 201
Striated muscle 103–*105*
Strength 173
Strip mining *475*
Stroma of chloroplasts *49*
Style, of flowers 210, 212
Suberin 178
Subsoil *439*
Substrate feeders 118
Substrates *438–439*
Subtidal (sublittoral) 429–*430*
Sucrose *25*

Suction curettage 241
Sugars *24–26*
  deoxyribose *316*
  ribose *316, 321*
Sulfur
  dioxide 478, 480
  trioxide 478
Sulfuric acid 480
Summation, principle of 163
Suntan 98
Surrogate motherhood 250
Suspension 13
Sweat glands 148, 153
Symbiosis 205, *456–459*
Sympathetic nervous system 158
Sympatric speciation *381*
Synapse *162*
  transmission across 162, *163*
Synapsis 83, 299
Synaptic networks *163*
Synaptic vesicles 162
Syphilis 245–246
Systemic circulation 130, 134–135
Systole 134

Taiga 426, 440
Tadpole *89*
Tapeworms 204
Tarsier *392*
Tar sands 475
Taste 164
  bud 164
Tatum, Edward 287
Taung child 403
Taxonomist(s) 7
Tay-Sachs disease 335–336
T cells 140–142
Teeth 174
Telophase 80
Temperature, human body 148
Template, DNA 319
Tendons 101
Tennis elbow 102
Tenting, vagina 233
Teratology 272, 274
Teratogens 272, 274
Terrestrial environments 421, *424–427*
Territory *499–502*
  fixed *500*
  floating 500–*501*
  functions 501
  human 502
  marking *500*
Tertiary sewage treatment 482–*483*
Test-tube babies 250
Testes *224–225*
Testicles *224–225*
Testosterone 228
Tetrads 83
Tetracycline 274

Tetrapods 362
Thalidomide 272, *274*
Theory 5
Thermal pollution 486
Thermocline 431
Thermodynamics, Laws of 57
Third trimester, human *264–267*
Threat *503*
  responses to *505*
Three Mile Island 32
Thymine 316
Thyroid gland *169*, 170
Tides 428
Tilapia 460
Tissues
  animal *94–96*
  plant *175–181*
Topsoil *439*
Trachea
  arthropod 128
  human 125
Tracheids *180*
Tracheophyta 518
Transcription *322*
Translation *322–325*, 326
Translocation 190
Translocation, chromosomal 326
Transmitter substance *162*
Transpiration 186, *188–189*
Treading 204
Tree shrew, Asian *392–394*
Trees
  deciduous 183, 426
  evergreen 114, *215*, 424–426
Tribolium 454
Trick knee 102
Tricuspid valve 131–132
Trimesters, human *262–267*
Trisomics, autosomal 338, *340–342*
  sex chromosomal *342–344*
Trisomy 338
Trisomy-13 342
Trisomy-18 342
Trisomy-X 342
Trophic levels 440
Trophic relationships *454–459*
Trophoblast *259*
Tropical rainforests *424–425*, 440, 450
Tuatara 373
Tubal ligation *240*
Tube nucleus 210
Tundra 426, 440
  Alpine 426
Turgor pressure *70*, 71
Turner's syndrome *343*–344
Twins *270–271*
  dizygotic *270*
  fraternal *270*

identical *270*
  monozygotic *270*
  Siamese *271*
Tympanic membrane *167*
Typhoid fever 458
Tyrosine 310, 336

Ulcers 116
Ultraviolet radiation 29, 59, 98, 482
Umbilical arteries 261
Umbilical cord 262
Umbilical vein 261
Umwelt *498–499*
Unit membrane *40–41*
United States population 488
Universal
  donors 308
  recipients 308
Unsaturated fats *26–27*
Upwelling *432–433*
Uracil (U) 321
Uranium 238, 360, 476
Urea 149, 155
Ureter 150
Urethra 150, 226
Urinary bladder *149*, 150
Urine 153
  formation *152*
Urogenital pore 378
Urogenital sinus 264
Uterus *232*

Vacuoles
  plant *47*, 50
  contractile *71*, 153
Vagina *231–233*
Valves, heart *131–133*
Vas deferens *224–225*
Vascular bundles 182, 186
Vascular tissue, plants 180–181, 190
Vasectomy *240*
Vegetables 213
Vegetal pole 255
Vegetation
  emergent 433
  floating 433
  submerged 433
Vegetative reproduction *198–201*
Vegetative structures 175
Vein(s) 130
Vena cava
  inferior 132, 150
  superior 132
Venereal diseases 244
Ventricles *131–132*
Venules 130
Vertebral column *98*
Vertebrates 362

embryo development *253–278*
Vessel elements *180*
Vestibule 167, 230
Vestigial structures *365–366*
Villi *116–117*
Viral replication *52*
Viruses *51–53*, 312
  symbiotic 313
Vision *165–166*
  color 396
  stereoscopic 166, *396*
Visual releasers *492–495*
Vitamin D 399
Vitamins *122–123*
Vitreous humor 166
Viviparous 257
Vocal cords 125
Vulva 228

Waistline 173
Wallace, Alfred Russel 353
Warblers *454–455*
Water 68–69
  cycle *444–445*
  density *432*
  fresh *432–434*
  in and out of cells *69–71*
  molecule *17–19*
  ocean 431
  ocean vs. fresh 429
  pollution *482–484*
  transport in plants 186–189
Watson, James 287
Watson-Crick model *318*
Weather 422
Weight, human 173
Wheat 380
White blood cells *138–139*
Wild type 292
Wind 422
  power 477
Windmills 477
Wood 180, *183–184*

Xylem *180*, 190
XYY males *342–343*

Yolk 257
Yolk sac *258*, 260

Zero population growth 488
Zone, growth of root
  elongation 184
  maturation 184
Zoomastigina 516
Zooplankton 428, 437
Zooxanthellae 459
Zygote 201, 255

†